Clinical Cardiovascular
Physiology

CLINICAL CARDIOLOGY MONOGRAPHS

SERIES CONSULTANTS

J. Willis Hurst, M.D.
Dean T. Mason, M.D.

Advances in Electrocardiography, Volume 1
Robert C. Schlant and J. Willis Hurst

Myocardial Diseases
Noble O. Fowler

Cardiac Pacing
Philip Samet

Advances in Cardiovascular Surgery
John W. Kirklin

Noninvasive Cardiology
Arnold M. Weissler

Shock in Myocardial Infarction
Rolf M. Gunnar, Henry S. Loeb, and Shahbudin H. Rahimtoola

Rheumatic Fever and Streptococcal Infection
Gene H. Stollerman

The Peripheral Circulations
Robert Zelis

Clinical Cardiovascular Physiology

Edited by

Herbert J. Levine, M.D.

Professor of Medicine, Tufts University School of Medicine and Chief, Cardiology Service, New England Medical Center Hospital, Boston, Massachusetts.

Grune & Stratton
New York San Francisco London
A Subsidiary of Harcourt Brace Jovanovich, Publishers

Library of Congress Cataloging in Publication Data
Main entry under title:

Clinical cardiovascular physiology.
(Clinical cardiology monograph)
Includes bibliographical references and index.
1. Cardiovascular system—Diseases. 2. Physiology, Pathological. I. Levine, Herbert Jerome, 1928- [DNLM: 1. Cardiovascular system—Physiology. 2. Cardiovascular system—Physiopathology. WG102 C641]
RC669.C54 616.1 75-35563
ISBN 0-8089-0914-2

Grune & Stratton, Inc.
111 Fifth Avenue
New York, New York 10003

Library of Congress Catalog Card Number 75-35563
International Standard Book Number 0-8089-0914-2

Printed in the United States of America

This book is dedicated to my wife, Sandra,
my son, Andrew, and my daughter, Rachel

Contents

Contributors

ABDUL S. ABBASI, M.D., Associate Clinical Professor of Medicine, UCLA School of Medicine (Formerly Director of Non-invasive Laboratory, UCLA Center for the Health Sciences)

FRANCOIS M. ABBOUD, M.D., Professor and Director, Cardiovascular Division, Department of Internal Medicine, Director Cardiovascular Center, University of Iowa

ZALMAN S. AGUS, M.D., Assistant Professor of Medicine, University of Pennsylvania School of Medicine, Veteran's Administration Hospital, and Hospital of the University of Pennsylvania, Philadelphia, Pennsylvania

JOHN S. BANAS, JR., M.D., Associate Professor of Medicine, Tufts University School of Medicine, Director, Cardiac Catheterization Laboratory, New England Medical Center Hospital, Boston, Massachusetts

ROBERT BARNES, M.D., Associate Professor of Surgery, Director, Peripheral Vascular Disease Laboratory, College of Medicine, University of Iowa

ARNOLD H. BLAUFUSS, Exchange Senior Cardiac Fellow, University of Cape Town, Groote Schuur Hospital, Cape Town, South Africa (Formerly Fellow in Cardiology, Harbor General Hospital)

NORMAN BRACHFELD, M.D., Associate Professor of Medicine, The New York Hospital—Cornell Medical Center, New York, New York

ARAM V. CHOBANIAN, M.D., Professor of Medicine, Boston University Medical Center, Boston, Massachusetts

J. MICHAEL CRILEY, M.D., Professor of Medicine and Radiology, University of California School of Medicine; Chief, Division of Cardiology, Harbor General Hospital, Torrance, California

MODESTINO G. CRISCITIELLO, M.D., Associate Professor of Medicine, Tufts University School of Medicine, Physician, New England Medical Center Hospital, Boston, Massachusetts

MARTIN GOLDBERG, M.D., Professor of Medicine, University of Pennsylvania School of Medicine, Chief, Renal Electrolyte Section, Hospital of the University of Pennsylvania, Philadelphia, Pennsylvania

RICHARD GORLIN, M.D., Professor and Chairman, Department of Medicine, Mt. Sinai School of Medicine, New York, New York

ROLF M. GUNNAR, M.D., Professor of Medicine, Chief, Section of Cardiology, Loyola University Stritch School of Medicine, Maywood, Illinois

J. WARREN HARTHORNE, M.D., Assistant Professor of Medicine, Harvard Medical School, Associate Physician, Massachusetts General Hospital, Boston, Massachusetts

DONALD D. HEISTAD, M.D., Associate Professor of Medicine, University of Iowa Hospitals and Clinics, Iowa City, Iowa

MICHAEL V. HERMAN, M.D., Professor of Medicine, Chief, Division of Cardiology, Mt. Sinai School of Medicine, New York, New York

MICHAEL A. HEYMANN, M.D., Associate Professor of Pediatrics, Cardiovascular Research Institute, University of California, San Francisco, California

WILLIAM P. HOOD, JR., M.D., Associate Professor of Medicine, Director, Cardiac Catheterization Laboratory, University of Alabama Medical Center, Birmingham, Alabama

SARAH ANN JOHNSON, M.D., Assistant Professor of Medicine, Assistant Director, Cardiopulmonary Laboratory, Loyola University Stritch School of Medicine, Maywood, Illinois

PAUL A. LENNON, M.D., Cardiologist, Memorial Hospital, Long Beach, California (Formerly Fellow in Cardiology, Harbor General Hospital)

HERBERT J. LEVINE, M.D., Professor of Medicine, Tufts University School of Medicine, Chief, Cardiology Service, New England Medical Center Hospital, Boston, Massachusetts

HENRY S. LOEB, M.D., Associate Professor of Medicine, Program Director, Section of Cardiology, Veterans Administration Hospital, Hines, Illinois

KEVIN M. McINTYRE, M.D., Assistant Professor of Medicine, Harvard Medical School, Assistant Chief, Cardiology Section, West Roxbury Veterans Administration Hospital, West Roxbury, Massachusetts

ALLYN L. MARK, M.D., Professor of Medicine, Assistant Director Cardiovascular Center, University of Iowa

SHAPUR NAIMI, M.D., Associate Professor of Medicine, Tufts University School of Medicine and Director, Intensive Cardiac Care Unit, New England Medical Center Hospital, Boston, Massachusetts

WILLIAM A. NEILL, M.D., Chief, Cardiology Section, University of Oregon Medical School, Veterans Administration Hospital, Sam Jackson Park, Portland, Oregon

GERALD POHOST, M.D., Instructor in Medicine, Harvard Medical School, Clinical Assistant in Medicine, Massachusetts General Hospital, Boston, Massachusetts

CHARLES E. RACKLEY, M.D., Professor of Medicine, Director, MIRU Program, University of Alabama Medical Center, Birmingham, Alabama

LEON RESNEKOV, M.D., F.R.C.P., Professor of Medicine (Cardiology), The University of Chicago, Chicago, Illinois

WILLIAM C. ROBERTS, M.D., Chief, Section of Pathology, National Heart and Lung Institute, National Institutes of Health, Bethesda, Maryland and Clinical Professor of Pathology and Medicine (Cardiology), Georgetown University, Washington, D.C.

ABRAHAM M. RUDOLPH, M.D., Professor of Pediatrics, Physiology, Obstetrics, and Gynecology, University of California, San Francisco, California

THOMAS J. RYAN, M.D., Professor of Medicine, Boston University School of Medicine, Head, Section of Cardiology, University Hospital, Boston, Massachusetts

ARTHUR A. SASAHARA, M.D., Associate Professor of Medicine, Harvard Medical School, Chief, Medical Services, West Roxbury Veterans Administration Hospital, West Roxbury, Massachusetts

PHILLIP G. SCHMID, M.D., Professor of Medicine, University of Iowa Hospitals and Clinics, Iowa City, Iowa

BERNARD F. SCHREINER, JR., M.D., Professor of Medicine, Director, Cardiac Catheterization Laboratory, University of Rochester Medical Center, Rochester, New York

C. LYNN SKELTON, M.D., Assistant Professor of Medicine, Department of Internal Medicine, University of Texas Health Science Center at Dallas, Southwestern Medical School, Dallas, Texas

THOMAS W. SMITH, M.D., Associate Professor of Medicine, Harvard Medical School, Chief, Cardiovascular Division, Peter Bent Brigham Hospital, Boston, Massachusetts

EDMUND SONNENBLICK, M.D., Professor of Medicine, Chief, Division of Cardiology, Albert Einstein College of Medicine, Bronx, New York

DAVID H. SPODICK, M.D., D.S.C., Professor of Medicine, Tufts University School of Medicine, Lecturer in Medicine, Boston University School of Medicine, Chief, Cardiology Division, Lemuel Shattuck Hospital, Jamaica Plain, Massachusetts

PAUL N. YU, M.D., Sarah M. Ward Professor of Medicine, Head, Cardiology Unit, University of Rochester Medical Center, Rochester, New York

Preface

With each new decade, the challenge of taking inventory in a broad medical field has become more difficult. Not only are we faced with a geometric increase in our data base, but the vulnerable truths of yesterday are being dispelled with increasing frequency. The subject matter which is appropriate for a text on clinical cardiovascular physiology is vast and, by necessity, many relevant aspects have been omitted in this book. Whole chapters could easily have been devoted to electrocardiography, cardiac echography, angiography, radionuclide imaging, and non-invasive technics. However, it was considered wise to avoid detailed review of these rapidly developing technical procedures and rather to present our current understanding of the pathophysiology of diseases of the heart and circulation, in their clinical setting when possible. In order to provide a basic understanding of the fundamental control systems of the heart and circulation, separate chapters are devoted to reviews such as cardiac muscle physiology, control of cardiac output, the venous system, and myocardial metabolism. While major emphasis has not been placed on the treatment of cardiovascular disorders, the preeminent role of digitalis glycosides, antiarrhythmic agents, modification of the adrenergic nervous system, and electrical therapy of arrhythmias are each discussed separately. Although portions of the text are quite detailed, it is hoped that the book will be found useful not only for cardiologists and cardiologists-in-training, but for medical students during their course in cardiovascular disease and for internists with a special interest in cardiology.

I am grateful to Dr. J. Willis Hurst and Dr. Dean T. Mason for inviting me to participate as an editor in their Clinical Cardiology Monographs series. Special thanks are also due to Grune & Stratton, Inc. and to the capable assistance of my secretary, Miss Sandra Bryan.

William C. Roberts

1

The Structural Basis of Abnormal Cardiac Function: A Look at Coronary, Hypertensive, Valvular, Idiopathic Myocardial, and Pericardial Heart Diseases

Cardiac dysfunction, with few exceptions, is the result of cardiac structural abnormality. The two basic structural abnormalities are destruction of normal tissue, with or without replacement by abnormal tissue, and proliferation of abnormal tissue, with or without destruction of the underlying normal tissue. Destruction of normal tissue most often results in altered myocardial contraction or valvular regurgitation; proliferation of abnormal tissue most often results in narrowing of coronary arteries or of valvular orifices. This chapter describes various structural alterations in certain cardiac diseases and emphasizes their functional counterparts.

CORONARY (ISCHEMIC) HEART DISEASE

The Coronary Arteries

Symptoms and signs of ischemic heart disease (IHD) are due to abnormal functioning of portions of left ventricular myocardium. The myocardial alterations, in turn, are consequences of narrowing of the lumens of the major coronary arteries. Certain characteristic changes in these arteries are present at necropsy in patients with fatal IHD:[1–6]

1. The coronary arteries are diffusely involved by atherosclerotic plaques. Virtually no segments of any of the major (right, left anterior descending, and left circumflex) extramural coronary arteries are free of atherosclerotic plaques. Although the lumens of some segments are narrowed more severely than others, all portions of the extramural coronary tree are involved by the atherosclerotic process.

2. In fatal IHD, with rare exception,[7] at least one and usually two of the three major coronary arteries are narrowed more than 75% by old atherosclerotic plaques (Fig. 1-1). The 75% demarcation point is useful because it separates normal from abnormal flow. Flow of a fluid (blood) through a tube (coronary artery) is not decreased until at least 75% of the lumen is obliterated. Thus the major challenge is not elimination of coronary atherosclerosis but simply the limiting of coronary atherosclerosis to less than 75% luminal narrowing. Not only is coronary arterial luminal narrowing of more than 75% observed in patients with fatal IHD, this same degree of narrowing is probably also present in patients with symptomatic IHD (Fig. 1-1). Studies of coronary arteries in patients dying during or shortly after aorto-coronary bypass procedures for severe angina pectoris have shown just as much narrowing of the major coronary arteries as in patients with fatal IHD dying naturally.

3. The atherosclerotic process is limited to the epicardial coronary arteries (i.e., the major trunks and their near right-angle branches) and spares the intramural (intramyocardial) coronary arteries.

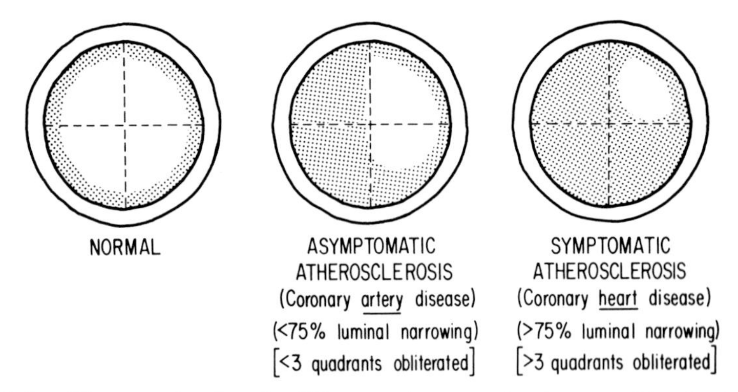

Fig. 1-1. Diagram showing the differences between the degrees of luminal narrowing of the coronary arteries in patients with symptomatic and asymptomatic ischemic heart disease. By age 20 years in all population groups worldwide there is fibrous intimal proliferation in the coronary arteries about equal in thickness to the thickness of the media of the coronary artery. In population groups with hypercholesterolemia (serum cholesterol levels > 200 mg/100 ml) this intimal proliferative process generally continues. However, symptoms of myocardial ischemia, with rare exception, do not occur until the intimal proliferative process obliterates more than 75% of the lumen.

4. Certain portions of the coronary tree tend to develop larger atherosclerotic plaques, and therefore more narrowed lumens, than other portions. The most severe narrowing of the left anterior descending and left circumflex branches is usually within 2 cm of the bifurcation of the left main; the distal third of the right coronary artery, in contrast, is usually more narrowed than is the proximal or middle third. The left main coronary artery is narrowed more than 75% in about 25% of patients with fatal IHD, and its narrowing to this degree is indicative of particularly severe diffuse coronary atherosclerosis (all three major coronary arteries more than 75% narrowed).[8]

5. Of the three types of atherosclerotic plaques, namely lipid, fibrous, and complicated,[9] only the complicated plaque is responsible for causing significant (more than 75%) narrowing of the lumens of the coronary arteries. The lipid and fibrous plaques are worldwide in their distribution. The complicated plaques (i.e., those containing calcific deposits, cholesterol clefts, pultaceous debris, etc.) are found only in populations that develop IHD. The major component of even the complicated atherosclerotic plaque is fibrous tissue (collagen). It is likely that the fibrous component of the atherosclerotic plaque is irreversible or nondissolvable. In contrast, the lipid component, at least experimentally, is reversible (dissolvable).[10] Whether low-lipid diets or lipid-lowering drugs will cause depletion of lipids in complicated symptom-producing coronary atherosclerotic plaques is uncertain. Although fat stores in the body consist predominantly of triglycerides and atherosclerotic plaques predominantly of cholesterol esters, nevertheless the latter might decrease in size when caloric intake is low enough to cause a decrease in the size of fat deposits in readily visible portions of the body (anterior panniculus, for example). Emaciated prisoners in World War II apparently had little coronary arterial luminal narrowing. Similar observations have been made in victims of malignant neoplasms. The amount of lipid in atherosclerotic plaques of emaciated necropsy patients appears to be less than that observed in nonemaciated necropsy patients.

Transient starvation may well be a neglected but beneficial form of therapy for symptomatic IHD (possibly also a form of prevention). If, for example, the lumen of a coronary artery is 90% obliterated by atherosclerotic plaques, generally about 20% of the plaques consist of lipid deposits. Thus depletion of these lipid deposits in the plaque would allow the lumen to be less than 75% narrowed, and there is no decrease in the amount of flow through a tube (artery) until the tube is more than 75% narrowed (Fig. 1-2). As long as the lumen of a coronary artery is less than 75% narrowed, symptoms of myocardial ischemia rarely occur.

Following are some observations on acute or recent lesions observed in major coronary arteries in patients with fatal IHD:

1. Among patients with fatal IHD, thrombi are infrequent (~ 10%)

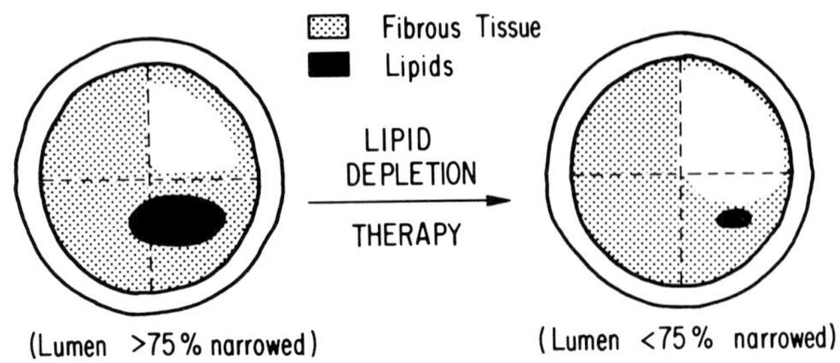

Fig. 1-2. Composition of atherosclerotic plaque in symptomatic coronary disease and the possible effect of lipid lowering or lipid withdrawal on its composition and luminal size. Although the dominant component of atherosclerotic plaque is usually fibrous tissue, lipid deposits (usually extracellular) form a portion of most plaques. It is likely that the fibrous component of the plaque is nonreversible. In contrast, the lipid component may well be reversible. Possibly, transient starvation may cause a diminution in the size of the lipid component and a decrease in luminal narrowing from more than 75% (the amount associated with symptoms, Fig. 1-1) to less than 75% (an amount rarely associated with symptoms).

in patients dying suddenly and in those in whom the necrosis is limited to subendocardium.[4] (Sudden death is defined herein as that occurring within 6 hr after onset of symptoms of myocardial ischemia unassociated with histologic evidence of myocardial necrosis. Subendocardium is the inner half of the myocardial wall.)

2. Thrombus is found in a coronary artery in about 60% of patients with fatal transmural acute myocardial infarction (AMI).[4]

3. Among patients with transmural myocardial necrosis, the major determinant of the presence of coronary thrombosis appears to the presence of cardiogenic shock. At necropsy, more than 70% of patients with fatal AMI with cardiogenic shock have coronary thrombi, whereas only about 15% of patients without the power-failure syndrome associated with fatal AMI have coronary thrombi.[11] Tissue necrosis itself, especially in a shock situation, also appears to increase the coagulability of blood.

4. The larger the area of myocardial necrosis, the greater the likelihood of coronary thrombosis. The larger the infarcted area, however, the greater the likelihood of cardiogenic shock. The latter generally indicates that more than 40% of the left ventricular wall is either necrotic or fibrotic or both, whereas shock is infrequently associated with infarcts or scars involving less than 40% of the ventricular wall.[12]

5. When coronary thrombosis is associated with AMI, the thrombus is always located in the artery responsible for perfusing the area of myocardial necrosis.[4] Thus in anterior wall infarction, a

thrombus, if present, will be located in the left anterior descending coronary artery.

6. Thrombi occur in fatal IHD in coronary arteries that already are severely narrowed by old atherosclerotic plaques (Fig. 1-3). At the distal site of attachment of the thrombus, or just distal to this site, the lumen of the coronary artery is nearly always more than 75% narrowed by old atherosclerotic plaques.[4] Not infrequently a thrombus may occur in an area between two sites of severe narrowing, as if in a valley between two mountains.

7. Coronary thrombi in fatal AMI are usually single (90%), occlusive (80%) (as opposed to mural or nonocclusive), short (< 2 cm long), and located entirely in the major trunks (as opposed to their near right-angle branches or intramural coronary arteries). The thrombus, when only a few hours old, may consist almost entirely of platelets, but thereafter is composed primarily of fibrin. By definition, the thrombus is adherent to the surface of the arterial wall bordering the lumen.

Since 1912, when Herrick first described the often dramatic clinical event characterized at necropsy by necrosis of portions of left ventricular

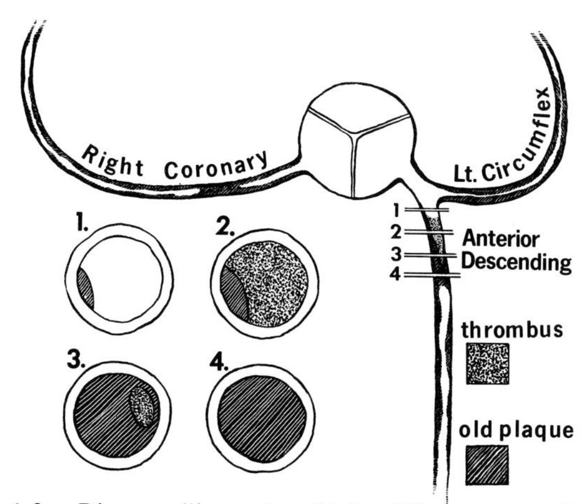

Fig. 1-3. Diagram illustrating (1) the diffuse nature of coronary atherosclerosis in fatal ischemic heart disease and (2) the usual status of a coronary artery at and distal to a thrombus. At level 2 in the anterior descending coronary artery the lumen is obstructed primarily by a thrombus. At level 3, however, the major percentage of narrowing is the result of old atherosclerotic plaquing and is just distal to the thrombus; the lumen is severely narrowed (> 75%) by old plaque. This situation occurred in 37 of 39 patients with fatal acute myocardial infarction and coronary thrombosis studied by the author.

wall, it has been assumed that the usual cause of AMI is coronary thrombosis.[13,14] Two factors implicate coronary thrombosis as the precipitating cause of AMI: (1) the occurrence of coronary arterial thrombi in many patients with fatal AMI; (2) the location of the thrombus in the coronary artery responsible for supplying the area of myocardial necrosis. Five factors, however, tend to indicate that coronary thrombosis is a consequence rather than the precipitating cause of AMI: (1) the very low frequency of thrombi in patients dying suddenly with or without previous evidence of cardiac disease; (2) the increasing frequency of thrombi with increasing intervals between onset of symptoms of AMI and death; (3) the absence of thrombi in fatal transmural AMI nearly as often as they are present; (4) the near absence of thrombi in fatal subendocardial AMI; (5) the occurrence of thrombi in high percentages only in patients with cardiogenic shock, most of whom have large transmural infarcts.

The key to coronary thrombosis, as well as to thrombosis occurring anywhere in the body, is slow blood flow, or relative stasis, and sufficient time for the thrombus to form. The absence of these two factors may explain the absence of coronary thrombosis in the sudden-death cases and the increasing frequency of thrombosis as the interval from onset of symptoms of myocardial ischemia to death increases.[15,16] There is marked reduction in blood flow in the coronary artery responsible for supplying the area of myocardial infarction. This observation was made in dogs after inducing AMI, and they had normal (i.e., widely patent) vessels.[17] In fatal AMI in humans the thrombus is always located in an artery already containing considerable atherosclerotic plaques, and therefore the infarct-induced relative coronary stasis is probably even greater. Cardiogenic shock must further diminish coronary flow.

The type of activity being engaged in by patients at the time of onset of AMI may reflect slowed blood flow. Nearly 75% of patients with AMI have the onset of chest pains while sleeping, resting, or performing mild activity.[18] Although inactivity may cause slight diminution in coronary blood flow, considerable stasis of blood (infarction-induced plus cardiogenic shock) is usually necessary for a thrombus to form. In contrast to fatal AMI, coronary thrombosis is rarely observed in fatal angina pectoris, although the coronary luminal narrowing by atherosclerotic plaques is similar in degree to that observed in AMI.[19] Evidence of thrombus formation is nearly always observed in arteries implanted into left ventricular myocardium but allowed to drain into the right ventricular cavity.[20] Thus it appears that a period of diminished coronary blood flow is necessary for a thrombus to form in a coronary artery. Shock, congestive cardiac failure, and inactivity all decrease coronary flow and with time may allow thrombosis.

Further support for the concept that coronary thrombosis is a consequence rather than a precipitating cause of AMI was supplied recently by Erhardt and associates,[21] who observed radioactivity at necropsy in coronary arterial thrombi in patients who had been given radioactive ^{125}I-labeled fibrinogen shortly after admission because of AMI. This finding implicates coronary thrombosis as a secondary event occurring sometime after the infarction.

Thus there is substantial evidence that acute thrombus formation does not precipitate acute fatal IHD. The major problem is a diffuse generalized coronary atherosclerosis with severe ($> 75\%$) luminal narrowing (at least two of the three major coronary arteries).

Now, acute lesions in coronary arteries other than thrombi:

1. Hemorrhage into an old atherosclerotic plaque: Hemorrhages into coronary atherosclerotic plaques are observed in about 25% of patients with fatal IHD.[4] Even when plaque hemorrhages occur, however, the lumen of the coronary artery is not further narrowed. Plaque hemorrhages have no relationship to the site of myocardial necrosis. At least when a coronary thrombus is observed, its location corresponds to the site of necrosis. When an anterior wall infarction occurs accompanied by a thrombus located in the left anterior descending coronary artery, other thrombi are not found (except rarely) in the right or left circumflex coronary arteries. In contrast, when hemorrhages into atherosclerotic plaques occur they bear no relation to the site of myocardial necrosis. The anterior wall may be the site of infarction, but the hemorrhage may involve a plaque in the right coronary artery, or plaques in all three coronary arteries, or plaques at multiple sites in a single coronary artery. The plaque hemorrhages may be occurring all through adult life in patients with and without symptomatic IHD. It appears unlikely that they are responsible for precipitating acute myocardial ischemia, because they do not narrow the lumen and they are not related to sites of myocardial necrosis.

2. Coronary arterial embolism (Fig. 1-4): Coronary arterial embolism is a rare cause of fatal IHD. Diagnosis of embolism requires identification of the site of dislodgment of the embolus or at least a condition predisposing to development of embolism, such as infective endocarditis, intracardiac mural thrombus, or a coagulopathy. Embolism is extremely difficult to recognize when superimposed on an extensively atherosclerotic coronary arterial tree. Thus diagnosis of embolism usually requires the occurrence of a clot in a coronary tree devoid of heavy atherosclerotic plaques. Furthermore, in contrast to coronary thrombosis, which never involves the intramural coronary arteries and infrequently the distal portions of the extramural coronary arteries, embolism generally involves both the intramural and extramural arteries, usually the distal portions of the latter.

In recent years a number of publications have described patients

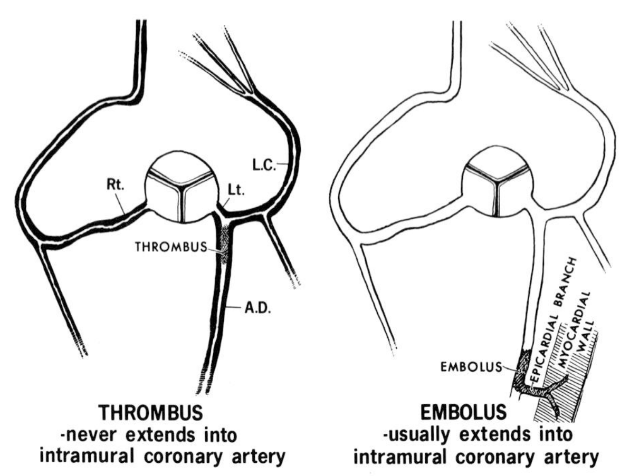

Fig. 1-4. Diagram depicting differences between coronary arterial thrombosis and embolism. The thrombus is usually proximal to and superimposed on old atherosclerotic plaque. The thrombus does not extend into intramural coronary arteries. The embolus is distal to and usually extends into an intramural artery. The embolus usually occurs in a coronary tree devoid of significant old atherosclerotic plaque.

with "myocardial infarction and angiographically normal coronary arteries."[22,23] It is essential to keep in mind that the term normal coronary arteries in this circumstance means angiographically normal, not necessarily anatomically normal, and that the angiographic studies were not done at the time of the AMI but usually several months later, at which time the patient was usually asymptomatic. It appears unlikely that all the major coronary arteries were "normal" at the time of the AMI. It is debatable whether or not spasm of a normal coronary artery can produce sufficient narrowing to cause myocardial necrosis. The most reasonable explanation for the occurrence of "myocardial infarction with angiographically normal coronary arteries" is coronary embolism. Such occlusion of a previously normal extramural coronary artery by a clot virtually always produces AMI. The embolus most likely organizes by developing large channels within it (recanalized channels) that on angiography some time later appear as normal vessels.

2. Dissecting aneurysm (hematoma) of a coronary artery with and without associated dissection of aorta: Dissection of one or both coronary arteries with resulting luminal narrowing is commonly associated with dissection of the aorta. The resulting myocardial ischemia in this circumstance may be fatal. Virtually all patients with aortic dissecting aneurysm with or without associated dissection of the coronary arteries have systemic hypertension.

Dissection of one or more major coronary arteries may nevertheless occur, although rarely, in the absence of dissection of the aorta.[24] When isolated to the coronary artery, systemic hypertension is infrequent, but other underlying precipitating causes have yet to be identified for this idiopathic dissection. Women are more often affected than men; death is usually sudden; and there is no histologic evidence of myocardial necrosis. In addition to the idiopathic variety, isolated coronary dissection may be iatrogenic in origin—a result of coronary angiography, coronary bypass operations,[25] or cardiac resuscitation.

The Myocardium

Symptoms of IHD are the result of myocardial dysfunction, and they take the form of angina pectoris (AP), acute myocardial infarction (AMI), or sudden coronary death (SCD). Arrhythmias, conduction disturbances, and congestive cardiac failure as isolated manifestations of IHD are infrequent. Most patients with the latter probably have had "silent" myocardial infarcts. What determines which of the three major coronary events (AP, AMI, or SCD) is manifest in a particular patient is uncertain. The degrees of luminal narrowing of the coronary arteries in all three appear similar.[4] Any of the three may be the first or the last manifestation of IHD. AP may precede or follow an AMI, or AMI may never occur in a patient with AP. About 50% of the patients with SCD have healed myocardial infarcts.

Sudden Cardiac Death Secondary to Coronary Atherosclerosis

Proof that sudden death is of cardiac origin, and specifically of coronary origin, requires necropsy examination. This procedure prevents inclusion under the heading of SCD the noncardiac causes of sudden death and the cardiac but noncoronary causes of sudden death, and it rules out previously clinically silent coronary events that might have occurred many hours or several days earlier and would otherwise have been considered SCD. On occasion, the first clinical manifestation of AMI may not be chest pain from myocardial ischemia or necrosis but a manifestation of a complication of myocardial necrosis. Rupture of the left ventricular free wall, for example, may be the first manifestation of AMI, but histologic examination of the area of myocardial necrosis may indicate that the infarct is several days old, and therefore that the AMI itself was clinically silent.

A good definition of sudden coronary death (SCD) is still debated. The definition must include a time interval, and this requires that the event be witnessed. If myocardial necrosis detectable by routine hematoxylin–eosin staining is present, a diagnosis of SCD, in my opinion, is not proper. The clinical time interval may be proper, but detectable necrosis cannot have occurred within a 6-hr period. The time interval in most definitions of SCD has ranged from instantaneous to up to 24 hr. Instantaneous, however, may be defined as a few seconds or several minutes. I have settled on an interval of up to 6 hr after onset of symptoms because it allows a great range of error when determining the time interval, and this interval provides a good separation between patients with and without histologic evidence of myocardial necrosis.

The coronary arteries of patients who die suddenly are strikingly similar to those of patients who die of AMI. Diffuse, severe coronary arterial luminal narrowing is present, with an average of 2.4 of the three major coronary arteries narrowed more than 75% by old atherosclerotic plaques.[4] Over 75% of these patients have cardiomegaly (heart weight > 400 g). Left ventricular myocardial scars, either subendocardial or transmural, are present in about 50%. Coronary arterial thrombi are infrequent (~ 10%). The mechanism of SCD in virtually all of these patients is ventricular fibrillation. No abnormality of the sinus node, atrioventricular node or bundle, or proximal portions of the left or right bundle branches is detectable by histologic examination of these structures.

Angina Pectoris

There are few necropsy studies on patients who during life had AP as the only manifestation of IHD. Patients with AP generally either die suddenly, in which case they are included in the category of SCD, or die of AMI, and consequently are included in the category of AMI. Of personally studied necropsy patients with AP as the only clinical manifestation of IHD, the hearts in most were hypertrophied (weight ≧ 400 g) and the ventricular cavities were of normal size. Focal scars were common in the left ventricular papillary muscles and in the subendocardial areas of the left ventricular free wall, but transmural left ventricular scarring was uncommon. None had left ventricular foci of necrosis. The lumens of at least two of the three major coronary arteries were narrowed more than 75% by old atherosclerotic plaques, and none had coronary arterial thrombi.

Acute Myocardial Infarction

Morphologic criteria for dating myocardial infarcts were first described by Mallory and associates[26] in 1939: polymorphonuclear leukocytes appear in 18 to 24 hr, the necrotic muscle is dissolved by 4 to 10 days, collagen is laid down after 10 days, and final healing is achieved by 60 days. Such a time course is probably an oversimplification because it is based on the assumption that AMI is precipitated by a thrombus suddenly obstructing a coronary artery and that all myocardial necrosis observed is the consequence of this sudden event. Myocardial necrosis, however, is on ongoing process over a period of hours to days, and within an area of infarction several different stages of an evolving infarct are present.

Cardiogenic shock, now the most common cause of death from AMI among hospitalized patients, is associated with considerably more myocardial necrosis or fibrosis or both than is present in patients with fatal AMI not associated with shock.[12] Also, the frequency of coronary thrombosis is much higher in the group with (> 80%) than in the group without (< 40%) shock.

HYPERTENSIVE HEART DISEASE

Other than age and sex, systemic hypertension, hyperlipidemia, and cigarette smoking are the three major factors increasing the risk of development of symptomatic cardiovascular disease. Of the three, two are correctable, smoking by abstinence and hypertension by certain drugs. It is uncertain as yet whether or not reduction in the levels of serum lipids will decrease the frequency of symptomatic cardiovascular disease. Among patients with symptomatic cardiovascular disease, however, hypertension is more common than hyperlipidemia, and reduction in the level of the blood pressure clearly decreases the frequency of complications of hypertension.[27-30] This section examines several cardiovascular disorders, termed *the hypertensive diseases,* as consequences, in whole or in part, of systemic hypertension* (Table 1-1). If cardiomegaly (heart weight > 400 g in men, > 350 g in women) is utilized as an indicator of hypertension, in addition to or in place of the blood pressure measurement, then hypertension is probably an even more common precursor of symptomatic vascular disease than previously realized. Therefore, cardiomegaly at necropsy is recorded (Table 1-1) where this information is available. (To be sure, ischemic heart disease, particularly when associated with congestive cardiac failure, appears capable of enlarging myocardial fibers and in some patients undoubtedly does so; but it ap-

*At what level of blood pressure normotension becomes hypertension is debatable. Certain studies[31] have emphasized that the frequency of complication increases with increasing increments of blood pressure. Although both levels might be considered normal, the person with a systolic pressure of 130 mm Hg is at greater risk of fatal or nonfatal complications than is the person with a systolic pressure of 110 mm Hg. The World Health Organization has defined systemic hypertension as systolic pressure > 160 mm Hg or diastolic pressure > 95 mm Hg or both. This study defines hypertension as systolic pressure > 140 mm Hg or diastolic pressure > 90 mm Hg or both. Because the pressure tends to increase with age, persons over the age of 65 are excluded from this analysis.

Table 1-1
Frequencies of Systemic Hypertension and Cardiomegaly in Various Cardiovascular Conditions (The Hypertensive Diseases)

Condition	% with Hypertension*	% with Cardiomegaly or, LVH @ Necropsy†
I. Sudden death	50	70
II. Angina pectoris	50	75
III. Acute myocardial infarction‡	50	75
IV. Complications of acute myocardial infarction		
A. Rupture of left ventricular free wall	70	85
B. Rupture of ventricular septum	70	85
C. Rupture of papillary muscle	70	85
D. Left ventricular aneurysm (healed)	70	80
V. Aneurysm of aorta		
A. Fusiform, sacular or cylindric	60	80
B. Dissecting	80	95
VI. Atherothrombotic obstruction of abdominal aorta or of its branches	70	?
VII. Cerebrovascular accident (stroke)		
A. Atherothrombotic cerebral infarction	50	70
B. Intracerebral hemorrhage	> 90	> 90
C. Lacunar softenings	> 90	> 95
D. Charcot-Bouchard aneurysm	> 90	> 95
VIII. Renal failure	> 90	> 90
IX. Miscellaneous		
A. Calcific aortic valve disease of the elderly	50	70
B. Calcification of the mitral anulus	40	60

* Blood pressure = systolic > 140 or diastolic > 90 mm. Hg. Limited to persons < age 66 years.
† Heart weight = women > 350 and men > 400 gms.
‡ Transmural type. Nearly all patients with fatal subendocardial necrosis have cardiomegaly.

pears more reasonable in most cases to attribute the cardiomegaly to hypertension.)

Sudden Coronary Death

About 50% of deaths from ischemic heart disease are sudden (ventricular fibrillation),[32] and sudden death is the initial coronary event in about 25% of patients with fatal ischemic heart disease.[33-42] Of 526 sudden coronary deaths (< 24 hr from onset of chest pain) reported by Kuller and associates,[42] 38% had had histories of hypertension and 40% had hearts weighing over 500 g. Thus, probably well over 50% had hearts weighing over 400 g. Among the blacks, 55% of the hearts weighed more than 600 g. Among 24 patients with SCD studied by the author (< 6 hr from onset of chest pain), at least 55% had had histories of hypertension and 70% had cardiomegaly at necropsy.[4]

Angina Pectoris

Most investigators have found that angina pectoris is more frequent in hypertensive than in normotensive persons,[43–52] and this association increases continuously as blood pressure levels rise.[52] Hypertension is present in about 50% of men with angina, and this percentage is much higher in women.[43] Indeed, Eppinger and Levine[43] in 1934 considered "the rarity of normal blood pressure in women with angina pectoris helpful in diagnosis of the presence of angina." The percentage of patients with fatal angina (without previously clinically diagnosed myocardial infarction) who have cardiomegaly at necropsy is uncertain, but it is probably about 75%. The major reason for this uncertainty is that most patients with angina pectoris usually either die suddenly, and are therefore placed in the "sudden coronary death" category, or die after an AMI, and are included in the "myocardial infarction" category. Of 72 patients with angina preceding the first diagnosed AMI reported by Francis and associates,[47] about 50% had histories of hypertension, and at necropsy about 70% had cardiomegaly.

Acute Myocardial Infarction

Overall, about 60% of patients with fatal AMI have positive histories of hypertension, and about 75% have cardiomegaly at necropsy.[53–63] These percentages are higher in women than in men. Not only does hypertension increase the risk of developing an AMI, it also increases the immediate and long-term mortality after AMI.[64–69] The reason hypertension worsens prognosis following AMI is uncertain, but it is probably because hypertension increases the frequency of major complications (see below). Although hypertension is a common precursor of AMI, the latter is the most common natural cure of hypertension. (Other natural cures include constrictive pericarditis and extensive cardiac amyloidosis.[70])

Cardiac Complications of AMI

Rupture of the heart during AMI, be it of left ventricular free wall, ventricular septum (acquired ventricular septal defect), or papillary muscle, is rare in the absence of left ventricular hypertrophy, which is infrequent among these patients in the absence of systemic hypertension.[71] When rupture occurs it is nearly always during the first AMI and in patients without previous congestive cardiac failure.[71] Therefore myocardial scars are rare in patients who die shortly after rupture of a papillary muscle or after perforation of ventricular septum or left ventricular free wall. Of 41 patients with rupture of the ventricular septum (18 patients) or left ventricular free wall (23 patients) studied at necropsy,[71] at least 33 had had histories of hypertension, none had previous evidence of congestive cardiac failure, and only 3 had evidence of previous AMI, and in them it was questionable. At necropsy, all had hypertrophied left ventricular free walls, and the hearts in 37 (90%) weighed more than 400 g. Only 3 had left ventricular scars, and each was small and subendocardial in location. Only 11 of the 41 patients had dilated left ventricular cavities, and in each the dilatation almost certainly was absent before the fatal AMI. Rupture may occur more frequently in patients in whom hypertension persists during the AMI, but its presence immediately preceding the rupture is not necessary. The recent study by Shell and Sobel[72] indicates that lowering of the blood pressure during AMI in the hypertensive patient improves prognosis.

It is also likely that left ventricular aneurysm after AMI is more frequent in patients with previous hypertension than in those with previous normotension. Healed anatomic left ventricular aneurysm, for example, is rarely observed at necropsy in hearts of normal weight. The left ventricle may be viewed as being capable of accommodating a systolic pressure of less than 140 mm Hg; when the pressure is higher and AMI occurs, the chances of aneurysmal formation are greater than when this pressure is lower.

Aneurysm of Aorta

Just as left ventricular aneurysm from AMI is more frequent in hypertensive than in normotensive patients, abdominal aortic aneurysm (etiology atherosclerosis)[73–78] and dissecting aortic aneurysm[79–95] (other than the type associated with the Marfan and Marfan-like syndromes) are far more frequent in the hypertensive subject. Indeed, both types of aneurysm are infrequent in patients with hearts of normal weight. The association of hypertension with aortic dissecting aneurysm has been appreciated for years, but the high frequency of hypertension in patients with abdominal fusiform aneurysms is not commonly appreciated. Burchell[90] stated nearly 20 years ago that "if Marfan's syndrome and pregnancy are excluded . . . aortic dissection is always associated with hypertension." Dissection of the ascending and transverse aorta also is far more frequent in patients with untreated coarctation of the aorta than in persons without aortic coarctation.[96] The higher the blood pressure, the greater the likelihood of aortic dissection.[90] Thus the frequency of aortic dissection in patients with accelerated or malignant hypertension is higher than in patients with "benign" (one of medicine's worse terms) hypertension.[90]

It is also likely that generalized aortic dilatation (senile ectasia) is more frequent in the hypertensive than in the normotensive subject. The numbers of elastic fibers in the aorta normally decrease with age, with resultant weakening of its wall.[97] Consequently, the higher the intraluminal pressure, the more the aortic expansion. Precordial diastolic, basal, blowing murmurs in the elderly person may result from this aortic dilatation, and these murmurs are more frequent in the hypertensive than in the normotensive person.[98]

Atherothrombotic Obstruction of the Abdominal Aorta or of Its Branches

The syndrome of thrombotic obliteration of the aortic bifurcation, described by Leriche in 1940,[99] is often associated with hypertension.[100–104] Hypertension was present in each of the 4 patients he described in 1948,[102] in 7 of 11 patients described by Kekwick and associates,[103] and in 9 of 13 patients described by Gunning and associates.[104] Atherosclerotic obstruction of arteries in the legs (arteriosclerosis obliterans) with intermittent claudication also is more frequent in the hypertensive than in the normotensive person.[105–110] Information regarding the weight of the heart in a large series of patients with obstructed abdominal aortas or more peripheral arteries is unavailable, but it seems highly likely that most patients with this condition also have cardiomegaly at necropsy.

Cerebrovascular Accidents

Atherothrombotic (nonembolic) cerebral infarction. Following age, hypertension is the most common and potent precursor of atherothrombotic cerebral infarction (ACI), which accounts for 60% to 80% of all strokes.[111–126] Its contribution is direct.[118] Asymptomatic, casual hypertension is associated with risk of ACI about four times as great as that in normotensives.[118] The frequency of hypertension in patients with ACI has been reported to be from 50% to 75%.[111,112,114,116–119] Heart weight is increased in the majority of patients with fatal ACI, and a finding of left ventricular hypertrophy on electrocardiogram increases the risk of cerebral infarction ninefold (only a threefold increase, however, after adjustment for coexisting hypertension).[120] Of patients with hypertensive heart disease, one-half will have one or more cerebral infarcts at necropsy.[114,126] As with AMI, the presence of hypertension increases the mortality rate at the time of the ACI.[112,117,118,120] Furthermore, although a person surviving a cerebral infarction more than a month is more likely to die of cardiac disease than of a subsequent stroke,[121,125] the frequency of a new ACI after recovery from the first one is greater in the hypertensive than in the normotensive person.[117] Treatment of hypertension clearly diminishes the frequency of ACI.[27–30] The frequency of hypertension and of electrocardiographic evidence of left ventricular hypertrophy in patients with transient ischemic attacks,* a frequent precursor of ACI, is probably similar to that in patients with ACI alone.[120,127–129]

Primary intracerebral hemorrhage. Primary intracerebral hemorrhage is defined as significant hemorrhage into the brain parenchyma, unassociated with trauma, arterial aneurysms, inflammatory vascular lesions, arteriovenous or angiomatous formations, neoplasms, or blood dyscrasias. Primary intracerebral hemorrhage almost always is preceded by hypertension,[118] and some authors define primary intracerebral hemorrhage as a disease exclusively of hypertensives. Millikan and associates[130] include the presence of systemic hypertension as one of their four criteria for the clinical diagnosis of nontraumatic intracerebral hemorrhage. The hearts of patients with primary intracerebral hemorrhage are nearly always increased in weight. Of 347 patients (193 males, 154 females) with primary intracerebral hemorrhage, 89% at necropsy had left ventricular hypertrophy: the hearts in 87% of the women weighed more than 400 g, and the hearts in 91% of the men weighed more than 420 g.[131]

Lacunar softenings. Lacunar strokes or infarcts are small infarcts, usually multiple and cystic, of the basal ganglia or pons resulting from occlusion of penetrating branches chiefly from the middle or posterior cerebral or basal arteries.[132,134] At their peak they cause a relatively minor neurologic deficit, which, although it appears submaximal, actually represents the total maximal stroke and not the initial stage of a devastating paralysis. Prognosis is good. Four lacunar strokes have been identified:[133] (1) pure motor hemiplegia and hemiparesis; (2) pure sensory stroke; (3) a syndrome in which cerebellar ataxia is combined with weakness of the corresponding leg; (4) dysarthria and clumsiness of one hand. Systemic hypertension nearly always is a precursor of the lucunar softening, the most common cause of stroke in persons over 60 years of age.[132–134]

*Transient (10–30 min) disturbances of neurologic function of 24-hr duration occurring in the territory of supply of the carotid or verterebrobasilar arteries and believed on clinical grounds to be associated with atherosclerotic disease in those arteries or in their branches.[127]

Charcot-Bouchard aneurysm. Miliary aneurysms of very small cerebral arteries were described in 1868 by Charcot and Bouchard in 60 elderly patients dying of cerebral hemorrhage.[135] These aneurysms occur on small perforating arteries of less than 1-mm diameter, especially in basal ganglia and subcortical regions. The media disappears at the neck of the aneurysm, whose walls consist of intima and adventitia. Cole and Yates[136] found these intracerebral microaneurysms at necropsy in 46% of 100 hypertensive patients and in only 7% of 100 normotensive patients matched for age and sex. Massive intracerebral hemorrhage occurred in 20 of the 100 hypertensive patients, 18 of whom had microaneurysms. In contrast, massive intracerebral hemorrhage occurred in only 1 of the 100 normotensive patients, and no microaneurysm was found in that patient. The microaneurysms are age-dependent: they occurred in 71% of the hypertensive patients age 65 to 69 years and in only 10% of the hypertensive patients less than 50 years of age. All 18 patients with microaneurysms and normotension were older than 65 years.

It is likely that these microaneurysms are responsible for most primary intracerebral hemorrhages in patients over 40 years of age with hypertension.[135–137] These aneurysms clearly are unrelated to atherosclerosis, and they provide the explanation for the much closer relationship between cerebral hemorrhage and hypertension than that between cerebral infarction and hypertension. The latter obviously is the result of atherosclerosis and is more nearly analogous to myocardial infarction. Cerebral hemorrhage involves small cerebral arteries, the medial weakening leading to aneurysm and rupture; therefore it is more nearly analogous to dissecting aortic aneurysm.

Renal Failure

Renal disease is both a cause and an effect of hypertension. Hypertension can be produced by parenchymal renal diseases (glomerulonephritis, pyelonephritis, tumor, cysts) and by renal arterial and renal ostial disease, the most common being atherosclerosis. Although the pathologic features of the parenchymal renal diseases are often distinctive, their separation from nephrosclerosis, the most common renal consequence of hypertension, is often not easy. Thus, determining whether renal disease is the cause or the consequence of hypertension may be difficult. Classically, in nephrosclerosis the kidneys are involved uniformly, the weights are usually normal, their surfaces are diffusely finely granular, and histologically some tubules are atrophied; the interstitium is focally fibrotic, and the lumens of the vascular channels, particularly the arterioles, are narrowed.[138] There appears to be some relationship between the severity of the hypertension, especially the diastolic pressure, and the arterial and arteriolar damage. In patients with "benign" essential (primary) hypertension, about 10% have renal failure, and then relatively late. In the accelerated or malignant phase of hypertension, the renal arteriolar changes appear soon after the striking elevation of pressure, and the resulting damage to the kidneys is usually severe. The vascular lesions are accompanied by severe parenchymal damage, including hemorrhagic infarction of glomeruli, tubular degeneration, and cellular infiltration of the interstitium. In contrast to patients with mild essential hypertension, those with accelerated hypertension frequently present initially with azotemia, and if untreated, they usually die within a year with evidence of advanced renal failure.

It is difficult to arrive at a precise figure regarding the percentage of patients with parenchymal renal disease who have hypertension. If the disease is one primarily of glomeruli (glomerulonephritis), the figure is high. If it is one involving primarily the tubules and interstitium (pyelonephritis), the figure is low. If both kidneys are involved uniformly, this percentage tends to be lower than when only one kidney is involved but that one severely. It is clear, however, that nearly all patients dying in chronic renal failure have hypertension and left ventricular hypertrophy.

Miscellaneous Consequences of Hypertension

Calcific deposits. Calcific deposits in the aortic valve and mitral anulus are extremely common in elderly patients, and in these patients they have been attributed to the consequences of wear and tear.[139,140] Their frequency is higher in patients with hypertension than in those with normal pressures.[140] Furthermore, hypertension tends to cause these calcific deposits to occur at younger ages than is the case in normotensive persons.[140] Hypertension greatly increases the closing pressures on both mitral and aortic valves, and thus the amount of resulting wear and tear. The calcific deposits are located on the aortic aspects of the aortic valve cusps, and if extensive enough they may produce stenosis from the immobility imparted to the cusps by the heavy calcific deposits.[139] Similar calcific deposits are located on the undersurface of the posterior mitral leaflet, and if extensive enough they can cause the mitral valve to be incompetent[140] and rarely stenotic.[141]

Not only does it increase or accelerate the development of calcific deposits in the aortic valve and mitral anulus, hypertension also is associated with a marked increased frequency of calcific deposits in the coronary arteries. Nearly 90% of patients with hypertension studied by Frink and associates[142] showed calcific deposits in their coronary arteries by roentgenographic examination of the heart at necropsy, whereas less than 50% of those with normal pressures had coronary calcific deposits.

Although usually asymptomatic (indeed, some physicians believe that hypertensive persons may feel better than normotensive persons—supernormal), headaches, pounding precordia and arterial pulses, local bleeding (nosebleed), increased risk from operations and other trauma (because of more rapid blood loss), increased rubor of skin (a cosmetic irritant), etc., may be consequences of elevated blood pressure. Because of their tendency to raise the blood pressure acutely, certain types of exercises [calisthenics (push-ups) or weightlifting], and sexual intercourse may be more dangerous in the hypertensive person.

Systemic hypertension. Systemic hypertension acts as a major risk factor to development of cardiovascular disease in two ways: (1) by increasing the deposition of atherosclerotic plaques (intimal lesions) in major arteries most commonly causing luminal narrowing with resulting organ ischemia or infarction or both; (2) by weakening the media of certain arteries causing aneurysms that may or may not rupture. Its effect on the arterial media is direct, whereas its effect on the arterial intima is indirect. Hypertension is the only known major underlying factor in two conditions: intracerebral microaneurysm (with or without rupture) and dissecting aortic aneurysm (excluding patients with the Marfan and Marfan-like syndrome),[135] both of which are associated with medial weakening or disruption. Reduction in blood pressure clearly leads to a reduction in the incidence of nontraumatic intracerebral hemorrhage and of dissecting aortic aneurysm.[28–30] Thus each of these conditions is virtually preventable simply by maintaining a normal blood pressure.

The more common manifestation of hypertension, however, is its ability to increase the amount of atherosclerotic plaquing in various arteries. This effect, however, is not direct, because hypertension in population groups with normal (< 180 mg/100 ml) serum cholesterol levels is not accompanied by more or larger atherosclerotic plaques than in the normotensive persons in those populations. Hypertension, for example, is by far the most common cardiovascular disease in the black African.[143] Yet complicated atherosclerotic plaques (as opposed to fatty and fibrous plaques), the only ones capable of producing significant luminal narrowing, have not been reported in either the hypertensive or the normotensive black African. Experimentally, however, lipid-induced atherosclerosis clearly is accelerated by hypertension.[144–146] Furthermore, certain complications of atherosclerosis—angina pectoris, myocardial infarction, sudden coronary death, atherothrombotic cerebral infarction, intermittent claudication—clearly are more frequent in hypertensive persons in the affluent or developed countries than in normotensive persons residing in the same countries.[40–43,107,118,120] Thus hypertension accelerates or increases atherosclerosis only in hypercholesterolemic (serum cholesterol > 200 mg/100 ml) population groups and not in those with normal serum cholesterol levels.

Other human examples of hypertension's increasing atherosclerosis are provided by patients with coarctation of the aorta and by those with pulmonary hypertension. Among patients with aortic isthmic coarctation, atherosclerotic plaques are always more frequent and more extensive proximal to the narrowing than distal to it. Pulmonary atherosclerotic plaques develop in young individuals with severely elevated pulmonary arterial pressures, whereas plaques in these arteries are virtually never observed in young individuals with normal pulmonary arterial pressures. Indeed, pulmonary hypertension may accelerate the development of atherosclerosis in the pulmonary circulation to a greater extent than systemic hypertension accelerates the development of atherosclerosis in the systemic circulation. Among patients with balanced pulmonary and systemic systolic pressures, for example, the amount of atherosclerosis is usually considerably more extensive in the pulmonary circuit than in the systemic circuit.

Although hypercholesterolemia (serum cholesterol > 200 mg/100 ml) may be necessary for hypertension to accelerate atherosclerosis, a certain level of blood pressure is required for atherosclerosis to develop, no matter what the level of serum cholesterol. The serum cholesterol levels are similar in both venous and arterial blood in the same person. Yet atherosclerotic plaques never develop in the veins no matter what the serum cholesterol level.

The mechanism by which hypertension accelerates atherosclerosis is uncertain. Possibly the higher pressure increases intimal damage, which provides a nidus for thrombus or lipid deposition with subsequent plaque formation. Likewise, the mechanism by which hypertension causes medial weakening is unknown. Although the media of aorta normally contain acid mucopolysaccharide (AMP) material, the amount of AMP (and cystic medial necrosis) in the aortic media is increased in persons with systemic hypertension as compared to normotensive subjects of similar age.[147] This small quantitative increase in AMP, however, does not provide, in my view, an adequate explanation for the occurrence of dissecting aortic aneurysm. Indeed, the amount of AMP is not increased at sites of aortic rupture as compared to nonruptured sites. The mechanism of cerebral aneurysm formation is not understood either. Possibly, focal thinning of the media is a common occurrence at birth, and the thinned media thereafter progressively bulge simply as a response to the elevated intraarterial pressure. Certainly there is no increased amount of AMP in the cerebral arteries, which are purely muscular arteries. The presence of aortic medial dissection and of cerebral microaneurysm is unrelated to the occurrence of atherosclerosis. Indeed, the greater the amount of atherosclerosis, the less the likelihood of aortic dissection or cerebral aneurysm rupture.

Although recognized for some time as a major risk factor for development of ischemic heart disease, systemic hypertension is the only risk factor (other than advanced age) that increases mortality during an acute myocardial infarction and during the first 6 years after recovery from an acute myocardial infarction.[148–151] The explanation for the increased mortality during AMI is uncertain, but it probably is the increased frequency of rupture of left ventricular free wall or ventricular septum or papillary muscle, or the increased propensity to develop left ventricular aneurysm, or both, in the hypertensive as compared to the normotensive person. The explanation for the poorer long-term prognosis after recovery from an AMI in the hypertensive is less certain. It is likely that the hypertensive patient develops a larger area of myocardial necrosis at the time of AMI than does the normotensive patient and thus is left with a larger myocardial scar. The larger the amount of scarring the greater the likelihood of persistent left ventricular failure, and congestive heart failure is considerably more frequent in the hypertensive than in the normotensive patient.[152] Kannel and associates[152] considered hypertension a causative factor in 75% of their patients with congestive cardiac failure. The incidence of congestive cardiac failure increased in both their female and male patients at all ages with increasing systolic or diastolic pressure. Even in the absence of overt congestive cardiac failure the function of the hypertrophied, hypertensive left ventricle is often impaired, with a decreased ejection rate and prolonged tension–time index.[153]

VALVULAR HEART DISEASE

Basic Concepts

Cardiac valve disease is present when there is structural alteration of a valve or when a valve functions abnormally, whether or not its structure is normal or abnormal (Fig. 1-5). A valve may be anatomically abnormal and function normally, it may function normally when anatomically abnormal, and it may be both structurally and functionally abnormal. A structurally abnormal valve is usually a thickened valve.

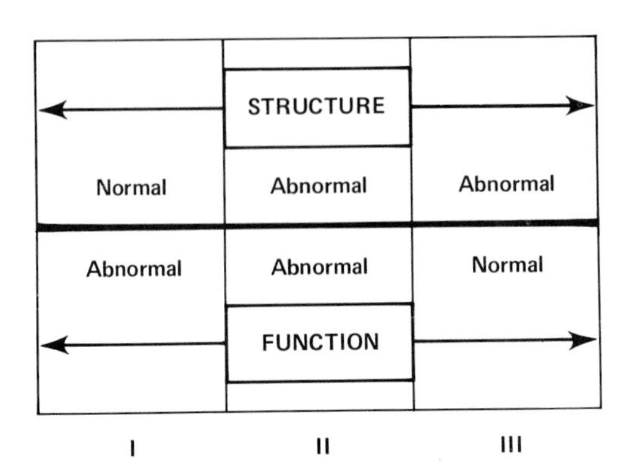

Fig. 1-5. Diagram showing the possible structural-functional relationships in valvular cardiac disease.

The thickening most commonly is the result of fibrous tissue proliferation with or without calcific deposits. The distribution of these structural abnormalities is variable, but knowledge of this distribution (that is, whether or not it is diffuse or focal) is helpful in determining the cause of the valvular abnormality.

The two types of valvular dysfunction are stenosis and regurgitation. Stenosis of a valvular orifice always indicates anatomic abnormality of the valve. Most stenotic valves also are incompetent, but any degree of stenosis indicates that the valve is thickened (by fibrous tissue with or without calcific deposits) and in addition that the thickening is diffuse. The purely incompetent or regurgitant valve may or may not be anatomically abnormal; if it is anatomically abnormal the lesions causing the incompetency may be focal rather than diffuse, whereas all stenotic valves are diffusely involved. Calcific deposits are absent or minimal in purely incompetent valves, whereas they are nearly always present and often heavy in stenotic valves. Thus it is important to establish whether or not a valve has some degree of stenosis or whether it is purely incompetent. A valve may have some degree of stenosis and still be predominantly incompetent; likewise it may be predominantly stenotic and have some degree of regurgitation. Knowing which lesion is predominant is helpful in explaining various clinical and laboratory findings and in determining proper therapy, but it is less helpful in determining the etiology or the structure of the valvular abnormality. If it can be established, for example, that a cardiac valve is stenotic (to any degree), then the number of causes for the valve dysfunction are, for practical purposes, only three, namely rheumatic, congenital, and degenerative (Table 1-2). If, however, a dysfunctioning valve can be demonstrated to be purely incompetent (no element of stenosis), then the possible etiologies are considerably increased (Table 1-2).

Table 1-2
Etiology of Severe Valvular
Heart Disease

I. STENOSIS (with or without regurgitation)
- A. Rheumatic
- B. Congenital
- C. Degenerative (A)

II. Pure REGURGITATION (No stenosis at all)
- A. Myocardial dysfunction (M +T)
- B. Floppy valve (M +T)
- C. Infective endocarditis
- D. Rheumatic
- E. Congenital
- F. Syphilis (A)
- G. Marfan and Marfanoid (A)
- H. Ankylosing spondylitis (A)
- I. Trauma
- J. Dissecting aneurysm (A)

Of the four cardiac valves, the mitral most frequently functions abnormally, but the aortic is most frequently anatomically abnormal. Any condition causing dilatation of the left ventricle can cause mitral regurgitation. Thus patients with congestive cardiac failure from ischemic or hypertensive heart diseases or cardiomyopathy often have mitral regurgitation even though the mitral leaflets and chordae tendineae are normal. Likewise, conditions causing dilatation of the right ventricle frequently lead to tricuspid regurgitation. Indeed, most patients with tricuspid regurgitation have normal tricuspid valve leaflets and chordae tendineae. The usual cause of tricuspid regurgitation in patients with dilated right ventricles is dilatation of the tricupsid valve anulus (the tricupsid valve is the only valve with a true, complete, 360° anulus). In contrast, papillary muscle malposition rather than mitral anular dilatation is the most common cause of mitral regurgitation in patients with dilated left ventricles and anatomically normal mitral leaflets and chordae tendineae.[154–156] Aortic or pulmonic regurgitation does not occur as a result of dilatation of the ventricular cavities. Dilatation of the aorta or of the pulmonary trunk, however, may be associated with aortic or pulmonic regurgitation; but these functional lesions are far less common than either functional mitral or tricuspid regurgitation, and the degree of semilunar valve regurgitation in this circumstance is always minimal or mild.

Just as anatomic changes in a cardiac valve may vary from minimal to severe, the degree of valvular dysfunction may also be mild, moderate, or severe. Milder forms of valvular dysfunction are more frequent than the more severe forms. Minimal or mild regurgitation of any valve is often functional in nature, i.e., the valve structure itself is anatomically normal. Even minimal or mild valvular stenosis, however, is indicative of diffuse structural alterations of the valve itself. Mild forms of valvular dysfunction usually are not associated with symptoms of cardiac dysfunction. Severe valvular dysfunction may or may not be associated with clinical evidence of cardiac dysfunction. A valve may be made purely incompetent acutely (by infective endocarditis or trauma), or more commonly the incompetence gradually progresses over many years. A patient may be well compensated despite severe regurgitation over many of those years. The sudden development of valvular destruction (by infective endocarditis or trauma, for example) is nearly always associated with the acute appearance of signs and symptoms of cardiac dysfunction.

Obstruction of a valve orifice, in contrast to pure incompetency, nearly always takes years to develop. Whether or not symptoms of cardiac dysfunction are present over many years or for only a few months in patients with valvular stenosis is dependent primarily on which valve is involved. Persons with tricuspid or mitral stenosis usually have symptoms of cardiac dysfunction for many years, whereas symptoms in patients with pulmonic or aortic stenosis do not appear until late.

The frequencies and types of functional valve lesions severe enough to be fatal and due to structural changes in the valve itself are listed in Table 1-3. Cases of papillary muscle dysfunction were excluded. The etiology of the valve condition when more than one valve is involved is usually rheumatic. Thus the etiology with rare exception in groups 3, 6, 7, 8, and 9 is rheumatic (Table 1-3). Clinically isolated mitral stenosis is rheumatic in origin in nearly 100% of patients, but in only about 10% of patients is clinically isolated aortic stenosis rheumatic in origin.[157] In about 50% of patients with pure mitral regurgitation and in about 10% of patients with pure aortic regurgitation, the origin is rheumatic.

Stenotic lesions are more frequent than purely incompetent lesions. In 703 valves functioning abnormally due to structural changes in the valve itself in 544 patients studied by the author, the valve lesion in 70% was stenosis, and in the other 30% the lesion was pure regurgitation.[158,159] Although multivalvular disease is common, univalvular disease is more common. Of 544 patients with fatal valvular heart disease, 392 (72%) had only one valve functioning abnormally and 152 (28%) had more than one valve functioning abnormally. Of 392 patients with univalvular disease, 71% had stenotic lesions (with or without regurgitation) and 29% had purely regurgitant lesions. Of the 152 patients with multivalvular disease, 53% had only stenotic lesions, 15% had only purely regurgitant lesions, and 32% had mixed lesions (that is, one valve was stenotic and another was purely incompetent). Thus in two-thirds of

Table 1-3
Frequencies of Valvular Lesions Causing Fatal Cardiac Dysfunction*
(data in patients $>$ age 14 years)

	Functional Valve Lesion(s)†	% of Patients
1.	Aortic stenosis (AS)	34
2.	Mitral stenosis (MS)	17
3.	AS + MS	12
4.	Pure aortic regurgitation (AR)	11
5.	Pure mitral regurgitation (MR)	10
6.	MS + AR	6
7.	MR + AR	5
8.	AS + MR	3
9.	Tricuspid stenosis + MS ± AS	2
	Total	100%

*Based on necropsy study of 543 cases of severe valvular heart disease (157).
†Any of the 9 lesions may have associated tricuspid regurgitation

subjects in whom more than one valve functioned abnormally, either all involved valves were stenotic or all involved valves were purely incompetent.

The alterations in size of the cardiac chambers and the great arteries in the various valvular lesions are shown in Table 1-4. Functional

Table 1-4
Size of Cardiac Chambers and Great Arteries in Valvular Heart Disease

	LA cavity	LA wall	LV cavity	LV wall	Asc. Aorta	PT	RV cavity	RV wall	RA cavity
1. Mitral stenosis (MS)	↑↑	↑↑	nl	nl	nl	↑	↑	↑↑	↑
2. Aortic stenosis (AS)	↑	↑↑	nl	↑↑	↑	nl	nl	↑	nl
3. MS + AS	↑	↑↑	nl	↑↑	nl	↑	nl	↑	↑
4. Mitral regurgitation (MR)	↑↑	↑	↑	↑	nl	↑	↑	↑↑	↑
5. Aortic regurgitation (AR)	↑	↑	↑↑	↑	↑↑	nl	↑	↑	nl
6. MR + AR	↑↑	↑	↑↑	↑	nl	↑	↑	↑	↑
7. MS + AR	↑↑	↑	↑	↑	nl	↑	↑	↑	↑
8. AS + MR	↑↑	nl	↑	↑↑	nl	↑	↑	↑	↑
9. Tricuspid stenosis	nl	nl	nl	nl	nl	nl	nl	nl	↑
10. TS + MS + AS	↑	↑	nl	↑	nl	↑	↑	↑	↑

and anatomic features of the various valvular lesions and considerations of their causes are discussed below.

Aortic Stenosis

Valvular aortic stenosis (AS) is the most common fatal cardiac valvular lesion (Table 1-3). Its cause, when associated with anatomic mitral disease, is usually rheumatic. The cause of anatomically isolated AS, with or without aortic regurgitation (AR), is always nonrheumatic.[157]

Four factors indicate that anatomically isolated aortic valve disease—either pure AS (rare), AS plus AR, or pure AR—is nonrheumatic in etiology: (1) The low frequency (about 10%) of a history of acute rheumatic fever or chorea. In contrast, positive histories are present in about 70% of patients with mitral stenosis and in those with multivalve disease. (2) The absence of Aschoff bodies. To my knowledge, no one has observed these structures in the heart of a patient with anatomically isolated aortic valvular disease. These lesions, however, which are pathognomonic of rheumatic heart disease, were observed in 5% of patients with mitral stenosis or multivalvular heart disease. (3) The normal atrial walls histologically. Rheumatic heart disease affects myocardium, particularly atrial myocardium, in addition to affecting at least the mitral valve. The consequences of this involvement are degeneration of atrial musculature and interstitial myocardial fibrosis. The latter is the usual in patients with rheumatic mitral valvular disease, but it does not occur in patients with anatomically isolated aortic valve disease. (4) The frequency of an underlying congenital malformation of the aortic valve. In patients with AS combined with mitral stenosis or regurgitation, the aortic valve practically always consists of three cusps. In patients with anatomically isolated aortic valve disease causing stenosis, the aortic valve is often bicuspid or unicuspid rather than tricuspid.

The structure of the stenotic aortic valve varies to a great extent with the age of the patient (Table 1-5). In subjects less than 15 years old

Table 1-5
Configuration of Aortic Valve in Isolated Aortic Stenosis in Relation to Age

	No. valve cusps	<age 15 years	Age 15-65	>age 65 years
1	1	60%	10%	0
2	2	20%	60%	10%
3	3	15%	25%	90%
4	Uncertain	5%	5%	0

the stenotic aortic valve is usually either unicuspid or bicuspid. In the age group 15 to 65 years, about 60% of the stenotic valves are either bicuspid or unicuspid and the rest are tricuspid. In the age group over 65 years, over 90% of the stenotic aortic valves are tricuspid.

At least two kinds of unicuspid aortic valves exist[158] (Fig. 1-6). In one, the orifice is central and there are no lateral attachments to the wall of the aorta (acommissural valve). This type of valve is often seen in "pure" pulmonic stenosis. Often three remnants of commissures are observed on these valves, however. A second type of unicuspid valve, the unicommissural, has an eccentric orifice.[160] Remnants of undeveloped commissures also are present on this type of valve.[160]

The congenitally bicuspid aortic valve is basically of two types (Fig. 1-7). In one type the cusps are located right and left, the commissures anteriorly and posteriorly; a raphe or false commissure, if present, is always in the right cusp, and a coronary artery arises from behind each cusp. In the second type the cusps are located anteriorly and posteriorly, the commissures right and left; a raphe, if present, is always in the anterior cusp, and both coronary arteries arise in front of the anterior cusp.[161]

Function of the malformed aortic valve at birth is variable. The fewer the number of cusps and commissures, the greater the likelihood that the valve is stenotic at birth.[162] The acommissural and unicommissural valves are almost surely stenotic at birth, but the degree of stenosis increases with age.[160]

The congenitally bicuspid valve is the most frequent major congenital malformation of the heart, occurring in possibly as many as 2% of births.[161] The percentage of congenitally bicuspid valves that develop complications, (namely stenosis, pure incompetency, and infection) is

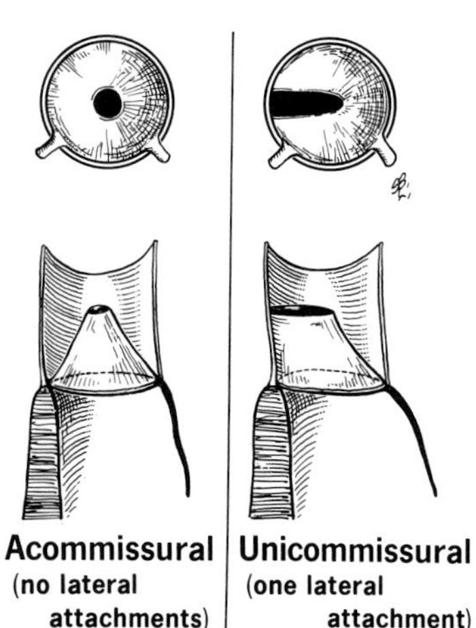

Fig. 1-6. Diagram illustrating the two types of unicuspid aortic valves. In the acommissural variety the orifice is centrally located and there are no lateral attachments. In the unicommissural variety the orifice is eccentrically located and there is only one lateral attachment to the wall of aorta.

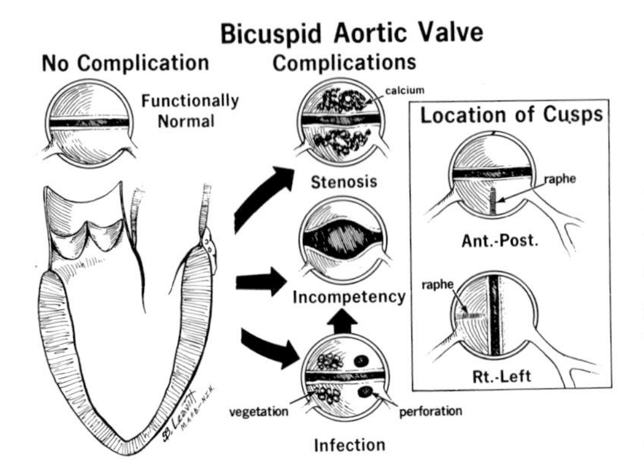

Fig. 1-7. Diagram illustrating the three major complications of the congenitally bicuspid aortic valve. The inset shows the two basic locations of the cusps in the bicuspid condition.

uncertain, but is probably more than 50%. The remainder may function more or less normally during an entire lifetime.

Among the complications of the bicuspid condition of the aortic valve, stenosis is by far the most frequent. Pure regurgitation is next, and the latter lesion is most frequently the result of the valve's being the site of infective endocarditis. The reason one congenitally bicuspid aortic valve becomes stenotic, another incompetent, and another the site of infective endocarditis is unknown. It is clear, however, that stenosis, regurgitation, and infection are complications of the bicuspid condition, and none are present at birth. The bicuspid valve becomes stenotic only as its cusps become fibrotic and calcified; neither complication exists at birth.

The mechanism by which a congenitally bicuspid aortic valve becomes stenotic or incompetent is uncertain. The explanation advanced by Edwards[163] is attractive. He proposed that it is not mechanically possible for a congenitally bicuspid aortic valve to open and close properly. The distances between lateral attachments of normal aortic valvular cusps along their free margins are curved lines. The extra length allows the cusps to move freely during opening and closing of the valve. In contrast, the distances between lateral attachments of congenitally bicuspid aortic valves along their free margins approach straight lines. If these distances were exactly straight lines the valves could not open during ventricular systole. Consequently, one cusp is larger than the other. The excessive length of one or both cusps of a congenitally bicuspid aortic valve produces abnormal contact between the cusps. This abnormal contact in turn causes focal fibrous thickening that with time becomes diffuse, and dystrophic calcification thereafter occurs. Thus stenosis of a congenitally bicuspid aortic valve may be the result of trauma to these cusps produced by their abnormal contact with each other. Although this explanation is appealing, it does not explain why one

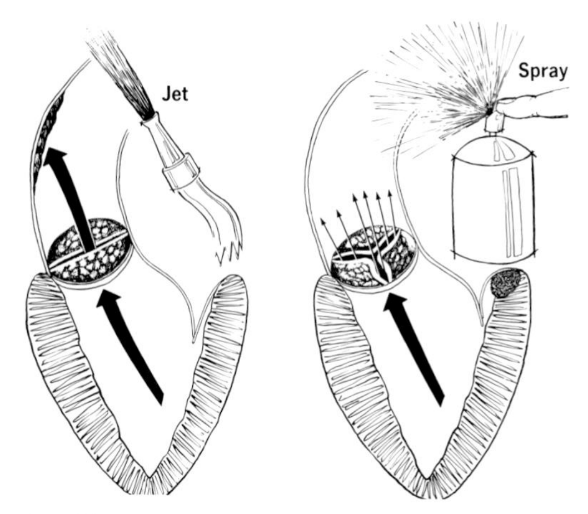

Fig. 1-8. Diagram illustrating the auscultatory differences between fixed orifice (left) stenosis of patients less than age 65 years and nonfixed orifice valvular aortic stenosis in patients over age 65 years. A jet lesion appears to occur when severe fixed stenosis is present, but is not apparent when there is a spray effect in nonfixed stenosis. The latter is also contributed to by the lack of commissural fusion. Calcium beneath the posterior mitral leaflet (anulus) is common (75%) in patients over age 65 with valvular aortic stenosis.

congenitally bicuspid aortic valve becomes stenotic, another becomes entirely incompetent, and another remains free of complications.

Stenosis of the aortic valve in patients over 65 years of age is characteristic[139] (Fig. 1-8). The valves are usually (90%) tricuspid, nodular calcific deposits are rather uniformly distributed on the aortic aspects of each cusp, and the commissures generally are not fused. Obstruction is due almost entirely to the presence of large calcific deposits that prevent the cusps from retracting adequately during ventricular systole. Because the commissures are usually not fused, associated aortic regurgitation is infrequent.

Clinically, aortic stenosis in the elderly is characterized by several features that are either absent or less prominent in younger subjects with this valvular lesion (Fig. 1-8): The systemic arterial pulse pressure may be increased due to systolic hypertension. A harsh right basal systolic murmur may be relatively inconspicuous, while a pure musical systolic murmur may be prominent at the apex. A coexisting apical systolic murmur of mitral regurgitation may be related to mitral anular calcification and not left ventricular failure. Systemic systolic hypertension can result in a comparatively small pressure gradient despite severe left ventricular outflow obstruction.

The systolic murmur of valvular aortic stenosis in patients less than 65 years old is classically rough, harsh, grunting, and loudest at the right base, with radiation upward and to the right. In patients over 65 years old with aortic stenosis, the harsh murmur at the right base may persist but be less intense, and auscultation may be dominated by a loud pure-frequency musical systolic murmur maximal at or near the cardiac apex. At least three reasons account for these differences. First, older persons tend to have increased upper thoracic chest dimensions, which soften basal murmurs. Second, the morphology of the stenotic aortic valve is different. A rigid, calcified valve with fused commissures generally provides a suitable mechanism for a high-velocity jet into the ascending aorta; the noisy murmur at the right base appears to originate within the ascending aorta because of turbulence caused by the high-velocity jet. A calcified, stenotic, trileaflet aortic valve without fused commissures, the most common type in the elderly, may produce a spray rather than a jet, so that the harsh murmur at the right base may be comparatively inconspicuous. Because the cusps are not fused, they may vibrate during ventricular ejection, and the latter may account for the pure-frequency, musical cooing murmur recordable within the left ventricle. Third, calcification of the mitral anulus, itself a producer of precordial murmurs, in elderly patients frequently accompanies calcification of the aortic valve. Mitral regurgitation can result from inability of the calcified anulus to decrease its circumference during ventricular contraction.[140] Accordingly, the holosystolic murmur of mitral regurgitation may be present at the apex and may be difficult to distinguish from the coexisting pure-frequency aortic stenotic apical murmur. Although mitral anular calcification tends to complicate the auscultatory diagnosis of aortic stenosis, the radiologic identification of mitral calcific deposits implies that coexisting aortic stenosis is likely to be caused by calcific deposits on a trileaflet valve without fused commissures.

The degree of left ventricular systolic hypertension in elderly patients may be considerably higher than that occurring in younger subjects with similar degrees of aortic stenosis, because the peripheral systolic systemic pressure usually rises with age. The degree of systolic hypertension that the elderly left ventricle is capable of achieving can be astonishingly high, particularly because this chamber seldom maintains a systolic pressure in excess of 250 mm Hg, even in younger subjects. Systemic hypertension may reduce the aortic gradient, and the degree of obstruction may be underestimated in the elderly patient if the peak systolic pressure difference is taken as the only index of severity. As the gradient diminishes in the presence of systemic systolic hypertension, the murmur at the right base may soften and shorten. Even if this auscultatory change does not occur with systemic hypertension, the modification of the systemic atrial pulse may be misleading, because the rate of rise can be relatively normal and the pulse pressure increased despite severe obstruction to left ventricular outflow.

Another important clinical feature in elderly patients with aortic stenosis is the presence of atrial fibrillation in about one-third of them.[139] This arrhythmia, however, is common in elderly persons without valvular cardiac disease. The left atrial cavities are mildly dilated in patients with aortic stenosis, but left atrial dilatation commonly accompanies old age in the absence of functionally significant valvular lesions. The left ventricular wall becomes less compliant with age, and the superimposed left ventricular hypertrophy (without left ventricular dilatation) of aortic stenosis may further impede left atrial emptying. The only common anatomic feature in elderly patients with atrial fibrillation is left atrial dilatation.[164] The combination of atrial fibrillation, a long apical systolic murmur, and left atrial dilatation can obscure the diagnosis of aortic stenosis and lead to a mistaken diagnosis of isolated mitral regurgitation. Atrial fibrillation is especially undesirable in aortic stenosis because the left ventricle requires the help of augmented left atrial contraction, which increases end-diastolic fiber length and permits the left ventricle to contract with greater force.

The cause of the aortic stenosis in elderly patients is uncertain, but simple wear and tear or degeneration appears to be the most reasonable explanation. The presence in the same hearts of other wear-and-tear lesions such as calcified mitral anuli, calcified coronary arteries, and focal thickening of other valve leaflets supports this contention. Certainly there is nothing to support a rheumatic etiology.

Although relatively few patients with isolated valvular aortic stenosis (with or without regurgitation) have positive histories of acute rheumatic fever, the occurrence of a positive history in a patient presenting clinically with apparently pure aortic stenosis (with or without regurgitation) should alert the examining physician to the possibility of accompanying mitral valvular disease.[159] Usually in this situation the mitral leaflets are diffusely, although mildly, thickened; but mitral valvular function is normal. On rare occasion, however, mitral stenosis may be an accompanying lesion and of such severity that it is clinically silent. Mitral stenosis should be ruled out, however, before operative therapy is undertaken in any patient presenting with apparently isolated valvular aortic stenosis (with or without regurgitation) and a positive history of acute rheumatic fever.

MITRAL STENOSIS

Of the nine single or combined valve lesions (excluding tricuspid regurgitation and papillary muscle dysfunction) severe enough to be fatal, mitral stenosis (MS) ranks second (Table 1-3). Among patients with fatal MS, in 45% the mitral valve is the only valve functioning abnormally because of anatomic abnormality. In the other 55%, either the aortic or tricuspid valves or both also function abnormally because of anatomic abnormality (Table 1-3). Mitral stenosis with relatively few exceptions is rheumatic in origin.[140] Indeed, rheumatic heart disease may be viewed as a disease of the mitral valve; other valves as well as myocardium also may be involved, but anatomically the mitral valve is always involved. In rheumatic MS, both mitral leaflets are always diffusely thickened, either by fibrous tissue or calcific deposits or both. One or both commissures are usually fused, and the chordae tendineae are shortened and usually fused to some degree. The greatest obstruction to this funnel-shaped valve occurs at its apex; the primary orifice, located at the level of the anulus, is far less narrowed. Occasionally in rheumatic MS the chordae tendineae are so retracted that the leaflets appear to insert directly into the papillary muscles. When this occurs the stenosis is always severe because the interchordal spaces are virtually entirely obliterated. The amount of calcium in the leaflets of stenotic mitral valves varies considerably. The most frequent site of calcific deposition is the posteromedial commissure. Men nearly always have more calcium than women, and older patients have more than younger patients. Gross calcific deposits generally preclude mitral commissurotomy and necessitate valve replacement. Previous commissurotomy, however, does not necessarily preclude a second commissurotomy, since the first procedure does not necessarily predispose the leaflets to the deposition of calcium.

The essential hemodynamic fault caused by obstruction at the mitral orifice is the inability of the left atrium to empty normally. As a consequence, not only is the pressure in the left atrium increased, in addition the pulmonary venous pressure is increased. A number of anatomic changes in the lung result from the elevation of the pulmonary venous pressure. These include changes in the veins themselves (intimal, medial, and adventitial fibrous proliferation), arteries (medial hypertrophy, intimal fibrous proliferation), lymphatics (marked dilatation), interlobular septa (considerable thickening either by edema fluid, dilated lymphatic channels, or fibrous tissue), alveolar septa (thickening initially by dilatation of alveolar capillaries and later by fibrous tissue), alveolar sacs (accumulation of serum and erythrocytes, the latter resulting in the deposition of collections of hemosiderin-laden macrophages), and bronchi (focal calcification of the bronchial cartilages). The pulmonary changes are far more prominent in MS than in any of the other valvular lesions. Many of the above-cited pulmonary changes are visible on roentgenogram.

Tricuspid Stenosis

Other than congenital and carcinoid etiologies, both of which are rare, tricuspid stenosis (TS) is rheumatic in origin.[158] Rheumatic TS, however, in contrast to rheumatic MS, never occurs as an isolated lesion. All patients with rheumatic TS always have MS, and often AS is present as well. To my knowledge, pure mitral regurgitation has never coexisted with rheumatic TS. Stenosis of the tricuspid valve, just like stenosis of any of the other cardiac valves, indicates diffuse thickening of the leaflets. The thickening virtually always is the result entirely of fibrous tissue proliferation; calcific deposits are absent. The tricuspid orifice never becomes as stenotic as the mitral orifice. Although mitral regurgitation coexists in most patients with mitral stenosis, the amount of regurgitation may be slight or trace. Not so with the tricuspid valve; a stenotic tricuspid valve is nearly always considerably incompetent.

Even though MS is always present in patients with rheumatic TS, the pulmonary changes delineated previously are always strikingly less than in the patient with isolated MS. The TS serves to some extent to "protect" the lung from the consequences of pulmonary venous and arterial hypertension.

Pulmonic Stenosis

This lesion is nearly always of congenital origin, and when unassociated with a defect in the ventricular septum, it produces the severest degrees of right ventricular hypertrophy. The valve usually has a central orifice with a dome shape.[158] In older patients it may calcify.[165] About one-half of patients with the carcinoid syndrome develop characteristic pulmonic valve lesions usually resulting in some degree of stenosis.[166]

Mitral Regurgitation

Mitral valve competence requires coordinated function of the mitral anulus, leaflets, chordae tendineae, papillary muscles, and left ventricular wall. Abnormalities of any of these structures may lead to mitral

Table 1-6
Spectrum of Left Ventricular Papillary Muscle Dysfunction

I. RUPTURE of papillary muscle
   A. Total ("Belly" or trunk)
      1. Rapid death
      2. Survival—not described
   B. Partial ("Head")
      1. Rapid death
      2. Survival but chronic congestive cardiac failure

II. FIBROSIS or NECROSIS of papillary muscle without rupture
   A. With coronary arterial narrowing
      1. Acute myocardial infarct
      2. Healed myocardial infarct
   B. Without coronary arterial narrowing
      1. Acute
      2. Chronic
         a. Valvular, subvalvular (discrete and diffuse) and supravalvular aortic stenosis
         b. Systemic hypertension
         c. Origin of left coronary artery from pulmonary trunk
         d. Primary endocardial fibroelastosis
         e. Endomyocardial fibrosis
         f. Löffler's fibroplastic parietal endocarditis
         g. Diffuse (primary) myocardial disease
         h. Focal myocardial disease
            1) Idiopathic
            2) "Neurogenic" heart disease
            3) Infectious

III. NORMAL or nearly-normal papillary muscles but severe fibrosis or necrosis in adjacent left ventricular free wall with and without coronary arterial narrowing

regurgitation (MR), the most common functional valve disorder. In contrast to mitral stenosis, which essentially has only two major etiologies (rheumatic and congenital), there are numerous causes of MR. Of them, the most frequent is papillary muscle dysfunction, most commonly the result of severe coronary atherosclerosis.[154,167,168] The papillary muscles, however, are fibrotic or necrotic or both in many conditions unassociated with coronary arterial luminal narrowing, because they are the last structures in the heart to be perfused by arterial blood (Table 1-6).[154] Of the two left ventricular papillary muscles, the posteromedial one has the poorer blood supply, and consequently lesions are more frequent in it than in the anterolateral muscle. Although common in one or both papillary muscles, scarring or necrosis also must be present in the left ventricular wall beneath or adjacent to the bases of the papillary muscles for MR to result. In addition, dilatation of the left ventricular cavity is probably also necessary for MR to result from papillary muscle dysfunction. Indeed, dilatation of the left ventricle from any cause may be associated with MR without any necrosis or fibrosis of myocardium or dilatation of the mitral anulus. Presumably the mechanism is altered spatial relationships between the attachments of the papillary muscles to the underlying left ventricular free wall.[154,155]

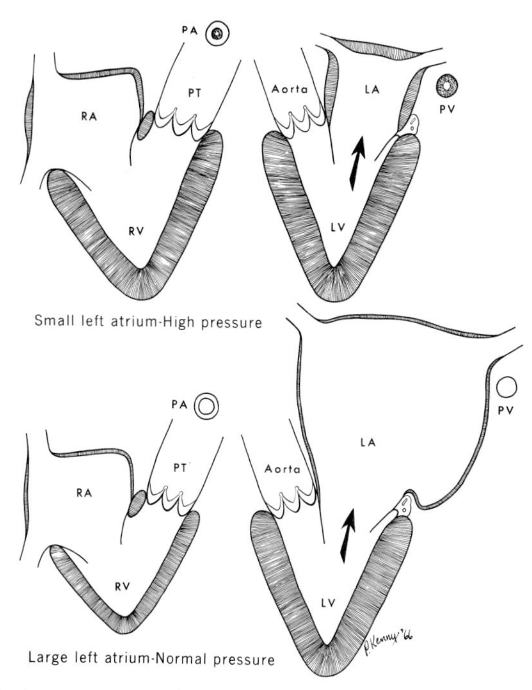

Fig. 1-9. The two extremes of the spectrum of pure mitral regurgitation. When severe mitral regurgitation appears suddenly in individuals with previously normal or near-normal hearts, the left atrium (LA) is relatively small, and the high pressure within it is reflected back into the pulmonary vessels and right ventricle (RV). The anatomic indicator of this latter physiologic event is severe hypertrophy of the left atrial and right ventricular walls and marked intimal proliferation and medial hypertrophy of the pulmonary arteries (PA), arterioles, and veins (PV). In the other extreme, the left atrial cavity is of giant size and its wall is thin. It is thus able to "absorb" the left ventricular (LV) pressure without reflecting it back into the pulmonary vessels or right ventricle. As a consequence, the pulmonary vessels remain normal, and the right ventricular wall does not thicken. PT = pulmonary trunk; RA = right atrium.

Rupture of chordae tendineae is one of the major causes of severe MR.[169] The normal chordae are the ones most likely to rupture, not the previously thickened ones.[170,171] Of the causes of chordal rupture, infective endocarditis in the most common cause of acute rupture and gives rise to the syndrome of acute severe MR associated with normal or near normal size of left atria and severe elevations of left atrial, pulmonary venous, pulmonary arterial, and right ventricular pressures (Fig. 1-9).[169,172] The left atrial myocardial fibers in these patients, as contrasted to those in patients with chronic rheumatic MR, have not been previously damaged by rheumatic fever; therefore the left atrial wall is capable of responding to the sudden elevation of left atrial pressure by vigorous contractions, leading to fiber hypertrophy. The left ventricular systolic pressure is reflected quickly to the vasculature of the lungs, and severe medial hypertrophy and intimal proliferation of the pulmonary arteries and veins progress rapidly.[169] The already diseased left atrium, in contrast, responds to volume overload by dilating, not contracting. Thus the smaller the left atrial cavity in pure MR, the thicker the right ventricular wall; conversely, the larger the left atrial cavity, the thinner the right ventricular wall.

Although it is the most common cause of acute rupture of mitral chordae, infective endocarditis may not be the most common cause of chronic MR associated with rupture of chordae. Of 42 patients with ruptured mitral chordae unassociated with an active process, only 14 had had histories of infective endocarditis. So-called floppy valves were common in these patients.

Of the leaflet abnormalities associated with MR, the most common is the floppy valve.[158] The common denominator of the floppy valve syndrome is an abnormal prolapse of one or both mitral valve leaflets toward left atrium during ventricular systole. Whether or not MR is a consequence of leaflet prolapse is determined by the extent of the prolapse and also by the degree of mitral anular dilatation.[156] Abnormal prolapse without MR generally results in a systolic click without an accompanying murmur (stage I). Additional prolapse generally produces mild MR as manifested by a late systolic murmur with retention of the systolic click (stage II) (Barlow syndrome).[173] Severe prolapse produces moderate to severe MR as manifested by prolongation of the murmur throughout systole (pansystolic) and disappearance of the systolic click (stage III). Thus MR represents a consequence of the prolapsing (floppy) mitral valve. Other complications, namely infective endocarditis, rupture of chordae tendineae (with or without infection), rupture of a valvular leaflet, mitral anular calcification, arrhythmias, chest pain, and sudden death may occur at any of the three stages. Although little information is available regarding progression from stage I to stage III, it appears reasonable to believe that progression is in this order even though one may remain only briefly at one stage or skip stage II entirely.[158] Furthermore, recent echocardiographic studies have demonstrated that the click or the late systolic murmur or both may disappear entirely during some cardiac cycles.

Fortunately patients with systolic clicks with or without late systolic murmurs infrequently have MR severe enough to be fatal. As a consequence, few mitral valves have been examined at necropsy in patients known to have systolic clicks and late systolic murmurs unequivocally documented during life. With the click alone, there are one or more areas of leaflet overlap during simulated ventricular systole. With the appearance of the late systolic murmur, multiple scallops, a severe exaggeration of the normal, are present in the leaflets. The scallops simply indicate that there is too much leaflet tissue (redundancy of leaflet), and the leakage probably occurs between the scallops. Although more prominent on the posterior leaflet, the scalloping usually involves both leaflets, leading to thickening of at least the distal halves of each. The appearance of the holosystolic murmur is probably indicative of considerable mitral anular dilation superimposed on the leaflet floppiness.[156] The floppy valve is the only type of valve associated with severe (> 60%) dilatation of the mitral anulus, and the anular dilation appears responsible for a great deal of the MR when it becomes severe in these patients. The mitral chordae are most commonly elongated, but they may be shorter than normal or normal in length.[158]

Why the mitral valve affected by acute rheumatic fever becomes stenotic in some patients and purely incompetent in others is unknown. Although the cause may be similar (rheumatic), the structural alterations in the purely incompetent mitral valve are quite different from those in the stenotic valve. The purely incompetent rheumatic valve infrequently contains calcific deposits; its commissures are not fused, its chordae are only mildly thickened, and chordal fusion is infrequent. It is difficult, actually, to derive the mechanism of pure MR from rheumatic fever. The mitral anuli in these patients also are usually dilated less than 25%, and consequently this mechanism appears inadequate, as a rule, to explain the MR.[156]

Congenital MR is usually associated with other anomalies.[140,174] "Isolated" congenital MR is usually even accompanied by diffuse left ventricular endocardial fibroelastosis. As with pure rheumatic MR, the mechanism of the MR is often uncertain in these patients.

Other causes of congenital MR include partial atrioventricular canal (with a cleft in the anterior mitral leaflet)[175] and the Ebstein-type anomaly involving the left-sided atrioventricular valve in corrected transposition.[176,177] The former is infrequently the cause of severe MR, the latter may cause severe MR.

Aortic Regurgitation

In contrast to valvular AS, which basically has only three etiologies (congenital, degenerative, and rheumatic), there are numerous causes of pure aortic regurgitation (AR).[158,159] The causes may be subdivided into those resulting from conditions affecting primarily the valvular cusps and those affecting primarily the aorta. Those potentially fatal conditions affecting primarily the cusps include rheumatic disease, infective endocarditis, and congenital malformations. Those causing fatal AR primarily by affecting the aorta include syphilis, "medial cystic necrosis" with or without the Marfan syndrome, and ankylosing spondylitis.

Rheumatic heart disease, at least anatomically, is never isolated to the aortic valve—the mitral valve is always involved anatomically.[157] The aortic valve, however, may be made seriously incompetent (or stenotic) by rheumatic involvement even while the mitral valve shows only mild fibrous scarring but no dysfunction. Most patients with serious rheumatic aortic valve disease have some aortic stenosis as well as incompetency. Less often the aortic valve is made purely incompetent by rheumatic involvement.

Among the causes of pure AR, rheumatic disease is still important. Clinically, the clue to a rheumatic etiology is a positive history of acute rheumatic fever. The purely incompetent rheumatic valve is diffusely thickened by fibrous tissue, and one or more of the three cusps is severely retracted. The commissures are only mildly fused, if at all. Calcific deposits are either absent or present in only small amounts.

Infective endocarditis (IE) is the most common cause of fatal isolated pure AR, and the aortic valve is the valve most commonly affected by fatal IE.[170,171] Although it is well appreciated that the congenitally bicuspid aortic valve has a particular propensity to become the site of vegetations, the aortic valve in the majority of patients with IE consists of three cusps.[170] Nevertheless, among persons with three-cuspid aortic valves, IE is infrequent. Fatal IE is more often observed when the infection involves a previously normal rather than a previously abnormal aortic valve. Because infection of previously anatomically normal valves is nearly always by so-called virulent organisms, and because the infective process can more easily destroy a thin, delicate (normal) valve as opposed to a thick, tough (abnormal) valve, the frequency of involvement by IE of previously anatomically normal valves in a necropsy study might be abnormally high. The frequency of IE is clearly higher in persons with underlying cardiac disease than in persons with previously normal hearts. Among patients with IE, however, there is more involvement of previously normal than of previously abnormal valves. This fact may be explained simply by the larger number of normal than abnormal valves.

Among patients with pure AR of congenital origin, the degree of AR is usually relatively mild unless the valve has previously been the site of infection. If AR is associated with ventricular septal defect, however, it may be severe. Although AR is due to prolapse of one or more of the cusps into the defect, the cause of the prolapse in these patients is uncertain. Poor support of the valve is not an explanation, because this valve appears to be less well supported in patients with the tetralogy of Fallot, and yet aortic cusp prolapse simply does not occur in them.[178] The prolapsed aortic cusp in patients with associated ventricular septal defect also has an increased propensity to develop IE. The largest hearts in humans occur in patients with AR superimposed on ventricular septal defect.

Trauma is a rare cause of AR, but AR is nevertheless the most common clinically recognized valvular lesion resulting from blunt trauma.[158] The lesion, limited almost entirely to men, generally is due to detachment of one or more cusps with prolapse or to tearing or fenestration of one or more cusps without detachment, or both. Cuspal detachment similar to that produced by trauma, however, may occur without historical evidence of trauma.

Among patients with AR secondary to disease of the aorta rather than to cuspal disease, syphilis continues to head the list of causes.[158] For AR to result from syphilis, however, the luetic process must involve the wall of aorta behind the sinuses. More often than not in patients with cardiovascular syphilis, the process involves only the tubular portion of ascending aorta (i.e., that portion above the sinuses), and without sinus wall involvement AR does not occur. Aortic wall thickening is virtually a sine qua non for the diagnosis of cardiovascular syphilis. Although thick, the wall is weak, and dilation is an inevitable consequence. The thickening results from adventitial scarring, which may cause the adventitia to be much thicker than the media, and from intimal proliferation, indistinguishable from atherosclerotic plaques. The dilatation results from the medial scarring with loss of both medial elastic fibers and smooth-muscle cells. The process appears to begin with inflammation in the walls of the adventitial vasa vasora, leading to severe intimal proliferation with luminal narrowing.

Aortic regurgitation occurs in about 30% of patients with cardiovascular syphilis. It results from dilation of the sinuses due to weakening of the sinus walls by the luetic process. Stretched by the dilated aortic root, the aortic valve cusps are unable to meet in ventricular diastole, and a central regurgitant stream results. The free margins of the aortic valve cusps thicken mildly, and this appears to be secondary to the AR and not to valvulitis per se. Thus whether AR occurs in cardiovascular syphilis is dependent on the site of the disease in the aorta. If the sinus wall is involved, AR occurs; if the process involves only the tubular portion of aorta, no AR results.

Ankylosing spondylitis, a chronic and usually progressive disease of the sacroiliac and apophyseal joints and adjacent soft tissues, also affects the root of aorta to cause AR.[158,179] The frequency of anatomic involvement of the ascending aorta appears to be about 20%, but only about 3% of individuals with ankylosing spondylitis have clinically detectable AR. Although the murmur of AR often appears within a year after onset of joint symptoms, evidence of congestive cardiac failure usually does not appear until 10 years or more after onset of joint symptoms. The AR is usually severe, with the systemic arterial diastolic pressure often equaling the left ventricular end-diastolic pressure.

The cardiovascular lesion in ankylosing spondylitis is distinctive morphologically.[158,179] The process always involves the wall of aorta behind the sinuses and often the proximal 1 cm or so of the tubular portion of aorta. The process also always involves the proximal or basal portions of the aortic valve cusps, and it always extends below the bases of the aortic valve cusps causing thickening of the membranous ventricular septum and the basal portion of anterior mitral leaflet. The latter often results in the formation of a fibrous bump in the left ventricular outflow tract, and occasionally it may be observed on left ventricular angiogram.[180] The process in the aorta is histologically similar to syphilis, but it differs from syphilis by involving primarily the sinus wall rather than the tubular wall of aorta, by involving the basal as well as distal portions of the aortic valve cusps, and by extending below the aortic valve cusps. The latter accounts for the frequent occurrence of conduction disturbances in patients with ankylosing spondylitis and the rarity of these disturbances in patients with syphilis. As in syphilis, however, the involved aortic wall is thickened and the sinuses are dilated.

The Marfan syndrome, either classic or forme fruste, is often associated with severe AR (and usually some degree of MR as well).[158,159] As in ankylosing spondylitis, the major site of involvement is the wall of aorta behind the sinus, but usually the proximal half of the tubular portion of ascending aorta is also involved. In contrast to syphilis and ankylosing spondylitis, however, the involved aorta is thinner than normal, not thicker. The thinning is due to severe loss of medial elastic fibers and to transverse tears in the aorta. The sinuses of Valsalva often are aneurysmally dilated, and the medial tears result in focal outpouchings. The interior lining over the outpouchings have an orange-peel appearance, whereas the other areas of intima usually appear grossly normal. Elastic fibers in the grossly normal areas of aorta are also usually severely decreased in number, and the amount of acid mucopolysaccharide material is focally increased. Few medial scars are present, and neither the adventitia nor intima is thickened. Since foci of medial necrosis are absent and medial cysts are infrequent, the older term "medial cystic necrosis" is inappropriate for this condition. The major abnormality is a striking loss of medial elastic fibers. Histologically, the aortic valve cusps contain an increased amount of acid mucopolysaccharide material, but only their distal margins are thickened, probably the result of AR.

Other causes of AR include dissecting aneurysm of aorta, rupture of an aortic sinus of Valsalva aneurysm, and aortico–left ventricular tunnel.[181] The first two produce AR acutely.

IDIOPATHIC CARDIOMYOPATHY

Abnormal cardiac function that cannot be attributed to coronary, hypertensive, pulmonary, congenital, or pericardial heart diseases is usually the result of a cardiomyopathy. The latter may be due to or be associated with a specific condition, such as amyloidosis, hemosiderosis, neoplasm, etc., and therefore would be in the category of secondary cardiomyopathies. Or the cardiomyopathy may be primary or idiopathic, in which case there are no known causes and no known associated conditions that in themselves would be capable of producing functionally significant cardiac dysfunction.

The idiopathic cardiomyopathies may be divided into two major types[182–185] (Fig. 1-10): (1) the ventricular dilated type (called by some "primary myocardial disease" or "congestive cardiomyopathy") and (2) the nonventricular dilated type.

The major feature of the dilated type of idiopathic cardiomyopathy is dilatation, usually of considerable proportions, of both ventricular cavities. Both atrial cavities are usually also dilated, but not to as great an extent. The ventricular dilatation appears to result from the poor ventricular contractions, resulting in low ejection fractions and high end-systolic volumes. The latter appear to limit atrial emptying, which in turn leads to abnormal atrial end-diastolic volumes that lead to atrial dilatation. The large end-systolic ventricular volumes lead to relative stasis of blood in the apical portions of the ventricular cavities, and this results in intracavitary thrombosis. Thrombi are also frequent in one or both atrial appendages, presumably also the consequence of poor atrial emptying and relative stasis of blood in the appendages.

The paradox of the dilated type of idiopathic cardiomyopathy is that the myocardium, at least grossly, looks normal and yet contracts poorly. Thus, this is one of the few cardiac conditions in which the anatomic changes in the heart are not adequate to explain the functional

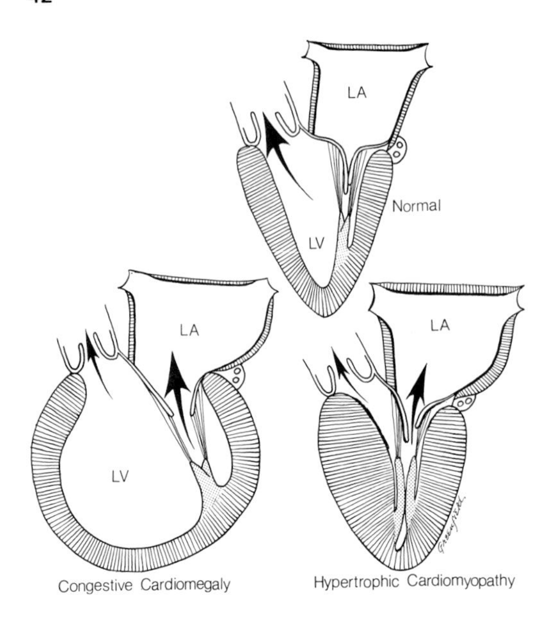

Fig. 1-10. Diagram illustrating the differences between the ventricular dilated (congestive) and nonventricular dilated (hypertrophic) types of idiopathic cardiomyopathy. Mitral regurgitation may occur in both forms. Focal scarring in the papillary muscles is also common in both forms of cardiomyopathy.

disturbances. The dilated cardiomyopathy thus may be considered the myasthenia gravis of heart disease—the muscle looks good, but function is terrible. Scarring of the ventricular myocardial walls in dilated cardiomyopathy is visible grossly in only about 20% of patients, and foci of myocardial coagulation necrosis are nearly always absent except in the apices of the papillary muscles.

The nondilated type of idiopathic cardiomyopathy is strikingly different both functionally and morphologically from the dilated type. The major problem is a hypercontracting left ventricle, not a hypocontracting ventricle as in the dilated type. The morphologic counterpart of this hypercontracting ventricle is a massive ventricular muscle mass and small ventricular cavities. The nondilated idiopathic cardiomyopathy is the classic example of the musclebound heart; the dilated type, in contrast, is the classic example of the flabby heart—mostly cavity with comparatively little myocardial mass.

In the nondilated type, in addition to there being relatively little ventricular cavity in comparison to the thick myocardial walls, the ventricular septum is usually (95% of cases) thicker than the left ventricular free wall.[158,159,162,182–188] Normally, at all ages, including birth, the ventricular septum is equal in thickness to the left ventricular free wall. With the exception of the occasional patient with right ventricular systolic hypertension (of any etiology) and a right ventricular free wall thicker than the left ventricular free wall, the nondilated idiopathic cardiomyopathy is the only known condition in which the ventricular septum is thicker than the left ventricular free wall. This fact gave rise to the name asymmetric septal hypertrophy (ASH), which was used in 1958 by Teare in his original description of this condition.[189]

In contrast to those with the dilated type, some patients with the nondilated idiopathic cardiomyopathy (ASH) have obstruction of left ventricular outflow. The occurrence of outflow obstruction gave rise to the names "idiopathic hypertrophic subaortic stenosis," "diffuse muscular subaortic stenosis," and "obstructive cardiomyopathy," among others. Subsequently it has become clear that the nonobstructive variety of ASH is more common than the obstructive variety.[189] For a number of years, anatomic features distinguishing between the nonobstructive and obstructive varieties were not recognized.[186,187] The extensive use of the echocardiogram indicated that the portion of left ventricular free wall behind the posterior mitral leaflet was of normal thickness in patients with nonobstructive ASH, but not in patients with obstructive ASH.[190] Then, reexamination of the heart specimens themselves proved the interpretation by echocardiogram to be correct (Fig. 1-11). Thus it is now possible to distinguish the obstructive from the nonobstructive variety: in the patient with nonobstructive ASH the most basal portion of the left ventricular free wall (i.e., the portion behind the posterior mitral leaflet) is pointed and thinner than normal, whereas this portion is rounded and thicker than normal in the obstructive variety.

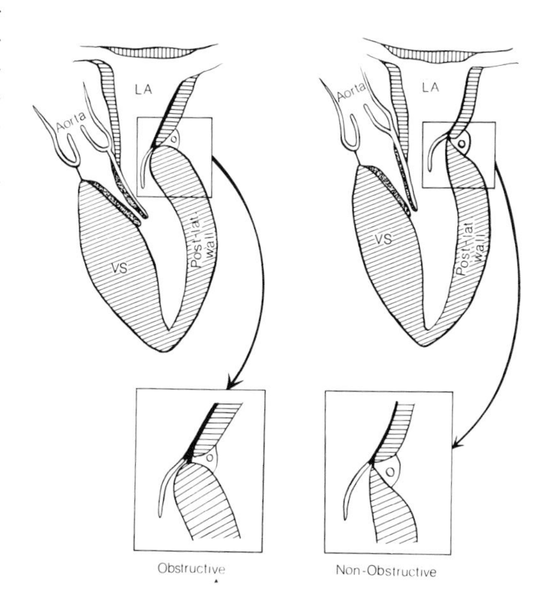

Fig. 1-11. Diagram illustrating anatomic differences between the obstructive and nonobstructive forms of hypertrophic cardiomyopathy. In both forms the ventricular septum (VS) is thicker than the left ventricular free wall. In the obstructive form the most basal portion of posterior left ventricular free wall is thick and rounded. In the nonobstructive form this portion of left ventricular free wall is thinned and more or less pointed.

Another distinguishing feature between the obstructive and nonobstructive varieties of ASH is in the orientation of the myocardial fibers in the left ventricular free wall. There is marked disorientation of the myocardial fibers in the ventricular septum in ASH.[191] The disorientation is present in the septum irrespective of whether outflow tract obstruction is present. The orientations of the myocardial fibers in the left (and also right) ventricular free wall, however, are different in the two varieties.[192] In patients without obstruction, the myocardial fiber disorientation, as observed in the ventricular septum, is also present in the left ventricular free wall, suggesting that this variety is a diffuse cardiomyopathy. In patients with the obstructive variety, in contrast, the orientation of the myocardial fibers in the free wall is normal (the fibers have their usual parallel relationship to one another), suggesting that the obstructive variety is fundamentally a cardiomyopathy localized to the ventricular septum. It is suspected, however, that in an occasional patient the obstructive variety may convert with time into the nonobstructive variety or vice versa.

Other characteristic features of ASH include thickening of the mural endocardium in the left ventricular outflow tract in apposition to the anterior mitral leaflet, which also is thickened.[187] In addition, the posterior mitral leaflet is focally thickened by fibrous tissue, possibly because there is inadequate space to accommodate it freely within the small left ventricle. It is suspected that the site of the outflow obstruction is at the most caudal extension of the mural endocardial plaque, a level that corresponds to the distal margin of the anterior mitral leaflet. The thickening of the endocardium (both mural and valvular) in the left ventricular outflow tract appears to result from contact between the anterior mitral leaflet and the mural endocardium. Interestingly, the frequencies of this endocardial thickening are similar in patients with and those without outflow tract obstruction.

PERICARDIAL HEART DISEASE

The normal parietal pericardium, the bag that encloses the heart, consists of a layer of dense fibrous tissue (collagen) 1 mm thick devoid of elastic fibrils and covered by a layer of mesothelial cells. Although present in the attached adipose tissue on its mediastinal side, vascular channels are absent in the dense collagen of the parietal pericardium itself. The visceral pericardium (epicardium) is the surface of the heart itself, and it also consists of a thin (< 1 mm) layer of fibrous tissue covered by mesothelial cells. Beneath this membrane is either myocardium or adipose tissue. The latter normally occurs in the areas of the extramural coronary arteries, generally completely surrounding them and consequently cushioning them from the underlying moving myocardium. In addition to containing the large coronary vessels, the subepicardial adipose tissue also contains numerous small vascular channels.

Myocardial malfunction from pericardial disease results either from large quantities of fluid in the pericardial sac or from marked thickening of the pericardium, or both. The amount of pericardial fluid necessary to cause ventricular constriction is highly variable and is dependent on several factors: (1) the time period required to accumulate the fluid, (2) the amount of thickening of the underlying or overlying pericardia, and (3) the amount of muscle mass (weight) of the cardiac ventricles. Obviously the quicker the fluid is accumulated the greater the chance that myocardial dysfunction (constriction) will result. The normal pericardial sac contains up to about 50 ml of serous fluid (similar in chemical composition to serum), but many times that amount may be present without myocardial constriction if the accumulation is slow. The thicker the parietal or visceral pericardia, or both, the less fluid is necessary to produce signs or symptoms of constriction. Lastly, the thicker the ventricular walls the more fluid required for constriction. More fluid is required, for example, to constrict the hypertensive ventricle than the normotensive ventricle.

In patients with chronic constrictive pericarditis the pericardium itself is thickened, and there may or may not be fluid (in normal, decreased, or increased amounts) in the space between the thickened parietal and visceral pericardia. Usually the parietal pericardium is much thicker than the epicardium, but it is possible for marked thickening of the latter to cause constriction without significant thickening of the parietal pericardium. In most patients with constriction secondary to pericardial thickening, the parietal pericardium is severely thickened (> 10 times normal thickness).

In some patients with pericardial disease with or without clinical evidence of constriction, the pericardial space may be totally devoid of fluid and may be replaced by fibrous adhesions. Probably obliterative pericarditis (i.e., the parietal and visceral pericardia are everywhere adherent to one another) is most often clinically silent. In nearly 50% of patients with rheumatoid arthritis the pericardial sac at necropsy is obliterated by fibrous adhesions, and signs of pericardial involvement in life in these patients are rare. After all cardiac operations the two layers of pericardium become adherent to one another, but clinical evidence of pericardial involvement is uncommon. For clinical evidence of myocardial constriction to occur in patients with obliterative pericarditis, either the parietal or visceral pericardium, usually both, must be quite thick; significant thickening of these layers does not occur in the usual patient with obliterative pericarditis.

Whether fluid is present in the pericardial sac in patients with severely thickened pericardia is probably most dependent on the length of time required to cause the thickening. In patients with thickened epicardium or parietal pericardium secondary to neoplastic infiltration, residual pericardial fluid is usually present. In patients with myocardial constriction secondary to tuberculous pericarditis, however, the pericardial space may be totally obliterated by fibrous adhesions.

An ideal classification of pericardial diseases is difficult. Anatomic and etiologic classifications often do not separate patients with and those without clinical evidences of constriction or tamponade, and functional classifications may ignore important anatomic and etiologic considerations. The presence or absence of the following should be taken into account: clinical evidence of myocardial constriction and nature of onset of symptoms (acute or chronic); condition primary to pericardium or pericardial involvement as simply one manifestation of a diffuse or systemic condition (primary or secondary); occurrence of pericardial fluid and composition of fluid (serous or bloody); nature of deposits on pericardial surfaces (fibrin, blood products, fibrous tissue, neoplastic cells, purulent material, etc.).

REFERENCES

1. Roberts WC: The pathology of acute myocardial infarction. Hospital Practice 6:89, 1971
2. Roberts WC: Coronary arteries in fatal acute myocardial infarction. Circulation 45:215, 1972
3. Roberts WC: Relationship between coronary thrombosis and myocardial infarction. Mod Concepts Cardiovasc Dis 41:7, 1972
4. Roberts WC, Buja LM: The frequency and significance of coronary arterial thrombi and other observations in fatal acute myocardial infarction. A study of 107 necropsy patients. Am J Med 52:425, 1972
5. Roberts WC: Does thrombosis play a major role in the development of symptom-producing atherosclerotic plaques? Circulation 48:1161, 1973
6. Roberts WC: Coronary thrombosis and fatal myocardial ischemia. Circulation 49:1, 1974
7. Eliot RS, Baroldi G, Leone A: Necropsy studies in myocardial infarction with minimal or no coronary luminal reduction due to atherosclerosis. Circulation 49:1127, 1974
8. Bulkley BH, Roberts WC: Severe (> 75%) narrowing of left main coronary artery: An indication of generalized severe coronary atherosclerosis. Circulation [Suppl] 49,50:III–150, 1974

9. Classification of atherosclerotic lesions. Report of a study group. WHO Tech Rep Ser (No 143), 1958
10. Armstrong ML, Megan MB: Lipid depletion in atheromatous coronary arteries in Rhesus monkeys after regression diets. Circ Res 30:675, 1972
11. Walston A, Hackel DB, Estes EH: Acute coronary occlusion and the "power failure" syndrome. Am Heart J 79:613, 1970
12. Page DL, Caulfield JB, Kastor JA, et al: Myocardial changes associated with cardiogenic shock. N Engl J Med 285:133, 1971
13. Herrick JB: Clinical features of sudden obstruction of the coronary arteries. JAMA 59:2015, 1912
14. Herrick JB: Thrombosis of the coronary arteries. JAMA 72:387, 1919
15. Spain DM, Bradess VA: The relationship of coronary thrombosis to coronary atherosclerosis and ischemic heart disease. (A necropsy study covering a period of 25 years.) Am J Med Sci 240:701, 1960
16. Spain DM, Bradess VA: Sudden death from coronary heart disease. Survival time, frequency of thrombi, and cigarette smoking. Chest 58:107, 1970
17. Hellstrom HR: Coronary artery stasis after induced myocardial infarction in the dog. Cardiovasc Res 5:371, 1971
18. Master AM, Dack S, Jaffe HL: Activities associated with the onset of acute coronary artery occlusion. Am Heart J 18:434, 1939
19. Roberts WC, Buja LM, Bulkley BH, Ferrans VJ: Congestive heart failure and angina pectoris. Opposite ends of the spectrum of symptomatic ischemic heart disease. Am J Cardiol 34:870, 1974
20. Yarbrough JW, Roberts WC, Abel RM, Reis RL: The cause of luminal narrowing in internal mammary arteries implanted into canine myocardium. Am Heart J 84:507, 1972
21. Erhardt LR, Lundman T, Mellstedt H: Incorporation of ^{125}I-labelled fibrinogen into coronary arterial thrombi in acute myocardial infarction in man. Lancet 1:387, 1973
22. Glancy DL, Marcus ML, Epstein SE: Myocardial infarction in young women with normal coronary arteriograms. Circulation 44:495, 1971
23. O'Reilly RJ, Spellberg RD: Rapid resolution of coronary arterial emboli. Myocardial infarction and subsequent normal coronary arteriograms. Ann Intern Med 81:348, 1974
24. Bulkley BH, Roberts WC: Dissecting aneurysm (hematoma) limited to coronary artery. A clinicopathologic study of six patients. Am J Med 55:747, 1973
25. Bulkley BH, Roberts WC: Isolated coronary arterial dissection. A complication of cardiac operations. J Thorac Cardiovasc Surg 67:148, 1974
26. Mallory GK, White PD, Salcedo-Salgar J: The speed of healing of myocardial infarction. A study of the pathologic anatomy in seventy-two cases. Am Heart J 18:647, 1939
27. Hamilton M, Thompson EN, Wisniewski TKM: Role of blood pressure control in preventing complications of hypertension. Lancet 1:235, 1964
28. Veterans Administration cooperative study group on antihypertensive agents: Effects of treatment on morbidity in hypertension. Results in

patients with diastolic blood pressures averaging 115 through 129 mm Hg. JAMA 202:1028, 1967
29. Veterans Administration cooperative study group on antihypertensive agents: Effects of treatment on morbidity in hypertension. II. Results in patients with diastolic blood pressure averaging 90 through 114 mm Hg. JAMA 213:1143, 1970
30. Freis ED: Effectiveness of drug therapy in hypertension: Present status. A review. Circ Res [Suppl] 28,29:II-70–II-75, 1971
31. Kannel WB, Gordon T, Schwartz MJ: Systolic versus diastolic blood pressure and risk of coronary heart disease. The Framingham study. Am J Cardiol 27:335, 1971
32. Lown B, Kosowsky BD, Klein MD: Pathogenesis, prevention, and treatment of arrhythmias in myocardial infarction. Circulation [Suppl] 39, 40:IV-261–IV-270, 1969
33. Moritz AR, Zamcheck N: Sudden and unexpected deaths in young soldiers: Diseases responsible for such deaths during World War II. Arch Pathol 42:459, 1946
34. Spain DM, Bradess VA, Mohr C: Coronary atherosclerosis as a cause of unexpected and unexplained death. JAMA 174:384, 1960
35. Kuller L: Sudden and unexpected nontraumatic deaths in adults: A review of epidemiological and clinical studies. J Chronic Dis 19:1165, 1966
36. Kuller L, Lilienfeld A, Fisher R: Epidemiological study of sudden and unexpected deaths due to arteriosclerotic heart disease. Circulation 34:1056, 1966
37. Kuller L, Lilienfeld A, Fisher R: An epidemiological study of sudden and unexpected deaths in adults. Medicine 46:341, 1967
38. Shapire S, Weinblatt E, Frank CW, Sager RV: Incidence of coronary heart disease in a population insured for medical care (HIP). Myocardial infarction, angina pectoris, and possible myocardial infarction. Am J Public Health 59:1 (Part II), 1969
39. Fulton M, Julian DG, Oliver MF: Sudden death and myocardial infarction. Circulation [Suppl] 39,40:IV-182–IV-193, 1969
40. Chiang BN, Perlman LV, Fulton M, et al: Predisposing factors in sudden cardiac deaths in Tecumseh, Michigan. Circulation 41:31, 1970
41. Gordon T, Kannel WB: Premature mortality from coronary heart disease. The Framingham study. JAMA 215:1617, 1971
42. Kuller LH, Cooper M, Perper J, Fisher R: Myocardial infarction and sudden death in an urban community. Bull NY Acad Med (2nd series) 49:532, 1973
43. Eppinger EC, Levine SA: Angina pectoris. Some clinical considerations, with special reference to prognosis. Arch Intern Med 53:120, 1934
44. Block WJ Jr, Crumpacher EL, Dry TJ: Prognosis of angina pectoris, observations in 6882 cases. JAMA 150:259, 1952
45. Richards DW, Bland EF, White PD: Completed 25 year follow-up study of 456 patients with angina pectoris. J Chronic Dis 4:423, 1956
46. Seim S: Angina pectoris: A prognostic study. Acta Med Scand 166:255, 1960

47. Francis RL, Achor RWP, Brown AL Jr: Angina pectoris preceding initial myocardial infarction. A clinicopathologic study. Arch Intern Med 112:226, 1963
48. Shurtleff D: The Framingham study. An epidemiological investigation of cardiovascular disease, Sect 26. Kannel WB, Gordon T (eds): Washington, DC, US Government Printing Office, 1970
49. Rosenman RH, Friedman M, Straus R, et al: Coronary heart disease in the Western collaborative group study. J Chronic Dis 23:173, 1970
50. Chapman JM, Coulson AH, Clark VA, Borun ER: The differential effect of serum cholesterol, blood pressure and weight on the incidence of myocardial infarction and angina pectoris. J Chronic Dis 23:631, 1971
51. Kannel WB, Feinleib M: Natural history of angina pectoris in the Framingham study. Prognosis and survival. Am J Cardiol 29:154, 1972
52. Medalie JH, Snyder M, Groen JJ, et al: Angina pectoris among 10,000 men. 5 year incidence and univariate analysis. Am J Med 55:583, 1973
53. Nathanson MH: Disease of the coronary arteries. Clinical and pathologic features. Am J Med Sci 170:240, 1925
54. Parkinson J, Bedford DE: Cardiac infarction and coronary thrombosis. Lancet 1:4, 1928
55. Levine SA: Coronary thrombosis. Its various clinical features. Medicine 8:245, 1929
56. Bartels EC, Smith HL: Gross cardiac hypertrophy in myocardial infarction. Am J Med Sci 184:452, 1932
57. Master AM, Jaffe HL, Dack S: The treatment and the immediate prognosis of coronary artery thrombosis (267 attacks). Am Heart J 12:549, 1936
58. Levy H, Boas EP: Coronary artery disease in women. JAMA 107:97, 1936
59. Rosenbaum FF, Levine SA: Prognostic value of various clinical and electrocardiographic features of acute myocardial infarction. I. Immediate prognosis. Arch Intern Med 68:913, 1941
60. Levine SA, Rosenbaum FF: Prognostic value of various clinical and electrocardiographic features of acute myocardial infarction. II. Ultimate prognosis. Arch Intern Med 68:1215, 1941
61. Billings FT, Kalstone BM, Spencer JL, et al: Prognosis of acute myocardial infarction. Am J Med 7:356, 1949
62. Zinn WJ, Crosby RS: Myocardial infarction. I. Statistical analysis of 679 autopsy-proven cases. Am J Med 8:169, 1950
63. Mulcahy R, Hickey N, Maurer B: Coronary heart disease in women. Study of risk factors in 100 patients less than 60 years of age. Circulation 36:577, 1967
64. Bland EF, White PD: Coronary thrombosis (with myocardial infarction) ten years later. JAMA 117:1171, 1941
65. Cole DR, Singian EB, Katz LN: The long term prognosis following myocardial infarction, and some factors which affect it. Circulation 9:321, 1954
66. Biörck G, Sievers J, Blomquist G: Studies of myocardial infarction in

Malmö 1935–1954. III. Follow-up studies from a hospital material. Acta Med Scand 162:81, 1958
67. Beard OW, Hipp HR, Robins M, et al: Initial myocardial infarction among 503 veterans. Five-year survival. Am J Med 28:871, 1960
68. Pell S, D'Alonzo CA: Immediate mortality and five-year survival of employed men with a first myocardial infarction. N Engl J Med 270:915, 1964
69. Kannel WB, Gordon T, Castelli WP, Margolis JR: Electrocardiographic left ventricular hypertrophy and risk of coronary heart disease. The Framingham study. Ann Intern Med 72:813, 1970
70. Buja LM, Khoi NB, Roberts WC: Clinically significant cardiac amyloidosis. Clinicopathologic findings in 15 patients. Am J Cardiol 26:394, 1970
71. Roberts WC, Ronan JA Jr, Harvey WP: Rupture of the left ventricular free wall (LVFW) or ventricular septum (VS) secondary to acute myocardial infarction (AMI): An occurrence virtually limited to the first transmural AMI in a hypertensive individual. Am J Cardiol 77:166, 1975
72. Shell WE, Sobel BE: Protection of jeopardized ischemic myocardium by reduction of ventricular afterload. N Engl J Med 291:481, 1974
73. Estes JE Jr: Abdominal aortic aneurysm: A study of one hundred and two cases. Circulation 2:258, 1950
74. Crane C: Arteriosclerotic aneurysm of the abdominal aorta. Some pathological and clinical correlations. N Engl J Med 253:954, 1955
75. Wheelock F, Shaw RS: Aneurysm of the abdominal aorta and iliac arteries. N Engl J Med 255:72, 1956
76. Brindley P, Stembridge VA: Aneurysms of the aorta. A clinicopathologic study of 369 necropsy cases. Am J Pathol 32:67, 1956
77. Sommerville RL, Allen EV, Edwards JE: Bland and infected arteriosclerotic abdominal aortic aneurysms: A clinicopathologic study. Medicine 38:207, 1959
78. Halpert B, Willms RK: Aneurysms of the aorta. An analysis of 249 necropsies. Arch Pathol 74:163, 1962
79. MacCallum WG: Dissecting aneurysm. Bull Johns Hopkins Hosp 20:9, 1909
80. Tyson MD: Dissecting aneurysms. Am J Pathol 7:581, 1931
81. Shennan T: Dissecting aneurysms. Medical Research Council. Special Report Series, No 193. London, His Majesty's Stationery Office, 1934
82. Weiss S: The clinical course of spontaneous dissecting aneurysm of the aorta. Med Clin North Am 18:1117, 1935
83. McGeachy TE, Paullin JE: Dissecting aneurysm of the aorta. JAMA 108:1690, 1937
84. Glendy RE, Castleman B, White PD: Dissecting aneurysm of aorta: Clinical and anatomic analysis of 19 cases (13 acute) with notes on differential diagnosis. Am Heart J 13:129, 1937
85. David P, McPeak EM, Vivas-Salas E, White PD: Dissecting aneurysm of the aorta: A review of 17 autopsied cases of acute dissecting aneurysm of the aorta encountered at the Massachusetts General Hospital from 1937 to

1946 inclusive, eight of which were correctly diagnosed ante mortem. Ann Intern Med 27:405, 1947

86. Levinson DC, Edmeades DT, Griffith GC: Dissecting aneurysm of the aorta: Its clinical, electrocardiographic and laboratory features. Report of fifty-eight autopsied cases. Circulation 1:360, 1950
87. Gore I, Seiwert VJ: Dissecting aneurysm of the aorta; Pathologic aspects; An analysis of eighty-five fatal cases. Arch Pathol 53:121, 1952
88. Gore I: Pathogenesis of dissecting aneurysm of the aorta. AMA Arch Pathol 53:142, 1952
89. Gore I: Dissecting aneurysms of the aorta in persons under forty years of age. AMA Arch Pathol 55:1, 1953
90. Burchell HB: Aortic dissection (dissecting hematoma; dissecting aneurysm of the aorta). Circulation 12:1068, 1955
91. Halpert B, Brown CA: Dissecting aneurysms of the aorta. Study of 12 cases. AMA Arch Pathol 60:378, 1955
92. Hirst AE Jr, Jones VJ Jr, Kime SW Jr: Dissecting aneurysm of aorta: A review of 505 cases. Medicine 37:217, 1958
93. De Bakey ME, Henly WS, Cooley DA, et al: Surgical treatment of dissecting aneurysm of the aorta. Analysis of seventy-two cases. Circulation 24:290, 1961
94. Lindsay J Jr, Hurst JW: Clinical features and prognosis in dissecting aneurysm of the aorta. A re-appraisal. Circulation 35:880, 1967
95. Gore I, Hirst AE Jr: Dissecting aneurysm of the aorta. Cardiovasc Clin 5:239, 1973
96. Reifenstein GH, Levine SA, Gross RE: Coarctation of the aorta: Review of 104 autopsied cases of "adult type" two years of age or older. Am Heart J 33:146, 1947
97. Ashworth CT, Haynes DM: Lesions in elastic arteries associated with hypertension. Am J Pathol 24:195, 1948
98. Puchner TC, Huston JH, Hellmuth GA: Aortic valve insufficiency in arterial hypertension. Am J Cardiol 5:758, 1960
99. Leriche R: De la résection du carrefour aortoiliague avec double sympathectomie lombaire pour thrombose artéritique de l'aorte: Le syndrome de l'oblitération terminoaortique par artérite. Presse Méd 48:601, 1940
100. Gross H, Philips B: Complete occlusion of the abdominal aorta: A review of seven cases. Am J Med Sci 200:203, 1940
101. Lueth HC: Thrombosis of the abdominal aorta: A report of four cases showing the variability of symptoms. Ann Intern Med 13:1167, 1940
102. Leriche R, Morel A: The syndrome of thrombotic obliteration of the aortic bifurcation. Ann Surg 127:193, 1948
103. Kekwick A, McDonald L, Semple R: Obliterative disease of the abdominal aorta and iliac arteries with intermittent claudication. Q J Med 21:185, 1952
104. Gunning AJ, Hackett MEJ, MacKenzie JR, et al: A clinicopathological study of aorto-iliac thrombosis (the Leriche syndrome). Q J Med 35(NS):475, 1966

105. Juergens JL, Barker NW, Hines EA Jr: Arteriosclerosis obliterans: Review of 520 cases with special reference to pathogenic and prognostic factors. Circulation 21:118, 1960
106. Tillgren C: Obliterative arterial disease of the lower limbs. II. A study of the cause of disease. Acta Med Scand 178:103, 1965
107. Kannel WB, Skinner JJ Jr, Schwartz MJ, Shurtleff D: Intermittent claudication. Incidence in the Framingham study. Circulation 41:875, 1970
108. Kannel WB, Shurtleff D: The natural history of arteriosclerosis obliterans. Cardiovasc Clin 3:37, 1971
109. Schenk EA: Pathology of occlusive disease of the lower extremities. Cardiovasc Clin 5:287, 1973
110. Lippmann HI: Peripheral vascular disease today. Bull NY Acad Med 49:674, 1973
111. Hicks SP, Warren S: Infarction of the brain without thrombosis. An analysis of one hundred cases with autopsy. Arch Pathol 52:403, 1951
112. David NJ, Heyman A: Factors influencing the prognosis of cerebral thrombosis and infarction due to atherosclerosis. J Chronic Dis 11:394, 1960
113. Low-Beer T, Phear D: Cerebral infarction and hypertension. Lancet 1:1303, 1961
114. Vost A, Wolochow DA, Howell DA: Incidence of infarcts of the brain in heart disease. J Pathol Bacteriol 88:463, 1964
115. Prineas J, Marshall J: Hypertension and cerebral infarction. Br Med J 1:14, 1966
116. Louis S, McDowell F: Stroke in young adults. Ann Intern Med 66:932, 1967
117. Baker RN, Schwartz WS, Ramseyer JC: Prognosis among survivors of ischemic stroke. Neurology 18:933, 1968
118. Kannel WB, Wolf PA, Verter J, McNamara PM: Epidemiologic assessment of the role of blood pressure in stroke. The Framingham study. JAMA 214:301, 1970
119. Heyden S, Heyman A, Goree JA: Nonembolic occlusion of the middle cerebral and carotid arteries—A comparison of predisposing factors. Stroke 1:363, 1970
120. Kannel WB: Current status of the epidemiology of brain infarction associated with occlusive arterial disease. Stroke 2:295, 1971
121. Whisnant JP, Fitzgibbons JP, Kurland LT, Sayre GP: Natural history of stroke in Rochester, Minnesota, 1945 through 1954. Stroke 2:11, 1971
122. Kannel WB, Gordon T, Wolf PA, McNamara P: Hemoglobin and the risk of cerebral infarction: The Framingham study. Stroke 3:409, 1972
123. Wolf PA, Kannel WB, McNamara PM, Gordon T: The role of impaired cardiac function in atherothrombotic brain infarction: The Framingham study. Am J Public Health 63:52, 1973
124. Siekert RG, Reagan TJ: Clinical and pathologic correlations in ischemic cerebrovascular disease. Cardiovasc Clin 5:311, 1973
125. Matsumoto N, Whisnant JP, Kurland LT, Okazaki H: Natural history of stroke in Rochester, Minnesota, 1955 through 1969: An extension of a previous study, 1945 through 1954. Stroke 4:20, 1973

126. Hudson AJ, Hyland HH: Hypertensive cerebrovascular disease: A clinical and pathologic review of 100 cases. Ann Intern Med 49:1049, 1958
127. Marshall J: The natural history of transient ischaemic cerebrovascular attacks. Q J Med 33:309, 1964
128. Acheson J, Hutchinson EC: Observations on the natural history of transient cerebral ischaemia. Lancet 2:871, 1964
129. Whisnant JP, Matsumoto N, Elveback LR: Transient cerebral ischemic attacks in a community: Rochester, Minnesota, 1955 through 1969. Mayo Clin Proc 48:194, 1973
130. Millikan CH, Siekert RG, Whisnant JP: The clinical pattern in certain types of occlusive cerebrovascular disease. Circulation 22:1002, 1960
131. Stehbens WE: Pathology of the Cerebral Blood Vessels. St Louis, CV Mosby, 1972, p 284
132. Fisher CM: Lacunes: Small deep cerebral infarcts. Neurology 15:774, 1965
133. Fisher CM: A lacunar stroke. The dysarthria–clumsy hand syndrome. Neurology 17:614, 1967
134. Fisher CM: The arterial lesions underlying lacunes. Acta Neuropathol (Berl) 12:1, 1969
135. Pickering G: Hypertension. Definitions, natural histories and consequences. Am J Med 52:570, 1972
136. Cole FM, Yates PO: The occurrence and significance of intracerebral microaneurysms. J Pathol Bacteriol 93:393, 1967
137. Fisher CM: Cerebral miliary aneurysms in hypertension. Am J Pathol 66:313, 1972
138. Bell ET, Clawson BJ: Primary (essential) hypertension. A study of four hundred and twenty cases. Arch Pathol 5:939, 1928
139. Roberts WC, Perloff JK, Costantino T: Severe valvular aortic stenosis in patients over 65 years of age. A clinicopathologic study. Am J Cardiol 27:497, 1971
140. Roberts WC, Perloff JK: Mitral valvular disease. A clinicopathologic survey of the conditions causing the mitral valve to function abnormally. Ann Intern Med 77:939, 1972
141. deLeon AC, Hammer WJ, Roberts WC: Mitral stenosis resulting from calcification of the mitral anulus. A rare occurrence. (in preparation)
142. Frink RJ, Achor RWP, Brown AL Jr, et al: Significance of calcification of the coronary arteries. Am J Cardiol 26:241, 1970
143. Hutt MSR: Pathology of African cardiomyopathies. Pathol Microbiol (Basel) 35:37, 1970
144. Bronte-Stewart B, Heptinstall RH: The relationship between experimental hypertension and cholesterol-induced atheroma in rabbits. J Pathol Bacteriol 68:407, 1954
145. Moses C: Development of atherosclerosis in dogs with hypercholesterolemia and chronic hypertension. Circ Res 2:243, 1954
146. Wakerlin GE, Moss WG, Kiely JP: Effect of experimental renal hypertension on experimental thiouracil-cholesterol atherosclerosis in dogs. Circ Res 5:426, 1957
147. Carlson RG, Lillehei CW, Edwards JE: Cystic medial necrosis of the as-

cending aorta in relation to age and hypertension. Am J Cardiol 25:411, 1970

148. Norris RM, Bensley KE, Caughey DE, Scott PJ: Hospital mortality in acute myocardial infarction. Br Med J 3:143, 1968
149. Norris RM, Brandt PWT, Caughey DE, et al: A new coronary prognostic index. Lancet 1:274, 1969
150. Norris RM, Caughey DE, Deeming LW, et al: Coronary prognostic index for predicting survival after recovery from acute myocardial infarction. Lancet 2:485, 1970
151. Norris RM, Caughey DE, Mercer CJ, Scott PJ: Prognosis after myocardial infarction. Six-year follow-up. Br Heart J 36:786, 1974
152. Kannel WB, Castelli WP, McNamara PM: Role of blood pressure in the development of congestive heart failure. N Engl J Med 287:781, 1972
153. Frohlich ED, Tarazi RC, Dustan HP: Clinical–physiological correlations in the development of hypertensive heart disease. Circulation 44:446, 1971
154. Roberts WC, Cohen LS: The left ventricular papillary muscles. Description of the normal and a survey of conditions causing them to be abnormal. Circulation 46:138, 1972
155. Perloff JK, Roberts WC: The mitral apparatus. Functional anatomy of mitral regurgitation. Circulation 46:227, 1972
156. Bulkley BH, Roberts WC: Dilatation of the mitral anulus. A cause of mitral regurgitation? Am J Med (1975) (in press)
157. Roberts WC: Anatomically isolated aortic valvular disease. The case against its being of rheumatic etiology. Am J Med 49:151, 1970
158. Roberts WC, Dangel JC, Bulkley BH: Nonrheumatic valvular cardiac disease: A clinicopathologic survey of 27 different conditions causing valvular dysfunction. Cardiovasc Clin 5:333, 1973
159. Roberts WC: Left ventricular outflow tract obstruction and aortic regurgitation, in: The Heart. International Academy of Pathology Monograph No 15 Edwards JE, Lev M, Abell MR (eds). Baltimore, Williams & Wilkins, 1974, p 110
160. Falcone MW, Roberts WC, Morrow AG, Perloff JK: Congenital aortic stenosis resulting from unicommissural valve. Clinical and anatomic features in twenty-one adult patients. Circulation 44:272, 1971
161. Roberts WC: The congenitally bicuspid aortic valve. A study of 85 autopsy cases. Am J Cardiol 26:72, 1970
162. Roberts WC: Valvular, subvalvular, and supravalvular aortic stenosis: Morphologic features. Cardiovasc Clin 5:97, 1973
163. Edwards JE: The congenital bicuspid aortic valve. Circulation 23:485, 1961
164. Neufeld HN, Wagenvoort CA, Burchell HB, Edwards JE: Idiopathic atrial fibrillation. Am J Cardiol 8:193, 1961
165. Roberts WC, Mason DT, Morrow AG, Braunwald E: Calcific pulmonic stenosis. Circulation 37:973, 1968
166. Roberts WC, Sjoerdsma A: The cardiac disease associated with the carcinoid syndrome (carcinoid heart disease). Am J Med 36:5, 1964
167. Morrow AG, Cohen LS, Roberts WC, et al: Severe mitral regurgitation

following acute myocardial infarction and ruptured papillary muscle: Hemodynamic findings and results of operative treatment in four patients. Circulation 38(Suppl II):II-124–II-132, 1968
168. Glancy DL, Stinson EB, Shepherd RL, et al: Results of valve replacement for severe mitral regurgitation due to papillary muscle rupture or fibrosis. Am J Cardiol 32:313, 1973
169. Roberts WC, Braunwald E, Morrow AG: Acute severe mitral regurgitation secondary to ruptured chordae tendineae. Clinical, hemodynamic, and pathologic considerations. Circulation 33:58, 1966
170. Buchbinder NA, Roberts WC: Left-sided valvular active infective endocarditis. A study of forty-five necropsy patients. Am J Med 53:20, 1972
171. Roberts WC, Buchbinder NA: Right-sided valvular infective endocarditis. A clinicopathologic study of twelve necropsy patients. Am J Med 53:7, 1972
172. Roberts WC, Ross RS, Eggleston JC, Massumi RA: Chronic mitral regurgitation of unresolved etiology in the elderly. A clinicopathologic study of two patients. Johns Hopkins Med J 122:26, 1968
173. Barlow JB, Pocock WA, Marchand P, Denny M: The significance of late systolic murmurs. Am Heart J 66:443, 1963
174. Carney EK, Braunwald E, Roberts WC, et al: Congenital mitral regurgitation, clinical, hemodynamic and angiographic findings in nine patients. Am J Med 33:223, 1962
175. Brockenbrough EC, Braunwald E, Roberts WC, Morrow AG: Partial persistent atrioventricular canal simulating pure mitral regurgitation. Am Heart J 63:9, 1962
176. Berry WB, Roberts WC, Morrow AG, Braunwald E: Corrected transposition of the aorta and pulmonary trunk. Clinical, hemodynamic and pathologic findings. Am J Med 36:35, 1964
177. Roberts WC, Ross RS, Davis FW Jr: Congenital corrected transposition of the great vessels in adulthood simulating rheumatic valvular disease. Bull Johns Hopkins Hosp 114:157, 1964
178. Glancy DL, Morrow AG, Roberts WC: Malformations of the aortic valve in patients with the tetralogy of Fallot. Am Heart J 76:755, 1968
179. Bulkley BH, Roberts WC: Ankylosing spondylitis and aortic regurgitation. Description of the characteristic cardiovascular lesion from study of eight necropsy patients. Circulation 48:1014, 1973
180. Roberts WC, Hollingsworth JF, Bulkley BH, et al: Combined mitral and aortic regurgitation in ankylosing spondylitis. Angiographic and anatomic features. Am J Med 56:237, 1974
181. Roberts WC, Morrow AG: Aortico–left ventricular tunnel. A cause of massive aortic regurgitation and of intracardiac aneurysm. Am J Med 39:662, 1965
182. Roberts WC, Ferrans VJ: Morphologic observations in the cardiomyopathies, in Fowler NO (ed): Myocardial Diseases. New York, Grune & Stratton, 1973, p 59
183. Roberts WC, Ferrans VJ: Pathological aspects of certain cardiomyopathies. Circ Res [Suppl] 34,35:II-128–II-144, 1974

184. Roberts WC, Ferrans VJ, Buja LM: Pathologic aspects of the idiopathic cardiomyopathies. Adv Cardiol 13:349, 1974
185. Roberts WC, Ferrans VJ: Pathologic anatomy of the cardiomyopathies Idiopathic dilated and hypertrophic types, infiltrative types, and endomyocardial disease with and without eosinophilia. Hum Pathol 6:287, 1975
186. Ewy GA, Marcus FI, Bohajalian O, et al: Muscular subaortic stenosis. Clinical and pathologic observations in an elderly patient. Am J Cardiol 22:126, 1968
187. Morrow AG, Roberts WC, Ross J Jr, et al: Obstruction to left ventricular outflow. Current concepts of management and operative treatment. Ann Intern Med 69:1255, 1968
188. Henry WL, Clarke CE, Epstein SE: Asymmetric septal hypertrophy. Echocardiographic identification of the pathognomonic anatomic abnormality of IHSS. Circulation 47:225, 1973
189. Epstein SE, Henry WL, Clark CE, et al: Asymmetric septal hypertrophy. Ann Intern Med 81:650, 1974
190. Henry WL, Clark CE, Roberts WC, et al: Differences in distribution of myocardial abnormalities in patients with obstructive and nonobstructive asymmetric septal hypertrophy (ASH). Echocardiographic and gross anatomic findings. Circulation 50:447, 1974
191. Ferrans VJ, Morrow AG, Roberts WC: Myocardial ultrastructure in idiopathic hypertrophic subaortic stenosis. A study of operatively excised left ventricular outflow tract muscle in 14 patients. Circulation 45:769, 1972
192. Maron BJ, Ferrans VJ, Henry WL, et al: Differences in distribution of myocardial abnormalities in patients with obstructive and nonobstructive asymmetric septal hypertrophy (ASH). Light and electron microscopic findings. Circulation 50:436, 1974

C. Lynn Skelton
Edmund H. Sonnenblick

2

Physiology of Cardiac Muscle

The rapid expansion of biomedical investigation, particularly during the past 15 years, has vastly increased our understanding of the basic mechanisms underlying the contractile process of cardiac muscle. This knowledge has forced a reappraisal of the traditional concepts regarding the characterization of the performance of the intact heart. It is now realized that the properties of the heart as a pump are largely dependent on its potentialities as muscle. The purpose of this presentation is to discuss some of these recent advances in our knowledge of the structural, biochemical, and mechanical properties of cardiac muscle and to integrate this information into a rational framework for the evaluation of myocardial performance.

ULTRASTRUCTURE OF THE MYOCARDIUM

General Organization

The ventricular myocardium is composed of interconnecting cells or fibers that normally are 30 to 60 μ in length and 5 to 15 μ in diameter (Fig. 2-1A). The individual muscle fibers are surrounded by a surface-limiting membrane, the sarcolemma. Under the light microscope, numerous cross-banded strands or bundles, termed fibrils (myofibrils), that run the length of the fiber are seen (Fig. 2-1B). In rapidly contracting or "fast" skeletal muscle, these fibrils are quite discrete. However, in slower skeletal muscles and heart muscle these fibrils are nonuniform in shape and diameter and tend to merge into one another along their

Supported in part by USPHS Grant HL-11306 and Training Grant 5T01-HL-0589003.

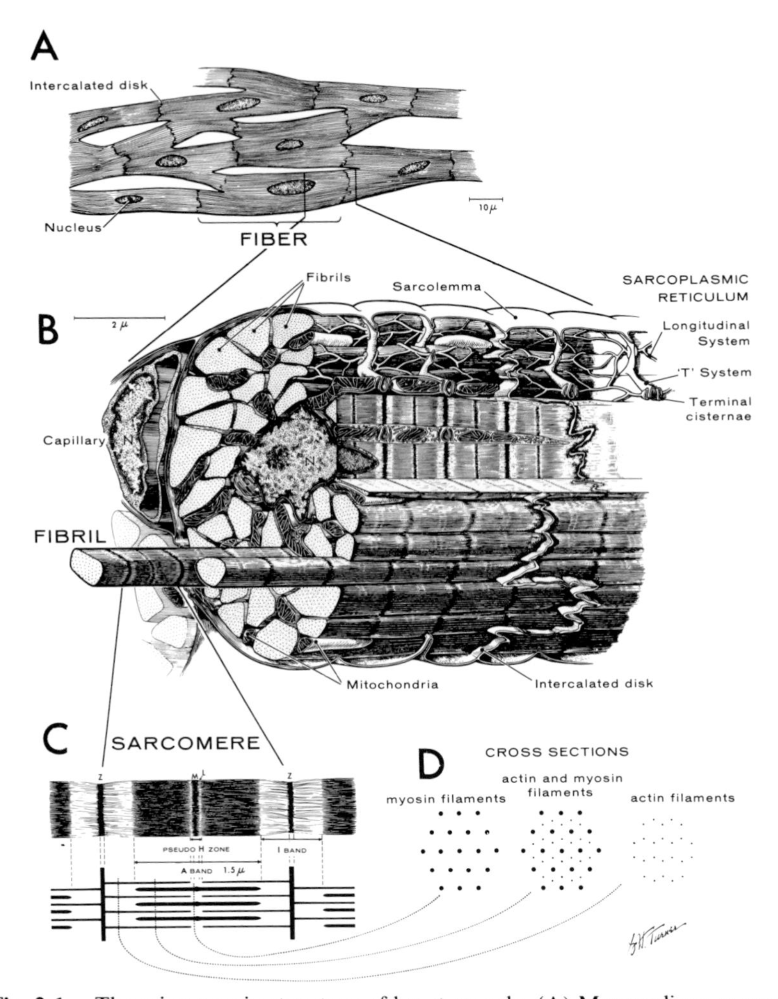

Fig. 2-1. The microscopic structure of heart muscle. (A) Myocardium as seen under the light microscope. Branching of fibers is evident, with each containing a centrally located nucleus. Fibers or cells are connected across intercalated disks. (B) A myocardial cell or fiber reconstructed from electronmicrographs, showing the arrangement of multiple parallel fibrils that compose the cell and serially connected sarcomeres that compose the individual fibril (N = nucleus). The sarco-tubular system that mediates activation, including the sarcolemma and sarcoplasmic reticulum, is also shown. An intercalated disk in the center of the reconstruction serves to connect two cells. (C) An individual sarcomere from a myofibril. A diagrammatic representation of the arrangement of myofilaments that make up the sarcomere is shown. Thick filaments approximately 1.5 μ in length composed of myosin are localized to the A band, while thin filaments 1.0 μ

length. Fibrils tend to be distinguished by clefts of cytoplasm that contain mitochondria and other membranous tubules, but this separation is not as complete in cardiac as in fast skeletal muscle.[1–4]

The myofibrils are composed of a serially repeating structure, the sarcomere (Fig. 2-1B), which is composed of the contractile proteins arranged in a very specific manner (Fig. 2-1C). The sarcomeres, which occupy about 50% of the mass of the myocardial cell (in contrast to 90% in skeletal muscle), are aligned so that the ends of sarcomeres in adjacent myofibrils are in register, giving the entire fiber its banded or striated appearance. It is the interaction between myofilaments in the sarcomere that ultimately generates force and shortening of the myocardium. The nature of this interaction and its control will be discussed in detail later in this chapter.

The remainder of the intracellular cytoplasm is occupied by membrane systems that modulate activity, mitochondria that serve as centers of aerobic metabolism, and a centrally located nucleus that controls protein synthesis in the cell (Fig. 2-1B). The nuclei of cardiac cells resemble those from other tissues that exhibit active protein synthesis; this is consistent with the continued turnover of myofilaments that occurs even in adult heart muscle. Although two nuclei may be found occasionally in normal adult myocardium, little or no mitotic activity occurs after the neonatal period. At the pole of the nucleus, a Golgi complex is commonly observed. This poorly understood organelle is characterized by an orderly array of membranes whose function is poorly understood. Lysosomes and lipofuscin bodies are present in the same area near the pole of the nucleus. Lysosomes are seen as membrane-limited vesicles, about 0.1 μ in diameter, which contain latent hydrolytic enzymes capable of lysing cellular membranes as well as other cell components. Lipofuscin bodies are thought to be dense deposits of insoluble lipoprotein surrounded by a discrete membrane. These bodies may represent the remains of autophagic lysosomal activity, and they tend to increase with age.

---

in length composed primarily of actin extend from the Z line through the I band into the A band, ending at the edges of the central pseudo H zone. The pseudo H zone is the central area of the A band where thin filaments are absent. Thick and thin filaments overlap only in the A band. (D) A diagrammatic cross section of the sarcomere showing the specific lattice arrangements of the myofilaments. In the center of the sarcomere (left) only the thick (myosin) filaments arranged in a hexagonal array are seen. In the distal portions of the A band (center) both the thick and thin (actin) filaments are found, with each thick filament surrounded by six thin filaments. In the I band only thin filaments are present.

Mitochondria

Mitochondria are the major sites of oxidative phosphorylation in the cell whereby energy released from the aerobic oxidation of fats and two carbon fragments is converted to adenosine triphosphate (ATP), the ultimate source of energy for cellular function.[5] Consonant with the high requirements for aerobic metabolism in heart muscle, mitochondria are present in great profusion, comprising 25% to 30% of the entire mass of the cell.[6] Under the electron microscope these structures appear as cylindrically shaped bodies, approximately 2 to 5 μ by 0.5 μ in size, that are distributed in the cell primarily between and in close apposition to the myofibrils (Fig. 2-1B). The close proximity of the mitochondria to the contractile filaments may facilitate the transfer of ATP from its site of production to its site of utilization in the contractile process. However, the process by which ATP is transferred to various sites within the cell is unknown.

The mitochondrion is a highly structured organelle. Projecting inward from the double membrane of the surface are found numerous platelike foldings or cristae, about 200 to 250 Å in thickness. These dense and closely packed cristae contain the enzymes of the tricarboxylic acid cycle orderly arranged in minute particles along the cristae membranes. Electron-dense bodies approximately 300 to 400 A in diameter, which are thought to be a site of calcium deposition, are also seen within the cristae.

The mitochondria have also been shown to have other capabilities in addition to the synthesis of ATP, including a capacity to swell reversibly and to accumulate calcium actively from the cytoplasm.[7] Under normal conditions this accumulation of calcium is relatively slow and appears to be insignificant at the low levels of free calcium that normally exist within the cell.[8] Thus this function appears to have little role in cellular control when the mechanisms that handle calcium are normally operative.[8,9] Nevertheless, mitochondrial calcium uptake may play a role when the handling of intracellular calcium is altered, as may occur in myocardial failure.

Membranes That Mediate the Control of Contraction: Sarcolemma, Intercalated Disks, and Sarcoplasmic Reticulum

Individual myocardial fibers are invested with a surface plasma membrane and its basement membrane that together comprise the sarcolemma. The plasma membrane is a thin (75–90 Å) electron-dense unit that is covered with an amorphous, granular basement membrane about 500 Å thick. The basement membrane, composed of glycoprotein, may play an important role in ion exchange, since it contains negatively charged sites where Ca^{++} may be bound.[9] The plasma membrane, which is the site of electrical polarization across the cell wall and is the major semipermeable membrane between the intracellular cytoplasm and extracellular space, is also the site of the enzymatic processes for active transport across the cell membrane.[9] The plasma membrane is comprised of a bimolecular phospholipid layer with its nonpolar portions directed inward and its polar heads outward. Monolayers of protein cover the surfaces of this bilayer.

Myocardial cells characteristically branch and interdigitate through a modification of the sarcolemma at the ends of the cells, termed the intercalated disk.[1–4] Since the intercalated disks are true cell membranes, the myocardium is not a true syncytium. However, since there is little electrical impedance between cells and since force can be generated across serially disposed cells, the myocardium has been considered a "functional" syncytium.

Three distinct variations in the structure of the intercalated disk have been described,[3,10] each of which appears to subserve a separate function. The fascia adherens consists of a dense cytoplasmic deposit into which the myofilaments of the terminal sarcomeres of a fibril are inserted. The fascia adherens serves to fix the ends of a myofibril within a given cell and to transmit force to the next cell.

Another variation of the intercalated disk consists of a desmosome-like structure, the macula adherens. This structure is composed of a dense set of paired arcuate cytoplasmic bodies 0.2 to 0.5 μ in diameter placed in apposition to the cell membranes. This structure, which is also noted in many epidermal tissues, is thought to hold cells together and allow the transmission of force between cells arranged in series.

The third variation of the disk, the nexus (or gap junction) is seen where the intercalated disk turns in direction to run parallel to the long axis of the two cells. It is composed of two closely apposed, but unfused, plasma membranes.[10] Between the two plasma membranes, a gap of 2 mμ is found that is accessible to small molecules coming from the extracellular space.[11] The nexus takes on functional importance as the probable site of low impedance and transmittal of electrical activity between cells.[12] Recently attempts have been made to reconstruct the structure of the nexus using freeze-etching techniques[13,14] in which tissues are broken after freezing and their fractured surfaces viewed under the electron microscope. These studies have supported previous observations that the nexus is composed of globular subunits arranged in a hexagonal array with 90- to 100-Å center spacings. Narrow channels may exist at these connected sites in the nexus that may permit small ions to pass from one cell to the next.[15]

The sarcolemma with its modifications serves multiple functions, including (1) maintenance of the intracellular milieu with its high [K^+] and low [Na^+], (2) a capacity to depolarize with rapid influx of Na^+ into the cell accompanied by efflux of K^+, and by a small inward flux of Ca^{++}, (3) propagation of electrical depolarization along adjacent fibers producing generalized activation, and (4) exclusion of extracellular Ca^{++} from the cell in diastole when the myoplasmic [Ca^{++}] level is very low.

The sarcolemma, like other surface membranes of cells, has an enzyme system for pumping Na^+ out of the cell.[16] This system, as studied in detail in the surface membrane of the red cell, utilizes ATP for energy, has an ATPase activity that is stimulated by Na^+ and K^+, is dependent on Mg^{++}, and is inhibited by Ca^{++} or digitalis glycosides. This Na^+ pump is located asymetrically in the sarcolemma, since it is stimulated by Na^+ only at the inner surface and inhibited by digitalis glycosides only on the outer surface.[17] Considerable evidence suggests that this Na^+–K^+-stimulated ATPase is responsible for active transport of Na^+ out of the cardiac cell and that this Na^+ extrusion is linked to K^+ uptake. A pump that extrudes Ca^{++} from the cell has also been described in association with the sarcolemma,[16] but its relation to the Na^+–K^+-stimulated ATPase is poorly understood.

In ventricular cells the sarcolemma displays large tubular invaginations 1000 to 2000 Å in diameter, termed transverse or T tubules, which form a ramifying network of tubules extending into the cell,[18–20] reaching the Z line of the sarcomere. The T system does not open into the cell but is rather an extension of the sarcolemmal plasma membrane with its basement membrane deep within the cell. The T tubules branch and change direction to run longitudinally along the fibril, becoming much smaller and losing their basement membrane. At this point they may be difficult to distinguish from truly intracellular tubules, except by the use of markers (e.g., ferric oxide or peroxidase) that are too large to cross membranes and thus remain extracellular.[21] These markers of the extracellular space have been used to demonstrate the connection of the T tubules with the extracellular fluid in both heart and skeletal muscle. The T system appears to play a vital role in the transfer of surface electrical depolarization deep into the cell.[18] Of interest is the finding that small cardiac cells such as occur in the atria, conduction tissue, and newborn ventricular myocardium generally do not have T tubules. This is also true of the cardiac cells of amphibia and reptiles.

The Sarcoplasmic Reticulum

In both cardiac[1,2,4] and skeletal muscle[18,22] the sarcoplasmic reticulum (Fig. 2-1B) consists of a complex network of anastamosing membrane-lined intracellular channels 200 to 400 Å in diameter that invest the myofibrils. These tubules extend longitudinally along the fiber without interruption from sarcomere to sarcomere but are not in continuity with the extracellular fluid. Specialized junctions or couplings are found where the sarcoplasmic reticulum comes in contact with the T tubules at the Z lines of the sarcomeres or at points of apposition to the sarcolemmal membrane.

While structurally similar in both cardiac and skeletal muscle,[18,19] the sarcoplasmic reticulum in heart muscle differs in certain important respects. In heart muscle it is much less profuse than in skeletal muscle. Further, the tubules of the cardiac sarcoplasmic reticulum extend from one sarcomere to the next and demonstrate junctions along the sarcolemma. One "functional" portion of the sarcoplasmic reticulum, termed the cisterna, is characterized by a flattened saclike enlargement that abuts closely to the sarcolemma or T tubules. The cisternae are about 150 to 200 Å from the T tubule and the sarcolemma, with the intervening space being filled by poorly defined globular densities. In skeletal muscle the cisternae from two adjoining sarcomeres form junctions on either side of the T tubule. The two organized terminal cisternae, together with the T tubule, have been called a triad. In cardiac muscle the cisternae tend to be flattened against the larger T tubule. Since these cisternae are less prominent and junctions do not occur as often as in skeletal muscle, only the T tubule and a single cisterna are generally visualized. This structure has been called a diad. Just as the T tubules are thought not to open into the cell,[21] the sarcoplasmic reticulum is probably closed to the extracellular space. Nevertheless, intermittent communications have not been ruled out. Connections to mitochondria are also unclear. In both cardiac and skeletal muscle the sarcoplasmic reticulum has been shown to contain an ATP-dependent capacity to bind calcium to its surface and pump it from the cytoplasm into its lumen against a large gradient.[23–25] In all striated muscle the overall speed of contraction is directly related to the profuseness and organization of the sarcoplasmic reticulum. As will be discussed subsequently, this sytem undoubtedly plays a vital role in the control of calcium movements within the cell, and thus the coupling of cellular excitation to contraction.

Detailed Organization of the Myofibril

Sarcomere

The sarcomere, which is the fundamental structural and functional unit of contraction in both heart and skeletal muscle,[26–29] is a repeating structure along the myofibril set off by two adjacent dark lines that are termed Z lines (Figs. 2-1B and 2-1C). Under physiologic conditions the distance between Z lines varies with overall muscle length and measures between 1.6 and 2.2 μ. Within the confines of the sarcomere, alternating light and dark bands are seen, giving the myocardium its striated appearance. At the center of the sarcomere is a broad dark area of constant width (1.6 μ) termed the A band. The A band is flanked by two lighter bands termed I bands, which are variable in width. The midportion of the A band is bisected by a dark line, the M line, immediately lateral to which are two lighter areas termed L lines. Taken together, this central area of the A band comprises the M-L complex, or the pseudo-H zone. In the longer sarcomeres of skeletal muscle ($> 2.2\ \mu$) there is a central light zone lateral to the M-L complex that is termed the H zone. The M line is further characterized by five sets of fixed bridges between the thick filaments that may serve to hold the A-band structure of the thick filaments intact.

On the basis of a series of classical x-ray diffraction studies and electron-microscopic studies of skeletal muscle[30,31] it was proposed that the contractile substance within the sarcomere is arranged in an ordered array of partially overlapping rodlike myofilaments that are fixed in length, both at rest and during contraction. This same basic structure was subsequently confirmed for heart muscle.[1,28] The myofilaments in turn consist of ordered macroaggregates of contractile proteins. The dimensions of the banding pattern of the sarcomere are of some practical importance, since the size reflects the relative disposition of these myofilaments and thus the localization of contractile proteins within the sarcomere.

The sarcomere of heart muscle, like that of skeletal muscle, is composed of two sets of myofilaments (Fig. 2-1C). Thicker myofilaments, composed of myosin molecules, are limited to the A band, where they are arranged in a hexagonal array. The thick filaments are about 100 Å in diameter, with tapered ends measuring 1.5 to 1.6 μ in length. Thinner myofilaments composed largely of actin course from the Z line through the I band into the A band (Fig. 2-1C). The thin filaments are approximately 50 Å in diameter and 1.0 μ in length. Thus there is an overlapping of thick and thin filaments only within the A band, the I band containing only thin filaments.

The well-ordered distribution of these two sets of myofilament in the sarcomere is even better viewed on cross section (Fig. 2-1D). Sections through the M line reveal only thick filaments arranged in a hexagonal array, with a separation of about 400 to 450 Å. More laterally in the A band, both thick and thin filaments are seen, with a thin filament located at the midpoint (trigonal point) between three thick filaments. Thus each thick filament is surrounded by six thin filaments (Fig. 2-1D). The I band contains only thin filaments terminating at the Z line, a complex matlike structure.[31]

The Thick Filament

The myofilaments themselves also have a substructure, and the elucidation and understanding of the interactions between the two sets of filaments have done much to provide an explanation of the contractile process.[32] The thick filament is composed of an orderly aggregation of myosin molecules that are stacked longitudinally, portions of which form bridgelike outcroppings from the filament that can interact with actin filaments. The myosin molecule is a fibrous protein with a molecular weight of about 500,000 d and dimensions of about 1500 by 30 Å (Fig. 2-2A). Its structure consists of a rodlike tail approximately 1300 A in length and 20 Å in diameter that terminates in a globular head, a bilobed structure 200 Å in diameter.[33,34]

In a purified state myosin can split ATP when activated by small

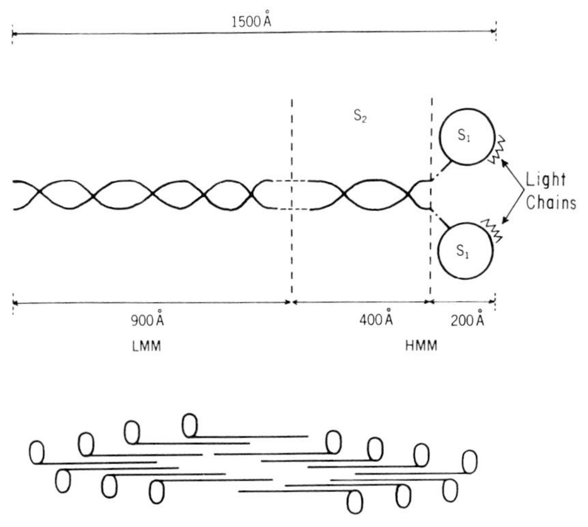

Fig. 2-2. The structure of the myosin molecule and its aggregation into a thick filament. At the top is a diagrammatic representation of the myosin molecule (after Lowey et al.[34]). The two-stranded portions of the molecule show a point of cleavage between the light meromyosin (LMM) and heavy meromyosin (HMM). The HMM can be cleaved into two portions, an S_2 fragment that is similar to the LMM portion of the molecule and an S_1 fragment that contains the ATPase portion of the molecule. Light chains are associated with the S_1 fragment. Below that is a diagram of the aggregation of myosin filaments into a thick filament. The long portion of the molecule tends to be oriented toward the center of the filament, with the enzymatically active heads of the molecule oriented laterally. The center of this aggregation will thus contain no active enzyme sites.

amounts of Ca^{++}, i.e., it is an ATPase. Under physiologic conditions this ATPase requires Mg^{++}, and it is activated only in the presence of actin to serve as a cofactor. Purified myosin is termed myosin A. When combined with actin it has been termed actomyosin or myosin B.[35]

While myosins extracted from red (slow) and white (fast) skeletal and heart muscle are quite similar in size, shape, and physical properties, differences in enzymatic behavior and composition have been noted. Specifically differences in ATPase activity have been found that are directly related to muscle speed.[36–40] Cardiac myosin has a significantly lower intrinsic ATPase activity than does myosin from fast skeletal muscle, but the activity is similar to that of slow skeletal muscle.[38] Further, cardiac myosin is more resistant to enzymatic (trypsin) digestion[40] and does not contain 3-methylhistidine, as is found in fast skeletal muscle.[41] The biochemical composition of cardiac myosin more closely resembles that of myosin from slow skeletal muscle.

The myosin molecule can be broken with the proteolytic enzyme trypsin into two fragments, light meromyosin (LMM) and heavy meromyosin (HMM) (Fig. 2-2A). Skeletal muscle LMM, with a molecular weight of 150,000 d, appears to be a double-stranded coil in the form of an α helix. The inert rodlike structure of LMM serves a structural purpose when myosin is reconstructed into a thick filament.[33] HMM, with a molecular weight of approximately 350,000 d, contains a short rodlike component and a globular head that contains the ATPase activity of the molecule. Further digestion of HMM with either trypsin or papain yields smaller fragments: a subfragment 1 (mol. wt. 120,000 d) that retains the ATPase activity and an ability to bind to actin[42] and a subfragment 2 (mol. wt. 60,000 d), which is the remainder of the rodlike component. Thus each subfragment 1 of HMM comprises one of the two globular heads of myosin. The reconstructed molecule contains the tail of LMM (mol. wt. 150,000 d), two intertwined subfragments 2 (mol. wt. 60,000 d each), and two subfragments 1 (mol. wt. 120,000 d each), for a molecular weight of about 500,000 d (Fig. 2-2A).

Thus the myosin molecule is comprised of an axial core of two heavy (mol. wt. 250,000 d) polypeptide chains that end in a double globular head. Each head contains one heavy and two light polypeptide chains.[43] One of the two light chains appears integral to ATPase activity and is responsible for the differences in activity noted between different types of striated muscle.[44] Further, Mg^{++} and Ca^{++} in association with ATP also appear to interact with this important portion of the myosin molecule. Myosin, and specifically the globular subfragment 1, has a site for combining with ATP and actin on each subfragment 1.[45] Whether these globular heads are identical has yet to be established.

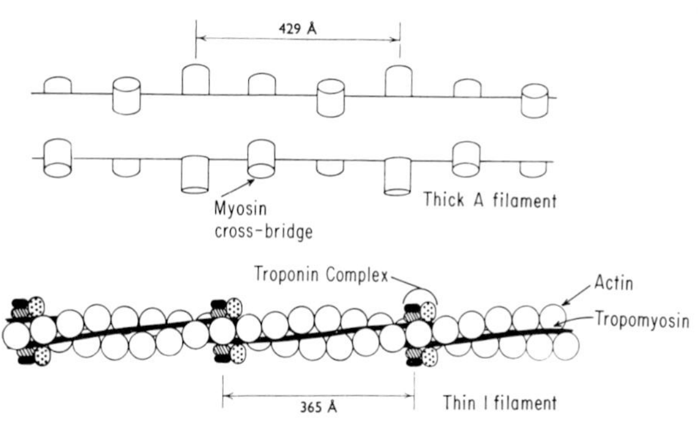

Fig. 2-3. The structure and relations of the thick (top) and thin (bottom) filaments of muscle. The active myosin cross-bridges turn progressively as one moves across the filament with a turnover point at 429 Å. The thin filament is composed of a double-chain α helix of actin molecules. The troponin complex is located at every seventh actin site, while the tropomyosin molecule lies close to the ridge between the two strands of actin. Thus each strand of seven molecules of actin is associated with one troponin complex and one long molecule of tropomyosin, which reaches the length of these seven molecules of actin.

When myosin molecules are placed into the proper solution, they will reaggregate to form myosin filaments.[33] The rodlike LMM portion of the molecule becomes oriented toward the center of the filament, while the globular head is directed outward (Fig. 2-2B). With this in mind, the structure of the thick filament within the sarcomere can be rationalized: the center 0.2 μ of the thick filament contains only the rodlike (LMM) portion of the myosin molecules. This creates the edge of the L lines and the lighter center of the thick filament, interrupted by the dark M-line bridges in the center. Farther laterally along the thick filaments, globular portions of myosin extend outward at intervals of 140 Å, with a complete rotation around the thick filament every 429 Å (Fig. 2-3).

The Thin Filament

The thin filament, which is 1.0 μ in length, is composed of a double helix of F (fibrous) subunits of actin. These F subunits of actin are polymers of the monomer form, G (globular) actin, which has a molecular weight of 47,000 d and a diameter of 55 Å.[40] The double-helical structure of F actin has been substantiated by both electron microscopy and x-ray diffraction studies.[32] Crossover points of the helix occur at 365 Å, with seven actins recurring at 55-Å intervals along each chain (Fig. 2-3). Unlike myosin, actin from different types of muscle is the same in amino acid composition, physical properties, molecular size, and shape.[40]

Tropomyosin

A rodlike protein with a molecular weight of 70,000 d is also associated with the thin filament.[46–48] Tropomyosins from both heart and skeletal muscle appear to be identical. Like actin, tropomyosin occurs as a two-stranded helical coil with each subunit having a molecular weight of 35,000 d. Tropomyosin is intimately associated with the thin filament, two strands of tropomyosin lying just slightly off the groove between the actin chains[49] (Fig. 2-3). Indeed, the 400-Å periodicity noted along the thin filament by either x-ray diffraction studies[50] or electron microscopy[48] appears to be due to tropomyosin rather than actin. Recent findings that tropomyosin has an important effect on the ATPase activity of actomyosin[51] has led to the suggestion that it plays a significant regulatory role in the contraction process.[24]

In addition to the contractile proteins, the sarcomere contains proteins that influence the contractile process and hence are termed regulatory proteins (Table 2-1). These constitute 10% to 15% of total muscle protein and are associated with the thin filament. The contraction-relaxation cycle of muscle is regulated by the intracellular concentrations of free Ca^{++}, which is controlled in turn by the surface membrane and sarcoplasmic reticulum.[24,52] The Ca^{++} responsiveness of the interaction of myosin and actin in which ATP is split depends on the presence of a protein system involving tropomyosin and a more recently recognized protein complex (troponin).[52,53] In the absence of troponin, actomyosin ATPase is fully activated, requiring the presence of only Mg^{++} and ATP. When troponin is present, actomyosin ATPase is inhibited. When small amounts of Ca^{++} are bound to the troponin, its inhibition of the actomyosin ATPase is removed. Thus the Ca^{++} ion does not participate directly in the enzymatic reaction, but may be considered to be a "derepressor" since it removes an inhibitor of the reaction between actin and myosin.[40] Whether Ca^{++} can also affect the bridge-forming component of myosin (HMM) has not been determined.

Troponin has been shown to be located exclusively along the thin filament in association with tropomyosin.[53,54] Based on this and related observations, a model of the thin filaments has been proposed in which tropomyosin forms a continuing strand through the center of the actin filaments with the troponin complex located at each 365-Å interval (Fig. 2-3).[25]

Troponin was first thought to be a single protein;[53] however, more recent studies[54–56] have shown that it can be separated into three distinct

Table 2-1
Myofibrillar Proteins

	Location	Molecular Weight (daltons)	Components
I Contractile			
Myosin	thick filament	500,000	LMM HMM {SF1, SF2}
Actin	thin filament	47,000	
II Regulatory			
Tropomyosin	thin filament	70,000	2 chains
Troponin*	thin filament		3 proteins
III Structural			
α actinin	thin filament		
β actinin	thin filament	62,000	
Fibrillin	?		
M-line protein	thick filament		
Z-line protein	Z line		
C protein	thin filament		

*Components of Troponin Complex	Molecular Weight (daltons)
Calcium-binding protein (TN-C)	17,000–20,000
ATPase-inhibiting protein (TN-I)	22,000–24,000
Tropomyosin-binding protein	35,000–40,000

components: (1) a "calcium sensitizing factor" with a molecular weight of 18,000 d that binds Ca^{++}, termed troponin C (TN-C);[55] (2) an "inhibitory factor" with a molecular weight of 23,000 d that inhibits the Mg^{++}-stimulated ATPase of actomyosin, termed troponin I (TN-I);[55] and (3) a component with a molecular weight of 37,000 d whose function has not as yet been determined. Whether the third component is necessary for the entire complex to function or merely serves a structural purpose is a matter of current study. Identical components of the troponin system have been noted for cardiac and skeletal muscle,[25,57] although some evidence suggests that the activity of this system in red (slow) skeletal muscle and cardiac muscle is somewhat reduced relative to white (fast) muscle.[25,58]

Since troponin is located only at intervals of 365 Å along the thin filament, i.e., with two actin units intervening (Fig. 2-3), it has been suggested that the inhibition of the interaction of actin with myosin is mediated by an ability of troponin to alter the steriochemistry of tropomyosin, which in turn changes the position of active sites all along the thin filaments.[52] In resting muscle, rods of tropomyosin lie toward the edge of the groove between the two strands of actin, each molecule of tropomyosin being in contact with seven actin monomers. In this position tropomyosin blocks the sites on actin that react to form cross-bridges with myosin.[49] When Ca^{++} is bound to troponin the rods of tropomyosin are drawn toward the center of the groove, allowing actin to bind to myosin and activate the splitting of ATP. An important corollary to this concept is the observation that the change in the configuration of the thin filament depends only on Ca^{++} binding and not on myofilament overlap.[59] Since tropomyosin is 400 Å in length, it could easily fulfill this role in amplifying the activity of troponin.[52,56] Thus one may think of a unit along the thin filament as one troponin complex, one tropomyosin, and seven monomers of actin.

Other proteins,[58] all of which appear to play a structural role, have also been isolated from the sarcomere. These include α actinin, which is localized at or near the Z line; β actinin, which is located around the A–I junction of the myofibril; and the M-line protein, which is intimately related to the fixed structural bridges in the area of the M line. β actinin, which is bound to the terminal actin molecules on the thin filament, is thought to limit its length.[60] α actinin has three components termed 6S, 10S, and 25S.[61] As the thin filament approaches the Z line it separates into four strands that extend to form the corners of a square. The 6S component of α actinin may provide the fixation for the thin filament in the Z line. The α actinin has also been localized part way along the I band and in a faint line termed the N line, where it may hold the thin filaments in relatively fixed position relative to the A band.[62] Fibrillin, which is composed of long strands of elastin, has also been isolated from skeletal muscle.[63] Since strands of fibrillin extend beyond the length of a single sarcomere, the possibility has been raised that they may provide a structural backbone for the sarcomere. In view of some of the special properties of the sarcomere in heart muscle, such as its resistance to stretch, the possible existence of fibrillin in heart muscle requires exploration.

EXCITATION-CONTRACTION COUPLING

At rest, the interior of the cardiac cell is negatively polarized relative to the outside of the cell to a resting transmembrane potential of −80 to −100 mV (Fig. 2-4). This resting potential is created by the sarco-

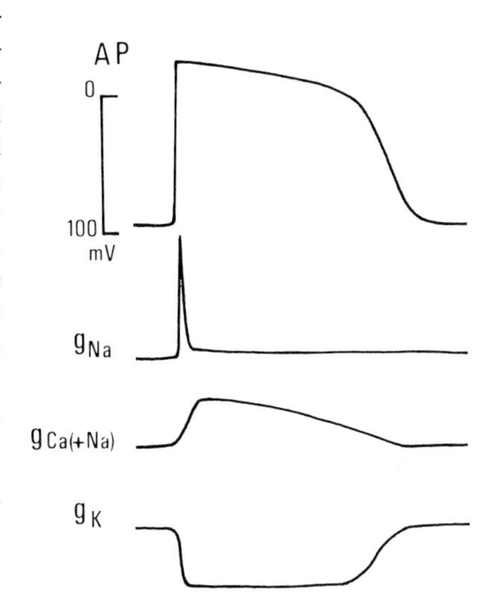

Fig. 2-4. The relation of the myocardial action potential to conductances of sodium, calcium, and potassium. The conductance of sodium (g_{Na}) rises abruptly during the rapid upstroke or spike of the action potential and then comes back nearly to control levels within a very short period of time. At the same time the conductance of potassium (g_K) falls and remains at a low level during most of the plateau of the action potential. The conductance of calcium (g_{Ca}) with a possible component of sodium rises more slowly and is maintained at an increased level throughout the plateau of the action potential. This has been termed the slow calcium current.

lemma, which is largely impermeable to Na^+ and also actively extrudes Na^+ from the cell.[64] Thus the inside of the cell contains mainly K^+, with little Na^+, while the extracellular milieu is just the opposite. Reduction of the transmembrane resting potential to the critical level (threshold) of approximately −50 mV causes a rapid depolarization of the cell membrane (Fig. 2-4). When this diastolic depolarization is spontaneous, the tissue exhibits automaticity, a characteristic of pacemaker tissue. Voltage-clamp techniques have demonstrated that during the rapid upstroke phase of depolarization (phase 0) there is a rapid inward current of Na^+ into the cell[64,65] (Fig. 2-4). This rapid depolarization not only returns the resting potential to zero but produces a reversal of the potential to about +30 mV (spike). The return of potential to zero is thought to be due to an egress of Cl− from the cell (phase 1). Subsequently a plateau of the action potential (phase 2) occurs during which there is a relatively slow inward current of Na^+ as well as Ca^{++}.[64,66–69] This slow phase of inward Ca^{++} current may actually be triggered by the prior rapid inward current of Na^+. The inward Ca^{++} current ceases at the end of the plateau of the action potential. Repolarization is characterized by an outward current of K^+. The late egress of K^+ from the cell, which ends the action potential, is termed delayed rectification. During the plateau of the action potential, the conductance for K^+ may even be less than at rest, a process known as anomalous rectification. These ionic currents are mediated through separate channels or pores in the surface membrane, eight of which have been identified.[64]

During the action potential, passive movements of ions result in an increase in the intracellular [Na^+] and [Ca^{++}] and a decrease in intracellular [K^+]. The ion levels in the cell are subsequently readjusted via the active "sodium pump" located in the sarcolemma. The sodium pump in the surface membrane extrudes Na^+ from the cell in exchange for K^+ while the resting potential remains unchanged (phase 4 of the action potential). The Na^+ pump has been envisioned in terms of a carrier molecule that attaches Na^+ on the interior and K^+ on the exterior of the membrane.[16,17] Energized by ATP, which is split in the process, the carrier rotates 180°, placing each molecule on the opposite side of the sarcolemma.[70] Four factors are known to stimulate the Na^+ pump: a return of the membrane potential toward the isoelectric point, a decrease in [Ca^{++}] at the surface membrane, an increase in [Na^+] on the inner surface of the membrane, and an increase of [K^+] on the outer side of the surface membrane.

The resting potential remains stable in ventricular cells during diastole, and small resting fluxes of Na^+ and K^+ are balanced. In pacemaker cells the situation is different: The resting potential falls slowly due to a decline in the conductivity of the membrane for K^+ (g_K) and a later increase in permeability to Na^+ as the threshold for rapid depolarization is reached. Chloride flows out of the cell as well. These spontaneous changes during phase 4 of the action potential create the pacemaker function of specific cells in the area of the sino-atrial node in the atrium.

The critical role of Ca^{++} in linking excitation to contraction has long been recognized.[71] The slow inward current during the plateau of the action potential (phase 2) reflects a movement of Ca^{++} into the myoplasm. However, the absolute amount of Ca^{++} that crosses the surface membrane is relatively small and appears to be incapable of completely initiating full activation of the contractile apparatus.[68] In mammalian myocardium this net influx of Ca^{++} during each action potential varies between 0.5 and 2.7 μmole/kg wet weight. It varies in different species and is dependent on the absolute extracellular concentration of Ca^{++}, the relative extracellular concentrations of Na^+ and Ca^+, and the frequency of contraction.[72–75] In general, the influx of Ca^{++} is matched by an equal efflux, so that there is no net intracellular gain of Ca^{++} over an extended period of time.

From recent studies a reasonable if incomplete synthesis of the role of Ca^{++} in activating muscle contraction can be presented.[76] An inward current of Ca^{++} takes place during the plateau of the action potential. In order to fully activate the contractile apparatus of the muscle, approximately 60 μmole/kg of Ca^{++} are required to saturate troponin sites. The Ca^{++} current, except under very special conditions such as low $[Na^+]$ in the frog, provides much less Ca^{++} than this. In addition, when stimulation is begun the inward Ca^{++} current is constant with each depolarization. Nevertheless, contractile force, which is small initially, continues to rise with each succeeding contraction. Therefore it has been proposed that the Ca^{++} flowing into the cell does not directly stimulate the contractile system but is stored in membrane sites within the cell, presumably the sarcoplasmic reticulum. The Ca^{++} influx associated with each action potential may lead to augmentation of these internal Ca^{++} stores, which may help explain the relation between developed tension and frequency of contraction (staircase or treppe phenomenon).[73] These observations support the suggestion that the Ca^{++} that actually activates the contractile system is stored in the cisternae of the sarcoplasmic reticulum, which have the capacity to actively bind calcium and store it in a bound form.

How then can surface events influence release of Ca^{++} within the cardiac cell? Studies in skeletal muscle have shown that contraction can be initiated by the application of a depolarizing current to the area of the T tubules.[77] Thus it is proposed that during the action potential a depolarizing current extends across the surface membrane and penetrates deep into the cell by the way of the ramifying T system. A flux of Ca^{++} as well as Na^+ takes place across the sarcolemma and T tubules as noted above. This current may in some way result in depolarization of the sarcoplasmic reticulum with a graded release of Ca^{++}.[78–82]

In both cardiac and skeletal muscle[83–86] a minute amount of Ca^{++} that cannot activate the contractile apparatus (2.5×10^{-7} mole) may be capable of releasing a much larger amount of Ca^{++} from the sarcoplasmic reticulum, which can then activate the contractile system. This Ca^{++}-mediated release of Ca^{++} has been termed a regenerative calcium release.[85] The physiologic significance of this finding is as yet unknown. What is known is that tension development by the contractile proteins is governed by the extracellular $[Ca^{++}]$, its relation to external $[Na^+]$,[87] and the character and duration of the action potential.[80] Each of these factors will determine the inward flux of Ca^{++} into the cell. Thus the action potential not only triggers but also determines to some extent the force of contraction.[80,88]

The sarcoplasmic reticulum also plays a vital role in relaxation of contraction. During repolarization, the sarcoplasmic reticulum avidly binds and stores myoplasmic Ca^{++} with an affinity much greater than troponin.[89] The removal of Ca^{++} from troponin results in an inhibition of actin–myosin interactions, and relaxation ensues.

In summary, the action potential is generated by movement of ions across the cell surface membrane. These movements are controlled by membrane potentials in a manner that is similar to that in nerves. Although the detailed manner of how this surface current leads to an increase in myoplasmic [Ca^{++}] sufficient to activate the contractile system (10^{-8} to 10^{-5} mole) still remains to be defined, it is clear that the major portion of the Ca^{++} used to activate contraction is stored within the cell, presumably in the sarcoplasmic reticulum. This Ca^{++} diffuses toward the myofibrils and binds to troponin, thereby removing its inhibition of the interaction of the head of myosin with actin. The number of sites so activated and the resultant force are directly related to the amount of Ca^{++} present, which in turn is influenced by the influx of Ca^{++} that accompanies the action potential.[69,90] Relaxation of contraction occurs during repolarization via a reduction in myoplasmic [Ca^{++}] mediated by active Ca^{++} binding to the sarcoplasmic reticulum.

SLIDING-FILAMENT MECHANISM OF MUSCLE CONTRACTION

Given the biochemical properties and physical structure of the thick and thin filaments of the sarcomere, a reasonable synthesis can now be made of how the fundamental unit of the contractile system may function.[91–93] These concepts are based largely on observations made on skeletal muscle, but they appear to be equally applicable to cardiac muscle. At rest, actin and myosin filaments do not interact due to the inhibition of actin molecules by tropomyosin, which in turn is controlled by the complexes of troponin arranged periodically along the thin filament. As discussed above, excitation increases myoplasmic [Ca^{++}] from a pCa of about 10^{-8} to about 10^{-6}. Two molecules of Ca^{++} bind to the TN-C component of the troponin complex with a very high affinity, triggering the TN-I component of troponin to cease its inhibition of actin. This inhibition appears to be mediated through motion of tropomyosin lying near the grooves of the actin helix. In some unspecified manner the attachment of Ca^{++} to one troponin triggers tropomyosin to leave seven units of actin capable of interacting with myosin.

In the presence of Mg^{++} and ATP, which are attached in a complexed form to the globular heads of myosin (HMM subfragment 1), cross-bridge links are formed between myosin and actin. This cross-bridge, which may be considered a stiff rod, is composed of the subfragment-2 portion of the HMM component of myosin. The LMM portion of the myosin molecule is embedded along the thick filament. The point between LMM and HMM functions like a hinge, allowing the

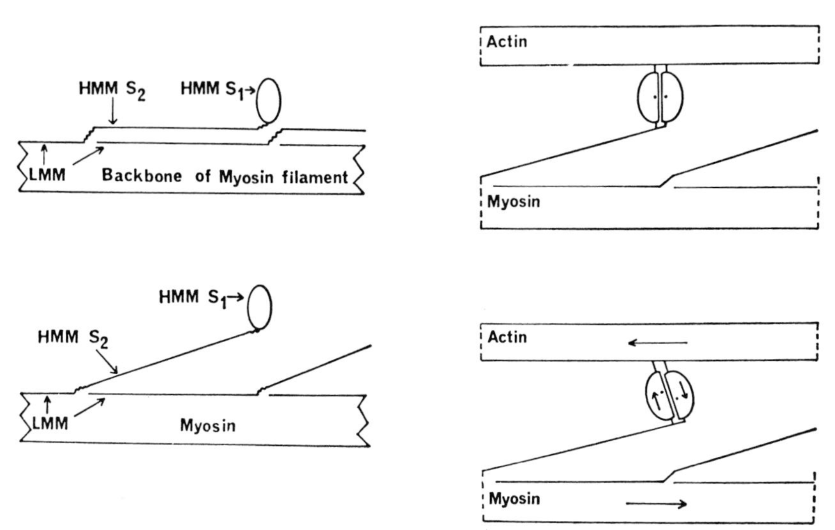

Fig. 2-5. A model for the behavior of myosin molecules in the thick filament and how they can interact with actin for the generation of shortening (after Huxley[27]). To the left is shown a diagram of myosin as it lies along the thick filament. The light-meromyosin (LMM) part of the molecule forms part of the backbone of the filament, while the linear part of the heavy-meromyosin (HMM) component can tilt out from the filament by bending of the junction between the HMM and LMM portions of the molecule. This allows the globular or enzymatically active portion of HMM (the S_1 fragment) to attach to actin over a very long distance. To the right is shown a diagram of how the HMM portion of myosin can bind to sites on actin. A change in the structure of the head of myosin, attached to actin, can then generate motion of these filaments. See text for further details.

myosin head to extend toward the thin filament (Fig. 2-5). Within the globular subfragment-1 portion of the myosin head resides the myosin ATPase. Once attached to actin, the globular head of myosin undergoes a change in conformation, exerting a pull on the myosin cross-bridge. Once the head moves, the link between myosin and actin is broken. ATP, which has been split to provide the energy for the motion of the cross-bridge link, is resynthesized from ADP, and the myosin head is repositioned. With sliding of the filaments relative to one another, the globular head becomes positioned to reattach and to repeat the contraction process. In this manner a directional force is created that when repeated at multiple sites propels the filaments in opposite directions. ATP, bound to HMM and converted to an activated complex, is split to ADP and Pi via the activity of myosin ATPase, leading to detachment of the bridge.

The freed HMM head is then rephosphorylated to reconstitute the ATP, and the bridge is thus reset to make contact farther along the actin filament and repeat the cycle. In this scheme, force and shortening result from a change in orientation of the myosin head that stretches the elastic structure of the cross-bridge link. The overall length of the myosin and actin filaments does not change during contraction. Contraction involves a relative change in position of these two sets of filaments, i.e., actin filaments are displaced by the force-generating process at many bridge sites, forming the basis of the sliding-filament hypothesis of contraction.[91,93]

The sliding-filament hypothesis is now generally accepted as the mechanism of contraction of striated muscle (i.e., skeletal and cardiac muscle). According to this hypothesis, if the filaments are prevented from sliding by an external restraint, the formation of cross-bridge links will result in a rise in force or tension in the muscle to a level that will be determined by the number of cross-bridge links per sarcomere. The latter largely depends on the degree of overlap between the interdigitating arrays of filaments. Thus the study of length–tension relationships in striated muscle may be expected to provide information about the number of cross-bridge links formed at various muscle lengths.

When the loading conditions permit sliding of the filaments, the muscle shortens at a velocity that is inversely related to the force opposing motion. The underlying mechanism for this force–velocity relationship might be conceptualized as follows. Tension or force in a muscle depends on the number of cross-bridges attached at a given time. During muscle shortening, cross-bridge links must be detached and reformed as the filaments slide past one another. The process of bridge formation and detachment must require a finite time. Thus the faster muscle shortens, the fewer will be the number of cross-bridges attached at any given time, and the lower will be the tension in the muscle. The maximal velocity of shortening (V_{max}) will occur when there is no force opposing shortening. Theoretically, V_{max} reflects the maximal rate of turnover of cross-bridge links, and according to some workers[93] V_{max} should be independent of the number of cross-bridges participating so long as internal forces opposing shortening do not arise. Thus the force–velocity relationship of a muscle should give potentially useful information concerning cross-bridge properties.

MECHANICS OF CARDIAC MUSCLE CONTRACTION

During the past 15 years there has been considerable interest in characterizing the mechanical properties of cardiac muscle. The reason for this interest is twofold. First, insight has been sought into the basic mechanisms of cardiac muscle contraction. Second, a conceptual basis for evaluating a "good" versus a "bad" heart has been sought. The first aim seeks biophysical and physiologic truths, while the second aim seeks to find whatever is useful. Unfortunately, the various controversies related to muscle mechanics, whether concerned with concept or application, are commonly confused with one another. This serves to further cloud the basic issues.

Analog Models of Muscle

Although one would wish to think in terms of cross-bridge activity when considering cardiac muscle mechanics, it is not possible to make measurements at anything near that level in cardiac muscle. The best studies have utilized isolated papillary muscles that are much more complex than a single sarcomere or even a single cell. Such studies have relied heavily on the classical approach to skeletal muscle mechanics developed by Hill[94] for the study of the mechanical properties of whole frog muscles. This approach depends heavily on a conceptual model of muscle postulating a contractile component and one or more passive elastic components (Fig. 2-6). The contractile element (CE) represents the actively contracting portion of the muscle that generates force and shortening. A passive series-elastic (SE) element with the properties of an undamped spring transmits the force generated by the contractile component to the exterior. At rest the CE is quite extensible, so that resting tension is sustained by another elastic component arranged in parallel—the parallel-elastic (PE) element. The exact arrangement of these elements has been the subject of much speculation. No single configuration has been uniformly successful in accounting for the mechanical properties of cardiac muscle. The most commonly used models differ only in the relation of the PE element to the other elements (Fig. 2-6). In the Maxwell (Model I) configuration the PE element is in parallel with both the CE and the SE element. The Voigt model (Model II) is similar except that the PE element is in parallel with only the CE. In both models the activated CE shortens and stretches the SE element during isometric contraction, and force develops in accordance with the stress–strain properties of the SE element.

As will be detailed subsequently, the analysis of cardiac muscle mechanics in terms of the Hill model has been extremely useful in providing a framework for viewing experimental findings in intact muscle in a number of complex settings as well as interrelating mechanical and energetic phenomena. However, the applicability of this simple model to skeletal muscle has recently been challenged by experiments[95] that support and extend the view that the SE element in skeletal muscle resides in

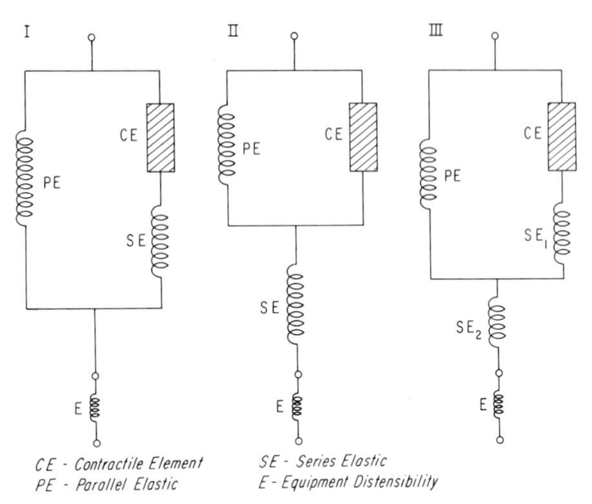

Fig. 2-6. Mechanical models of cardiac muscle. Models of muscle with various arrangements of the specific components. Model I represents the Maxwell arrangement in which the parallel-elastic (PE) component is in parallel with both the contractile element (CE) and the series-elastic (SE) component. Model II represents the Voigt configuration in which the PE element is parallel to the CE alone. In this model resting tension is supported by both the PE and the SE elements. In model I the PE alone supports the resting tension. In model III, models I and II have been combined so that the SE element is split into two components. The SE element is in series with the CE and in parallel with the PE element. SE_2 is in series with the PE element as well as with SE_1.

the cross-bridge links between actin and myosin. Such a mechanism would imply, at least in skeletal muscle, that the rise in force development is largely due to the progressive formation of more cross-bridge links rather than to extension of a true SE element. Noble and Else[96] and Pollack et al.[97] have recently reported that the compliance of the SE element in cardiac muscle varies with time after activation, suggesting that cross-bridge elasticity may be an important component of the series-elastic compliance measured in heart muscle. Their inability to define a unique series-elastic compliance with the properties of a passive nonlinear spring suggests that the traditional Hill model for cardiac muscle may not be useful for quantitative characterization of CE properties. These findings disagree with earlier[98] reports as well as with more recent studies from our laboratory[99,100] in which the series-elastic compliance of cardiac muscle varied in a linear fashion with force and was independent of time. The reasons for the differences between these studies probably relate to differences in experimental technique and to the extrapolation procedure used to obtain data in the studies by Noble and Else[96] and Pollack et al.[97] Such extrapolations were avoided in our laboratory by directly measuring muscle force and length continuously throughout the release interval. However, because of the exponential shape of the stress–strain relationship for cardiac muscle series elasticity derived from real-time measurements, it was not possible to exclude appreciable series-elastic compliance residing in attached myofilament cross-bridges.

An additional consideration is the finding in cardiac muscle of a very compliant SE element amounting to an extension of 5%–8% of muscle length for peak isometric force, compared to less than 1% extension in skeletal muscle. Assuming that the SE element residing in the cross-bridges is the same for both heart and skeletal muscle, an additional 4%–7% of series elasticity remains to be explained in cardiac muscle. The site of this large elasticity is as yet unknown, but could reflect such structural factors as branching of cardiac fibers.[100] Moreover, shortening of cardiac sarcomeres during so-called isometric contraction has been observed,[101] so that some of the compliance of the cardiac series elasticity may be thought of in terms of sliding of filaments even during isometric contraction. With these considerations in mind, a literal interpretation of models is not warranted, and series elasticity might best be considered to be a passive spring whose site is widely distributed in cardiac muscle. While this does not deny that some of this compliance may be in the cross-bridges per se, the use of a Hill type of model, at least as a working model for the analysis of cardiac muscle contraction, appears warranted.

Length–Tension Relationships

When cardiac muscle is stimulated it contracts in different ways depending on the loading conditions. The muscle can be permitted to shorten isotonically with just a preload (e.g., Fig. 2-7, panel I, points *A* to *I*). Alternatively, if the ends of the muscle are fixed to prevent shortening, by definition isometric force is developed (Fig. 2-7, panel I, points *A* to *F*). Under physiologic conditions (i.e., in the beating heart in vivo) contraction involves a sequential combination of an isometric and an isotonic contraction, termed an afterloaded isotonic contraction, i.e., an afterloaded contraction in which shortening occurs at a constant force. Thus, in such a contraction, starting from point *A* in Figure 2-7 panel I, force is generated until it equals an imposed load (afterload) at

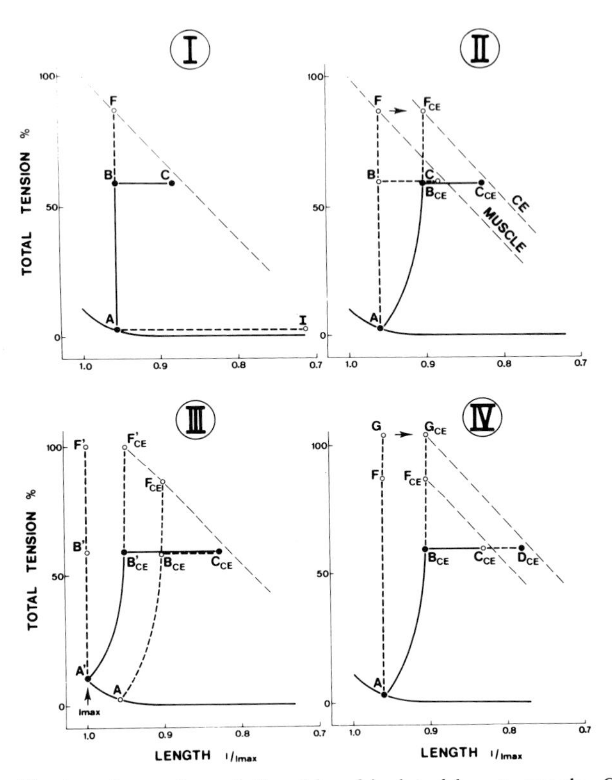

Fig. 2-7. The length–tension relationship of isolated heart muscle. On the ordinate is shown total tension with 100% corresponding to the total tension that is developed by the muscle at a length where developed tension is maximum (L_{max}). (I) The lower solid line portrays the passive or resting tension curve of the non-stimulated muscle. The diagonal dash line represents the total isometric force (resting + developed force) that can be generated from each resting muscle length. With the muscle starting its contraction at point A with a small resting tension (or preload), A to I represents the extent of isotonic shortening with this preload alone. A to F represents the total force generated if the ends of the muscle are fixed. With an afterload amounting to 65% of total tension the muscle will develop isometric force from point A to point B and then shorten with this load between point B and point C. (II) The length–tension curve of panel I is shown as a vertical dash line. Starting from point A, the generation of force by the CE in the muscle stretches the SE element so that force rises while the CE shortens to B_{CE}. Were the muscle isometric it would shorten to F_{CE}. The dash lines to the upper right represent the length–tension curve for the muscle and the CE that is generated from the shortening of the CE at the expense of the SE element. Under afterloaded conditions the CE shortens between B_{CE} and C_{CE} rather than from B to C (see text). (III) The effects of increasing muscle length from A to A' on isometric force development when represented in terms of CE

point *B*. The muscle then shortens with the load (afterload + preload) until the length–active-tension curve is reached at point *C*. Relaxation then begins, isotonic lengthening (i.e., lengthening at a constant force) occurs, and then force itself falls. The preload then stretches the muscle back to its starting length at point *A*.

In all forms of striated muscle, including cardiac muscle, the force of contraction (i.e., the active tension developed) depends on initial muscle length (Fig. 2-8). By adjusting the initial degree of stretch of the resting muscle, a length of the resting muscle is found at which the resultant isometric force is maximal; this resting muscle length is termed L_{max}. The force or load placed on the muscle necessary to stretch the resting muscle to its initial length is termed the preload or resting tension, and the relationship between preload and the length of the resting muscle is the length–resting tension relationship. The relationship between developed isometric tension (i.e., the increment in tension during contraction) and initial muscle length constitutes the length–active tension relationship. If the initial muscle length is either increased or decreased on either side of L_{max}, actively developed isometric force and rate of force development (df/dt) decline. The length–active tension relationship at lengths below L_{max} is termed the ascending limb of the curve; the portion beyond L_{max} is the descending limb of the curve.

In the presence of high concentrations of Ca^{++} in the medium perfusing or surrounding the muscle, or following the addition of agents such as catecholamines or digitalis glycosides, the force and rate of contraction at any muscle length are augmented[102,103] (Fig. 2-9). This shifts the length–active tension curve upward, so that the isometric force of contraction and df/dt at any length are higher. Maximum force is still reached at the same length (i.e., L_{max} is not shifted). In addition, the length–resting tension curve is not altered (Fig. 2-9).

In this description, the whole muscle has been considered. In terms of the Hill model, development of force can be reinterpreted in terms of a

---

shortening. Substantial shortening of the CE takes place during isometric force development. With the increase in the initial muscle length, isometric force is increased from F_{CE} to F'_{CE}. Nevertheless, isotonic afterloaded shortening with an afterload still proceeds from B_{CE} to C_{CE}, which is the same final length that is reached when the muscle shortens from B'_{CE} to C_{CE}. (IV) The effects of norepinephrine on the length–tension curve in the course of afterloaded isotonic shortening. The addition of norepinephrine increases the force of the CE from F_{CE} to G_{CE} and augments the extent of isotonic shortening from B_{CE} to C_{CE} to D_{CE}. The curve of the SE component or the resting-length–tension curve is not altered by this inotropic intervention.

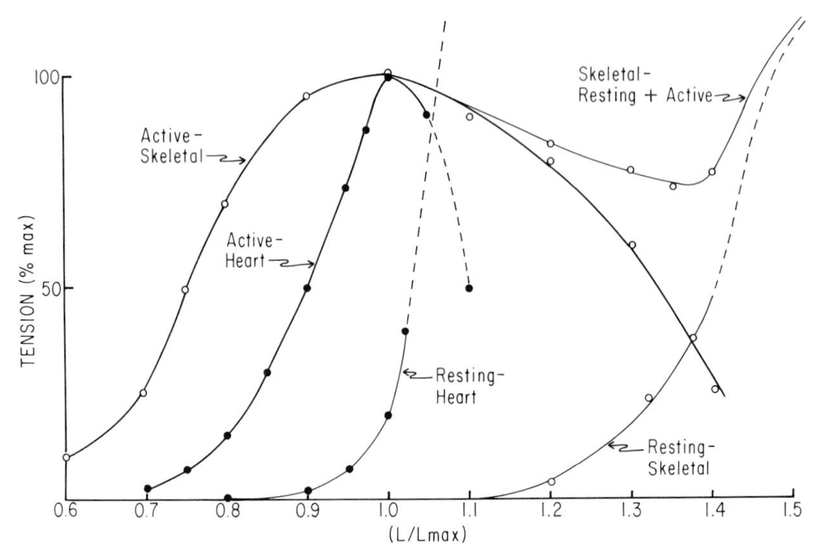

Fig. 2-8. The relation of resting tension and actively developed isometric tension to initial muscle length for both heart and skeletal muscle. The curves for heart muscle are shown in solid circles, while the curves for skeletal muscle are shown in open circles. Length has been normalized relative to L_{max}, that length where actively developed isometric tension is maximum. Tension has been normalized considering this maximum tension as 100%, and changes with initial muscle length related to this starting length. At L_{max} skeletal muscle demonstrates little or no resting tension, while cardiac muscle demonstrates substantial resting tension. At L_{max} a minor change in either shortening or lengthening cardiac muscle leads to a substantial decline in actively developed isometric tension. Changes in developed tension in skeletal muscle are much less marked for similar changes in initial muscle length.

CE and SE element (Fig. 2-10). Resting tension, which is small relative to actively developed force, is borne by the PE element. Following activation (Fig. 2-7, panel II) the CE shortens at the expense of the springlike SE element, and force is developed. Due to this shortening of the CE, which is substantial in heart muscle, isometric force is associated with a decrease in CE length to F_{CE}, rather than remaining at F. In the afterloaded isotonic contraction the CE shortens to C_{CE}, rather than to C. Thus the length–active tension curve for the CE lies to the right of the length–active tension curve of the muscle as a whole. Moreover, given the complaint character of the SE elements, the CE is always shortening while force is being generated, even though overall muscle length is kept constant.

When initial muscle length is increased (Fig. 2-7, panel III, A to A'),

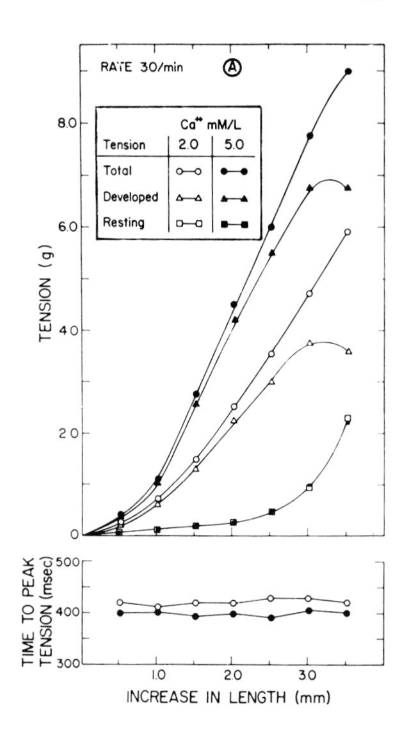

Fig. 2-9. The effects of increased calcium on the relations between muscle length and tension in an isolated cat papillary muscle. The level of calcium in the perfusing medium has been increased from 2.0 mmole to 5.0 mmole. The relationship between resting muscle length and tension is not altered. However, the tension development at any given muscle length is increased, although that length where maximum tension exists (L_{max}) is not altered. Total tension equals the developed tension plus the resting tension.

developed force is increased from F_{CE} to F'_{CE}, and the extent of afterloaded isotonic shortening occurs from B'_{CE} to C_{CE} rather than from B_{CE} to C_{CE}. It should be noted that not only does substantial shortening of the CE occur during isometric force development, the CE also shortens to the same final length (i.e., C_{CE}) despite differing starting lengths.

When the contractile state of the muscle is augmented, e.g., by the addition of norepinephrine or other inotropic agents (Fig. 2-7, panel IV), isometric force development is augmented (F_{CE} to G_{CE}). Again, internal shortening of the CE at the expense of the SE element produces isometric force at the shorter CE length (G_{CE}) rather than at G. Neither the length–resting tension curve nor the load–extension curve of the SE element is altered by the inotropic intervention.

Given the properties of the CE and the SE element of heart muscle, the rate of force development (df/dt) can readily be understood. As already noted, the load–extension curve of the SE element is exponential in form, and its stiffness or modulus of elasticity (df/dl) equals $Kf + c$ (where K is the slope of the curve and c is a constant). It is axiomatic that $df/dt = dl/dt_{CE} \times df/dl_{SE}$, where dl/dt_{CE} is the velocity of shortening of the CE. Substituting Kf for df/dl and disregarding the c constant, which is small, the equation can be rewritten as as $dl/dt = (df/dt/Kf$.

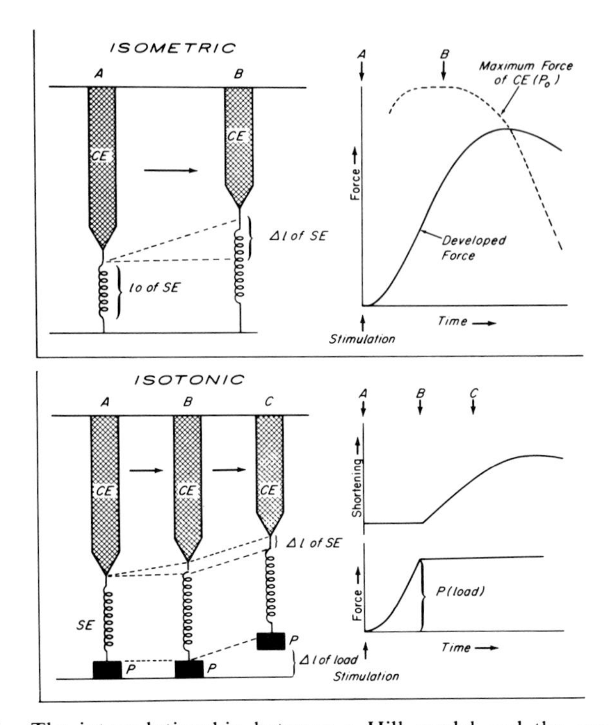

Fig. 2-10. The interrelationship between a Hill model and the generation of isometric force and afterloaded isotonic shortening. Isometric contraction is shown (top), and an afterloaded isotonic contraction is portrayed (bottom). In the isometric model the resting contractile element (CE) is in series with the series elastic (SE) element. With activation the CE shortens and stretches the SE element, with the development of force across the system. To the right the dash line represents the course of activation or the capacity to develop force by the CE. It is apparent that the CE develops a capacity for force generation prior to its manifestation of external force. This is due to the fact that time is required to stretch the SE element. Even with very rapid full activation of the CE, a delay in the development of force would be due to this time-dependent stretching of the SE element. In panel A (bottom) the CE is attached to a load *P* through the SE element. When contraction begins the CE shortens and lengthens the SE element, with the development of force. When this force equals the load *P* the load will move off of the base. From this point on the SE element is stretched only by the *P* and thus will remain constant length between *B* and *C*. Thus all shortening of the muscle between *B* and *C* will represent shortening of the CE. For this reason, afterloaded isotonic shortening has been used to study the course of shortening of the CE independent of the SE element. This simplification requires that the contribution from resting tension and the parallel-elastic (PE) element be trivial.

Accordingly, one may predict the rate of force development from the velocity of shortening of the CE, or *vice versa,* depending on what information is given as long as K is known.

Velocity–Length Relationships

In Figure 2-7 panel I the shortening of the muscle between points B and C represents the physiologic course of contraction. As shown in Figures 2-10 and 2-7, isometric force is generated until the load to be carried (afterload) is reached, and then shortening ensues. The muscle shortens with a velocity that depends to varying degrees on the instant during contraction that it is measured, on the instantaneous muscle length, and on the load that is being carried. Thus the velocity of shortening rises to a peak quite early in the course of contraction and then declines as shortening ensues. The course of relaxation depends on the afterload. As the afterload is increased the time to peak shortening is prolonged and the rate of isometric relaxation is progressively delayed.[104]

The relationship between velocity of shortening and muscle length at every instant during shortening constitutes the *velocity–length relationship.* When the afterload is increased the velocity of shortening and the extent of shortening are decreased (Fig. 2-11, panel III). The velocity of shortening at any given muscle length is also reduced. When initial muscle length is increased (Fig. 2-11, panel II) the resting load (preload) is increased. With the onset of contraction the rate of force development (df/dt) is increased. However, when the total load (preload plus afterload) is unchanged but the preload is increased, velocity tends to rise to about the same level and then follows the same course as the muscle beginning to shorten from the shorter length.[105,106] Moreover, with the same total load, shortening proceeds to about the same final length, although total shortening is greater from the longer length. Following an inotropic intervention such as the addition of norepinephrine, the dl/dt, the peak velocity of shortening at any load and muscle length, and the extent of shortening are all augmented (Fig. 2-11, panel IV).[105,106]

In terms of the length–tension curve (Fig. 2-7), an increase in initial muscle length (panel III) leads to increased force development. However, although the muscle can contract a longer distance with the same load (i.e., B'_{CE} to C_{CE} rather than B_{CE} to C_{CE}), the contractile-element length at the end of shortening is approximately the same for the same afterload. When contractility is augmented (panel IV), the contractile units shorten to a shorter length (i.e., from B_{CE} to D_{CE} rather than C_{CE}).

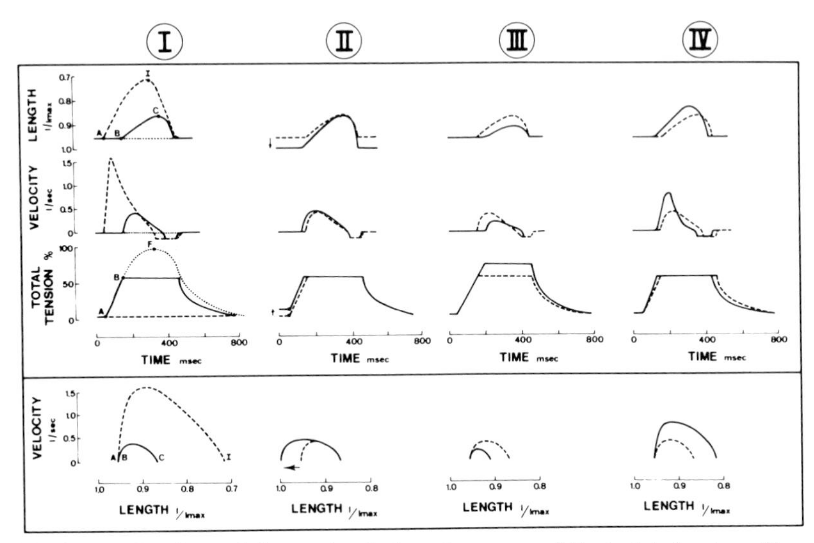

Fig. 2-11. Tracings of the mechanical performance of the isolated cat papillary muscle. From above downward are shown muscle length, velocity of shortening obtained as a derivative of the length tracing, and tension development relative to time. In the lower panel velocity of shortening has been plotted as a function of muscle length during the course of contraction as shown in the length and velocity tracings given above. The velocity–length tracing represents the phase of shortening between the start at point *A* or *B* and ending at points *I* and *C* in panel I. The phase of relaxation is not portrayed. (I) The dash line depicts the course of isotonic shortening with the preload alone. The solid line depicts an afterloaded isotonic contraction. On the tension tracing, point *A* is where force development begins, rising until point *B* is reached, which is equivalent to the load. Shortening then ensues and continues between *B* and *C*, as shown on the shortening trace, following which relaxation occurs. The relationship between the length change and shortening as shown in the upper two panels is plotted in the lower section giving the velocity–length relationship. On the force tracing, the course that force would have followed had shortening not occurred is shown by the dotted line. (II) The effects of increasing initial muscle length on the afterloaded isotonic contraction. With an increase in length, resting tension (preload) is increased as shown in the tension trace. The muscle now generates tension at a faster rate (i.e., df/dt is increased) and shortens a greater distance. Nevertheless the muscle shortens to approximately the same final length for the same total load. Plotted below is the relationship of velocity of shortening to muscle length during the course of shortening. When initial muscle length is increased the velocity trace starts from a longer length. Nevertheless, shortening tends to follow the same common pathway to reach the same final length of shortening. (III) The effects of increasing afterload. The initial course of tension development remains the same, but the course rises to a higher level when the afterload is increased. Shortening then begins slightly later in time and at a slower rate, i.e., the velocity

Force–Velocity Relationships

A fundamental property of active muscle is the inverse relationship between peak velocity of shortening and load that is carried, i.e., the force–velocity relationship. This relationship is obtained by setting by setting the initial muscle length with a small preload and by plotting the peak velocity of isotonic shortening with various afterloads (Fig. 2-12). When considered in terms of the Hill model (Fig. 2-10), the course of isotonic shortening reflects the characteristics of the CE, since the length of the SE element is constant at any level of afterload. As the afterload is increased, the peak velocity of isotonic shortening is reduced (Figs. 2-11 and 2-12). When the load is so great that no shortening occurs, isometric force is produced. Further, when the afterload is increased, contraction is prolonged and the isometric phase of relaxation is prolonged.[107] When the curve relating load and velocity is extrapolated back to zero load, the maximum velocity of shortening (V_{max}) is obtained.

This inverse relationship between force and speed can be fitted by a displaced hyperbola, at least with smaller afterloads, and is described by a well-known equation* of Hill.[94] In some species, such as the rabbit,[111] or in the cat papillary muscle in the presence of caffeine,[112–114] the force–peak-velocity curve is nonhyperbolic, even at low loads. This fact is responsible in part for some of the controversy that has developed relative to the use of force–velocity relationships in evaluating cardiac

*Using the frog sartorius muscle, Hill[94] demonstrated that the relationship between force or load (P) and velocity (V) could be described as a displaced hyperbola: $(P + a)V = b(P_0 - P)$. The constants a and b provide the displacement of the hyperbola from the axes, while P_0 represents isometric force. When P is zero, $V = V_{max} P_0 b/a$. Hill utilized thermopile techniques to study the heat flux from muscle and demonstrated that a could be correlated with the extent of shortening, independent of velocity, while b was correlated with the rate at which heat was produced. Thus a was termed the heat of shortening, while b was related to V_{max}. More recently, Hill[108] revised this conclusion by showing that the constant for the heat of shortening, now termed α, depended on load. Nevertheless, a and α could be correlated relative to load. The complexities of relating heat measurement to changes in chemical energy have been reviewed,[109] as have studies on heat measurements in heart muscle.[110]

---

of shortening is slower when the afterload is increased. Furthermore, the muscle does not shorten as far. The relationship between velocity at any length as plotted below also shows that velocity at any length is decreased. (IV) The effects of the addition of norepinephrine to the afterloaded isotonic contraction. The rate of tension development (df/dt) is increased. Shortening begins slightly earlier in time and at a higher velocity. Moreover, the extent of shortening is increased. The relationship between velocity and length shows that the velocity at any given length is increased.

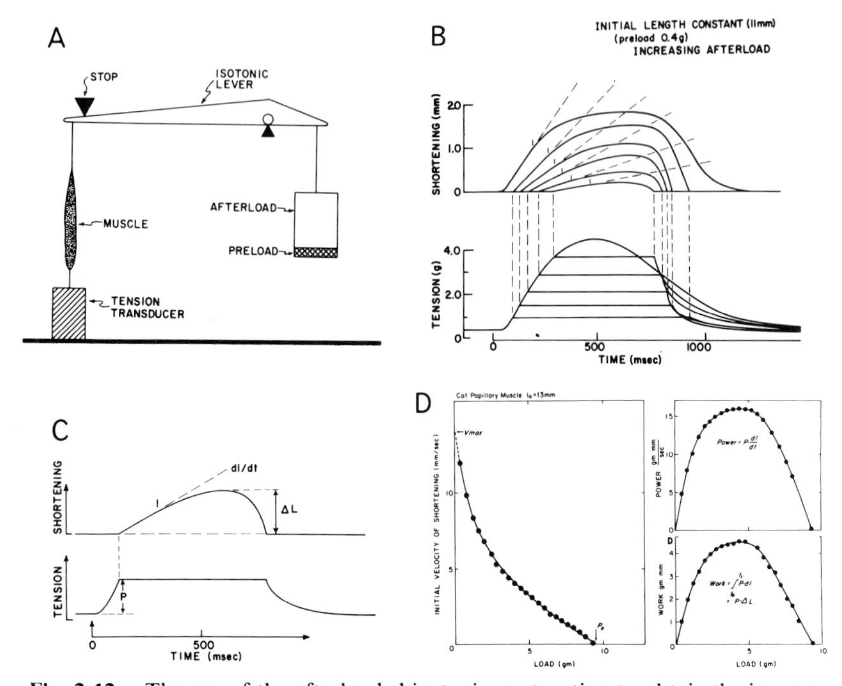

Fig. 2-12. The use of the afterloaded isotonic contraction to obtain the inverse force–velocity relationship in the papillary muscle preparation. (A) Diagrammatic representation of an isotonic lever system. A papillary muscle is placed in a bath (not shown) of Krebs bicarbonate solution and stimulated by electrodes along its lateral aspect. The lower end of the muscle is attached to an extension from a tension transducer, while the upper free end is attached to the end of a lever system that is free to move. The fulcrum of the lever system is shown toward the right. Initially the stop is not above the tip of the lever which is above the muscle. A small weight termed a preload is placed on the opposite end of the lever, and this preload will stretch the muscle to a length consistent with its resting length–tension relationship. The stop is then fixed above the tip of the lever, so that any added weight over and above the preload will not be sensed by the muscle until it contracts. Additional loads can then be added to the preload (afterloads). Total load equals the preload plus the afterload. (B) Tracings of the afterloaded isotonic contraction. Below is shown tension development and shortening as would be recorded on the tension transducer shown above in panel A with the isotonic lever. The contraction is shown relative to time on the abscissa. Following stimulation at time zero, there is a short latency period followed by the generation of isometric force. When the force p equals the load, shortening begins as shown in the upper half of the panel. The maximum velocity of shortening is reached shortly after shortening ensues, and the tangent to this slope (dl/dt) approximates the maximum velocity of shortening with this particular load. ΔL denotes the extent of shortening. Subsequently the muscle

performance. In skeletal muscle the force–velocity curve is generally obtained during the steady state of tetanic contractions; under physiologic conditions cardiac muscle cannot be tetanized,[115] and in general the analysis of cardiac contractions is limited to a single twitch. Thus, in some circumstances, force–velocity curves in heart muscle may be subject to such factors as changing levels of activation.

Two clear alternative explanations have been developed for nonhyperbolic force–velocity curves of isolated cardiac muscle at low loads. Recognizing the fact that peak velocities occur at progressively later times during contraction and at progressively shorter contractile-element lengths where velocity and force development may be falling, nonhyperbolic curves would occur if the muscle were activated very slowly and if velocity of shortening were highly dependent on instantaneous muscle length. In heart muscle the onset of the "active state" or mechanical activity is slower than in skeletal muscle.[104,116] Moreover, the onset of the active state appears to be substantially slower in the rabbit and frog heart than in the cat heart.[114,117] When expressed in terms of the time required to reach peak isometric tension (TTP), maximum active state appears to be achieved earlier in cat heart muscle than in rabbit myocardium. Thus in the cat heart most isotonic velocity measures are made when the muscle is fully or nearly fully activated, while in the rabbit the active state is still rising at the point where measurements are being made. Accordingly, velocities from afterloaded isotonic contractions obtained from rabbit heart muscle are too low, and the resultant force–velocity curve may actually appear linear. Furthermore, when the preload is removed by electronic means from the activated cat papillary

---

elongates, following which isometric relaxation ensues. (C) The effects of increasing afterloads on the course of tension development and subsequent shortening. Tension and shortening are displayed on the ordinate, and time is given on the abscissa. Several superimposed contractions are displayed. The muscle develops a force equal to the afterload, followed by shortening. One notes that as the afterload is increased, velocity of shortening (as represented by the dashed lines) and the extent of shortening decrease. (D) Velocity of shortening plotted relative to load: the force–velocity relationship. On the abscissa the slope of the initial portion of the shortening curve as shown above has been plotted relative to several total loads. As the load is increased the velocity of shortening decreases. When the load is such that no external shortening is recorded, velocity is zero and the force is equivalent to the isometric contraction (P_0). When the curve is extrapolated back to zero load, V_{max} is obtained. Also shown are power and work relative to increasing afterloads. Both power and work are zero when the load is zero, or when only isometric force is developed. Both power and work peak at an intermediate load.

muscle, maximum velocity of shortening is rapidly obtained[118] and is maintained until the muscle has shortened to a length less than 87% of L_{max}.[119] The level of this plateau of velocity of shortening with virtually zero load correlates well with the V_{max} that is obtained by extrapolation of afterload isotonic contractions.[120,121] In addition, studies in which muscle loads have been altered during shortening have supported the view that activation of the contracting heart muscle is not changing greatly between shortly after the onset of contraction to just before peak tension development.[119] Thus instances exist where a very slow onset of mechanical activation may preclude the interpretation of force–peak-velocity curves; this is usually the case with rabbit heart muscle or in the cat following the addition of caffeine.[114,116] However, in most species and in experimental circumstances these phenomena do not significantly alter force–peak velocity curves, since the mechanical activity is established early in contraction, is maintained for the major time during which the measurements are made, and is not highly dependent on length along the critical lengths where it is being measured.

At higher loads the force–peak velocity curve may fall off and intersect the force axis at lower forces than the hyperbolic curve would predict. This could be explained by the limited duration of the active state. Also, in the process of force generation the CE shortens at the expense of the SE element and thus moves down the length–active tension curve where it produces less force,[122] preventing the force–velocity curve from assuming a true hyperbolic shape.[105]

The force–velocity relationship obtained from afterloaded isotonic contractions has been of special interest, since it helps to distinguish the two major ways in which cardiac performance is altered, i.e., changing initial muscle length (the basis of the Frank-Starling phenomenon) and changing contractile state brought about by inotropic interventions.[102] When initial muscle length is increased (Fig. 2-13), increases in the rate of force development (df/dt) and peak isometric force (P_0) are noted with little change in the time to peak tension. The force–peak velocity curves are shifted to the right (Fig. 2-13), so that both P_0 and the velocity of shortening with a given load are augmented. However, when the force–velocity curves are extrapolated back to zero load, V_{max} shows little or no change when initial muscle lengths are varied within a physiologic range, although at shorter initial lengths V_{max} tends to fall.

When the contractile state is altered by an inotropic intervention such as a catecholamine, digitalis glycosides, or increased levels of calcium, the force–velocity curve is shifted upward and to the right with increases in P_0 and velocity of shortening at any load (Fig. 2-14). However, in contrast to the effects of changes in initial muscle length, an increase

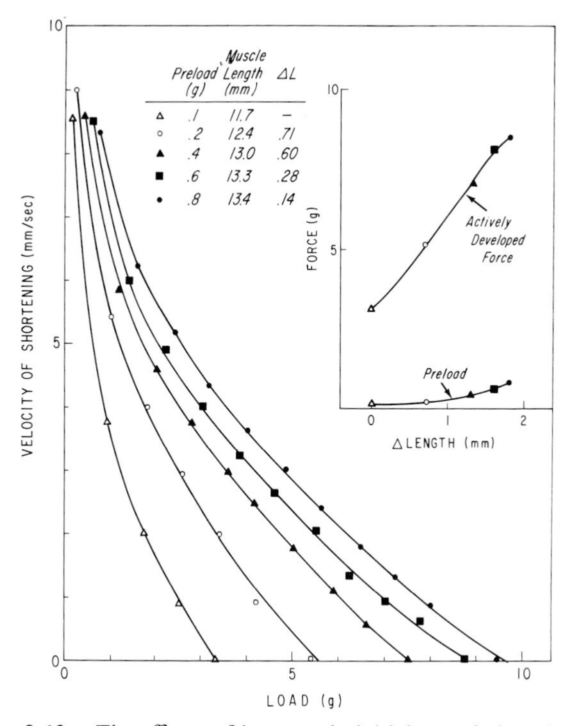

Fig. 2-13. The effects of increase in initial muscle length on the force–velocity relationship. The relationship between peak velocity of afterloaded isotonic shortening and total load is shown for several initial muscle lengths of a cat papillary muscle. The inset to the right shows the resting and developed active forces at these various lengths. When initial muscle length is increased, the actively developed force is augmented, as is the velocity of shortening with any individual load. Nevertheless the maximum velocity of shortening with the preload alone is little altered. Moreover, were these curves to be extrapolated back to zero load (V_{max}) this value would also show little if any change.

in V_{max} is also observed. At the same time the rate of isometric force development (df/dt) is increased, while the time to peak tension (TTP) tends to be shortened. Both the extent of shortening and P_0 are increased. The TTP is most markedly shortened by catecholamines and less so by augmenting [Ca^{++}] or digitalis glycosides. Increases in initial muscle length produce little change or even a slight prolongation of TTP.

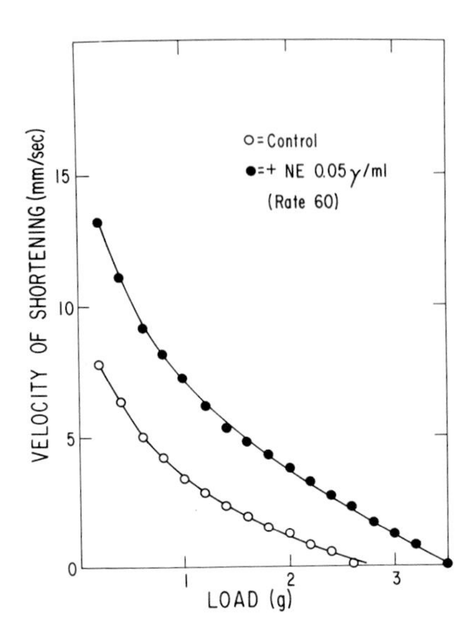

Fig. 2-14. The effects of the addition of norepinephrine on the force–velocity relationship of the cat papillary muscle. Norepinephrine induces an increase in the velocity of shortening at any load. Furthermore, the maximum force of contraction on the abscissa is also increased, as is the maximum velocity of unloaded shortening (V_{max}).

Force–Velocity–Length Relationships

From the foregoing, the interdependence among force, velocity, and length is apparent. Thus these variables can be plotted as a force–velocity–length diagram, which provides an integrated view of the way in which the muscle can perform.[105,106,123] Moreover, with corrections for series elasticity, contractile element motion can be viewed during both the isometric and isotonic phases of contraction (Fig. 2-15).

In Figure 2-15, panel I, the force–velocity–length relationship has been plotted using data derived from recordings of shortening relative to muscle length in a series of afterloaded contractions. The base of this diagram shows a length–tension relationship; the projection to the rear illustrates the length–velocity relationship for the unloaded muscle; the projection to the left shows the force–velocity relationship. Point *A* on the base corresponds to the preload starting point. With activation, the contractile elements rise onto a hypothetical force–velocity curve, with force increasing until afterload is reached at point *B*, after which shortening proceeds across the surface until point *C* is reached. In any given contraction the muscle moves in a predictable manner across the surface describing this relationship among force, length, and velocity.

In Figure 2-15, panel II, the force–velocity–length relationship of the CE (rather than the muscle illustrated in panel I) is shown. The relationship has now been corrected for series elasticity, with the CE shortening at the expense of the SE element as isometric force is generated. The force–velocity curve is shown as a dotted line superim-

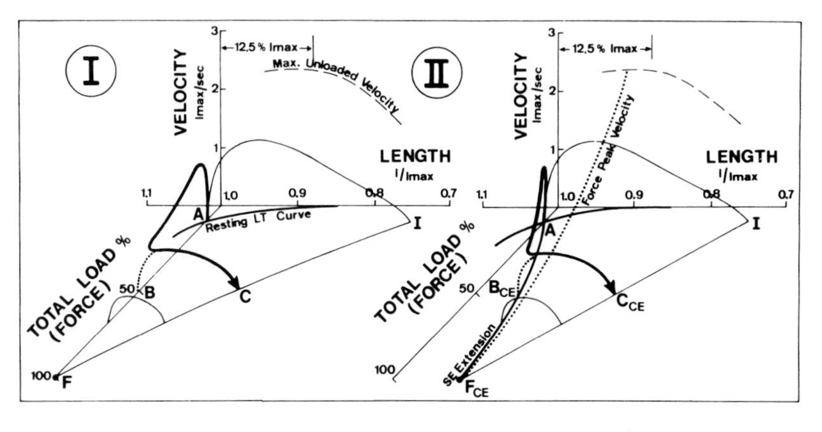

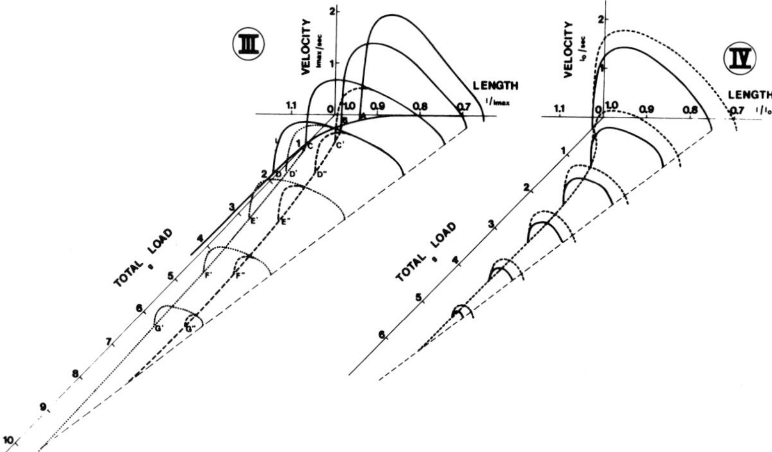

Fig. 2-15. The three-dimensional representation of the force–velocity–length relationships of the cat papillary muscle. (I) The velocity–length relationships of isotonic contractions obtained at $L_{\max}$ have been replotted as a function of total load. Such a reconstruction can be obtained by plotting the velocity–length relationships as obtained in panel I of Figure 2-7 at various loads. The course of velocity of a hypothetical afterloaded isotonic contraction is superimposed (thick line). The velocity of shortening during the isometric phase of the contraction has been theoretically derived from a two-component model. Velocity rapidly rises to that appropriate for the plane of this three-dimensional composite. During isometric contraction the CE velocity will fall as force rises. This velocity is not seen, but is expressed in terms of the rate of force development (i.e., df/dt). At point B the force development equals the load, and external shortening can then proceed between points B and C. The velocity of shortening between B and C will depend on the level of the force–velocity–length plane. On the right is portrayed the velocity–length relationship, and the maximum unloaded velocity of shortening ($V_{\max}$) is shown. Reflection of the plane of the force–velocity–length

posed on the surface. When initial muscle length is increased, contraction starts from a point farther to the left along the base. Velocity of shortening rises and meets the same surface that it did when shortening from the initially shorter length. The surface relating velocity of shortening to load and length stays the same. Moreover, although shortening begins from a longer initial length, it proceeds to approximately the same end-systolic length.

Despite some reservations,[124–126] the bulk of available evidence suggests that for a given state of contractility the surface created by the force, velocity, and length relationship is unique. Two facts have been clearly demonstrated that support the validity of such a surface in characterizing contractility. First, strong support has been lent to the view that the development of full contractile activity in heart muscle is relatively rapid in onset when considered in terms of velocity[118] or in terms of onset of the interrelationship between force, velocity, and length.[119] The unique surface of force–velocity–length can actually be viewed as maximum activity for that state of contractility. This surface is reached quite rapidly and is maintained for a considerable time, approaching the point where the maximal extent of shortening occurs.[119] This conclusion is in contrast to the findings of previous studies that have claimed a very slow onset and nonsteady state for the maximal activity of heart muscle.[116,127–129] This latter conclusion may be best explained by limitations of methodology.[117] Second, as illustrated in Figure 2-15 panel III, the velocity of shortening is the same at any given length during contraction for the same total load and is independent of the length from

---

relationship to the right defines the force–velocity relationship, while on the base the length–tension curve is reflected. (II) The force–velocity–length relationships of the same muscle as shown in panel I after correction for extension of the SE component. The entire curve is moved over and to the right. The dash line shown on the plane created by the force–velocity–length relationship represents the force–velocity curve as obtained from afterloaded contractions. (III) The effect of a positive inotropic intervention (dash line) on the force–velocity–length relationship obtained from a series of afterloaded isotonic contractions. The velocity of shortening at any given muscle length is augmented, so that the entire surface relating force–velocity and length is increased. Furthermore, the extent of shortening is augmented. Such an elevation in the force–velocity–length relationship and the surface that it creates characterizes an increase in contractility. The projection of this surface to the right would be characterized by an increase in V_{max}. (IV) Influence of a positive inotropic intervention (dotted line) on the force–velocity–length surface described by a series of afterloaded contractions starting from an initial muscle length (L_0) somewhere between A and B in panel III. The entire surface is raised at all loads and lengths.

where the contraction began or the time at which it attained that length.[119] Therefore the time-independent portion of the force–velocity–length relationship serves to characterize a given state of contractility.

Although the force–velocity–length surface is independent of time and initial length, it is substantially shifted by inotropic interventions such as catecholamines (Fig. 2-15, panel IV). For any given load and length the velocity of shortening is markedly altered. A new surface is created among force–velocity–length, indicating an altered contractility.

While the midportion of the force–velocity–length relationship characterizes the range of normal physiologic performance, the analysis of this portion of the surface is quite complex. Therefore the extremities of the surface take on both theoretical and practical interest. When shortening of muscle is prevented, isometric force is registered, which when corrected for series elasticity provides us with the length–active-tension curve of the CE (Fig. 2-15, panel II, projection on the basal plane). When the load is progressively reduced, velocity of shortening increases, and the inverse curve forming the force–velocity relationship is described (Fig. 2-15, panel II, dotted line and projection on the left plane). When the force approaches zero, the velocity of shortening reaches a maximum (V_{max}) (Fig. 2-15, panel II, dash line on the right plane). Thus V_{max} is merely one aspect of the force–velocity–length relationship.

While afterloaded force–velocity curves appear valid in a wide variety of circumstances, such measurements are made at different times in contraction and at progressively shorter contractile-element lengths, and an extrapolated V_{max} might therefore be misleading. An attempt to overcome this difficulty has been to unload the preloaded muscle to near zero external load once contraction has begun.[120] In this manner it has been suggested that V_{max} can be measured directly. In Figure 2-16 the course of velocity of such an "unloaded" muscle relative to length is shown in the upper panel, while in the lower panel the relationship between length and developed tension is depicted.[130] Contraction has been initiated at L_{max} and at four shorter muscle lengths starting with different preloads, which are then removed. The unloaded velocity of shortening rises rapidly to a maximum and remains on a plateau where unloaded velocity remains relatively constant for the range of muscle length between L_{max} and 87% of L_{max}. When the initial muscle length is reduced, velocity climbs onto the same plateau and then falls along the same trajectory. This plateau corresponds closely to the value of V_{max} as determined from extrapolation of force–peak-velocity curves. Only at muscle lengths less than 87% of L_{max} does the unloaded velocity fail to reach this plateau. Thus, over the range of muscle lengths where the heart normally functions, V_{max} is independent of muscle length.

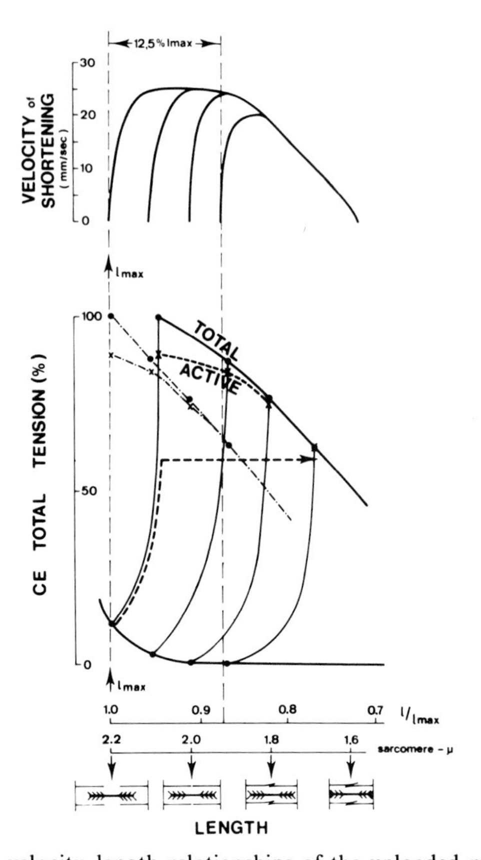

Fig. 2-16. The velocity–length relationships of the unloaded muscle (top) and the length–tension relationships of the CE (bottom) in the papillary muscle of the cat. These relationships have been shown for contractions arising from four different initial muscle lengths. L_{max} is the length where actively developed tension is maximum. Below are shown the approximate sarcomere lengths that relate to the diastolic muscle lengths as shown on the resting tension curve. When a contraction is initiated with the preload removed, velocity rises to its peak level very rapidly, as shown above; then a constant velocity is maintained, following which velocity of shortening decreases. When initial muscle length is decreased, velocity rises onto this same plateau and then follows this common pathway, with velocity ultimately declining. Only with the shortest initial muscle length does the velocity fail to rise onto this plateau of velocity of shortening. When the preload is removed this plateau approximates V_{max}. Below, the CE length-tension curve has been portrayed to show a shift in the force curve over to the right. The dashed line shows a hypothetical afterloaded isotonic contraction. Clearly, unloaded velocity of shortening occurs at longer CE lengths than does the ultimate isometric force development, due to shortening of the CE when isometric force is generated.

Relationship of Myofilament Disposition to Force–Velocity–Length in Heart Muscle

The mechanical characteristics of contraction are based ultimately on the ultrastructure of cardiac muscle. The relationship of diastolic sarcomere length to either force development or velocity of shortening remains of central importance. According to the sliding-filament theory of muscle contraction,[91,93] the total force on any one filament is the sum of the forces contributed by each cross-bridge interaction with actin, which in turn is proportional to the number of cross-bridge links formed and therefore to the width of the overlap zone between actin and myosin filaments. In skeletal muscle the relationship between muscle length and active tension development can be reasonably well explained at lengths greater than L_{max} by the degree of myofilament overlap, which in turn determines the number of force-generating cross-bridge attachments (Fig. 2-17).[131] Below L_{max} other factors in addition to myofilament overlap become important determinants of measured force; these factors include internal resistive forces, restoring forces, and deactivation induced by shortening.[131–133] This relationship is not so clear-cut in cardiac muscle (Fig. 2-17). Diastolic sarcomere lengths at L_{max} are about 2.2 μ in both cardiac and skeletal muscle. In skeletal muscle a plateau of maximal force is seen between initial sarcomere lengths of 2.0–2.2 μ, corresponding to the area of optimal myofilament overlap.[131] However, in cardiac muscle a decrease in sarcomere length to 2.0 μ results in a 30%–40% decline in actively developed tension.[122] Why then is developed tension or force dependent on muscle length in the physiologic range of cardiac sarcomere lengths?

It has been suggested that some, if not all, of the length dependence of force development in heart muscle contracting from diastolic sarcomere lengths where myofilament overlap is optimal (2.0–2.2 μ) may result from the compliant series elasticity that is characteristic of heart muscle.[111,122,130] Thus when cardiac muscle is activated at L_{max} with diastolic sarcomere lengths at 2.2 μ, the development of force is accompanied by an internal shortening of the contractile unit, so that when full isometric force is reached sarcomeres in systole are no longer at the initial 2.2 μ (Fig. 2-18); rather, end-systolic sarcomeres are 2.0 μ, which represents about a 10% decrease in initial muscle length. When the muscle's initial length is shortened such that diastolic sarcomeres are only 2.0 μ, sarcomeres during isometric contraction shorten to approximately 1.8 μ. Thus the true sarcomere length–tension curve during the development of force is shifted, so that sarcomeres are shorter and a decline in force relative to length would be anticipated (Fig. 2-18). Recent studies with living muscles using laser techniques have supported the

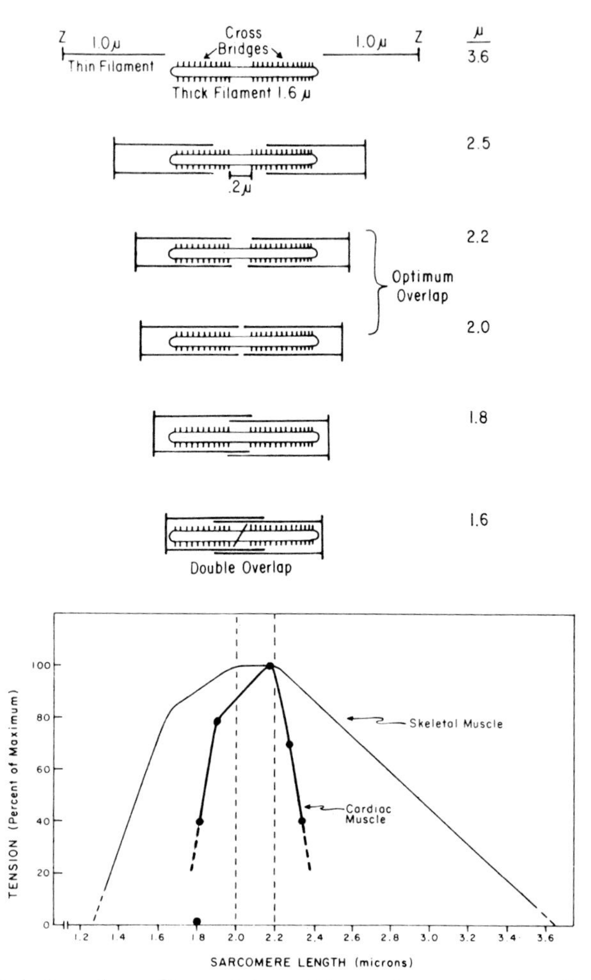

Fig. 2-17. Relation of myofilament disposition and sarcomere length to active tension development in striated muscle. In the upper panel is shown a schematic representation of the relative positions of the thick myosin filaments and the thin actin filaments in vertebrate striated muscle at six selected sarcomere lengths ranging from 1.6 to 3.6 μ. The area of overlap between actin filaments and portions of the myosin filaments containing force-generating cross-bridges increases as the sarcomere is lengthened from 1.6 to 2.0 μ and remains constant between 2.0 and 2.2 μ. In the lower panel the relationship between actively developed isometric tension and sarcomere length for frog skeletal muscle fibers and cat right ventricular papillary muscles is depicted. The curve for skeletal muscle fibers is derived from the data of Gordon et al.[131] In both tissues, peak developed tension is attained at a sarcomere length of 2.2 μ. At a sarcomere length of 2.0 μ, tension development is substantially decreased in cardiac muscle but remains

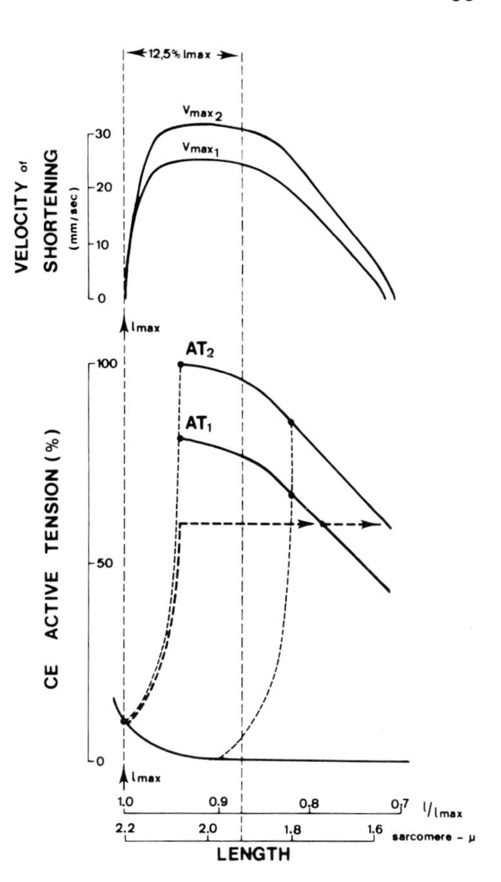

Fig. 2-18. The effects of increased calcium concentration on the unloaded velocity of shortening (top) and the length–tension curve of the CE corrected for series-elastic extension (bottom). With increased calcium, active tension generation is augmented from AT_1 to AT_2, while the extent of afterloaded isotonic shortening is increased. The series-elastic extension is not altered. In addition, the velocity of shortening of the unloaded muscle is also increased onto a new elevated plateau. Nonetheless the width of this plateau is not altered.

view that sarcomeres do indeed shorten substantially during apparent isometric contraction of heart muscle.[101,122] Thus, since isotonic velocity measurements with very small loads do not involve force development and resultant "internal" sarcomere shortening, the velocity of unloaded shortening should and does remain constant between sarcomere lengths of 2.0 and 2.2 μ. Accordingly, the length dependence of force in heart muscle can be attributed to a compliant SE element that places the sar-

---

maximal in skeletal muscle. At sarcomere lengths less than 2.0 μ, developed force falls in both types of muscle, but more precipitously in cardiac tissue. At sarcomere lengths longer than 2.0 μ, cardiac muscle is resistant to further extension and developed force declines precipitously when compared with the linear fall determined for skeletal muscle. In cardiac muscle this decrease in force cannot be entirely explained by sarcomere elongation and a decrease in myofilament overlap.

comeres of isometrically contracting heart muscle at shorter lengths than had previously been appreciated.[111,122,130] Therefore a dissociation between the effects of increasing muscle length on force development and velocity of shortening occurs. Were it not for this effect, both force and V_{max} would be constant in the physiologic range and would change in parallel. Moreover, despite arguments based on abstract models, direct measurements of V_{max} support the view that it is reasonably constant over the range of diastolic sarcomere and initial muscle lengths cardiac muscle would encounter under physiologic conditions (Fig. 2-16). At shorter muscle lengths obtained whether following a quick release or during the process of shortening, V_{max} does indeed fall as theory would predict (Fig. 2-16).

CONTRACTILITY IN THE ISOLATED MUSCLE

A rigorous definition of myocardial contractility is not possible even in isolated cardiac muscle. One can only describe the contractile state of the heart in terms of force, length, velocity, and time. In the previous discussion it has been suggested that contractility of heart muscle can be best characterized by the time-independent portion of the force–velocity–length relationship. Moreover, the extension of this surface to zero load is a measure of the unloaded maximal velocity of shortening (V_{max}). V_{max} is independent of initial muscle length and remains constant during shortening over the muscle lengths of physiologic importance. Thus V_{max} can be used to characterize a given surface of contractility. A change in contractility can be characterized by a level shift in the time-independent portion of the force–velocity–length relationship that is paralled by a change in V_{max}.

Since a three-dimensional construction of force–velocity–length relationships is difficult to obtain, the V_{max} concept can be used as an index of contractility. However, basic questions have been raised concerning the validity of such a generalization.[124–126]

Theoretically, it has been claimed that V_{max} cannot be measured directly. However, the unloaded velocity of shortening (Fig. 2-16) approaches such a direct measure and correlates well with extrapolated values.[120] That the force–velocity curve is nonhyperbolic and thus not subject to extrapolation[111] has been considered. However, it has been shown that the curve is indeed hyperbolic at lower loads, and deviations from this hyperbolic relationship at high loads are largely explained by a compliant SE element.[107] Accordingly, the curve can be corrected for this deviation.

While a dependence of V_{max} on muscle length has warranted careful study,[124–126] initial conclusions that V_{max} was independent of initial muscle length[103] have been supported at least along physiologic portions of the length–tension curve by recent studies utilizing unloading techniques.[120] Conclusions from quick-release experiments[115,126] that V_{max} is dependent on length have been readily explained[120] by the fact that following quick releases from peak isometric force, measurements can only be made at relatively short CE lengths (i.e., 12.5% below L_{max}). At these shorter (nonphysiologic?) diastolic lengths unloading studies clearly show that V_{max} is falling with length. Furthermore, only toward the latter portion of a normal contraction does the muscle pass through these lengths.

Were the active state (i.e., the activation of the muscle) to be turned on very slowly as has been claimed,[104,111,116] V_{max} measurements would be time-dependent. However, recent studies have shown that the capacity of the muscle to shorten maximally with light loads is established very early in contraction and that the course of shortening is largely independent of the time between onset of shortening and peak shortening, as well as the length from which shortening began.[117–120]

While the derivation of V_{max} from isometric data requires assumption of a model that will greatly affect derived values for CE velocity, the basic measurements utilized for the three-dimensional force–velocity–length relationships as well as the unloading and zero-load clamping experiments involve "shortening" muscle where the models are of minor importance when measurements are made. In physiologically competent preparations resting tensions are relatively very small, permitting the use of a simple two-component system. However, these conclusions do not detract from the pragmatic problems inherent in the application of models to the isovolumic portion of contraction in the intact heart to obtain a measure of V_{max}, as will as discussed subsequently.

Experimentally, extrapolation of a force–velocity curve to zero load to obtain V_{max} at very high resting tensions is admittedly difficult, if not impossible. However, we have found that the resting tension at L_{max} is less than 15% of peak isometric force and substantially less than that with minor muscle shortening, e.g., at peak velocities.[120] A small resting tension in relation to peak isometric tension allows extrapolation to zero load to be made with reasonable precision and accuracy.

Serious reservations concerning the use of V_{max} as an index of contractility have also arisen on a conceptual basis. It has been theorized that since troponin may inhibit the interaction of sites between actin and myosin filaments on an all-or-none basis, and since ATPase activity of "pure" actin and myosin is not calcium-dependent, the true V_{max} of the CE reflecting this chemical turnover rate at each contractile site should

FORCE (Po)	Number of Active Sites Internal Elasticity (SE)	Myofilament Overlap Ca^{++} Activation Restoring Forces
VELOCITY with LOAD	Number of Active Sites and Turnover Rate	
VELOCITY with ZERO LOAD (V max)	Turnover Rate at Active Sites	Actomyosin AT Pase Ca^{++} Activation (?)

Fig. 2-19. Mechanisms of the force–velocity relationship relative to proposed biochemical models of contraction.

be constant, independent of contractility changes.[40] Force, on the other hand, would depend on the number of cross-bridge attachments permitted between actin and myosin when calcium ions inhibit troponin[89,93] (Fig. 2-19). Thus if calcium is only a trigger to permit the interaction of actin and myosin, and does not itself affect this interaction, then increased availability of calcium would only create more force-generating sites and would not alter V_{max}.[40] The increase in V_{max} obtained from muscle studies might then reflect an unrecognized internal load, which is carried by more activated sites so that the load per active site is reduced, and not by a faster intrinsic speed of the enzymatic reaction at each site.[133] Were this the case, cardiac force–velocity relationships would tend to converge at zero load when normalized for force; this does not occur.[134] Furthermore, Julian[135] has demonstrated that V_{max} is dependent of $[Ca^{++}]$ in isolated skeletal fibers and that this calcium dependence cannot be explained by an internal load. Although these results have been contested by others,[136,137] we feel that in cardiac muscle the true V_{max}, and thus the rate of turnover at individual contractile sites, appears to be altered by changes in contractility. Indeed, the recent model of the CE proposed by Huxley and Simmons[95] suggests a multistep interaction between myosin and actin, which may be more complex than isolated disorganized protein solutions where stearic considerations may be absent. Furthermore, recent studies indicate that the biochemistry of troponin is more complex than was previously appreciated.[54–56] One troponin moiety inhibits actomyosin ATPase, while another contributes calcium sensitivity. These considerations, combined with the observation of Huxley[138] that myosin bridges appear to move in the absence of overlap with actin, leave the important question of whether calcium alters the intrinsic actomyosin ATPase reaction (or the rate of turnover of one cross-bridge) or merely alters the rate of active and inactive enzyme site activation (or the total number of cross-bridges activated) unresolved. Experimental mechanical data suggest that both phenomena are occurring.

CONTRACTILITY IN THE INTACT HEART

Pressure–Volume Relationships

In systole, the ventricle generates pressure that permits ejection of a volume of blood into the aorta. For this to occur the myocardium develops force in order to increase intraventricular pressure and then shortens with a load determined by the aortic pressure and the ventricular geometry (Fig. 2-20, panel I). Physiologic requirements necessitate remarkable flexibility for this pump to vary its volume output on a beat-to-beat basis in relation to venous return and changing peripheral resistance. Under any given set of conditions the pump has the ability to put out a given stroke volume (SV) from a given end-diastolic ventricular volume (EDV) (Fig. 2-20, panel II). The SV can be augmented by increasing the EDV, the Frank-Starling phenomenon (Fig. 2-20, panel I). The relation of SV to EDV is relatively linear until filling pressures at the upper limits of normal are reached, and thus the ejection fraction (SV/EDV) is actually the slope to this line (Fig. 2-20, panels II and IV). If the aortic pressure is increased (point *b''*) the SV will be reduced (*c''*), but it can be restored in subsequent beats by augmenting the EDV (*a* to *a'*). Nevertheless the SV/EDV would be reduced to a lesser or greater degree by the increase in pressure (Fig. 2-20, panel II). Additionally, catecholamines released from sympathetic nerves drugs such as digitalis glycosides, calcium, or other potentiating inotropic influences such as increased heart rate or postextrasystolic contractions can alter the performance of the pump. These inotropic influences are characterized by an increased SV from any given EDV, or alternatively by an augmented ejection fraction (SV/EDV) (Fig. 2-20, panels III and IV). This change in SV/EDV has been termed a change in contractility. A major problem for the cardiovascular physiologist as well as the clinician has been to evaluate the basic performance capacity of this pump and distinguish changes in performance due to altered initial volume (Frank-Starling) from alterations in contractility. Unfortunately, as noted above, SV is dependent on systolic pressure. While increasing systolic pressure will reduce the ejection fraction, reduction in impedance to ventricular emptying (e.g., mitral regurgitation) would tend to augment the SV at any EDV and increase the ejection fraction.

The relation of SV to EDV has also been restated in terms of stroke work (the product of the SV and the aortic pressure) and plotted relative to end-diastolic pressure, which has been used as an index of EDV. This derived relation has been termed a ventricular function curve.[139] Unfortunately, the stroke work at any given end-diastolic pressure is also altered by changes in aortic pressure per se,[140] as well as by inotropic

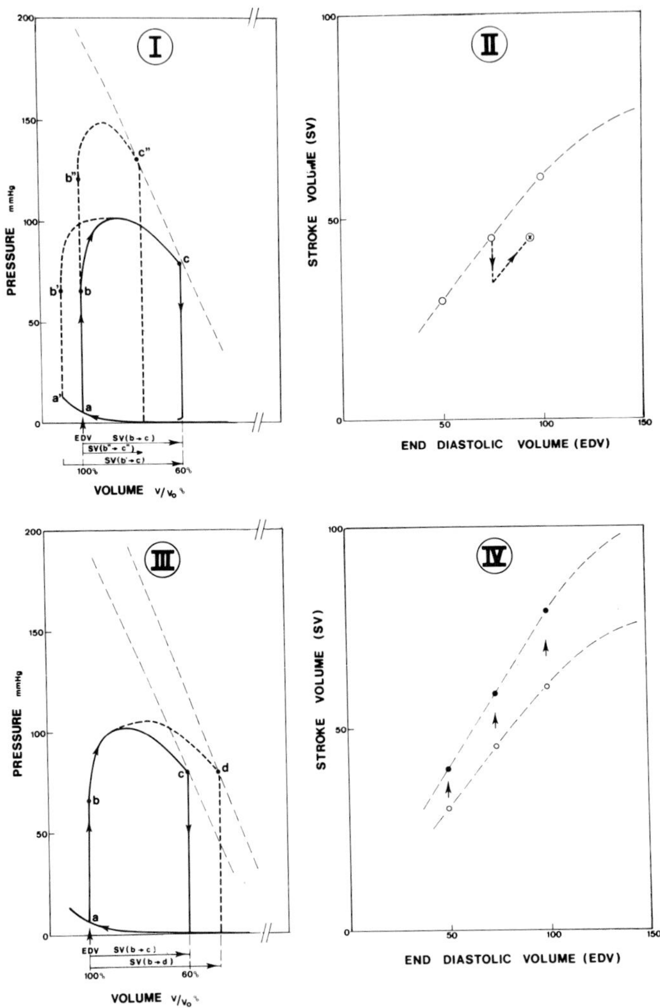

Fig. 2-20. Pressure–volume relationships of the intact ventricle. (I) Volume is expressed as a percentage of an arbitrary volume. Points *a* and *a′* lie along the passive relation between filling pressure and intraventricular volume. Point *b* represents the pressure at which the aortic valve is opened, with point *c* where it closes at the end of systole. Point *a′* represents an increase in end-diastolic volume (EDV), while point *b″* represents an increase in arterial pressure. This latter intervention leads to a reduction in the stroke volume (SV) of the first beat after the increased loading (*b″* to *c″*). In subsequent beats this decrement in SV could be restored by a small increase in EDV, e.g., *a* to *a′*. The broad-dashed line represents the maximum isovolumic pressure that could be developed were no ejection to occur. (II) The relationship of SV to EDV. The effects of the initial and subsequent events following an abrupt increase in aortic pressure are shown

agents. Furthermore, on a chronic basis, changes in wall thickness may increase the end-diastolic pressure for any volume.[141] Thus an increase in end-diastolic pressure need not indicate ventricular dilatation and associated "heart failure."

For simplification, these properties of the heart as a pump can be expressed in terms of segments of muscle in the ventricular wall. The diastolic volume of the ventricle prior to the onset of systole (EDV), established by a diastolic filling pressure (EDP), is reflected in initial fiber length, which is determined in turn by a resting tension or preload. The pressure developed by the stimulated left ventricle, which leads to opening of the aortic valve and ejection of blood, is directly related to afterload. The course of ejection is reflected by the shortening of muscle fibers in the ventricular wall. Of course, the afterload is not the pressure itself but is directly related to the pressure and the size of the ventricle. Thus the force in the wall is a function of pressure P and radius r and is inversely related to wall thickness h. Hence the smaller heart requires less force or afterload to generate the same pressure (law of LaPlace): $T \simeq P(r/h)$. Moreover, the force in the wall falls during normal ejection as the volume (or r) is reduced, even though pressure P itself may continue to rise.

Pressure–Volume–Flow Relationships

In addition to ejecting a SV from a given EDV, the ventricle delivers this SV with a velocity determined by the instantaneous volume of the ventricle and resistance to ejection (pressure) (Fig. 2-21). Thus when the end-diastolic volume is augmented (Fig. 2-21, panel II) there is an increase in stroke volume (b' to c') as well as in the rate at which the blood is ejected. When the arterial pressure is increased (Fig. 2-21, panel III) the velocity of ejection of blood and extent of ventricular emptying (b'' to c'') are reduced. Furthermore, an inotropic intervention (Fig. 2-21, panel IV) will increase both the SV (b to d) and the velocity of ejection. The stroke power, which is the product of the SV and aortic pressure divided

---

by the dotted line and ⓧ. The initial effect of increasing aortic pressure is to reduce the SV, which is restored to the control level by increasing EDV. (III) The effect of augmenting contractility on the pressure–volume relationships of the intact left ventricle. Starting from the same EDV, the SV is increased. The dashed slope, the maximum isovolumic pressure that could be developed, is moved upward and to the right. (IV) The relationship between SV and EDV following the addition of an inotropic agent (●). The control is as in panel II (○). The SV at any EDV is increased, so that SV/EDV is augmented.

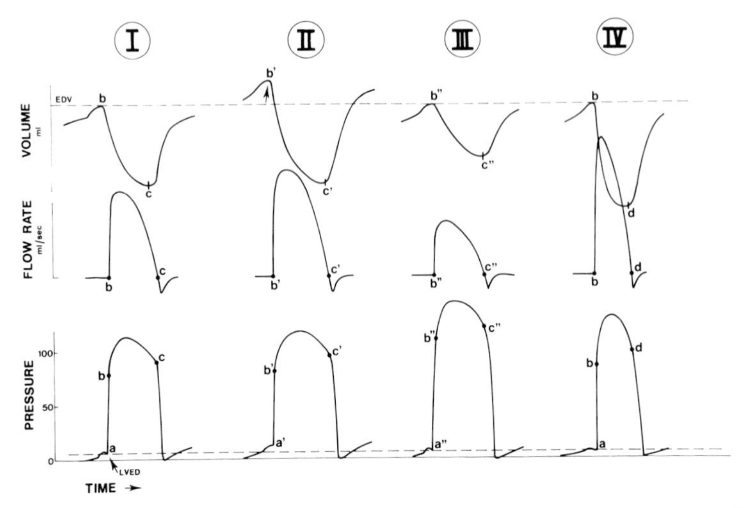

Fig. 2-21. Effects of altered ventricular volume, pressure, and contractility. Intraventricular volume, aortic flow rate, and intraventricular pressure have been illustrated relative to time. (I) Control state. (II) Effects of an augmented EDV. (III) Effects of increased intraventricular pressure. Diastolic volume has been maintained constant. (IV) Effects of administering isoproterenol. Diastolic volume has been held constant.

by the duration of ejection (i.e., stroke work per unit time), is also increased at any EDV; but as with stroke work, stroke power remains pressure-dependent.

Muscle Mechanics in the Intake Heart

Ultimately, all of these properties of the functioning pump mirror the properties of the underlying myocardium, which leads to indices of performance derived from the principles of muscle mechanisms as determined from studies in isolated cardiac muscle. The hemodynamic parameters of pressure, flow rate, and volume reflect to a degree the activity of the underlying myocardium as expressed by force, velocity, and length. In Figure 2-22 these three variables have been portrayed with the measurements that are made during the preejection and ejection phases of systole. As noted, the hemodynamic measurements reflect one or more of these variables, but no single measurement encompasses them all, or one reflects only two of the three, independent of the third. In general, indices for contractility in the intact heart need to be measurable, reproducible, and predictive of non-volume-dependent function. However, it should be appreciated that these indices are only

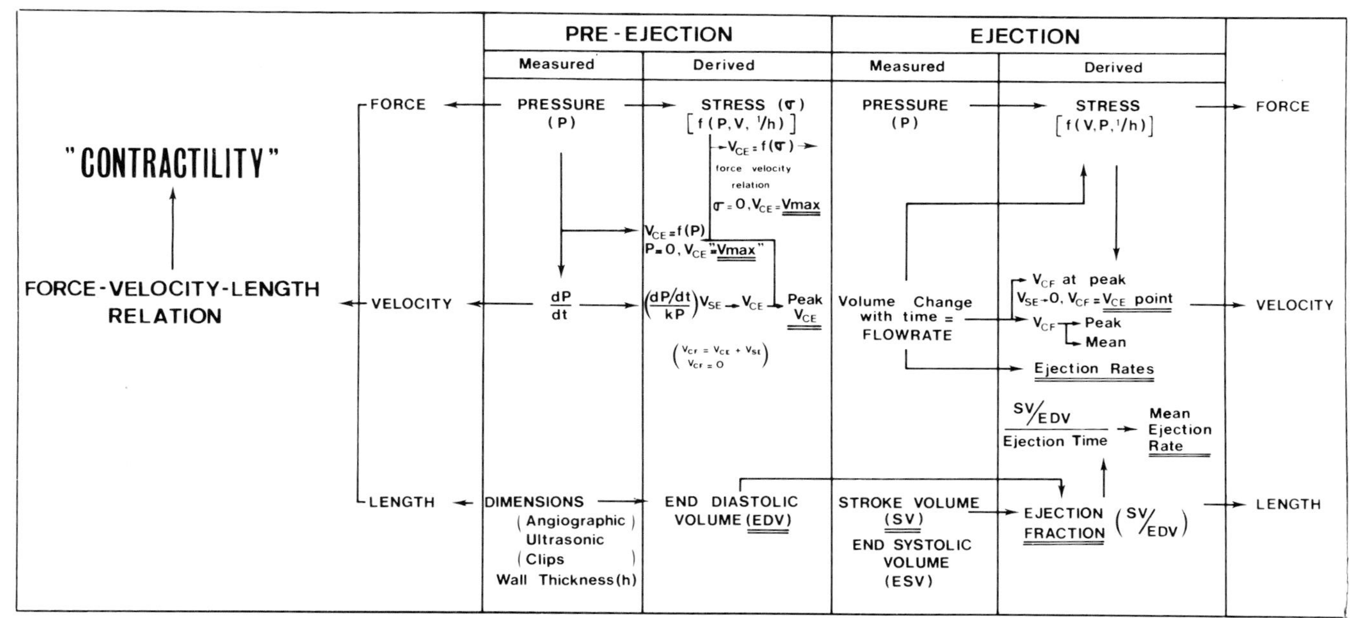

Fig. 2-22. Measured and derived indices of contractility in the intact heart during preejection and during ejection. For the derivation of V_{CE} during the preejection phase, a two-component model is assumed in which $dP/dt = dl/dt$ (velocity of shortening of CE or V_{CE}) × (dP/dl) (stiffness of SE). Since $dP/dl = kP + c$ (where k and c are constants), $V_{CE} = (dP/dt)/kP = c$ or (since c is negligible) $V_{CE} = (dP/dt)/kP$. As an approximation for this value, $(dP/dt)/P$ or $(d \log P)/dt$ has also been used. Notice that peak V_{CE} represents only one single velocity point on the force–velocity–length relationship.

approximations dependent on admittedly oversimplified assumptions and difficult measurements. While the principles may be clear from isolated muscle, the direct translation into pragmatic tools in the intact heart is difficult.

Isovolumic Indices

During ejection, measurements of changes in length (SV and SV/EDV) and velocity are dependent on load (aortic pressure). Accordingly, attempts have been made to utilize measurements recorded during the preejection phase of systole such as the rate of pressure development (dP/dt) and its derivatives to evaluate myocardial contractility. Furthermore, left ventricular dP/dt can be measured readily with appropriate care, while accurate ejection volumes may pose somewhat greater difficulties. Indeed, dP/dt itself is directly related to contractility and would be a useful index of contractility except that its dependence on volume (length) as well as contractility tends to limit its quantitative usefulness.[142] Nevertheless, an increase in dP/dt accompanied by a decrease in left ventricular end-diastolic pressure (and thus volume) can only be construed as an increase in contractility.

Utilizing a model for muscle in which a CE is arranged in series with a SE element, contractile-element velocity (V_{CE}) can be calculated.[143–151] This V_{CE} is derived from the dP/dt, the instantaneous pressure, and an elasticity constant k that has been taken from studies with isolated cardiac muscle.[98] Actually, V_{SE} is being calculated from dP/dt, but during isovolumic contraction $V_{SE} = V_{CE}$. From pressure, volume, and wall thickness, stress can be calculated and a portion of the force–velocity relationship derived. By extrapolating the relationship between derived V_{CE} and pressure (or stress) to zero, a figure for V_{max} can be obtained.[143,144,146–148] Since the relationship between pressure and stress is linear, a scaling factor is involved along the abscissa only if pressure alone is utilized rather than stress, and extrapolations to zero should be the same for both stress and pressure.[147,148]

Despite the potential advantages of such an approach, including its independence of the effects of aortic pressure and volume, important limitations need to be recognized before this technique can be accepted and applied to disease states. Some of these limitations have been noted in Figure 2-23. Certain of these are trivial, while others are potentially major and limiting. First, dP/dt must be measured from high-fidelity catheters; the frequency response of fluid-filled catheters may be great enough if very special care is taken, but even then it may be subject to unpredictable changes relative to time. Second, values for derived V_{CE} and thus extrapolated V_{max} are highly dependent on the model that is utilized,[152] especially at high preloads beyond L_{max}.[153] If total (i.e., diastolic plus systolic) pressure is used for the calculations, a Voigt model and its variations are assumed; in this situation a long and uncertain extrapolation to zero pressure is required, and linear plots of V_{CE} relative to pressure are obtained rather than hyperbolae.[147,148] Peak measured V_{CE} from this approach has been used without extrapolation, but it is reduced when preload is increased.[153,154] If developed pressure (systolic less diastolic) is utilized, a Maxwell model is assumed. In this model the c constant approaches zero, and thus the start of the calculation requires a finite pressure. In order to avoid an infinite value for V_{max}, calculations with this model assume a small initial load. Such calculations yield relatively hyperbolic curves with short extrapolations and a V_{max} that is relatively independent of preload, at least within physiologic limits.[153,154] Other limitations to this approach are more speculative. For example, in considering disease states one must assume similar values for the k constant for the SE. While k may be the same under a number of conditions, asynchronous or akinetic areas of the myocardium might alter it. Moreover, analysis of ventricular performance in terms of a lumped model assumes synchronous ventricular contraction, which is commonly absent in coronary artery disease and cardiomyopathies. In addition, massive valvular regurgitation might vitiate the measurement of dP/dt.

While recognizing all of these problems and criticisms, the measurement of indices derived from the isovolumic portion of contraction may still be useful if subsequent study supports the view that they can help in a pragmatic way to differentiate the "good" from the "bad" heart, and the gradations that lie in between.[148,155] Should this be the case, practical measurements are warranted for this purpose to complete the entire profile of the heart's performance. As a unique measurement, V_{max} or peak V_{CE} derived from dP/dt during the preejection phase must be viewed with caution, but should provide supplementary information when combined with other data.

Indices from the Ejection Phase of Systole

During ejection the measurement of indices reflecting force, velocity, and length requires the measurement of changes in pressure and volume relative to time. From pressure and volume, wall stress can be derived, but it will depend on the model assumed for ventricular shape, configuration, fiber direction, and distribution of stress in the wall.[156] From changes in intraventricular volume relative to time, velocity of

Table 2-2
Criticisms of Derived Velocity Measurements in the Intact Heart (dP/dt = rate of rise of left ventricular pressure; P = intraventricular pressure; k = a constant of SE compliance; C = a constant for the intercept of the SE curve)

Derived Fiber Shortening Velocities (V_{CF}) and Contractile Element Velocity (V_{CE}).

A. <u>PRE-EJECTION (ISOVOLUMIC) PHASE:</u> V_{CE} derived from $\frac{dP/dt}{kP+C}$ (or simplified $\frac{dP/dt}{kP}$; $\frac{dP/dt}{P}$ or $\frac{d \log P}{dt}$)

1. dP/dt must be high fidelity measurement.
2. Model dependent results:
   a. Use of <u>total</u> or <u>developed</u> pressure curve and extrapolation.
   b. <u>C</u> : dependent on model.
   c. Long extrapolations uncertain with mode dependent on model.
3. <u>k</u> (stiffness constant) : could vary with disease.
4. Peak V_{CE} preload dependent.
5. Asynchrony may limit <u>lumped</u> model.
6. Severe valvular regurgitation may alter dP/dt.

B. <u>EJECTION PHASE:</u> V_{CF} measured. V_{CE} derived from $V_{CF} - V_{SE} = V_{CE}$, where $dP/dt = 0$.

1. V_{CF} load dependent. Applies to peak and mean V_{CF}.
2. V_{CE} at $dP/dt = 0$, load dependent ; forms one point on a force-velocity-length relation.

Table 2-3
Indices of Contractility in the Intact Heart (columns 4 and 5 note the dependence on volume and load of each of these indices)

"CONTRACTILITY" INDEX in the INTACT HEART	FORCE	VELOCITY	DISPLACE-MENT	Dependent on:	
				VOLUME	LOAD
End Diastolic Volume (EDV)	0	0	0	+	0
Stoke Volume (SV)	+	0	+	+	+
Ejection Volume (SV/EDV)	+	0	+	+	+
Ejection Time	0	+	0	+	+
Ejection Rate $\frac{SV/EDV}{time}$	+	+	+	+	+
Rate of Pressure Development (dP/dt)	+	+	0	+	±
Fiber Shortening Rate	+	+	0	+	+
V max	0	++	0	0	0
Peak V ce	+	+	0	+	0

volume change and thus velocity of fiber shortening (V_{CF}) can be derived. Using dP/dt and P, the rate of stretch of the elastic elements in the wall (V_{SE}) can then be calculated. When stress reaches a maximum (i.e., is not changing) V_{SE} becomes zero and $V_{CF} = V_{CE}$.[157,158] This provides a single value for V_{CE} on a force–velocity curve. It should be emphasized that this V_{CE} point will be dependent on the assumptions used to derive stress. Given these limitations, peak or mean V_{CF} may also be useful indices of speed,[151] but these measurements will also depend on the pressure (Table 2-2). The same may be said for indices of change in length such as SV, ejection fraction (SV/EDV), or ejection rates (SV/EDV/time). A major reason why correlations between different variables are obtained despite afterload dependence is that they all reflect one or more of the variables we have noted, and with altered contractility they all tend to change in a parallel manner. Accordingly, a reasonable correlation between all of the velocity-related measurements should be anticipated.

In summary, indices of contractility in the intact heart (Table 2-3) are all partial measurements of the force–velocity–length relationship. None is complete. All have limitations of either methodology, practicality, or concept. In an attempt to exclude the effects of volume (fiber length) and loading from consideration of performance, indices of velocity have been derived. While V_{max} may be estimated from isovolumic measurements, practical and modeling problems bear consideration. Ultimately, a useful index will require the ability to distinguish a "good" from a "bad" heart, independent of load and volume, which is measurable in a reliable and reproducible manner. At present this is best done by combining methods rather than seeking a unique measurement. In addition, problems of segmental performance (e.g., asynchrony or akinesis) require broader analysis rather than unified oversimplifications.

REFERENCES

1. Stenger RJ, Spiro D: The ultrastructure of mammalian cardiac muscle. J Biophys Biochem Cytol 9:325–351, 1961
2. Sommer JR, Johnson EA: Cardiac muscle. A comparative study of purkinje fibers and ventricular fibers. J Cell Biol 36:497–526, 1968
3. McNutt NS, Weinstein RS: Membrane ultrastructure at mammalian intercellular junctions. Prog Biophys Mol Biol 26:45–101, 1972
4. Fawcett DW, McNutt NS: The ultrastructure of the cat myocardium. I. Ventricular papillary muscle. J Cell Biol 42:1–45, 1969
5. Lehninger AL: Bioenergetics (ed 2). Menlo Park, Calif, W. A. Benjamin, 1971, pp 91–98

6. Page E, Polimeni PI, Zak R, et al: Myofibrillar mass in rat and rabbit heart muscle. Circ Res 30:430–439, 1972
7. Patriarca P, Carafoli E: A study of the intracellular transport of calcium in rat heart. J Cell Physiol 72:29–37, 1968
8. Kübler W, Shinebourne EH: Calcium and the mitochondria, in Harris P, Opie LH (eds): Calcium and the Heart. London, Academic Press, 1971, pp 93–123
9. Langer GA: Ion fluxes in cardiac excitation and contraction and their relation to myocardial contractility. Physiol Rev 48:708–757, 1968
10. Dewey MM, Barr L: A study of the structure and distribution of the nexus. J Cell Biol 23:553–585, 1964
11. Revel JP, Karnofsky MJ: Hexagonal array of subunits in intercellular junctions of the mouse heart and liver. J Cell Biol 33:C7–C12, 1967
12. Weidmann S: The diffusion of radiopotassium across intercalated disks of mammalian cardiac muscle. J Physiol 187:323–342, 1966
13. Sommer JR, Johnson EA: Comparative ultrastructure of cardiac cell membrane specializations. Am J Cardiol 25:184–194, 1970
14. McNutt NS, Weinstein RS: The ultrastructure of the nexus. A correlated thin section and freeze-cleave study. J Cell Biol 47:666–688, 1970
15. Sommer JR, Johnson EA: Cardiac muscle: A comparative ultrastructural study with special reference to frog and chicken hearts. Z Zellforsch Mikrosk Anat 98:437–468, 1969
16. Schwartz A: Active transport in mammalian myocardium, in Langer GA, Brady AJ (eds): The Mammalian Myocardium. New York, John Wiley & Sons, 1974, pp 81–104
17. Skou JC: Enzymatic basis for active transport of Na^+ and K^+ across the cell membrane. Physiol Rev 45:596–617, 1965
18. Smith DS: Organization and function of sarcoplasmic reticulum and T-system of muscle cells. Prog Biophys Mol Biol 16:107–142, 1966
19. Nelson DA, Benson ES: On the structural continuities of the transverse tubular system of rabbit and human myocardial cells. J Cell Biol 16:297–313, 1963
20. Simpson FO: The transverse tubular system in mammalian myocardial cell. Am J Anat 117:1–18, 1965
21. Forssmann WG, Girardier L: A study of the T system in rat heart. J Cell Biol 44:1–19, 1970
22. Franzini-Armstrong C: Studies of the triad. I. Structure of the junction in frog twitch fibers. J Cell Biol 47:488–499, 1970
23. Weber A, Herz R, Reiss I: The nature of the cardiac relaxing factor. Biochem Biophys Acta 131:188–194, 1967
24. Ebashi S, Endo M: Calcium ion and muscle contraction. Prog Biophys Mol Biol 18:123–183, 1968
25. Ebashi S, Endo M, Ohtsuki I: Control of muscle contraction. Q Rev Biophys 4:351–384, 1969
26. Hanson J, Lowy J: Molecular basis of contractility in muscle. Br Med Bull 21:264–271, 1965

27. Huxley HE: The structural basis of muscular contraction. Proc R Soc Lond [Biol] 178:131–149, 1971
28. Spiro D, Sonnenblick EH: Comparison of the ultrastructural basis of the contractile process in heart and skeletal muscle. Circ Res [Suppl 2] 15:14–37, 1964
29. Huxley HE: The fine structure of striated muscle and its functional significance. Harvey Lect 60:85–118, 1966
30. Huxley HE, Hanson J: Changes in the cross-striations of muscle during contraction and stretch and their structural interpretation. Nature 173:973–976, 1954
31. Huxley HE: The double array of filaments in cross-striated muscle. J Biophys Biochem Cytol 3:631–647, 1957
32. Huxley HE: The mechanism of muscular contraction. Science 164:1356–1366, 1969
33. Huxley HE: Electron microscope studies on the structure of natural and synthetic protein filaments from striated muscle. J Mol Biol 7:281–308, 1963
34. Lowey S, Slayter HS, Weeds AG, Baker H: Substructure of the myosin molecule. I. Subfragments of myosin by enzymic degradation. J Mol Biol 42:1–29, 1969
35. Englehardt A, Ljubimova J: Myosin adenosine-triphosphatase. Nature 144:668–669, 1939
36. Barany M, Conover TE, Schlisselfeld LH, et al: Relation of properties of isolated myosin to those of intact muscle of the cat and sloth. J Biochem 2:156–164, 1967
37. Barany M: ATPase activity of myosin correlated with speed of muscle shortening. J Gen Physiol 50:197–216, 1967
38. Mueller H, Franzen J, Rice RV, Olson RE: Characterization of cardiac myosin from the dog. J Biol Chem 239:1447–1456, 1964
39. Muir JR, Weber A, Olson RE: Cardiac myofibrillar ATPase: A comparison with that of fast actomyosin in its native and in an altered conformation. Biochim Biophys Acta 234:199–209, 1971
40. Katz AM: Contractile proteins of the heart. Physiol Rev 50:63–158, 1970
41. Johnson P, Harris CI, Perry SV: 3-Methylhistidine in actin and other muscle proteins. Biochem J 105:361–370, 1967
42. Mueller H, Perry SV: The degradation of heavy meromyosin by trypsin. Biochem J 85:431–439, 1962
43. Dreizen P: Structure and function of myofibrillar contractile proteins. Ann Rev Med 22:365–390, 1971
44. Dreizen P, Richards DH: Studies on the role of light and heavy chains in myosin adenosine triphosphatase. Cold Spring Harbor Symp Quant Biol 37:29–45, 1972
45. Weeds AG, Frank G: Structural studies on the light chains of myosin. Cold Spring Harbor Symp Quant Biol 37:9–14, 1972
46. Perry SV, Corsi A: Extraction of proteins other than myosin from isolated rabbit myofibril. Biochem J 68:5–12, 1958

47. Endo M, Nonomura Y, Masaki T, et al: Localization of native tropomyosin in relation to striation patterns. J Biochem (Tokyo) 60:605–609, 1966
48. Hanson J, Lowy U: The structure of F-actin and of actin filaments isolated from muscle. J Mol Biol 6:46–60, 1963
49. Spudich JA, Huxley HE, Finch JF: Regulation of skeletal muscle contraction. II. Structural studies of the interaction of the tropomyosin–troponin complex with actin. J Mol Biol 72:619–632, 1972
50. Worthington CR: Large axial spacings in striated muscle. J Mol Biol 1:398–401, 1959
51. Ebashi S, Ebashi F: A new protein component participating in the super-precipitation of myosin B. J Biochem (Tokyo) 55:604–613, 1964
52. Weber A, Murray JM: Molecular control mechanisms in muscle contraction. Physiol Rev 53:612–673, 1973
53. Ebashi S, Kodoma A: A new protein factor promoting aggregation of tropomyosin. J Biochem (Tokyo) 58:107–108, 1965
54. Wilkinson JM, Perry SV, Role HA, Trayer LP: The regulatory proteins of the myofibril. Separation and biological activity of the components of inhibitory factor preparation. Biochem J 127:215–228, 1972
55. Greaser ML, Yamaguchi M, Brekke C, et al: Troponin subunits and their interactions. Cold Spring Harbor Symp Quant Biol 37:235–244, 1972
56. Potter JD, Gergely J: Troponin, tropomyosin and actin interactions in the Ca^{++} regulation of muscle contraction. Biochemistry 13:2697–2703, 1974
57. McCubbin WD, Kouba RF, Kay CM: Physicochemical studies on bovine cardiac tropomyosin. Biochemistry 6:2417–2425, 1967
58. Ebashi S, Nonomura Y: Proteins of the myofibril, in Bourne GH (ed): The Structure and Function of Muscle, vol 3 (ed 2). New York, Academic Press, 1973, pp 286–362
59. Kawaura M, Maruyama K: Electron microscopic particle length of F-actin polymerized in vitro. J Biochem (Tokyo) 67:437–443, 1970
60. Vibert PJ, Haselgrove JC, Lowry J, Roulsen FR: Structural changes in actin-containing filaments of muscle. J Mol Biol 71:757–767, 1972
61. Masaki T, Endo M, Ebashi S: Localization of 6S component of α-actinin at Z band. J Biochem (Tokyo) 62:630–642, 1967
62. Franzini-Armstrong C: Details of the I band structure as revealed by the localization of ferritin. Tissue Cell 2:327–338, 1970
63. Guba F, Harsanyi V, Vajda E: The muscle protein fibrillin. Acta Biochem Biophys 3:353–363, 1968
64. Trautwein W: Membrane currents in cardiac muscle fibers. Physiol Rev 53:793–835, 1973
65. Weidmann S: The effect of the cardiac membrane potential on the rapid availability of the sodium carrying system. J Physiol 127:213–224, 1955
66. Beeler GW, Reuter H: The relation between membrane potential, membrane currents, and activation of contraction in ventricular myocardial fibers. J Physiol 207:211–229, 1970
67. Reuter H: Divalent cations as charge carriers in excitable membranes. Prog Biophys Mol Biol 26:1–43, 1972

68. Morad M, Goldman Y: Excitation-contraction coupling in heart muscle: Membrane control of development of tension. Prog Biophys Mol Biol 27:257–313, 1973
69. Mascher D, Peper K: Two components of inward current in myocardial muscle fibers. Pfluegers Arch 307:190–203, 1969
70. Fozzard HA, Gibbons WR: Action potential and contraction of heart muscle. Am J Cardiol 31:182–192, 1973
71. Ringer S: A further contribution regarding the influence of the different constituents of the blood on the contraction of the heart. J Physiol 4:29–42, 1883
72. Winegrad S, Shanes AM: Calcium flux and contractility in guinea pig atria. J Gen Physiol 45:371–394, 1962
73. Grossman A, Furchgott RF: The effects of frequency of stimulation and calcium concentration on Ca^{45} exchange and contractility in the isolated guinea pig auricle. J Pharmacol Exp Ther 143:120–130, 1964
74. Langer GA: Kinetic studies of calcium distribution in ventricular muscle of the dog. Circ Res 15:393–405, 1964
75. Langer GA, Brady AJ: Calcium flux in the mammalian ventricular myocardium. J Gen Physiol 46:703–720, 1963
76. Langer GA: Heart: Excitation-contraction coupling. Ann Rev Physiol 35:55–86, 1973
77. Huxley AF, Taylor RE: Local activation of striated muscle fibers. J Physiol 144:426–441, 1958
78. Costantin LL, Taylor SR: Graded activation in frog muscle fibers. J Gen Physiol 61:424–443, 1973
79. Schneider MF, Chandler WK: Voltage dependent charge movement in skeletal muscle: A possible step in excitation-contraction coupling. Nature 242:244–246, 1973
80. Morad M, Trautwein W: The effect of the duration of the action potential on contraction in the mammalian heart muscle. Pfluegers Arch 299:66–82, 1968
81. Fabiato A, Fabiato F, Sonnenblick EH: Electrical depolarization of intracellular membranes and "regenerative calcium release" in single cardiac cells with disrupted sarcolemmal. Circulation 44:(Suppl II):II-90, 1971
82. Nakajuna Y, Endo M: Release of calcium induced by "depolarization" of the sarcoplasmic reticulum membrane. Nature 246:216–218, 1973
83. Fabiato A, Fabiato F: Excitation-contraction coupling of isolated cardiac fibers with disrupted or closed sarcolemmas. Circ Res 31:293–307, 1972
84. Fabiato A, Fabiato F: Activation of skinned cardiac cells. Eur J Cardiol 1:143–155, 1973
85. Ford LE, Podolsky RJ: Regenerative calcium release within muscle cells. Science 67:58–59, 1970
86. Endo M, Tanaka M, Ogawa Y: Calcium induced release of calcium from the sarcoplasmic reticulum of skinned skeletal muscle fibers. Nature 228:34–36, 1970
87. Luttgau HC, Niedergerke R: Antagonism between Ca and Na ions on the frog's heart. J Physiol 143:486–505, 1958

88. Wood EH, Heppner RL, Weidmann S: Inotropic effects of electrical currents. Circ Res 24:409–445, 1969
89. Ebashi S, Ohtsuki I, Mihashi K: Regulatory proteins of muscle with special reference to troponin. Cold Spring Harbor Symp Quant Biol 37:245–252, 1973
90. New W, Trautwein W: The ionic nature of slow inward current and its relation to contraction. Pfluegers Arch 334:24–38, 1972
91. Huxley HE: The mechanism of muscular contraction. Sci Am 213:18–27, 1965
92. Fuchs F: Striated muscle. Ann Rev Physiol 36:461–502, 1974
93. Huxley AF: Muscle structure and theories of contraction. Prog Biophys 7:255–318, 1957
94. Hill AV: The heat of shortening and the dynamic constants of muscle. Proc R Soc Lond [Biol] 126:136–195, 1938
95. Huxley AF, Simmons RM: Proposed mechanism of force generation in striated muscle. Nature 233:533–538, 1971
96. Noble MIM, Else W: Reexamination of the applicability of the Hill model of muscle to cat myocardium. Circ Res 31:580–589, 1972
97. Pollack GH, Huntsman LL, Verdugo P: Cardiac muscle models: An overextension of series elasticity? Circ Res 31:569–579, 1972
98. Parmley WW, Sonnenblick EH: Series elasticity in heart muscle: Its relation to contractile element velocity and proposed muscle models. Circ Res 20:112–123, 1967
99. McLaughlin RJ, Sonnenblick EH: Time behavior of series elasticity in cardiac muscle: Real-time measurement by controlled length techniques. Circ Res 34:798–811, 1974
100. Meiss RA, Sonnenblick EH: Dynamic elasticity of cardiac muscle as measured by controlled length changes. Am J Physiol 226:1370–1381, 1974
101. Pollack G: Discussion on light diffraction of cardiac muscle, in: The Physiologic Basis of Starling's Law of the Heart. Amsterdam, Elsevier, 1974, p 90
102. Abbott BC, Mommaerts WFHM: A study of inotropic mechanisms in the papillary muscle preparation. J Gen Physiol 42:533–551, 1959
103. Sonnenblick EH: Force–velocity relations in mammalian heart muscle. Am J Physiol 202:931–939, 1962
104. Sonnenblick EH: Active state in heart muscle: Its delayed onset and modification by inotropic agents. J Gen Physiol 50:661–676, 1967
105. Brutsaert DL, Sonnenblick EH: Force–velocity–length–time relation of the contractile elements in heart muscle of the cat. Circ Res 24:137–149, 1969
106. Sonnenblick EH: Instantaneous force–velocity–length determinants in the contraction of heart muscle. Circ Res 16:441–451, 1965
107. Brutsaert DL, Sonnenblick EH: Nature of force–velocity relation in heart muscle. Cardiovasc Res 1[Suppl 1]:18–33, 1971
108. Hill AV: The effect of load on the heat of shortening of muscle. Proc R Soc Lond [Biol] 159:297–318, 1964
109. Mommaerts WFHM: Energetics of muscular contraction. Physiol Rev 49:427–508, 1969

110. Gibbs CL: Cardiac energetics, in Langer G, Brady AJ (eds): The Mammalian Myocardium. New York, John Wiley & Sons, 1974, pp 105–133
111. Brady AJ: Active state in cardiac muscle. Physiol Rev 48:570–600, 1968
112. Henderson AH, Claes VA, Brutsaert DL: Influence of caffeine and other inotropic interventions on the onset of unloaded shortening velocity in mammalian heart muscle. Circ Res 33:291–302, 1973
113. Bodem R, Sonnenblick EH: Deactivation of contraction by quick release in the isolated papillary muscle of the cat: Effects of lever damping, caffeine and tetanization. Circ Res 34:214–225, 1974
114. Bodem R, Sonnenblick EH: Mechanical activity of mammalian heart muscle: variable onset, species differences, and effect of caffeine. Am J Physiol 228:250–261, 1975
115. Forman R, Ford LE, Sonnenblick EH: Effect of muscle length on the force–velocity relationship of tetanized cardiac muscle. Circ Res 31:195–206, 1972
116. Brady AJ: Onset of contractility in cardiac muscle. J Physiol 184:560–580, 1966
117. Brutsaert DL, Henderson AH: Time course of mechanical activation in cardiac muscle. Eur J Cardiol 1:201–208, 1973
118. Brutsaert DL, Sonnenblick EH: The early onset of maximum velocity of shortening in heart muscle of the cat. Pfluegers Arch 324:91–99, 1971
119. Brutsaert DL, Claes VA, Sonnenblick EH: Effects of abrupt load alterations on force-velocity-length and time relations during isotonic contractions of heart muscle: Load clamping. J Physiol 216:319–330, 1971
120. Brutsaert DL, Claes VA, Sonnenblick EH: Velocity of shortening of unloaded heart muscle and the length–tension relation. Circ Res 29:63–75, 1971
121. Henderson AH, van Ocken E, Brutsaert DL: A reappraisal of force–velocity measurements in isolated heart muscle preparation. Eur J Cardiol 1:105–118, 1973
122. Sonnenblick EH, Skelton CL: Reconsideration of the ultrastructural basis of cardiac length–tension relations. Circ Res 35:517–526, 1974
123. Brutsaert DL, Parmley WW, Sonnenblick EH: Effects of various inotropic interventions on the dynamic properties of the contractile elements in heart muscle of the cat. Circ Res 27:513–522, 1970
124. Blinks JR, Jewell BR: The meaning and measurement of myocardial contractility, in Bergel DH (ed): Cardiovascular Fluid Dynamics, vol I. New York, Academic Press, 1970, pp 225–260
125. Pollack GH: Maximum velocity as an index of contractility in cardiac muscle. Circ Res 26:111–127, 1970
126. Noble MIM: Problems in the definition of contractility in terms of myocardial mechanics. Eur J Cardiol 1:209–216, 1973
127. Brady AJ: Time and displacement dependence of cardiac contractility: Problems in defining the active state and force–velocity relations. Fed Proc 24:1410–1420, 1965
128. Sonnenblick EH: Determinants of active state in heart muscle: Force, velocity, instantaneous muscle length, time. Fed Proc 24:1396–1409, 1965

129. Edman KAP, Nilsson E: Mechanical parameters of myocardial contraction studied at a constant length of the contractile element. Acta Physiol Scand 72:205–219, 1968
130. Brutsaert DL, Sonnenblick EH: Cardiac muscle mechanics in the evaluation of myocardial contractility and pump function: Problems, concepts and directions. Prog Cardiovasc Dis 16:337–361, 1973
131. Gordon AM, Huxley AF, Julian FJ: The variation in isometric tension with sarcomere length in vertebrate muscle fibers. J Physiol 184:170–192, 1966
132. Taylor SR, Rüdel R: Striated muscle fibers: Inactivation of contraction induced by shortening. Science 167:882–884, 1970
133. Wise RM, Rondinone JF, Briggs FN: Effect of calcium on force–velocity characteristics of glycerinated skeletal muscle. Am J Physiol 221:973–979, 1971
134. Brutsaert DL, Claes VA, Goethals M: Effect of calcium on force–velocity–length relations of heart muscle of the cat. Circ Res 32:385–392, 1973
135. Julian FJ: The effect of calcium on the force–velocity relation of briefly glycerinated frog muscle fibers. J Physiol 218:117–145, 1971
136. Podolsky RJ, Teichholz LE: The relation between calcium and contraction kinetics in skinned fibers. J Physiol 211:19–35, 1970
137. Thames MD, Teichholz LE, Podolsky RJ: Ionic strength and contraction kinetics of skinned muscle fibers. J Gen Physiol 63:509–530, 1973
138. Huxley HE: Structural changes in the actin and myosin containing filaments during contraction. Cold Spring Harbor Symp Quant Biol 37:361–376, 1972
139. Sarnoff SJ, Mitchell JH: The control of the function of the heart, in: Handbook of Physiology, vol 1. Washington D.C., American Physiological Society, 1962, pp 489–532
140. Sonnenblick EH, Downing SE: Afterload as a primary determinant of ventricular performance. Am J Physiol 204:604–610, 1963
141. Braunwald E, Ross J Jr: The ventricular end-diastolic pressure; Appraisal of its value in the recognition of ventricular failure in man. Am J Med 34:147–150, 1963
142. Wallace AG, Skinner NS Jr, Mitchell JH: Hemodynamic determinants of the maximal rate of rise of left ventricular pressure. Am J Physiol 205:30–36, 1963
143. Levine HJ, Britman NA: Force–velocity relation in intact dog heart. J Clin Invest 43:1383–1396, 1964
144. Covell JW, Ross J, Sonnenblick EH, Braunwald E: Comparison of the force–velocity relation and the ventricular function curve as measures of the contractile state of the heart. Circ Res 19:364–372, 1966
145. Fry DL, Griggs DM, Greenfield JC: Myocardial mechanics: Tension–velocity–length relationships of heart muscle. Circ Res 14:73–85, 1964
146. Falsetti HL, Mates RE, Greene DG, Bunnell IL: V_{max} as an index of contractile state in man. Circulation 43:467–479, 1971
147. Hugenholtz PG, Ellison RC, Urschel CW, et al: Myocardial force–velocity relationships in clinical heart disease. Circulation 41:191–202, 1970
148. Mason DT: Usefulness and limitations of the rate of rise of intraventricular

pressure (*dP/dt*) in the evaluation of myocardial contractility in man. Am J Cardiol 23:516–527, 1969

149. Graham TP, Jarmakani JM, Canent RV, Anderson PAW: Evaluation of left ventricular contractile state in childhood. Circulation 44:1043–1053, 1971
150. Mehmel H, Krayenbuhl HP, Rutishauser W: Peak measured velocity of shortening in the canine left ventricle. J Appl Physiol 29:637–645, 1970
151. Karliner JS, Gault JH, Eckberg D, et al: Mean velocity of fiber shortening. Circulation 44:323–333, 1971
152. Urschel CW, Henderson AH, Sonnenblick EH: Model dependency of ventricular force–velocity relations: Importance of developed pressure. Fed Proc 29:719, 1970
153. Parmley WW, Chuck L, Sonnenblick EH: Relation of V_{max} to different models of cardiac muscle. Circ Res 34:34–43, 1972
154. Nejad NS, Klein MD, Mirsky I, Lown B: Assessment of myocardial contractility from ventricular pressure recordings. Cardiovasc Res 5:15–23, 1971
155. Mirsky I, Ellison RC, Hugenholtz PG: Assessment of myocardial contractility in children and young adults from ventricular pressure recordings. Am J Cardiol 27:359–367, 1971
156. Streeter DD, Vaishnav RN, Patel DJ, et al: Stress distribution in the canine left ventricle during diastole and systole. Biophys J 10:345–363, 1970
157. Gault JH, Ross J, Braunwald E: Contractile state of the left ventricle in man: Instantaneous tension–velocity–length relations in patients with and without diseases of the left ventricular myocardium. Circ Res 22:451–463, 1968
158. Hefner LL, Bowen TE: Elastic components of cat papillary muscle. Am J Physiol 212:1221–1227, 1967

William A. Neill

3

Regulation of Cardiac Output

PURPOSE OF CARDIAC OUTPUT

Cardiac output is the rate at which the heart delivers blood to the body. The need for blood flow by the body to maintain the proper interstitial environment for its cellular constituents provides the basis for cardiac output regulation. A human or any other large organism requires a means by which substances can be transported over long distances and brought into intimate contact with remote individual cells. Circulating blood is an exchange medium linking cells to cells, and the cells to the environment outside the organism. The blood supplies the cells with needed raw materials from the lungs, splanchnic region, and fat depots, and removes their undesired metabolic end products, which are then carried to the lungs, kidneys, and liver. Hormones, when dispersed by the blood, can influence the function of remote cells.

The transport function of blood flow must be capable of meeting (1) basal needs to maintain cell homeostasis over an extended period and (2) specific needs that arise when activity and metabolic requirements of different tissues of the organism change in diverse ways. Blood flow is not maintained at a constant level that can satisfy all possible ranges of needs for each tissue in the body. Instead, it is responsive to changes in local tissue needs. With the body at rest, blood flow is regulated to a relatively low rate that is appropriate to the low transport requirements. Therefore, when requirements vary due to increased activity, blood flow must change appropriately in order to continue to serve its purpose.

CARDIOVASCULAR ORGANIZATION AND MECHANISMS OF ADJUSTMENT

Organization

Heart muscle contraction provides the impetus for blood flow. The blood vessels conduct the blood and by varying their resistance to flow locally determine the distribution of blood flow.

The pulmonary and systemic circulations are arranged in series. Both circulations are composed of enormous numbers of parallel circuits. Individual control of each parallel circuit permits independent control of blood flow through different regions. In the pulmonary circulation independent local changes in vascular resistance shunt blood flow away from poorly ventilated alveoli and ensure oxygenation of mixed pulmonary venous blood. But it is the systemic circulation, serving different regions of widely varying blood flow needs, that fully utilizes the parallel-circuit arrangement in keeping local blood flow appropriate for local needs.

The large arteries and veins that conduct blood over long distances offer little resistance to blood flow and are relatively inactive in blood flow regulation. The arterioles and precapillary sphincters are major points of resistance, and they contain smooth muscle; therefore, they are suitable sites for blood flow control. The veins also contain smooth muscle and serve as a reservoir of blood that can be redistributed elsewhere in the circulatory system. Exchange of substances between the blood and tissue occurs as the blood traverses thin-walled capillaries linking the arteries and veins.

Mechanisms of Adjustment

Mechanisms for control of cardiac output operate through two general pathways: peripheral vasomotion and force of cardiac contraction, as represented in Figure 3-1.

Peripheral Vasomotion

Resistance to blood flow is regulated mainly by intravascular smooth-muscle tension. The muscle tension in arterioles determines their caliber, and in the precapillary sphincters muscle tension determines which capillaries will remain open to receive the blood flow. This selectivity of capillaries exerts an important influence on diffusion distance between the capillary blood and tissue. The arteriole and precapillary muscle tensions are controlled (1) locally by tissue and (2) remotely by the sympathetic nervous system and by chemical substances

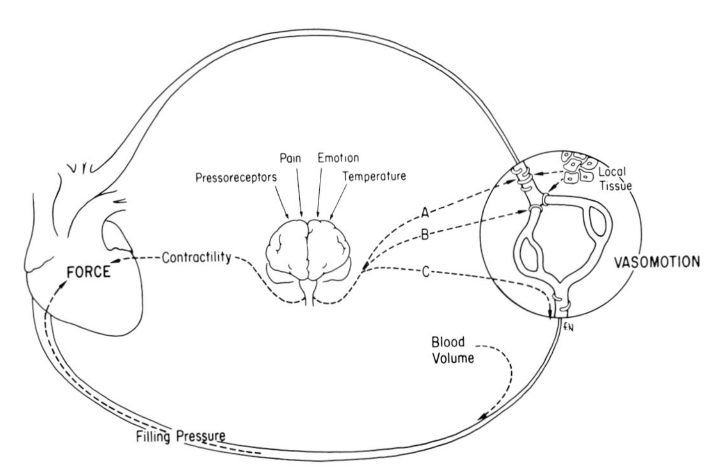

Fig. 3-1. A scheme representing the control of cardiac output by peripheral vasomotion and regulation of cardiac contraction force. Varied information reaching the brain modulates the sympathetic nerve output to the heart and peripheral blood vessels. Peripheral vasomotion is determined by the smooth-muscle tension in the arterioles, precapillary sphincters, and veins. All three types of vessels respond to sympathetic nerve impulses (A, B, and C, respectively), and the arterioles and precapillary sphincters also respond to local tissue metabolites. Peripheral vasomotion and blood volume together determine cardiac filling pressure. The force of cardiac contraction is regulated by filling pressure (intrinsic response) and sympathetic nerve impulses (contractility). The influences of humoral substances (other than local tissue metabolites) are not represented.

secreted by the adrenals, kidneys, and spleen and delivered to the peripheral vessels via the blood. Smooth-muscle tension in the walls of the veins is controlled by sympathetic nerves and determines their capacity for containing blood, which influences venous pressure, cardiac filling pressure, and the force of cardiac contraction, as will be described subsequently.

There is much evidence that tissue itself judges the adequacy of its own blood supply and institutes the necessary arteriole and precapillary sphincter adjustments to keep blood flow appropriate for current needs. Local autonomous control of blood flow is most suitably designed for and is most used by tissues in which the purpose of blood flow is metabolic support (e.g., heart, brain, working skeletal muscle). Local control is the most powerful factor in long-term blood flow regulation (homeostasis).

Remote autonomic nervous control over peripheral vasomotion is organized in the brain. Sympathetic adrenergic vasoconstrictor impulses exiting the brain travel to the arteriole, precapillary sphincter, and venous smooth muscles (Fig. 3-1, A, B, and C, respectively). Sympathetic vasodilation occurs principally by diminution of the adrenergic impulses, whereas sympathetic vasodilator impulses, which are cholinergic, appear to play a relatively unimportant role in humans. The vasoconstrictor impulse transmission from the brain is modulated by nerve connections from a variety of sources, including arterial pressoreceptors and chemoreceptors, somatic afferents appreciating pain and muscle proprioception, hypothalamic connections involved in emotions and temperature regulation, and cerebral connections. Sympathetic control in some instances is diffuse. For example, afferent stimuli arising from pressoreceptors in the carotid artery, which participate in maintenance of normal systemic arterial blood pressure, inhibit sympathetic impulses and cause widespread vasodilation in the systemic circulation. The sympathetic vasomotor response to stimulation of the hypothalamus, however, may be highly selective anatomically and may comprise part of a complex performance of the organism such as body temperature regulation or preparation for combat or escape, as will be seen later when acute regulation of the circulation is discussed.

The vascular beds of some organs, including heart muscle and brain, are unresponsive to remote control. In others (e.g., skeletal muscle), reflex vasoconstriction designed to maintain arterial blood pressure may compete with local vasodilation aimed at maintaining tissue blood flow. For example, sympathetic vasoconstriction may occur during exercise and oppose the vasodilation induced by increased metabolic rate of the muscles,[1] and during hemorrhagic hypotension, basal nutritional needs of muscle may interfere with the pressoreceptor reflex vasoconstriction.[2] Under these conditions local vasomotor control dominates in the arterioles and especially in the precapillary sphincters, while reflex vasomotion is essentially unopposed in the veins. In some diving animals, however, reflex vasoconstriction of peripheral arteries overrides local control, and skeletal muscle blood flow practically ceases during submersion despite marked muscular exertion.[3]

Force of Cardiac Contraction

Intrinsic control. The heart adjusts its contraction force so that it pumps blood as fast as it is received from the veins. This fundamental intrinsic property of heart muscle enables peripheral arterial and venous vasomotion and changes in blood volume to influence cardiac output (Fig. 3-1). The filling pressure of the heart is responsible for diastolic stretch of the heart muscle fibers, and the increased stretch augments the strength of cardiac contraction, increasing the stroke volume and cardiac output.

Filling pressure of the heart depends upon the volume of blood contained in the circulatory system relative to its capacity. Venous vasomotor tone controls the capacity and is utilized especially for acute adjustments of cardiac filling pressure. Blood volume is regulated by fluid exchange with the extravascular interstitial space and with the external environment via the kidneys. Fluid exchange between the capillary blood and interstitial space normally is balanced by opposing hydrostatic and osmotic pressure gradients. When a fall in blood volume activates the arterial pressoreceptor mechanism, constriction of arterioles, which are located proximal to the capillaries, decreases capillary hydrostatic pressure. The opposing forces are unbalanced until fluid moves from the interstitial space into the capillaries. Fluid excretion externally via the kidneys is controlled by glomerular filtration and tubular reabsorption. Systemic arterial hypotension reflexly constricts renal arteries proximal to the glomeruli, which decreases glomerular fluid filtration, and induces secretion of renin and aldosterone, which increases renal tubular fluid reabsorption. The consequent expansion in blood volume restores cardiac filling pressure and cardiac output, withdrawing the arterial pressoreceptor signals for renal vasoconstriction and renin secretion.[4]

External control. Sympathetic stimulation of the heart markedly increases its force of contraction without an increase in filling pressure. (In the intact organism, cardiac volume and filling pressure decrease.) A change in contraction strength that does not depend upon filling pressure is referred to as a change in contractility (Fig. 3-1). Even when contractility increases, the intrinsic property of the heart to adjust to filling pressure persists; the cardiac output at each level of filling pressure is higher. Cardiac contractility is increased also by catecholamines secreted by the adrenal medulla in response to sympathetic nerve stimulation. Recently reported experiments in dogs suggest that sympathetic stimulation of the spleen induces the elaboration of a different chemical capable of profound augmentation in cardiac contractility.[5] Sympathetic stimulation also increases heart rate, whereas the vagal parasympathetics slow the rate.

The relatively stable distribution of volume and pressure of blood in the different regions of the circulatory system, even in the face of marked alterations in blood flow, is evidence that the peripheral vasomotor and cardiac adjustments work smoothly together. Major characteristics of the circulatory system that assist in the coordination are (1) the intrinsic property of heart muscle to pump the amount of venous blood presented, (2) the renal mechanism for preservation of an appropriate blood volume, and (3) the arterial pressoreceptors, which through their sympathetic connections in the medulla modulate cardiac activity and peripheral vasomotion to keep arterial pressure within the range necessary to perfuse the vital organs of the body.

HOMEOSTASIS

Homeostasis refers to the regulation of blood flow over an extended period to maintain an appropriate chemical and physical environment for cells throughout the body in the basal resting state.

Local Control of Blood Flow

Cardiac output is not uniformly distributed. Table 3-1 shows the variation in blood flow per unit mass of tissue in different organs (see also additional data[6,7]). The variation can be explained partly by different O_2 consumption rates in different organs; however, the diverse values for blood A-V O_2 show that the relationship between O_2 consumption and blood flow is not consistent. Organs with relatively low or high blood flow in the intact body maintain similar blood flow when they are separated from the body and perfused artificially at physiologic pressure, indicating that the pattern of cardiac output distribution is determined mainly by

Table 3-1
Blood Flows in Different Organs

	Blood Flow		O_2 Consumption		Blood A-V O_2
Organ	(ml/100 g/min)	(ml/min)	(ml/100 g/min)	(ml/min)	(ml/liter)
Kidneys	314	1100	4.6	16	13
Heart	89	250	10.0	27	114
Cerebrum	54	750	3.3	46	63
Splanchnic	29	1400	1.2	58	41
Skin	14	500	0.14	5	10
Skeletal muscle	4.9	1200	0.29	70	80
Other	1.7	600	0.034	12	30
Total	8.3	5800	0.33	234	40

Data adapted from (1) Schaeffer, J Parsons (ed): Morris' Human Anatomy (ed 11). New York, McGraw-Hill, 1953, p 43; (2) Wade OL, Bishop JM: Cardiac Output and Regional Blood Flow, Oxford, Blackwell Scientific Publications, 1962, p 93.

the individual organs themselves rather than by a centrally organized plan.

Blood flow is actively regulated to the level that normally exists in the basal state. For example: (1) an increase or a decrease in the arterial perfusion pressure is met by an immediate opposing change in local vascular resistance, which maintains the flow near normal (autoregulation); (2) temporary interruption of blood flow is followed by a transient increase in flow above the original level (postocclusion reactive hyperemia). The magnitude of the reactive hyperemia is proportional to the duration of occlusion and appears to be related to delinquent transport needs. Autoregulation and postocclusion reactive hyperemia are manifestations of local vasomotor regulation; they operate in organs isolated from the body and therefore cannot depend upon extrinsic neural or humoral mechanisms.

By manipulating independently the concentration of O_2 and the rate of blood flow, it has been shown that local vasomotor regulation involves mainly local regulation of O_2 delivery. Decrease or increase in O_2 concentration in the blood perfusing an organ causes a compensatory change in vascular resistance, which tends to preserve the amount of O_2 delivered to the organ and to maintain a constant tissue P_{O_2}. Preservation of O_2 delivery dominates blood flow regulation in skeletal and cardiac muscle, whose variable metabolic rates depend upon a mechanism responsive to changes in O_2 requirements. On the other hand, O_2 delivery is not the major regulating factor for all tissues. In the kidney, blood flow is high in proportion to O_2 requirements and is more responsive to changes in fluid and electrolyte excretory demands. Blood flow to the skin is influenced less by the skin's metabolic needs than by the body's variable requirements for heat loss needed to maintain a constant body temperature. Although metabolic requirements of the brain are relatively high, they are also rather stable, and brain blood flow is regulated mainly to keep brain tissue P_{CO_2} and pH constant rather than to compensate for variations in O_2 need.

The local regulation of blood flow according to the supply and demand of tissue for O_2 suggests that arteriole and precapillary sphincter vasomotion is linked to the availability of O_2 to tissue for its metabolism. Probably some metabolic consequence of O_2 availability, rather than O_2 itself, is responsible for the local vasomotion. As evidence, cyanide administration causes all of the metabolic changes of O_2 deficiency and results in vasodilation; yet O_2 concentration itself increases.[8]

Local metabolic regulation of blood flow requires (1) a means of judging the adequacy of O_2 availability in the tissue and (2) a mechanism for inducing appropriate vasomotor adjustments in O_2 delivery when needed to restore its adequacy. Tissue hypoxia decreases the redox balance of the respiratory enzymes[9] and decreases the phosphorylation of adenine nucleotides.[10] Both chemical changes can be expected to occur when O_2 availability falls. A system that monitored either one would provide a suitable means of judging adequacy of O_2 supply and would, moreover, link regulation of O_2 supply to regulation of O_2 utilization rate. In the first case, ATP utilization rate governs O_2 utilization rate, and the redox balance would be kept constant by adjustment of blood flow and O_2 delivery, thus ensuring a satisfactory energy flow for reconstitution of ATP. In the second instance, the breakdown products of ATP utilization would govern both O_2 utilization rate and blood flow directly, thus regulating O_2 use and O_2 delivery in parallel.

The second requirement concerns the mechanism for transforming a recognition of inadequate O_2 availability into an appropriate vasomotor adjustment. Since the purpose of local regulation supposedly is to preserve O_2 supply to the tissue, adequacy should be judged within the tissue cells themselves, rather than in the vascular smooth-muscle cells serving the tissue. Yet this arrangement will demand transmitting information from the cell to the neighboring arteriole or precapillary sphincter. There is no evidence to suggest the existence of nerve fibers to carry out this function, although some direct pathway remains a theoretical possibility. A vasodilator substance released into the adjacent capillary could not easily reach the upstream resistance sites. Diffusion via interstitial fluid seems the most likely possibility, and considerable experimental evidence is accumulating that adenosine, which is a breakdown product of ATP utilization and a potent vasodilator, diffuses from cells and is responsible for metabolically induced local vasomotor regulation.[10]

Cardiac Output in Homeostasis

Normal Control

Cardiac output at rest is the sum of blood flows locally regulated throughout the body. In the normal adult human lying supine at rest, cardiac output ranges from approximately 4 to 8 liters/min. The variability is less if corrected for body size, and since there is a closer relationship to body surface area than to body weight, cardiac output often is expressed as cardiac index, which normally lies between 2.5 and 4.5 liters/min per square meter of estimated body surface area. In Table 3-1 the quantitative relationship between blood flow and O_2 consumption, expressed as A-V O_2 difference, is shown to vary between different organs.

The P_{O_2} of venous blood draining different organs ranges from 20 to 60 mm Hg. The A-V O_2 difference for the body as a whole is approximately 40 ml per liter of blood (one-fifth of the O_2 contained in the arterial blood), and the P_{O_2} of mixed systemic venous blood is approximately 40 mm Hg.

The narrow ranges of cardiac output and mixed venous blood P_{O_2} might suggest that these variables are monitored and regulated in the intact organism. It is highly unlikely, however, that systems exist for monitoring pulmonary artery or aorta blood flow or mixed venous blood P_{O_2}. The blood flow and P_{O_2} in these structures have no biologic importance to the organism, and their control would have only the most crude influence on maintaining blood supply to individual cells at appropriate levels. Cardiac output and mixed venous blood P_{O_2} remain within narrow limits because blood flows in many of the diverse components of the systemic circulation are regulated by the local tissue needs for O_2.

Since a major purpose of blood flow is to deliver O_2 to the tissues, the large amount of O_2 that is not normally extracted by the tissues and thus remains in the venous blood at rest provides a reserve that can be utilized when cardiac output fails to keep pace with body O_2 needs, e.g., when O_2 needs rise acutely during muscular exercise or when cardiac function is impaired. The venous blood O_2 reserve can be recruited either by a decrease in capillary blood P_{O_2} or by a decrease in the affinity of hemoglobin for O_2.[11]

The heart assumes a passive role in homeostasis and takes no initiative in cardiac output control. To consider the cardiac output as the product of heart rate and stroke volume, although mathematically valid, is likely to lead to biological misunderstanding. For example, a primary increase in heart rate, which can be induced by electric pacing or atropine administration, is compensated by a reciprocal fall in stroke volume, while cardiac output remains unchanged.[12–14] Even when heart rate is below normal, as in congenital atrioventricular conduction block, cardiac output is adjusted to normal levels by increased stroke volume.[15] Moreover, the increase in cardiac output during exercise, although normally accompanied by increased heart rate, does not depend on the increased heart rate. Cardiac output still increases appropriately during exercise even when the heart rate is kept constant by electric pacing.[14]

Pathologic Conditions

Certain pathologic conditions in which cardiac output is outside of the normal range illustrate, paradoxically, local homeostatic control mechanisms. To be considered are systemic arteriovenous fistula, anemia and polycythemia, hyperthyroidism, and heart disease.

A systemic arteriovenous fistula can be viewed as a low-resistance circuit parallel to the other systemic circuits of the body. If the fistula is of small or moderate size, cardiac output rises by an amount equal to the fistula flow, since blood flows through the other circuits are controlled independently.[16] Mixed venous blood O_2 concentration and P_{O_2} are above normal, illustrating that neither cardiac output nor mixed venous blood O_2 is under active regulation. If the fistula is extremely large the increase in cardiac output does not keep pace with the fistula flow; blood flows through selected parallel circuits are decreased by increased arteriole resistance in order to prevent an unacceptably large fall in arterial blood pressure (e.g., by the pressoreceptor reflex initiating sympathetic vasoconstriction). The basis of the increase in cardiac output is the intrinsic response of the heart to the increased venous filling pressure and decreased peripheral resistance, which are caused by the fistula. Over a more prolonged period, reduced blood flow through the renal circuit may lead to fluid retention and expanded blood volume with further increases in cardiac filling pressure and output.

In regard to a major transport responsibility of the cardiovascular system, anemia and polycythemia represent a decrease and an increase, respectively, in the quantity of O_2 that can be carried by the blood. Although the P_{O_2} of arterial blood is normal in anemia, the amount of O_2 contained in the blood is low, and each increment of O_2 molecules removed from the blood by the tissues produces a greater than normal fall in capillary blood P_{O_2}. Capillary blood P_{O_2}, not blood O_2 concentration, is responsible for O_2 diffusion into the tissue; therefore, an increase in blood flow during anemia that would restore capillary (and venous) blood P_{O_2} to normal would be a perfect adjustment, at least as far as tissue O_2 supply at rest is concerned. In fact, the blood flow increase is less, and venous blood P_{O_2} and O_2 concentration are both below normal in anemia.[17] Since venous blood O_2 reserve is low, any increase in tissue O_2 needs (e.g., during exercise) depends to an abnormal degree upon an increase in cardiac output.[18]

An inverse relationship exists between cardiac output and blood O_2 capacity that extends above as well as below normal values for O_2 capacity[19] and appears to be related to preservation of normal tissue O_2 supply. The distribution of the increased cardiac output is consistent with regulation by O_2 supply. Blood flow does not increase uniformly in all parallel circuits, as one would expect from a simple change in viscosity, but increases most in the tissues where local regulation by metabolism dominates (e.g., heart muscle[20]) and increases less or not at all where nutrition of tissue is not the major consideration (e.g., kidney[21]). Although local blood flow control appears to be involved in the peripheral vasodilation in anemia, other aspects of the mechanism of the increase in cardiac output are less clear. Increased cardiac filling pressure or low systemic arterial pressure have not been found during anemia, and surgical or pharmacologic interruption of the sympathetic control of cardiac contractility has little effect on the high cardiac output.[22,23]

In anemia an increase in cardiac output compensates for the low O_2 concentration in the blood. In hyperthyroidism an increase in cardiac output compensates for the elevated requirement of the tissue for O_2. Although the increase in anemia does not prevent a slight fall in mixed venous blood P_{O_2}, the increase in cardiac output in hyperthyroidism is proportionately greater than the increase in metabolic rate. Thus the systemic A-V O_2 in hyperthyroidism at rest or during a given level of exercise is slightly less than normal.[17,24] This observation would be difficult to explain solely on the basis of peripheral blood flow regulation related to tissue metabolic needs. Excessive sympathetic stimulation of cardiac contractility does not appear to be involved, since sympathetic inhibition in hyperthyroid patients does not decrease their resting cardiac output.[25] There is evidence that thyroid hormone has a direct effect on cardiac contractility,[26] which could be responsible for the increase in cardiac output beyond the augmented nutritional requirements of the tissue.

Pathologic limitations in the functional capabilities of the heart or blood vessels also demonstrate cardiac output homeostasis. The underlying causes of dysfunction of the heart as a pump may be categorized as either impairment of the contractile property of the heart muscle itself or abnormal performance of the associated mechanical apparatus that forces the heart muscle to achieve the necessary systemic cardiac output at a physical disadvantage. In either case the cardiac output for any combination of filling pressure and aortic pressure is below normal. The heart muscle, however, retains, even if to a reduced degree, both its intrinsic property to adjust its contraction strength to changes in filling pressure and its capacity for altered contractility under the influence of various factors, including sympathetic nerve stimulation. If the cardiac dysfunction is slight, cardiac output can be maintained in the normal range by an increase in its filling pressure[27] due to augmented blood volume and venous constriction.[28] When the resting cardiac output falls below normal, its distribution shifts: blood flows to the heart muscle and brain are maintained at the expense of the rest of the body.[29] Tachycardia, decreased myocardial norepinephrine concentration, and increased urinary excretion of norepinephrine[30] are signs that the circulatory system is operating under intensified activity of the sympathetic nervous system as a means of compensating for the low cardiac output.

When these mechanisms that are normally used to increase cardiac output during stress are partially exploited in the resting state, cardiac reserve falls. Therefore even when the output of the diseased heart is normal or only slightly low at rest the increase in its output for a given amount of exercise is much less than normal and is achieved only by excessive increases in heart rate and filling pressure.[27,31] The homeostatic circulatory adjustments brought on by cardiac dysfunction may result in signs and symptoms related to diminished skeletal muscle and skin blood flow (weakness, fatigue, cold extremities), increased cardiac filling pressure (dyspnea, edema), and augmented sympathetic nervous system activity (sweating, palpitations, angina).

Pathologic conditions of the peripheral blood vessels, because of their focal nature, are more often considered from the standpoint of the organs they directly affect. Nevertheless the vasodilation of downstream arterioles and the progressive growth of collateral blood vessels in response to arterial stenosis are examples of local homeostatic control of blood flow.

ACUTE REGULATION

Humans engage in abrupt changes in their activities that can be met only by circulatory responses that are both prompt and suitable for the specific activity undertaken. The sympathetic nervous system is a prominent factor in acute circulatory adjustments. Since acute circulatory needs extend beyond satisfying basal requirements of the organism, in the presence of cardiovascular disease they may precipitate signs and symptoms of impaired circulatory function not present under basal resting conditions. The responses to alterations in effective blood volume, exercise, emotion, and temperature will be considered.

Effective Blood Volume (Hemorrhage, Transfusion, Posture)

Loss of blood decreases the filling pressure of the heart. The intrinsic cardiac response to the lower filling pressure is a decrease in its contraction strength, resulting in a fall in cardiac output and systemic arterial blood pressure. This fall in blood pressure, however, diminishes arterial pressoreceptor stimulation, which augments sympathetic vasoconstriction and cardiac contractility, thus turning systemic arterial blood pressure and cardiac output back toward normal. Atrial pressoreceptors[32] and vasoconstriction due to vasopressin secretion[33] may also participate significantly. The magnitude of the fall in cardiac output depends upon the volume of blood lost, the responsiveness of the sympathetic system, and the ability of the heart to increase its contraction force under sympathetic stimulation. Carotid sinus denervation or pharmacologic sympathetic inhibition weakens the pressoreceptor reflex, resulting in a larger drop in blood pressure and cardiac output for a given volume of blood loss.[34] Since the reflex arteriole vasoconstriction released by the pressoreceptor stimulation is not uniformly distributed and is antagonized variably by local blood flow regulation in different organs, the decrease in cardiac output during hemorrhage is very unevenly distributed. For example, an acute bleed of 30% of the blood volume in conscious monkeys results in a marked fall in skin and skeletal muscle blood flow, a moderate fall in visceral blood flow, and no change in blood flow to the brain and heart muscle.[35] Since the arterial perfusion pressure is low after the hemorrhage, the normal blood flow to the brain and heart muscle implies that vascular resistance in these organs must decrease, presumably due to dominant local metabolic regulation of blood flow.

Acute increase in blood volume produces direct hemodynamic effects and secondary reflex adjustments that are opposite to those of hemorrhage. Infusion of 1500 ml of blood into a normal adult in 1.5 hr causes no detectable increase in cardiac output. The increased blood volume apparently is largely accommodated by an increase in venous capacity due to reflex venous vasodilation. On the other hand, when the sympathetic reflex adjustments are inhibited the same infusion increases the cardiac filling pressure and causes a substantial increase in cardiac output.[36]

Gravitational forces along the long axis of the body have a profound effect on blood volume distribution when humans stand up. The pressure in the veins and capillaries in the feet is greater than the filling pressure of the heart by an amount equal to the pressure exerted by a column of water extending from the feet to the heart. In the standing position an extra 500 ml of blood accumulates in the distensible blood vessels of the legs and thus is effectively removed from circulation. If this posture persists, transcapillary filtration leads to interstitial edema and further reduction in blood volume. The decrease in cardiac filling pressure in the upright position leads to cardiovascular sympathetic stimulation via the arterial pressoreceptor mechanism, as in hemorrhage. Despite the resulting sympathetic stimulation of cardiac contractility, however, the stroke volume and cardiac output decrease.[37,38] The accompanying increases in heart rate and diastolic arterial blood pressure are related to the sympathetic stimulation. As evidence that the circulatory effects of the upright posture, including the compensatory adjustments, are consequences of reduced effective blood volume due to venous pooling, the circulatory changes with posture disappear when the increased transmural pressure gradient of the blood vessels in the dependent part of the body is nullified by immersing the subject in water. When exercise is performed in the upright position, rhythmic contraction of leg muscles counteracts the gravitational effects by extravascular venous compression. As with hemorrhage, inhibition of the sympathetic system diminishes the effectiveness of the compensatory circulatory adjustments to upright posture, and patients receiving sympatholytic agents may become hypotensive when they stand. Cardiac disease further limits the response of the heart to compensatory sympathetic stimulation. The effects of decreased blood volume and upright posture are additive. For example, a patient with heart disease who is capable of adjusting to a decrease in blood volume due to dehydration, gastrointestinal bleeding, or excessive diuresis as long as he remains in the horizontal position may be incapable of the additional compensation needed to assume the upright position and may therefore develop hypotension and faint when he stands.

Longer term adjustments to hemorrhage or posture involve regulation of blood volume. Low capillary pressure due to intravascular volume loss plus arteriole vasoconstriction withdraws interstitial fluid into the capillaries. Blood loss or the upright standing position causes renal arteriole constriction and decreased renal blood flow, glomerular filtration, and fluid excretion. Effective blood volume is increased toward normal. On the other hand, prolonged horizontal position increases renal fluid excretion and decreases the blood volume,[39] which may be a factor in the postural sensitivity of convalescing patients following prolonged bedrest.

Exercise

Muscular exercise poses the greatest challenge for acute circulatory adjustment, due mainly to the need for transport of metabolites to support the enormous increase in metabolic rate of the exercising skeletal muscles and to a lesser extent the heart muscle. The increase in cardiac output during exercise is proportional to but quantitatively less than the increase in systemic O_2 consumption, i.e., the increase in O_2 consumption is met partly by increase in blood flow and partly by increase in extraction of O_2 from the blood. Stroke volume and cardiac output at any level of O_2 consumption are higher for supine exercise than for upright exercise due to the higher cardiac filling pressure in the supine position.[38] During upright exercise the increase in cardiac output is accomplished partly by increased heart rate and partly by increased stroke volume.

When exercise is performed lying down the increase in cardiac output is due entirely to increased heart rate, whereas stroke volume, which is higher at rest in this position, does not change consistently during exercise. The limits of cardiovascular adjustment to upright exercise in untrained normal young adults are: heart rate 2.5× (to 190 beats/min), cardiac output 4× (to 25 liters/min), stroke volume 2× (to 120 ml), and systemic A-V O_2 2.5× (to 150 ml/liter).[40] The heart rate is lower and stroke volume higher at any level of O_2 consumption in physically trained subjects. Training increases the maximum cardiac output but does not alter its relationship to O_2 consumption.[41] Serial observations made over 1-min intervals in exercising humans[42] and data from studies utilizing electromagnetic flowmeters in conscious exercising dogs show that the major cardiovascular adjustments reach a nearly steady state within the first minute of exercise.[43]

Exercise profoundly alters the distribution of cardiac output. Vascular resistance in the exercising muscles falls markedly, and most of the increase in systemic blood flow traverses these circuits. Presumably the vasodilation in the exercising muscles is induced locally as a response to their increased metabolic rate, as occurs when muscles isolated from the body are stimulated to contract. Vasoconstriction in the splanchnic region[44] and in skeletal muscles[45] that are not involved in the physical activity may increase resistance in these circuits, thereby helping to sustain aortic blood pressure and as a result enhancing the effect of the vasodilation where the augmented transport for metabolism is needed.

Mean aortic pressure rises, and the volume and filling pressure of the heart diminish during upright exercise of mild or moderate intensity.[46,47] Therefore the increase in cardiac output cannot be attributed simply to the intrinsic response of the heart to an increased cardiac filling pressure resulting from decreased systemic arterial resistance, as occurs with a systemic arteriovenous fistula. Instead, contractility of the heart increases and provides a counterpart, equal in importance, to the vasodilation in the exercising muscles. Stimulation of the diencephalon in the dog increases its heart rate, contractility, and cardiac output, changes that resemble the cardiovascular response to exercise and suggest that the sympathetic outflow from the central nervous system might be responsible for the increased contractility in exercise.[48] Surgical denervation of the heart[49,50] or inhibition of sympathetic β-adrenergic receptors by administration of propranolol,[46,50,51] when employed alone, markedly attenuates the tachycardia but only modestly decreases the cardiac output response to exercise. On the other hand, cardiac denervation and propranolol together profoundly limit the cardiac output response.[50] These findings imply that the effect of cardiac nerve stimulation is not confined to sympathetic β-adrenergic activation and that sympathetic activation during exercise influences cardiac contractility not only via the cardiac nerves. The latter proposal is supported by the observation that adrenalectomy further reduces the exercise cardiac output in dogs already subjected to cardiac denervation.[52] These experiments suggest that the sympathetic system, operating normally through cardiac nerve stimulation as well as release of catecholamines and perhaps other humoral substances remote from the heart, is responsible in a major way for the increase in cardiac contractility during muscular exercise. When either the cardiac nerve stimulation or humoral release is blocked, the alternative surviving mechanism operates almost as effectively alone. When they are both blocked, the sympathetic system is effectively paralyzed, and any increase in cardiac output must depend upon other, less effective mechanisms.

Even with the sympathetic system intact, when the limit of its employment is reached a further increase in cardiac output ultimately depends upon the intrinsic strength of the heart itself. Although cardiac volume and filling pressure decrease during moderate exercise in dogs, the further increase in cardiac output during extreme exercise depends upon increase in filling pressure.[47,53] As discussed previously, when cardiac disease is present sympathetic stimulation is utilized to satisfy even basal requirements, and an increase in cardiac output even for mild exercise depends upon increase in filling pressure. Unfortunately, the abnormality of the cardiac muscle compromises its capacity to respond to the higher filling pressure. Ancillary mechanisms (e.g., vasoconstriction in skin and splanchnic and renal circuits) are brought into play during mild exertion.[45] The cardiac output response to a given increase in O_2 consumption is low, and the systemic A-V O_2 during exercise widens abnormally.

Emotion

Emotional excitement is second only to muscular exercise in imposing acute stress on the circulatory system in humans. Excitement increases the heart rate, cardiac output, and systemic arterial pressure.[54–56] Changes in systemic vascular resistance are inconstant. Blood flow to heart muscle and skeletal muscle increases, and splanchnic and skin blood flow decreases.[54] The origins of the emotional state and accompanying circulatory phenomena appear to be closely associated in the brain, since electrical stimulation of regions in the lateral hypothalamus in conscious cats that produces overt anxiety also produces the same redistribution of blood flow typical of anxiety: increased skeletal muscle blood flow and decreased splanchnic blood flow.[57] These circulatory changes do not depend upon increased muscular activity and still occur when the cat is anesthetized.

The cardiovascular response to muscular exercise is characterized by a coordinated balance between sympathetic stimulation of the heart and metabolically induced vasodilation in muscles. During excitement the sympathetic cardiac stimulation exceeds by far the relatively minor need for increased systemic transport of metabolites. As a result the increase in cardiac output during excitement is proportionately greater than the increase in O_2 consumption, and systemic A-V O_2 narrows.[58] Sympathetic β-adrenergic inhibition attenuates the increases in cardiac output and heart rate to a much greater extent during excitement than during exercise,[55,56] and the circulatory changes of exciting stimuli can even be markedly reduced by tranquilization.[56]

Temperature Regulation

Blood flow through the skin is regulated according to the needs of the organism as a whole for heat conservation, rather than to satisfy provincial metabolic requirements of the skin. Exposure of the body to a cold environment causes vasoconstriction, decrease in cutaneous blood flow, and increase in systemic arterial blood pressure. Since skin blood flow normally constitutes such a minor component of cardiac output, the influence of a decrease in skin blood flow on cardiac output is negligible.[59] The effect on body heat conservation, however, which is the objective, is considerable. Exposure to a hot environment causes cutaneous vasodilation, and the resulting increase in cutaneous blood flow can be large enough to cause a significant increase in cardiac output.[60] Since the increase in cardiac output is not brought about by increased metabolic rate, systemic A-V O_2 narrows or remains unchanged. If the peripheral vasodilation is not balanced by a proportional increase in cardiac output in patients with cardiac disease, hypotension may occur. Hypotension is especially apt to occur if muscular exercise is superimposed on the hot environment, further augmenting cardiac requirements due to simultaneous increases in muscular and cutaneous blood flow.

The mechanism of cutaneous vasoconstriction due to cold exposure is sympathetic nerve stimulation of arterioles, which may be induced either reflexly by cooling the skin or by cooling the hypothalamus directly.[61] Heating the skin or the hypothalamus causes cutaneous vasodilation by sympathetic inhibition. The cutaneous vasomotion constitutes only a part of a coordinated response of the organism that is designed to maintain a constant body temperature. Cooling the hypothalamus of dogs causes not only cutaneous vasoconstriction, but stimultaneously shivering, which increases heat production.[62] Warming the hypothalamus provokes cutaneous vasodilation and panting, which in dogs is an additional means of dissipating heat from the body by evaporation.[63] The hypothalamic locus responsible for promoting heat loss appears to be distinct from that promoting heat conservation. The specialized vasomotor responses integrated with other facets of regulation of body temperature are examples of the ability of the central nervous system to exert not just diffuse responses but discrete control over the cardiovascular system.

SUMMARY

The cardiovascular system is responsible for maintaining a viable environment for the body's cellular constituents, giving priority to cells vital for survival of the organism as a whole. Blood flow is responsive to changing needs, including blood flow regulation according to local metabolic requirements, as well as regulation in one region designed to meet homeostatic needs elsewhere. Cardiac output control operates through peripheral vasomotion coordinated with adjustment of the strength of cardiac contraction. Long-term control is dominated by metabolic vasomotor regulation combined with the intrinsic response of cardiac contraction strength to changes in filling pressure. For acute circulatory adjustments, sympathetic control of vasomotion and especially of cardiac contraction force is superimposed over these mechanisms.

REFERENCES

1. Kjellmer I: On the competition between metabolic vasodilation and neurogenic vasoconstriction in skeletal muscle. Acta Physiol Scand 63:450–459, 1965
2. Lundgren O, Lundvall J, Mallander S: Range of sympathetic discharge and reflex vascular adjustments in skeletal muscle during hemorrhagic hypotension. Acta Physiol Scand 62:380–390, 1964
3. Folkow B, Fuxe K, Sonnenschein RR: Responses of skeletal musculature and its vasculature during "diving" in the duck: Peculiarities of the adrenergic vasoconstrictor innervation. Acta Physiol Scand 67:327–342, 1966
4. Folkow B, Neil E: Circulation. New York, Oxford University Press, 1971, p 364

5. Liang C, Huckabee WE: Effects of sympathetic stimulation of the spleen on cardiac output. Am J Physiol 224:1099–1103, 1973
6. Forsyth RP, Nies AS, Wyler F, et al: Normal distribution of cardiac output in the unanesthetized, restrained rhesus monkey. J Appl Physiol 25:736–741, 1968
7. Kaihara S, Rutherford RB, Schwentker EP, Wagner HN Jr: Distribution of cardiac output in experimental hemorrhagic shock in dogs. J Appl Physiol 27:218, 1969
8. Liang C, Huckabee WE: Mechanisms regulating the cardiac output response to cyanide infusion, a model of hypoxia. J Clin Invest 52:3115–3128, 1973
9. Huckabee WE: Metabolic regulation of cardiac output. J Clin Invest 39:998, 1960
10. Rubio R, Berne RM: Release of adenosine by the normal myocardium in dogs and its relationship to the regulation of coronary resistance. Circ Res 25:407–415, 1969
11. Metcalfe J, Dhindsa DS: The physiological effects of displacements of the oxygen dissociation curve, in Rørth M, Astrup P (eds): Oxygen Affinity of Hemoglobin and Red Cell Acid Base Status. Copenhagen, Munksgaard, 1972, pp 613–628
12. Bristow JD, Ferguson RE, Mintz F, Rapaport E: The influence of heart rate on left ventricular volume in dogs. J Clin Invest 42:649–655, 1963
13. Ross J Jr, Linhart JW, Braunwald E: Effects of changing heart rate in man by electrical stimulation of the right atrium. Circulation 32:549–558, 1965
14. Stein ES, Damato AN, Kosowsky BD, et al: The relation of heart rate to cardiovascular dynamics. Pacing by atrial electrodes. Circulation 33:925–932, 1966
15. Holmgren A, Karlberg P, Pernow B: Circulatory adaptation at rest and during muscular work in patients with complete heart block. Acta Med Scand 164:119–130, 1959
16. Frank CW, Wang HH, Lammerant J, et al: An experimental study of the immediate hemodynamic adjustments to acute arteriovenous fistulae of various sizes. J Clin Invest 34:722–731, 1955
17. Bishop JM, Donald KW, Wade OL: Circulatory dynamics at rest and on exercise in the hyperkinetic states. Clin Sci 14:329–360, 1955
18. Sproule BJ, Mitchell JA, Miller WF: Cardiopulmonary physiological responses to heavy exercise in patients with anemia. J Clin Invest 39:378–388, 1960
19. Richardson TQ, Guyton AC: Effects of polycythemia and anemia on cardiac output and other circulatory factors. Am J Physiol 197:1167–1170, 1959
20. Regan TJ, Frank MJ, Lehan PH, et al: Myocardial blood flow and oxygen uptake during acute red cell volume increments. Circ Res 13:172–181, 1963
21. Aperia AC, Liebow AA, Roberts LE: Renal adaptation to anemia. Circ Res 22:489–500, 1968
22. Glick G, Plauth WH, Braunwald E: Role of the autonomic nervous system in the circulatory response to acutely induced anemia in unanesthetized dogs. J Clin Invest 43:2112–2124, 1964

23. Neill WA, Oxendine JO, Moore SC: Acute and chronic cardiovascular adjustments to induced anemia in dogs. Am J Physiol 217:710–714, 1969
24. Massey DG, Becklake MR, McKenzie JM, Bates DV: Circulatory and ventilatory response to exercise in thyrotoxicosis. N Engl J Med 276:1104–1112, 1967
25. Wilson WR, Thielen EO, Hege JH, Valence MR: Effects of beta-adrenergic receptor blockade in normal subjects before, during, and after triiodothyronine-induced hypermetabolism. J Clin Invest 45:1159–1169, 1966
26. Buccino RA, Spann JF Jr, Pool PE, et al: Influence of the thyroid state on the intrinsic contractile properties and energy stores of the myocardium. J Clin Invest 46:1669–1682, 1967
27. Ross J Jr, Gault JH, Mason DT, et al: Left ventricular performance during muscular exercise in patients with and without cardiac dysfunction. Circulation 34:597–608, 1966
28. Wood JE: The mechanism of the increased venous pressure with exercise in congestive heart failure. J Clin Invest 41:2020–2024, 1962
29. Wade OL, Bishop JM: The distribution of the cardiac output in patients with heart disease, in: Cardiac Output and Regional Blood Flow. Oxford, Blackwell Scientific Publications, 1962, pp 134–148
30. Chidsey CA, Braunwald E, Morrow AG: Catecholamine excretion and cardiac stores of norepinephrine in congestive heart failure. Am J Med 39:442–451, 1965
31. Epstein SE, Beiser GD, Stampfer M, et al: Characterization of the circulatory response to maximal upright exercise in normal subjects and patients with heart disease. Circulation 35:1049–1062, 1967
32. Johnson JM, Rowell LB, Niederberger M, Eisman MM: Human splanchnic and forearm vasoconstrictor responses to reductions of right atrial and aortic pressures. Circ Res 34:515–524, 1974
33. Rocha e Silva M Jr, Rosenberg M: The release of vasopressin in response to haemorrhage and its role in the mechanism of blood pressure regulation. J Physiol 202:535–557, 1969
34. Chalmers JP, Korner PI, White SW: The effects of hemorrhage in the unanesthetized rabbit. J Physiol 189:367–391, 1967
35. Forsyth RP, Hoffbrand BI, Melmon KL: Redistribution of cardiac output during hemorrhage in the unanesthetized monkey. Circ Res 27:311–320, 1970
36. Fry RL, Braunwald E: Studies on Starling's law of the heart. I. The circulatory response to acute hypervolemia and its modification by ganglionic blockade. J Clin Invest 39:1043–1050, 1960
37. Rushmer RF: Postural effects on the baselines of ventricular performance. Circulation 20:897–905, 1959
38. Bevegard S, Holmgren A, Jonsson B: The effect of body position on the circulation at rest and during exercise, with special reference to the influence on the stroke volume. Acta Physiol Scand 49:279–298, 1960
39. Saltin B, Blomqvist G, Mitchell JH, et al: Response to exercise after bed rest and after training. Circulation 37[Suppl 7]:VII–1, 1968

40. Mitchell JH, Sproule BJ, Chapman CB: The physiological meaning of the maximal oxygen intake test. J Clin Invest 37:538–547, 1958
41. Hartley LH, Grimby G, Kilbom A, et al: Physical training in sedentary middle-aged and older men. 3. Cardiac output and gas exchange in submaximal and maximal exercise. Scand J Clin Lab Invest 24:335–344, 1969
42. Donald KW, Bishop JM, Cummings G, Wade OL: The effect of exercise on the cardiac output and circulatory dynamics of normal subjects. Clin Sci 14:37–73, 1955
43. Van Citters RL, Franklin DL: Cardiovascular performance of Alaska sled dogs during exercise. Circ Res 24:33–42, 1969
44. Wade OL, Combes B, Childs AW, et al: The effect of exercise on the splanchnic blood flow and splanchnic blood volume in normal man. Clin Sci. 15:457–463, 1956
45. Zelis R, Mason DT, Braunwald E: Partition of blood flow to the cutaneous and muscular beds of the forearm at rest and during leg exercise in normal subjects and in patients with heart failure. Circ Res 24:799–806, 1969
46. Keroes J, Ecker RR, Rapaport E: Ventricular function curves in the exercising dog. Circ Res 25:557–568, 1969
47. Horwitz LD, Atkins JM, Leshin SJ: Role of the Frank-Starling mechanism in exercise. Circ Res 31:868–875, 1972
48. Smith OA Jr, Rushmer RF, Lasher EP: Similarity of cardiovascular responses to exercise and to diencephalic stimulation. Am J Physiol 198:1139–1142, 1960
49. Donald DE, Shepherd JT: Response to exercise in dogs with cardiac denervation. Am J Physiol 205:393–400, 1963
50. Donald DE, Ferguson DA, Milburn SE: Effect of beta-adrenergic receptor blockade on racing performance of greyhounds with normal and with denervated hearts. Circ Res 22:127–134, 1968
51. Epstein SE, Robinson BF, Kahler RL, Braunwald E: Effects of beta-adrenergic blockade on the cardiac response to maximal and submaximal exercise in man. J Clin Invest 44:1745–1753, 1965
52. Ashkar E, Stevens JT, Houssay BA: Role of the sympathicoadrenal system in the thermodynamic response to exercise in dogs. Am J Physiol 214:22–27, 1968
53. Vatner SF, Franklin D, Higgins CB, et al: Left ventricular response to severe exertion in untethered dogs. J Clin Invest 51:3052–3060, 1972
54. Brod J, Fencl V, Hejl Z, Jirka J: Circulatory changes underlying blood pressure elevation during acute emotional stress (mental arithmetic) in normotensive and hypertensive subjects. Clin Sci 18:269–279, 1959
55. Eliasch H, Rosen A, Scott HM: Systemic circulatory response to stress of stimulated flight and to physical exercise before and after propranolol blockade. Br Heart J 29:671–683, 1967
56. Bergamaschi M, Longoni AM: Cardiovascular events in anxiety: Experimental studies in the conscious dog. Am Heart J 86:385–394, 1973
57. Folkow B, Rubinstein E: Behavioral and autonomic patterns evoked by

stimulation of the lateral hypothalamic area in the cat. Acta Physiol Scand 65:292–299, 1965

58. Hickman JB, Cargill WH, Golden A: Cardiovascular reactions to emotional stimuli. Effect on the cardiac output, arteriovenous oxygen difference, arterial pressure, and peripheral resistance. J Clin Invest 27:290–298, 1948
59. Leon DF, Amidi M, Leonard JJ: Left heart work and temperature responses to cold exposure in man. Am J Cardiol 26:38–45, 1970
60. Carlsten A, Gustafson A, Werko L: Hemodynamic influence of warm and dry environment in man with and without rheumatic heart disease. Acta Med Scand 169:411–417, 1961
61. Abboud FM, Eckstein JW: Reflex vasoconstrictor and vasodilator responses in man. Circ Res 18[Suppl I]:96–103, 1966
62. Hammel HT, Hardy JD, Fusco MM: Thermoregulatory responses to hypothalamic cooling in unanesthetized dogs. Am J Physiol 198:481–486, 1960
63. Phillips HH, Jennings DB: Cardiorespiratory effects of hypothalamic heating in conscious dogs. Am J Physiol 225:700–705, 1973

Francois M. Abboud, Phillip G. Schmid
Donald D. Heistad, Allyn L. Mark

4

Regulation of Peripheral and Coronary Circulation

The integrity of cellular function is dependent on (1) the delivery of blood to organs and (2) the availability of an adequate capillary surface area for an appropriate exchange of substrates for metabolic activity and for removal of waste products. The regulation of the circulation would therefore require the adequate maintenance of these functions in physiologic states and the development of compensatory mechanisms that allow for the preservation of these functions in pathologic states.

The flow of blood to an organ is dependent upon the systemic arterial pressure and the vascular resistance of that organ (Table 4-1): flow = pressure/resistance. Systemic arterial pressure is a force generated by two biologic functions, cardiac output and total vascular resistance: systemic arterial pressure = cardiac output × total vascular resistance.

In an earlier chapter the factors responsible for the regulation of cardiac output were discussed. The emphasis in this chapter is on the regulation of vascular resistance and its effect on the circulation. Vascular resistance is primarily a function of vascular caliber, or the radii of blood vessels. Selective changes in arterial and arteriolar caliber or resistance in different vascular beds will determine the distribution of blood flow to various organs at a given cardiac output. Changes in vascular tone of precapillary sphincters will determine the patency of nu-

The more recent studies of the authors and their colleagues that are cited in this chapter were supported by USPHS Grants HL-14388, HL-16149, and HL-16066. Parts of this review were published by the authors elsewhere (Arch Intern Med 133:935, 1974, and Anesthesiology 41:139, 1974) and are reprinted here by permission of those publishers.

Table 4-1
Distribution of Blood Flow Relative to Organ Size and Metabolic Needs

	Total Organ Flow (Maximal Dilatation) (ml/min)	Flow/100 g* (ml/min/100 g)	O_2 Used/100 g (ml/min/100 g)	Flow/ml of O_2† (ml)
Brain	750 (1,500)	55	3	18
Heart	250 (1,200)	80	8	10
Liver	1,300 (5,000)	85	2	31
GI tract	1,000 (4,000)	40		
Kidneys	1,200 (1,800)	400	5	80
Muscle	1,000 (20,000)	3	0.15	20
Skin	200 (3,000)	10	0.2	50
Other (skeleton, fat, connective tissue)	800 (4,000)	3	0.15	20

Reproduced by permission from Folkow B, Neil E: Circulation. New York, Oxford University Press, 1971, p 12.

*Flow per 100 g of tissue may provide an indication of relative "vascular density" in each organ. Flow at maximal dilation is shown in parentheses.

The distribution of flow to organs is determined by the relative vascular resistances. The maximal dilator capacity is in the muscle and skin (20- and 15-fold, respectively).

†Overperfusion with respect to oxygen requirements and metabolic needs is seen in kidney and skin. The flow per unit O_2 delivered is least in the heart.

tritional capillaries and transcapillary exchange. Changes in the precapillary and postcapillary resistance ratio will determine capillary pressure, transcapillary fluid movement, and consequently total blood volume, which in turn affects cardiac output. Finally, changes in venous tone or tone of the capacitance vessels will determine the intravascular distribution of blood volume and can modify cardiac filling pressure and cardiac output. Thus the factors that determine vascular caliber in the arteries, arterioles, precapillary sphincters, venules, and veins are major determinants of circulatory homeostasis.

There are two determinants of vascular caliber: the tone of vascular smooth muscle and the structural changes in the vessel walls. The factors that determine the tone of vascular smooth muscle are related primarily to the contractile properties of smooth muscle, to the neurogenic and humoral control of smooth muscle, and to the determinants of autoregulatory adjustment of vascular resistance in response to changes in blood flow, arterial pressure, and metabolic demands.

CONTRACTILE PROPERTIES OF VASCULAR SMOOTH MUSCLE

Membrane Potential

It now appears that the major features of vascular smooth-muscle contraction are similar to those of cardiac and skeletal muscle. In certain types of vascular smooth muscle, contraction may be initiated by a process of membrane depolarization triggered by spontaneous action potentials whereby a vascular pacemaker cell would generate a spike that would then propagate in a manner such as is seen in portal mesenteric veins.[1,2] This spontaneous excitability may be accentuated with stretch of the smooth muscle or with neural activity.[3,4] In other types of smooth muscle, particularly in some larger vessels, there appears to be a more graded depolarization that may be present simultaneously with the spike potentials during contraction.[5,6] The process of depolarization is associated with an increased cellular calcium influx.[7] As in other types of muscle, changes in transmembrane cation gradients of sodium and potassium may alter membrane potential, excitability, influx of calcium, and smooth-muscle contraction.[8] Reductions in extracellular potassium will cause contraction of vascular smooth muscle associated with depolarization, increased spike frequency, and increased free ionic calcium concentration in the cytoplasm[9–13] (Fig. 4-1). We have found that acute isotonic changes in extracellular sodium concentration in man and dog do not change vascular resistance, but alter responsiveness to catecholamines.[14] Increases in intracellular concentration of sodium may alter resting membrane potential, and this in turn may underlie increased reactivity to pharmacologic agents in hypertensive vessels.[15]

Recent observations by Jones[16] and by Hermsmeyer[16a] suggest that there is genetically linked involvement of ion transport in the aorta of rats with spontaneous hypertension (SHR), which results in a reduced ability to accumulate potassium and to extrude sodium; in these studies ^{42}K washout from smooth muscle with norepinephrine was greater in SHR than in control rats. This altered ionic permeability may play an important role in vascular regulation in hypertension, since several vasoactive substances cause their effects on vascular smooth-muscle contraction by altering transmembrane cation gradient.

Excitation-Contraction Coupling

It is now obvious that calcium is the rate-limiting factor in contraction of vascular smooth muscle. Calcium may be derived from extracellular spaces or intracellularly from a loosely bound fraction in

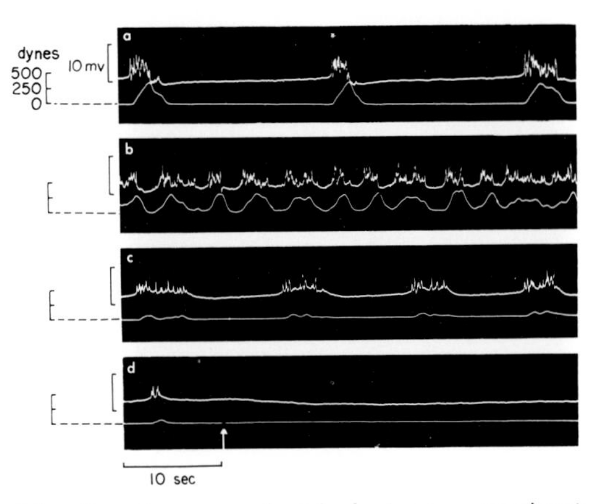

Fig. 4-1. Spontaneous electrical (upper tracings) and mechanical (lower tracings) activity of rat portal vein and its response to K^+-free solution. (c): After 10 min in K^+-free solution. (d): Return to normal solution (arrow) after 15 min in K^+-free medium. The zero level of the tension scale, which is the same in each panel, represents a resting tension of approximately 300 dynes applied to the muscle by passive stretch. The 10-mV marker for the electrical recording is shown at the same potential level in each panel. (Reproduced by permission of the American Medical Association, Inc. from Axelsson J, et al: Circ Res 21:609, 1967.)

the sarcolemma or a less labile fraction located in the sarcoplasmic reticulum.[17] As in cardiac muscle, calcium will activate actomyosin by binding with troponin of the troponin–tropomyosin complex, thereby removing the restraint on actomyosin ATPase activity caused by troponin.[18] ATPase activity splits ATP, providing the energy for contraction. The contractile protein complex actomyosin has been extracted from vascular smooth muscle and appears to have properties similar to those of actomyosin from skeletal muscle; specifically, its ATPase activity is increased by the presence of calcium, and it is precipitated when ATP is added in the presence of calcium.[18–20] The troponin–tropomyosin complex also has been extracted.[18] The uptake of calcium by the sarcoplasmic reticulum or other sarcolemmal binding sites or by mitochondria will result in relaxation.[21]

Vascular smooth-muscle contraction may be elicited in the absence of a membrane action potential; thus in a depolarizing solution norepinephrine may cause contraction of vascular smooth muscles that have alpha receptors.[22] Vasoactive agents can excite the release of calcium from intracellular binding sites independently of an action potential, a process that has been referred to as pharmacomechanical coupling[23] or nonelectrical activation.[24]

The relative affinities of the sarcoplasmic reticulum and of troponin for calcium will determine to a significant extent the processes of smooth-muscle relaxation and contraction. Drugs that activate vascular beta receptors and cause smooth-muscle relaxation may do so by increasing cyclic AMP, which appears to increase the calcium binding by microsomal fractions of vascular smooth muscle, making less calcium available to the contractile protein. The competition between hydrogen and calcium for troponin binding sites that was demonstrated in cardiac muscle[25] may explain the vasoconstricting effect of alkalosis and the inhibition of vasoconstriction in acidotic states. A decreased pH will tend to reduce the availability of calcium to contractile protein.

Generation of ATP

The work of Detar and Bohr[26] suggests that the substrate level necessary for generation of sufficient amounts of high-energy phosphates to subserve vascular smooth-muscle contraction is rarely a rate-limiting factor. When aortic strips were maintained for 4 hr or more in a nitrogen environment, their ability to contract anaerobically was enhanced. This so-called hypoxic adaptation may represent the capacity of vascular smooth muscle for anaerobic synthesis of ATP for contraction. Work in our laboratories indicates that hypoxemia will not prevent vasoconstrictor response to norepinephrine or angiotensin in the perfused gracilis muscle unless the muscle is contracting.[27] This would suggest that when the demands of skeletal muscle for ATP increase, its limited supply from anaerobic pathways during hypoxemia is reflected by a reduction in vasoconstrictor responses (Fig. 4-2). In resting man,[28] however, the vasoconstrictor responses to norepinephrine are reduced in the presence of systemic hypoxemia, possibly because of systemic acidosis resulting from a generalized hypoxic state. In a simple experiment on the portal vein designed to elucidate the relative vulnerability of the contractile apparatus to the metabolic and ionic supply, Johansson[29] demonstrated that spontaneous activity, as well as response to norepinephrine, is abruptly eliminated in a calcium-free medium and promptly restored by the addition of calcium. On the other hand, a glucose-free medium bubbled with nitrogen causes a gradual deterioration of spontaneous activity and weak response to norepinephrine, both of which are restored on readministration of glucose and oxygen (Fig. 4-3). It would thus appear that the intracellular stores of metabolic substrate can apparently support contractile responses under anaerobic conditions

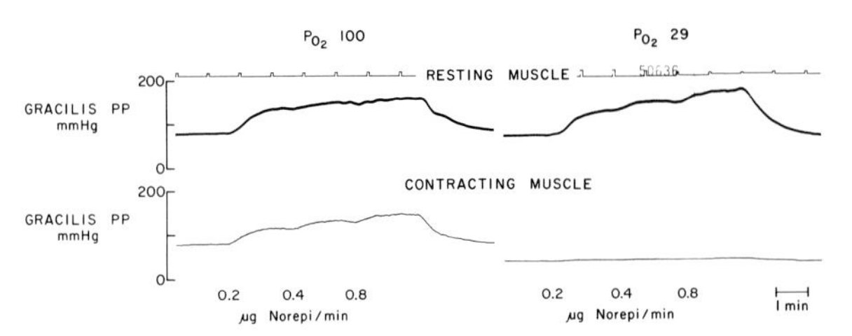

Fig. 4-2. Effect of hypoxemia in resting and contracting muscle gracilis at constant blood flow. Responses to three doses of norepinephrine were compared when the Po_2 of the blood perfusing the resting and contracting gracilis muscle was 100 mm Hg (lest panel) and 29 mm Hg (right panel). Contraction at normal Po_2 or hypoxemia in the absence of contraction had a slight effect on the constrictor response to norepinephrine, whereas a combination of hypoxemia and contraction antagonized the vasoconstriction. (Reproduced by permission from Heistad DD, Abboud FM, Mark AL, Schmid PG: J Pharmacol Exp Ther 193:941, 1975)

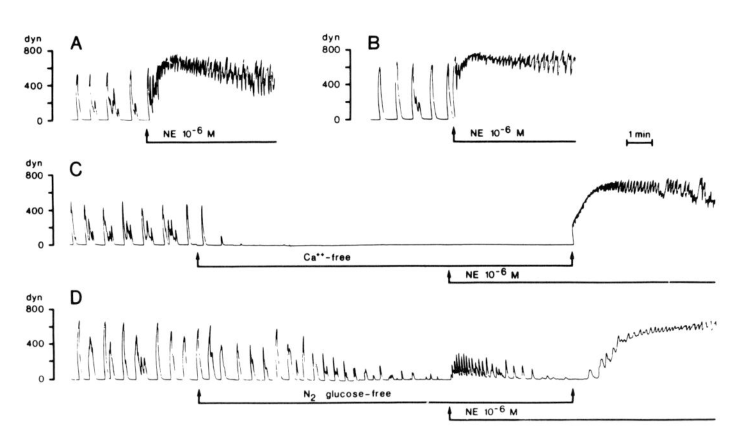

Fig. 4-3. Relative vulnerability of the contractile system to metabolic and ionic support. Panels A and B indicate the spontaneous activity and responses to norepinephrine (NE) in two portal veins bathed in physiologic salt solution; the responses to the addition of NE were similar in both veins. When Vein A was exposed to a Ca^{++}-free solution (tracing C) the spontaneous activity was eliminated and the response to NE was abolished; addition of Ca^{++} restored the response to NE immediately. When vein B was exposed to a glucose-free medium bubbled with nitrogen (tracing D) there was a gradual deterioration of spontaneous activity and a minimal response to NE; but the response to NE recovered slowly on readministration of glucose and oxygen. Intracellular stores of metabolic substrate can support contractile responses under anaerobic conditions to some extent, but not for long periods of activity. (Reproduced by permission from Johansson B: Fed Proc 33:121, 1974.)

to some extent, but the supply of calcium is critically needed for the contractile system.

NEUROGENIC CONTROL OF THE CIRCULATION

The immediate circulatory adjustment to stress is mediated through the autonomic system with its sympathetic and parasympathetic components. More sustained adjustment is brought about through both humoral and neural factors, as well as structural changes in the vessel walls. The sympathetic arm of the autonomic system supplies the heart and blood vessels, and the parasympathetic provides innervation to the coronaries, the conduction system, and the atria and ventricles.

Medullary Vasomotor Centers and Efferent Pathways

Medullary Centers

Pressor responses are elicited by electrical stimulation of neurons in the rostrolateral areas of the medulla, whereas depressor responses occur when stimuli are delivered to the mediocaudal parts of the medulla. The pressor areas are tonically active, and their rate of discharge is modulated by an inhibitory input from the depressor area.[30,31]

The output from the pressor area is carried along efferent fibers that synapse one or more times in passing through the spinal cord, reach the sympathetic preganglionic cells in the lateral horn, exit with motor nerves in the ventral root, and synapse in the paravertebral ganglia. Postganglionic sympathetic fibers either array in bundles identifiable as discrete nerves such as the cardiac or splanchnic nerves or rejoin the motor roots and are distributed to blood vessels along with the sensory motor nerves in the peripheral nerve trunks.

A third medullary cardiovascular control center is the dorsal motor nucleus of the vagus, which provides the parasympathetic outflow and exerts an inhibitory effect on the heart through the vagus nerves.

Efferent Sympathetic Adrenergic Pathways

The sympathetic adrenergic fibers provide the neural mechanism that mediates reflex vasoconstriction. The extent of innervation and the catecholamine content of vessels correlate with their responsiveness to stimulation of the sympathetic nerve,[32] but not necessarily with the level of resting sympathetic tone. Sympathetic tone is a function of neural discharge rate and is best evaluated by measurement of norepinephrine turnover or circulating dopamine-β-hydroxylase.[33] Norepinephrine, the neurotransmitter, is located in adrenergic nerve endings in dense core vesicles (Fig. 4-4). An action potential generated in sympathetic neurons triggers the release of norepinephrine from these vesicles.[34] The release of the neurotransmitter may be related to calcium influx in the axon membrane in a manner similar to events occurring at motor nerve terminals and involving the release of acetylcholine. The released norepinephrine may activate vasoconstrictor alpha receptors in vascular smooth muscle or excitatory adrenergic beta receptors in the heart and sinoatrial node. The action of norepinephrine is terminated by the uptake of excess hormone by the adrenergic nerves or by its metabolism through the catechol-*o*-methyltransferase and monoamine oxidase enzyme (MAO) systems. Drugs that block the reuptake of norepinephrine by the adrenergic terminal such as the tricarboxylic antidepressants (e.g., imipramine) and guanethidine may produce exaggerated pressor responses, particularly in the presence of a stimulus causing the release of endogenous norepinephrine.

Serious interactions between drugs that have different effects on storage, uptake, and metabolism of norepinephrine may occur clinically, such as during the simultaneous administration of an MAO inhibitor and imipramine, which causes coma, hyperpyrexia, and convulsions, or the administration of imipramine to a hypertensive patient on guanethidine, which may cause severe hypertension.

Catecholamines activate the enzyme adenylcyclase and catalyze the formation of cyclic 3′- and 5′-AMP, which are thought to mediate cardiac contractile responses and vasodilator responses, as well as metabolic effects such as activation of lipase or phosphorylase (Fig. 4-5). These effects can be blocked by beta-receptor blockers such as propranolol. Cyclic guanosine 3′- and 5′-monophosphate may mediate alpha-receptor responses such as vasoconstriction. The biochemical link between cyclic AMP or GMP, activation of kinases, and ensuing muscular contraction or relaxation requires further studies that may relate to the effects on Ca^{++} binding to subcellular fractions.

There is a good deal of selectivity in the distribution of adrenergic pathways to various vascular segments, as well as to the various regions of the heart. We have found that electrical stimulation of the sympathetic nerves in the forelimb of the dog causes constriction of large arteries and veins and cutaneous venules with minimal or no constriction of small veins draining skeletal muscle.[35,36] The magnitude of constriction of the large arteries indicates that they are not merely conduit vessels; they are innervated and may constrict not only with electrical

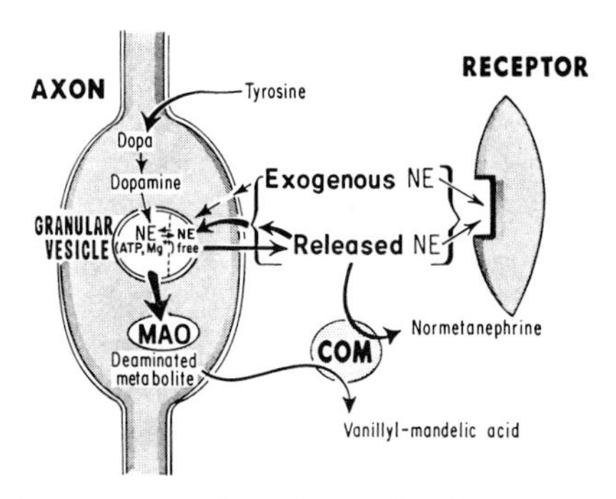

Fig. 4-4. Schematic representation of a varicosity on a terminal axon and a receptor on an effector cell. The major steps involved in the synthesis, storage, release, and metabolism of norepinephrine are depicted. NE = norepinephrine; MAO = monoamine oxidase; COM = catechol-*o*-methyltransferase. The granular vesicle contains most of the endogenous norepinephrine in an ATP-Mg^{++}-dependent storage site (site I), which can be released by reserpine, and in a more loosely bound site (site II), which can be released by nerve action potential. NE released from site I intraneuronally is deaminated as it is exposed to MAO, whereas NE released from Site II directly extraneuronally is only *o*-methylated. Urinary excretion of VMA has been used as an index of storage, tissue content, or intraneuronal release; whereas normetanephrine, which is the *o*-methylated metabolite, has been used as an index of the level of sympathetic activity, extraneuronal release and turnover rate [Hertting G, Axelrod J: Nature (Lond) 192:172, 1961]. The action of exogenous NE or NE released from the vesicle during neural activation is terminated in part by its uptake by the neuronal membrane and storage in an intraneuronal but extravesicular site (site III). Drugs that compete with this membrane uptake will increase the availability of norepinephrine to receptor sites and augment its effects. These include the indirect-acting sympathomimetic amines (tyramine, amphetamine, and ephedrine) that act primarily by releasing NE from site III, the tricyclic antidepressants (desipramine and imipramine), some antihypertensive drugs (guanethidine, bethanidine), and cocaine. The tricyclic antidepressants compete with uptake of guanethidine by the neuron and reverse its antihypertensive effect. The uptake of injected or infused exogenous norepinephrine by extraneuronal cells (site IV) may determine also its availability to receptors and the response. This uptake may represent a specific transport mechanism that is blocked by α-receptor blocking (Eisenfeld AJ, Axelrod J, Krakoff L: J Pharmacol Exp Ther 156:107, 1967), is reduced by lack of extracellular sodium,[14] and is unaffected by sympathetic tone (Abboud FM, Mayer HE: Fed Proc 27:281, 1968). [Reproduced by permission from Abboud FM: in Stollerman G (ed): Advances in Internal Medicine, Vol. 15. Chicago, Year Book Medical Publishers, 1969).]

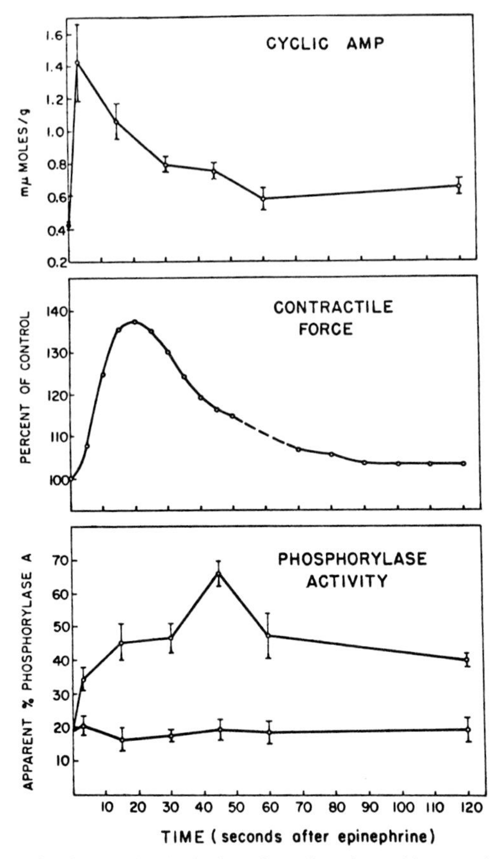

Fig. 4-5. Effect of epinephrine in isolated perfused working rat heart. Activation of adenyl cyclase was extremely rapid, and an increase in cyclic AMP occurred within seconds and preceded both the inotropic response and the activation of phosphorylase. The lower curve in the phosphorylase panel shows lack of response to saline. The administration of the beta blocker pronethalol prevented the effects of the epinephrine on contractility and cyclic AMP levels. Changes in cyclic AMP may mediate the actions of many other hormones besides the catecholamines. ACTH and glucagon will activate adenyl cyclase and increase cyclic AMP, but this action is not antagonized by β-adrenergic blockers. Insulin and prostaglandins lower cyclic AMP levels and have an antilipolytic effect and antagonize the lipolytic action of catecholamines. The mechanisms for these actions are not known, but they do not appear to involve beta receptor activity. (Reproduced by permission from Robison GA, Sutherland EW, et al: Mol Pharmacol 1:168, 1965.)

stimulation of sympathetic nerves but also in response to the neurogenic stimulus induced by the "diving reflex" in both experimental animals and man.[37,38] Stimulation of sympathetic nerves to various vascular beds will elicit vasoconstrictor responses that differ widely in magnitude; for example, vasoconstriction is greater in the kidney than in the forelimb of the dog and is negligible in coronaries.[39]

Peripheral sympathetic nerves are also distributed selectively to specific regions of the heart. Randall and his co-workers have shown that in both the dog and primate[40] the epicardial and endocardial surfaces are capable of highly localized alterations in contractile force. A separate influence of sympathetic nerve activity on the control of right and left ventricle contractility has been shown during exercise and emotional responses.[41] Furthermore, Armour et al.[42] have made significant contributions to the understanding of the potential importance of sympathetic activity in the peripheral adrenergic nerve supply to nodal tissue in the genesis of cardiac arrhythmias.

In addition to the release of norepinephrine from nerve terminals, sympathetic activity induces the release of epinephrine and norepinephrine from the adrenal medulla and causes the release of renin from the juxtaglomerular apparatus, not only through an effect on vascular tone of preglomerular arterioles, but also through a direct effect independent of the vascular influence.[43,44] Low-level sympathetic stimulation of the renal nerves may induce sodium retention in the absence of an effect on intrarenal blood flow distribution.[45]

Efferent Parasympathetic Pathways

It is now fairly well established that the parasympathetic innervation supplies not only the SA node, atrial structures, and conduction system, but also the ventricles. Several workers have characterized the extent of parasympathetic innervation of the ventricles through the determination of cholinesterase activity and choline acetyltransferase activity or the demonstration of nerve fibers associated with ganglia in the canine ventricular septum.[46–48] Since the parasympathetic supply of the heart is essentially preganglionic, the persistence of ganglia in the ventricle after total surgical denervation of the heart also supports the presence of ventricular parasympathetic innervation. There is sufficient evidence to indicate that the parasympathetic control of the ventricle exerts a negative inotropic effect[49,50] through a muscarinic action blocked by atropine. The inhibitory action appears to be related to a decrease in the duration of action potential and associated decrease in cellular influx of calcium ions.[51] Furthermore, acetylcholine decreases cyclic AMP accumulation and cyclase activity in atria and ventricles.[52]

In general, efferent sympathetic adrenergic activity to the heart is excitatory and parasympathetic cholinergic activity is inhibitory. However, several studies have indicated that these two components of the autonomic system are very closely interrelated anatomically and functionally.

Although the muscarinic action of acetylcholine is inhibitory, this neuromediator is known to have a specific excitatory effect mediated through the release of norepinephrine, an action that can be blocked by hexamethonium and not by atropine and may represent a cholinergic "nicotinic receptor."[53] This interaction is best demonstrated with respect to the chronotropic effect. Levy[54] has summarized the paradoxical effects of cardioacceleration seen after vagal cholinergic-type interventions. There appears to be an excitatory effect of acetylcholine on the cells of the SA node existing concurrently with the deceleratory effect.[55] The reason the vagus is a cardiac decelerator nerve is that the inhibitory effect is the dominant one. The findings of Scher et al.[56] indicate that the predominant control of heart rate in response to changes in arterial pressure is vagal in the dog heart; in the human it is predominantly vagal and slightly sympathetic, and in the baboon it is predominantly sympathetic. The work of Eckberg et al.[57,58] also points to a predominant vagal control of heart rate in man and suggests the presence of a cholinergically mediated adrenergic response with respect to heart rate during carotid sinus stimulation by stretch. The vagal parasympathetic supply to the heart is not restricted to cardiac tissue, but appears also to innervate the coronary vessels and may mediate a coronary vasodilatation during stimulation of chemoreceptors.[59,60]

Sympathetic Vasodilator Pathways

Vasodilatation of peripheral blood vessels may occur from passive withdrawal of sympathetic vasoconstrictor tone or active vasodilatation through activation of sympathetic cholinergic pathways. Active vasodilatation in skeletal muscle mediated through sympathetic cholinergic fibers has been demonstrated in experimental animals during the defense reaction in response to hypothalamic stimulation.[61]

This cholinergic vasodilator pathway[62] originates in the frontal cortex, passes through synaptic relays in hypothalamic and collicular levels, and runs through the ventral medulla to the spinal cord.[63] These fibers innervate predominantly larger arteriolar resistance vessels of skeletal muscle. During the defense reaction this cholinergic vasodilatation of arterioles may occur simultaneously with withdrawal of sympathetic vasoconstrictor tone from precapillary sphincters and thereby improve the perfusion of skeletal muscle and increase muscle blood flow.

The work of Takeuchi and Manning[64] indicates that the cholinergic vasodilator fibers to skeletal muscle may be activated during the carotid sinus baroreceptor response in cats. Although it is generally accepted that the cholinergic vasodilator system is present in skeletal muscle of cats and dogs, there has been some question regarding its importance in the subhuman primate or in man. In two papers by Schramm et al.[65,66] the cholinergic sympathetic dilator pathways in the cat have been compared with the noncholinergic vasodilatation produced in the squirrel monkey by stimulating homologous foci primarily along the route of the lateral spinothalamic tract. These authors concluded that there is a primate homologue of the cholinergic sympathetic vasodilator system that can be defined anatomically instead of pharmacologically.

In man, active vasodilatation occurs during emotional stress,[67] during the Valsalva maneuver or application of ice to the forehead after adrenergic blockade with guanethidine, and in diabetic patients with postural hypotension.[68] This suggests that the defense reaction of hypothalamic stimulation in animals may have important physiologic counterparts in man (Fig. 4-6).

In addition to our work in man, Mason and Braunwald have shown that chronic oral administration of an adrenergic blocking agent not only abolishes the reflex increase in forearm vascular resistance and in venous tone that normally occur during the cold pressor tests and during leg exercise, but also produces a paradoxical decline in venous tone and vascular resistance.[69] Atropine either abolishes or reduces this dilatation.

Other Dilator Systems

In addition to the cholinergic dilator system, there appear to be other dilator efferent pathways. These can be identified by direct stimulation of efferent nerves following adrenergic blockade.[70] Pathways that are not blocked by atropine, tripelennamine, propranolol, or carboxypeptidase may be activated in the hindpaw of the dog during chemoreceptor stimulation.[71] Histaminergic sympathetic efferent pathways may be reflexly activated by stimulation of baroreceptors.[72,73] Inhibition of this active dilator system may occur following short-term intravertebral infusions of angiotensin and may contribute a neurogenic element to renal hypertension.[74,75]

A sympathetic vasodilator response may also occur through activation of β-adrenergic receptors[76] or through the release of bradykinin.[77]

Afferent Pathways

Strategically located sensors in the carotids, aorta, and cardiopulmonary areas or in skeletal muscle spindles are activated by changes in pressure, by stretch, or by changes in the chemical com-

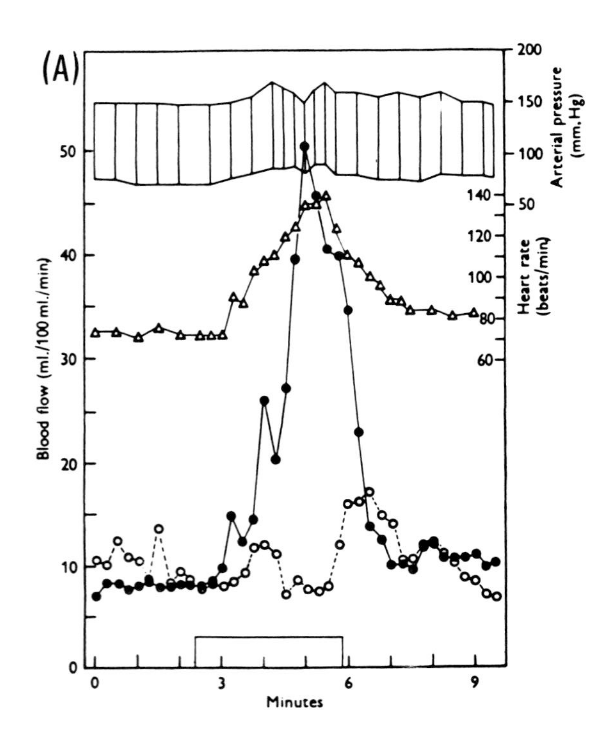
(A)
Blood flow (ml./100 ml./min)
Arterial pressure (mm.Hg)
Heart rate (beats/min)
0
10
20
30
40
50
200
150
100
50
140
120
100
80
60
0
3
6
9
Minutes

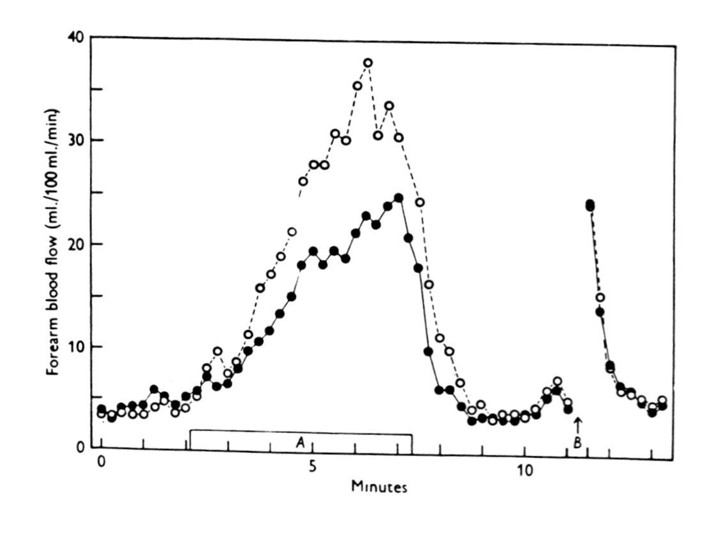
Forearm blood flow (ml./100 ml./min)
0
10
20
30
40
0
5
10
A
B
Minutes

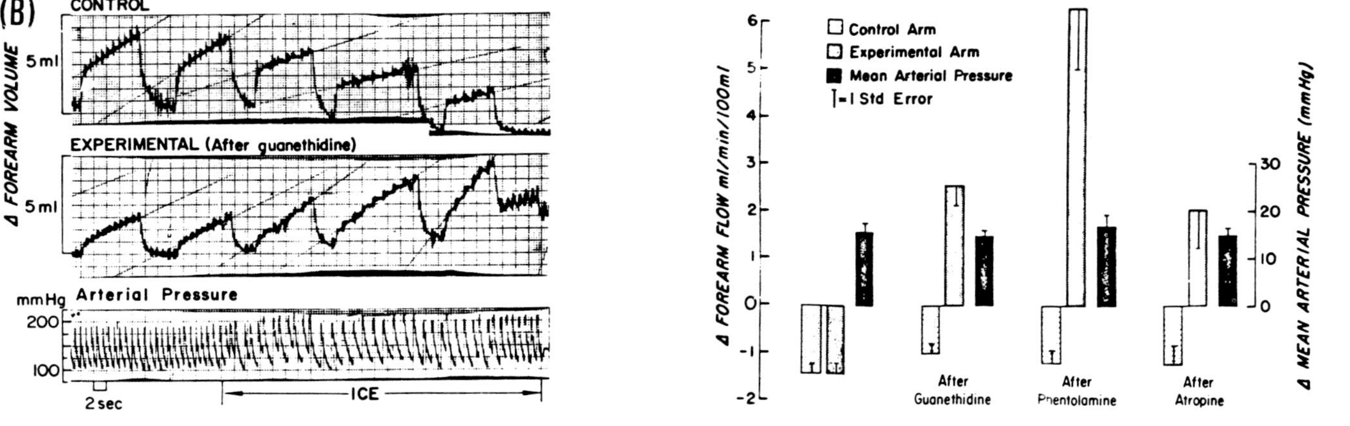

Fig. 4-6. Cholinergic vasodilatation in the forearm of man. (A) The left panel shows that severe emotional stress induced by the suggestion to the subject that he was suffering from severe blood loss caused tachycardia and marked vasodilatation in the forearm (black dots) and not in the hand (circles). The right panel indicates that this dilatation was in part cholinergic: A = period of stress; circles = flow to normal forearm; black dots = flow to atropinized forearm; B = vasodilator response after arrest of circulation to both forearms for 2 min. (Reproduced by permission from Blair DA, et al.[67]) (B) Left panels: The tracings represent plethysmographic flow curves. Application of ice to the forehead caused vasodilatation in the forearm after intraarterial guanethidine, which blocked adrenergic neural vasoconstriction; the response in the opposite control arm was the expected vasoconstriction. Right panel: The reflex vasodilatation in the "experimental" arm was blocked in part by intraarterial atropine. (Reproduced by permission of the American Heart Association, Inc. from Abboud FM, Eckstein JW: Suppl I Circ Res 18, 19:96, 1966.)

position of the blood perfusing them. The activity initiated in these sensors or receptors is transmitted to the central nervous system and modulates the output from the medullary centers over the efferent sympathetic and parasympathetic pathways.[78] Afferent impulses originating in arterial and cardiopulmonary baroreceptors exert primarily a restraining influence on medullary excitatory or pressor centers, whereas afferent impulses originating in chemoreceptors or somatic receptors are predominantly stimulating.

Arterial Baroreceptors

Mechanoreceptors or stretch receptors are located in the adventitia and media of the arch of the aorta and in the adventitia of the carotid arteries in the carotid sinus regions. Afferent impulses in these regions travel along the vagus nerve and the glossopharyngeal nerve. In 1932 Bronk and Stella[79] demonstrated that impulses in single fibers from carotid mechanoreceptors varied rhythmically with pulse pressure and had a maximum frequency early in systole. Increases in afferent nerve activity relate clearly to stretch or distension of baroreceptors. Recent observations[80] suggest that the aortic and carotid baroreceptors have a different pressure threshold; afferent nerve activity in the carotid sinus appears at a lower level of arterial pressure than in the aortic buffer nerve.

Chronic hypertension is associated with disturbances in baroreceptor function that may result from structural changes in the aorta and carotid vessels.[81,82] The term "resetting of baroreceptors" has been used to identify this peculiar characteristic in mechanoreceptors. It is doubtful that the abnormality represents a causative factor in hypertension; it probably represents an adaptation to a sustained level of elevated arterial pressure.

A rise in arterial pressure causes an increase in afferent nerve activity from carotid and aortic baroreceptors and triggers a reflex reduction in peripheral resistance, in heart rate, and in myocardial contractility. This is achieved primarily through withdrawal of sympathetic vasoconstrictor activity and increases in activity of parasympathetic vagal efferent pathways to the heart. Activation of sympathetic vasodilator pathways and inhibition of cardiac sympathetic activity may also participate in this circulatory adjustment, but to a lesser degree. In contrast, a decrease in arterial pressure would trigger reflex vasoconstriction, tachycardia, and an increase in cardiac output and myocardial contractility through activation of sympathetic excitatory efferent pathways.

It is essential to realize that the activation of sympathetic and parasympathetic efferent pathways, as well as their inhibition, are not generalized phenomena; that is to say, not all components of the cardiovascular system respond in a uniform fashion to the autonomic drive or restraint. For example, during carotid sinus nerve stimulation in man, reflex bradycardia and hypotension were demonstrated, but there was no evidence of venodilatation.[83] In studies using lower-body negative pressure to simulate blood loss and produce systemic hypotension, we have demonstrated reflex tachycardia and vasoconstriction in the forearm, but there was little evidence of an increase in venous tone[84] (Fig. 4-7). Thus, although the veins are innervated with adrenergic vasoconstricting fibers, they do not appear to participate in the baroreceptor reflex.

This selective modulation of efferent autonomic pathways to parts of the cardiovascular system was also demonstrated by our group.[85,86] We stretched the carotid sinus mechanoreceptors by applying negative pressure to the neck in man to simulate hypertension.[87] Reflex bradycardia and hypotension were induced without decreasing vascular re-

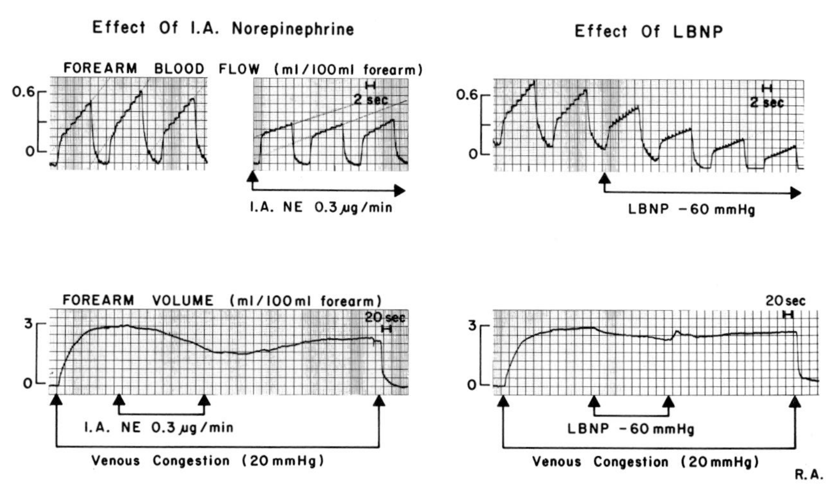

Fig. 4-7. Effect of pooling blood in the lower extremities by lower-body negative pressure (LBNP, right panels) on forearm blood flow (upper tracings) and venous volume at a distending pressure of 20 mm Hg (lower tracings). Both blood flow and venous volume were measured plethysmographically. Reduction in the slope of the flow curves indicates arteriolar vasoconstriction, and reduction in venous volume indicates venoconstriction. LBNP caused marked reflex arteriolar vasoconstriction with minimal venoconstriction (right panels). When contrasted with the effect of intraarterial norepinephrine in the same subject, the latter caused marked arteriolar as well as marked venous constriction. (Reproduced by permission from Abboud FM, et al.[86])

sistance of the forearm.[85,86] On the other hand, a withdrawal of sympathetic vasoconstrictor tone from forearm vessels and forearm vasodilatation can be demonstrated during elevation of the legs, a maneuver that increases central venous pressure and activates cardiopulmonary mechanoreceptors.[86]

The carotid sinus reflex had been considered to exert a significant restraint on heart rate and myocardial contractility,[88,89] but more recent observations in conscious dogs by Vatner and his collaborators indicate that the suppression of myocardial contractility by carotid sinus stimulation is a relatively minor factor in the circulatory adjustments to this reflex.[90] It is important in this context to appreciate the fact that the parasympathetic inhibitory influences on rate, as well as contractility during a rise in arterial pressure, will depend to some extent on preexisting sympathetic tone.[91] Thus stimulation of the vagus nerve, which by itself may have a minimal depressing effect on myocardial contractility, would in the presence of exaggerated sympathetic tone produce a dominant depressing action. In contrast, in the presence of high-level parasympathetic tone, sympathetic stimulation will have little effect.

There are several factors that should be taken into consideration in evaluating the circulatory adjustments to baroreceptor reflex mechanisms in man:

1. There is a selectivity with respect to the component of the efferent sympathetic limb activated during hypotension or inhibited during hypertension exemplified by the fact that veins or capacitance vessels in the extremities do not seem to participate in this reflex, whereas the arteriolar resistance vessels and the heart do.

2. There is not a uniform response of resistance vessels to activation of arterial baroreceptors. The forearm vessels do not appear to participate significantly in the circulatory adjustment to carotid sinus stretch, whereas the same vessels might respond to activation of sympathetic efferent pathways triggered from other mechanoreceptor sites in the cardiovascular system, such as the low-pressure cardiopulmonary receptors. In contrast to vessels of the forearm, neural regulation of other vessels such as the splanchnic or renal may be modulated significantly by arterial baroreceptors.

3. The parasympathetic influence on heart rate and myocardial contractility appears to predominate during a rise in arterial pressure, and the magnitude of the inhibitory action will depend to a significant degree on the level of cardiac sympathetic tone existing at the time of the rise in arterial pressure.

Cardiopulmonary Receptors

Afferent impulses originating in the cardiopulmonary area and arising predominantly from atrial and ventricular receptors are relayed centrally in the vagi along relatively large medullated fibers at high frequency and with fast conduction velocity (8–29 m/sec).[92] These fibers may carry impulses that are rhythmic with atrial or ventricular contraction, such as those originating in type-A atrial receptors and in some ventricular receptors.[93,94] The fibers may also transmit impulses that are more closely related to stretch, such as those originating in type-B atrial receptors.[95]

Impulses may also travel in nonmedullated small fibers at lower frequency and at slow conduction velocity (< 2.5 m/sec). Such fibers characteristically transmit receptor activity generated by chemical substances, but in physiologic states they respond to mechanical activation; such receptors are in the epicardium and in the left ventricular wall.[94,96,96a] A similar pattern of receptor activation by mechanical stimuli with transmission of impulses along slow-conducting nonmedullated afferents occurs also in carotid baroreceptors.[97] The cardiac nonmedullated afferents that transmit the mechanical activation of these receptors under physiologic states and also transmit their chemical activation with nicotine or veratridine or phenyl diguanide in experimental situations are referred to as the C fibers. They are more widely distributed in the cardiac region than the medullated fibers and appear to have more profound circulatory effects. Cooling the vagi will block first the medullated (fast conduction, high frequency) afferent impulses, but the relatively low frequency impulses in nonmedullated fibers will be intact until very low temperatures are reached (< 8° C). Cardiac afferent impulses may also be transmitted along the sympathetic nerves and may elicit a spinal reflex with efferent sympathetic activity transmitted through the stellate ganglia. The receptors are probably located in the coronaries and in the ventricle and are activated by stretch, by myocardial ischemia, or by accumulation of metabolites.[98,99]

Electrical stimulation of cardiopulmonary afferents. In general, one may conclude from the work of Öberg and Thorén[100] that electrical stimulation of cardiac vagal afferents at parameters that activate large medullated fibers (high frequency, short duration) triggers reflex sympathetic excitatory activity with tachycardia and vasoconstriction in kidney and muscle. It is proposed that these afferents originate predominantly in the atria and play a relatively minor role in circulatory adjustment. In contrast, electrical stimulations of vagal afferents at parameters that activate small nonmedullated fibers (low frequency, longer duration) thought to originate predominantly in ventricles are inhibitory, causing hypotension vasodilatation, bradycardia, and powerful hemodynamic responses such as described in 1948 by Jarisch and Zotterman.[101] Öberg and Thorén believe that the more significant cardiac reflexes such as the Bezold-Jarisch reflex are inhibitory and are transmitted along nonmedullated C fibers originating in ventricular receptors; these responses have erroneously been ascribed by others to afferent activity in medullated fibers.[93]

Activation of sympathetic afferent fibers by electrical stimulation of the central end of the cut left inferior cardiac nerve or pericoronary nerves elicits an excitatory response consisting of reflex increases in heart rate, contractility, and pressure.[102,103]

Mechanical or chemical activation of cardiopulmonary receptors. The exact correlation between responses to electrical stimulation of specific cardiopulmonary afferent fibers and the hemodynamic response obtained with atrial, ventricular, or pulmonary stretch or with denervation has been difficult and has led to some confusion. One way to approach the role of cardiopulmonary receptors is to describe what circulatory changes occur during their mechanical or chemical activation, without attempting to ascribe them rigidly to medullated or nonmedullated afferents.

1. Atrial receptors. The work of Ledsome, Linden, Carswell, and Hainsworth[104–106] indicates that stretch of atrial receptors or receptors at pulmonary vein–left atrial junctions will trigger a reflex tachycardia, confirming that the Bainbridge reflex does indeed exist. The magnitude of the tachycardia, however, is relatively small; and the total circulatory response, despite significant stretch of these receptors, is also small. More recently, the work of Karim et al.[107] has indicated that stretch of left atrial receptors may cause an increase in cardiac sympathetic activity and a selective reduction in sympathetic activity to the kidney; others have found no significant alteration in sympathetic efferent activity to the hindlimb.[106] The combination of tachycardia and renal vasodilatation during left atrial stretch might be an effective mechanism by which the ventricular volume becomes smaller and the intravascular volume is reduced in response to cardiac decompensation and dilatation.[105] Greenberg et al.[108] suggest that there is an impaired atrial receptor response in dogs with heart failure; one might be tempted to hypothesize that one of the actions of digitalis is to restore the responsiveness of atrial receptors to stretch and decrease the sympathetic efferent activity to the kidneys. Sympathetic activity to kidneys is excessive in heart failure and may contribute to sodium retention.

Recent studies by Mancia et al. indicate that excessive cooling of the vagi (thereby impeding all afferent impulses) is accompanied by an increase in sympathetic nerve activity to the renal nerve,[109] as well as an increase in renin release.[109a] Although the afferent activity may originate in atria or ventricles, the work of Karim et al.[107] would support an atrial contribution.

In man we have observed that pooling of blood in the lower extremities with low levels of lower-body negative pressure sufficient to decrease right atrial pressure without altering systemic arterial pressure triggers reflex vasoconstriction in the forearm without a significant change in heart rate, and on occasion with slight bradycardia[110] (Fig. 4-8). Although we are not certain whether the reflex is triggered by a decrease in stretch of the atria, ventricles, or pulmonary vessels, it is clear that in unanesthetized man reduction in stretch of low-pressure cardiopulmonary receptors may remove a sympathetic inhibition imposed by stretch of these receptors and activate sympathetic vasoconstriction at least with respect to resistance vessels of the forearm.

Atrial receptors may also have a modulating influence on humoral substances; left atrial stretch reduces levels of vasopressin,[111] and right atrial stretch reduces renin release.[112]

2. *Ventricular receptors.* Electrical stimulation of nonmedullated vagal afferent C fibers from the heart causes a fall in blood pressure, slowing of the heart rate, and vasodilatation.[101,113] A similar response is also observed when cardiac receptors are chemically excited with veratrum alkaloid[114] or mechanically stimulated when pressure elevation is induced in isolated heart chambers.[115,116]

The characteristic cardiocirculatory inhibition described as the Bezold-Jarisch reflex was ascribed by Dawes and Comroe[114] to a coronary "chemoreflex" rather than to a coronary "chemoreceptor reflex" because of the lack of convincing evidence that there are chemoreceptors in the coronaries. Recent observations by our group[117] would indicate that a reflex vasodilatation can be triggered by intracoronary injection of nicotine, but not by perfusion of coronaries with hypoxemic, hypercapnic, and acidotic blood, indicating that although chemoreflexes that are inhibitory in nature can be demonstrated the physiologic role of the receptors that mediate this reflex does not appear to be one of adjustment to changes in the chemical composition of blood.[117]

Our studies in humans would also indicate that such a coronary chemoreflex might be triggered during intracoronary injections of

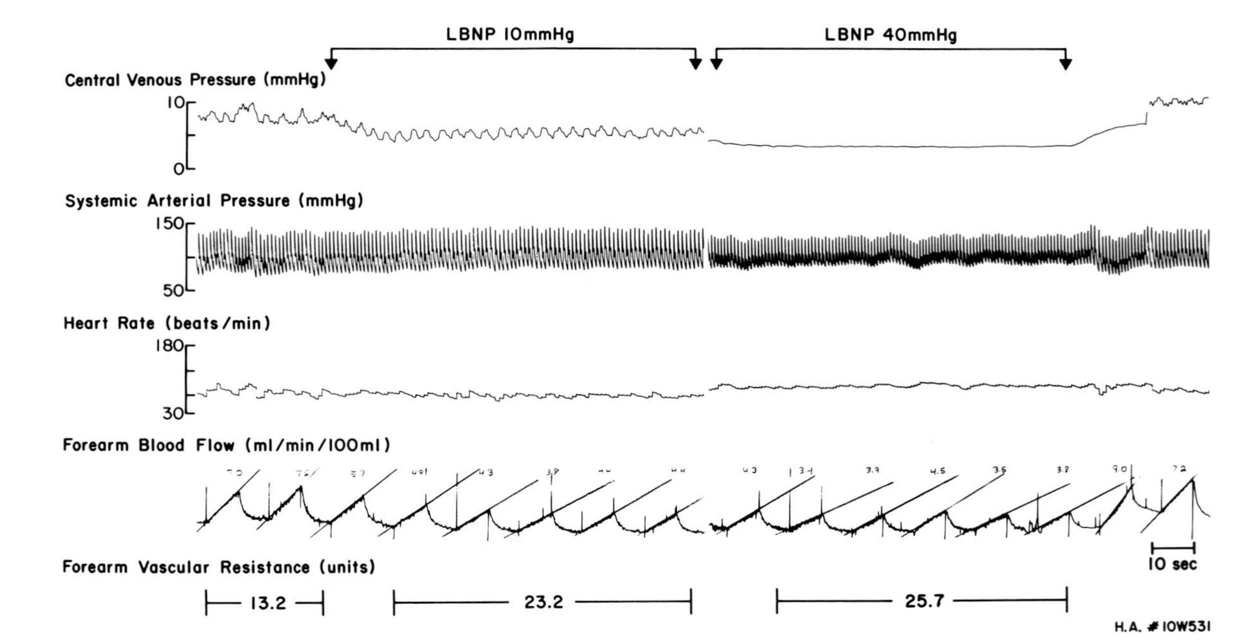

Fig. 4-8. Activation of cardiopulmonary (low-pressure) baroreceptor reflex in man. A vasoconstrictor response in the forearm was seen during minimal pooling of blood in lower extremities by lower-body negative pressure (LBNP, 10 mm Hg); there was no reduction in arterial pressure, pulse pressure, or *dp/dt;* there was no tachycardia, and at times bradycardia was noted. This reflex (left half of figure), which is thought to be mediated through low-pressure cardiopulmonary baroreceptors as indicated by the fall in central venous pressure, differs from the reflex induced during excessive pooling at LBNP of −40 mm Hg, which causes systemic hypotension and tachycardia but no additional vasoconstriction (right half of figure).

contrast medium for coronary arteriography and may represent a clinical counterpart of the Bezold-Jarisch reflex.[118] In addition, cardiocirculatory inhibition similar to that induced by stimulation of C fibers occurs during stimulation of epicardial receptors with topical application of nicotine[96] or acetylstrophanthidin[119] or with ventricular stretch.[96,115,116] Despite the fact that under resting physiologic states the rate of discharge along the nonmedullated C fibers is minimal and most of the evidence for their activation has involved unphysiologic chemical or mechanical stimulation, their role in the regulation of the circulation under normal physiologic conditions may be important. Under pathologic states such as following coronary occlusion and myocardial infarction in man[120] or in animals[121,122] these fibers may mediate bradycardia and hypotension. Furthermore, these afferent impulses may also prevent reflex efferent sympathetic activity associated with hypotension and shock if the hypotension is secondary to myocardial infarction.[123] Recent findings in man by Mark et al.[124] would indicate that leg exercise that ordinarily triggers reflex vasoconstriction in the nonexercising forearm fails to do so in aortic stenosis; this inhibition of the expected reflex sympathetic vasoconstriction might be caused by an undue stretch of the left ventricle in aortic stenosis and activation of ventricular stretch receptors.

Although the predominant afferent activity from the ventricles appears to be vagally mediated and inhibitory, there is evidence for some afferent sympathetic activity that is mostly excitatory and is activated by a rise in coronary artery pressure or by myocardial bulging secondary to ischemia. This spinal reflex causes tachycardia and hypertension.

3. Pulmonary receptors. Just as epicardial and ventricular receptors may generate impulses that impose a restraint on the excitatory medullary centers, there is some suggestion that afferent impulses from the lung may also mediate a similar restraint. It seems, however, that the magnitude of that influence is relatively small.[125] According to Paintal, it has been difficult to isolate endings that satisfy requirements for a pulmonary artery baroreceptor.[92] Pulmonary congestion during exercise or in pulmonary edema may activate pulmonary receptors of type J through an increase in interstitial pressure or volume and cause a reflex tachypnea and dyspnea.[126] Other structures in the lung besides the pulmonary vasculature might mediate afferent impulses that are inhibitory in nature. A reflex from the lungs affecting systemic blood vessels was first demonstrated by Brodie and Russell[127] in 1900 and has been more thoroughly analyzed recently by Daly and Robinson.[128] Inflation of the lungs will activate bronchopulmonary stretch receptors and inhibit the vasomotor center, causing a reflex vasodilatation.[129]

Chemoreceptors

Afferent impulses originating in peripheral vascular chemoreceptors and in skeletal muscle (somatic afferents) can induce reflex excitatory responses, in contrast to the reflex inhibitory responses described with stimulation of baroreceptors in the aortic, carotid, and left ventricular regions. The chemoreceptors provide a control system for oxygen conservation and for increasing oxygen delivery, primarily in hypoxic states. The sensors are present in the carotid and aortic bodies and are particularly sensitive to reduction in arterial Po_2, increases in Pco_2, and decreases in pH. Sensitivity to these chemical changes in the blood may be a function of the relatively very high blood flow to the carotid and aortic bodies or may be due to the presence of an enzyme system with an unusual oxygen affinity. These receptors may also be activated by nicotine or cyanide.

The adjustment to stimulation of the chemoreceptors involves not only a respiratory response with hyperventilation to increase oxygen delivery and reduce hypercapnia, but also a circulatory response that might effectively redistribute blood flow away from less vital organs, which can withstand anaerobic metabolism, toward the cerebral and coronary circulations, which are critically dependent on oxygen delivery. The circulatory response, therefore, includes reflex vasoconstriction, which appears to predominate in skeletal muscle and is mediated through activation of sympathetic efferent pathways and a bradycardia resulting from activation of parasympathetic vagal efferents to the heart (Fig. 4-9). Bradycardia is effective in reducing myocardial oxygen requirements. Activation of carotid chemoreceptors may decrease cardiac output and contractility, whereas aortic chemoreceptors will increase cardiac output.[130] We have demonstrated also that there is a vagally mediated coronary vasodilatation during stimulation of chemoreceptors that would tend to favor greater oxygen delivery to the heart.[60] The work of Ponte and Purves[131] indicates that hypercapnia may trigger a reflex cerebral vasodilatation by activation of chemoreceptors; such a dilatation appears to be mediated through cholinergic sympathetic fibers.* We have reported that reflex vasodilatation in the hindpaw and vasoconstriction in skeletal muscle occur simultaneously during stimulation of chemoreceptors.[71,132] The dilatation is noncholinergic. A similar differential vascular response to chemoreceptor stimulation has also been noted by Pelletier and Shepherd.[133] All these findings indicate that there is a fine process of selectivity in the activation of the efferent sympathetic

*Recent studies by Heistad *et al.*[131a] indicate that stimulation of chemoreceptors has no effect on total or regional cerebral blood flow in the dog.

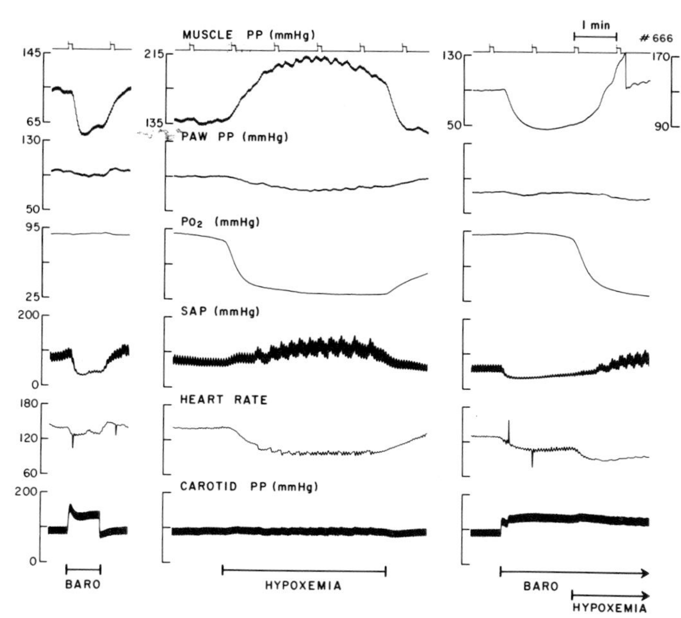

Fig. 4-9. Effect of stimulation of carotid chemoreceptors by perfusing hypoxemic blood at constant pressure into both carotids (middle panels). The decrease in arterial Po_2 was associated with reflex vasoconstriction in muscle (PP = perfusion pressure at constant flow), vasodilatation in the hindpaw, and bradycardia. The vasoconstrictor response in muscle will override and reverse the dilator effect of carotid hypertension (right panel). (Reproduced by permission from Heistad DD, et al.[132])

pathways to the heart and to various vascular beds controlled at the level of the central nervous system.

Somatic Afferents

The circulatory adjustments to exercise are important in increasing cardiac output and redistributing blood flow to the exercising muscle. A reflex tachycardia that occurs almost instantaneously with the beginning of muscle contraction, a rise in arterial pressure, and vasoconstriction in the nonexercising muscle are all parts of this response. Freyschuss[134] has shown that an abrupt tachycardia occurs when subjects are asked to perform an isometric contraction of the arm; the tachycardia is essentially unaltered in magnitude after neuromuscular blockade of the contracting arm. This finding would point toward a prepatterned output of the central nervous system in initiating at least the tachycardia of the cardiovascular response to isometric exercise. The work of Lind and his collaborators[135] indicates that the tachycardia may also in part be due to an inhibition of the baroreceptor reflex, which accounts for an increased rate despite the associated simultaneous rise in arterial pressure. The studies of Coote et al.[136] and McCloskey and Mitchell[137] and the more recent work of Clement et al.[138] indicate that the circulatory response is also in part the result of the contracting muscle. The receptors that appear to trigger that reflex, although originating in the contracting muscle, do not appear to be associated with the muscle spindles or tendons. A chemical factor may be important.

When these reflexes are coupled with other adjustments, such as increases in venous return, changes in activity of chemoreceptors, reactive hyperemia, changes in temperature, and release of humoral substances during exercise, the overall adjustment to the excessive demand of flow for exercising muscle is effectively satisfied.

Central Nervous System Modulation of Medullary Centers

In the foregoing paragraphs the modulation of medullary cardiovascular centers by peripheral afferent impulses originating in arterial and cardiopulmonary baroreceptors, in chemoreceptors, and in somatic receptors has been discussed. The medullary vasomotor centers are also under the influence of input from various parts of the central nervous system, such as the hypothalamus and the cerebellum. Integration of various inputs occurs in the paramedian reticular nucleus.

Hypothalamic Input

The association of emotional lability with fluctuations in blood pressure has long been known. Clinicians have recognized the entity of neurogenic or labile hypertension. The hypothalamus is regarded generally as having a key role in mediating these pressure changes. Recently, sustained hypertension has been induced experimentally with chronic stimulation of the pars posterior of the hypothalamus.[139] An earlier detailed analysis of the effects of hypothalamic stimulation on sympathetic discharge was provided by Bronk and co-workers.[140] Hypothalamic stimulation caused constriction of resistance vessels in skin, intestine, and kidneys and dilatation in skeletal muscle; this differential effect on the peripheral circulation causes a redistribution of blood flow that favors perfusion of skeletal muscle for the "flight or fight" defense reaction. Despite the rise in arterial pressure, the cardiac output and the heart rate remain increased during hypothalamic stimulation, a finding suggesting that hypothalamic activation might interrupt baroreceptor-mediated cardiac inhibition.[141] A similar mechanism may explain the high output and heart rate, despite the rise in arterial pressure, during emotional stress. This inhibition of the baroreceptor-mediated reflex bradycardia is not associated with an inhibition of the baroreceptor-mediated reflex vasodilatation in the kidney[142]; the reflex withdrawal of sympathetic drive to the kidney during the rise in arterial pressure may be offset in part by the renal vasoconstrictor effect of hypothalamic stimulation. It is also of interest that hypothalamic stimulation suppresses rather than augments renin release by the kidney.[143]

An important circulatory response during the defense reaction and hypothalamic stimulation in animals is the active vasodilatation in skeletal muscle that is mediated by sympathetic cholinergic pathways.[61] A similar cholinergic vasodilatation occurs in man during emotional stress,[67] suggesting that the "defense reaction" of hypothalamic stimulation in animals may have important physiologic counterparts in man. Other stimuli may also trigger a reflex vasodilatation in the forearm of man that is in part cholinergic.[68] Additional work is necessary for a better and more comprehensive understanding of the role of sympathetic cholinergic vasodilatation in man. The hypothalamus might play a role in the neurogenic regulation of the circulation in early labile hypertension. The high cardiac output, high arterial pressure, vasodilatation in skeletal muscle, and renal vasoconstriction that characterize a hypothalamic response are seen in these patients.

Recently Nathan and Reis[143a] demonstrated a fulminating arterial hypertension with pulmonary edema from release of adremedullary catecholamines after lesions of the anterior hypothalamus in the rat. Several years ago we noted that insulin causes a marked pressor response in the dog through a CNS effect independent of hypoglycemia.[143b] It is tempting to speculate that this effect is mediated through activation of hypothalamic nuclei.

Cerebellar Control

Miura and Reis[144] discovered that stimulation of the fastigial nucleus of the cerebellum produced a very consistent and large pressor effect. This pressor response was eliminated by section of the fastigiobulbar tract or the paramedian reticular nucleus in the medulla. It was also found that fastigial stimulation produced a pattern of responses similar to the arterial baroreceptor responses to a fall in systemic pressure[145] with vasoconstriction and tachycardia. It seems, therefore, that the circulatory adjustments to the upright posture might be triggered not only from arterial baroreceptors and cardiopulmonary baroreceptors[110] but also from afferent impulses originating in the vestibular apparatus, all of which may be acting in concert.

Paramedian Reticular Nucleus and Interaction of Reflexes

Baroreceptor–chemoreceptor interaction. Humphrey provided a description of the central projections of the carotid sinus nerve afferents to the nucleus tractus solitarius.[146] The more recent work of Miura and Reis[147] verified the presence of monosynaptic as well as polysynaptic pathways to the nucleus tractus solitarius, but localized more precisely the active cells to the paramedian reticular nucleus. Lesions of the paramedian reticular nucleus abolished or reversed the depressor response to electrical stimulation of myelinated fibers of the carotid sinus nerve (which are predominantly baroreceptor fibers) and attenuated the depressed response to carotid sinus stress. In contrast, these lesions tended to augment the pressor response to chemoreceptor stimulation by lobeline. Thus the baroreceptor projections into the paramedian reticular nucleus appear to buffer the pressor response to chemoreceptor stimulation, making a strong case for a true synaptic interaction between these two afferent inputs. These findings also provide the electrophysiologic explanation for our recent observations[148] of an interaction between chemoreceptor and baroreceptor reflexes that indicate that hypertension inhibits or buffers the vasoconstrictor response to stimulation of chemoreceptors.

Cerebellar–baroreceptor interactions. Since the paramedian reticular nucleus also relays impulses from the fastigial nucleus of the cerebellum, it was suggested that there might be an inhibitory interaction between fastigial and baroreceptor fibers on the cells of the paramedian reticular nucleus, but this was not found to a sufficient degree to make a strong case for a true synaptic interaction.[145,149] Nathan[150] showed, however, that there was an inhibitory interaction between these two stimuli with respect to the final efferent output as recorded over the splanchnic nerve. Cerebellar stimulation was also shown to suppress the defense reaction induced by hypothalamic stimulation.[151]

Cardiopulmonary–baroreceptor or -chemoreceptor interactions. Although several afferent inputs, both peripheral and central, appear to influence the vasomotor center by first relaying in the paramedian reticular nucleus, there is no a priori reason to postulate that excitatory and inhibitory afferents will have identical synaptic connections. Recent

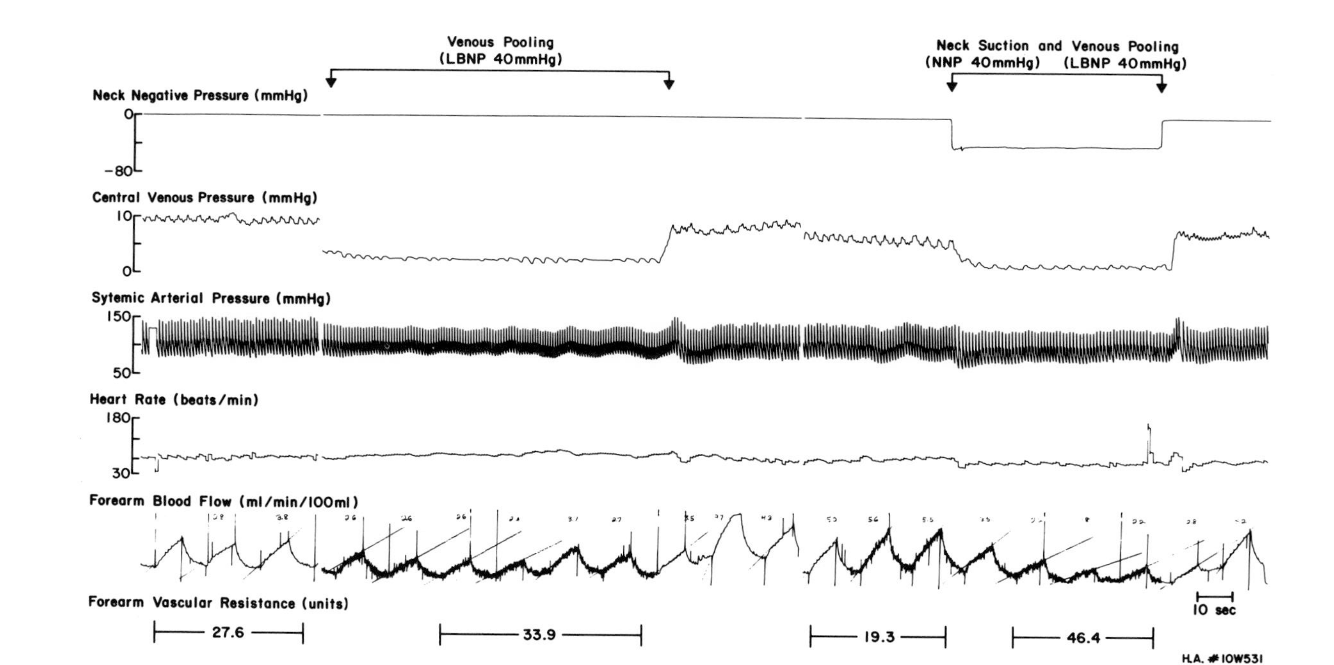

Fig. 4-10. Interaction between carotid baroreceptor reflex (neck suction) and cardiopulmonary reflex (venous pooling). Venous pooling caused a fall in central venous pressure, tachycardia, and forearm vasoconstriction (left half of figure). The simultaneous application of neck suction and venous pooling (right half of figure) abolished the reflex tachycardia but did not prevent vasoconstriction of the forearm. Not shown here is the vasoconstriction of the splanchnic bed, which was suppressed by neck suction. (Reproduced by permission from Abboud FM, et al.[86])

electrophysiologic studies on the organization of the central vasopressor pathways by Gebber et al.[152] indicate that there are two distinct systems of vasopressor pathways. One has a short latency and is rapidly conducting; the other has a longer latency and is slowly conducting. It has also been found that activation of the baroreceptor reflex blocks the more slowly conducting vasoconstrictor pathways without interfering with the more rapidly conducting system.[152] This finding could explain the difficulty in demonstrating an algebraic summation of effects between two reflex responses that may be acting in opposite directions, such as the fastigial and arterial baroreceptor reflexes. An interaction may be apparent only with respect to selective synaptic connections. A potentially exciting human correlary of the findings by Gebber et al. is the observation by Mark et al. that "carotid neck suction in man" (which activates the carotid baroreceptor reflex) prevents the reflex tachycardia and reflex splanchnic vasoconstriction induced by pooling of blood in the lower extremities without affecting the reflex vasoconstrictor response to the forearm[85,86] (Fig. 4-10). Pooling of blood in the lower extremities activates a cardiopulmonary reflex[110] that may relay in the paramedian reticular nucleus and may cause vasoconstriction through both the slow and rapid conducting vasopressor pathways of Gebber. If the reflex forearm vasoconstriction is mediated through the more rapidly conducting sympathetic efferent pathway and the splanchnic constriction and tachycardia are mediated through slow conducting pathways, the forearm response would not be blocked by simultaneous carotid baroreceptor stimulation. This would imply a failure of synaptic interaction between the cardiopulmonary and baroreceptor inputs with respect to the rapidly conducting sympathetic efferent fibers. The physiologic role of cardiopulmonary inhibitory afferents has been suggested from the work of Mancia et al., who by cooling the vagi demonstrated an increase in renal sympathetic efferent activity[109] and renin release.[109a] More recently, Koike, Mark, Heistad, et al.[152a] have extended these observations and demonstrated that the "resting" vagal cardiopulmonary afferent activity exerts a significant restraint on the reflex sympathetic activation induced by carotid hypotension or by stimulation of chemoreceptors.

HUMORAL FACTORS REGULATING THE CIRCULATION

Many substances with vasoactive properties are circulating in the bloodstream and exert important vascular effects modifying tissue perfusion. The effects of humoral substances on various vascular beds are often different. An understanding of the differential effects upon various parts of the circulation is important for a thorough appreciation of their effect on distribution of blood flow to various organs.

Catecholamines

The cardiovascular and metabolic effects of catecholamines are caused by activation of adrenergic receptors. The classification of these receptors is based primarily on identification of the relative potency of the different catecholamines that activate them and the demonstration of their antagonism by specific receptor blockers.[153] Adrenergic receptors are classified as follows:

1. Alpha receptors mediate the vasoconstrictor responses to norepinephrine, methoxamine, and epinephrine and are blocked by phentolamine or phenoxybenzamine. These receptors are not evenly distributed in all vascular beds nor in consecutive segments of the same vascular bed. The renal, cutaneous, splanchnic, and skeletal muscle arterioles and the cutaneous and splanchnic venules appear to be more generously supplied with alpha receptors. Their distribution in the coronary and cerebral vessels is sparse.
2. Beta-1 receptors mediate the positive inotropic and chronotropic effects on the myocardium, but do not mediate vascular effects; these are blocked by practolol.[154]
3. Beta-2 receptors activate vasodilator responses in peripheral vessels, but not in myocardium. Propranolol blocks both beta-1 and beta-2 receptors.[154]

Norepinephrine activates predominantly alpha and beta-1 adrenergic receptors and causes vasoconstriction that is relatively greater in cutaneous and renal beds. Epinephrine activates alpha, beta-1, and beta-2 receptors; it is a powerful vasodilator of skeletal muscle in man and a vasoconstrictor in skin and kidney. Norepinephrine is derived primarily from the adrenergic nerve terminals, and the major source of epinephrine is the adrenal medulla. Ordinarily the levels of catecholamines circulating in the blood exert minimal effects on tissue perfusion; however, under abnormal conditions increased blood levels may affect significantly the distribution of blood flow. These situations might arise in shock, in pheochromocytoma, or during vasovagal syncope, severe postural hypotension, exercise, and cardiac failure.

Renin–Angiotensin System

The release of renin by the juxtaglomerular cells of the afferent glomerular arteriole into the circulation is modulated by the level of renal sympathetic tone and also by the concentration of sodium in the macula densa. The release of renin by the kidney during sympathetic nerve stimulation may be blocked by alpha blockers, which prevent the vasoconstrictor effects of adrenergic stimulation from activating the juxtaglomerular baroreceptor mechanism for release of renin, or by beta blockers, which appear to antagonize an effect of sympathetic stimulation on renin release independent of the vascular response.

The pathophysiologic role of the renin–angiotensin system in hypertension is considered elsewhere. The circulating levels of renin or angiotensin appear to be too small to cause significant circulatory effect in physiologic states. It is possible, however, that the renin content of vascular wall[155] and the high concentrations found in the area postrema of the medulla[156] may be more important than the circulating levels in directly or indirectly adjusting vascular tone.[74] Angiotensin and renin may traverse the blood-brain barrier in the area postrema where they would influence medullary cardiovascular centers responsible for baroreceptor reflexes. It was shown by Ueda et al.[157] that microinjections of angiotensin in this area will cause systemic pressor responses.

Angiotensin is also known to cause endogenous release of norepinephrine from vascular walls and facilitate the release during nerve stimulation in renal and cutaneous beds.[158] When infused intravenously or intraarterially it has potent constrictor effects on precapillary resistance vessels, predominantly in renal, skeletal, splanchnic, and coronary circulation; its effects on postcapillary resistance vessels[36] and on capacitance vessels are negligible.[159,160] The effect of angiotensin on distribution of blood flow suggests that blood flow is diverted away from the kidneys during its intravenous infusion.[161]

Whelan et al.[162] have demonstrated in man that the vasoconstrictor effect of norepinephrine is augmented when the catecholamine is given intraarterially as an infusion superimposed on infusion of a small dose of angiotensin, and our findings[163] indicate a potentiating effect of small doses of angiotensin on the vasoconstrictor response to the reflex release of norepinephrine.

Angiotensin may also have effects on intrarenal distribution of blood flow through its intrarenal release, which may cause postglomerular arteriolar constriction and depress proximal tubular sodium reabsorption. This would maintain adequate sodium delivery to the distal tubular segments, thus permitting sodium and water absorption and preservation of volume during hypotensive states.

Vasopressin

Vasopressin is another potent constrictor hormone, but its circulating blood level under physiologic states is probably too low to cause sufficient direct effects on vascular tone to significantly alter distribution of blood flow. There is some evidence, however, that small doses, possibly close to the physiologic range, might influence certain vascular beds directly, such as the skeletal muscle or kidney, and produce vasoconstriction by indirectly augmenting the constrictor action of norepinephrine. Its release from the posterior pituitary gland is regulated primarily by changes in osmolality, and its most prominent physiologic role is in the control of water balance. The release of vasopressin may be altered by stretch of left atrial receptors that are sensitive to changes in circulating blood volume. Similarly, inhibition of the release of vasopressin may be achieved by activation of arterial mechanoreceptors, but these probably play a secondary role in relation to the effects of blood volume. Under conditions of significant circulatory stress, such as severe hemorrhage or reduction of left atrial stretch during cardiopulmonary bypass, high circulating levels of vasopressin may occur. In addition to vessels in skeletal muscle, the coronary and splanchnic vessels are particularly sensitive to the constrictor effect of vasopressin. The cutaneous vessels respond only minimally to this hormone.[164]

Corticosteroids

Steroids have minimal direct effects on blood vessels. However, they may cause accumulation of sodium and water in blood vessels and augment their responsiveness to constrictor stimuli.[165] The work of Schmid et al.[166] shows that the increased reactivity is not related to the effects of steroids on the uptake or storage of norepinephrine, and the fact that the response to norepinephrine is potentiated significantly after mineralocorticoids and not glucocorticoids suggests that a primary vascular change related to cation concentration might be responsible for the augmentation. It is not certain, however, whether this potentiation is a result of a change in geometry of the vessel with an increase in wall/lumen ratio or a change in contractility of vascular smooth muscle. Steroids have been advocated in the treatment of septic, and more recently cardiogenic, shock. Their effectiveness under these circumstances is probably unrelated to their direct cardiovascular effects, but may rather be through their effects on stabilization of cellular membranes and prevention of release of proteolytic enzymes. The possibility that they might facilitate oxygen transport across capillaries by reducing reactive tissue changes following ischemia or during shock is considered.

Kinins

The kinins are potent vasodilator polypeptides formed by the action of proteolytic enzymes (kallikrein) on plasma protein precursors (kininogens). Bradykinin is the substance that is essentially the prototype of this class of polypeptides. Their major physiologic role has been in the local regulation of blood flow in such organs as the salivary gland and the pancreas, where they are released in significant amounts. In these glands the kinins are activated by the secretion of the proteolytic enzymes. In response to inflammatory changes or during anaphylactic reactions and possibly during endotoxin shock these compounds appear to play a significant role as vasodilators and may increase capillary permeability to protein, which would decrease intravascular volume and exaggerate the hypotension. The most prominent pathophysiologic role of the kinins, however, appears to be in the carcinoid syndrome, where they characteristically induce cutaneous flushing. Patients afflicted with the carcinoid syndrome and liver metastasis will release the enzyme kallikrein from the metastic lesions under the influence of the administration of epinephrine or alcohol. Kallikrein liberates the decapeptide lysylbradykinin, which is converted to bradykinin. Serotonin and histamine may also be involved in the flushing attacks of the carcinoid syndrome or in the capillary permeability changes in inflammation or shock.

Recent evidence indicates that kallikreins and kinins liberated by the kidneys play an important role in regulating sodium excretion and may participate in the pathogenesis of certain types of hypertension.[166a,b]

Prostaglandins

The prostaglandins are unsaturated lipids derived from arachidonic acid and dihomolinolenic acid. Active synthesis of these compounds accounts for most of the prostaglandins in the tissues; storage is meager. They are either metabolized locally or immediately upon their release from the tissues. Thus their major effect occurs in situ in response to their generation as a result of biosynthesis. Release of prostaglandins into the circulation probably has minimal physiologic influence. Prostaglandins of the E and F series are metabolized in the lung, but prostaglandin A can pass through the lung intact and might possibly serve as circulating hormone, except that its concentration in blood is low.

Prostaglandins E cause vasodilatation in most vascular beds, whereas the F prostaglandins cause constriction of veins but not resistance vessels.[167–169] The concentration of prostaglandins E in the kidney is high, particularly in the inner cortex, and their rate of biosynthesis may influence vascular tone in this organ.[170] They apparently inhibit renal vasoconstrictor effects of angiotensin and norepinephrine and may inhibit renin secretion.[171] Thus it would appear that prostaglandins may have a protective role against hypertension, and they have recently been demonstrated in the renal venous blood in patients with proven renal vascular disease.[172] These compounds may also represent the hypotensive factor of renal medullary extracts.[173]

In addition to their effects on the kidney, prostaglandins may play a major role in the modulation of sympathetic nervous system function. Responses to nerve stimulation, norepinephrine, and angiotensin are inhibited by prostaglandins E, which have also been demonstrated to prevent the release of norepinephrine from sympathetic terminals.[167,168,174,175]

In the central nervous system prostaglandins inhibit the release of dopamine and norepinephrine from adrenergic neurones. They appear to play a role in peripheral as well as central neural control of the circulation and in the local control of vascular tone.[176]

Recent observations suggest that prostaglandins released in the coronary sinus during periods of ischemia may play an important role in coronary reactive hyperemia.

INTRINSIC TISSUE FACTORS

Autoregulation

Blood vessels have an intrinsic ability to regulate their vascular tone and thereby maintain blood flow to an organ independently of systemic neurogenic influences or humoral factors. This property has been referred to as the autoregulatory capacity for blood flow. Different vascular beds vary with respect to their ability to maintain blood flow over a wide range of perfusion pressures. The cerebral, coronary, and renal circulations appear to be most potent in this regard. Two of the three (cerebral and coronary) exhibit high metabolic activity with a significant oxygen requirement, whereas the kidney is essentially overperfused with respect to its oxygen demand. By its ability to autoregulate, the kidney can maintain renal blood flow and possibly glomerular filtration constant when arterial pressure is raised or decreased. Three theories have been advanced to explain the phenomenon. The first is that the activity of smooth-muscle pacemaker cells in vessel walls is increased during stretch, causing greater smooth-muscle contraction and an increase in resistance to flow when arterial pressure rises. This has been generally referred to as the Bayliss response, characterizing a contraction of smooth muscle during stretch. Another theory is that the rate of oxygen delivery to arterioles and precapillary sphincters, as well as the rate of accumulation of metabolites around these vessels, might regulate their smooth-muscle tone when flow is either increased or decreased. Hypoxia and accumulation of metabolites would tend to produce arteriolar vasodilatation and relaxation of precapillary sphincters and would favor restoration of flow and perfusion of the microcirculation. The third theory is that an increase in perfusion pressure might increase capillary filtration and tissue pressure, and secondarily increase resistance to flow by compressing arterioles and precapillary sphincters.

An autoregulatory response appearing in the total circulation has been proposed as a pathogenetic mechanism in essential hypertension.[177] Several investigators have suggested that the first phases of primary hypertension involve an increase in cardiac output without a change in total peripheral vascular resistance. The autoregulatory hypothesis[177] suggests that peripheral vessels, and in particular renal vessels, will constrict and increase their resistance in an effort to maintain blood flow constant in the face of an increase in arterial pressure and cardiac output. Renal vasoconstriction might lead to the elaboration of renin, sodium, and water retention and further acceleration of the development of hypertension by causing generalized vasoconstriction. The increase in resistance will be followed by a return of cardiac output to normal, but arterial pressure would be maintained. Although the hypothesis is intriguing, there has been little experimental evidence that correlates directly the levels of renin with the development of an increase in peripheral vascular resistance as hypertension progresses from the labile to the fixed state or from the high-output to the normal-output state.

Hypoxia, Ischemia, and Hyperemia

Numerous studies have sought to identify the factors responsible for local regulation of blood flow, particularly in skeletal muscle during the hyperemia of exercise or following ischemia. Release of adenosin,[178] hyperosmolality, hyperkalemia, hypoxemia, or a combination of these factors[179] have all been implicated. It seems likely that the hyperemic response in skeletal muscle represents an interaction of several of these factors.

Studies of acute systemic hypoxemia in man have been of interest, since they demonstrate vasodilatation in the forearm.[180] Hypoxemia activates a chemoreceptor reflex that would cause reflex vasoconstriction in animals.[132] In man, however, the vasoconstrictor response induced by chemoreceptor stimulation may be attenuated by two opposing factors: (1) hypoxemia may attenuate neurogenic constrictor reflex responses,[28,181] and (2) the activation of pulmonary stretch receptors during the hyperventilation caused by hypoxemia may also inhibit vasoconstrictor reflexes.[182]

The forearm vasodilator response during hypoxemia in man occurs in the presence of nerve block; it is not caused by the release of catecholamines and is not caused by hypocapnia.[183,184] It may well be due to the local effect of hypoxemia on arterioles or the release of a local vasodilator substance.[185]

Studies by Heistad et al.[27] indicate that local hypoxemia of skeletal muscle causes vasodilatation that is greater during muscular contraction. Hypoxemia alone does not inhibit vasoconstrictor responses to norepinephrine and angiotensin significantly, but when the skeletal muscle contracts the inhibition by hypoxemia is pronounced (Fig. 4-2). A synergistic inhibitory action of hypoxemia plus muscular contraction seems to take place. The release of inhibitory metabolites may be potentiated by a combination of hypoxemia and contraction. The insufficient generation of ATP from anaerobic pathways for both smooth and skeletal muscle contraction may be the limiting factor.

Failure of delivery of oxygen from anemia, from high altitude, or from deficiency of thiamine, which is necessary for oxidative decarboxylation of pyruvic acid, will all cause a vasodilatation. Conditions associated with uncoupling of oxidative phosphorylation, such as cyanide poisoning, idiopathic hyperkinetic states, nonshivering thermogenesis, or defects in chemicomechanical coupling (as in hyperthyroidism), may also cause vasodilatation. In these conditions the "tissue factors" responsible for vascular adaptation are not defined precisely.

STRUCTURAL CHANGES

The major cause of reduction in blood flow to the most vital organs in pathologic states is the arteriosclerotic process, which induces structural changes in large- and medium-size arteries of a magnitude sufficient to obstruct blood flow either at rest or during reactive hyperemia. The problem of ischemic heart disease and arteriosclerotic vascular changes is discussed elsewhere. It is already well appreciated that significant reduction in the lumen of blood vessels can take place without any reduction in resting blood flow. Lesions will produce critical narrowing when the lumen of major vessels is reduced by approximately 70%. It must be stressed, however, that the critical level of 70% narrowing is applicable only to the situation in which blood flow is at "resting" levels. When flow is increased with increased demand, a lesion that is not critical may become limiting and prevent adequate flow.

Structural changes in the vascular wall may cause the augmentation of vasoconstrictor responses to adrenergic stimuli seen in hypertensive states. The observations of Folkow et al.[186] and Sivertsson[187] suggest that there is greater vascular resistance at maximal vasodilatation in the blood vessels of the hand and forearm of hypertensive patients than in normals. This structural limitation manifested by an increase in resistance to flow at maximal vasodilatation may account for the augmented vasoconstrictor responses to humoral agents.

When one examines the responsiveness to norepinephrine in vessels with increased wall/lumen ratio, one finds that the threshold dose of norepinephrine is not increased, but the dose–response curve is much steeper (Fig. 4-11). Thus the increase in wall/lumen ratio caused an increase in vascular reactivity based on a structural wall change rather than on the reactivity of smooth muscle. Such a "hypertrophic" change in the vascular architecture would then be expected to sustain the elevated blood pressure in the face of normal neurohumoral constrictor influences. If it occurs in the carotid area, for example, it could contribute to the well-known resetting of the baroreceptors. What causes the "hypertrophic" change in arterioles? One could postulate that an initial excitatory trigger factor may be responsible for vasoconstriction. This could be an intermittent increase in neurogenic drive or a specific hormonal factor, or maybe even in some cases a primary increase in the intrinsic myogenic activity of smooth muscle in resistance vessels. Whichever is the cause, the presence of these excitatory influences on resistance vessels and the rise in arterial pressure will cause an adaptive increase in wall/lumen ratio. When this is established, a given degree of smooth-muscle shortening will cause a greater increase in resistance in hypertensives as compared to normals. The possibility also exists that a primary genetically linked factor in essential hypertension might be simply an increased tendency for the blood vessels to develop increased wall/lumen ratios as an expression of "increased mesenchymal endowment, as seen in subjects with essential hypertension with a mesomorphic body build."[187] The hypothesis ascribing major significance to structural changes in the blood vessels of hypertensive animals has been tested in the spontaneously hypertensive strain of rats (SHR) by Folkow et al.[186] By producing an arterial obstruction and reducing regional arterial pressure to about 50%–70% of the initial high level in the hindquarters of young SHR rats, there were obvious reductions in the wall/lumen ratios, and the resistance curves revealed considerable normalization. Even in old animals preliminary studies done by Folkow indicate that there is considerable regression of the hypertrophic vascular adaptation to high pressure within a period of less than 2 weeks. It appears from these animal models of essential hypertension that aggressive pharmacologic treatment of hypertension when intense enough and early enough to normalize the pressure load would lead

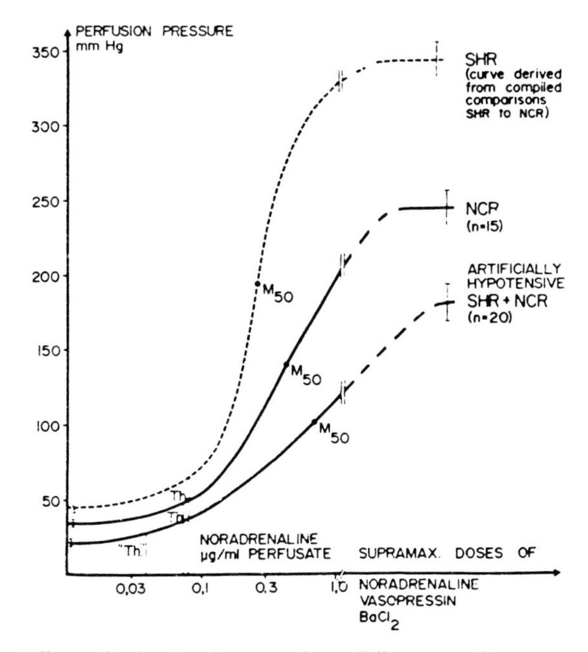

Fig. 4-11. Effect of adaptive increase in wall/lumen ratio on vasoconstrictor responses to norepinephrine. Average changes in perfusion pressure at constant flow in hindquarter vascular beds from spontaneously hypertensive rats (SHR), control rats (NCR), and SHR that had had aortic ligation were done to minimize or reverse the structural increases in wall/lumen ratio in the hindquarters exposed to low arterial pressure (artificially hypotensive SHR and NCR). Responses to norepinephrine are shown. The threshold dose is not decreased in SHR, whereas the slope of the dose-response curve is. In artifically hypertensive SHR and NCR the threshold and the slope were similar. These findings suggest that the increased reactivity in SHR is a function of the structural change rather than an increased "excitability" of vascular smooth muscle. Work by others, however,[188,189] suggests that there is a selective augmentation of vasoconstricting responses to different compounds, which supports a functional alteration in smooth muscle. (Reproduced by permission from Folkow B, et al.[186])

to rapid and considerable regression of the structural increase in wall/lumen ratio and vascular resistance. Thus, given a trigger factor, the morphologic vascular adaptation seems to be a key factor in maintaining the true hypertensive state in primary hypertension, and a very intense treatment might efficiently interrupt the tendency toward the vicious cycle that is inherent in these structural changes.

It is important, however, to appreciate the possibility that functional changes in vascular smooth muscle, which in the first place would make it more responsive to trigger influences, may be playing an important role in hypertension, either acquired renal or genetic. This is suggested by the findings of Hansen and Bohr[188] and Lais et al.,[189] who demonstrated that the augmented responses of vessels of hypertensive rats were specific to certain vasoconstricting interventions, indicating a selective sensitization of smooth muscle. A uniform sensitization to all compounds would be expected if structural changes accounted for the augmentation. There is recent evidence to indicate that in genetic hypertensive rats a specific membrane defect in vascular smooth muscle might account for an increased response.[16,16a]

RHEOLOGIC AND MECHANICAL FACTORS ALTERING RESISTANCE TO BLOOD FLOW

Poiseuille's formula provides a mathematical expression for the definition of the various factors that enter into the determination of vascular resistance to flow. It is strictly valid only for nonpulsatile streamline flow of a fluid of uniform viscosity. Although these requirements are not met in the circulation, the law can be applied in hemodynamics in an approximate fashion. The formula can also be written in a simplified manner as an analogue to Ohm's law for the relationship between voltage, current, and resistance: $Q = (P_1 - P_2)/R$, where $R = 8\,Ln/\pi r^4$, $P_1 - P_2$ = pressure head, R = resistance to flow, r = radius, L = length of cylinder, and n = viscosity of liquid. We have discussed in the preceding sections factors that determine the vascular caliber or the radius r; we shall now address ourselves to a consideration of the role of viscosity. It is apparent from the formula that the radius is the most important determinant of blood flow, since flow is proportional to the fourth power of r. The length of the vessel and viscosity of blood do not change significantly under ordinary circumstances.

Viscosity

Although blood is a non-Newtonian nonhomogenous fluid containing plasma with protein macromolecules and other solutes, it behaves very much like a Newtonian fluid, and the deformable erythrocytes can easily squeeze through capillaries with smaller diameters. At very low shear rates, however, the viscosity may increase, but only by about 30%–50%, provided the blood composition remains unaltered.[190] In abnormal pathologic states such as severe shock the average velocity of flow may be so slow, particularly in the wide postcapillary vessels as compared to the precapillary vessels, that the relative viscosity would be greater in the postcapillary compartment. The resultant reduction in the ratio of precapillary to postcapillary resistance will tend to favor the transcapillary filtration of fluid and loss of intravascular volume.

Viscosity varies relatively little in the range of hematocrits between zero and 40%, but the increase in viscosity is rather steep as the hematocrit increases between 40% and 60%; such an increase in hematocrit would double the relative viscosity and double resistance to blood flow. The effect of hematocrit on relative viscosity is important at the level of resistance vessels and arterioles, but becomes less critical in the microcirculation in capillaries of approximately 6 μ in diameter because the red cells are deformable and can be squeezed readily through pores of 3–5 μ with little additional force. The noncellular components of the blood have predictably similar effects on viscosity. For example, chylomicrons present in large quantities in hyperlipidemia and macroglobulins in Waldenström's macroglobulinemia could increase blood viscosity substantially. The high hematocrits found in polycythemia undoubtedly contribute to increased resistance to blood flow and thus to the increase in blood pressure seen in this condition.

Temperature has a significant influence on viscosity, but any increase would be most apparent in the peripheral parts of the body, which may be exposed to considerable cooling. If the fingers are cooled in ice water, regional viscosity may be increased three times, perhaps even more. If the effects of cooling on vascular resistance are studied in vivo, there will be in addition to the doubling or tripling of viscosity an increase in resistance to flow induced by vasoconstriction related to the release of endogenous catecholamines, most apparent in the cutaneous bed.[191]

Red-Cell Wall Rigidity

If for some reason the easy deformability of the red cells is hindered by an increase in rigidity of the red-cell wall, tissue perfusion could be altered drastically by this physical change alone. It is possible that hypoxia might alter the deformability of red cells and create a significant increase in apparent viscosity.

Clotting-Platelet Aggregation and Thrombosis

A physical obstruction of blood vessels may be induced by clotting or thrombosis. Major vessels such as the pulmonary, coronary, or cerebral vessels may occlude, leading to catastrophic consequences of severely compromised perfusion of these organs.

The role of altered coagulation or platelet aggregation in shock has received much attention. In irreversible stages of shock there may be damage to the endothelial lining of blood vessels and capillaries with subsequent fibrin deposition, accumulation of microthrombi, and intravascular coagulation. When this is coupled with diffuse vasospasm, it may severely restrict tissue perfusion and cause cellular death. Such a phenomenon occurs most dramatically in the disseminated intravascular coagulation associated with renal cortical necrosis caused by Gram-negative septicemia. The generalized reaction mimicks in some ways the Shwartzman reaction.

More recent evidence suggests that the presence of platelet aggregation may correlate with clinical situations associated with intravascular thrombosis, transient cerebral ischemic attacks, primary pulmonary hypertension, and myocardial infarction. The exact role that platelet aggregation may play in such disease states, the possible role of prostaglandin release as a factor in generating this platelet aggregation, and the potential therapeutic indication for the use of aspirin as a prostaglandin synthetase inhibitor all require further evaluation and study.

MECHANISMS INVOLVED IN THE REGULATION OF BLOOD VOLUME

Significant increases in blood volume occur in heart failure and may represent a compensatory mechanism in an attempt to increase filling pressure and ventricular stretch in diastole in order to enhance ventricular performance through the Frank-Starling mechanism and maintain cardiac output. Conversely, reductions in blood volume and filling pressure reduce cardiac output. A careful and tuned regulation of the intravascular volume is an important determinant of cardiac performance and is itself regulated by adjustments in blood vessels and by important neurohumoral factors.

Blood volume is determined by the rate of fluid exchange between the intravascular volume and extracellular fluid, the rate of loss of fluid to the external environment through excretory functions primarily in the urine, and finally the rate of formation of red cells.

The transcapillary fluid exchange depends on the hydrostatic and colloidal osmotic pressure across the capillary. According to the Starling hypothesis, hydrostatic pressure difference across the wall of the capillary favors filtration, and colloidal osmotic pressure difference favors absorption.

Capillary Hydrostatic Pressure

Capillary hydrostatic pressure P_c is dependent upon the ratio between precapillary resistance r_a and postcapillary resistance r_v; so $P_c = \{P_a[(r_v/r_a) + P_v]\}/[1 + (r_v/r_a)]$, where P_a and P_v are the central arterial and venous pressures, respectively. During hypotension or shock there is a reflex increase in precapillary resistance exceeding that seen in postcapillary resistance. This set of circumstances will favor a significant reduction in capillary pressure, absorption of fluid from the extracellular space, and augmentation of the intravascular volume to compensate for hypotension and shock. On the other hand, after exercise, when there is a marked precapillary vasodilatation because the effects of local metabolites on precapillary resistance vessels are more pronounced than those on postcapillary resistance vessels, there is a rise in capillary pressure and hemoconcentration as fluid is lost from the intravascular volume to the extracellular space. True capillary pressure may be 20 mm Hg, but the effective hydrostatic pressure of "exchange vessels," which include postcapillary venules, is lower. In hepatic and pulmonary capillaries the pressure is less than 10 mm Hg, whereas in glomerular capillaries it is 50–70 mm Hg.

Plasma Colloidal Osmotic Pressure

The crystalloids in plasma (300 mOsm/liter) are freely exchanged across the capillary wall and do not contribute significantly to the effective osmotic pressure across the wall. The osmotic pressure counteracts filtration and results primarily from plasma proteins (7 g/100 ml plasma), which exert an osmotic pressure of about 25 mm Hg. Albumin contributes the most to osmotic pressure because of its concentration and lower molecular weight. The interstitial fluid protein concentration is relatively low in skeletal muscle, i.e., 10%–30% of that in plasma, as compared to the intestine and liver with concentrations ranging from 50% to 80% of that in plasma. The loss of protein across the capillaries is returned to the circulation via the lymphatics, so that the plasma colloidal osmotic pressure is maintained. There appears to be a significant regulation of plasma protein synthesis by the liver, since blood loss is followed by an increased production of proteins; but the regulatory mechanisms of the production of proteins are not clear.

Regulation of Fluid Loss

We mentioned earlier that arterial high pressure or cardiopulmonary low pressure may regulate renal vascular resistance and consequently glomerular filtration, renin release, ADH secretion, or the ratio of precapillary to postcapillary resistance in several beds and particularly skeletal muscle. All these effects will regulate intravascular volume.

Gauer et al.[192] have shown that negative-pressure breathing will induce diuresis, and Henry and Pearce have reported that stretch of the left atrial–pulmonary venous junction increases the discharge rate of impulses from the left atrium and also causes a diuretic effect.[193] ADH secretion is inhibited during left atrial stretch, and renal vascular resistance decreases.

It is known that renin secretion will be affected by changes in sympathetic tone, which can result from changes in blood volume and cardiac output. Increases in renin secretion stimulate the secretion of aldosterone, which in turn modulates sodium and fluid retention and provides a compensatory mechanism for the preservation of intravascular volume.

Red-Cell Volume

Erythropoietin is secreted primarily by the kidney and is responsible for red-cell formation. Chronic hypoxia is known to stimulate the secretion of erythropoietin and is associated with polycythemia. The stimulus for erythropoietin formation may be arterial Po_2. A hormonal factor may also play a role.[194]

REGULATION OF THE CORONARY CIRCULATION

The regulation of the circulation in each organ has unique features characteristic of that organ. A discussion of factors that regulate flow to each bed is beyond the scope of this chapter; however, the specific features of the coronary circulation will be discussed because of their potential clinicophysiologic implications.

Neurohumoral Control

Neurohumoral interventions can change coronary resistance, either through vascular effects or indirectly by changing myocardial metabolic needs. Increases in myocardial metabolism resulting from increases in heart rate, arterial pressure, or contractility will cause coronary vasodilatation, whereas decreases in metabolic requirements will cause coronary vasoconstriction. These metabolic effects are often the overriding influence in the control of coronary circulation. More recent evidence has indicated that there are direct vascular effects of neurohumoral stimuli that may play a role in the physiologic regulation of coronary flow. The availability of new specific beta blocking agents such as practolol that antagonize the myocardial stimulating effects of catecholamines, but not their vascular effects, permits the assessment of the magnitude of the direct vascular component.

Neural Control

The stimulation of cardiac sympathetic nerves causes coronary vasodilatation; but after the administration of practolol, which blocks myocardial stimulation and its associated metabolic effects, sympathetic stimulation causes vasoconstriction.[195,196] (Fig. 4-12) Constriction of coronary vessels in response to sympathetic stimulation, however, is small compared to the responses observed in other vascular beds.[197]

There are coronary vasodilator beta receptors that can be demonstrated in the intact beating heart in response to isoproterenol after the administration of practolol.[196,197] These vascular beta receptors do not contribute to the coronary response to nerve stimulation.

The sympathetic cholinergic vasodilator fibers that produce vasodilatation in skeletal muscle during the defense reaction of hypothalamic stimulation and during stimulation of the sympathetic nerve following adrenergic blockade do not seem to have a counterpart in the coronary circulation.[198] On the other hand, there is a parasympathetic vagal in-

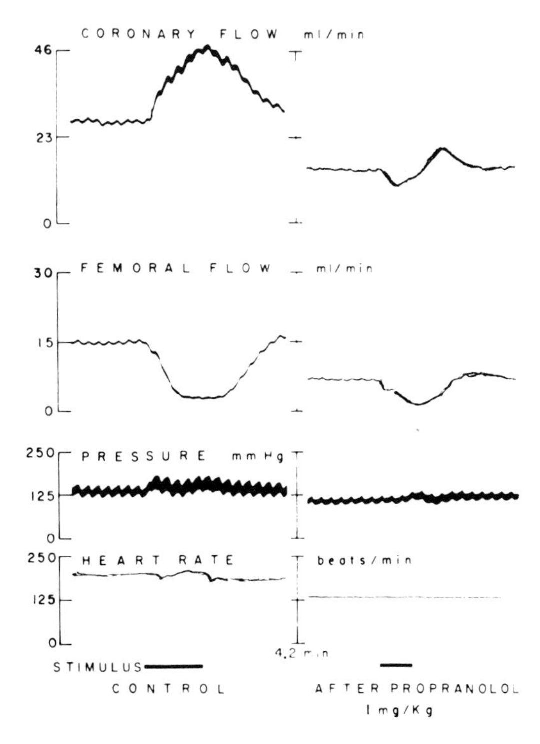

Fig. 4-12. Effects of sympathetic stimulation on femoral and coronary flow. A redistribution of flow away from the extremities and toward the coronary circulation is apparent. (Reproduced by permission of the American Heart Association, Inc. from Feigl EO.[198])

nervation of the coronary vessels that mediates a cholinergic vasodilator response.[199,200] Feigl[200] and Berne et al.[199] have demonstrated by direct vagal stimulation that there are coronary vasodilator responses during parasympathetic stimulation that cannot be accounted for on the basis of changes in myocardial contractility and heart rate.

We have been able to demonstrate in the beating heart that activation of carotid chemoreceptors with nicotine or cyanide produces cholinergic vasodilatation in the coronary vessels that can be blocked by vagotomy.[60] The dilatation resulted from direct neurogenic influences on the coronary vessels and not from metabolic effects. This vasodilator response was associated with a rise in coronary sinus Po_2, indicating that it was not secondary to an increased metabolic requirement for oxygen; furthermore, it was seen after the administration of propranolol.

In studying the effects of carotid sinus nerve stimulation on the coronary circulation in the conscious dog, Vatner and collaborators[201] have demonstrated a decrease in coronary vascular resistance. This vasodilatation results from withdrawal of adrenergic tone, as well as activation of vagal cholinergic fibers.[60,201] It appears from these studies that the direct neurogenic influence on the coronary vessels may predominate over the metabolic influence in the reflex control of the coronary circulation, since with reduction in metabolic requirements associated with carotid sinus nerve stimulation one would have expected coronary vasoconstriction.

Recently[201a] Feigl demonstrated a reflex parasympathetic cholinergic vasodilation in response to chemical activation of cardiac receptors during injection of veratridin into a coronary vessel.

Humoral and Therapeutic Factors

Adrenergic receptors. Studies on isolated vessels by Bohr et al.[202,203] suggest that beta receptors in the coronary vessels behave more like those in myocardium than like those in peripheral vessel and are of the beta-1 rather than the beta-2 type. In contrast, our results and those of others in the beating heart[196,204,205] indicate that the beta receptors in the coronaries do resemble those in other peripheral vessels and are not blocked by small doses of the beta-1 blocker practolol.

As with nerve stimulation (Fig. 4-13), the intracoronary injection of norepinephrine produces vasodilatation before practolol and vasoconstriction after blocade of the myocardial effects of norepinephrine. Epinephrine, on the other hand, produces vasodilatation, both indirectly and directly. After practolol, the intracoronary injection of epinephrine causes vasodilatation, which is subsequently blocked by administration

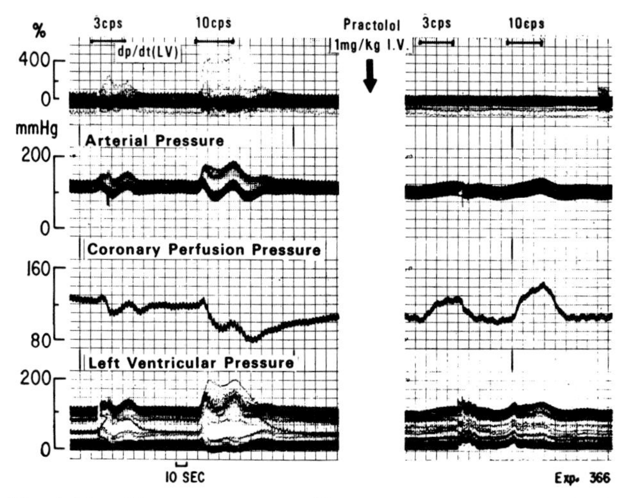

Fig. 4-13. The coronary vasodilator response to sympathetic nerve stimulation (left panel) is reversed following the administration of practolol (right panel), which blocks the myocardial stimulation and associated metabolic effects. The direct coronary vasoconstrictor effect is unmasked. (Reproduced by permission from McRaven DR, et al.[196])

of propranolol. These observations suggest that the predominant direct effect of epinephrine is dilatation mediated through beta receptors; the constictor effect may then be unmasked after the beta receptors are blocked by propranolol.

There are specific dopamine receptors in the coronaries that contribute to the coronary vasodilator effect of that compound.[206] Dopamine also activates myocardial beta receptors and causes a direct vasoconstrictor effect by activating vascular alpha receptors, but that effect is not a dominant one.

Propranolol. Propranolol is a commonly used drug for the treatment of angina. It is known to increase coronary vascular resistance by blocking myocardial beta receptors, thereby decreasing metabolic demands, and by blocking coronary dilator beta receptors. The magnitude of the increase in coronary resistance after propranolol will depend on the preexisting level of beta receptor activity in the heart, so that in resting unanesthetized animals the increase in resistance would not be noted.[207] The major effect of the drug is through blockade of myocardial rather than coronary beta receptors, although the latter may contribute to the effect. Theoretically a drug like practolol, which blocks myocardial but spares coronary beta receptors, might be associated with less vasoconstriction than propranolol during circulatory stresses associated with the release of endogenous catecholamines, such as exercise.

Because propranolol increases coronary vascular resistance, its beneficial effect in angina pectoris must be through a reduction in myocardial oxygen consumption and not an increase in myocardial oxygen delivery. The expected coronary vasoconstriction with a reduction in blood flow after propranolol does not take place in the ischemic myocardium.[208,209] Furthermore, it has been reported that propranolol increases the transmural distribution of flow to ischemic subendocardium.[209] A possible explanation for these effects is that vasoconstriction occurs after propranolol in the normally perfused myocardium, with redistribution of flow favoring the ischemic region.

Vasopressin and angiotensin. These are very potent nonadrenergic vasoconstrictor agents and may contribute to myocardial ischemia in pathophysiologic states. An understanding of the relative effects of a stimulus or a drug on different vascular beds is important, because a quantitatively different effect on various beds may induce a redistribution of blood flow to various organs. For example, the constrictor effect of adrenergic stimuli on coronary vessels is considerably less than that in the extremities; therefore the differential effect would favor redistribution of flow toward the coronaries and away from the extremities. In contrast, the coronary vessels are more sensitive to angiotensin and vasopressin than are vessels in skin and skeletal muscle; these stimuli might redistribute blood flow away from the heart.[164]

Prostaglandins. The pathophysiologic significance of the coronary effects of prostaglandins is not known and may be minor in relation to the indirect effects on the coronary circulation that accompany systemic and hemodynamic effects of prostaglandins. The elaboration of prostaglandins from the myocardium, however, may be an important determinant of the coronary vasodilator response to local ischemia and hypoxia and may play an important role in the pathogenesis of angina.

Ouabain. The administration of ouabain causes direct coronary vasoconstriction, as well as constriction in peripheral vessels.[210] Increases in coronary and systemic vascular resistance may precede the beneficial inotropic effect of this compound. The intravenous administration of rapidly acting digitalis preparations may carry some hazard by exaggerating myocardial ischemia or increasing afterload, at least transiently.

Nitroglycerin. Although nitroglycerin is a potent coronary dilator, it does not consistently increase total coronary flow in patients with coronary artery disease.[211,212] It is now well known that one of its most potent actions is to reduce myocardial oxygen demand by decreasing arterial pressure and ventricular volume through its effect on peripheral resistance and capacitance vessels. The drug also produces a favorable redistribution of flow within the myocardium by increasing nutritional flow[213] and by redistributing blood flow from epicardium to ischemic endocardium.[209,214]

Ganz and Marcus[215] have reported that the direct effects of nigroglycerin on coronary vessels play little role in the relief of angina pectoris. In their study the administration of nitroglycerin into the obstructed artery or the artery supplying collateral vessels did not relieve angina pectoris, despite the fact that coronary blood flow increased. In contrast, intravenous administration consistently relieved angina in association with a fall in arterial pressure and reduction in coronary flow. The possibility that decreases in wall tension with intravenous or sublingual nitroglycerin may redistribute flow and increase perfusion of ischemic myocardium in addition to decreasing metabolic demand should still be considered.

Relationship of Neurohumoral and Metabolic Determinants of Coronary Blood Flow

Hypoxia

Severe hypoxia is a potent coronary vasodilator. In a recent study Heistad et al.[27] demonstrated that moderate hypoxia (Po_2 40 mm Hg), which produced only minimal coronary vasodilatation, inhibited vasoconstrictor responses to angiotensin and norepinephrine, whereas the same degree of moderate hypoxia did not inhibit these responses in skeletal muscle. When skeletal muscle was made to contract by direct electrical stimulation, the same level of moderate hypoxia caused an inhibition of the vasoconstrictor responses. This inhibition seen when hypoxemia was accompanied by muscular contraction did not appear to be caused by the release of prostaglandins. The selective inhibition of coronary vasoconstrictor responses seen in heart but not in skeletal muscle at rest would favor redistribution of blood flow to the heart during the release of endogenous vasoconstrictor hormones and during the administration of vasoconstrictor drugs to hypoxemic patients.

When severe hypoxemia is induced, coronary vasodilatation is striking, and inhibition of vasoconstrictor responses is also maximal.

Therefore, during hypoxemia or ischemia maximal dilatation of coronary resistance vessels is maintained regardless of constricting neurohumoral influences, so that coronary flow becomes regulated primarily by aortic diastolic pressure. This concept has important clinical implications. Mueller et al.[216,217] have demonstrated that norepinephrine increases coronary flow in patients with cardiogenic shock, presumably by increasing arterial pressure. Although isoproterenol would produce greater increases in cardiac output than norepinephrine under such circumstances, it would also have a variable effect on coronary flow because it decreases aortic diastolic pressures through its potent vasodilating effect in other vascular beds. Thus small increases in coronary flow with isoproterenol are not adequate to meet the increased metabolic demands of the heart. Myocardial lactate production increases, indicating an increase in anaerobic metabolism. In contrast, during administration of norepinephrine, myocardial lactate metabolism improves as coronary perfusion improves consonant with the rise in arterial pressure.

Heart Failure

Myocardial stores of norepinephrine are depleted and cardiac responses to sympathetic nerve stimulation are reduced in heart failure, but histochemical studies suggest that adrenergic innervation of the coronary vessels is preserved.[218,219] If myocardial and indirect coronary vasodilator effects of sympathetic nerve stimulation are attenuated in heart failure, while direct constrictor effects are preserved, then stimuli such as carotid sinus hypotension could produce reflex coronary vasoconstriction instead of reflex vasodilatation. The indirect coronary dilator responses to circulating catecholamines, however, might be preserved, since myocardial beta receptors are intact in heart failure. However, the contractile response might be limited. Additional work is necessary to define the relative contribution of the direct vascular effect in heart failure.

Summary

Several concepts that should be kept in mind when thinking about the regulation of coronary circulation are as follows:

1. The effect of neurohumoral stimuli should be considered in terms of the direct and indirect effects and in terms of the balance between oxygen delivery and oxygen demand. A decrease in coronary blood flow and oxygen delivery may under certain circumstances be associated with greater reductions in myocardial oxygen requirement through effects on rate and contractility of blood pressure and may result in a favorable balance between oxygen delivery and demand.

2. The coronary vessels are exquisitely sensitive to their metabolic environment, so that stimuli that alter the chronotropic and inotropic states, as well as the metabolic requirements of the myocardium, will influence vasomotor tone markedly. Increases in metabolic requirements decrease coronary vascular resistance; conversely, decreases in metabolic demands increase vascular coronary resistance. In general, the indirect and metabolic effects override the direct vascular effects, but there are several exceptions that may be critical in certain pathophysiologic states.

3. There are striking contrasts between the effects of various neurohumoral stimuli on coronary vessels and other vessel. The differential responses of the vascular beds determine to a large extent the distribution of cardiac output and coronary flow. A stimulus that constricts the coronary vessels may still redistribute cardiac output in such a way as to increase coronary flow if pronounced constriction occurs in other vascular beds. Similarly, a stimulus that causes coronary vasodilatation might reduce coronary flow if the vasodilator responses of other parallel vascular beds are of greater magnitude, thus causing a fall in arterial pressure.

4. Diastolic pressure is a critical determinant of coronary blood flow, particularly in the presence of reactive hyperemia in ischemic segments. In the presence of myocardial ischemia the effects of various interventions on myocardial flow will depend predominantly on their effects on arterial blood pressure.

5. It is becoming evident that neurohumoral stimuli may alter intramyocardial distribution of blood flow from nonischemic to ischemic myocardium and from epicardium to endocardium. These effects on transmural distribution may result from changes in ventricular volume and tension, as well as from direct effects on different parallel segments of the coronary vasculature.

REFERENCES

1. Johansson B: Relationship between electrical and mechanical activity in vascular smooth muscle. Proceedings of the 25th International Congress of the International Union of Physiological Scientists. Munich, German Physiological Society 8:51, 1971
2. Steedman WM: Micro-electrode studies of mammalian vascular muscle. J Physiol 186:382, 1966
3. Speden RN: Electrical activity of single smooth muscle cells of the mesenteric artery produced by splanchnic nerve stimulation in the guinea pig. Nature 202:193, 1964

4. Holman ME: Electrophysiology of vascular smooth muscle. Ergeb Physiol 61:137, 1969
5. Somlyo AP, Somlyo AV: Vascular smooth muscle: I. Normal structure, pathology, biochemistry, and biophysics. Pharmacol Rev 20:197, 1968
6. Somlyo AV, Vinallp P, Somlyo AP: Excitation-contraction coupling and electrical events in two types of vascular smooth muscle. Microvasc Res 1:354, 1969
7. van Breeman C, Farinas BR, Gerba P, et al: Excitation-contraction coupling in rabbit aorta studied by the lanthanium method for measuring cellular calcium influx. Circ Res 30:44, 1972
8. Haddy FJ: Local control of vascular resistance as related to hypertension. Arch Intern Med 133:916, 1974
9. Haddy FJ, Scott JB, Florio MA, et al: Local vascular effects of hypokalemia, alkalosis, hypercalcemia, and hypomagnesemia. Am J Physiol 204:202, 1963
10. Konald P, Gebert G, Brecht K: The effect of potassium on the tone of isolated arteries. Pfluegers Arch 301:285, 1968
11. Scott JB, Daugherty RM, Overbeck HW, et al: Vascular effects of ions. Fed Proc 27:1403, 1968
12. Kuriyama H, Ohshima F, Sakamoto Y: The membrane properties of the smooth muscle of the guinea-pig portal vein in isotonic and hypertonic solutions. J Physiol 217:179, 1971
13. Axelsson J, Wahlström B, Johansson B, et al: Influence of the ionic environment on spontaneous electrical and mechanical activity of the rat portal vein. Circ Res 21:609, 1967
14. Heistad DD, Abboud FM, Ballard DR: Relationship between plasma sodium concentration and vascular reactivity in man. J Clin Invest 50:2022, 1971
15. Somlyo AP, Somlyo AV: Vascular smooth muscle: II. Pharmacology of normal and hypertensive vessels. Pharmacol Rev 22:249, 1970
16. Jones AW: Reactivity of ion fluxes in rat aorta during hypertension and circulatory control. Fed Proc 33:133, 1974
16a. Hermsmeyer K: Electrogenesis of increased norepinephrine sensitivity of arterial vascular muscle in hypertension. Circ Res 1975 (in press)
17. Somlyo AV, Somlyo AP: Strontium accumulation by sacroplasmic reticulum and mitochondria in vascular smooth muscle. Science 174:955, 1971
18. Sparrow MP, Maxwell LC, Ruegg JC, et al: Preparation and properties of a calcium ion-sensitive actomyosin from arteries. Am J Physiol 219:1366, 1970
19. Murphy RA: Arterial actomyosin: Effects of pH and temperature on solubility and ATPase activity. Am J Physiol 220:1494, 1971
20. Maxwell LC, Bohr DF, Murphy RA: Arterial actomyosin: Effects of ionic strength on ATPase activity and solubility. Am J Physiol 220:1871, 1971
21. Hess ML, Ford GD, Griggs FN: Calcium uptake in subcellular fractions of vascular smooth muscle. Physiologist 15:165, 1972
22. Bohr DF, Sitrin M: Regulation of vascular smooth muscle contraction: Changes in experimental hypertension. Circ Res 26, 27 [Suppl II]:1970

23. Somlyo AV, Somlyo AP: Electromechanical and pharmacomechanical coupling in vascular smooth muscle. J Pharmacol Exp Ther 159:129, 1968
24. Bohr DF, Uchida E: Activation of vascular smooth muscle, in Fishman A, Hecht H (eds): The Pulmonary Circulation and Interstitial Space. Chicago, University of Chicago Press, 1969, pp 133–145
25. Katz AM, Hecht HH: The early "pump" failure of the ischemic heart. Am J Med 47:497, 1969
26. Detar R, Bohr DF: Adaptation to hypoxia in vascular smooth muscle. Fed Proc 27:1416, 1968
27. Heistad DD, Abboud FM, Mark AL, Schmid PG: Effect of hypoxemia on responses to norepinephrine and angiotensin in coronary and muscular vessels. J Pharmacol Exp Ther 193:941, 1975
28. Heistad DD, Wheeler RC: Effect of acute hypoxia on vascular responsiveness in man. J Clin Invest 49:1252, 1970
29. Johansson B: Determinants of vascular reactivity. Fed Proc 33:121, 1974
30. Alexander RS: Tonic and reflex functions of medullary sympathetic cardiovascular centers. J Neurophysiol 9:205, 1946
31. Folkow B, Neil E: Circulation. New York, Oxford University Press, 1971, pp 307–319, 320–340
32. Mayer HE, Abboud FM, Ballard DR, et al: Catecholamines in arteries and veins of the foreleg of the dog. Circ Res 23:653, 1968
33. Weinshilboum R, Axelrod J: Serum dopamine-beta-hydroxylase activity. Circ Res 28:307, 1971
34. Rolett ES: Adrenergic mechanisms in mammalian myocardium, in Langer GA, Brady AJ (eds): The Mammalian Myocardium. New York, John Wiley & Sons, 1974, pp 29–250
35. Abboud FM, Eckstein JW: Comparative changes in segmental vascular resistance in response to nerve stimulation and to norepinephrine. Circ Res 18:263, 1966
36. Abdel-Sayed WA, Abboud FM, Ballard DR: Contribution of venous resistance to total vascular resistance in skeletal muscle. Am J Physiol 218:1291, 1970
37. Irving L, Scholander PF, Grinnell SW: The regulation of arterial blood pressure in the seal during diving. Am J Physiol 135:557, 1941
38. Heistad DD, Abboud FM, Eckstein JW: Vasoconstrictor response to simulated diving in man. J Appl Physiol 25:542, 1968
39. Abboud FM: Control of the various components of the peripheral vasculature. Fed Proc 31:1226, 1972
40. Randall WC, Armour JA, Geis WP, Lippincott DB: Regional cardiac distribution of the sympathetic nerves. Fed Proc 31:1199, 1972
41. Randall DC, Smith OA: Heart rate, pressure, and myocardial contractility responses to exercise and emotional conditioning in the nonhuman primate. Physiologist 14:213, 1971
42. Armour JA, Hageman GR, Randall WC: Arrhythmias induced by local cardiac nerve stimulation. Am J Physiol 223:1068, 1972
43. Davis JO: The control of renin release. Am J Med 55:333, 1973
44. Winer N, Chokshi DS, Yoon MS, et al: Adrenergic receptor mediation of renin secretion. J Clin Endocrinol Metab 29:1168, 1969

45. Slick GL, Aguilera AJ, Zambraski EJ, et al: Renal neuroadrenergic transmission. Am J Physiol (in press)
46. Hamlin RL, Smith CR: Effect of vagal stimulation on S-A and A-V nodes. Am J Physiol 215:560, 1968
47. Roskoski R Jr, Schmid PG, Mayer HE, Abboud FM: In vitro acetylcholine biosynthesis in vitro in normal and failing guinea pig hearts. Circ Res 36:547, 1975
48. Hirsch EF, Kaiser GC, Cooper T: Experimental heart block in the dog: I. Distribution of nerves, their ganglia and terminals in septal myocardium of dog and human hearts. Arch Pathol 78:522, 1964
49. Levy MN, Ng ML, Zieske H: Sympathetic and parasympathetic interactions upon the left ventricle of the dog. Circ Res 19:5, 1966
50. Widenthal K, Mierzwiak DS, Mitchell JH: Influence of efferent vagal stimulation on left ventricular function in the dog. Am J Physiol 216:577, 1969
51 Burgen ASV, Terroux KG: Membrane potentials of the cat auricle and their response to parasympathetic agents. Fed Proc 10:21, 1951
52. La Rai PJ, Sonnenblick EH: Autonomic control of cardiac C-AMP. Circ Res 28:377, 1971
53. Volle RL: Pharmacology of the autonomic nervous system. Ann Rev Pharmacol 3:129, 1963
54. Levy MN: Brief reviews: Sympathetic–parasympathetic interactions in the heart. Circ Res 29:437, 1971
55. Iano TL, Levy MN, Lee MH: An acceleratory component of the parasympathetic control of heart rate. Am J Physiol 224:997, 1973
56. Scher AM, Ohm WW, Bumgarner K, et al: Sympathetic and parasympathetic control of heart rate in the dog, baboon and man. Fed Proc 31:1219, 1972
57. Eckberg DL, Fletcher GF, Braunwald E: Mechanism of prolongation of the R-R interval with electrical stimulation of the carotid sinus nerves in man. Circ Res 30:131, 1972
58. Eckberg DL, Abboud FM, Mark AL: Baroreflex augmentation by upright posture after β-adrenergic blockade in man. Circ 52 (Suppl II): 52, 1975
59. Feigl EO: Parasympathetic control of coronary blood flow in dogs. Circ Res 25:509, 1969
60. Hackett JG, Abboud FM, Mark AL, et al: Coronary vascular responses to stimulation of chemoreceptors and baroreceptors; Evidence for reflex activation of vagal cholinergic innervation. Circ Res 31:8, 1972
61. Abrahams VC, Hilton SM, Zbrozyna AW: The role of active muscle vasodilatation in the alerting stage of the defence reaction. J Physiol (Lond) 171:189, 1964
62. Uvnas B: Cholinergic vasodilator nerves. Fed Proc 25:1618, 1966
63. Lindgren P: Mesencephalon and the vasomotor system. Acta Physiol Scand 35 [Suppl 121]:1, 1955
64. Takeuchi T, Manning JW: Muscle cholinergic dilators in the sinus baroreceptor response in cats. Circ Res 29:350, 1971

65. Schramm LP, Honig CR, Bignall KE: Active muscle vasodilatation in primates homologous with sympathetic vasodilation in carnivores. Am J Physiol 221:768, 1971
66. Schramm LP, Bignall KE: Central neural pathways mediating active sympathetic muscle vasodilation in cats. Am J Physiol 221:754, 1971
67. Blair DA, Glover WE, Greenfield ADM, et al: Excitation of cholinergic vasodilator nerves to human skeletal muscles during emotional stress. J Physiol (Lond) 148:633, 1959
68. Abboud FM, Eckstein JW: Active reflex vasodilatation in man. Fed Proc 25:1611, 1966
69. Mason DT, Braunwald E: Effects of guanethidine, reserpine and methyldopa on reflex venous and arterial constriction in man. J Clin Invest 43:1449, 1964
70. Ballard DR, Abboud FM, Mayer HE: Release of a humoral vasodilator substance during neurogenic vasodilatation. Am J Physiol 219:1451, 1970
71. Calvelo MG, Abboud FM, Ballard DR, Abdel-Sayed WA: Reflex vascular responses to stimulation of chemoreceptors with nicotine and cyanide. Circ Res 27:259, 1970.
72. Brody MJ: Neurohumoral mediation of active reflex vasodilatation. Fed Proc 25:1583, 1966
73. Beck L: Active reflex dilatation in the innervated perfused hind leg of the dog. Am J Physiol 201:123, 1961
74. Sweet CS, Brody MJ: Central inhibition of reflex vasodilatation by angiotensin and reduced renal pressure. Am J Physiol 219:1751, 1970
75. Brody, MJ, Dorr LD, Shaffner RA: Reflex vasodilatation and sympathetic transmission in the renal hypertensive dog. Am J Physiol 219:1746, 1970
76. Viveros OH, Garlick DG, Renkin EM: Sympathetic beta adrenergic vasodilatation in skeletal muscle of the dog. Am J Physiol 215:1218, 1968
77. Fox RH, Hilton SM: Bradykinin formation in human skin as a factor in heat vasodilatation. J Physiol (Lond) 142:219, 1958
78. Heymans C, Neil E: Reflexogenic Areas of the Cardiovascular System. London, J & A Churchill, 1958
79. Bronk DW, Stella G: Afferent impulses in carotid sinus nerve: Relation of discharge from single end organs to arterial blood pressure. J Cell Physiol 1:113, 1932
80. Pelletier C, Shepherd JT: Circulatory reflexes from mechanoreceptors in the cardio-aortic area. Circ Res 33:131, 1973
81. McCubbin JW, Green JH, Page IH: Baroreceptor function in chronic renal hypertension. Circ Res 4:205, 1956
82. Kezdi P, Wennemark JR: Baroreceptor and sympathetic activity in experimental renal hypertension in the dog. Circulation 17:785, 1958
83. Epstein SE, Beiser GD, Goldstein RE, et al: Circulatory effects of electrical stimulation of the carotid sinus nerves in man. Circulation 40:269, 1969
84. Abboud FM, Heistad DD, Mark AL, Schmid PG: Effect of lower body negative pressure on capacitance vessels of forearm and calf. Fed Proc 32:396, 1973

85. Mark AL, Eckberg D, Abboud FM: Selective contribution of cardiopulmonary and carotid baroreceptors to forearm and splanchnic vasoconstrictor responses during venous pooling in man. Physiologist 18:305, 1975
86. Abboud FM, Mark AL, Heistad DD, Schmid PG: Selectivity of autonomic control of the peripheral circulation in man. Trans Am Clin Climatol Assoc 86:184, 1974
87. Eckberg DL, Cavanaugh MS, Mark AL, Abboud FM: A simplified neck suction device for activation of carotid baroreceptors. J Lab Clin Med 85:167, 1975
88. DeGeest H, Levy MN, Zieske H Jr: Carotid sinus baroreceptor reflex effects upon myocardial contractility. Circ Res 15:327, 1964
89. Glick G: Importance of the carotid sinus baroreceptors in the regulation of myocardial performance. J Clin Invest 50:1116, 1971
90. Vatner SF, Higgins CB, Franklin D, Braunwald E: Extent of carotid sinus regulation of the myocardial contractile state in conscious dogs. J Clin Invest 51:995, 1972
91. Levy MN, Zieske H Jr: Effect of enhanced contractility on the left ventricular response to vagus nerve stimulation in dogs. Circ Res 24:303, 1969
92. Paintal AS; Cardiovascular receptors, in Neil E (ed): Handbook of Sensory Physiology, vol 3 Enteroceptors. New York, Springer-Verlag, 1972, pp 1–45
93. Paintal AS: A study of ventricular pressure receptors and their role in the Bezold reflex. Q J Exp Physiol 40:348, 1955
94. Coleridge HM, Coleridge JCG, Kidd C: Cardiac receptors in the dog, with particular reference to two types of afferent ending in the ventricular wall. J Physiol 174:323, 1964
95. Paintal AS: A study of right and left atrial receptors. J Physiol 120:596, 1953
96. Sleight P, Widdicombe JG: Action potentials in fibres from receptors in the epicardium and myocardium of the dog's left ventricle. J Physiol 181:235, 1965
96a. Öberg B, Thorén P: Studies on left ventricular receptors, signaling in nonmedullated vagal afferents. Acta Physiol Scand 85:164, 1972
97. Sato A, Fidone S, Eyzaguirre C: Presence of chemoreceptor and baroreceptor C-fibers in the carotid nerve of the cat. Brain Res 11:459, 1968
98. Uchida Y, Murao S: Excitation of afferent cardiac sympathetic nerve fibers during coronary occlusion. Am J Physiol 226:1094, 1974
99. Brown AM, Malliani A: Spinal sympathetic reflexes initiated by coronary receptors. J Physiol 212:685, 1971
100. Öberg B, Thorén P: Circulatory responses to stimulation of medullated and non-medullated afferents in the cardial nerve in the cat. Acta Physiol Scand 87:121, 1973
101. Jarisch A, Zotterman Y: Depressor reflexes from the heart. Acta Physiol Scand 16:31, 1948

102. Peterson DF, Brown AM: Pressor reflexes produced by stimulation of afferent fibers in the cardiac sympathetic nerves of the cat. Circ Res 28:605, 1971
103. Malliani A, Parks M, Tuckett RP, Brown AM: Reflex increases in heart rate elicited by stimulation of afferent cardiac sympathetic nerve fibers in the cat. Circ Res 32:9, 1973
104. Ledsome JR, Linden RJ: A reflex increase in heart rate from distension of the pulmonary-vein-atrial junctions. J Physiol 170:456, 1974
105. Linden RJ: Function of cardiac receptors. Circulation 48:463, 1973
106. Carswell F, Hainsworth R, Ledsome JR: The effects of distension upon peripheral vascular resistance. J Physiol 207:1, 1970
107. Karim F, Kidd C, Malpus CM, Penna PE: The effects of stimulation of the left atrial receptors on sympathetic efferent nerve activity. J Physiol 227:243, 1972
108. Greenberg TT, Richmond WH, Stocking RA, et al: Impaired atrial receptor responses in dogs with heart failure due to tricuspid insufficiency and pulmonary artery stenosis. Circ Res 32:424, 1973
109. Mancia G, Donald DE, Shepherd JT: Inhibition of adrenergic outflow to peripheral blood vessels by vagal afferents from the cardiopulmonary region in the dog. Circ Res 33:713, 1973
109a. Mancia G, Romero JC, and Shepherd JT: Continuous inhibition of renin release in dogs by vagally innervated receptors in the cardiopulmonary region. Circ Res 36:529, 1975
110. Zoller RP, Mark AL, Abboud, FM, et al: The role of low pressure baroreceptors in reflex vasoconstrictor responses in man. J Clin Invest 51:2967, 1972
111. Johnson JA, Moore WW, Segar WE: Small changes in left atrial pressure and plasma antidiuretic hormone sites in dogs. Am J Physiol 217:210, 1969
112. Brennan LA Jr, Malvin RL, Jochim KE, et al: Influence of right and left atrial receptors in plasma concentrations of ADH and renin. Am J Physiol 221:273, 1971
113. Öberg B, White S: Circulatory effects of interruption and stimulation of cardiac vagal afferents. Acta Physiol Scand 80:383, 1970
114. Dawes GS, Comroe JH: Chemoreflexes from the heart and lungs. Physiol Rev 34:167, 1954
115. Salisbury PF, Cross CE, Rieben PA: Reflex effects of left ventricular distension. Circ Res 8:530, 1960
116. Mark AL, Abboud FM, Schmid PG, Heistad DD: Reflex vascular responses to left ventricular outflow obstruction and activation of ventricular baroreceptors in dogs. J Clin Invest 52:1147, 1973
117. Mark AL, Abboud FM, Heistad DD, et al: Evidence against the presence of ventricular chemoreceptors activated by hypoxia and hypercapnia. Am J Physiol 227:178, 1974
118. Eckberg DL, White CW, Kioschos JM, Abboud FM: Mechanisms mediating bradycardia during coronary arteriography. J Clin Invest 54:1455, 1974

119. Sleight P, Lall A, Muers M: Reflex cardiovascular effects of epicardial stimulation by acety strophanthidin in dogs. Circ Res 25:705, 1969
120. Adgey AAJ, Geddes JS, Mulholland HC, et al: Incidence, significance and management of early bradyarrhythmia complicating acute myocardial infarction. Lancet 23:1097, 1968
121. Kolatat R, Ascanio G, Tallarida RJ, Oppenheimer MJ: Action potentials in the sensory vagus at the time of coronary infarction. Am J Physiol 213:71, 1967
122. Thoren P: Left ventricular receptors activated by severe asphyxia and by coronary artery occlusion. Acta Physiol Scand 85:455, 1972
123. Toubes DB, Brody MJ: Inhibition of reflex vasoconstriction after experimental coronary embolization in the dog. Circ Res 26:211, 1970
124. Mark AL, Kioschos JM, Abboud FM, et al: Abnormal vascular responses to exercise in patients with aortic stenosis. J Clin Invest 52:1138, 1973
125. Shepherd JT: The cardiac catheter and the American Heart Association. Circulation 50:418, 1974
126. Paintal AS: Mechanism of stimulation of type J pulmonary receptors. J Physiol 203:511, 1969
127. Brodie TG, Russell AE: On reflex cardiac inhibition. J Physiol 26:92, 1900
128. Daly M De B, Robinson BH: An analysis of the reflex systemic vasodilator response elicited by lung inflation in the dog. J Physiol 195:387, 1968
129. Ott, NT, Shepherd JT: Vasodepressor reflex from lung inflation in the rabbit. Am J Physiol 221:889, 1971
130. Stern S, Rapaport E: Comparison of the reflexes elicited from combined or separate stimulation of the aortic and carotid chemoreceptors on myocardial contractility, cardiac output and systemic resistance. Circ Res 20:214, 1967
131. Ponte J, Purves MJ: The role of the carotid body chemoreceptors and carotid sinus baroreceptors in the control of cerebral blood vessels. J Physiol 237:315, 1974
131a. Heistad DD, Marcus ML, Ehrhardt JC, Abboud FM: Effect of stimulation of carotid chemoreceptors on total and regional cerebral blood flow. Circ Res 1976 (in press)
132. Heistad DD, Abboud FM, Mark AL, Schmid PG: Response of muscular and cutaneous vessels to physiologic stimulation of chemoreceptors. Proc Soc Exp Biol Med 148:198, 1975
133. Pelletier CL, Shepherd JT: Venous responses to stimulation of carotid chemoreceptors by hypoxia and hypercapnia. Am J Physiol 223:97, 1972
134. Freyschuss U: Cardiovascular adjustment to somatomotor activation. Acta Physiol Scand 79:63, 1970
135. Lind AR, Taylor SH, Humphreys PW, et al: The circulatory effects of sustained voluntary muscle contraction. Clin Sci 27:229, 1964
136. Coote JH, Hilton SM, Perez-Gonzalez JF: The reflex nature of the pressor response to muscular exercise. J Physiol 215:789, 1971
137. McCloskey DI, Mitchell JH: Reflex cardiovascular and respiratory responses originating in exercising muscle. J Physiol 224:173, 1972

138. Clement DL, Pelletier CL, Shepherd JT: Role of muscular contraction in the reflex vascular responses to stimulation of muscle afferents in the dog. Circ Res 33:386, 1973
139. Folkow B, Rubinskin EH: Cardiovascular effects of acute and chronic stimulation of the hypothalamic defense area in the rat. Acta Physiol Scand 68:48, 1966
140. Bronk DW, Pitts, RF, Larrabee MG: Role of hypothalamus in cardiovascular regulation. Assoc Res Nerv Ment Dis Proc 20:323, 1940
141. Gebber GL, Snyder DW: Hypothalamic control of baroreceptor reflexes. Am J Physiol 218:124, 1970
142. Bagshaw RJ, Iizuka M, Peterson LN: Effect of interaction of the hypothalamus and the carotid sinus mechanoreceptor system on renal hemodynamics in the anesthetized dog. Circ Res 29:569, 1971
143. Zehr JE, Feigl EO: Suppression of renin activity by hypothalamic stimulation. Circ Res 32, 33:1–17, 1973
143a. Nathan MA, Reis DJ: Fulminating arterial hypertension with pulmonary edema from release of adrenomedullary catecholamines after lesions of the anterior hypothalamus in the rat. Circ Res 37:226, 1975
143b. Pereda S, Eckstein JW, Abboud FM: Cardiovascular responses to insulin in the absence of hypoglycemia. Am J Physiol 202:249, 1962
144. Miura M, Reis DJ: A blood pressure response from fastigial nucleus and its relay pathway in brainstem. Am J Physiol 219:1330, 1970
145. Lisander B, Martner J: Interaction between the fastigial pressor response and the baroreceptor reflex. Acta Physiol Scand 83:505, 1971
146. Humphrey DR: Neuronal activity in the medulla oblongata of cat evoked by stimulation of the carotid sinus nerve, in Kezdi P (ed): Baroreceptors and Hypertension. New York, Pergamon, p 131, 1965
147. Miura M, Reis DJ: The role of the solitary and paramedian reticular nucleii in mediating cardiovascular reflex responses from carotid baro- and chemoreceptors. J Physiol 223:525, 1972
148. Heistad DD, Abboud FM, Mark AL, Schmid PG: Interaction of baroreceptor and chemoreceptor reflexes. J Clin Invest 53:1226, 1974
149. Miura M, Reis DJ: The paramedian reticular nucleus: A site of inhibitory interaction between projections from fastigial nucleus and carotid sinus nerve acting on blood pressure. J Physiol 216:441, 1971
150. Nathan MA: Vasomotor projection of the nucleus fastigii to the medulla. Brain Res 41:194, 1972
151. Lisander B, Martner J: Cerebellar suppression of the autonomic components of the defence reaction. Acta Physiol Scand 81:84, 1971
152. Gebber GL, Taylor DG, Weaver LC: Electrophysiological studies on organization of central vasopressor pathways. Am J Physiol 224:470, 1973
152a. Koike H, Mark AL, Heistad DD, et al: Influence of cardiopulmonary vagal afferent activity on carotid chemoreceptor and baroreceptor reflexes in the dog. Circ Res 37:422, 1975
153. Furchgott RF: The pharmacological difference of adrenergic receptors. Ann NY Acad Sci 139:553, 1967

154. Mark AL, Abboud FM, Schmid PG, et al: Differences in direct effects of adrenergic stimuli on coronary, cutaneous, and muscular vessels. J Clin Invest 51:279, 1972
155. Genest J, Simard S, Rosenthal J, et al: Norepinephrine and renin content in arterial tissue from different vascular beds. Can J Physiol Pharmacol 47:87, 1969
156. Scroop GS, Lowe RD: Efferent pathways of the cardiovascular response to vertebral artery infusions of angiotensin in the dog. Clin Sci 37:605, 1969
157. Ueda H, Uchida Y, Ueda K, et al: Centrally mediated vasopressor effect of angiotensin II in man. Jap Heart J 10:243, 1969
158. Zimmerman BG: Evaluation of peripheral and central components of action of angiotensin on the sympathetic nervous system. J Pharmacol Exp Ther 158:1, 1967
159. Folkow B, Johansson B, Mellander S: The comparative effect of angiotensin and noradrenaline on consecutive vascular sections. Acta Physiol Scand 53:99, 1961
160. Abboud FM: Vascular responses to norepinephrine, angiotensin, vasopressin and serotonin. Fed Proc 27:1391, 1968
161. Mandel MJ, Sapirstein LA: Effect of angiotensin infusion on regional blood flow and regional vascular resistance in the rat. Circ Res 10:807, 1962
162. Whelan RF, Scroop GC, Walsh JA: Cardiovascular action of angiotensin in man. Am Heart J 77:546, 1969
163. Abboud FM: Effect of sodium, angiotensin, and steroids on vascular reactivity in man. Fed Proc 33:143, 1974
164. Schmid PG, Abboud FM, Wendling MG, et al: Vascular effects of vasopressin: Plasma levels and circulatory responses. Am J Physiol 227:998, 1974
165. Schmid PG, Eckstein JW, Abboud FM: Effect of 9-α-fluorohydrocortisone on forearm vascular responses to norepinephrine. Circulation 36:620, 1966
166. Schmid PG, Eckstein JW, Abboud FM: Comparison of effects of deoxycorticosterone and dexamethasone on cardiovascular responses to norepinephrine. J Clin Invest 46:590, 1967

166a. Adetuyibi A, Mills IH: Relation between urinary kallikrein and renal function, hypertension and excretion of sodium and water in man. Lancet 2:203, 1972

166b. Margolius HS, Horwitz D, Geller RG, Alexander RW, Gill JR, Pisano JJ, Keiser HR: Urinary kallikrein excretion in normal man: Relationships to sodium intake and sodium-retaining steroids. Circ Res 35:812, 1974

167. Hedwall PR, Abdel-Sayed WA, Schmid PG, et al: Vascular responses to prostaglandin E_1 in gracilis muscle and hindpaw of dog. Am J Physiol 221:42, 1971
168. Mark AL, Schmid PG, Eckstein JW, et al: Venous responses to prostaglandin $F_{2\alpha}$ Am J Physiol 220:222, 1971
169. Brody MJ, Kadowitz PJ: Prostaglandins as modulators of the autonomic nervous system. Fed Proc 33:48, 1974

170. McGiff JC, Itskovitz HD: Prostaglandins and the kidney. Circ Res 33:479, 1973
171. Smeby RR, Sen S, Bumpus FM: Naturally occurring renin inhibitor. Circ Res 21:II–129, 1967
172. Edwards WG Jr, Strong CG, Hunt JC: Vasodepressor lipid resembling prostaglandin E_2 (PGE_2) in the renal venous blood of hypertensive patients. J Lab Clin Med 74:389, 1969
173. Muirhead EE, Jones F, Stirman JA: Antihypertensive property in renoprival hypertension of extract from renal medulla. J Lab Clin Med 56:167, 1960
174. Kadowitz PJ, Sweet CS, Brody MJ: Influence of prostaglandin on adrenergic transmission to vascular smooth muscle. Circ Res 31:II–36, 1972
175. Hedqvist P: Studies on the effect of prostaglandins E_1 and E_2 on the sympathetic neuromuscular transmission in some animal tissues. Acta Physiol Scand 345:1, 1970
176. Bergstrom S, Farnebo LO, Fuxe K: Effects of prostaglandin E_2 on central and peripheral catecholamine neurones. Eur J Pharmacol 21:362, 1973
177. Guyton AC, Coleman TG, Cowley AW, et al: A systems analysis approach to understanding long-range arterial blood pressure control and hypertension. Circ Res 35:159, 1974
178. Dobson JG, Jr, Rubio R, Berne RM: Role of adenine nucleotides, adenosine, and inorganic phosphate in the regulation of skeletal muscle blood flow. Circ Res 29:375, 1971
179. Skinner NS, Costin JC: Interactions of vasoactive substances in exercise hyperemia: O_2, K^+, and osmolality. Am J Physiol 219:1386, 1970
180. Black JE, Roddie IC: The mechanism of the changes in forearm vascular resistance during hypoxia. J Physiol 143:226, 1958
181. Heistad DD, Abboud FM, Mark AL, Schmid PG: Impaired cardiovascular reflexes in chronically hypoxic patients. J Clin Invest 51:331, 1972
182. Daly MD, Scott MJ: An analysis of the primary cardiovascular reflex effects of stimulation of the carotid body chemoreceptors in the dog. J Physiol 162:555, 1962
183. Richardson DW, Kontos HA, Raper AJ, et al: Modification by beta-adrenergic blockade of the circulatory responses to acute hypoxia in man. J Clin Invest 46:77, 1967
184. Richardson DW, Kontos HA, Shapiro W, et al: Role of hypocapnia in the circulatory responses to acute hypoxia in man. J Appl Physiol 21:22, 1966
185. Hutchins PM, Bond RF, Green HD: Participation of oxygen in the local control of skeletal muscle microvasculature. Circ Res 34:93, 1974
186. Folkow B, Gurevich M, Hallback M, et al: The hemodynamic consequences of regional hypotension in spontaneously hypertensive and normotensive rats. Acta Physiol Scand 83:532, 1971
187. Silvertsson R, Olander R: Aspects of the nature of the increased vascular resistance and increased "reactivity" to noradrenaline in hypertensive subjects. Life Sci 7:1291, 1968

188. Hansen TR, Bohr DF: Hypertension, transmural pressure, and vascular smooth muscle response in rats. Circ Res (in press)
189. Lais LT, Shaffer RA, Brody MJ: Neurogenic and humoral factors controlling vascular resistance in the spontaneously hypertensive rat. Circ Res 35:764, 1974
190. Djojosugito AM, Folkow B, Oberg B, White S: A comparison of blood viscosity measured in vitro and in a vascular bed. Acta Physiol Scand 78:70, 1970
191. Abdel-Sayed WA, Abboud FM, Calvelo MG: Effect of local cooling on responsiveness of muscular and cutaneous vessels Am J Physiol 219:1772, 1970
192. Gauer OH, Henry JP, Sieker HO, Wendt WE: The effect of negative pressure breathing on urine flow. J Clin Invest 33:287, 1954
193. Henry JP, Pearce JW: The possible role of cardiac atrial stretch receptors in the induction of changes in urine flow. J Physiol 131:572, 1956
194. Linman JW, Bethell FH: Factors Controlling Erythropoiesis. Springfield, Ill, Charles C Thomas, p 3, 1960
195. Feigl EO: Control of myocardial oxygen tension by sympathetic coronary vasoconstriction in the dog. Circ Res 37:88, 1975
196. McRaven DR, Mark AL, Abboud FM, Mayer HE: Responses of coronary vessels to adrenergic stimuli. J Clin Invest 50:773, 1971
197. Mark AL, Abboud FM, Schmid PG, et al: Differences in direct effects of adrenergic stimuli on coronary cutaneous and muscular vessels. J Clin Invest 51:279, 1972
198. Feigl EO: Sympathetic control of coronary circulation. Circ Res 20:262, 1967
199. Berne RM, DeGeest H, Levy MN: Influence of the cardiac nerves on coronary resistance. Am J Physiol 208:763, 1965
200. Feigl EO: Parasympathetic control of coronary blood flow in dogs. Circ Res 25:509, 1969
201. Vatner SF, Franklin D, Van Citters RL, Braunwald E: Effects of carotid sinus nerve stimulation on the coronary circulation of the conscious dog. Circ Res 27:11, 1970
201a. Feigl EO: Reflex parasympathetic coronary vasodilation elicited from cardiac receptors in the dog. Circ Res 37:175, 1975
202. Bohr DR: Adrenergic receptors in coronary arteries. Ann NY Acad Sci 139:799, 1967
203. Baron GD, Speden RN, Bohr DR: Beta-adrenergic receptors in coronary and skeletal muscle arteries. Am J Physiol 223:878, 1972
204. Adam KR, Boyles S, Scholfield PD: Cardio-selective beta-adrenoceptor blockade and the coronary criculation. Br J Pharmacol 40:534, 1970
205. Ross G, Jorgensen CR: Effects of a cardio-selective beta-adrenergic blocking agent on the heart and coronary circulation. Cardiovasc Res 4:148, 1969
206. Schuelke DM, Mark AL, Schmid PG, Eckstein JW: Coronary vasodila-

tation produced by dopamine after adrenergic blockade. J Pharmacol Exp Ther 176:320, 1971

207. Pitt B, Green HL, Sugishita Y: Effect of beta-adrenergic receptor blockade on coronary haemodynamics in the resting unanesthetized dog. Cardiovasc Res 4:89, 1970
208. Pitt B, Craven P: Effect of propranolol on regional myocardial blood flow in acute ischaemia. Cardiovasc Res 4:176, 1970
209. Becker LC, Fortuin NJ, Pitt B: Effect of ischemia and antianginal drugs on the distribution of radioactive microspheres in the canine left ventricle. Circ Res 28:263, 1971
210. Vatner SF, Higgins CB, Franklin D, Braunwald E: Effects of a digitalis glycoside on coronary and systemic dynamics in conscious dogs. Circ Res 28:470, 1971
211. Gorlin R, Brachfeld N, MacLeod C, Bopp P: Effect of nitroglycerin on the coronary circulation in patients with coronary artery disease or increased left ventricular work. Circulation 19:705, 1959
212. Bernstein L, Friesinger GC, Lichtlen PR, Ross RS: The effect of nitroglycerin on the systemic and coronary circulation in man and dogs. Circulation 33:107, 1966
213. Knoebel SB, McHenry PL, Bonner AJ, Phillips JF: Myocardial blood flow in coronary artery disease. Circulation 47:690, 1973
214. Winbury MM: Redistribution of left ventricular blood flow produced by nitroglycerin. Circ Res 28, 29:I–140, 1971
215. Ganz W, Marcus HS: Failure of intracoronary nitroglycerin to alleviate pacing-induced angina. Circulation 46:880, 1972
216. Mueller H, Ayres SM, Gregory JJ, et al: Hemodynamics, coronary blood flow and myocardial metabolism in coronary shock; Response to l-norepinephrine and isoproterenol. J Clin Invest 49:1185, 1970
217. Mueller H, Ayres SM, Giannelli S, et al: Effects of isoproterenol l-norepinephrine, and intra-aortic counterpulsation on hemodynamics and myocardial metabolism in shock following acute myocardial infarction. Circulation 45:335, 1972
218. Vogel JHK, Jacobowitz D, Chidsey CA: Distribution of norepinephrine in the failing bovine heart. Circ Res 24:71, 1969
219. Schmid PG, Brody MJ, Mayer HE, et al: Innervation of the coronary and peripheral vasculature in heart failure. Circulation 46:11, 1972

Francois M. Abboud, Allyn L. Mark
Donald D. Heistad, Phillip G. Schmid, (Part I)
Robert W. Barnes (Part II)

5

The Venous System

In this chapter we describe the role of the venous system in circulatory control in health and disease under two parts: I. Pathophysiology of the Venous System: In this section the emphasis is on the basic and applied physiologic and pharmacologic concepts as they relate to the role of the venous system in circulatory control in physiologic and pathologic states that do not involve primarily the venous system. Several excellent reviews on this subject have been published in recent years,[1-4] and they provided an important base of information for this chapter. II. Pathophysiology of Venous Disease of the Lower Extremities: In this section the emphasis is predominantly on the utilization of diagnostic techniques for the precise definition of the abnormal venous dynamics in venous disease of the lower extremities.

I. PATHOPHYSIOLOGY OF THE VENOUS SYSTEM

The venous system is often thought of as a large low-pressure reservoir of blood that serves as a conduit for the venous return from the tissues to the heart. This notion fails to underscore the facts that the venous system is a determinant of circulatory homeostasis and that it participates in circulatory adjustments to stresses.

Functions of the Venous System

Whereas the arterial system serves primarily a resistance function that regulates the distribution of blood flow to organs and maintains per-

The more recent studies of the authors and their colleagues that are cited in this review were supported by U.S. Public Health Service Grants HL-14388, HL-16149, and HL-16066.

fusion pressure, the venous system serves three important functions: (1) a capacitance function that regulates cardiac output, (2) a resistance function that regulates hydrostatic capillary pressure, and (3) a temperature-regulating function.

Venous Capacitance Function and Regulation of Cardiac Output

The cardiac stroke volume is determined by the contractile property of the ventricles and by the force for cardiac filling, or the filling pressure. The pressure within the venous system and right atrium, or the central venous pressure, represents the cardiac filling force, determines the magnitude of diastolic filling of the right ventricle, and according to Starling's law of the heart regulates the vigor of contraction during the subsequent systole.

Central Venous Pressure

Since the cardiac filling force is a function of pressure in the large veins in proximity to the right atrium, it is important to identify the determinants of central venous pressure. These are primarily the total blood volume, the capacity of venous compartments in different organs, the venous tone, the ventricular compliance, and the duration of diastolic filling, or heart rate.

Normal venous pressure in a supine resting individual is approximately 10 mm Hg when measured in the large veins with reference to the level of the right atrium. At the venular end of the capillaries and in the venules it is approximately 15–25 mm Hg, whereas in the right atrium it is 5 mm Hg. The driving force for venous return between the capillaries and the right atrium is approximately 15 mm Hg. Despite the fact that the elastic modulus of veins is not significantly different from that of arteries and the total arterial blood flow is equivalent to venous flow, venous pressure is significantly lower than arterial pressure. The reasons are the greater cross-sectional area in the venous system than in the arterial system and the minimal resistance to venous outflow, which is determined by ventricular compliance as compared to arteriolar resistance.[5] Under certain experimental circumstances small-vein pressure might increase significantly to levels approaching 50–100 mm Hg in certain vascular beds where venular constriction is intense. Such a rise in small-vein pressure has a significant effect on capillary hydrostatic pressure and on blood flow to the organ involved, but it would have little effect on the pressure in the large veins or cardiac filling pressure unless a significant loss of intravascular volume took place or the larger capacitance veins also constricted.

1. *Passive change in venous pressure–volume relationship.* Changes in central venous pressure of only 2–3 mm Hg can cause

significant changes in cardiac output in the absence of heart failure. A change in filling pressure is brought about passively during postural changes. In the Trendelenburg position the distending pressure in the veins of the elevated lower extremities is reduced, and their venous volume decreases because of venous viscoelastic properties. Blood volume from the lower extremities is displaced centrally toward the heart and lung and contributes to an elevation in central venous pressure and in stroke volume (Fig. 5-1). Conversely, a head-up tilt position increases the hydrostatic distending pressure in the veins of the lower extremities. The distribution of blood volume changes, so that the capacity of the venous system in the lower extremities increases, whereas the capacities of the central veins, the heart, and the lungs decrease and filling pressure falls.[6]

Although we are referring to shifts of venous blood volume as if we were dealing with a static system in which blood is translocated from one part of the venous system to another, the total blood volume is in fact circulating, but the capacities of the various venous compartments (central versus peripheral) are changing dynamically as a result of changes in distending pressure resulting from changes in hydrostatic pressure.

A reduction in blood volume from hemorrhage results also in a reduction in central venous pressure and stroke volume. Thus the capacitance function of the venous system is best described as a pressure–volume relationship of the veins, in contrast to the resistance function of the arterioles or venules, which is described in terms of a pressure–flow relationship.

At the normal venous distending pressure the veins accommodate approximately 60%–70% of the total blood volume. As the venous pressure increases from 5 to 15 mm Hg, venous volume in the extremities may double as a result of venous distensibility and the known pressure–volume relationship in the limbs of man (Fig. 5-2). An increase in hydrostatic pressure in the veins of the lower extremities to 70 or 80 mm Hg may occur if the lower limbs are dependent, resulting in pooling of approximately 600–700 ml of blood with simultaneous reduction of capacity of the upper venous compartment and a fall in central venous pressure. Fortunately, however, in the standing position hydrostatic pressure is not fully transmitted to all veins of the lower extremities because of contraction of skeletal muscle, which tends to compress the veins of the lower limbs and interrupt the hydrostatic pressure column.[6,7]

2. *Active change in venous pressure–volume relationship.* There can be significant changes in the capacity of the venous system in different organs through active constriction of the venous smooth muscle. The constriction may occur either in response to a neurogenic

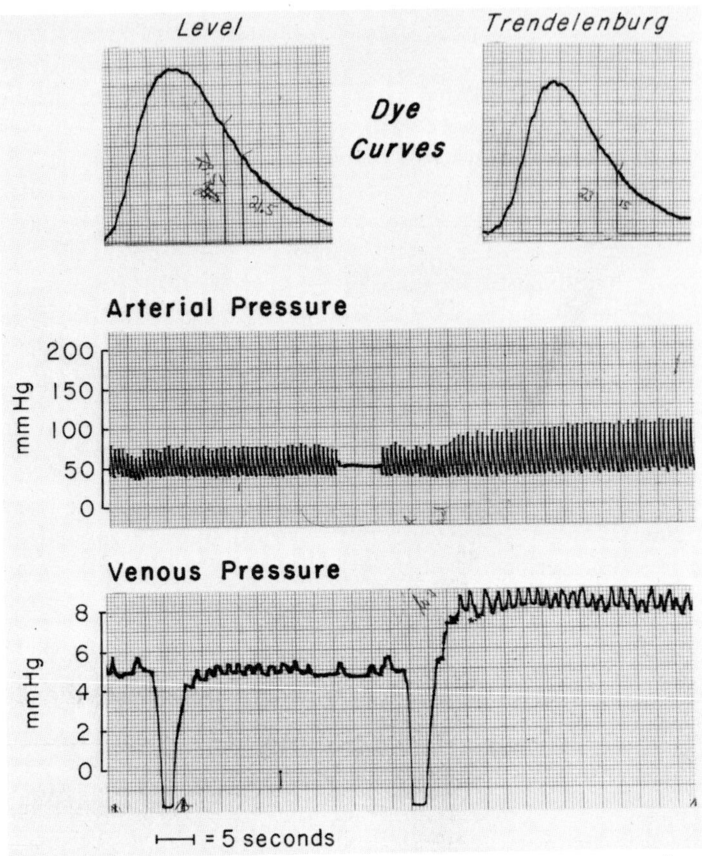

Fig. 5-1. A small increase in central venous pressure or cardiac filling pressure causes a significant increase in cardiac output and arterial pressure. This effect of filling pressure on cardiac output is very evident here because of prior administration of a ganglion blocker. In the absence of ganglion blockers, neurogenic circulatory adjustments are such that cardiac output does not increase in the Trendelenburg position.

stimulus mediated through the adrenergic sympathetic supply or through a humoral stimulus. The responsiveness of the veins in the different organs is not uniform. Some may react negligibly to a stimulus that causes intense venoconstriction in other beds. For example, the capacity in the splanchnic vascular bed may change significantly in response to activation of arterial baroreceptor and chemoreceptor reflexes. Particularly during hemorrhage the contraction of the splanchnic bed might maintain temporarily an adequate filling pressure by shifting blood volume centrally. On the other hand, the cutaneous

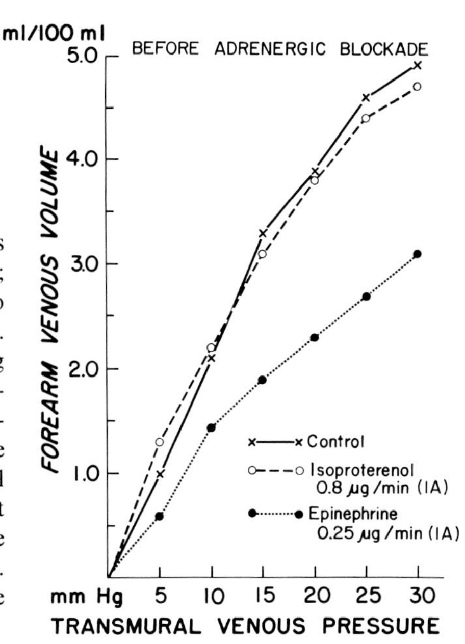

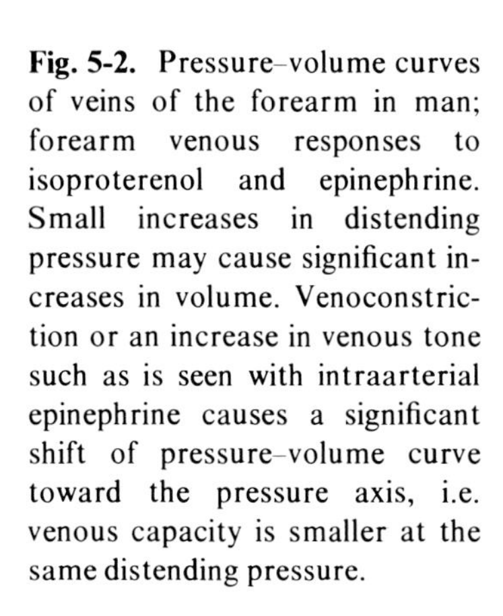
Fig. 5-2. Pressure–volume curves of veins of the forearm in man; forearm venous responses to isoproterenol and epinephrine. Small increases in distending pressure may cause significant increases in volume. Venoconstriction or an increase in venous tone such as is seen with intraarterial epinephrine causes a significant shift of pressure–volume curve toward the pressure axis, i.e. venous capacity is smaller at the same distending pressure.

venous bed is much less responsive to baroreceptor and chemoreceptor reflexes, but more sensitive to changes in temperature and to the constricting effect of circulating catecholamines (Fig. 5-2). In contrast to the splanchnic and cutaneous veins, those of skeletal muscle have minimal innervation and are much less responsive to humoral substances. They contribute little to blood volume shifts in the circulatory system in an active way, but they may constrict passively during arteriolar constriction and reduction in blood flow to skeletal muscle.

3. *Compliance of the ventricle.* Ventricular hypertrophy caused by pulmonary hypertension or pulmonic stenosis causes a rise in atrial or venous pressure, particularly in the venous A wave. Changes in compliance occur also during myocardial ischemia, but these are most apparent in the left ventricle and are reflected in an elevation of the pulmonary venous pressure. Significant elevations in pulmonary venous congesting pressure are seen during coronary arteriography in patients with coronary artery disease. There are other factors related to diastolic filling that can cause significant increases in venous pressure; these include tricuspid stenosis, constrictive pericarditis, and pericardial tamponade. In the latter states venous pressure is high, but transmural ventricular pressure, which is the distending pressure, may not be increased and the anticipated increase in stroke volume expected from stretch of ventricular muscle by the diastolic filling force does not occur.

4. *Duration of diastolic filling.* A rapid tachycardia may result in a reduction in diastolic filling time and a fall in venous pressure, such as is seen sometimes with infusions of isoproterenol. This may not take place if the tachycardia is associated with myocardial ischemia and a loss of ventricular compliance, such as is sometimes seen during pacing to induce angina.

Central Venous Pressure versus Pulmonary Venous Congesting Pressure

The monitoring of central venous pressure has been exceedingly helpful in the management of shock, as well as in the treatment of acute pulmonary edema and heart failure.[8] In shock, a low central venous pressure is used as a guide for blood volume replacement, and in acute pulmonary edema it is used as a guide for continued diuretic therapy and for evaluation of progress of heart failure. Although the normal central venous pressure averages 7 mm Hg, it can vary widely from 2 to 10 mm Hg. A single measurement of central venous pressure may be of relatively little value in the management of hypotensive patients, unless the value is extreme. For example, a pressure of 5 mm Hg may represent hypovolemia in a subject whose normal venous pressure is 10 mm Hg, and conversely it would represent hypervolemia if the venous pressure were normally 2 mm Hg. When one is confronted with such a situation clinically, a therapeutic trial of volume replacement with careful monitoring of changes in both central venous and arterial pressures is a reasonable course to follow. A favorable response would be reflected by a small increase in central venous pressure and a significant rise in arterial pressure; conversely, a rapid rise in venous pressure without a significant change in arterial pressure would suggest myocardial failure and contraindicate further volume expansion. Patients in heart failure may have borderline elevations of venous pressure that become significant during mild exercise; thus a normal venous pressure by itself should not exclude the diagnosis of congestive heart failure.

Measurement of central venous pressure reflects right ventricular filling pressure, but it has been used as a guide for left ventricular filling pressure or pulmonary venous congesting pressure. When ventricular function is normal, changes in central venous pressure are paralleled by changes in pulmonary venous or congesting pressure during volume expansion or during blood loss. This fact and the relative ease with which central venous pressure can be measured have led to the regular monitoring of central venous pressure in many precarious hemodynamic states. The practice is useful and should be continued. One should keep in mind, however, that the absolute level of central venous pressure does not equal pulmonary venous pressure; the latter usually exceeds central venous pressure by approximately 5–10 mm Hg. Furthermore, although the directional changes in pressure in both systems may be the same, the magnitude of the changes may be different in different disease states. For example, during cardiogenic shock from left ventricular failure, normal central venous pressure may be associated with significant elevation of left ventricular end-diastolic pressure and pulmonary congesting pressure. Although increases in central venous pressure will generally accompany increases in pulmonary venous pressure, the latter might be of much greater magnitude and represent a significant hazard.[9] Conversely, in patients with pulmonary disease increases in central venous pressure may be greater than corresponding increases in pulmonary venous pressure (Fig. 5-3).

During treatment of cardiogenic shock it is preferable to measure pulmonary artery wedge pressure, which reflects pulmonary venous congesting pressure. If this is not possible, pulmonary artery diastolic

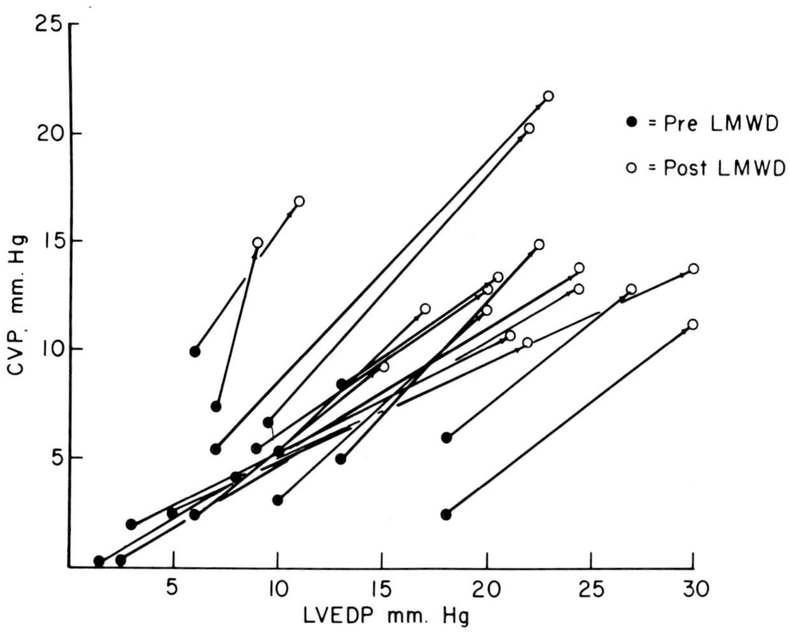

Fig. 5-3. Comparison of central venous pressure (CVP) and left ventricular end-diastolic pressure (LVEDP) before and after infusion of low molecular—weight dextran (LMWD) in patients with various types of shock. Although the absolute levels differed, the increases in pressure after volume overload were often comparable. (Reproduced from Cohn JN, Tristani FE, Khatri IM. J Clin Invest 48:2013, 1969.)

pressure would reflect pulmonary venous pressure. The availability of Swan-Ganz catheters has made the measurement of pulmonary artery diastolic and wedge pressures relatively easy at the bedside without the need for fluoroscopy. When neither of these measurements is obtainable, the use of central venous pressure can provide a useful guide.

Cardiopulmonary Volume

In addition to the regulation of cardiac output through a change in filling pressure, the capacitance function of veins modifies cardiopulmonary blood volume, which influences afferent neural activity and myocardial oxygen consumption.

1. *Afferent cardiopulmonary neural activity.* The atria, ventricles, and lungs contain stretch receptors that trigger reflexes mediated through afferents in the vagi or in the sympathetic nerves. The role of these afferent impulses in circulatory regulation has been discussed in the chapter on the regulation of the peripheral and coronary circulation. Stretch of the left atrial receptors may cause an increase in cardiac sympathetic activity and a selective reduction in sympathetic activity to the kidney.[10] The resulting combination of tachycardia and renal vasodilatation due to left atrial stretch might be an effective mechanism by which diuresis may be evoked, and thus a reduction in intravascular volume as well as ventricular volume would result.[11] In man, pooling of blood in veins of the lower extremities with mild lower body negative pressure (LBNP) sufficient to decrease right atrial pressure without altering systemic arterial pressure triggers reflex vasoconstriction in the forearm without a significant change in heart rate.[12] Conversely, elevation of the lower limbs increases filling pressure and causes reflex vasodilatation in the forearm, presumably by stretching atrial receptors. Furthermore, atrial receptors may have a modulating influence on release of humoral substances; left atrial stretch reduces levels of vasopressin[13] and right atrial stretch reduces renin release.[14] Interruption of afferent impulses from the cardiopulmonary area along the vagi by cooling them tends to increase sympathetic nerve activity in the renal nerve and increase renin release.[15,16] Removal of cardiopulmonary afferent activity by vagotomy augments the vasoconstrictor and respiratory responses to chemoreceptor activation, as well as the responses to carotid baroreceptor reflexes.[17]

Thus, although the veins of the extremities respond rather poorly to chemoreceptor and baroreceptor reflexes, they can indirectly influence the circulatory adjustments to these reflexes by either increasing or decreasing cardiopulmonary volume.

2. *Myocardial oxygen consumption.* The veins fulfill another important indirect role through their influence on cardiac volume. Venous dilatation in response to nitroglycerin, nitroprusside, or theophylline in patients with angina and heart failure decreases cardiac size, pulmonary blood volume, and pulmonary venous congesting pressure. Reduction in cardiac size and in ventricular wall tension reduces myocardial oxygen consumption.

3. *Reduction of venous pressure in heart failure.* Diuretics decrease blood volume, relax peripheral veins, and thus reduce cardiopulmonary volume and central venous pressure. It might seem paradoxical that beneficial results are obtained in heart failure when the cardiac filling pressures are reduced. The paradox relates to the fact that according to Starling's law of the heart a reduction in cardiac filling force decreases stroke volume. The situation in heart failure is particular in that the ventricular function curve relating filling pressure to stroke volume is relatively flat at high levels of filling pressure, and despite a considerable increase in filling pressure there is minimal increase in stroke volume. Therefore, reduction in pulmonary venous congesting pressure from levels of, for example, 35 to 20 mm Hg may be achieved with little change in stroke volume. Furthermore, reduction in venous congesting pressure may be induced with relatively small reductions in blood volume, since in heart failure the venous pressure–volume curve is shifted toward the pressure axis as venous tone increases. Once the pulmonary venous congesting pressure is reduced significantly, pulmonary capillary filtration will be reduced and pulmonary edema reversed. The coincidental improvement in ventilation and oxygenation, as well as the reduction in myocardial oxygen demand as a result of the decrease in cardiac size, will improve ventricular performance (Fig. 5-4).

4. *Effect of rotating tourniquets.* Elevation of congesting cuff pressures to levels close to diastolic causes pooling of blood in the extremities, but the decreased venous compliance in heart failure limits the maximal capacity for pooling. Increased capillary filtration during congestion contributes to loss of intravascular volume, but it is surprising that only minimal reductions in left ventricular end-diastolic pressure occur in patients with left ventricular failure during application of rotating tourniquets.[18]

Venular Resistance Function and Capillary Filtration

Capillary hydrostatic pressure is dependent upon the ratio of postcapillary to precapillary resistance, as well as the levels of venous and arterial pressures. It has been estimated that 80% of an increase in

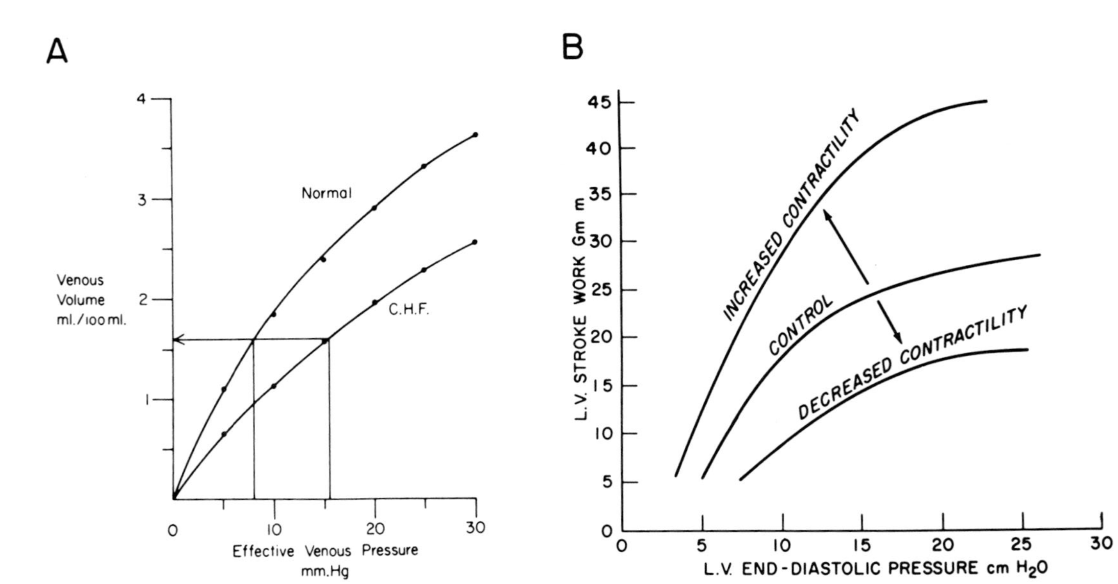

Fig. 5-4. Relationship of venous pressure and cardiac filling pressure to venous volume and stroke work. In heart failure, a small reduction in blood volume through diuresis causes a significant fall in venous pressure (left panel). A significant fall in cardiac filling pressure (L.V. End-Diastolic Pressure) causes only a negligible reduction in cardiac stroke volume and work (right panel), yet pulmonary congestion and cardiac size decrease wherein the beneficial effect of diuresis. (Left panel: reproduced from Wood JE (ed): The Veins: Normal and Abnormal Function. Boston, Little, Brown, 1965, p 153.) (Right panel: reprinted by permission from Braunwald, Ross, Sonnenblick. N Engl J Med 277:911, 1967.)

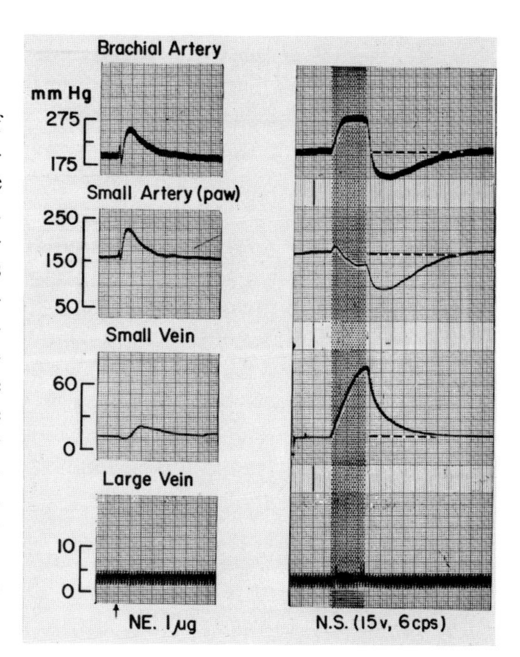

Fig. 5-5. Measurement of vascular pressure in consecutive segments in the perfused forelimb of dog. During electrical stimulation of sympathetic nerves (N.S.) there are significant increases in both arterial and small vein pressures. The increase in the latter is pronounced and the ratio of venous to arteriolar resistance or post to precapillary resistances increases favoring capillary filtration. Reproduced from Abboud FM. Fed Proc 31:1227, 1972.)

venous pressure is transmitted to the capillaries, whereas only 10% of an increase in arterial pressure is reflected as an increase in capillary hydrostatic pressure. A small increase in the absolute value of venous resistance may cause a significant increase in the ratio of venous to arteriolar resistance and consequently in capillary hydrostatic pressure (P_c) (Fig. 5-5).

$$P_c = \frac{\left(Pa\,\dfrac{Rv}{Ra}\right) + Pv}{1 + \dfrac{Rv}{Ra}}$$

where P_a and P_v represent central arterial and venous pressures, respectively, and R_a and R_v represent precapillary and postcapillary resistances.[19]

During hypovolemic shock there is a reflex increase in precapillary resistance that exceeds postcapillary resistance and favors absorption of fluid from the extracellular space. The fall in arterial and venous pressures also decreases capillary hydrostatic pressure and favors intravascular fluid absorption. In late or irreversible shock, precapillary resistance tends to decrease because of progressive acidosis, whereas postcapillary resistance continues to rise; this increases hydrostatic capillary pressure and favors the loss of intravascular volume.[20]

Venules in certain vascular beds may be very sensitive to neurogenic or humoral stimuli, but they do not appear to be as sensitive to local metabolites or to hypoxia as the precapillary resistance vessels. Because of their resistance function, small veins may play an important role in the regulation of blood volume through capillary filtration.

Viscosity. Although blood is non-Newtonian and nonhomogenous, it behaves very much like a Newtonian fluid, because erythrocytes can easily be squeezed through capillaries. At very low shear rates, however, viscosity may increase significantly, and this factor becomes apparent in the postcapillary vessels and venules where the average velocity of flow is very low, particularly in hypotension or shock. This relative increase in viscosity would increase the ratio of postcapillary/precapillary resistance and favor transcapillary fluid filtration and loss of intravascular volume.[21]

Capillary filtration coefficient. It has been estimated that an increase in venous pressure of approximately 10 mm Hg will cause a loss of intravascular volume of approximately 600 ml/hr. This calculation is based on an estimate of capillary filtration amounting to 0.0033 ml/min/mm Hg venous pressure rise per 100 ml of forearm volume.[22] It assumes the absence of any compensatory mechanism that could result in closure of precapillary sphincters and reduction in capillary surface area. A venoarteriolar reflex takes place, however, during the rise in venous pressure and may explain the lack of dependent edema when hydrostatic pressure increases in dependent capillaries.[23]

Temperature-Regulating Function

Cutaneous veins. Pronounced responses to changes in total-body or blood temperature occur primarily in the cutaneous veins and are minimal or absent in splanchnic veins or in veins draining skeletal muscle.[24–26] Cutaneous venoconstriction during cooling decreases heat dissipation, particularly when coupled with arteriolar constriction and reduction in cutaneous blood flow. When most of the venous return takes place through the deeper veins of the extremities that travel in close proximity to the arteries, heat conservation through a contercurrent system is more efficient. Heating dilates cutaneous veins favoring dissipation of heat through an increase in venous surface area and a greater flow of blood, possibly at reduced velocity.

Local cooling by perfusing a vascular bed with cold blood or by exposing the skin to cold increases venous resistance either by increasing viscosity, by causing the release of endogenous catecholamines, by aug-

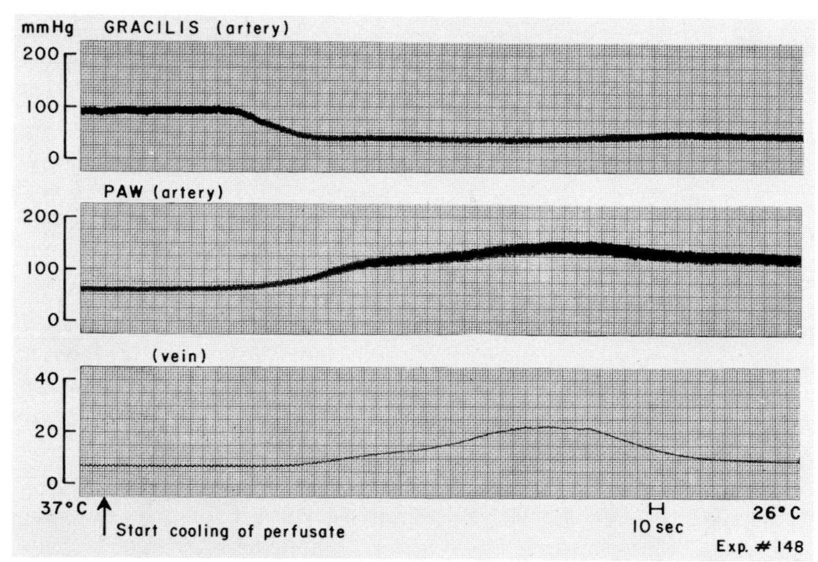

Fig. 5-6. Effect of cooling—denervated hind limb; cooling of blood perfusing the paw and gracilis muscle causes an increase in arteriolar and venous resistances in the paw which is a predominantly cutaneous bed and arteriolar dilatation with negligible venous responses in muscle. These opposite responses in skin and muscle of the extremity should result in efficient heat conservation through both a change in distribution of blood flow and in the vascular area for dissipation of heat. (Reproduced from Abdel-Sayed WA, Abboud FM, Calvelo MG. Am J Physiol 219:1773, 1970.)

menting responses to sympathetic impulses, or by a direct effect on veins (Fig. 5-6). Furthermore, responsiveness to various venoconstrictor agents such as serotonin, angiotensin, and vasopressin is augmented, but the augmentation is not uniform in magnitude, suggesting some specific interactions between the various constrictor agonists and the thermal stimulus.[26]

The tone of cutaneous veins is affected not only through local thermoregulatory mechanisms but also through central neural mechanisms. Changes in central body temperature affect the venous volume reflexly; with heating the venous effect is predominantly passive as a result of reflex changes in arteriolar resistance, whereas with cooling it is partly passive and partly active resulting from both arteriolar vasoconstriction and an increase in sympathetic venous tone. In addition, cooling of the skin on one part of the body activates thermal receptors that reflexly alter the caliber of resistance, as well as capacitance vessels in other parts of the circulation.[27] Cooling of large areas of skin on one limb, for example, induces reflex venoconstriction and exaggerates reflex cutaneous responses in the opposite limb to baroreceptor reflexes, exercise, or deep breath.[28] Conversely, heating of one limb opposes reflex cutaneous arteriolar or venous constriction in the opposite limb. This interaction appears selective to the cutaneous bed. Thus thermal receptors may induce a specific central interaction that can modify reflex responses of vessels in the skin in other portions of the body not exposed to the thermal stimulus.[29,30]

Splanchnic veins. In the isolated mesenteric vein the response to adrenergic stimulation is augmented in the cold.[31] This in vitro response, however, does not reflect the behavior of splanchnic veins in the intact circulation. Müller demonstrated that in man heating tends to produce a shift of blood volume away from abdominal-visceral capacitance segments, whereas cooling produces abdominal pooling.[32] It was also shown that heating causes a decrease in liver size by x-ray.[33] One might explain these changes in splanchnic blood volume that seem to be opposite and reciprocal to those occurring in the cutaneous veins as a result of passive changes in splanchnic volume in response to changes in distending pressure. During heating, cutaneous venodilatation causes a reduction in systemic venous pressure that could passively cause a reduction in splanchnic volume.[34,35] The converse may occur during cooling—as venous pressure rises splanchnic veins may distend passively. There may also be an element of active increase in sympathetic nerve activity to the splanchnic circulation during heat stress in man that causes reduction in splanchnic blood flow; this increase in sympathetic activity may also affect splanchnic volume.[36] In the dog, raising central body temperature does not cause a reflex venomotor response in the isolated spleen or loop of ileum, however.[25]

Skeletal muscle veins. Local cooling of isolated femoral veins that drain muscular tissues causes venodilatation.[31] Arteriolar resistance vessels of the perfused gracilis muscle also dilate during perfusion with cold blood.[26] This is in contrast to the arteriolar and venous resistance vessels of cutaneous beds, which constrict during cooling. The differential effect on resistance as well as capacitance vessels in muscle and skin results in efficient heat conservation.

Methods of Studying Venous Responses

As indicated earlier, one might think of the venous system as performing both capacitance and resistance functions. The capacitance function, predominant in the large veins, can be evaluated by studying venous compliance, i.e., pressure–volume relationships. To evaluate the resistance function of veins, an estimation of the pressure–flow relationship is needed. Several methods will be described briefly.

Venous Pressure Measurement

Central venous pressure may be used as an overall index of the change in compliance of the total venous system. Assuming there is no change in venous volume or in ventricular compliance, the rise in central venous pressure would reflect an increase in venous tone in some venous compartment. Since contraction of one venous compartment may occur simultaneously with relaxation of another, the absence of a change in venous pressure would not necessarily reflect a lack of change in venous tone.

Translocation of Volume in an Extracorporeal Reservoir

If the venous system is connected to an extracorporeal reservoir and blood is allowed to move freely in and out of the reservoir, the change in blood volume in the reservoir may reflect changes in venous capacity. A source of error in these studies, for example, would be the segregation of large volumes of blood into the splanchnic circulation as a result of hepatic venous constriction or portal constriction. The passive accumulation of volume in the splanchnic veins and spleen reduces venous pressure in the more central veins and causes a shift of volume out of the reservoir, which might be interpreted as a venodilator response. Another source of error is the small contribution of arterial vessels to the total capacity for blood.

Pressure–Volume Relationship

Equilibration technique. The most commonly used method to assess changes in venous tone is that of the measurement of changes in limb volume as a result of changes in venous pressure. This method has been used extensively by Litter, by Wood, and by Eckstein and has provided the tool upon which a very large part of our information concerning the venous system in man has accumulated over the past couple of decades.[37,38] The advantages of the method are that it is noninvasive and relatively simple and provides accurate information on the pressure–volume relationship in the extremities. Its disadvantages are that the behavior of the venous system in the extremities may not necessarily reflect the behavior of other venous compartments in the body, and transient changes in venous tone may be missed. In this technique, forearm volume is measured by plethysmography using either the conventional water-filled plethysmograph or a mercury strain gauge wrapped around the limb that permits calculation of the changes in forearm volume from the changes in forearm circumference.[39] Progressive increases in venous pressure are caused by inflating a congesting cuff placed around the arm just above the elbow. A catheter inserted into a forearm vein allows measurement of changes in venous pressure. By careful placement of the congesting cuff around the arm, reasonably accurate estimation of changes in venous pressure can be reached from measurement of cuff pressure without intravenous cannulation. The cuff is inflated in increments of 5 mm Hg to a pressure of 30 mm Hg. The increase in venous pressure is accompanied by an increase in forearm volume detected by the plethysmograph. After each cuff inflation, time is allowed for equilibration of forearm volume. The same technique may be used to measure changes in venous tone of the calf, hand, or foot by applying a congesting cuff above the knee, the wrist, or the ankle. Several precautions are necessary in using this technique. It is important that the veins be collapsed at the initiation of venous congestion; this can be achieved either by elevation of the limb or by external compression with water in the plethysmograph. The temperature of the water and the temperature of the room in which the studies are carried out should be comfortably warm to minimize changes in venous tone or venous reflexes induced by cold.[38] This equilibration technique requires time for the stabilization of the venous volume at each level of venous pressure. At congesting levels between 10 and 30 mm Hg there is little increase in capillary filtration, so that the increase in volume reflects primarily venous volume. One can thus plot the pressure–volume relationship and determine the shift in this pressure-volume relationship with the interventions desired (Fig. 5-7A).

A modification of the technique involves inflation of the venous occluding cuff to a pressure of 30 mm Hg, allowing the venous volume to stabilize, and then introducing the intervention at a constant distending pressure.[40,41] Under these circumstances the change in volume at a constant distending pressure reflects the change in venous tone (Fig. 5-7B).

Occluded-vein technique. Another modification involves the use of the constant-volume technique in which blood is trapped in the extremity or in a venous segment, with the consequent change in venous pressure at a constant venous volume reflecting the change in venous tone (Fig. 5-7C). This technique is referred to as the occluded-vein or occluded-limb technique.[42,43] In the occluded-limb technique the circulation to the limb is arrested by inflating a pneumatic cuff to suprasystolic pressures, and the pressure in one of the large veins is measured continuously. The assumption is that in the occluded-limb technique the veins will contain a constant volume, and constriction of arterioles will not result in a shift in volume into the venous compartment. The venous valves may prevent any translocation of volume from the venous segments into the capillary or arterial segments coupled in series. One also has to assume that the venous wall does not permit extravasation of fluid at high distending pressure.

Changes in splanchnic capacitance may also be determined from changes in venous pressure in the isovolumetric spleen. The splenic artery and vein are occluded; after ligation of all other vessels, changes in venous pressure in the organ reflect changes in tone of the splenic vein and capsule and may represent changes in the splanchnic venous bed. In the occluded-vein or occluded-limb technique the response to neurogenic stimuli can be studied, but the method does not lend itself to studies of humoral responses.

Intermittent occlusion technique. Another estimate of the venous pressure–volume relationship in an extremity is the ratio of the rate of increase in pressure to the rate of increase in volume during intermittent and transient periods of venous occlusion. This method has been popularized by Sharpey-Schafer.[44] It has limitations. The measurement of venous pressure in one vein of the forearm during sudden venous congestion does not necessarily reflect the change in venous pressure in all other veins of the arm that contribute to the change in volume measured. Furthermore, the rate of change in venous volume will vary with blood flow, and at high rates of blood flow the viscoelastic properties of veins may modify compliance independently of an active change in venous smooth-muscle tone (Fig. 5-7D).

Pressure–Flow Relationship

Some investigators have studied the behavior of large veins such as the saphenous or colic veins by maintaining blood flow through them at a high rate sufficient to cause significant elevation in perfusion pressure outside the physiologic range (between 20 and 60 mm Hg) and then determining the effect of interventions on the pressure gradient along the length of the perfused vein.[45,46] All tributaries between the upstream and downstream sites of pressure measurement are ligated (Fig. 5-8).

Measurement of Small-Vein Pressure

A small vein is cannulated retrogradely, and blood flow is maintained constant. From changes in small-vein pressure one can estimate changes in venous resistance downstream from the point of pressure measurement. One assumes the presence of several connections between

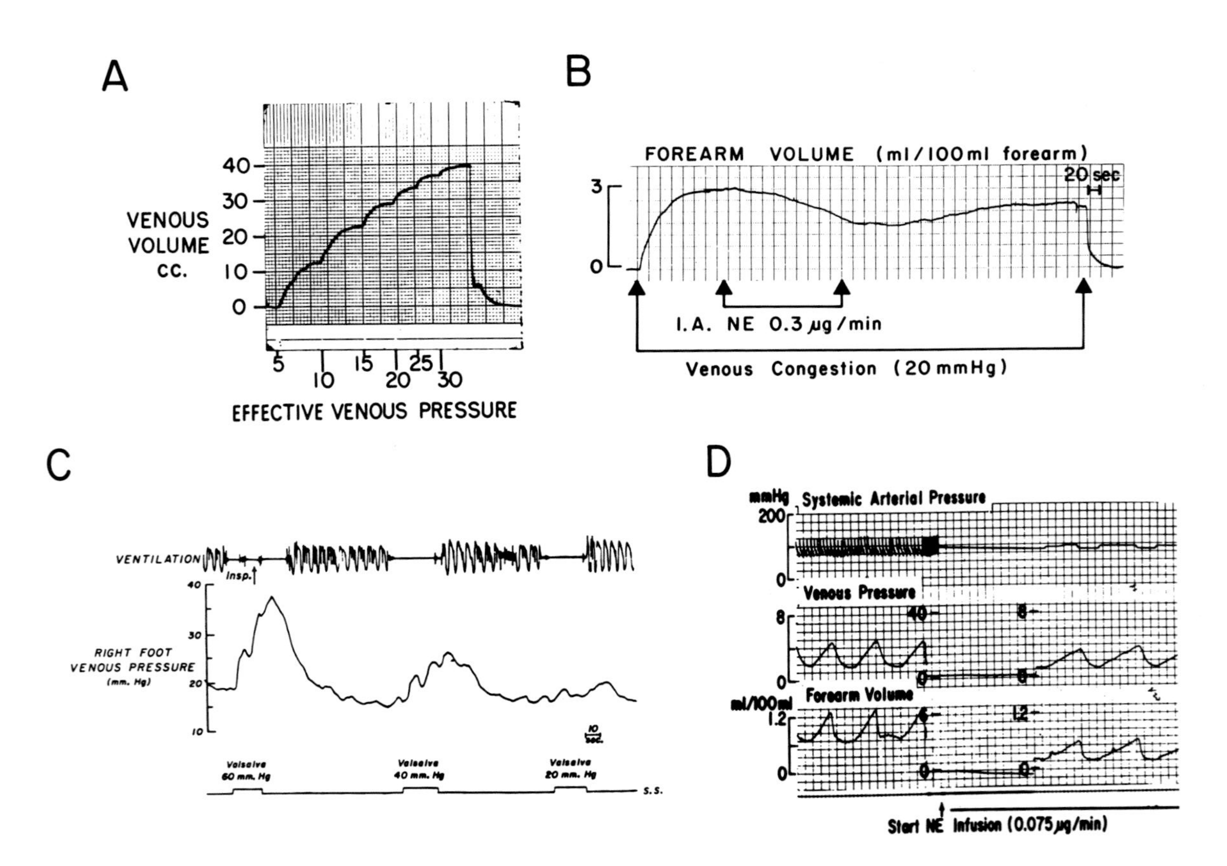
A
VENOUS VOLUME CC.
40
30
20
10
0
5
10
15
20
25
30
EFFECTIVE VENOUS PRESSURE
B
FOREARM VOLUME (ml/100ml forearm)
3
0
20 sec
I.A. NE 0.3 μg/min
Venous Congestion (20 mmHg)
C
VENTILATION
Insp.
RIGHT FOOT VENOUS PRESSURE (mm. Hg)
40
30
20
10
Valsalva 60 mm. Hg
Valsalva 40 mm. Hg
Valsalva 20 mm. Hg
S.S.
D
mmHg
200
0
Systemic Arterial Pressure
Venous Pressure
8
0
40
8
ml/100ml
1.2
0
Forearm Volume
1.2
Start NE Infusion (0.075 μg/min)

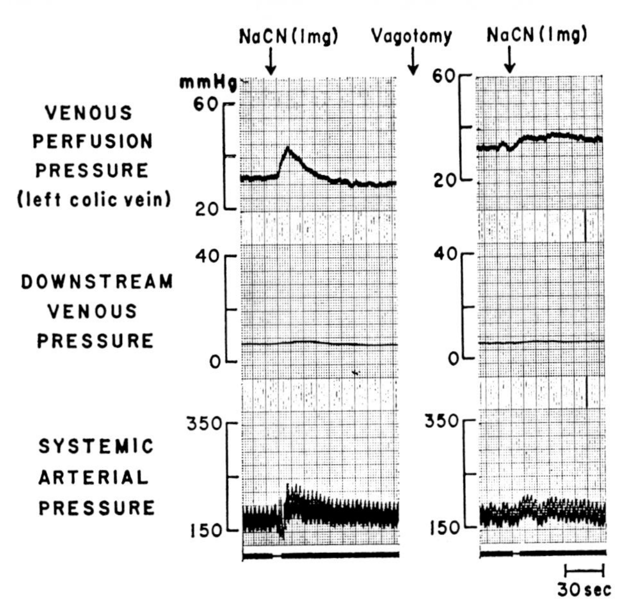

Fig. 5-8. An increase in perfusion pressure of the colic vein with a minimal change in downstream pressure at constant flow indicates venoconstriction. The response was triggered by activation of the aortic chemoreceptors with sodium cyanide and abolished by vagotomy. (Reproduced from Eckstein JW, Mark AL, Schmid PG, et al. Trans Am Clin Clim Assoc 81:60, 1969.)

Fig. 5-7. Four techniques for measurement of venous tone are shown: (A) The equilibration technique—stepwise increases in congesting cuff pressure are associated with gradual increases in venous volume which reach a new plateau at each level of distending pressure; (B) The constant distending pressure technique—a single level of congesting cuff pressure increases venous volume which reaches a plateau. A decrease in volume at a constant distending pressure means venoconstriction; (C) The occluded-vein technique—the circulation to the organ or limb is arrested and venous volume or capacity is unchanged. A rise in venous pressure indicates venoconstriction. (Reproduced from Zitnick R, Shepherd J. Prog in Cardiol. Philadelphia, Lea and Febiger, 1972, pp 185–203); (D) Intermittent-occlusion technique—rapid transient and intermittent increases in venous pressure and volume are recorded simultaneously. An increase in the ratio of the slope of venous pressure to venous volume indicates venoconstriction. (Reproduced from Abboud FM, Schmid PG, Eckstein JW. J Clin Invest 47:1–9, 1968.)

the small veins that allow equilibration of venous pressure and permit the assumption that the pressure measurement in one vein reflects the pressure in all parallel venous segments. This method has been used to determine changes in venous resistance in the paw, in muscle, and in splanchnic veins.[47,48]

Measurement of Capillary Pressure and Estimation of Postcapillary Venular Resistance

Using the isogravimetric technique of Pappenheimer and Soto-Rivera[19] one can estimate the effective capillary hydrostatic pressure, and with the additional measurement of large-artery and large-vein pressure and flow through the organ one can calculate precapillary and postcapillary resistances. These techniques have been applied in animals to the limbs, to the isolated muscle, and to the splanchnic bed and have provided reasonably accurate estimations of venular resistances in response to humoral as well as neurogenic interventions.[49]

Venomotor Responses

Innervation

The cutaneous and splanchnic veins contain several layers of smooth muscle and are innervated with adrenergic sympathetic fibers, in contrast to the vein draining skeletal muscle.[50] We have shown a good correlation between the responsiveness of cutaneous and muscular veins to sympathetic nerve stimulation and their norepinephrine content; the cutaneous veins respond more vigorously and contain more norepinephrine.[51,52] (Fig. 5-9). Innervation is not restricted to venules. Nerves are distributed along the larger veins, which may also constrict.[48] Neurogenic venoconstriction occurs during activation of sympathetic vasoconstrictor fibers, and neurogenic venodilatation occurs during withdrawal of sympathetic tone.

It has been suggested that in man activation of sympathetic cholinergic venodilatation may occur after adrenergic blockade.[53] It has also been shown that active venous constriction of the perfused saphenous vein may be triggered by injection of acetylcholine.[54] The role of cholinergic mechanisms in the regulation of venous tone is not clear at this time.

Electrical stimulation of splanchnic nerves at low frequencies may cause maximum decrements in splanchnic blood volume. When one considers that approximately 20%–25% of the total blood volume is in the splanchnic bed, a translocation of about half of that volume from the splanchnic circulation into the more central veins represents a significant

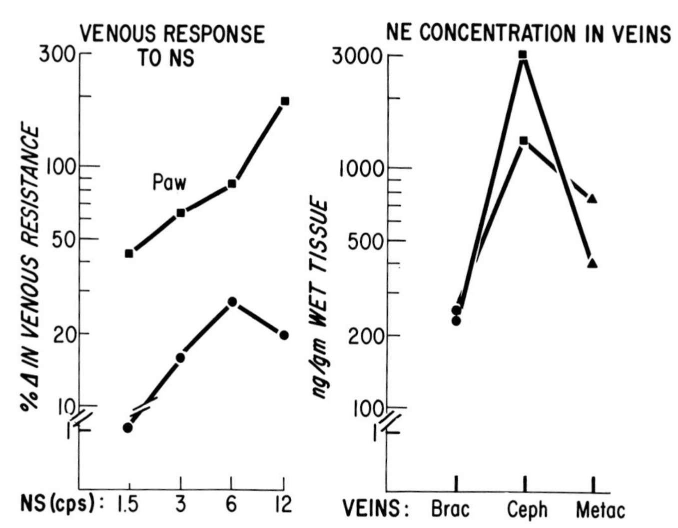

Fig. 5-9. Correlation between responsiveness to sympathetic nerve stimulation (NS) and the catecholamine content of various venous segments. ■—■ indicates responses of veins draining the paw (i.e., cephalic vein: Ceph; and metacarpal veins: Metac.). ●—● indicates responses of veins draining more proximal muscular parts of the forelimb (i.e., brachial vein: Brac.). (Reproduced by permission of the American Heart Association, Inc. from Mayer HE, Abboud FM, Ballard DR, *et al.*: Catecholamines in arteries and veins of the foreleg of the dog. Circ Res 23:653, 1968.)

transfusion of blood, particularly after blood loss.[55] Venous responses are generally more gradual and more sustained than arterial responses.[56,57] It has been estimated that part of the reduction in splanchnic volume is secondary to reduction in distending pressure of veins as arteriolar constriction takes place.[58] More specifically, Oberg has suggested that in the splanchnic bed passive changes in transmural pressure have a significant effect on splanchnic blood volume as venous geometry changes at low pressure levels from a rounded vein to a more elliptical one at pressures approximating 5–10 mm Hg.[59]

Reflexes

Baroreceptor reflexes. A fall in arterial pressure triggers the baroreceptor reflex response that activates excitatory sympathetic efferent pathways. This activation is selective in that there are several parts of the circulatory system that do not respond to the stimulus. It is now apparent that cutaneous veins and the veins of the extremities in animals and in man do not respond significantly or in a sustained fashion.[53,60,61] Splanchnic veins, however, respond promptly and significantly to changes in carotid sinus pressure in the cat.[61a] If reductions in systemic arterial pressure are prolonged, humoral factors induce venous constriction in vascular beds that may not be responsive to the sympathetic reflex.

Hemorrhage, pooling of blood in the lower extremities, and postural changes. Hemorrhage and pooling reduce cardiopulmonary volume and systemic arterial pressure and thereby activate the arterial high-pressure baroreceptor as well as the cardiopulmonary low-pressure baroreceptor reflexes. Hemorrhage has been shown to reduce splanchnic blood volume significantly in man[61b]; in animals this reduction is in part caused by the carotid sinus reflex, since it is reduced after carotid sinus denervation.[62] In the dog, hemorrhage causes a rise in small-vein resistance in the paw, but not in the muscular portion of the extremities. This increase is small compared to the corresponding rise in precapillary arteriolar resistance, but it is more sustained as local metabolic changes reverse arteriolar but not venular constriction.[20]

One may assume that in the upright position as in hemorrhage there is also reduction in the stretch of pulmonary arterial receptors. When these receptors are activated in the dog with capsaicin, bradycardia, hypotension, and relaxation of hindlimb resistance vessels as well as splanchnic capacitance vessels occur.[63] Vagotomy abolishes these responses. Opposite responses would be expected when stretch of pulmonary arterial receptors is reduced such as in hemorrhage, i.e., tachycardia and constriction of resistance and capacitance vessels.

In man it has been found that pooling of blood in the lower extremities triggers only minimal changes in venous capacitance of the forearm, even when attempts are made to augment the reflex response by cooling the opposite forearm and activating thermal receptors.[64–66] This negligible capacitance response was not restricted to the upper limb. During upright tilt, active changes in venous tone in the lower extremities also were negligible.[66] There are transient increases in cutaneous venous resistance; these do not represent a baroreceptor reflex but rather an emotional response or the increase in depth of respiration associated with upright tilt.[65] There may be a reduction in splanchnic blood volume in the upright position. The major compensatory adjustments, however, include the neurogenic increase in arteriolar resistance and heart rate, the compressive force of skeletal muscle contraction in the standing position, which prevents venous distension and in the venoconstrictor effect of humoral factors, rather than neurogenic factors.

Chemoreceptor reflexes. Activation of the carotid chemoreceptors with nicotine, cyanide, or hypoxic hypercapnic blood causes shifts in blood volume which suggest that there is a decrease in vascular capacitance and venoconstriction.[67] Experiments in the dog with perfused colic and saphenous veins, as well as with measurements of small-vein pressure in the hindlimb, indicate that the splanchnic veins are the most responsive to chemoreceptor stimulation.[68,69] The response in cutaneous veins is negligible and inconsistent, and in some experiments a venodilatation was described.[70]

Hyperventilation. A single breath or hyperventilation consistently produces reflex venous constriction in man. This is evident in the small isolated cutaneous veins, as well as in the capacitance vessels of the forearm.[71] Positive-pressure breathing also increases venous tone, and the increase may be of such a magnitude as to oppose the venous pooling coincident with the rise in venous pressure during an increase in airway pressure.[72] It has also been shown by Wood[73] that if expansion of the chest is prevented during pressure breathing the reflex venoconstrictor response is prevented, suggesting that stretching the chest wall initiates reflex venoconstriction during pressure breathing, hyperventilation, or deep breathing.

Hypoxemia. Hypoxemia produces modest increases in venous tone that are in part caused by hyperventilation (Fig. 5-10).[74] Inhalation of carbon dioxide also induces marked venoconstriction by both hyperventilation and a possible central nervous system action. It is doubtful that the chemoreceptors are involved in the increase in venous tone in the extremities during either hypoxia or hypercapnia, since the experimental work in animals indicates little if any change in venous tone or resistance in the extremities during chemoreceptor stimulation.

There is a paucity of information concerning the effects of changes in ventilation on the capacity of the splanchnic venous bed in man. In animals, however, hypoxia as well as stimulation of chemoreceptors causes a reflex constriction of splanchnic veins.[68,75]

Diving reflex. Activation of afferents in the trigemminal area either by the simulated diving reflex[76] or by stimulation of the nasal mucosa during inhalation of noxious gases or "smelling salts" causes reflex venoconstriction in the extremities of man.[77] Although this venoconstrictor response may be demonstrated at a constant distending venous pressure and therefore represents an active venoconstriction, another element of passive constriction must exist, since during the diving reflex there is marked arteriolar constriction and a decrease in peripheral blood flow[76] (Fig. 5-11).

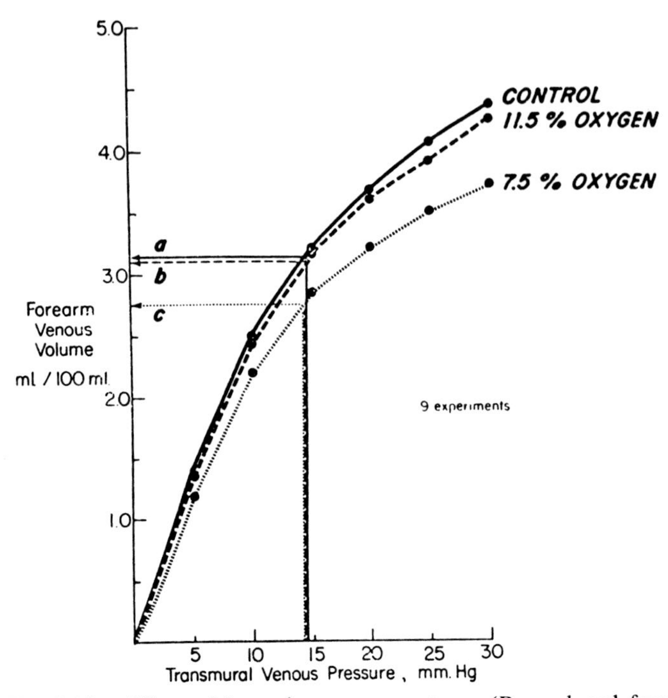

Fig. 5-10. Effect of hypoxia on venous tone. (Reproduced from Eckstein JW, Horsley AW: Effects of hypoxia on peripheral venous tone in man. J Lab Clin Med 56:847, 1960.)

Exercise. The primary response to exercise in both animals and man is venoconstriction.[78,79] This response is of physiologic importance since it maintains cardiac filling pressure and a high output. In the exercising limb the venoconstriction is in part passive, because of compression of veins by contracting skeletal muscle. The reflexes that trigger the circulatory adjustments during exercise may originate in the arterial or cardiopulmonary baroreceptors or in mechanoreceptors or chemoreceptors in the exercising muscle. It has been estimated that in man splanchnic blood volume decreases by one-third during mild supine exercise, and since splanchnic blood flow decreases only slightly and central venous pressure does not fall the decrease in splanchnic volume is attributed predominantly to an active venoconstriction.[80] In animals, stimulation of afferent nerves from skeletal muscle causes constriction of the splanchnic capacitance vessels.

In the cold environment, exercise causes constriction of cutaneous veins, but with continued exercise the reflex venoconstriction decreases because of the rise in central body temperature and the attempt to

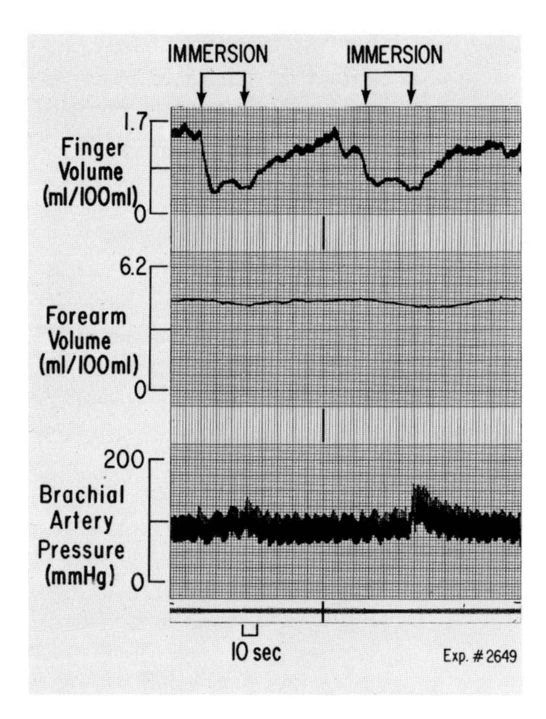

Fig. 5-11. Simulated diving in normal subjects causes marked reductions in blood flow to the forearm and finger as well as reductions in venous volume or capacity at a constant congesting cuff pressure. (Reproduced from Heistad DD, Abboud FM, Eckstein JW. J Appl Physiol 25:547, 1968.)

dissipate heat. Thus in a hot environment reflex dilatation of cutaneous vein occurs during exercise because of local as well as central inhibitory effects of the rise in temperature on the sympathetic response of cutaneous veins to exercise.[79,81] Wood and Bass demonstrated that there may be a heat adaptation of venomotor responses to exercise.[82] When exercise was carried out regularly in a hot environment the venomotor responses were improved, as well as the tolerance to exercise in the heat—a finding that indicates the importance of venous responses to exercise.[81]

Humoral Factors

Venous responses to humoral stimuli vary markedly, from one vascular bed to the other. In the same vascular bed, venous responses may differ from responses of arteriolar resistance vessels coupled in series. The differences may be related in part to distribution of adrenergic or other receptors and the density of smooth-muscle layers in the venous wall.

Alpha and beta adrenergic receptors. The predominant response to sympathomimetic amines in veins, whether cutaneous, muscular, or

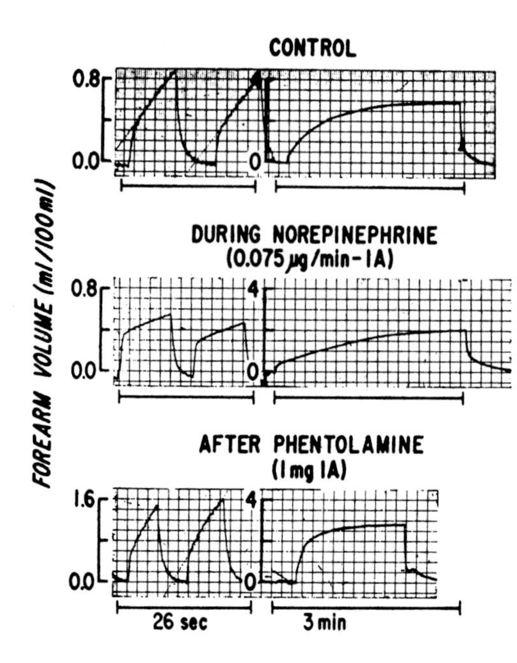

Fig. 5-12. Effect of intraarterial infusion of norepinephrine on forearm blood flow and venous capacity measured in man with plethysmographic technique. Reductions in both flow and capacity reflect the arteriolar and venous constrictor effects of activation of alpha adrenergic receptors. Phentolamine antagonizes these adrenergic constrictor effects. (Reproduced from Abboud FM, Schmid PG, Eckstein JW. J Clin Invest 47:2, 1968.)

splanchnic, is a constrictor response mediated through alpha adrenergic receptors.[40,47,51,55,58,83,84]

Venoconstrictor responses occur with norepinephrine, metaraminol, phenylephrine, and epinephrine and are blocked by alpha blockers such as phenoxybenzamine or phentolamine (Fig. 5-12).[40,41,85] There appears to be a greater sensitivity to alpha receptor blockade in the capacitance vessels of the forearm of man and venous segments in the dog, as compared to the arteriolar resistance vessels.[41,85] Furthermore, a drug like epinephrine that may activate vasodilator beta receptors in arteriolar resistance vessels of skeletal muscle of the forearm uniformly causes venoconstriction in the forearm (Fig. 5-2).[86,87] All experimental evidence in both animals and man indicates that there is a predominance of alpha constrictor receptors in veins and a paucity or absence of dilator beta receptors.[40,88] Extensive studies on forearm responses to brachial artery infusions of isoproterenol have indicated that whereas the resistance vessels dilate markedly in response to isoproterenol through activation of beta receptors to increase blood flow, there are no changes in venous distensibility.[40] Systemic administration of isoproterenol, however, induces hypotension and reflex constriction of capacitance vessels of the forearm, but this venoconstriction is neurogenic and results from activation of sympathetic vasoconstrictor fibers rather than through activation of constrictor beta receptors, as had been suggested in one study. In the small veins of the perfused forelimb of the dog minimal venodilatation may be demonstrated with beta receptor stimulation, but this amounts to less than 5% of the total vasodilator response to isoproterenol, and its physiologic significance is probably limited.[88]

Different sympathomimetic amines vary in their ability to constrict veins.[83,89] Metaraminol (Aramine) and phenylephrine (Neosynephrine) produce venous effects similar to those of norepinephrine, whereas methoxamine (Vasoxyl), a potent arteriolar constrictor, causes only slight venoconstriction. Other sympathomimetic amines such as mephentermine (Wyamine) and ephedrine cause smaller venoconstrictor responses.

Responses of venous segments in different vascular beds. Large limb veins do not seem to constrict uniformly in response to adrenergic stimuli; the veins draining superficial parts of the limb are more reactive than those draining deeper structures.[51,90] Similarly, small veins do not constrict uniformly, and pressor responses to sympathetic stimulation, norepinephrine, and serotonin are greater in small veins of the paw than in small veins of the more proximal and more muscular portions of the limb[47,84] (Fig. 5-13). Norepinephrine, acetylcholine, and serotonin, which constrict markedly the perfused saphenous vein of the dog, have little effect on the perfused colic vein. For example, the amount of drug required to produce a response in the colic vein equivalent to that seen in the saphenous vein is approximately three times greater with acetylcholine, 18 times greater with norepinephrine, and 400 times greater with serotonin.[46] The variability in the relative potency of the three drugs in the two venous segments suggests that the greater response in the saphenous vein cannot be explained only on the basis of structural or geometric differences between the two vessels. Specific characteristics of the smooth muscle in these two systems must be responsible for differences in responsiveness. Similarly, when one contrasts responses of small veins in the hindpaw of the dog to those in the perfused gracilis muscle, several differences become apparent.[84] Sympathetic stimulation and several vasoconstrictor agents, most of which are naturally occurring potent vasoactive substances, have little or no effect on venous segments in the isolated gracilis; their major constrictor effect in the muscle is on arteriolar resistance vessels. We have found that norepinephrine, dopamine, and vasopressin cause some venoconstriction in the perfused gracilis, but this contributed only 4%–8% of the average increase in vascular resistance in that muscle. Nerve stimulation, angiotensin, and serotonin had no effect on the venous

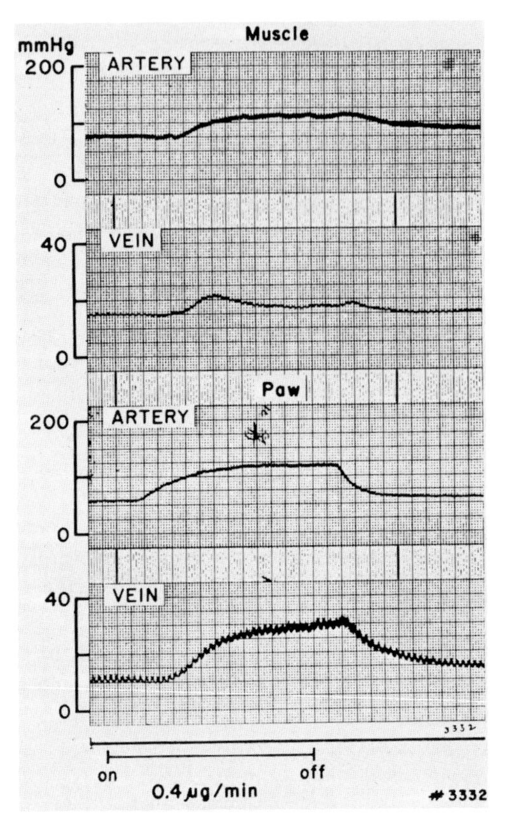

Fig. 5-13. Contrast between the magnitude of venous constrictor responses to the same drug in different vascular beds. Norepinephrine causes marked venoconstriction in the paw and minimal venoconstriction in the muscle of the hindlimb of dogs. (Reproduced from Abdel-Sayed WA, Abboud FM, Ballard DR. Am J Physiol 218:1293, 1970.)

segment in muscle. In contrast, the small veins draining the hindpaw constricted intensely and contributed up to 50% of the total increase in resistance. The responses of the venous segment to norepinephrine, angiotensin, serotonin, and dopamine in the paw were between 10 and several hundred times greater than those of the small veins of the muscle. Vasopressin is rather unique in its action in that its effect on venous as well as on arteriolar resistance is negligible in the paw as compared to the muscle. In contrast to vasopressin, responses to serotonin are negligible in the muscle and very marked in the paw. These differential responses also suggest that vessel geometry and wall structure do not account for differential effects.

Venous versus arteriolar responses. There are several instances in which the venous responses are different from those of arterioles coupled in series, not only quantitatively but also qualitatively. An understanding of these differences allows the appreciation of the effects of various humoral agents on capillary filtration, since the capillary hydrostatic pressure is determined predominantly by the ratio of postcapillary to precapillary resistance. Several examples will be cited.

Although norepinephrine constricts both arterioles and veins, epinephrine given into the brachial artery in man dilates arteriolar resistance vessels to skeletal muscle but constricts forearm veins. Bradykinin also increases forearm blood flow by dilating arteriolar resistance vessels, but it increases venous tone.[91]

Isoproterenol dilates arteriolar resistance vessels markedly, but its effect on veins is negligible.[84] This set of circumstances causes an increase in hydrostatic pressure and capillary filtration by increasing the ratio of postcapillary to precapillary resistance. Isoproterenol, therefore, would not favor volume conservation in any shock situation in which hypovolemia is a factor.

Acetylcholine dilates arterial resistance vessels in both cutaneous and muscular beds; however, in the perfused cutaneous veins it causes significant venoconstriction[54] (Fig. 5-14). Although it is commonly suggested that histamine causes arteriolar dilatation and venular constriction, recent studies[49] indicate that both precapillary and postcapillary resistances decrease during its intraarterial administration into the hindlimb of dog.

Subcutaneous injections of histamine (3 mg base/kg) in the dog cause a fall in arterial prssure, a reduction in splanchnic blood flow, and a rise in hepatic venous wedge pressure caused by hepatic venoconstriction.[92]

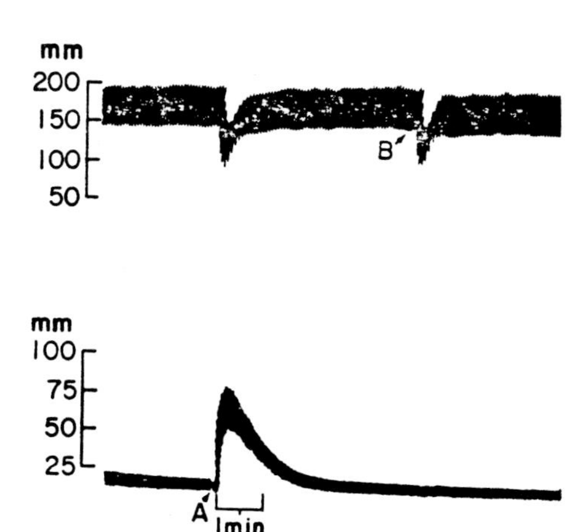

Fig. 5-14. Acetylcholine which dilates arterioles may cause venous constriction. Systemic arterial blood pressure measured in the carotid artery (top tracing); perfusion pressure measured in the accessory cephalic vein (bottom tracing). (A) 40 μg of ACh given by close venous injection; (B) 40 μg of ACh given systemically through the femoral vein. (Reproduced from Rice AJ, Long JP. J Pharmacol Exp Ther 151:425, 1966.)

The vasoconstrictor effects of serotonin in the hindlimb are predominant in the cutaneous paw, and the venous segment constricts markedly and to a much greater degree than the arterial segment. Angiotensin, on the other hand, a very potent arteriolar constrictor agent particularly in skeletal muscle beds, produces only minimal venoconstriction.[93,94]

There are few drugs that produce venodilatation. Nitroglycerin causes relaxation of smooth muscle in the venous wall and increases the volume of blood in capacitance vessels of limbs. Venous relaxation with nitroglycerin and pooling of blood in the extremeties may alter cardiac dynamics in such a way as to decrease the cardiopulmonary volume and reduce myocardial oxygen consumption—a mechanism of action that might explain in part its beneficial effect in angina pectoris.[87,95] In contrast, amyl nitrite, which also causes arteriolar vasodilation and a decrease in vascular resistance as does nitroglycerin, will cause reflex constriction of the limb veins, probably as a result of the fall in arterial pressure in a manner analogous to the indirect effect of isoproterenol.[95] Its salutary effect in angina pectoris might be more related to the decrease in arterial pressure and afterload.

Prostaglandins of the A and E groups are potent vasodilator compounds, but their effect is predominantly on arteriolar resistance.[96] Their venous effects differ in muscle and skin; they cause venodilatation in muscle and venoconstriction in the cutaneous paw.[97] Prostaglandin $F_{2\alpha}$ also may cause arteriolar dilatation, but it is a potent venoconstrictor.[98] In isolated vascular strips contraction has been reported with the vasodilator prostaglandins, unless the strip is partially contracted with catecholamines or KCl.[99]

Since cutaneous veins are most responsive to cold, to humoral as well as to neurogenic stimuli, we examined the effect of cooling on their responsiveness.[26] Cooling reduced the arteriolar constrictor responses to sympathetic stimulation and to norepinephrine, but the venoconstrictor responses to these interventions were augmented. The difference in behavior of arterial and venous segments negates the possibility that cooling alters responsiveness through its effect on the metabolism of norepinephrine or its binding or degradation. If this were the case one would have expected parallel changes in both arteries and veins. In order to explain the selectively augmented venous responses we postulate a selective distribution of thermal receptors in cutaneous veins, but not in arteries. The thermal receptors may sensitize veins, and the direct depressing effect of cooling on physicochemical reactions necessary for contraction and relaxation may decrease arteriolar responses and delay contraction and prolong relaxation.

Venous Adjustments in Pathologic States

Hemorrhage and Shock

In hemorrhage or hypovolemia the reduction in blood volume is compensated for by contraction of capacitance vessels, predominantly in the splanchnic bed. The neurogenic response of veins of the extremities is negligible, but the humoral factors released during sustained hypotension and shock such as vasopressin, angiotensin, norepinephrine, and epinephrine released from the adrenal gland would all tend to increase venous tone and cause reduction in the capacitance of the forearm vessels, particularly the cutaneous veins in shock. One may postulate that warming subjects in hemorrhagic shock would eliminate the venoconstrictor adaptation and might have a detrimental effect.

It has also been suggested that in prolonged shock and during the associated acidosis an increase in postcapillary resistance might be sustained and eventually attain a relatively greater magnitude than the increase in precapillary resistance, thereby increasing the loss of intravascular fluid by capillary filtration.[100] The administration of an alpha blocker will eliminate the venoconstrictor response to adrenergic stimuli more completely than the arteriolar constrictor responses and thereby reverse the loss of intravascular volume.[41,85,100] There is a paucity of information concerning venous responses in man during hemorrhagic or cardiogenic shock.

Orthostatic Hypotension

A dramatic fall in arterial pressure occurs during upright tilt in patients who have orthostatic hypotension on the basis of autonomic denervation, such as patients with peripheral neuropathy, diabetes, and the Shy-Drager syndrome. This is caused primarily by an absence of reflex vasoconstriction. As indicated earlier, the veins of the extremities do not participate significantly in the reflex reduction of venous capacitance, but the splanchnic veins may. There is no information on responses of splanchnic capacitance vessels during orthostatic hypotension. Although all indications are that the arteriolar resistance vessels lack adrenergic vasoconstrictor innervation, the state of innervation of the venous system in this disease is still equivocal. We have failed to demonstrate venoconstrictor responses to deep breath in such an individual (Fig. 5-15), whereas others have.[65] Additional information is necessary on the venous innervation in this disease state.

It has been noted that vasovagal syncope is associated with marked increase in capacitance of veins of the forearm, suggesting that a venodi-

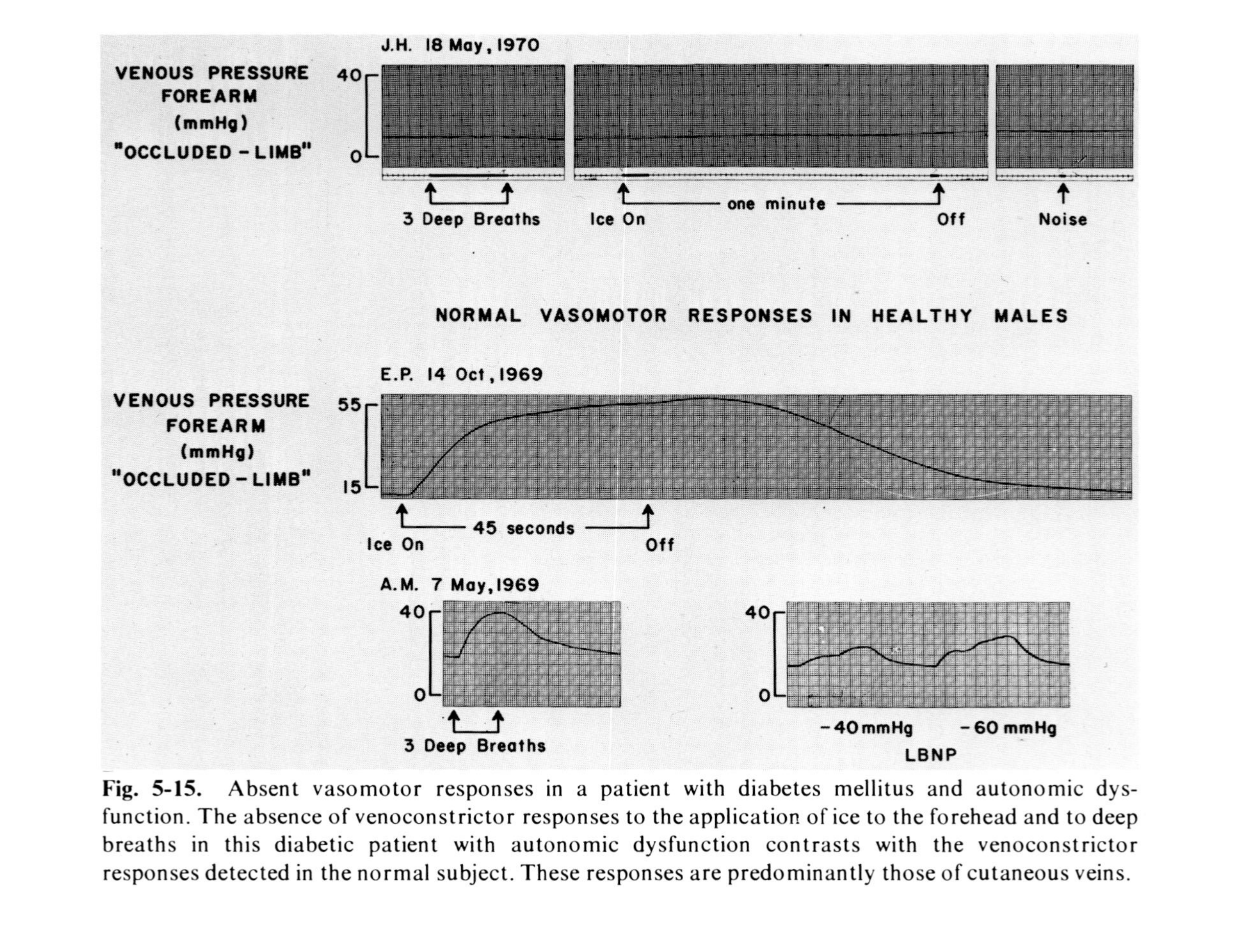

Fig. 5-15. Absent vasomotor responses in a patient with diabetes mellitus and autonomic dysfunction. The absence of venoconstrictor responses to the application of ice to the forehead and to deep breaths in this diabetic patient with autonomic dysfunction contrasts with the venoconstrictor responses detected in the normal subject. These responses are predominantly those of cutaneous veins.

lator cholinergic innervation might contribute to fainting. On the other hand, recent evidence by Epstein et al.[101] suggests that the veins retain their ability to constrict during such an episode, the constriction, however, may be a delayed response to humoral factors.

Congestive Heart Failure

Several factors should be considered in evaluating the behavior of the venous system in congestive heart failure.

1. The elevated venous pressure is caused in part by hypervolemia and some reduction in ventricular compliance. Splanchnic blood volume increases. This increase may be entirely passive and may be related to the high central venous pressure.

2. Wood et al.[102] found that resting venous tone in the forearm of patients with cardiac failure is greater than in normal subjects and in patients with compensated heart failure. Furthermore, Wood has demonstrated that in cardiac failure the venoconstrictor response to exercise is exaggerated.[103] There is a question whether these increases in venous tone reflect greater sympathetic discharge to the veins or an augmented response to circulating catecholamines or simply higher levels of circulating catecholamines in heart failure. Recent studies utilizing the occluded-limb technique and ganglionic blockade have demonstrated that sympathetic adrenergic activity to the limb veins is minimal in patients in heart failure, as it is in normal patients provided the subjects are relaxed and warm and not dyspneic.

3. After blockade of alpha adrenergic receptors and presumably the removal of all adrenergic activity in patients in congestive heart failure, there is a residual increase in tone of forearm veins that has been ascribed at least in part to structural changes in the venous walls, possibly related to sodium and water retention.[104]

4. The action of digitalis or ouabain on veins is of interest in that its direct effects induce venoconstriction, probably through the release of endogenous catecholamines or the facilitation of responses to adrenergic stimulation.[105] Digitalis glycosides may also increase cutaneous venous tone in normal man with or without alpha receptor blockade, suggesting a direct action. In contrast to the venous constrictor responses in the normal state, the effect in congestive heart failure is venodilatation.[106] This may be related to the fact that improvement in cardiac output coincidental with the administration of digitalis in heart failure reduces the prevailing sympathoadrenal stimulation.

5. With respect to the effects of morphine in acute pulmonary edema, the possible role of venodilatation is still inconclusive. In one study intravenous administration of morphine was associated with

decreases in resting venous tone. Sharpey-Schafer has also reported that the intravenous administration of theophylline, another drug used in the treatment of acute pulmonary edema, causes dilatation of veins and sharp reductions in central venous pressure, suggesting a possible mechanism for its beneficial effect.[107]

6. Vasodilator therapy has been advocated for the treatment of intractable heart failure.[108] The mechanisms of action may include, in addition to the decrease in afterload, arterial pressure and aortic impedance, an increase in venous capacitance, and reduction in cardiac size.

Hypertension

The role of the venous system in hypertension is still equivocal. Studies in man have shown either no change in pressure–volume relationship in the forearm or a slight increase that is reduced after antihypertensive therapy.[109] The direct effect of angiotensin on veins of the forearm is negligible.[94] There has been some suggestion that it can cause constriction of isolated venous segments when given intravenously and that this effect may be reflex in nature.[110] In pheochromocytoma, circulating catecholamines may increase venous tone, and this in turn may cause a high output and arterial pressure.

Studies in animals indicate that there is an increase in contractility of portal veins from spontaneously hypertensive rats[111] and reduced compliance of femoral and mesenteric veins in renal hypertension in dogs.[111a,b] The evaluation of the contribution of the venous system to the hemodynamic state in different types of hypertension in man requires more study.

Pregnancy, Oral Contraceptives, and Varicose Veins

McCausland suggests from clinical observation that there is greater venous distensibility during pregnancy, associated with a decrease in venous tone and velocity of blood flow. Progesterone, rather than estrogen, is thought to mediate this loss of venous tone, which may contribute to stasis and to increased incidence of venous thrombosis during pregnancy and also in women taking oral contraceptives.[112,113]

Although varicose veins have been attributed to mechanical factors, such as incompetent valves, several investigators have demonstrated that there is a generalized loss of venous tone in patients with varicose veins that can be demonstrated in the veins of the upper extremities. In these patients venous distensibility is increased in the forearm, as well as in the leg, suggesting a more systemic venous disorder despite the predominance of the manifestations in the lower extremities.[114]

II. PATHOPHYSIOLOGY OF VENOUS DISEASE OF THE LOWER EXTREMITIES

Peripheral venous disease may result in two mechanisms of abnormal venous dynamics: obstruction to venous flow and retrograde flow (reflux) through damaged venous valves. Obstruction to venous flow is usually a consequence of acute venous thrombosis. Venous valvular damage with reflux may occur as a late sequela of deep vein thrombosis and is the predominant abnormality in varicose veins. These three diseases (acute venous thrombosis, postphlebitic stasis syndrome, and varicose veins) are common conditions of the lower extremities, and their frequency may surpass that of arterial occlusive disease. The clinical diagnosis and management of these conditions have been greatly aided by recent developments in noninvasive techniques that detect the abnormal venous dynamics. This section will briefly review the principles of two useful diagnostic methods (Doppler ultrasound and strain-gauge plethysmography) and will describe their application to the study of acute venous thrombosis, postphlebitic syndrome, and varicose veins.

Techniques

Doppler Ultrasound

The Doppler ultrasonic velocity detector[115] contains an oscillator that causes a piezoelectric crystal in a hand-held probe to emit a beam of ultrasound of 5–10 MHz (megacycles). The sound is transmitted into the tissues through an acoustic gel on the skin. Sound reflected from moving blood cells is shifted in frequency by an amount proportional to the blood flow velocity. The reflected sound is received by a second crystal, and the frequency shift (Doppler effect) is detected and amplified by the instrument as an audible signal or a recordable analogue waveform. All of the major veins of the extremities may be directly examined with the instrument.[116] Information is indirectly obtained about the patency of the major intraabdominal and thoracic veins. A normal venous velocity signal sounds like a cyclic windstorm that waxes and wanes with respiration. With inspiration and descent of the diaphragm, intraabdominal pressure is raised and venous outflow from the lower extremity diminishes or ceases completely. With expiration, venous flow is augmented. In the upper extremities and neck, venous velocity may be maximal with inspiration and diminish with expiration. The patency of the venous system is further assessed by augmenting venous flow by compression of muscle masses distal or upstream to the point of auscultation or by release of compression proximal or downstream to the probe. In the presence of venous thrombosis the venous signal may be completely absent. Thrombosis cardiad or downstream to the vein being examined may result in sufficient impedance to flow to result in a more continuous velocity signal little affected by respiration. Thrombosis may also attenuate the augmented velocity response to limb compression maneuvers. Compression of the limb cardiad to the vein being examined results in no retrograde flow if the venous valves are competent. In the presence of the postphlebitic syndrome or varicose veins, such proximal limb compression will result in audible retrograde venous flow that may be heard with the Doppler detector. Doppler ultrasound will thus provide a qualitative detection of both venous outflow obstruction and venous reflux through incompetent valves in various peripheral venous disorders.

Strain-Gauge Plethysmography

A variety of plethysmographs have been employed to study arterial and venous disease in human limbs. Strain-gauge plethysmography[117] has proved to be a simple and accurate method of quantitating abnormalities of venous outflow obstruction[118] and venous reflux[119] in peripheral venous diseases. The patients are studied supine with the lower extremity elevated above atrial level. A mercury-in-silastic strain gauge is placed around the maximal girth of the calf. A pneumatic cuff is placed on the thigh above the knee and is inflated to 50 mm Hg to temporarily occlude venous return. Continued arterial inflow into the calf results in an increase in calf circumference, as evidenced by an upslope on the recording. When the calf circumference stabilizes in 1 or 2 min the thigh cuff is rapidly deflated, decompressing the calf with a resultant downslope on the recording. From a tangent line drawn to the initial steepest downslope one may calculate maximum venous outflow of the calf in milliliters per minute per 100 ml of calf tissue (Fig. 5-16A). To measure maximum venous reflux flow, a proximal thigh tourniquet is applied and inflated above systemic arterial pressure (250 mm Hg). With the limb temporarily isolated from the circulation, a thigh cuff is then inflated to 50 mm Hg. The underlying thigh venous blood can only be translocated in an upstream or distal direction toward the calf at a rate and magnitude proportional to the incompetence of the venous valves. The resultant rate of increase in calf circumference permits calculation of maximum venous reflux flow in milliliters per minute per 100 ml of calf tissue.

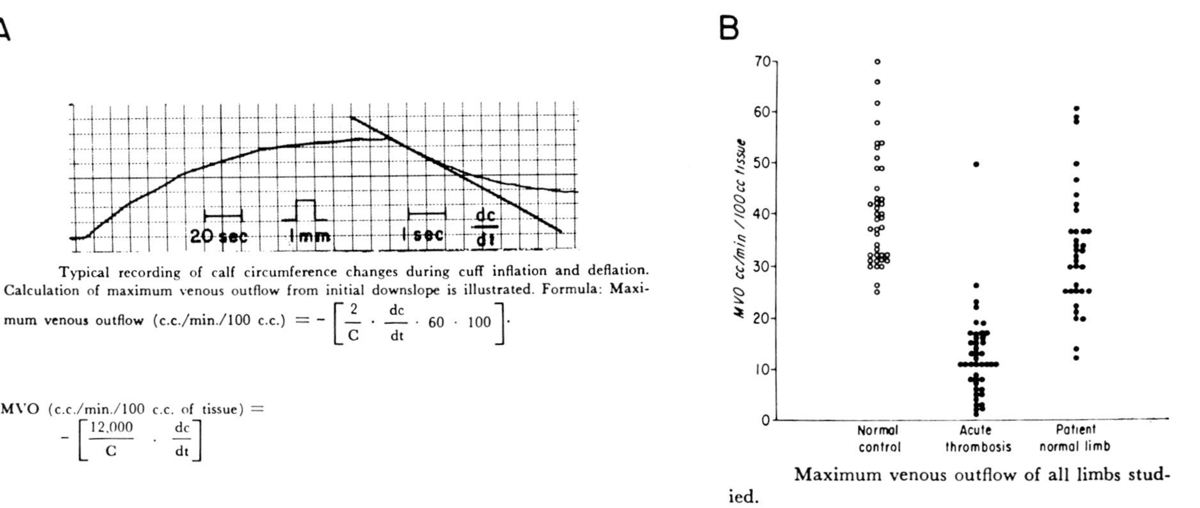

Typical recording of calf circumference changes during cuff inflation and deflation. Calculation of maximum venous outflow from initial downslope is illustrated. Formula: Maximum venous outflow (c.c./min./100 c.c.) $= -\left[\frac{2}{C} \cdot \frac{dc}{dt} \cdot 60 \cdot 100\right]$.

MVO (c.c./min./100 c.c. of tissue) $= -\left[\frac{12{,}000}{C} \cdot \frac{dc}{dt}\right]$

Maximum venous outflow of all limbs studied.

Fig. 5-16. Obstruction of venous outflow. The tracing represents the change in calf circumference during inflation and deflation. Maximum venous outflow (MVO) can be calculated from the slope (dc/dt) of the change in calf circumference (or volume) with time. Values in the graph are obtained from normal subjects and patients with acute thrombosis of one limb. (Reproduced from Barnes RW, Collicott PE, Mozersky DJ, et al.: Noninvasive quantitation of maximum venous outflow in acute thrombophlebitis. Surgery 73:971, 1972.)

Acute Venous Thrombosis

Deep Vein Thrombosis

Acute deep vein thrombosis most often occurs in the veins draining the lower extremities, which are considered to be the source of most pulmonary emboli. The clinical manifestations of deep vein thrombosis are nonspecific and are an inaccurate guide to the presence or absence of venous disease. Objective demonstration of deep vein thrombosis has been traditionally carried out by contrast phlebography. The outline of the thrombus as a radiologic defect in contrast material injected into a peripheral vein of the foot remains the most accurate index of the presence of venous thrombosis. In experienced hands the abnormalities of venous outflow obstruction detectable by Doppler ultrasound permit identification of major venous thrombosis with an accuracy of 95%, as compared to contrast phlebography.[120] Most errors of diagnosis on Doppler examination relate to the presence or absence of venous disease in small muscle veins of the calf. During acute venous thrombosis venous reflux is usually not demonstrable on Doppler examination.

The effect of acute venous thrombosis on obstruction to venous outflow from the leg as quantitated by strain-gauge plethysmography is depicted in Fig. 5-16. The mean maximum venous outflow of both limbs of 20 normal subjects was 41 ± 11 ml/min/100 ml of tissue. No normal subject had a maximum venous outflow less than 25 ml/min/100 ml, while all but 2 of the patients' affected limbs had an outflow below this value. The maximum venous outflow was normal in most of the patients' limbs that were free of disease. Five of the 6 patients who had low venous outflows in undiseased limbs had been at bedrest for prolonged periods. Whether a reduction in maximum capacity of venous outflow from the leg in such patients predisposes to thrombosis is problematical.

Superficial Thrombophlebitis

Acute venous thrombosis of the superficial veins commonly occurs on the leg, especially in varicose veins. In the absence of predisposing varicose veins, the diagnosis of superficial thrombophlebitis may be confused with lymphangitis or cellulitis. Inasmuch as the superficial lymphatics may parallel the saphenous veins of the leg, diagnostic differentiation may be aided by Doppler ultrasound. In the presence of acute superficial thrombophlebitis, no venous velocity is heard in the area of inflammation although prominent arterial signals may be heard associated with the inflammation. In contrast, acute lymphangitis is characterized by the presence of high velocity venous signals in the saphenous vein coursing through the affected area. The Doppler detector is particularly useful in such inflammatory conditions which are difficult to examine by invasive phlebographic techniques. The patency of the major deep veins results in normal venous outflow of the extremity as measured by plethysmography.

Postphlebitic Syndrome

The clinical postphlebitic syndrome consists of lower extremity pain, edema, hyperpigmentation, stasis dermatitis, and ulceration. This syndrome usually develops only after several months or years following an episode of deep vein thrombosis that may or may not have been clinically recognized. The condition is considered to be the result of persistent ambulatory venous hypertension in the lower leg. While either venous outflow obstruction or venous reflux through incompetent venous valves may contribute to postphlebitic stasis, the latter mechanism is felt to be the most important.

In the normal standing individual the saphenous venous pressure at the ankle is approximately 100 mm Hg and is equal to the hydrostatic column of blood. During ambulation this pressure should fall to approximately 20 mm Hg as the superficial veins are decompressed through the action of a competent leg muscular venous pump. In the patient with primary varicose veins and a normal deep venous system the ambulatory venous pressure will fall to approximately 40–60 mm Hg, as the superficial veins are incompletely decompressed because of retrograde venous flow through incompetent valves of the superficial venous system. In the patient with a postphlebitic syndrome with deep venous valvular incompetence the ambulatory venous pressure at the ankle may show little fall and occasionally may increase as muscle compression drives blood upstream distally through incompetent deep and perforating venous valves. This ambulatory venous hypertension is the pathophysiologic mechanism that results in the postphlebitic stasis syndrome.

The Doppler ultrasonic velocity detector permits recognition of retrograde venous flow in the deep venous system in response to proximal limb compression. While deep venous obstruction is also noted in the postphlebitic syndrome, most abnormalities on Doppler examination relate to incompetent venous valves and retrograde venous flow. The localization of incompetent perforating veins can be readily carried out by Doppler examination.[121] With a tourniquet about the calf to prevent retrograde flow in superficial veins, compression above the tourniquet results in incompetent venous flow signals in the lower leg at the site of incompetent perforating veins.

Quantitation of abnormal venous hemodynamics in the postphlebitic syndrome has been carried out by strain-guage plethysmography.[122] Figure 5-17A shows the maximal venous outflow in postphlebitic limbs compared to normal legs. The mean maximum venous outflow of postphlebitic limbs was 34 ± 15 ml/min/100 ml, compared to 39 ± 7 ml /min/100 ml in the normal control limbs. This difference is not statistically significant. However one-third of postphlebitic legs had an outflow below the lower normal limit of 25 ml/min/100 ml. Seven of the 13 limbs were in patients with plication of the inferior vena cava. Figure 5-17B shows the results of maximum venous reflux flow in postphlebitic limbs compared to normal legs. The mean value of 12.7 ± 7.0 ml/min/100 ml of maximum venous reflux flow in postphlebitic legs was significantly greater than that of normal controls, 3.1 ± 1.1 ml/min/100 ml. All but 2 of the 39 postphlebitic legs had reflux flow greater than 6.0 ml/min/100 ml, the upper limit of normal values. Thus plethysmography confirms that the major pathophysiologic abnormality in the postphlebitic syndrome is abnormal venous reflux through incompetent venous valves. The majority of postphlebitic limbs have no impedance to maximum venous outflow.

Varicose Veins

Varicose veins are usually classified as primary or secondary. Primary varicose veins are usually familial, and abnormalities are limited to the superficial veins, with a normal deep venous system. Secondary varicose veins are associated with underlying deep venous disease, usually prior deep vein thrombosis that may or may not have been clinically evident. The postphlebitic syndrome and the so-called varicose ulcer at the ankle are almost always associated with secondary varicose veins. The natural histories of primary and secondary varicose veins may differ, inasmuch as secondary varices may be associated with recurrent episodes of deep vein thrombosis and pulmonary embolism. These problems are uncommon in primary varicose veins. The differentiation of these two conditions is thus of considerable clinical importance.

Primary varicose veins usually involve the greater or lesser saphenous venous systems. The dilated veins are associated with venous valvular incompetence. The exact etiologic mechanism remains unknown. The theory of progressive descending valvular incompetence is supported by studies with Doppler ultrasound.[123] Patients with greater saphenous varicosities almost always have incompetence of the common femoral vein, implying a lack or dysfunction of iliac venous valves. Increased

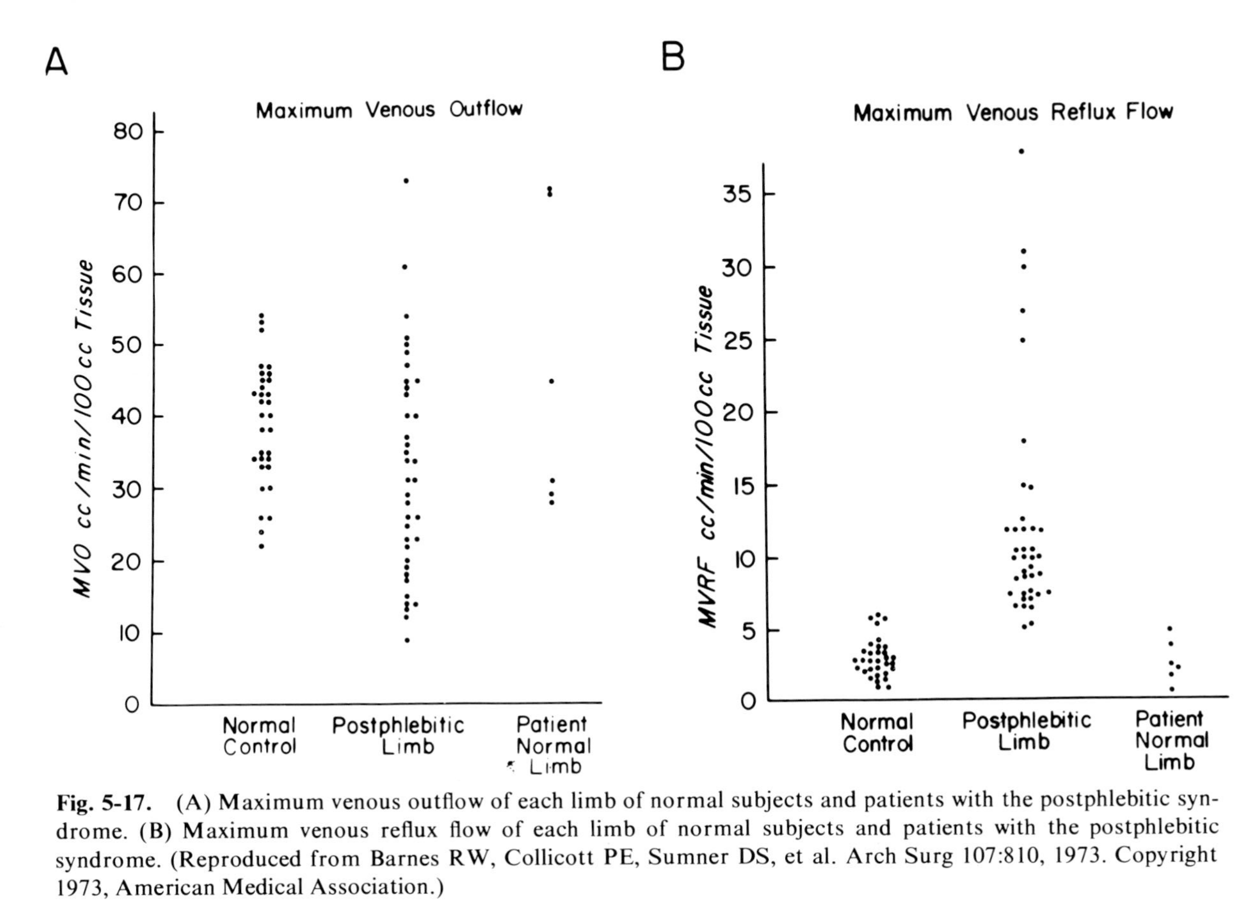

Fig. 5-17. (A) Maximum venous outflow of each limb of normal subjects and patients with the postphlebitic syndrome. (B) Maximum venous reflux flow of each limb of normal subjects and patients with the postphlebitic syndrome. (Reproduced from Barnes RW, Collicott PE, Sumner DS, et al. Arch Surg 107:810, 1973. Copyright 1973, American Medical Association.)

intraabdominal pressure results in retrograde venous flow into the saphenous venous system. Approximately 16% of normal subjects have absent iliac venous valves. The incidence of nonfunctional iliac venous valves is approximately 32% in children of patients with primary varicose veins who were surveyed by Doppler ultrasound.[124]

The differentiation of primary from secondary varicose veins can be qualitatively determined by Doppler ultrasound.[125] In primary varicose veins, venous valvular incompetence is detectable in the superficial veins by retrograde Doppler venous flow signals in response to proximal compression of the varices. However, examination of the deep venous system confirms valvular competence and patency of the veins. In secondary varicose veins, venous valvular incompetence is detectable in both the superficial and deep veins. In addition, deep venous obstruction may be detectable by Doppler ultrasound.

Quantitation of abnormal venous hemodynamics and varicose veins has been carried out by strain-gauge plethysmography.[125] The contribution of varicose veins to maximum venous outflow of the leg can be assessed by performing venous outflow plethysmography before and after application of a rubber tourniquet immediately below the knee between the calf strain gauge and the thigh cuff. The narrow rubber tourniquet compresses the superficial but not the deep veins. The results of determination of maximum venous outflow before and after tourniquet application are shown in Fig. 5-18A. The mean maximum venous outflow of normal legs (38 ± 7 ml/min/100 ml tissue) did not differ significantly from that of limbs with Doppler evidence of primary or secondary varicose veins (39 ± 12 or 39 ± 10 ml/min/100 ml, respectively). The small drop in maximum venous outflow resulting from application of the rubber tourniquet at the knee did not differ significantly between the normal subjects and the two groups of patients with varices. The results of maximum venous reflux flow in normal subjects and patients with primary and secondary varices are shown in Fig. 5-18B. The mean maximum venous reflux flow in limbs with Doppler evidence of primary or secondary varices (9 ± 6 or 13 ± 7 ml/min/100 ml, respectively) was significantly greater than that of normal legs (3 ± 2 ml/min/100 ml). Tourniquet compression in patients with primary varicose veins reduced the mean maximum venous reflux flow to normal (2 ± 1 ml/min/100 ml). However, tourniquet compression of the varices did not significantly reduce the maximum venous reflux flow in limbs with Doppler evidence of secondary varicose veins (11 ± 5 ml/min/100 ml). These findings suggest that although abnormal venous reflux is present in patients with both primary and secondary varicose veins the reflux can be obliterated by tourniquet compression of primary varicose veins. However, patients

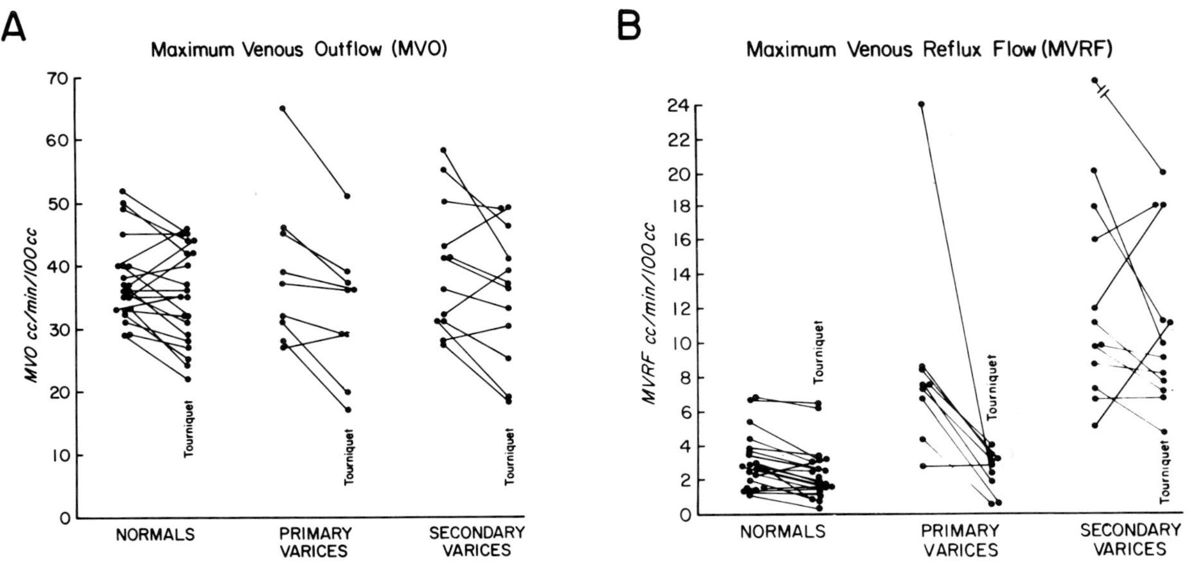

Fig. 5-18. (A) In both primary and secondary varices the maximum venous outflow was within the normal range before and after occlusion of drainage through the superficial veins with the tourniquet. (B) The maximum venous reflux was greater in the presence of both primary and secondary varices. Application of the tourniquet permitted the separation of patients with primary varices and incompetence of superficial veins from those with the postphlebitic secondary varices who have incompetence of the deep venous valves. (Reproduced from Barnes RW by permission of Surg Gyn and Obstet 141:207, 1975)

with secondary varicose veins continue to manifest abnormal deep venous reflux despite compression of the superficial varicosities. These findings may have considerable importance in proper selection of patients for surgical treatment of varicose veins.

Conclusions

While venous disease remains an important health problem, previous management of venous disease has primarily been based upon the bedside clinical manifestations of these problems. The inaccuracy of the clinical examination has been apparent in studies involving objective diagnostic techniques for the evaluation of venous disease. The traditional diagnostic assessment of venous disease involves the use of contrast phlebography. While this remains the diagnostic standard with which other techniques must be compared, recent years have led to many noninvasive approaches for objective assessment of venous abnormalities. Two very useful techniques, Doppler ultrasound and plethysmography, have permitted increased accuracy of diagnosis of venous problems based upon the pathophysiologic sequelae of these conditions. Further application of these unique clinical tools may permit improved understanding of the incidence, altered venous dynamics, and natural history of peripheral venous disease in man.

REFERENCES

1. Eckstein JW, Abboud FM: Peripheral venous mechanisms which influence arterial pressure, in Mills LC, Moyor JH (eds): Shock and Hypotension: Pathogenesis and Treatment. New York, Grune & Stratton, 1965
2. Mark AL, Eckstein JW: Venomotor tone and central venous pressure. Med Clin North Am 52:1077, 1968
3. Zitnik RS, Shepherd JT: Control of the venous system in man, in Yu PN, Goodwin JF (eds): Progress in Cardiology. Philadelphia, Lea & Febiger, 1972, pp 185–203
4. Rowell LB: The cutaneous circulation; Circulation to skeletal muscle; The splanchnic circulation, in Ruch TC, Patton HD (eds): Textbook of Physiology and Biophysics, vol II (ed 20). Philadelphia, WB Saunders, 1974
5. Burton AC: The vascular bed, in: Physiology and Biophysics of the Circulation. Chicago, Year Book Medical Publishers, 1968, pp 61–84
6. Rushmer RF: Effect of posture, in: Cardiovascular Dynamics. Philadelphia, W. B. Saunders, 1961, pp 171–192
7. Pollack AA, Wood EH: Venous pressure in the saphenous vein at the ankle in man during exercise and changes in posture. J Appl Physiol 1:649, 1949

8. Cohn JN: Central venous pressure as a guide to volume expansion. Ann Intern Med 66:1283, 1967
9. Tristani FE, Cohn JN: Masked heart failure in shock. Circulation 34 [Suppl III]:230, 1966
10. Karim F, Kidd C, Malpus CM, Penna PE: The effects of stimulation of the left atrial receptors on sympathetic efferent nerve activity. J Physiol (Lond) 227:243, 1972
11. Linden RJ: Function of cardiac receptors. Circulation 48:463, 1973
12. Zoller RP, Mark AL, Abboud FM, et al: The role of low pressure baroreceptors in reflex vasoconstrictor responses in man. J Clin Invest 51:2967, 1972
13. Johnson JA, Moore WW, Segar WE: Small changes in left atrial pressure and plasma antidiuretic hormone sites in dogs. Am J Physiol 217:210, 1969
14. Brennan LA Jr, Malvin RL, Jochim KE, et al: Influence of right and left atrial receptors in plasma concentrations of ADH and renin. Am J Physiol 221:273, 1971
15. Mancia G, Donald DE, Shepherd JT: Inhibition of adrenergic outflow to peripheral blood vessels by vagal afferents from the cardiopulmonary region in the dog. Circ Res 33:713, 1973
16. Mancia G, Romero C, Shepherd JT: Continuous inhibition of renin release in dogs by vagally innervated receptors in the cardiopulmonary region. Circ Res 36:529, 1975
17. Koike H, Mark AL, Heistad DD, et al: Influence of cardiopulmonary vagal afferent activity on carotid chemoreceptor and baroreceptor reflexes. Circ Res (in press)
18. Habak PA, Mark AL, Kioschos JM, et al: Effectiveness of congesting cuffs ("rotating tourniquets") in patients with left heart failure. Circulation 50:366, 1974
19. Pappenheimer JR, Soto-Rivera A: Effective osmotic pressure of the plasma proteins and other quantities associated with the capillary circulation in the hindlimbs of cats and dogs. Am J Physiol 152:471, 1948
20. Mellander S, Lewis DH: Effect of hemorrhagic shock on reactivity of resistance and capacitance vessels and on capillary filtration transfer in cat skeletal muscle. Circ Res 13:105, 1963
21. Djojosugito AM, Folkow B, Öberg B, White S: A comparison of blood viscosity measured in vitro and in a vascular bed. Acta Physiol Scand 78:70, 1970
22. Landis EH, Gibbon JH: The effects of temperature and of tissue pressure on the movement of fluid through the human capillary wall. J Clin Invest 21:105, 1933
23. Haddy FJ, Gilbert RP: The relation of a venous-arteriolar reflex to transmural pressure and resistance in small and large systemic vessels. Circ Res 4:25, 1956
24. Webb-Peploe MM, Shepherd JT: Response of dogs' cutaneous veins to local and central temperature changes. Circ Res 23:693, 1968
25. Webb-Peploe MM: Effect of changes in central body temperature on capacity elements of limb and spleen. Am J Physiol 216:643, 1969

26. Abdel-Sayed WA, Abboud FM, Calvelo MG: Effect of local cooling on responsiveness of muscular and cutaneous arteries and veins. Am J Physiol 219:1772, 1970
27. Cooper KE, Johnson RH, Spalding JMK: The effects of central body and trunk skin temperatures on reflex vasodilatation in the hand. J Physiol (Lond) 174:46, 1964
28. Zitnik RS, Ambrosioni E, Shepherd JT: Effect of temperature on cutaneous venomotor reflexes in man. J Appl Physiol 31:507, 1971
29. Rowell LB, Brengelmann GL, Detry JMR, Wyss C: Venomotor responses to local and remote thermal stimuli to skin in exercising man. J Appl Physiol 30:72, 1971
30. Heistad DD, Abboud FM, Mark AL, Schmid PG: Interaction of thermal and baroreceptor reflexes in man. J Appl Physiol 35:581, 1973
31. Vanhoutte PM, Lorenz RR: Effect of temperature on reactivity of saphenous, mesenteric and femoral veins of the dog. Am J Physiol 218:1746, 1970
32. Müller O: Uber die Blutverteslung im menschlichen Korper unter dem Ernfluss thermischer Reize. Dtsch Arch Klin Med 82:547, 1905
33. Glaser EM, Berridge FR, Prior KM: Effects of heat and cold on the distribution of blood within the human body. Clin Sci 9:181, 1950
34. Rowell LB, Brengelmann GL, Blackmon JR, et al: Splanchnic blood flow and metabolism in heat-stressed man. J Appl Physiol 24:475, 1968
35. Rowell LB, Brengelmann GL, Blackmon JR, Murray JA: Redistribution of blood flow during sustained high skin temperature in resting man. J Appl Physiol 28:415, 1970
36. Rowell LB, Detry JMR, Blackmon JR, Wyss C: Importance of the splanchnic vascular bed in human blood pressure regulation. J Appl Physiol 31:864, 1971
37. Litter J, Wood JE: The venous pressure–volume curve of the human leg measured in vivo. J Clin Invest 33:953, 1954
38. Wood JE, Eckstein JW: A tandem forearm plethysmograph for study of acute responses of the peripheral veins of man: The effect of environmental and local temperature change, and the effect of pooling blood in extremities. J Clin Invest 37:41, 1958
39. Whitney RJ: The measurement of volume changes in human limbs. J Physiol 121:1, 1953
40. Eckstein JW, Wendling MG, Abboud FM: Forearm venous responses to stimulation of adrenergic receptors. J Clin Invest 44:1151, 1965
41. Abboud FM, Schmid PG, Eckstein JW: Vascular responses after alpha adrenergic receptor blockade. I. Responses of capacitance and resistance vessels to norepinephrine in man. J Clin Invest 47:1, 1968
42. Samueloff SL, Bevegard BS, Shepherd JT: Temporary arrest of circulation to a limb for the study of venomotor reactions in man. J Appl Physiol 21:341, 1966
43. Burch GE, Murtadha M: A study of the venomotor tone in a short intact venous segment of the forearm of man. Am Heart J 51:807, 1956
44. Sharpey-Schafer EP: Venous tone: Effects of reflex changes, humoral agents and exercise. Br Med Bull 19:145, 1963

45. Webb-Peploe MM, Shepherd JT: Response of large hindlimb veins of the dog to sympathetic nerve stimulation. Am J Physiol 215:299, 1968
46. Eckstein JW, Mark AL, Schmid PG, et al.: Responses of capacitance vessels to physiologic stimuli. Trans Am Clin Climatol Assoc 81:57, 1969
47. Haddy FJ, Fleishman M, Emanuel DA: Effect of epinephrine, norepinephrine and serotonin upon systemic small and large vessel resistance. Circ Res 5:247, 1957
48. Abboud FM: Control of the various components of the peripheral vasculature. Fed Proc 31:1226, 1972
49. Diana JN, Kaiser RS: Pre- and post-capillary resistance during histamine infusion in isolated dog hindlimb. Am J Physiol 218:132, 1970
50. Fuxe K, Sedvall G: The distribution of adrenergic nerve fibers to the blood vessels in skeletal muscle. Acta Physiol Scand 64:75, 1965
51. Abboud FM, Eckstein JW: Comparative changes in segmental vascular resistance in response to nerve stimulation and to norepinephrine. Circ Res 18:263, 1966
52. Mayer HE, Abboud FM, Ballard DR, Eckstein JW: Catecholamines in arteries and veins of the foreleg of the dog. Circ Res 23:653, 1968
53. Epstein SE, Beiser GD, Stampfer M, Braunwald E: Role of the venous system in baroreceptor-mediated reflexes in man. J Clin Invest 47:139, 1968
54. Rice AJ, Long JP: Unusual venoconstriction induced by acetylcholine. J Pharmacol Exp Ther 151:423, 1966
55. Greenway CV, Stark RD, Lautt WW: Capacitance responses and fluid exchange in the cat liver during stimulation of the hepatic nerves. Circ Res 25:277, 1969
56. Greenway CV, Stark RD: Hepatic vascular bed. Physiol Rev 51:23, 1971
57. Mellander S, Johansson B: Control of resistance, exchange and capacitance functions in the peripheral circulation. Pharmacol Rev 20:117, 1968
58. Brooksby GA, Donald DE: Release of blood from the splanchnic circulation in dogs. Circ Res 31:105, 1972
59. Öberg B: The relationship between active constriction and passive recoil of the veins at various distending pressures. Acta Physiol Scand 71:233, 1967
60. Epstein SE, Beiser GD, Goldstein RE, et al: Circulatory effects of electrical stimulation of the carotid sinus nerves in man. Circulation 40:269, 1969
61. Browse NL, Donald DE, Shepherd JT: Role of the veins in the carotid sinus reflex. Am J Physiol 210:1424, 1966
61a. Brooksby GA, Donald DE: Dynamic changes in splanchnic blood flow and blood volume in dogs during activation of sympathetic nerves. Circ Res 29:227, 1971
61b. Price HL, Deutsch S, Marshall BE, *et al.*: Hemodynamic and metabolic effects of hemorrhage in man, with particular reference to the splanchnic circulation. Circ Res 18:469, 1966
62. Chien S: Role of the sympathetic nervous system in hemorrhage. Physiol Rev 47:214, 1967
63. Brender D, Webb-Peploe MM: Vascular responses to stimulation of pulmonary and carotid baroreceptors by capsaicin. Am J Physiol 217:1837, 1969

64. Brown E, Goei JS, Greenfield ADM, et al: Circulatory responses to simulated gravitational shifts of blood in man induced by exposure of the body below the iliac crests to sub-atmospheric pressure. J Physiol (Lond) 183:607, 1966
65. Samueloff SL, Browse NL, Shepherd JT: Response of capacity vessels in human limbs to head-up tilt and suction on lower body. J Appl Physiol 21:47, 1966
66. Abboud FM, Mark AL, Heistad DD, Schmid PG: Selectivity of autonomic control of the peripheral circulation in man. Trans Am Clin Climatol Assoc (in press)
67. Braunwald E, Ross J Jr, Kahler RL, et al: Reflex control of the systemic venous bed. Effect on venous tone of vasoactive drugs, and baroreceptor and chemoreceptor stimulation. Circ Res 12:539, 1963
68. Iizuka T, Mark AL, Wendling MG, et al: Differences in responses of saphenous and mesenteric veins to reflex stimuli. Am J Physiol 219:1066, 1970
69. Calvelo M, Abboud FM, Abdel-Sayed W: Reflex vascular responses to stimulation of chemoreceptors with nicotine and cyanide: Activation of adrenergic constriction in muscle and non-cholinergic dilatation in paw. Circ Res 27:259, 1970
70. Browse NL, Shepherd JT: Response of veins of canine limb to aortic and carotid chemoreceptor stimulation. Am J Physiol 210:1435, 1966
71. Eckstein JW, Hamilton WK, McCammond JM: Pressure–volume changes in forearm veins of man during hyperventilation. J Clin Invest 37:956, 1958
72. Eckstein JW, Horsley AW, Hamilton WK: Responses of the peripheral veins in man to continuous positive pressure breathing. J Clin Invest 40:1036, 1961
73. Wood JE: The Veins: Normal and Abnormal Function. Boston, Little, Brown, 1965
74. Eckstein JW, Horsley AW: Effects of hypoxia on peripheral venous tone in man. J Lab Clin Med 56:847, 1960
75. Webb-Peploe MM: The isovolumetric spleen: Index of reflex changes in splanchnic vascular capacity. Am J Physiol 216:407, 1969
76. Heistad DD, Abboud FM, Eckstein JW: Vasoconstrictor response to simulated diving in man. J Appl Physiol 25:542, 1968
77. Zitnik RS, Burchell HB, Shepherd JT: Hemodynamic effects of inhalation of ammonia in man. Am J Cardiol 24:187, 1969
78. Bevegard BS, Shepherd JT: Reaction in man of resistance and capacity vessels in forearm and hand to leg exercise. J Appl Physiol 21:123, 1966
79. Bevegard BS, Shepherd JT: Changes in tone of limb veins during supine exercise. J Appl Physiol 20:1, 1965
80. Wade OL, Combes B, Childs AW, et al: The effect of exercise on the splanchnic blood flow and splanchnic blood volume in normal man. Clin Sci 15:457, 1956
81. Eichna LW, Park CR, Nelson N, et al: Thermal regulation during acclimatization in a hot, dry (desert type) environment. Am J Physiol 163:585, 1950

82. Wood JE, Bass DE: Responses of the veins and arterioles of the forearm to walking during acclimatization to heat in man. J Clin Invest 39:825, 1960
83. Zimmerman BG, Abboud FM, Eckstein JW: Comparison of the effects of sympathomimetic amines upon venous and total vascular resistance in the foreleg of the dog. J Pharmacol Exp Ther 139:290, 1963
84. Abdel-Sayed WA, Abboud FM, Ballard DR: Contribution of venous resistance to total vascular resistance in skeletal muscle. Am J Physiol 218:1291, 1970
85. Abboud FM, Eckstein JW: Vascular responses after alpha adrenergic blockade. II. Responses of venous and arterial segments to adrenergic stimulation in the forelimb of dog. J Clin Invest 47:10, 1968
86. Eckstein JW, Hamilton WK: The pressure–volume responses of human forearm veins during epinephrine and norepinephrine infusions. J Clin Invest 36:1663, 1957
87. Sharpey-Schafer EP, Ginsburg J: Humoral agents and venous tone. Effects of catecholamines, 5-hydroxytryptamine, histamine, and nitrites. Lancet 2:1337, 1962
88. Abboud FM, Eckstein JW, Zimmerman BG: Venous and arterial responses to stimulation of beta adrenergic receptors. Am J Physiol 209:383, 1965
89. Schmid PG, Eckstein JW, Abboud FM: Comparison of the effects of several sympathomimetic amines on resistance and capacitance vessels in the forearm of man. Circulation 34 [Suppl III]:209, 1966
90. Zelis R, Mason DT: Comparison of the reflex reactivity of skin and muscle veins in the human forearm. J Clin Invest 48:1870, 1969
91. Mason DT, Melmon KL: Abnormal forearm vascular responses in the carcinoid syndrome: The role of kinins and kinin-generating system. J Clin Invest 45:1685, 1966
92. Chien S, Krakoff L: Hemodynamics of dogs in histamine shock, with special reference to splanchnic blood volume and flow. Circ Res 12:29, 1963
93. Abboud FM: Vascular responses to norepinephrine, angiotensin, vasopressin and serotonin. Fed Proc 27:1391, 1968
94. Wood JE: Peripheral venous and arteriolar responses to infusions of angiotensin in normal and hypertensive subjects. Circ Res 9:768, 1961
95. Mason DT, Braunwald E: The effects of nitroglycerin and amyl nitrite on arteriolar and venous tone in the human forearm. Circulation 32:755, 1965
96. Hedwall PR, Abdel-Sayed WA, Schmid PG, et al: Vascular responses to prostaglandin E_1 in gracilis muscle and hindpaw of the dog. Am J Physiol 221:42, 1971
97. Hedwall PR, Abdel-Sayed WA, Schmid PG, Abboud FM: Inhibition of venoconstrictor responses by prostaglandin E_1. Proc Soc Exp Biol Med 135:757, 1970
98. Mark AL, Schmid PG, Eckstein JW, Wendling MG: Venous responses to prostaglandin $F_{2\alpha}$. Am J Physiol 220:222, 1971
99. Strong CG, Bohr DF: Effects of prostaglandins E_1, E_2, A_1 and $F_{1\alpha}$ on isolated vascular smooth muscle. Am J Physiol 213:725, 1967

100. Nickerson M, Gourzis JT: Blockade of sympathetic vasoconstriction in the treatment of shock. J Trauma 2:399, 1962
101. Epstein SE, Stampfer M, Beiser GD: Role of the capacitance and resistance vessels in vasovagal syncope. Circulation 37:524, 1968
102. Wood JE, Litter J, Wilkins RW: Peripheral venoconstriction in human congestive heart failure. Circulation 13:524, 1956
103. Wood JE: The mechanism of the increased venous pressure with exercise in congestive heart failure. J Clin Invest 41:2020, 1962
104. Zelis R: The contribution of local factors to the elevated venous tone of congestive heart failure. J Clin Invest 53:315, 1974
105. Brender D, Vanhoutte PM, Shepherd JT: Potentiation of adrenergic venomotor responses in dogs by cardiac glycosides. Circ Res 25:257, 1969
106. Mason DT, Braunwald E: Studies on digitalis. X. Effects of ouabain on forearm vascular resistance and venous tone in normal subjects and patients in heart failure. J Clin Invest 43:532, 1964
107. Sharpey-Schafer EP: Venous tone. Br Med J 2:1589, 1961
108. Gucha NH, Cohn JN, Mikulic E, et al: Treatment of refractory heart failure with infusion of nitroprusside. N Engl J Med 291:587, 1974
109. Walsh JA, Hyman C, Maronade RF: Venous distensibility in essential hypertension. Cardiovasc Res 3:338, 1969
110. DePasquale NP, Burch GE: Effect of angiotensin II on the intact forearm veins of man. Circ Res 13:239, 1963
111. Greenberg S, Bohr DF: Venous smooth muscle in hypertension: enhanced contractility of portal veins from spontaneously hypertensive rats. Circ Res 36:791, 1975
111a. Overbeck HW: Hemodynamics of early experimental renal hypertension in dogs: Forelimb hemodynamics. Circ Res 31:653, 1972
111b. Simon G, Pamnani MB, Dunkel JF, Overbeck HW: Mesenteric hemodynamics in early experimental renal hypertension in dogs. Circ Res 36:791, 1975
112. McCausland AM, Holmes F, Trotter AD Jr: Venous distensibility during the menstrual cycle. Am J Obstet Gynecol 86:640, 1963
113. Goodrich SM, Wood JE: Peripheral venous distensibility and velocity of venous blood flow during pregnancy or during oral contraceptive therapy. Am J Obstet Gynecol 90:740, 1964
114. Wheeler RC, Kayan JB, Wood JE: Venous capacitance in varicose veins. Circulation 34[Suppl III]:239, 1966
115. Rushmer RF, Baker DW, Stegall HF: Transcutaneous Doppler flow detection as a nondestruction technique. J Appl Physiol 21:554, 1966
116. Sumner DS, Baker DW, Strandness DE Jr: The ultrasonic velocity detector in a clinical study of venous disease. Arch Surg 97:75, 1968
117. Whitney RJ: The measurement of changes in human limb volume by means of a mercury-in-rubber strain gauge. J Physiol 109:5P, 1949
118. Barnes RW, Collicott PE, Mozersky DJ, et al: Noninvasive quantitation of maximum venous outflow in acute thrombophlebitis. Surgery 72:971, 1972

119. Barnes RW, Collicott PE, Mozersky DJ, et al: Noninvasive quantitation of venous reflux in the postphlebitic syndrome. Surg Gynecol Obstet 136:767, 1973
120. Strandness DE Jr, Sumner DS: Ultrasonic velocity detector in the diagnosis of thrombophlebitis. Arch Surg 104:180, 1972
121. Folse R, Alexander RH: Directional flow detection for localizing venous valvular incompetency. Surgery 67:114, 1970
122. Barnes RW, Collicott PE, Sumner DS, et al: Noninvasive quantitation of venous hemodynamics in the postphlebitic syndrome. Arch Surg 107:807, 1973
123. Folse R: The influence of femoral vein dynamics on the development of varicose veins. Surgery 68:974, 1970
124. Reagan R, Folse R: Lower limb venous dynamics in normal persons and children of patients with varicose veins. Surg Gynecol Obstet 132:15, 1971
125. Barnes RW, Ross EA, Strandness DE Jr: Differentiation of primary from secondary varicose veins by Doppler ultrasound and strain gauge plethysmography. Surg Gynecol Obstet 141:207, 1975

Modestino G. Criscitiello

6

Pathophysiology of Heart Sounds and Murmurs

Some of the energy generated with each heartbeat is released in the form of sound, produced by vibrations occurring in the heart and great vessels and transmitted through surrounding soft tissues to the chest wall. Cardiovascular sound is divided into two categories: (1) brief transients referred to as *heart sounds* and (2) vibrations of longer duration, occupying approximately 25% or more of systole or diastole, termed *murmurs*. The mass of the heart tissue is large in relation to its elasticity; hence these vibrations are of low frequency, in the range of 30 to 500 Hz*. The human auditory range lies between 20 and 16,000 Hz, with maximal sensitivity between 1000 and 2000 Hz. Heart sounds and murmurs fall at the lower end of the audible range where the ear is less efficient and vibrations of low intensity may escape detection. Clinical auscultation involves listening for events that are near the threshold of hearing, and a quiet environment is essential when using the stethoscope.

The bell-shaped chestpiece, when applied lightly to the chest wall, utilizes the skin as a diaphragm, capable of responding to low frequencies (20–150 Hz) generated by ventricular filling sounds or by the diastolic murmur of mitral stenosis, for example. Murmurs of higher frequency (200–400 Hz), as in aortic insufficiency, are more easily heard with a rigid diaphragm that filters out the lower frequencies. (Note that a murmur in the high-frequency range of cardiovascular sound falls well within the lower end of the total human auditory range.) To avoid loss of intensity as sound is transmitted from the chest wall to the ear, the stethoscope

*The term *Hertz* (Hz) is used interchangeably with "cycles per second" (c.p.s.)

chestpiece should permit no air leakage, the tubing should be short and inelastic with a narrow bore (⅛ in.), and the earpieces should be properly angulated to fit snugly into the external auditory canals.

Phonocardiography produces a graphic display of acoustic events and greatly facilitates their interpretation. Simultaneous recordings of sound events, arterial and venous pulses, chest wall pulsations, intracardiac pressures, and the ultrasound echoes of heart valves and chamber walls have led to a greater understanding of the pathophysiology of heart sounds and murmurs.

HEART SOUNDS

Although there is no universal agreement as to the mechanism of production of heart sounds, it is apparent that they result from forces that initiate vibrations in the heart, the proximal great vessels, and the contained blood mass, described as the cardiohemic system.[1] These forces are produced by (1) rises in pressure and acceleration of blood flow induced by atrial and ventricular contraction and (2) deceleration and arrest of flow resulting from valve closure or from sudden changes in wall tension of relaxing ventricles. Although there have been attempts to demonstrate a single factor such as valve closure, muscle tension, or flow velocity change as the cause of a given heart sound, it is unlikely that an exclusively valvular, muscular, or flow origin exists, since no component of the cardiohemic system can vibrate independently of the others.

FIRST HEART SOUND

For over a century there has been lively debate about the origin of the first heart sound (S_1), and this controversy has sharpened in the last decade as increasingly precise means of studying mechanical and acoustic events of the heart have become available. One point of view relates S_1 primarily to closure of atrioventricular valves,[2,3] but this is challenged by those who attribute the sound to vibrations resulting from sudden acceleration of the blood mass induced by ventricular contraction.[4] A concept that incorporates both mechanisms explains S_1 as oscillations of a complex mass that includes blood, ventricular walls, and valves.[5,6]

The duration of the first sound is normally 60–90 msec, but the number of phonocardiographic components comprising S_1 varies according to the sound filters used and the enthusiasm of the observer.

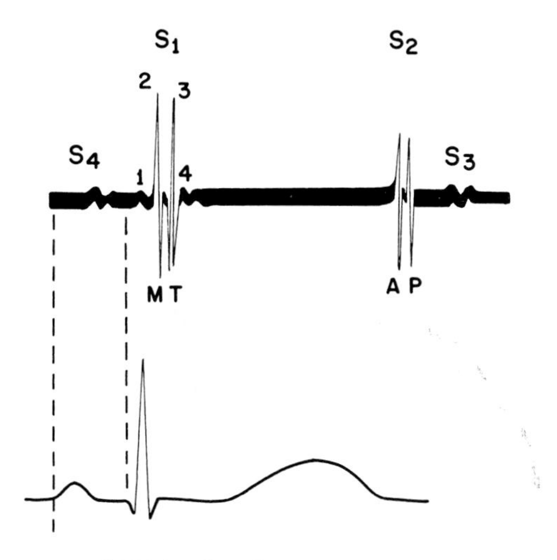

Fig. 6-1. Schematic representation of heart sounds. The vibrations of the first sound (S_1) begin after the onset of the QRS complex. S_1 contains four components including initial (1) and terminal (4) low-frequency vibrations that are inaudible. The major high-frequency deflections (2, 3) are in the audible range and are related to mitral (M) and tricuspid (T) closure, respectively. The second sound (S_2) consists of two deflections, the first related to aortic closure (A) and the second to pulmonic closure (P). The third heart sound (S_3) is a low-frequency event that occurs during rapid ventricular filling in early diastole. The fourth heart sound (S_4) or atrial sound is inscribed after the onset of the P wave and is attributed to ventricular filling during atrial systole.

Recordings employing a series of filters show that S_1 begins with an inaudible low-frequency vibration followed by a short sequence of high-frequency components of greater amplitude. The sound ends with a low-frequency aftervibration (Fig. 6-1).

The initial low-frequency component (25 Hz) of S_1 begins approximately 30 msec after the onset of the QRS complex of the electrocardiogram. It is inscribed during initial ventricular pressure rise and is attributed to apposition of mitral and tricuspid leaflets[7] or to the acceleration of blood toward the atrioventricular valves prior to their closure.[8] This vibration has been recorded in patients with atrial fibrillation;[8] hence an older theory of atrial origin seems untenable. It has also been recorded in patients who have undergone insertion of an artificial mitral valve. The low-frequency component is registered just prior to the sharply defined vibrations of prosthetic valve closure.[3] This presents a strong argument against leaflet apposition as the mode of origin and favors the view that the vibration arises from oscillations within the cardiohemic system produced by acceleration of ventricular blood during early pressure rise.

In phonocardiograms of many normal subjects there is an inaudible low-frequency vibration created in the ventricle during atrial systole prior to the onset of ventricular contraction. This may be continuous with the early vibrations of S_1 and is sometimes mistakenly interpreted as part of it. It can usually be separated from the initial vibration of S_1 by noting its relationship to the QRS. The low-frequency vibration of atrial origin is recorded shortly before the QRS or just at its onset and is hence too early to be considered as part of S_1. It disappears with the onset of atrial fibrillation.

The components of S_1 detectable with a stethoscope are those major high-frequency vibrations (50–200 Hz) that begin 35–70 msec after the onset of the QRS. They are usually divisible phonocardiographically into two components. The first of these is believed by most workers in the field to be due to tension in the mitral leaflets and chordae tendineae generated by rapidly rising ventricular pressure following valve closure. It is therefore commonly labeled M_1. It has been argued that this first major component actually follows the point of crossover of left ventricular and left atrial pressures.[9] There is evidence that valve closure does not actually begin until 20–40 msec after pressure crossover due to the inertia of mitral flow.[10] Also, this M_1 deflection is not coincident with initial coaptation of mitral leaflets as demonstrated echocardiographically.[11] The process of valve closure requires 30–50 msec,[12] and it is unlikely that vibrations are generated until the valve is completely closed and under rising tension. This concept is also supported by the demonstration that the M_1 deflection is simultaneous with the left atrial *c* wave.[13] This same high-frequency component also coincides with a point on the left ventricular pressure curve where there is a sudden rapid rise, and it is therefore also attributed to oscillations set up in the cardiohemic system as tension develops rapidly in the contracting myocardium.[9]

The second of the major high-frequency components of S_1, often audibly separate from the first, is ascribed to tricuspid valve tension following its closure. It is synchronous with the right atrial *c* wave[14] and is more widely separated from the M_1 component when right ventricular pressure rise is delayed by right bundle branch block. However, several investigators have expressed doubt that tricuspid closure is audible and argue that this second major component of S_1 is also left-sided in origin, inscribed as ventricular contraction raises ventricular pressure above aortic pressure and blood is accelerated toward the aortic valve.[1,4,5]

The first sound ends with a component that is low in both frequency and amplitude (and therefore inaudible). It coincides with ejection of blood into the great vessels. This component may be delayed and greatly accentuated in the presence of semilunar valve stenosis or with dilatation of the great vessels. Under these circumstances it becomes easily audible as an "ejection sound."

In clinical auscultation it is the two high-frequency components in the middle of S_1 that are heard in most normal subjects. They may be detected as distinctly separate components, but are often heard as a single event. The first sound is usually loudest on the chest surface at the point of the apical impulse, but it may be equally well heard in the third and fourth interspaces just to the left of the sternum. Inotropic agents that increase the rate of rise of pressure in the left ventricle accentuate S_1. These include sympathetic stimulation, administration of catecholamines, digitalis, and thyroid hormone, or an increase in heart rate. The amplitude of S_1 is reduced with impaired myocardial contractility, as in acute myocardial infarction, during angina pectoris, or with myocardial failure.

The intensity of S_1 correlates well with the position of the atrioventricular valve at the end of diastole. At the end of atrial contraction the leaflets move toward each other, and under normal circumstances the valve is nearly closed at the onset of ventricular contraction. However, the valve may remain open at the end of diastole if flow is increased due to a shunt, if diastole is short because of tachycardia, if flow is prolonged by stenosis of the valve, or if ventricular contraction follows early upon atrial contraction as in the case of a short PR interval. From this open position the leaflets move through a greater excursion to reach the point of full closure. This permits the ventricle to have reached a greater velocity of contraction at the moment of valve tension. This produces more intense oscillations in the cardiohemic system and an accentuated S_1.

The PR interval greatly influences the intensity of S_1, with loud sounds produced by intervals of 100 to 140 msec and softer sounds with intervals increasing beyond 200 msec. With the longer intervals the atrioventricular valve leaflets have drifted toward the closed position, and they reach maximum tension at a lower level of left ventricular pressure. In third-degree heart block the intensity of S_1 is variable, ranging from loud and booming to barely audible, according to whether the PR interval is very short or markedly prolonged, respectively.

In mitral stenosis an atrioventricular pressure gradient still exists at the end of diastole. The scarred, adherent cusps, moving together as a diaphragm, are held forward in the ventricle at the end of diastole, as demonstrated by echocardiographic or radiographic techniques. With ventricular contraction this diaphragmlike structure is therefore required to move through a considerable distance, and the interval between the onset of the QRS and the major components of S_1 (Q–S_1 interval) is prolonged. The normal Q–S_1 interval is 30 to 70 msec, but in mitral stenosis it may exceed 100 msec. If the process of fibrosis and calcification has rendered the leaflets adherent and immobile, S_1 will be soft.

Audible splitting of S_1 occurs when the two major high-frequency components are separated by 20 msec or more.[15] Splitting is frequently heard in normal subjects, most commonly at the lower left sternal border. In the presence of a left-to-right shunt at atrial level, high flow through the tricuspid valve maintains wide separation of the leaflets up to the point of end-diastole, and the second major component of S_1 is accentuated and delayed, with wide splitting heard at the lower left sternal margin.

Wide splitting may result from asynchronous activation and contraction of the ventricles as in bundle branch block, ventricular premature beats, idioventricular rhythm, or endocardial pacing. For example, delay in onset of rise of right ventricular pressure occurs in right bundle branch block (RBBB) or in ectopic beats originating in the left ventricle, and the second major component is late and well separated from the first. (This wide splitting in RBBB is offered as evidence that the second major component is related to tricuspid valve closure).[16] Wide splitting of S_1 is not a feature of left bundle branch block (LBBB) because the onset of left ventricular pressure rise is not delayed, although left ventricular systole is prolonged. This suggests that LBBB does not usually involve interruption of all fibers in the main left bundle, only in its subdivisions or more peripheral distribution.[16] Since audible splitting of S_1 occurs in many normal subjects, its presence cannot be construed as evidence of heart disease.

SECOND HEART SOUND

Analysis of the second heart sound (S_2) shows it to be divided into two sharply defined components (Fig. 6-2). These appear to be related in some manner to closure of the semilunar valves. The earlier of the two deflections (A_2) is associated with aortic closure, and the second (P_2) with pulmonic closure. As in the case of S_1 there is controversy as to whether it is valve closure itself[17] or the sudden deceleration of blood flow in the great vessels[4] that actually generates the sound. The deflection of A_2 on the phonocardiogram actually begins an average of 12 msec after approximation of the aortic cusps, and P_2 follows pulmonary closure by 56 msec as determined with echocardiographic techniques.[18] With the use of aortic flowmeters in dogs it has been determined that A_2 is recorded slightly before the nadir of aortic flow reversal at the moment of peak deceleration of flow.[19] Simultaneous pressure and sound tracings show A_2 to be synchronous with the aortic incisura and P_2 with the pulmonary incisura.[20] It is not clear to what extent it is the sudden

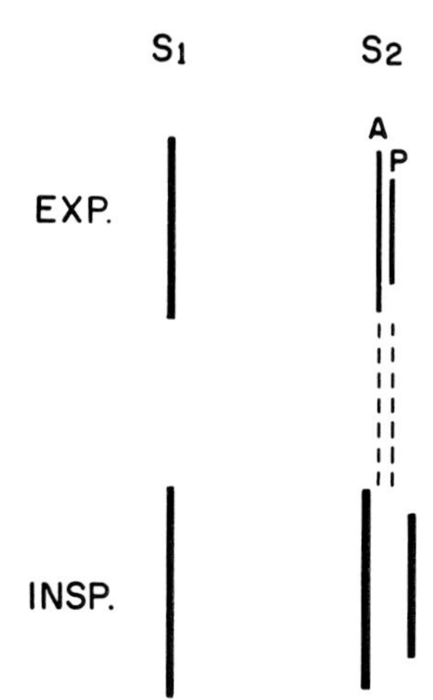

Fig. 6-2. Second heart sound normal splitting; during normal expiration the aortic (A) and pulmonic (P) components of the second sound (S_2) are close and may be heard as a single sound. Splitting of S_2 during inspiration is chiefly the result of later pulmonic closure.

tensing of the valve cusps following their closure or the oscillations of the blood column and vessel walls at the point of sudden flow reversal that account for these sound events.

During quiet expiration the components of S_2 close within 10 msec or less and become single, but during inspiration they separate into two distinct events (Fig. 6-2). This inspiratory splitting has been considered to be due to a transient disparity in stroke volume between the ventricles.[21] Animal studies have demonstrated that negative intrathoracic pressure during inspiration results in an increased inflow into the right heart from the systemic veins and augments right ventricular stroke volume.[22] This is believed to have the effect of prolonging right ventricular systole and delaying pulmonic closure. However, some observers find no delay in right ventricular ejection during inspiration. They argue that the high capacitance of the pulmonary vessels accommodates the increased flow without increased resistance.[23,24] They explain the late P_2 as the result of delay in the rebound of the pressure pulse from the pulmonary artery due to inspiratory dilatation of the pulmonary network. The negative pressure around the pulmonary venous bed increases its capacity also and theoretically allows some pooling of blood there with reduced left-sided filling. However, flowmeter studies in intact dogs actually demonstrated an increase in pulmonary venous flow within one or two beats of the inspiratory increase in right ventricular output.[22] The respiratory effect upon left ventricular stroke volume and ejection time is minimal, and aortic closure occurs only slightly earlier during inspiration as compared with expiration. The interval between A_2 and P_2 increases to the range of 30–60 msec during inspiration, the greater part of the change being due to delay in pulmonic closure. During expiration systemic venous return decreases and the A_2–P_2 interval narrows. A minimal degree of separation may be heard during expiration in the supine position, but normal subjects in the upright position rarely have detectable expiratory splitting.

The aortic component is the louder of the two even at the second left interspace over the pulmonary artery, and it can usually be heard well over the entire precordium and at the apex. The pulmonic component is heard well only at the second or third left interspaces, and audible splitting of S_2 is confined to the upper left parasternal area. In the second right interspace and over the lower precordium and apex, S_2 is single, consisting of A_2 alone. In pulmonary hypertension the vibrations of P_2 increase in amplitude and become audible at areas farther removed from the pulmonary area. If the right ventricle is dilated by a large stroke volume and extends leftward to the apex, as in atrial septal defect, P_2 may be heard widely over the precordium. Hence, splitting of S_2 which is easily heard at the apex suggests pulmonary hypertension or dilatation of the right ventricle.

Aneurysmal dilatation of either great vessel may increase the ease with which its respective semilunar valve closure is heard on the chest wall. Both components of S_2 seem louder with pectus excavatum, loss of dorsal kyphosis, or any chest configuration in which the heart is closely applied to the sternum. Obesity, emphysema, and advancing age lead to softening of S_2, and the sound may appear single due to fading of P_2 altogether.

Assessment of the presence and degree of splitting of S_2 is of great importance and provides clues to a number of forms of heart disease.[25] Splitting that is detectable during expiration (20–40 msec) and widens further during inspiration can be produced by (1) delayed electrical activation of the right ventricle, (2) prolongation of right ventricular ejection due to increase in stroke volume or to outflow obstruction, (3) increased capacitance of the pulmonary artery with loss of elastic recoil, and (4) early aortic closure due to shortened left ventricular ejection (Table 6-1, Fig. 6-3).

Right bundle branch block causes delay in the onset of right ventricular ejection and a corresponding delay in pulmonic closure with clear separation of A_2 and P_2 during expiration. The A_2–P_2 interval increases in normal fashion with inspiration (Fig. 6-3). Ectopic beats of left ventricular origin or left ventricular epicardial pacing will also lead to late right ventricular activation and delayed P_2.

In pulmonary stenosis ejection is prolonged by obstruction to outflow. Pulmonic closure is late and also soft because of low pulmonary

Table 6-1
Wide Splitting of Second Heart Sound

- Delayed electrical activation of right ventricle
  - Complete right bundle branch block
  - Ectopic beat initiated in left ventricle
  - W-P-W preexcitation (some forms)
- Prolongation of right ventricular ejection
  - Increased stroke volume
    - Atrial septal defect (fixed splitting)
    - Partial anomalous pulmonary venous drainage
    - Ventricular septal defect (left-to-right shunt)
    - Congenital pulmonary valve insufficiency
  - Outflow obstruction
    - Pulmonary valve stenosis
    - Infundibular stenosis
    - Pulmonary arterial branch stenosis
  - Right ventricular failure
- Increased capacitance of pulmonary artery and loss of recoil
  - Idiopathic dilatation of pulmonary artery
  - Poststenotic dilatation
- Early aortic valve closure
  - Mitral insufficiency
  - Ventricular septal defect (left-to-right shunt)
  - Constrictive pericarditis

arterial pressures and reduced cusp mobility. There is a good correlation between the A_2–P_2 interval and the level of right ventricular systolic pressure.[26] The higher the right ventricular pressure, the longer is the interval required for it to drop below pulmonary pressure during isovolumetric relaxation. Loss of elastic recoil due to poststenotic dilatation may contribute to the slow drop in pulmonary arterial pressure and delayed valve closure. A similar delay in P_2 occurs in idiopathic dilatation of the pulmonary artery in the absence of obstruction. Here the increased capacitance and slow recoil alone are responsible.

Right ventricular myocardial failure of any origin may be associated with prolonged ejection and delayed P_2, producing wide splitting. Inspiratory augmentation of right ventricular filling may or may not occur, and in the latter case the splitting will be fixed. Fixed splitting of S_2 also occurs in atrial septal defect where the stroke volume of the right ventricle exceeds that of the left in both phases of respiration. In the presence of a large communication between the atria the ventricles share a common

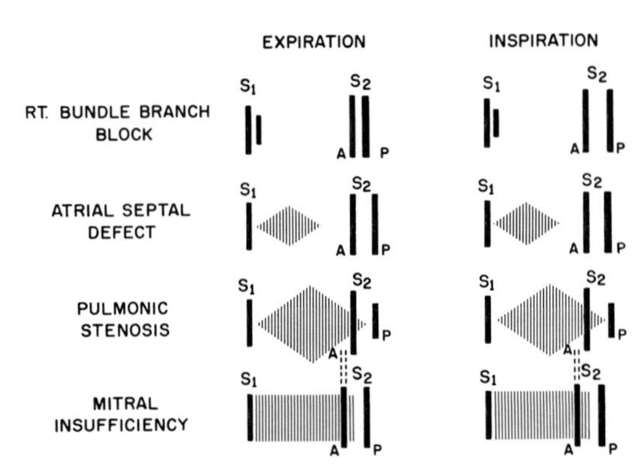

Fig. 6-3. Diagrammatic representation of four mechanisms of wide splitting of the second heart sound. In right bundle branch block onset of right ventricular systole is delayed and both the tricuspid component of S_1 and the pulmonic component of S_2 are late. In atrial septal defect pulmonary closure is delayed as the result of increased right ventricular stroke volume and delayed recoil of the dilated pulmonary artery. In pulmonic stenosis right ventricular systole is delayed because of outflow obstruction, and pulmonary closure is late and diminished in amplitude. The pulmonary ejection murmur may overlap aortic closure. In mitral insufficiency left ventricular systole is short and aortic closure early. A further increase in A-P interval occurs during inspiration in all lesions except atrial septal defect, where the splitting is fixed in degree.

venous reservoir and undergo the same inspiratory increase in venous return. Aortic and pulmonic closure are delayed to the same extent, and the interval between them remains unchanged, usually varying by less than 10 msec between inspiration and expiration (Fig. 6-3).

One might anticipate fixed splitting of S_2 in constrictive pericarditis since both ventricles are subjected to the same restraint during diastolic filling. Actually the normal pattern of inspiratory separation does occur, but as the result of earlier closure of the aortic valve.[27] This is probably due to inspiratory sequestration of blood in pulmonary veins. Aortic closure occurs early as left ventricular stroke volume falls, but P_2 fails to move as right ventricular stroke volume remains unchanged during inspiration.

In ventricular septal defect or in mitral insufficiency, despite the increase in left ventricular stroke volume, left ventricular ejection is shortened because two exit orifices are available. Therefore S_2 may be

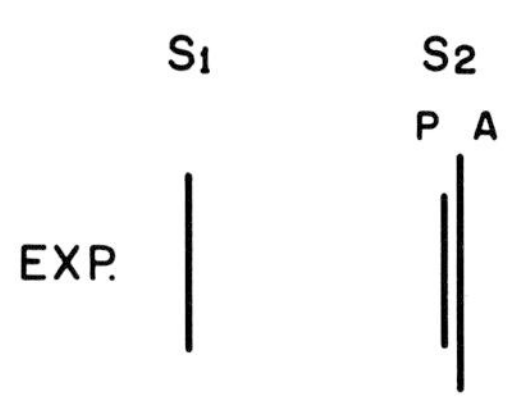

Fig. 6-4. Second heart sound paradoxical splitting; prolongation of left ventricular systole may cause aortic closure to follow pulmonic. The inspiratory delay in pulmonic closure will cause lessening of the split and S_2 will become paradoxically single on inspiration.

widely split because aortic closure is early. In either lesion the loud pansystolic murmur may mask S_2 and obscure the splitting.

Delay in aortic closure to the extent that it occurs later than pulmonic will result in a reversed sequence of splitting and yield a paradoxical shortening of the P_2–A_2 interval during inspiration (Fig. 6-4). This reverse sequence of semilunar valve closure may occur as a result of (1) delayed electrical activation of the left ventricle, (2) prolongation of left ventricular ejection, or (3) early pulmonary valve closure (Table 6-2).

The onset of left ventricular pressure rise is not delayed in complete LBBB, but the rate of rise is slower than normal and ejection is prolonged. This may delay aortic closure sufficiently to result in paradoxical splitting. Preexcitation of portions of the myocardium, as occurs in the Wolff-Parkinson-White syndrome, may lead to wide splitting of the second sound with normal respiratory variation, or to reversed sequence of closure and paradoxical splitting, according to whether the left or the right ventricle is activated earlier. Paradoxical splitting commonly occurs with transvenous pacing of the right ventricle. This is not always the case, since an electrode placed near the interventricular septum of the right ventricle may be close enough to the left bundle system to activate the left ventricle without delay.

Increase in left ventricular stroke volume of the sort produced by patent ductus arteriosus may lead to A_2 delay with paradoxical splitting, but the loud continuous murmur of this lesion usually obscures the second sound. In aortic regurgitation left ventricular ejection is sometimes prolonged sufficiently by the increase in stroke volume to delay aortic closure. In all forms of outflow obstruction of the left heart, A_2 may be late. In valvular stenosis, particularly with heavily calcified and

Table 6-2
Paradoxical Splitting of Second Heart Sound

Delayed electrical activation of left ventricle
Complete left bundle branch block
W-P-W preexcitation (some forms)
Right ventricular pacing or ectopic beats
Prolongation of left ventricular ejection
Increased stroke volume
Patent ductus arteriosus
Aortic regurgitation (occasionally)
Outflow obstruction
Aortic valve stenosis
Hypertrophic subaortic stenosis
Ischemic heart disease
Systemic hypertension
Early pulmonic valve closure
Tricuspid insufficiency

immobile cusps, A_2 is often inaudible and the second sound is faint, consisting only of pulmonary closure.

Paradoxical splitting has been recorded transiently during angina pectoris or following acute myocardial infarction as contractility of ischemic portions of the left ventricle is impaired. This is not a consistent finding, however, and it is uncommonly observed in patients with healed infarctions. Impaired left ventricular function in systemic hypertension may also delay A_2 enough to cause paradoxical splitting. In tricuspid insufficiency pulmonary valve closure is early due to shortened right ventricular ejection time, and the second sound in this lesion may also demonstrate paradoxical splitting.

The second sound is single when P_2 is inaudible because of obesity or hyperinflation of the lungs, when A_2 and P_2 are simultaneous as in pulmonary hypertensive ventricular septal defect or in a single ventricle, when A_2 is inaudible due to aortic stenosis, or when A_2 is obscured by the long murmurs of ventricular septal defect or pulmonary stenosis. In some normal subjects in whom detectable splitting does not occur during inspiration, the Valsalva maneuver may help to confirm the presence of two semilunar sounds. Systemic venous return to the heart is reduced by increased intrathoracic pressure during Valsalva strain. Following release the surge of blood to the right heart usually augments right ventricular stroke volume sufficiently to cause marked delay in P_2 and easily audible splitting.

ADDED SOUNDS IN SYSTOLE

Ejection Sound

The ejection sound (ES) is a sharply defined, high-frequency event that immediately follows S_1 at the onset of ventricular ejection. It does not occur in normal subjects but is likely to be present in patients with semilunar valve stenosis or with proximal great-vessel dilatation. Sound tracings recorded during angiographic study of patients with semilunar valve stenosis have registered an ejection sound at the point where the valve with its fused commissures "domes" upward into the open position. If the valve is heavily fibrotic and immobile an ejection sound is not recorded. The normal S_1 has an inaudible terminal vibration recorded during early ejection, and in the presence of great-vessel enlargement an ejection sound may merely represent some accentuation and delay of this component. In patients with dilatation of the aorta or pulmonary artery, an ejection sound is recorded in the absence of any valvular obstruction at a point when the vessel walls are suddenly distended at the onset of ejection.[28]

In relation to its site of origin, the ejection sound is usually best heard at the base of the heart at the upper sternal margins. The interval between S_1 and ES corresponds to the isovolumetric contraction time of the ventricle involved. This interval becomes longer as the pressure required to open the valve grows higher. For example the ejection sound is later in severe as compared with mild pulmonary hypertension because the right ventricular pressure must rise to a higher level before the valve opens. For the same reason, aortic ejection sounds are later in timing than pulmonic.

An *aortic ejection sound* is common in patients with congenital aortic stenosis or with a congenital bicuspid valve, but it is rarely produced in any form of subvalvular stenosis. It follows S_1 by approximately 60–80 msec. An ejection sound may be recorded in the aortic area whenever there is aortic dilatation, as in luetic or atherosclerotic aneurysms, cystic medial necrosis, dissection of the aorta, or systemic hypertension. The aortic ejection sound is rarely altered by respiration, either in its intensity or in its relationship to S_1. It may be heard at the apex as well as at the base, and for this reason may produce there a sequence of S_1–ES that resembles the combination of a fourth sound followed by the first sound (S_4–S_1) (see below). The ejection sound is a high-frequency event heard best with the diaphragm of the stethoscope and is not influenced by change in the patient's position. A fourth sound is low frequency, is best heard with the bell chestpiece, and tends to fade as venous return is reduced by assumption of the upright position.

Some ejection sounds may occur as early as 50 msec after the first high-frequency component of S_1 and therefore imitate normal splitting. Since splitting of S_1 is usually most distinct at the lower left sternal border in the tricuspid area, any reduplication heard well at the apex or base should raise the possibility of a first-sound–ejection-sound pairing. A loud ejection sound heard at the base can be mistaken for S_1 if the latter is quite faint in this area, and the presence of two events may not be appreciated. In this instance simultaneous recordings with two microphones will demonstrate that the loud ejection sound registered at the base actually follows the onset of S_1 recorded at the apex.

A *pulmonic ejection sound* is usually loudest in the second left interspace, and a double sound heard there at the onset of a systolic murmur should always raise the possibility of pulmonary stenosis or dilatation of the proximal pulmonary artery. The latter includes enlargement due to pulmonary hypertension of any origin (mitral stenosis, ventricular septal defect) or to idiopathic dilatation with normal pressures. Occasionally lesions associated with dilatation due to high flow (atrial septal defect) may produce an ejection sound.

In pulmonary stenosis the duration of the S_1–ES interval varies inversely with the severity of the obstruction.[29] Because the pulmonary arterial pressure is subnormal in pulmonary stenosis, the valve opening pressure is achieved early during right ventricular contraction. With more severe stenosis, the lower pulmonary arterial pressure permits earlier valve opening and closer approximation of S_1 and the ejection sound. In addition, as the right ventricle undergoes hypertrophy in response to valvular stenosis, its rate of rise of pressure during isovolumetric contraction is increased. A similar index of severity is provided by measurement of the interval between the onset of the QRS deflection on the EKG and the ejection sound on a phonotracing (Q–ES). This is in the range of 110–150 msec in mild pulmonary stenosis and drops to 60–100 msec in severe stenosis. Unfortunately there is not an equally good correlation between these intervals and the degree of obstruction in the assessment of aortic stenosis.[29,30]

The ejection sound in pulmonic stenosis decreases in intensity during inspiration and moves closer to the first sound, often fusing with it. The hypertrophied right ventricle has low compliance, and the inspiratory increase in venous return leads to an elevated diastolic pressure. This, together with the vigorous contraction of the hypertrophied right atrium, may lead to a right ventricular end-diastolic pressure that actually exceeds pulmonary arterial pressure prior to the onset of ventricular systole. This end-diastolic ventriculo-arterial gradient moves the pulmonary valve upward to its open position, and any vibrations produced in this structure by ventricular contraction are soft and merge with those of the first sound.[31]

Midsystolic Click

There are very sharply defined high-frequency sounds occurring in midsystole or late systole in many apparently healthy subjects. These sounds, often described as systolic clicks, are usually best heard at the apex and are frequently followed by a murmur that fills the remainder of systole. The term systolic click is an accurate description since these sounds are usually very brief transients of high pitch. They were formerly considered to be of extracardiac origin related in some way to healed pericardial or pleural disease.[32] In many patients studied with left ventricular cineangiography, the systolic click has been associated with prolapse of a portion of one of the mitral leaflets.[33] This prolapse may be due to aneurysmal ballooning of a segment of the valve and can involve either the anterior or posterior leaflet. This type of deformity usually occurs in the absence of other disease, but it may be present as a component of the Marfan syndrome or in conjunction with a congenital cardiac lesion, most commonly atrial septal defect. Valve prolapse may also be due to attenuation of chordae tendineae, to stretching of noncontractile papillary muscle, or to an asynchronous contraction pattern in the left ventricle. The click is recorded during systole when the lax leaflet structure is suddenly checked as it prolapses toward the atrium. Very commonly this prolapse permits a small amount of mitral regurgitation to develop in late systole, in which case the click is followed by a short murmur. These clicks may be multiple if there are a number of separate aneurysmal bulges of one or both mitral leaflets. A number of individuals initially free of systolic clicks or late systolic murmurs have developed these findings over a period of observation, and it has been concluded that mitral prolapse is an acquired change. Since there is a definite familial incidence of this deformity, there may be a genetically determined predisposition. The most common pathologic finding is myxomatous degeneration of the valve tissue.[34] Since some patients with mitral prolapse have recurrent chest pain, the possibility of an ischemic origin has been raised, and this has been supported by some coronary angiographic studies that have demonstrated small or absent circumflex arteries. In most instances, however, angiograms have demonstrated no detectable abnormality in coronary anatomy or blood flow.

Mitral prolapse represents a dynamic or variable form of mitral regurgitation that can be altered by changes in left ventricular size or systolic pressure. Any reduction in left ventricular size tends to permit earlier onset of prolapse. The systolic click moves closer to the first sound when heart size is reduced by standing upright or by the strain phase of the Valsalva maneuver. Both of these result in reduced ventricular filling and allow earlier prolapse during ejection. Amyl nitrite also leads to a reduction in heart size and usually produces an earlier click and murmur onset, but the drop in systolic pressure produced by this drug may result in a lower regurgitation gradient with some softening of the murmur intensity.[35]

Other features of the mitral prolapse syndrome include premature beats with runs of tachycardia, a nondescript form of recurrent chest pain, and T-wave changes in the electrocardiogram suggesting inferior and lateral ischemia.[34] The relationship of these changes to the mitral valve deformity is not clearly understood.

ADDED SOUNDS IN DIASTOLE

Opening Snap

The opening of atrioventricular valves at the end of isovolumetric relaxation is silent in the absence of disease. If the leaflets and chordae tendineae are fibrosed and contracted, as in rheumatic heart disease, valve opening produces a sharply defined sound termed the opening snap. In mitral stenosis left atrial pressure is elevated, and as left ventricular pressure falls rapidly during relaxation, a marked atrioventricular gradient develops. The leaflets that are shortened and fused at the commissures move abruptly toward the left ventricle at the point of pressure crossover, and it is likely that the sudden tensing of the valve structure produces the snapping sound. Vibrations may also be generated by the sharp tug of the relaxing ventricle on the shortened chordae tendineae. Mitral stenosis is the lesion most commonly involved in the production of an opening snap, although it may be heard in occasional patients with predominant or even pure mitral insufficiency. Rarely an opening snap has been recorded in the absence of organic valvular changes in disorders associated with very high atrioventricular flow (tricuspid opening snap in atrial septal defect,[36] mitral opening snap in patent ductus arteriosus[37] or ventricular septal defect[32]).

In mitral stenosis the presence of an opening snap indicates some mobility of the scarred leaflets, and left ventricular angiographic studies will show the negative shadow of the fused leaflets moving as a thickened diaphragmlike structure. The opening snap is recorded as this diaphragm shifts briskly toward the apex of the left ventricle. When fibrosis is excessive and the valve structure is rendered entirely immobile, as in severe calcific mitral stenosis, an opening snap is not present.

The opening snap is of brief duration with its frequency in the range of 75–100 Hz. It is best heard with the diaphragm at an area just medial to the apex or along the left lower sternal border, but it is often widely transmitted and frequently heard well at the base. It follows aortic closure by an interval of 50–120 msec. Since the snap has a sharp, discrete quality resembling loud pulmonic closure, the combination of the second sound and an opening snap (S_2–OS) may be interpreted as a widely split second sound. This is often the case when the opening snap is transmitted well toward the pulmonary area. Unlike the split second sound, however, the S_2–OS interval does not vary noticeably with respiration. In tight mitral stenosis pulmonary hypertension creates a pulmonary component loud enough to be heard widely, and splitting of S_2 may be audible at the apex mimicking an S_2–OS combination. In this circumstance careful auscultation at the upper left sternal border may resolve the problem. Here the two components of the second sound separate during inspiration and are followed by the opening snap. This triple sequence (A_2–P_2–OS) allows certain identification of the opening snap.

In general, the duration of the interval between the second sound and the opening snap varies inversely with the severity of the stenosis. With a higher left atrial pressure, the mitral valve will open earlier as left ventricular pressure falls, and the opening snap will occur closer to aortic closure. This interval is subject to influence by factors other than the degree of mitral stenosis; hence it is not the reliable noninvasive guide one might wish it to be. For example, a very low cardiac output reduces left atrial pressure and prolongs the S_2–OS interval. Administration of intravenous phenylephrine is a means of temporarily increasing systemic pressure and widening the S_2–OS interval.[38] This response differs from the split S_2, which is uninfluenced by pressure rise. In atrial fibrillation left atrial pressure varies according to the length of diastole. With a long cycle length, atrial pressure drops to a lower level at end-diastole, and the mitral valve opens later after the next ejection. This lengthens the S_2–OS interval. The interval becomes of little value, therefore, in assessing stenosis in the fibrillating heart.

The tricuspid opening snap is generally softer than that of the mitral valve and is often obscured by it, since tricuspid stenosis rarely occurs in the absence of associated mitral disease. The tricuspid snap is heard in an area localized to the lower sternum and is accentuated by inspiration, which momentarily increases the gradient across the valve.

In left atrial myxoma an opening snap may be imitated by the sound

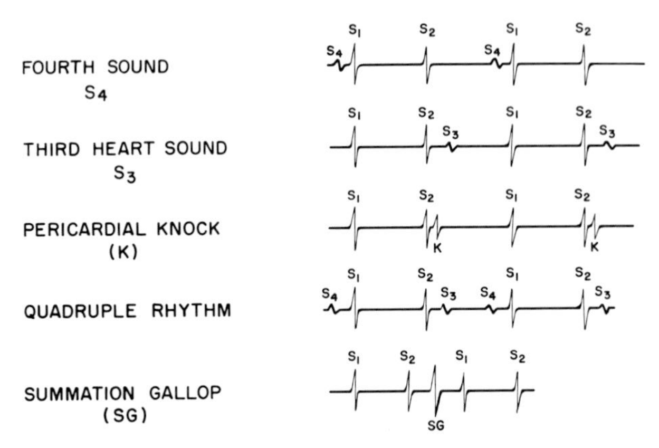

Fig. 6-5. Diagrammatic representation of ventricular filling sounds. The fourth heart sound (S_4) is a presystolic event that occurs in patients with ventricular hypertrophy due to hypertension, semilunar valve stenosis, ischemic heart disease, or acute-onset mitral insufficiency. The third heart sound (S_3) is a physiologic event occurring during early diastolic filling in healthy children and young adults. It disappears with age and when heard in later adult life is a pathologic sound indicative of myocardial failure. It may also occur in ventricular diastolic overload. In constrictive pericarditis a loud sound is heard in early diastole. Termed a pericardial knock, it follows S_2 by a shorter interval than S_3. If both a third sound and a fourth sound are present, a quadruple rhythm results. If the rate is rapid, these two events may fuse into a single mid-diastolic sound, the summation gallop (SG). (This figure and those which follow were adapted by permission from Leonard JJ, Kroetz FW: Examination of the Heart. Auscultation. New York, American Heart Association, 1967).

of the tumor as it moves forward into the mitral orifice during ventricular diastole. The impact of the tumor striking the mitral leaflets and suddenly impeding flow produces vibrations at the moment of mitral opening. If this sound is followed by a diastolic murmur, the resemblance to mitral stenosis may be striking.

Physiologic Third Heart Sound

A soft low-frequency sound is heard at the apex following S_2 in many healthy young subjects (Fig. 6-5). It is termed a third heart sound (S_3), and it is detected best with the bellpiece of the stethoscope applied with the lightest possible pressure. It follows aortic closure by 100–180 msec and is often variable in intensity to the extent that it may be inaudible after some beats. It is often referred to as a filling sound, since it occurs during the phase of rapid flow into the ventricle in early diastole. Since it is generated most often in the left ventricle and is of low intensity, it is usually confined to the apex.

There have been several proposals to explain the formation of the third heart sound: (1) vibrations in the wall of the ventricle itself as the limit of elastic distension is reached during rapid inflow,[39] (2) oscillations in the combined mass of blood and myocardium at the point of sudden deceleration of blood flow,[4] (3) tension and stretch of the atrioventricular valve apparatus during elongation of the ventricle and sudden ascent of the annulus fibrosis,[40,41] and (4) momentary closure of the atrioventricular cusps during rebound from sudden ventricular filling.[42] The last concept has been abandoned as a result of echocardiographic and angiographic demonstrations that the mitral leaflets are still some distance apart at the third sound. Those who favor the mechanism of sudden tension on elongating chordae tendineae and valve leaflets must explain the recorded presence of a typical third sound in a group of patients with prosthetic mitral valves (Starr-Edwards) and paravalvular leaks in whom valve cusps and chordae have been excised at surgery.[43] On the other hand, in vitro studies have suggested that forces much larger than those generated by physiologic filling pressures are required to elicit low-pitched sounds from strips of ventricular wall.[44]

The third sound is commonly heard in normal subjects up to the age of 35 years and is most easily audible in children in whom it may be quite loud, equaling the first and second sounds in intensity although somewhat lower in frequency (20–50 Hz). It fades momentarily during inspiration and can usually be obliterated altogether by having the subject stand or by application of tourniquets to the extremities—maneuvers that reduce venous return. It is best heard at the apex in the supine or left lateral decubitus positions and can be augmented by exercise or elevation of the legs.

Pathologic Third Sound

In the process of normal maturation following early adult years the third heart sound is lost, presumably because of alterations in the viscoelastic properties of ventricular myocardium. A low-frequency sound identical in character and timing with the physiologic S_3 may be recorded in subjects beyond the age of 35 years, but in this age range it must be regarded as an abnormal event (pathologic S_3) denoting myocardial disease. A triple rhythm imparted by this extra sound in the adult has long been regarded as an indication of congestive heart failure. This cadence, accelerated by the presence of sinus tachycardia in heart failure, has led to the terminology gallop rhythm. The added sound itself has been labeled protodiastolic gallop or ventricular diastolic gallop, but the terms third-sound gallop or pathologic third sound seem more descriptive.

A pathologic S_3 is also recorded when the ventricle is overloaded by increased flow (mitral regurgitation, hyperthyroidism, intracardiac shunts), but in this group the S_3 does not necessarily indicate myocardial failure and may disappear with correction of the lesion responsible for the excessive flow. For example, the pathologic S_3 generated in the left ventricle in a patient with a large flow ventricular septal defect will become inaudible after surgical closure of the defect.

A pathologic S_3 is heard in myocardial failure due to many causes, including coronary heart disease, hypertension, or congestive cardiomyopathy. Hemodynamic studies of patients with these disorders show that they have the following in common: left ventricular dilatation, an increase in left ventricular end-diastolic pressure and mean left atrial pressure, rapid filling of the ventricle in early diastole, but reduced cardiac output.[45] There may be clinical manifestations of cardiac decompensation such as dependent edema, hepatomegaly, and pulmonary congestion, but occasionally the presence of a pathologic S_3 is the only evidence of myocardial failure. Appropriate treatment with digitalis, diuretics, antihypertensive agents, etc., often leads to clearing of the pathologic S_3. Since the advent of more effective forms of treatment for failure, this sound is no longer the index of poor prognosis that it was formerly stated to be.[46]

The pathologic S_3 may be right-sided, as in failure due to pulmonary hypertension (mitral stenosis, obstructive lung disease), in which case it is usually best heard at the lower left sternal edge or over the xiphoid. The right-sided S_3 tends to increase on inspiration and fades when the patient assumes the upright position.

The increased diastolic inflow present in a large left-to-right shunt may produce a loud S_3 in the right ventricle when due to an atrial septal defect, and in the left ventricle when due to a patent ductus arteriosus or ventricular septal defect. In some shunts the low-frequency vibration of this S_3 is actually prolonged into a murmur of short duration (flow murmur).

Since its timing and frequency range are similar to those of the physiologic S_3, the diagnosis of pathologic S_3 may be difficult to make in the young subject. A loud, early diastolic filling sound in a young person with rheumatic myocarditis or a large left-to-right shunt need not signify myocardial failure. In general the pathologic third sound remains audible with maneuvers that reduce venous return (standing upright, application of tourniquets, strain phase of the Valsalva maneuver), whereas the physiologic third sound fades.[47] However, it is not until the subject is beyond the age of 35 that one can be certain a third sound indicates myocardial disease.

An early diastolic sound, often quite loud and sharply defined, is heard in constrictive pericarditis and may resemble the third sound in its mode of origin. Because of its forceful character, this is often referred to as a pericardial knock (Fig. 6-5). It appears to be related to the sudden arrest of ventricular filling as the expanding chamber abruptly reaches the limit imposed by thickened pericardium. It is recorded somewhat earlier in diastole than the third sound and follows aortic closure by an interval averaging 90–120 msec.

Fourth Heart Sound

In some patients with ventricular hypertrophy there is a low-frequency event immediately preceding the first sound, heard and recorded best near the apex. It is inscribed after the onset of the P wave but prior to the QRS in the EKG and is therefore related in some way to atrial contraction (Figs. 6-1 and 6-5). It occurs at the peak of the *a* wave of the apexcardiogram, coinciding with an outward bulge of the ventricle during the presystolic phase of rapid filling. Although it immediately precedes the first sound, it is clearly an event in late diastole following S_3, and it has therefore been designated as the fourth heart sound. Since it is generated during atrial systole, it is also described as the atrial sound. It seems clear, however, that there are no audible vibrations produced directly by contraction of atrial myocardium. The sound is generated in the ventricle, and the mechanism of origin is thought to be similar to that of the third sound described above. As in the case of the third sound, the relative importance of tension on the valve structures, deceleration of the transported blood mass, and vibrations in the ventricular wall in the production of the fourth sound has been a subject of lively controversy.[44,48] Although the S_4 occurs commonly in acute myocardial infarction and in disorders leading to ventricular hypertrophy, it also has been recorded in subjects with no evidence of heart disease,[49] and there is currently considerable debate as to its clinical significance. It is not an indicator of heart failure in the same manner that the pathologic S_3 is, and for this reason the terms atrial gallop and presystolic gallop should probably be avoided, since historically the word gallop has implied deteriorating myocardial function and poor prognosis.

The fourth sound is best heard at the apex, with the patient supine or in the left lateral decubitus position. Since its frequency is low (35–100 Hz), light pressure with the bell chestpiece provides the best means of detecting it. Firm pressure will tense the skin rendering it unresponsive to low-frequency vibrations and thereby damping out the S_4. Since this maneuver will not effect the high-frequency components of a split S_1, it provides one means of distinguishing this from the double event of S_4–S_1. In addition the S_4 is altered by respiration and is usually abolished by standing, whereas the split S_1 is not affected by respiration or position change.[50]

Low-amplitude vibrations have been recorded phonocardiographically at the apex ranging from 70 to as long as 280 msec after the onset of the P wave, but those more intense vibrations that produce an audible S_4 and that coincide with the *a* wave of the apexcardiogram fall in the narrower range of 120–170 msec. A fourth sound is more easily distinguished as separate from S_1 when there is prolongation of the PR interval in first-degree heart block. In complete heart block the S_4 has a constant relationship with the P wave and produces an especially loud filling sound when atrial contraction coincides with the rapid phase of ventricular filling in early diastole (summation gallop). The S_4 disappears with the onset of atrial fibrillation.

Although S_4 vibrations can be recorded in normal subjects, a sound loud enough to be heard easily usually implies decreased ventricular compliance. This assertion has been challenged by recent studies in which audible as well as graphically recorded fourth sounds have been detected in the absence of cardiac disease.[49,51] The S_4 is frequently present in ventricular hypertrophy secondary to systemic hypertension or aortic stenosis and in idiopathic hypertrophic cardiomyopathy. There is a correlation between the presence of the S_4 and the level of the left ventricular end-diastolic pressure in aortic stenosis, the sound being present in all patients with a pressure of 12 mm Hg or more.[52] A fourth sound may emerge during an episode of angina pectoris or following acute myocardial infarction, presumably due to changes in compliance of the ischemic or infarcted myocardium.[53] Acute-onset mitral insufficiency of the type related to ruptured chordae tendineae or to papillary muscle dysfunction is characterized by the presence of a fourth sound and minimal left ventricular or left atrial enlargement.[54] If the regurgitation is allowed to persist over a long period and dilatation and failure supervene, the S_4 is replaced by a pathologic S_3. In those forms of mitral insufficiency that develop slowly, as in rheumatic disease, left atrial and left ventricular dilatation are present and a fourth sound is uncommonly heard.

In most of the lesions described above early diastolic filling is limited by impaired ventricular distensibility in relation to the volume of available inflow. This necessitates a prominent atrial contribution with a vigorous "atrial boost" that achieves additional ventricular filling in presystole, causes elevated end-diastolic pressure, and generates an audible fourth sound. Hemodynamic studies show patients with these lesions to have normal atrial mean pressures with a tall atrial *a* wave, an elevated end-diastolic pressure in the ventricle, and a normal cardiac output. They may have no signs of cardiac decompensation for long periods of time, but the loss of atrial contraction, as in atrial fibrillation, may lead to rapid deterioration.[45]

The apexcardiogram demonstrates a prominent presystolic outward bulge (*a* wave) synchronous with the S_4 in most patients with left ventricular hypertrophy or with ischemic heart disease during angina or acute myocardial infarction. In subjects with a fourth sound who have no obvious evidence of heart disease, the presence of a prominent *a* wave in the apexcardiogram strongly suggests an abnormality of ventricular compliance.[55]

In aortic stenosis the fourth sound has been used as an index of severity of the obstruction, being associated usually with a peak systolic aortic gradient of 70 mm Hg or more. However, since subjects over the age of 40 may have a fourth sound from other causes, this is not a reliable indication of severe stenosis in the older age range.[56] A fourth sound is also commonly heard in hypertrophic cardiomyopathy (asymmetric septal hypertrophy) without obstruction and is not a reliable indication of the presence or absence of stenosis in this entity. In congenital pulmonary valvular stenosis, a right-sided S_4 does identify the more severe obstruction and indicates a right ventricular systolic pressure of 100 mm Hg or more.[57]

A fourth sound can be heard immediately after an acute myocardial infarction in almost all patients.[58] In fact, its absence should cast some doubt on the diagnosis. It tends to fade and move toward S_1 with recovery; its persistence is an unfavorable sign.[55] The temporal relationship of S_4 to the onset of the P wave is a clue to the state of ventricular function, the P–S_4 interval tending to shorten with deterioration and lengthen with improvement. With a short P–S_4 interval the S_4 is more clearly separated from S_1 and its presence easily detected. As ventricular function improves, the P–S_4 interval increases, the intensity of S_4 fades, and its vibrations become indistinguishable from those of the first low-frequency component of S_1.[58]

In some patients a quadruple rhythm is produced by the occurrence of both a fourth sound and a pathologic third sound (Fig. 6-5). This is likely to be the case, for example, in the individual with left ventricular hypertrophy due to hypertension who develops congestive failure, or in the patient with diffuse cardiomyopathy with deteriorating ventricular performance. If the heart rate is rapid, diastole may be short enough to allow the two rapid filling phases to run together. The third and fourth sounds become superimposed, producing a single, loud mid-diastolic sound termed a summation gallop (Fig. 6-5). Slowing of the sinus tachycardia with carotid sinus stimulation will usually permit separation of the two components with the characteristic quadruple pattern emerging for a moment.

HEART MURMURS

The preceding discussion has focused upon vibratory transients of brief duration that are identified as discrete heart sounds. Heart murmurs consist of a series of vibrations that are long enough to occupy approximately 25% of systole or diastole. This distinction between heart sounds and murmurs is a purely arbitrary one, not accepted by all observers. However, there are no clear-cut auscultatory or phono criteria upon which a sharp separation can be based. A heart sound usually consists of a single vibration or a short burst of vibrations discrete and intense enough to be easily distinguished from events that precede or follow. A murmur is a sequence of many vibrations occurring during periods of blood flow. Murmurs are generated when flow velocity reaches a point sufficient to convert silent laminar flow into turbulent flow, with the creation of vortices and eddies. These cause oscillations in blood and in the adjacent vessels or chamber walls that are manifested as audible vibrations.

Turbulence is generated in the thoracic aorta and in the right ventricular outflow tract during ejection in most normal resting subjects, but any murmur produced in this way is usually quite faint and is poorly transmitted to the chest wall. Although such murmurs may be recorded inside the heart with catheter-tip microphones, they are usually not detectable over the anterior chest wall in the adult.

Some degree of narrowing is present at normal valve orifices, and if flow velocity is accelerated, audible murmurs may be generated even in the absence of disease. Narrowing of valves resulting from congenital deformities or from fibrosis secondary to acquired disease may lead to murmur production even at normal flow levels. Murmurs may also be generated in any channel where there is sudden change in the cross-sectional area due to constriction or dilatation. This results in disruption of laminar flow with downstream turbulence and eddy formation. Finally, murmurs are formed by retrograde flow through incompetent valves, the turbulence being caused by the collision of the incoming and regurgitant streams.

The intensity (loudness) of murmurs is influenced by the velocity and volume of flow as well as orifice size. For example, a loud systolic murmur is produced in severe aortic stenosis when a left ventricle with unimpaired contractility ejects a normal stroke volume. With the onset of myocardial failure, ejection of a reduced stroke volume at diminished velocity through the same orifice will produce a less intense murmur. The frequency or pitch of a murmur seems more closely related to the pressure head across a valve and to its orifice size than to the amount of flow. In aortic insufficiency, even a small volume of backflow through a slender regurgitant orifice yields a high-frequency or blowing murmur in response to the high aorto-ventricular gradient (80 mm Hg). In mitral stenosis the transvalvular gradient is small (10 mm Hg), and diastolic flow of the entire filling volume produces only a low-frequency or rumbling murmur.

Use of intracardiac phonocardiography has demonstrated that murmurs are loudest in the cardiac chamber or vessel where turbulence is generated, but the vibrations may be transmitted widely through the blood-filled heart and great vessels to adjacent mediastinal tissue and to extensive portions of the chest surface. The transmission of a murmur is influenced by its intensity, the direction of turbulent flow, and the surface projection of the anatomic structures in which it is formed. For example, the murmur of aortic stenosis is generated in the ascending aorta and is transmitted to the chest wall at the level of the upper right sternal border, a point some distance removed from the aortic valve itself. The vibrations of this murmur are also transmitted for short distances along the course of arteries branching from the aortic arch, particularly the common carotids. The aortic murmur may also be transmitted quite well through the medium of the contracting ventricle itself and the contained volume of blood, so that it is easily heard on the chest wall overlying the apex of the heart.

In describing the intensity of murmurs as heard through the stethoscope, a grading scale of six has been devised by Levine as follows: I, very faint, audible only after a few moments of listening in quiet surroundings; II, faint but heard with first application of the stethoscope; V, very loud, but audible only with the stethoscope touching the chest; VI, very loud, heard with the stethoscope held off the chest. Grades III and IV are intermediate between II and V.[59]

Murmurs have been classified according to a number of characteristics including their location on the chest wall (apical, basilar), their musical qualities (harsh, "cooing") or their phonocardiographic configuration (diamond-shaped), but the most widely accepted classification is that of Leatham, which relates the murmur to its hemodynamic mode of production.[60] His concepts will be adhered to in the following discussion.

SYSTOLIC MURMURS

Systolic murmurs are divided into two main groups: (1) midsystolic ejection murmurs produced by forward flow across semilunar valves and (2) pansystolic murmurs resulting from regurgitant flow across atrioventricular valves or through an interventricular septal defect (VSD). (This second definition includes but does not adequately describe the VSD murmur that is limited to early systole or the midsystolic or late systolic murmur of dynamic mitral insufficiency.)

Systolic Ejection Murmurs

Systolic ejection murmurs may result from disorders involving (1) increased velocity or increased volume of ventricular ejection, (2) semilunar valve or subvalvular stenosis, or (3) dilatation of the aorta or pulmonary artery above a normal valve. The innocent or physiologic murmur heard in the absence of any disease is also of this variety.

The systolic ejection murmur does not begin until the semilunar valve is open at the end of isovolumetric contraction; hence there is a brief interval between the first sound and the first vibrations of the murmur. Its intensity rises as velocity of ejection increases and fades as ejection is completed. This accounts for the typical crescendo–decrescendo character, with the murmur terminating before the onset of the second sound (Fig. 6-6).

Innocent Murmur

A soft, early or midsystolic ejection murmur can be heard in the majority of healthy children and in many young adults. It is usually loudest over the third left interspace or slightly lower along the left sternal border and is probably generated in the right ventricular outflow tract and pulmonary artery.[61] There also may be some vibrations in the aorta or subaortic region.[62] Such murmurs are in part the reflection of increased velocity of ejection secondary to the heightened sympathetic tone of youth or excitement. They may be considerably amplified by the

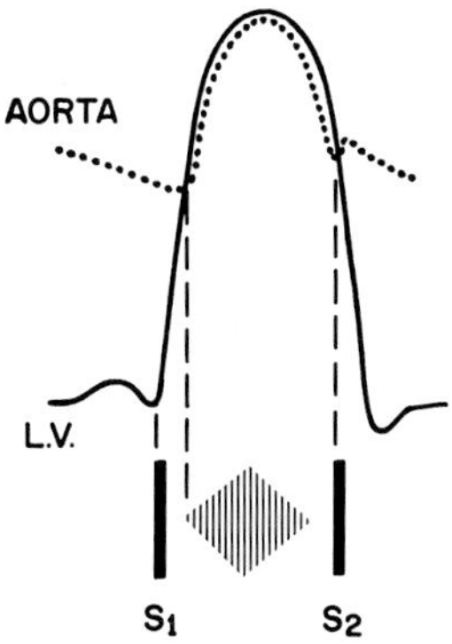

Fig. 6.6 Diagrammatic representation of a systolic ejection murmur. The murmur does not begin immediately with the first sound (S_1) but starts at the end of isovolumetric contraction at the onset of ejection. It reaches peak intensity during peak flow velocity and fades entirely before the second sound.

presence of a thin chest wall, a depressed sternum or a straightened dorsal spine leading to forward displacement of the heart. This innocent or physiologic murmur is usually grade I or II in intensity, but it may increase to grade III with enhancement of output by exercise or by use of amyl nitrite. Some innocent murmurs have a musical or buzzing quality as described by Still,[63] but they are all brief in duration and reach their peak intensity well before midsystole.

The innocent murmur itself may be indistinguishable from those of mild pulmonic stenosis or of atrial septal defect, but attention to the second heart sound is helpful. A single or very closely split second sound during inspiration is strong evidence against either of these lesions. Also, evidences of right ventricular enlargement should be absent. With reliance upon other features of the cardiac examination, the designation of a brief, soft ejection murmur as innocent can be made with confidence. Any murmur that is clearly pansystolic in duration or of grade IV intensity or louder indicates organic disease.

High-Flow Ejection Murmur

A large stroke volume ejected through a normal semilunar valve by a competent ventricle will produce a typical midsystolic ejection murmur up to grade III in loudness. The high-output states of pregnancy, anemia, beriberi, or thyrotoxicosis cause such a murmur. Also, the large stroke volume resulting from significant insufficiency of a semilunar valve will invariably lead to a systolic ejection murmur in addition to the murmur in diastole. In the case of aortic regurgitation the systolic murmur may be quite loud and associated with a thrill even in the absence of stenosis, particularly if the ascending aorta is dilated and contiguous with the chest wall. In left-to-right shunts at the atrial level (atrial septal defect, anomalous pulmonary venous return) the excessive right ventricular stroke volume leads to a moderately loud pulmonary ejection murmur accompanied by wide, fixed splitting of S_2 and vigorous pulsations over the right ventricle.

Since no obstruction is present, all of these high-flow ejection murmurs peak in early systole and end well before the onset of S_2. In many instances, particularly if there is dilatation of the pulmonary artery or aorta, an ejection sound is audible at the onset of the murmur. Physiologic maneuvers (inspiration, change in position, sudden squatting, etc.) that alter venous return and arterial pressure have little influence on the intensity or duration of these high-flow ejection murmurs.

Outflow Obstruction Murmurs

Obstruction of a semilunar valve results in a typical systolic ejection murmur, the duration of which is related to the degree of stenosis. In the case of severe obstruction ejection may be prolonged to the extent that the terminal vibrations of the murmur mask the sound of semilunar closure on the other side. In mild pulmonic stenosis the murmur peaks in midsystole and ends well before S_2 with both A_2 and P_2 audible. In more severe pulmonic stenosis right ventricular ejection is prolonged beyond left. The murmur peaks in late systole and obscures aortic closure. Pulmonic closure, which is late and soft, may be barely audible. Recognition of such a murmur as ejection rather than pansystolic may be difficult and is based primarily upon its crescendo–decrescendo configuration and its site of maximum intensity at the upper sternal margins.

In advanced aortic stenosis with the valve area reduced to 0.6–0.8 cm^2 the ejection murmur tends to peak beyond midsystole, and aortic closure may be prolonged past pulmonic closure, producing paradoxical splitting of S_2. However, this is often difficult to detect since P_2 is masked by the late vibrations of the murmur and aortic closure itself is faint due to immobility of the cusps.

Because it contains many low-frequency vibrations the murmur of valvular aortic stenosis has a "rough" character when heard over the aortic area. These may be damped out as the murmur is transmitted to the apex. Here it often sounds higher in frequency, similar to that of mitral insufficiency. If the second sound is inaudible the murmur may be mistaken for one of pansystolic duration. The presence of varying cycle lengths, as in atrial fibrillation or premature beats, may be helpful in differentiating between these lesions. Typically the intensity of the aortic ejection murmur varies with changes in stroke volume, increasing in a telltale manner after the prolonged diastolic filling following a premature beat. The regurgitant murmur of mitral insufficiency is little influenced by change in cycle length.

In obstructive cardiomyopathy (idiopathic hypertrophic subaortic stenosis) the anterior mitral leaflet is pulled during systole toward the greatly hypertrophied interventricular septum and into the path of left ventricular outflow. This narrows the left ventricular outlet and creates a systolic ejection murmur. Because of its subvalvular location the murmur is heard well over the lower precordium but is poorly transmitted to the ascending aorta. The abnormal stresses on the anterior mitral leaflet prevent its coaptation with the posterior leaflet, which leads to some degree of mitral insufficiency. The superimposed regurgitant murmur tends to prolong the murmur of outflow obstruction but rarely renders it pansystolic or obscures the second sound. The degree of obstruction is greatly influenced by left ventricular size and by the force of contraction. Sudden squatting, by simultaneously increasing venous return and raising arterial mean pressure, increases ventricular size and reduces obstruction. The systolic ejection murmur thus is made shorter and softer.[64] Reduction of heart size during the Valsalva strain phase increases the murmur intensity and duration. Accentuation of contractility, which occurs following a premature beat, may also augment the murmur.

In pulmonic stenosis analysis of the ejection murmur yields reliable criteria for the degree of obstruction.[29] The smaller the orifice the longer the duration of the murmur and the later the peak intensity. In severe stenosis, as noted above, right ventricular systole may be prolonged beyond aortic closure, with the pulmonary component of S_2 delayed, widening the degree of second-sound splitting. Infundibular stenosis produces a similar change, but its murmur is loudest at the third left interspace. If a large ventricular septal defect is also present, as in the tetralogy of Fallot, there is an escape of right ventricular blood into the aorta with reduced flow across the infundibulum. The murmur in this instance is shorter and quieter than that of infundibular stenosis with an intact septum and grows softer as the stenosis progresses in severity.[65]

Systolic Murmur in Dilated Great Vessels

The turbulent flow associated with dilatation of the aorta or pulmonary artery results from a sudden change in cross-sectional area between the level of the valve orifice and the distended vessel beyond. Idiopathic dilatation of the pulmonary artery, or persistent enlargement following closure of an atrial septal defect, produces a short systolic murmur in the second left interspace. An ejection sound is usually recorded at the onset of the murmur as the lax vessel walls are suddenly distended at the onset of ejection. The increased capacitance of the dilated pulmonary artery results in slow deceleration and reversal of flow, delayed closure of the pulmonary valve, and wide splitting of S_2.[24] A combination of obstruction and vessel enlargement is responsible for the murmur in patients with poststenotic dilatation of the aorta or pulmonary artery.

Pansystolic Regurgitant Murmurs

In mitral and tricuspid insufficiency retrograde flow generates a murmur that fills the entire systolic interval. A pressure gradient exists between ventricle and atrium as soon as valve closure occurs, and the regurgitant murmur begins immediately with the first sound (Fig. 6-7). It persists up to the second sound, at which point a significant ventriculo-atrial gradient remains. A few vibrations extend beyond this into the isovolumetric relaxation interval. A ventricular septal defect permits a similar pansystolic gradient to develop between left and right ventricles.

If in rheumatic mitral insufficiency the valvular regurgitant orifice is small and remains constant throughout systole, the murmur is usually high in frequency (200–300 Hz). This orifice may change in size or shape during systole, however, and the murmur may correspondingly vary in pitch or intensity. Although remaining pansystolic, the murmur may change in amplitude, starting softly and growing louder, or vice versa.[66] If the murmur reaches its maximal intensity late in systole, S_2 may be entirely obscured at the apex. The third sound often associated with mitral regurgitation may be confused as S_2 and the murmur erroneously labeled as ejection in character. Identification of the true S_2 may require phonocardiographic study with simultaneous recording at the apex and base.

In forms of mitral insufficiency of acute onset (infectious endocarditis, ruptured chordae tendineae) the left atrium does not rapidly enlarge to accommodate the regurgitant volume. This leads to a tall atrial *v* wave with a resultant sharp drop in ventriculo-atrial gradient and reduction in murmur intensity in late systole.[67]

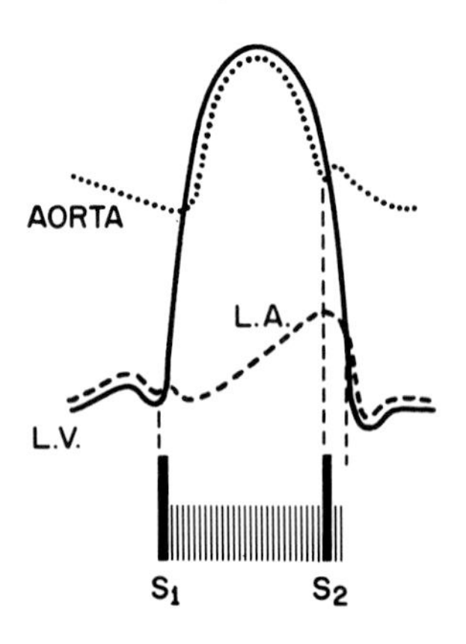

Fig. 6-7. The pansystolic regurgitant murmur of mitral insufficiency. The murmur begins promptly with the first sound when left ventricular pressure exceeds left atrial pressure. The murmur persists to and beyond the second sound, since a ventriculo-atrial gradient still exists at the moment of aortic valve closure.

In chronic mitral insufficiency a loud murmur can occur with a relatively small regurgitant jet; conversely, a grossly incompetent valve may produce only a soft murmur. The volume of regurgitation in long-standing mitral insufficiency is better judged by the degree of left atrial and left ventricular enlargement. Raising left ventricular systolic pressure acutely by administering phenylephrine or by having the patient perform isometric handgrip will lead to more severe regurgitation and a louder murmur. The administration of amyl nitrite leads to a fall in systemic resistance and a drop in left ventricular systolic pressure with resultant softening of the mitral regurgitation murmur.

Tricuspid regurgitation produces turbulence in the right atrium, with a pansystolic murmur transmitted to the region of the lower sternum. The *v* wave in the jugular venous pulse is a telltale accompanying sign, its amplitude being related directly to the volume of regurgitation and inversely to the size and compliance of the right atrium. Increase in venous return during inspiration augments the murmur selectively,[68] and it also tends to grow louder during inhalation of amyl nitrite, since this agent leads to an increase in venous return with no change in pulmonary arterial or right ventricular systolic pressures.[69]

The pansystolic murmur of a ventricular septal defect (VSD) may show midsystolic accentuation, particularly along the upper left sternal border, because of the associated ejection murmur produced by high pulmonary valve flow. If pulmonary hypertension is present a significant difference between left and right ventricular pressure levels may occur only during midsystole, as a result of which the murmur is correspondingly short and diamond-shaped. If the VSD is small and located in the muscular septum, the murmur may be cut off in midsystole by constriction of the defect during septal contraction (Fig. 6-8A).

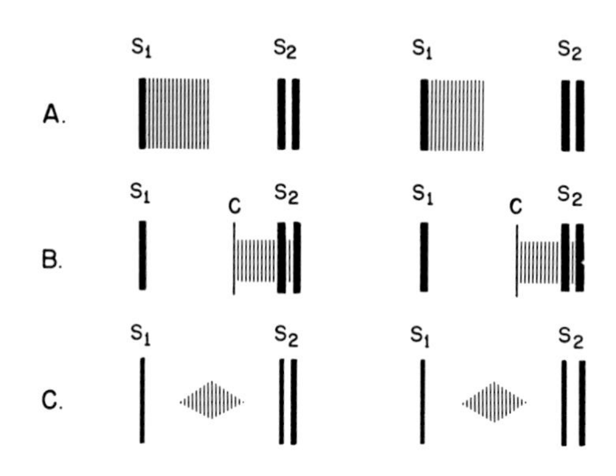

Fig. 6-8. Varieties of systolic regurgitant murmurs. (A) The murmur of a small muscular interventricular septal defect may be confined to early systole if shunt flow is terminated by septal muscular contraction (B) The murmur of mitral valve prolapse commonly is initiated by a systolic click and is confined to late systole. (C) The murmur of papillary muscle dysfunction is due to mitral regurgitation but may be confined to midsystole.

Amyl nitrite has its major dilating effect upon systemic vessels and usually reduces systemic arterial pressure without changing pulmonary pressure. In the presence of a VSD with normal pulmonary resistance, inhalation of amyl nitrite leads to a reduction in the gradient between the left and right ventricles, a decrease in the shunt magnitude, and softening of the pansystolic murmur. If vasoactive pulmonary hypertension is present, amyl nitrite may have a predominant vasodilatory effect upon the pulmonary circulation with a greater drop in pulmonary than systemic pressure and a resultant increase in left-to-right shunt and accentuation of the murmur.[70]

Late Systolic Murmur

Fitting into neither of the categories of midsystolic murmur and pansystolic murmur is the apical late systolic murmur of delayed-onset mitral regurgitation. Left ventricular angiography has defined this murmur as being most commonly the result of dynamic mitral valve incompetence, caused by a ballooning deformity of one or both mitral leaflets, stretching and attentuation of chordae tendineae, dysfunction of papillary muscle, or dyskinesis of subjacent myocardium.[33,34] The mitral leaflets remain coapted in early systole and become incompetent only when the ventricular pressure has risen sufficiently to force prolapse of a portion of the valve into the atrium. The previously described midsystolic click produced by this prolapse initiates the late systolic murmur (Fig. 6-8B). Alterations in ventricular size that affect the dynamics of the subvalvular apparatus may influence the onset, duration, and intensity of the murmur. A temporary reduction of venous return by sudden standing or by Valsalva strain leads to a decrease in heart size and allows prolapse of the mitral leaflets to take place earlier in systole. This results in earlier onset of the click and the murmur. Sudden squatting, by increasing venous return and raising left ventricular systolic pressure, increases ventricular dimensions and delays prolapse or eliminates it altogether. Amyl nitrite has a more complex effect. By reducing left ventricular systolic pressure it tends to reduce the gradient responsible for regurgitation and thereby soften the murmur. However, the reflex increase in contractility and the reduction in heart size lead to earlier prolapse and therefore earlier onset of the murmur.[35]

When papillary muscle dysfunction is the underlying mechanism permitting mitral prolapse, the murmur does not always have the typical late systolic configuration.[71] Passive elongation of these muscles sufficient to produce regurgitant flow may occur only during peak systolic pressures, with the murmur therefore confined to midsystole (Fig. 6-8C).

DIASTOLIC MURMURS

Based on hemodynamic considerations, all diastolic murmurs fall into one of two groups: (1) those generated during forward flow across atrioventricular valves and caused by obstruction or high flow and (2) those produced by retrograde flow across semilunar valves resulting from cusp deformity or dilatation of the entire valve ring. In both categories eddies and vortices are formed within a ventricular chamber, with low levels of energy released as sound. Hence these murmurs are often soft and are transmitted only to limited areas on the chest wall.

Diastolic Murmurs of Atrioventricular Valve Origin

Atrioventricular Valve Stenosis

The most common murmurs of this group are those resulting from fibrosing stenosis of the mitral and tricuspid valves. The murmur of mitral stenosis begins in early diastole, a moment after the second sound (Fig. 6-9). Ventricular filling begins with mitral opening, but there is often a very short gap between the opening snap and the onset of the murmur, for reasons that are not clear. Because the atrioventricular gradient is small, the murmur tends to be low frequency. The duration of the murmur is an index of severity of the stenosis. If the obstruction is mild the atrial and ventricular pressures equalize by middiastole and the murmur fades. If the stenosis is severe a gradient persists to end-diastole and the murmur is prolonged (Fig. 6-9). With atrial contraction there is presystolic increase in the valve gradient and accentuation of the

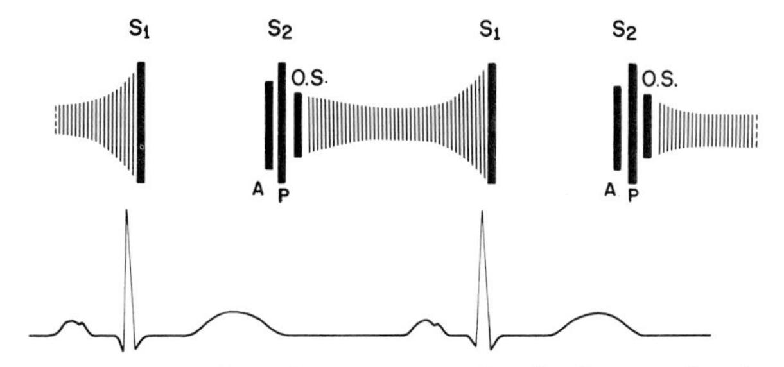

Fig. 6-9. The diastolic murmur of mitral stenosis. An opening snap (OS) is recorded at the moment of valve opening. There is a short gap between the OS and the onset of the low-frequency diastolic murmur. During atrial contraction flow through the stenotic valve is accelerated and the murmur is accentuated.

murmur. If the PR interval is normal this murmur of atrial systole runs into the first sound in typical crescendo fashion. Presystolic accentuation does not entirely depend upon atrial contraction, because it may persist with atrial fibrillation. Particularly if the cycle length is short there may still be a significant atrioventricular gradient when ventricular systole begins. Because atrial pressure remains high the interval between the onset of ventricular pressure rise and mitral closure (preisovolumetric period) is somewhat prolonged. During this interval the increasing ventricular pressure drives the mitral leaflets closer together. The atrioventricular gradient falls rapidly but is not obliterated until pressure crossover. Hence forward flow continues through the mitral valve, and as the orifice diminishes the murmur intensifies.[72]

The murmur of mitral stenosis is confined to the left ventricle and may be heard only in the localized area overlying the apex. In fact, the loud S_1 and opening snap are more readily detected and serve as clues leading to careful search for the murmur. Turning the patient left-side down brings the tip of the left ventricle closer to the chest wall, and the murmur may be audible only in this position with the bell chestpiece held lightly against the skin. Although the duration of the murmur relates to severity of stenosis, its intensity is a function of flow, and with lower flow levels the murmur may be inaudible. Increasing the flow by exercise, forced coughing, or inhalation of amyl nitrite increases murmur intensity.

The murmur of tricuspid stenosis is formed within the right ventricle and is best heard at the lower sternal borders. Though relatively low pitched, it is often higher in frequency than that of mitral stenosis, and its presystolic accentuation begins earlier because onset of right atrial systole precedes left. When sinus rhythm is present the jugular venous *a* wave may be an important clue to tricuspid obstruction. The transvalvular gradient, and hence the murmur, is augmented by inspiration in tricuspid stenosis but not in mitral stenosis. These points of differentiation between tricuspid and mitral stenosis are important because tricuspid obstruction rarely occurs in the absence of mitral stenosis.

Atrioventricular Flow Murmurs

Short diastolic murmurs may be heard as the result of excessive flow through normal atrioventricular valves. Increased mitral diastolic flow occurs in severe mitral insufficiency, in complete heart block with very slow ventricular rates, and in left-to-right shunt lesions such as ventricular septal defect or patent ductus arteriosus. Tricuspid diastolic flow is increased in tricuspid insufficiency, atrial septal defect, or anomalous pulmonary venous drainage. These flow murmurs are usually confined to early diastole and may be initiated by a third sound. They are distinguished from murmurs of organic valve disease by the absence of an opening snap and of presystolic accentuation.

Occasionally, as in cardiomyopathy with heart failure, the combination of a third sound shortly followed by a fourth sound may resemble a low-frequency diastolic murmur. Here carotid sinus stimulation may slow the heart rate and split the vibrations into two discrete filling sounds.

Diastolic Regurgitant Murmurs

Aortic insufficiency causes a high-frequency murmur that begins immediately upon aortic valve closure. The gradient producing retrograde flow increases rapidly during isovolumetric relaxation as ventricular pressure falls below aortic, and the murmur achieves peak intensity in very early diastole (Fig. 6-10A). As aortic diastolic pressure falls, the gradient is reduced and the murmur fades with a gradual decrescendo. It is usually best heard along the left sternal margin, indicating direct transmission from the left ventricular cavity, where it is formed, through the septum and right ventricle to the overlying chest wall. It is also usually heard well at the apex and in the second right interspace overlying the ascending aorta.

Because the frequency of this murmur is usually high and its intensity soft, it may be difficult to detect. Its audibility is enhanced by auscultation during held expiration with the diaphragm pressed firmly against the chest wall and with the patient in an upright position leaning forward. The murmur may be further increased by use of isometric handgrip,[73] by sudden squatting,[74] or by administration of phenylephrine

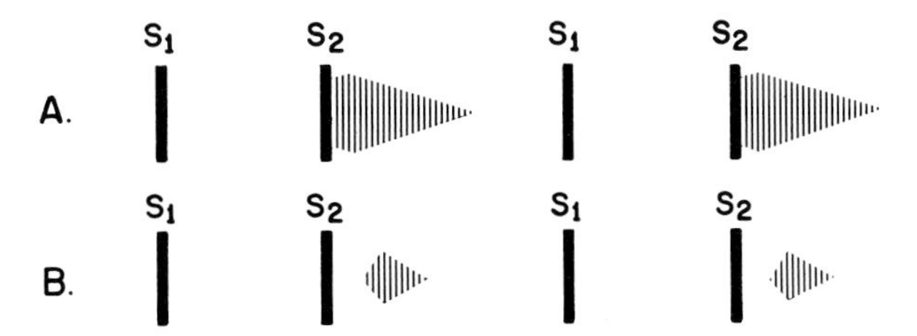

Fig. 6-10. Diastolic regurgitant murmurs. (A) The murmur of aortic insufficiency reaches maximum intensity shortly after the second sound and fades in decrescendo fashion. (B) The diastolic murmur of congenital pulmonary valve insufficiency is short in duration and its onset delayed until early or middiastole.

or other peripheral constricting agents.[69] The effect of all of these is to raise the blood pressure and increase the retrograde aortic gradient.

In higher grades of aortic insufficiency left ventricular stroke volume is considerably increased, and a loud systolic ejection murmur is added to the diastolic murmur producing a to-and-fro sequence. In this group a low-frequency apical diastolic murmur (Austin Flint) is also heard. It is believed to be generated by partial mitral obstruction resulting from posterior displacement of the anterior mitral leaflet as aortic regurgitant flow impinges upon it. Fluttering of this leaflet is seen echocardiographically as a hallmark of aortic insufficiency and may be involved in the genesis of the murmur.[75] Since the Austin Flint murmur is typically late diastolic or presystolic, it may resemble that of mitral stenosis and cannot be distinguished from it on the basis of its quality or timing. A loud first sound, an opening snap, right ventricular hypertrophy, and accentuation of the murmur with amyl nitrite indicate mitral stenosis. A quiet first sound, left ventricular enlargement, and diminution of the murmur with amyl nitrite suggest the Austin Flint murmur.

Dilatation of the valve ring and of the ascending aorta, as in Marfan's syndrome, luetic aortitis, or cystic medial necrosis, leads to diastolic separation of the commissures and creation of regurgitation. The murmur produced in this manner is often transmitted well to the right sternal margin in the third and fourth interspaces, and a diastolic decrescendo murmur loudest in this area is an indicator of enlargement of the aortic root.[76]

In severe aortic insufficiency S_1 may be absent or soft presumably because of premature mitral closure, which results from marked elevation of left ventricular end-diastolic pressure.[77] (This clinical observation supports a valve-closure origin for S_1.)

Organic disease of the pulmonary valve is rare, and pulmonic insufficiency is most commonly associated with dilatation of the valve ring due to pulmonary hypertension. The typical murmur is high frequency and decrescendo in character, similar to that of aortic insufficiency, but it is confined to the left sternal border and is introduced by a loud pulmonary component of S_2. Pulmonary insufficiency due to congenital valve deformity or to idiopathic dilatation without increased pulmonary arterial pressure generates a low-frequency murmur in a localized area at the lower left sternal border. This murmur is usually of short duration and often follows S_2 after considerable pause (Fig. 6-10B). It may be selectively accentuated during inspiration and is commonly associated with wide splitting of S_2 due to delayed pulmonary valve closure.[78]

CONTINUOUS MURMURS

When a pressure gradient exists throughout the cardiac cycle between a high- and low-pressure area, continuous flow and hence a continuous murmur may occur. Most commonly communications of this type are extracardiac and involve abnormal channels connecting the systemic and pulmonary circulation (patent ductus arteriosus, aortopulmonary septal defect, bronchopulmonary anastomoses, Pott or Blalock shunts) or direct passages between arteries and adjacent veins (pulmonary or systemic arteriovenous fistulas). More rarely there is an anomalous pathway between a systemic artery and a right heart chamber (sinus of Valsalva aneurysm ruptured into right ventricle, coronary-artery–right-atrial fistula). Occasionally a continuous gradient exists between proximal and distal segments of a large artery across a narrowed segment (branch stenosis of a pulmonary artery, incompletely occluding pulmonary emboli, coarctation of the aorta, severe subclavian or carotid arterial stenosis).

In each of these circumstances the murmur begins in systole and continues through the second sound into diastole. It may be high in frequency or have coarse vibrations of mixed frequency depending upon the dimensions of the channel involved. Peak intensity is usually achieved late in systole at the time of peak gradient and highest flow velocity. In patent ductus arteriosus, for example, the murmur is typically loudest at the time of the second sound and may actually obscure it (Fig. 6-11A). In

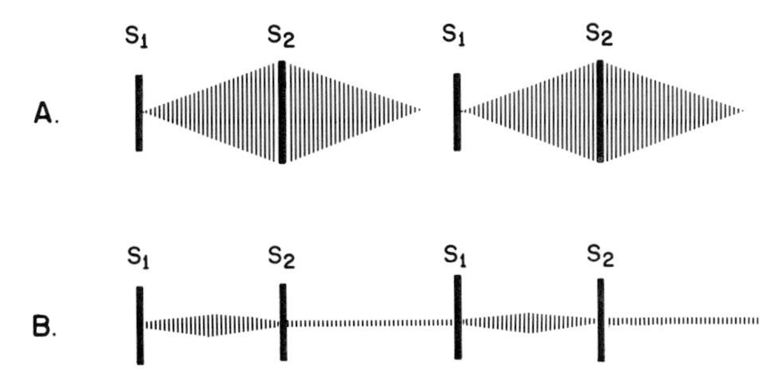

Fig. 6-11. (A) The murmur of the typical patent ductus arteriosus increases in late systole and spills over into diastole. It achieves a maximum intensity at the second sound and often obscures it. (B) An example of a continuous murmur that has a more nearly uniform intensity throughout systole and diastole. This type of murmur is recorded when the gradient across the channel where it is formed does not vary significantly during the cardiac cycle.

Table 6-3
Physiologic and Pharmacologic Alterations of Heart Sounds and Murmurs

Maneuver or Drug	Circulatory Change	Response
Inspiration	Increases volume and velocity of venous inflow to RV, augments RV stroke volume	Splits S_2; increases murmurs of TS, TI, and congenital PI; augments right-sided S_3 and S_4 and tricuspid OS; softens ES and decreases S_1–ES interval in PS
Valsalva	Strain phase: reduces venous return, heart size, arterial pressure; reflex tachycardia	Increases murmur of IHSS; earlier onset of click and LSM in MP; reduces murmur of AS, PS
	Release phase: sudden increase in venous return, first to right heart, then to left	Immediate increase in murmur of TS, TI, PS, right-sided S_3 and S_4; separation of A_2–P_2; later (6–8 beats) increase in murmur of MS, MI, AS
Passive leg raising	Increase in venous return, increase in heart size	Increases murmur of TI, TS, PS, AS; reduces murmur of IHSS; later onset of LSM in MP; enhances both right- and left-sided S_3, S_4
Sudden squatting	Increases venous return, heart size, systemic arterial pressure	Increases murmur of AI, MI; reduces murmur of IHSS

Sudden standing	Reduces venous return, heart size	Increases murmur of IHSS; earlier onset of LSM of MP; increases S_2–OS interval; reduces S_3, S_4
Isometric handgrip	Increases heart rate, cardiac output, systemic arterial pressure	Increases murmur of AI, MI, MS, TS; increases S_3, S_4
Carotid sinus stimulation	Slows heart rate	Separates SG into S_3 and S_4; permits accurate timing of S_3, S_4
Phenylephrine	Vasoconstrictor: raises systemic arterial pressure; reflex bradycardia; decrease in cardiac output	Increases murmurs of AI, MI, VSD, PDA; increases S_2–OS interval; decreases murmurs of IHSS
Amyl nitrite	Vasodilator: lowers systemic arterial pressure and LV systolic pressure; reflex increase in heart rate and velocity of ejection; increase in venous return, increase in cardiac output; PA pressure unchanged	Increases murmurs of MS, TS, TI, AS, IHSS, PS, pulmonary hypertensive VSD; earlier onset of LSM of MP; decreases murmurs of AI, MI, Austin Flint, VSD, tetralogy of Fallot

Key: S_2, S_3, S_4, second, third, and fourth heart sounds; OS, opening snap; AS, PS, MS, TS, aortic, pulmonic, mitral, and tricuspid stenosis; AI, PI, MI, TI, aortic, pulmonic, mitral and tricuspid insufficiency; IHSS, idiopathic hypertrophic subaortic stenosis; PDA, patent ductus arteriosus; VSD, ventricular septal defect; LSM of MP, late systolic murmur of mitral prolapse; ES, ejection sound; SG, summation gallop.

some cases, as in coronary-artery–right-atrial fistula, the gradient varies little during the entire cardiac cycle, and the murmur may be of fairly uniform intensity throughout (Fig. 6-11B). The location on the chest wall of the site of maximal intensity of these murmurs provides the chief clue to their site of origin, but angiographic study is almost always necessary in tracking down the specific lesion.

If pulmonary hypertension is present, flow through a patent ductus arteriosus or an aortopulmonary septal defect may occur only during systole, and the murmur may be short and fail to extend beyond the second sound. There are some intracardiac lesions that produce combinations of murmurs that fill both systole and diastole (e.g., aortic insufficiency associated with ventricular septal defect) and that therefore resemble a continuous murmur. There is usually a distinct difference in pitch between the systolic and diastolic murmurs or a telltale to-and-fro effect that distinguishes these from true continuous murmurs.

SPECIAL TECHNIQUES FOR DIFFERENTIATION OF HEART SOUNDS AND MURMURS

In the preceding discussion the effects of a number of physiologic maneuvers have been described in the analysis of heart sounds and murmurs. These include exercise, respiration, change in position, the Valsalva strain, and isometric handgrip. Also, the effects of pharmacologic agents such as phenylephrine and amyl nitrite have been considered. With the selective use of these maneuvers and drugs, one can momentarily alter venous return, heart size, ejection velocity, arterial pressure, and heart rate in a predictable way and obtain important diagnostic clues to differentiate between lesions that produce similar auscultatory findings.[79] Table 6-3 presents a summary of the circulatory changes and the responses of the heart sounds and murmurs in each case.

REFERENCES

1. Rushmer RF: Cardiovascular Dynamics (ed 3). Philadelphia, W. B. Saunders, 1970, p 305
2. Dock W: Mode of production of the first heart sound. Arch Intern Med 51:737, 1933
3. Dayem MKA, Raftery EB: Mechanism of production of heart sounds based on records of sounds after valve replacement. Am J Cardiol 18:837, 1966
4. Piemme TE, Barnett GO, Dexter L: Relationship of heart sounds to acceleration of blood flow. Circ Res 18:303, 1966

5. Luisada AA, MacCanon DM, Coleman B, Feigen LP: New studies on the first heart sound. Am J Cardiol 28:140, 1971
6. Van Bogaert A: Role of the valves in the genesis of normal heart sounds. Cardiologia 52:330, 1968
7. Lakier JB, Fritz VU, Pocock WA, Barlow JB: The left sided components of the first heart sound. S Afr J Med Sci 35:85, 1970
8. Armstrong TG, Gotsman MS: Initial low frequency vibrations of the first heart sound. Br Heart J 35:691, 1973
9. Shah PM, Mori M, MacCanon DM, Luisada AA: Hemodynamic correlates of the various components of the first heart sound. Circ Res 12:386, 1963
10. Laniado S, Yellin EL, Miller H, Frater RWM: Temporal relation of the first heart sound to closure of the mitral valve. Circulation 47:1006, 1973
11. Parisi AF, Milton BG: Relation of mitral valve closure to the first heart sound in man. Echocardiographic and phonocardiographic assessment. Am J Cardiol 32:779, 1973
12. Criley JM, Feldman IM, Meredith T: Mitral valve closure and the crescendo presystolic murmur. Am J Med 51:456, 1971
13. Lakier JB, Fritz VU, Pocock WA, Barlow JB: Mitral components of the first sound. Br Heart J 34:160, 1972
14. Lakier JB, Bloom KR, Pocock WA, Barlow JB: Tricuspid component of first heart sound. Br Heart J 35:1275, 1973
15. Heintzen P: The genesis of the normally split first heart sound. Am Heart J 62:332, 1961
16. Haber E, Leatham A: Splitting of heart sounds from ventricular asynchrony in bundle-branch block, ventricular ectopic beats, and artificial pacing. Br Heart J 27:691, 1965
17. Leatham A: Splitting of first and second heart sounds. Lancet 2:607, 1954
18. Chandraratna PAN, Lopez JM, Gindlesperger D: Echocardiographic correlates of the second heart sound. Circulation [Suppl 3] 50:85, 1974 (abstract)
19. Brough RD, Talley RC: Temporal relation of the second heart sound to aortic flow in various conditions. Am J Cardiol 30:237, 1972
20. Mori M, Shah PM, MacCanon DM, Luisada AA: Hemodynamic correlates of the various components of the second heart sound. Cardiologia 44:65, 1964
21. Tavel ME: Clinical Phonocardiography and External Pulse Recordings (ed 2). Chicago, Year Book Medical Publishers, 1972, p 38
22. Morgan BC, Abel FL, Mullins GL, Guntheroth WG: Flow patterns in cavae, pulmonary artery, pulmonary vein, and aorta in intact dogs. Am J Physiol 210:903, 1966
23. Luisada AA: The second heart sound in normal and abnormal conditions. Am J Cardiol 28:150, 1971
24. Shaver JA, Nadolny RA, O'Toole JD, et al: Sound pressure correlates of the second heart sound. Circulation 49:316, 1974
25. Leatham A: Auscultation of the Heart and Phonocardiography. London, J. & A. Churchill, 1970, p 52
26. Leatham A, Weitzman D: Auscultatory and phonocardiographic signs of pulmonary stenosis. Br Heart J 19:303, 1957

27. Beck W, Schrire V, Vogelpoel L: Splitting of the second heart sound in constrictive pericarditis, with observations of the mechanism of pulsus paradoxus. Am Heart J 64:765, 1962
28. Epstein EJ, Criley JM, Raftery EG, et al: Cineradiographic studies of the early systolic click in aortic valve stenosis. Circulation 31:842, 1965
29. Gamboa R, Hugenholtz PE, Nadas AS: Accuracy of the phonocardiogram in assessing severity of aortic and pulmonic stenosis. Circulation 30:35, 1964
30. Whittaker AV, Shaver JA, Gray S, Leonard J: Sound-pressure correlates of the aortic ejection sound. Circulation 39:475, 1969
31. Hultgren HN, Reeve R, Cohn K, McLeod R: Ejection click of valvular pulmonic stenosis. Circulation 40:631, 1969
32. McKusick VA: Cardiovascular Sound in Health and Disease. Baltimore, Williams & Wilkins, 1958
33. Barlow JB, Bosman CK: Aneurysmal protrusion of the posterior leaflet of the mitral valve. Am Heart J 71:166, 1966
34. Jeresaty RM: Mitral valve prolapse–click syndrome. Prog Cardiovasc Dis 15:623, 1973
35. Bittar N, Sosa J: Billowing mitral valve leaflet: Report on fourteen patients. Circulation 38:763, 1968
36. Tavel ME, Baugh D, Fisch C, Feigenbaum H: Opening snap of the tricuspid valve in atrial septal defect. Am Heart J 80:550, 1970
37. Neil P, Mounsey P: Auscultation in patent ductus arteriosus with description of two fistulae simulating patent ductus. Br Heart J 20:51, 1958
38. Tavel ME, Frazier WJ, Fisch C: Use of phenylephrine in the detection of the opening snap of mitral stenosis. Am Heart J 77:274, 1969
39. Potain PCE: Les bruits de galop. Semaine Med (Paris) 20:175, 1900
40. Dock W, Grandell F, Taubman F: The physiologic third heart sound: Its mechanism and relation to the protodiastolic gallop. Am Heart J 50:449, 1955
41. Fleming JS: Evidence for a mitral valve origin of the left ventricular third sound. Br Heart J 31:192, 1969
42. Thayer WS: Further observations on the third sound. Arch Intern Med 4:297, 1909
43. Coulshed N. Epstein EJ: Third sound after mitral valve replacement. Br Heart J 34:301, 1972
44. Dock W: The genesis of diastolic heart sounds. Am J Med 50:178, 1971
45. Shah P, Yu PN: Gallop rhythm. Hemodynamic and clinical correlation. Am Heart J 78:823, 1969
46. Harvey WP, Stapleton J: Clinical aspects of gallop rhythm with particular reference to diastolic gallop. Circulation 18:1017, 1958
47. Craige E: Gallop rhythm. Prog Cardiovasc Dis 10:246, 1967
48. Luisada AA, Shah PM: Controversial and changing aspects of auscultation. Am J Cardiol 13:243, 1964
49. Swistak M, Mushlin H, Spodick DH: Comparative prevalence of the fourth heart sound in hypertensive and matched normal persons. Am J Cardiol 33:614, 1974

50. Fowler NO, Adolph RF: Fourth sound gallop or split first sound? Am J Cardiol 30:441, 1972
51. Rectra EH, Khan AH, Pigott VM, Spodick DH: Audibility of the fourth heart sound. JAMA 221:36, 1972
52. Goldblatt A, Augen MM, Braunwald E: Hemodynamic phonocardiographic correlations of the fourth heart sound in aortic stenosis. Circulation 26:92, 1962
53. Kincaid-Smith P, Barlow J: The atrial sound in hypertension and ischemic heart disease. Br Heart J 21:479, 1959
54. Cohen LS, Mason DT, Braunwald E: Significance of an atrial gallop sound in mitral regurgitation. Circulation 35:112, 1967
55. Bethell HJN, Nixon PGF: Understanding the atrial sound. Br Heart J 35:229, 1973
56. Caulfield WH, DeLeon AC, Perloff JK, Steelman RB: The clinical significance of the fourth sound in aortic stenosis. Am J Cardiol 28:179, 1971
57. Yahini JH, Dulfano MJ, Toor M: Pulmonic stenosis: A clinical assessment of severity. Am J Cardiol 5:744, 1960
58. Hill JC, O'Rourke RA, Lewis RP, McGranahan GM: The diagnostic value of the atrial gallop in acute myocardial infarction. Am Heart J 78:194, 1969
59. Freeman AR, Levine SA: Clinical significance of systolic murmur: A study of 1000 consecutive "noncardiac" cases. Ann Intern Med 6:1371, 1933
60. Leatham A: A classification of systolic murmurs. Br Heart J 17:575, 1955
61. Lewis DH, Deitz GW, Wallace JD, Brown JR: Present status of intracardiac phonocardiography. Circulation 18:991, 1958
62. Wennevold A: The origin of the innocent vibratory murmur studied with intracardiac phonocardiography. Acta Med Scand 181:1, 1967
63. Still GF: Common Disorders and Diseases of Childhood. London, Oxford University Press, 1915, p 495
64. Nellen M, Gotsman MS, Vogelpoel L, et al: Effects of prompt squatting on the systolic murmur in idiopathic hypertrophic obstructive cardiomyopathy. Br Med J 3:140, 1967
65. Vogelpoel L, Schrire V: Auscultatory and phonocardiographic assessment of Fallot's tetralogy. Circulation 22:73, 1960
66. Perloff JK, Harvey WP: Auscultatory and phonocardiographic manifestations of pure mitral regurgitation. Prog Cardiovasc Dis 5:172, 1962
67. Sutton GC, Craige E: Clinical signs of severe acute mitral regurgitation. Am J Cardiol 20:141, 1967
68. Rivero Carvallo JM: Signo para el diagnóstico de las insufiencias tricuspideas. Arch Inst Cardiol Mex 16:531, 1946
69. Beck W, Schrire V, Vogelpoel L, et al: Hemodynamic effects of amyl nitrite and phenylephrine on the normal circulation and their relation to changes in cardiac murmurs. Am J Cardiol 8:341, 1961
70. Vogelpoel L, Schrire V, Beck W, et al: Variations in the response of the systolic murmur to vasoactive drugs in ventricular septal defect with special reference to the paradoxical response in large defects with pulmonary hypertension. Am Heart J 64:169, 1962

71. Phillips JH, Burch DE, DePasquale NE: The syndrome of papillary muscle dysfunction. Ann Intern Med 59:580, 1963
72. Criley JM, Hermer AJ: The crescendo presystolic murmur of mitral stenosis with atrial fibrillation. N Engl J Med 285:1284, 1971
73. McCraw DB, Siegel W, Stonecipher HK, et al: Response of heart murmur intensity to isometric (handgrip) exercise. Br Heart J 34:605, 1972
74. Vogelpoel L, Nellen M, Beck W, Schrire V: The value of squatting in the diagnosis of mild aortic regurgitation. Am Heart J 77:709, 1969
75. Joyner CR, Dyrda I, Reid JM: Behavior of the anterior leaflet of the mitral valve in patients with the Austin Flint murmur. Clin Res 14:251, 1966
76. Harvey WP, Corrado MA, Perloff JK: "Right-sided" murmurs of aortic insufficiency. Am J Med Sci 245:533, 1963
77. Meadows WR, Sharp JT, Zachariudakis S: Premature mitral valve closure: A hemodynamic explanation for the absence of the first sound in aortic insufficiency. Circulation 28:251, 1963
78. Criscitiello MG, Harvey WP: Clinical recognition of congenital pulmonary valve insufficiency. Am J Cardiol 20:765, 1967
79. Dohan MC, Criscitiello MG: Physiological and pharmacological manipulations of heart sounds and murmurs. Mod Concepts Cardiovasc Dis 39:121, 1970

Shapur Naimi

7

Electrophysiologic Basis for Arrhythmias and Their Therapy

Concepts relating to the electrophysiologic basis of cardiac arrhythmias are so intimately bound to the modern understanding of the electrophysiology of the myocardial cell that a brief review of the developments in this area is germane to the understanding of cardiac arrhythmias.

The landmark in our understanding of the basic electrophysiology of the single myocardial fiber dates from the introduction of the microelectrode by Ling and Gerard in 1949 for studies of skeletal muscle[1] and its subsequent adaptation to studies of cardiac transmembrane potential.[2,3] When a resting myocardial fiber is impaled by such a microelectrode (Fig. 7-1) the cell is found to be negative inside relative to the outside, with the potential difference generally around −90 mV. Upon activation of the cell an action potential is initiated with a rapid upstroke designated as phase 0, which usually results in reversal of polarity of the cell to about +30 or +40 mV. This is followed by the stage of repolarization, which goes through phases 1, 2 (also designated as the plateau), 3 (the period of rapid repolarization), and finally phase 4 (which is the period of electrical diastole).

It is noted in Fig. 7-1 that the surface electrogram inscribes a curve that roughly correlates with the rate of change of the transmembrane potential. Thus when the transmembrane potential undergoes a rapid change, such as during phase 0, a sharp spike or R wave is recorded by the surface electrogram. During phase 2, or the plateau, on the other hand, with the membrane potential relatively constant the surface electrogram registers an isoelectric baseline. Finally, during the rapid re-

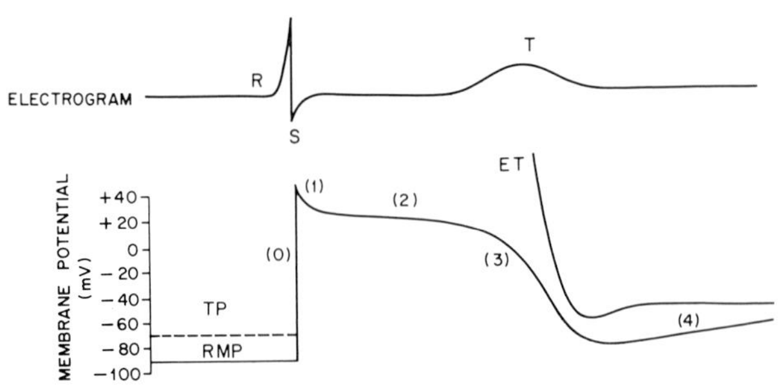

Fig. 7-1. A schematic diagram of the transmembrane action potential and unipolar electrogram of myocardial cell. Depolarization and sequence of repolarization phases are designated by numerals (0), (1), (2), (3), and (4). It is noted that the electrogram by and large correlates with the rate of change of transmembrane potential. It thus has a rapid upstroke during phase 0 and inscribes the T wave during rapid repolarization of phase 3. The threshold potential (TP) during diastole and return of membrane excitability with rapidly declining excitation thresholds (ET) following a period of absolute refractoriness is also illustrated. A brief period of supernormal excitability toward the end of phase 3 and the beginning of phase 4 is also noted. (Adapted from Hoffman, BF, Singer DH: Effects of digitalis on electrical activity of cardiac fibers. Prog Cardiovasc Dis 7:226, 1964.)

polarization of phase 3 the T wave of the surface electrogram is inscribed. Since the electrocardiogram is a composite of surface electrograms, it has a similar, albeit less exact, relationship to the action potential.

If a depolarizing current is passed through a cardiac fiber, it is noted that the membrane potential cannot be reduced beyond a certain limit without the development of a self-regenerative action potential. This critical level of membrane potential at which an action potential is triggered is called the threshold potential. This level generally lies in the vicinity of −65 to −70 mV (Fig. 7-1). Immediately upon excitation the cell becomes completely refractory to stimuli (absolute refractory period). This is followed during repolarization (starting at a membrane potential of about −55 mV) by the relative refractory period, during which stronger stimuli than normal are required for excitation. At the beginning of the relative refractory period stimulation of the cell results in local response without propagation. Later on during this period, stimulation causes a propagated response (effective refractory period). Most experimental measurements of the refractory period are based on observation of a propagated response and are therefore measurements of effective refractory period. It is noted in Fig. 7-1 that there is a rapid fall in excitation threshold during phase 3, leading to a brief period of supernormal excitability.

Myocardial fibers in general can be divided into two groups: pacemakers and nonpacemakers. In a pacemaker cell (Fig. 7-2, top panel A) during electrical diastole or phase 4 of the action potential there is a slow spontaneous depolarization. This is due to a time- and voltage-dependent diminution in potassium conductance. The membrane potential is thus depolarized to the level of threshold potential, and a self-regenerative action potential ensues. Pacemaker cells are found in the sinoatrial node, the cells of specialized atrial fiber tracts, the distal region of the atrioventricular node (NH), and the His-Purkinje system. Evidence points to the

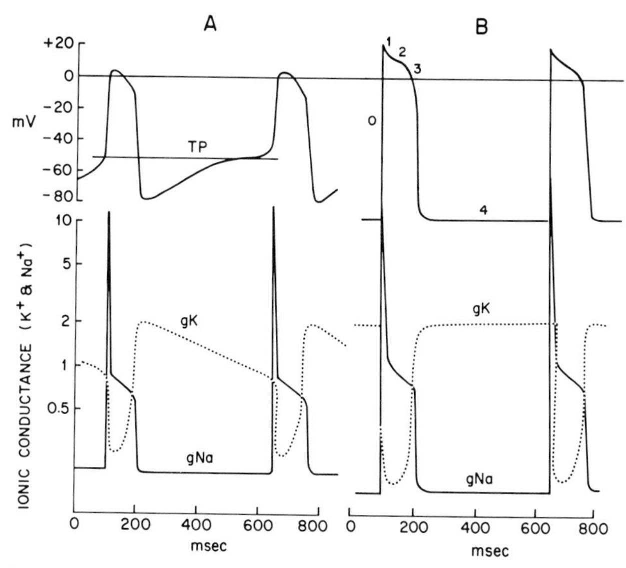

Fig. 7-2. Schematic diagram of pacemaker (panel A, top) and nonpacemaker (panel B, top) action potentials and corresponding changes in ionic conductances of the two types of action potential (g_K and g_{Na}) in the lower panels. It is noted that a pacemaker cell differs from a nonpacemaker in having a less negative resting membrane potential (based on a higher resting level of sodium conductance), a less steep upstroke of action potential, and a smaller amplitude. It is also noted that the slow diastolic depolarization of the pacemaker is based on a time- and voltage-dependent decline in potassium conductance together with a larger but steady sodium conductance of these cells. (Adapted from Trantwein W.[14])

absence of automaticity in the more proximal (AN and N) regions of the atrioventricular node,[4,5] except under extremely unphysiologic experimental conditions. In the nonpacemaker fibers that include atrial and ventricular muscle cells, on the other hand, membrane potential during electrical diastole remains stable (Fig. 7-2, top panel B), and therefore these fibers lack automaticity. As such, their depolarization normally depends on the propagation of an impulse from a pacemaker cell through the conducting system. In Fig. 7-2 it is also noted that the pacemaker fibers generally have a less negative transmembrane potential and a less steep phase 0 of the action potential. Furthermore, to a lesser or greater extent, they lack the overshoot or the reversal of polarity on activation and have a more blunt peak of the action potential, with the various phases of repolarization being less distinct from one another than in the nonpacemaker fibers.

The steepness of the slope of the slow diastolic depolarization differs among various pacemaker fibers. Thus as one moves down from the sinoatrial node to the A-V junctional area and through the His-Purkinje system to the ventricular junction there is a gradual diminution of this slope resulting in a decreasing automaticity. This provides a natural system of command in the heart, whereby the sinoatrial node having the steepest diastolic depolarization arrives at its threshold potential earlier and becomes the physiologic pacemaker. Furthermore, inherent in this system of decreasing automaticity down the conduction system is an efficient failsafe mechanism whereby, should the automaticity of the pacemaker fibers higher up in the system fail for physiologic or pathologic reasons, or should there be interruption of impulse propagation down the system of conduction, the pacemaker fibers immediately below the zone of failure or conduction block will take over the pacemaking function.

IONIC BASIS OF MEMBRANE POTENTIAL

Membrane potential is the result of the specific conductance of the membrane to certain ions (in particular, potassium and sodium), together with a specific transmembrane concentration gradient of these ions. Furthermore, the fundamental property of an excitable membrane is the ability to alter its permeability to various ions in a specific manner. The electrophysiologic basis of the ionic origin of membrane potential is firmly based on the classical work of Hodgkin and Huxley.[6,7] At rest (Fig. 7-2, bottom panels) the cell membrane is highly permeable to potassium and impermeable to sodium. The resting membrane potential is therefore largely determined by the potassium equilibrium potential given by the Nernst equation: $E_k = -RT/F \ln (K_i/K_0)$. In this equation E_k is the potassium equilibrium potential, R is the gas constant, T is absolute temperature, and F is Faraday's number. K_i and K_0 represent intracellular and extracellular concentrations of potassium, respectively, in millimoles. At 37°C RT/F has a numerical value of about 60. K_i is usually about 150 mM and K_0 about 5 mM. Thus $\ln (K_i/K_0)$ will have a value of about 1.5, and hence E_k equals about −90 mV, which is close to the physiologic resting membrane potential. It thus appears that the level of resting membrane potential is largely a function of the potassium conductance and concentration gradient across the membrane. As extracellular potassium level increases, there is a tendency for the resting membrane potential to shift to a less negative value and become partially depolarized. Conversely, with decrease in extracellular potassium there is initially a rise in membrane potential, as predicted by the Nernst equation. However, with more marked degrees of hypokalemia, observed membrane potential deviates from the theoretical value and again shifts to a less negative level. This phenomenon is at least partly accounted for by a reduction in potassium conductance at very low levels of extracellular potassium.[4]

Ionic conductance plays an extremely important role in the time course of action potential. Thus upon activation of the myocardial fiber there is a rapid rise in sodium conductance (Fig. 7-2, bottom panels) and an influx of sodium current through fast sodium channels. Hence the transmembrane potential approaches sodium equilibrium potential, which has a positive value. The main difference between pacemaker and nonpacemaker cells, as noted in Fig. 7-2, bottom panels, is that a pacemaker demonstrates a slow time- and voltage-dependent decay in potassium conductance in diastole, which is the electrophysiologic basis for the slow depolarization during phase 4 of pacemaker cells. It is also noted that the magnitude of the membrane potential in a pacemaker is less than that of the nonpacemakers. This is related to the presence of higher sodium conductance at rest and a larger coexisting steady inward sodium current in the pacemakers as compared to nonpacemakers.

Although potassium and sodium are the most important ions in determining membrane potential at rest, there is also some contribution from other ions such as calcium and chloride. A more elaborate equation for determining membrane potential has been worked out by Hodgkin and Katz based on theoretical considerations of Goldman.[8,9] Detailed discussion of this equation would be beyond the scope of this chapter.

Ionic mechanisms involved in the time course of the action potential have been the subject of intensive study in recent years. It appears that cardiac cells are of two types in terms of their ionic mechanisms for impulse generation—fast fibers with fast response and slow fibers with slow response.[5,10] The former include atrial and ventricular muscle cells as well as conducting systems in both chambers. The latter include sinoatrial and atrioventricular nodes as well as fibers in the mitral and tricuspid valve leaflets and A-V ring. Fast fibers have a high magnitude of resting potential (−80 to −90 mV) and a threshold potential of about −70 mV. The sharp upstroke of phase 0 in these fibers is the result of a strong influx of inward (deplorizing) sodium current through fast sodium channels that is rapidly inactivated and is followed by a weaker inward current that is triggered at membrane potentials less negative than −55 mV. This current, probably due to calcium, passes through slow channels distinct from sodium channels.[11] The slow channel is inactivated slowly, causing a long repolarization beyond the phase 0, and as such is probably responsible for the plateau phase of action potential. Fast fibers are rapid conductors of impulse (0.5–3 m/sec). This is related to their large resting potential and rapid rise of phase 0 (dv/dt). Slow cardiac fibers, on the other hand, are poor conductors (0.05 m/sec). This in turn is related to a relatively small resting potential (−60 to −70 mV), a slow self-regenerative depolarization, and a small amplitude of the action potential.[12] The slow depolarization phase in slow fibers is due to a weak inward current probably carried by calcium through slow channels[10,12,13] with characteristic slow channel responses to interventions. Thus the depolarization phase of these fibers is not affected by tetrodotoxin (sodium channel blocker), but is depressed by manganese and verapamil.[12,13]

Table 7-1
Differences between Fast and Slow Fibers

	Fast Fibers	Slow Fibers
Location	Atrial and ventricular muscle and conduction system	Sinus and A-V nodes and fibers in A-V valve ring
Magnitude of membrane potential	Large (−80 to −95 mV)	Small (−60 to −70 mV)
Rate of rise of phase 0	Fast	Slow
Amplitude of action potential	High (+30 to +40 mV)	Low (0 to +15 mV)
Threshold potential	−65 to −70 mV	Less than −55 mV
Conduction velocity	Fast (0.5 to 5 m/sec)	Slow (0.01 to 0.1 m/sec)
Fast sodium channels	Present	Absent
Slow channels (probably calcium)	Present	Present
Inhibition of depolarization phase	Tetrodotoxin (inactivating sodium channels)	Manganese and verapamil (inactivating calcium channels)

Table 7-1 illustrates the important characteristics of the two types of fibers and the differences between the two.

ELECTROPHYSIOLOGIC BASIS OF CARDIAC ARRHYTHMIAS

Excitation of cardiac muscle is the result of the processes of automaticity and conductivity. Hence cardiac arrhythmias are generally considered to be due to the abnormality of impulse formation or conduction. However, an alternative and perhaps more useful classification

Table 7-2
Electrophysiologic Basis of Cardiac Arrhythmias

Tachyarrhythmias: Essentially two mechanisms are involved in the pathogenesis of extrasystoles and tachyarrhythmias:
- Alteration of automaticity
  - Physiologic and environmental changes in normal automaticity (autonomic, electrolytic, etc.) of the sinus, NH region of A-V node, and specialized atrial and ventricular fibers
  - Abnormal automatic activity
    - Partial depolarization of fast fibers resulting in automaticity of the slow fiber type
    - Afterpotential oscillations
- Reentry
  - Circus type
    - At sinus node or A-V junction
    - Over large circuits, i.e., atrial flutter or anomalous tracts (W-P-W)
    - Microreentry due to local unidirectional block
  - Focal reexcitation due to boundary currents from local potential difference
    - Dispersion of refractoriness
    - Local partial depolarization

Bradyarrhythmias: Essentially two mechanisms may be involved in the pathogenesis of bradyarrhythmias:
- Alteration of automaticity
  - Physiologic and environmental depression in the normal automaticity of pacemaker fibers (autonomic, electrolytic, etc.)
  - Pathologic depression, i.e., sick sinus syndrome
- Depression of conductivity
  - Decremental conduction due to:
    - Refractory tissue in the path due to disparity of action potential duration
    - Failure of propagation due to partial depolarization
      - Poor repolarization
      - Enhanced depolarization

illustrated in Table 7-2 differentiates cardiac arrhythmias into tachyarrhythmias and bradyarrhythmias, the former arising from increased automaticity or reentry and the latter resulting from depressed automaticity or conductivity.

DISTURBANCES OF NORMAL AUTOMATIC MECHANISMS

Figure 7-3 illustrates factors that influence the degree of automaticity. These factors include the slope of phase-4 diastolic depolarization, the absolute level of resting membrane potential, and the magnitude of the threshold potential. If the slope of spontaneous depolarization is decreased (trace *c*), cycle length is prolonged. Similarly, when threshold potential is shifted to a less negative level (trace *b*) or when membrane is hyperpolarized, firing rate is reduced. Figure 7-3 (trace *d*) indicates the marked effect on prolongation of cycle length of a combination of all three factors.

There are a number of environmental factors that affect the slope of diastolic depolarization.[14] Increase in temperature or stretching of the

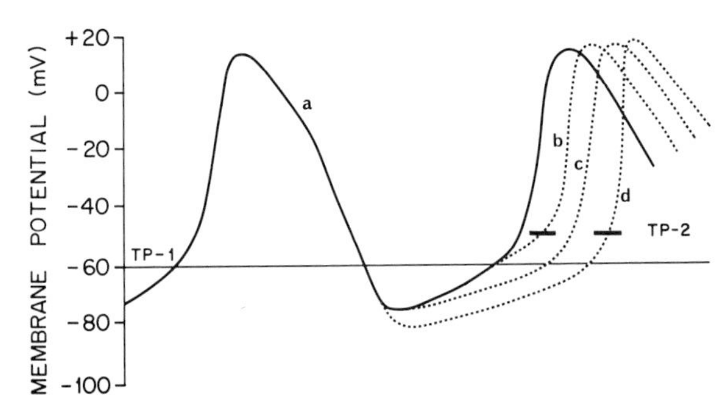

Fig. 7-3. Schematic representation of an action potential from a pacemaker cell illustrating the mechanisms involved in the alteration of normal automaticity. In *b* a shift in the threshold potential from TP-1 to TP-2 results in a longer cycle length. In *c* the reduced automaticity is due to a decrease in the slope of diastolic depolarization without a shift in threshold potential. In *d* a combination of hyperpolarization of the membrane, a decrease in the slope of diastolic depolarization, and a shift of the threshold potential to a less negative level causes a marked widening of the cycle length. (Adapted from Hoffman BF: the genesis of cardiac arrhythmias. Prog. Cardiovasc Dis 8:319, 1965.)

myocardial fibers[4,15,16] cause enhancement of automaticity. Thus sinus tachycardia in the context of fever or cardiac failure (resulting in the stretch of atrial fibers) is an expected consequence of the environmental factors affecting the pacemaker fibers. Similarly, hypoxia[17] and increases in pH and PCO_2[18] cause tachycardia through an increase in the slope of phase 4. Paramount among the factors that affect automaticity is the influence of the autonomic nervous system. Thus stimulation of the adrenergic nervous system or catecholamines increase the slope of pacemaker potential,[19] resulting in sinus tachycardia. Vagal stimulation or acetylcholine have the reverse effect, resulting in sinus bradycardia.[4] This effect of vagal stimulation is due to reduction in the slope of phase 4 as well as hyperpolarization of the membrane as a result of increased potassium permeability. Variations in vagal tone associated with respiration frequently causes sinus arrhythmia. Furthermore, parasympathetic activity has a more depressing effect on the sinoatrial node than the lower regions of the conducting system. This may result in the shifting of the pacemaker to lower atrial foci or the atrioventricular (junctional) area. Thus a number of relatively benign physiologic arrhythmias may result from modifications of the autonomic influences. The emergence of a junctional pacemaker from the upper reaches of the His-Purkinje system or the NH region of the A-V node is particularly likely when there is simultaneous increase in parasympathetic and sympathetic tone, with the former depressing the atrial pacemaking foci and the latter increasing the automaticity of the junctional area.[20] This may occasionally result in interference dissociation, with the automaticity of the junctional area exceeding that of the sinus node. Here the junctional pacemaker controls the ventricle at a higher rate, and since there is a retrograde (ventriculo-atrial) block the sinus node is protected and controls the atria at a slower rate; occasional capture of the ventricle by the sinus node occurs when the impulse arrives in the ventricular conducting system beyond its refractory period. Thus interference dissociation may at times result from modification of the balance between adrenergic and cholinergic activity.

Changes in the electrolytic environment may have a significant effect in the genesis of arrhythmias. Hyperkalemia has a complex effect on automaticity. It shifts the membrane potential to a less negative level, as predicted by the Nernst equation, and therefore should be expected to accelerate the firing rate. However, simultaneously it reduces the slope of spontaneous diastolic depolarization, particularly in the Purkinje fibers, and shifts the threshold potential to a less negative value.[21] Thus when the His-Purkinje system is the pacemaker of the heart (such as in complete heart block), spontaneous diastolic depolarization may not achieve the level-of-firing threshold. Hence the cell is locked in a polarized state,[22] and complete asystole results. Conversely, hypokalemia (less than 3 mM) increases the steepness of phase-4 depolarization and, when marked, shifts the membrane potential to a less negative level.[23] This enhanced automaticity in hypokalemic states may lead to ventricular tachyarrhythmias. Low extracellular sodium concentration decreases automaticity of phase 4, since a large inward sodium current is an important aspect of automaticity in fast fibers. Increased automaticity of the low calcium states, on the other hand, is the result of a shift in threshold potential toward the resting potential. Conversely, high extracellular calcium levels raise the threshold potential and reduce the firing rate.[4]

It should be noted that these changes of automaticity with alterations of electrolytic environment are a hallmark of fast fibers (i.e., conduction systems in the atria and ventricles) and probably do not apply to the slow fibers or to partially depolarized Purkinje cells that may assume slow response characteristics.[24] The ionic mechanism for spontaneous depolarization in slow fibers is not clear. However, it is clearly different from that of fast fibers, since there is evidence that hyperkalemia[25] and hypercalcemia[26] may enhance this automaticity, and hyponatremia has little effect on it. Catecholamines, on the other hand, increase the diastolic depolarization of both fast and slow fibers.[12]

Antiarrhythmic agents generally reduce the slope of phase 4, particularly in an ectopic focus,[21] and some may also shift the threshold potential to a less negative level (vide infra) and result in bradycardia. Digitalis glycosides at a toxic level enhance the automaticity of the His-Purkinje system,[27] which in the clinical setting may lead to the development of nonparoxysmal junctional tachycardia. More recently it has been demonstrated that digitalis toxicity through depolarization of the cell membrane may convert the normal fast fiber response activity of Purkinje fibers to slow fiber response. Under these circumstances, Cranefield and Aronson have demonstrated that electrical stimulation of Purkinje fibers results in a sustained rhythmic activity rather than the overdrive suppression of a normal fiber with fast response.[28] In this regard it is possible that the beneficial effect of hypocalcemia in controlling digitalis toxic arrhythmias may be based on its depressant effect on calcium-dependent automaticity of the slow fiber type. Thus Surawicz[29] has demonstrated that reduction in serum calcium level induced by chelating agents is very effective in suppressing ventricular arrhythmias of digitalis intoxication.

It should be noted that the presence of more than one factor working in the same direction may potentiate the effect on automaticity. Thus enhancement of automaticity in the His-Purkinje system caused by toxic levels of digitalis is compounded by the presence of a hypokalemic state.

The operation of environmental factors in the genesis of arrhythmias is through their effects on the normal automatic mechanism and therefore does not involve the atrial and ventricular muscles, which do not under physiologic conditions demonstrate slow diastolic depolarization.

ABNORMAL AUTOMATIC ACTIVITY IN ISCHEMIC HEART DISEASE

The influence of automaticity in the pathogenesis of the arrhythmias in the acute phase of ischemic disease is not completely settled. There are electrophysiologic reasons to believe that enhancement of automaticity may occur. Thus Trautwein and Kassebaum have demonstrated that the passage of a subthreshold current through the Purkinje cells increases their automaticity.[30] Hence the current of injury of acute infarction may in like manner trigger ectopic impulse formation in the Purkinje fibers in the vicinity of an infarct. Furthermore, acute ischemia results in leakage of potassium into the extracellular space,[31] resulting in partial depolarization of the adjacent Purkinje fibers, inactivation of their fast sodium channels, and conversion to slow fiber response. Simultaneously, there is release of catecholamines in the infarcted area,[32] which tends to enhance the slow response seen in depolarized Purkinje fibers.[10] In addition, Solberg and collaborators have demonstrated, by microelectrode studies, cells with low-amplitude action potential and diastolic depolarization within 1 hr of coronary ligation.[33] On the other hand, Scherlag and collaborators have produced evidence that in acute experimental myocardial ischemia the ventricular escape rate (a measure of inherent ventricular automaticity) as determined by vagally induced atrial arrest remains unchanged.[34,35] This would tend to argue against automaticity. In addition, these investigators found that overdrive, expected to suppress automaticity, increased the incidence of ventricular tachyarrhythmias in this setting; they used this evidence as a further indication of reentry being the cause of tachyarrhythmias of early acute ischemia. More recent evidence, however, would indicate that overdrive suppresses the automaticity of fast fibers and may on the contrary enhance automaticity of the slow fibers[12] or that of the Purkinje fibers that have assumed slow fiber characteristics.[28] Thus overdrive interventions may not necessarily help determine whether an arrhythmia is of automatic origin. The same group has found that in later phases of ischemia, within

24 hr after infarction, idioventricular rate is increased, denoting enhancement of automaticity.[35,36] Moreover, this increased automaticity is independent of catecholamine release and appears to relate to a primary increase in the automaticity of the His-Purkinje system. This increased automaticity in the later phases of myocardial ischemia has been found in surviving Purkinje fibers by other investigations.[37,38] It is thought that the mechanism for spontaneous depolarization in these partially depolarized (−60 mV or less) Purkinje cells is of the slow fiber response type.[12]

SPECIAL FORMS OF ABNORMAL AUTOMATIC ACTIVITY

Apart from the alterations of normal pacemaker potential leading to arrhythmias, microelectrode studies have demonstrated other possible mechanisms for abnormal automaticity. One such mechanism may be oscillation of afterpotentials, generally cited as a form of abnormal automatic activity that may be responsible for tachyarrhythmias. It has been argued that these afterpotentials, consisting of oscillations of membrane potential following repolarization, may attain sufficient magnitude to generate ectopic beats or repetitive firing. There is no doubt that under certain circumstances such as exposure of atrial or Purkinje fibers to aconitine[39] and veratrine[40] such oscillation of afterpotentials leading to tachyarrhythmias has been documented. Furthermore it has been demonstrated that exposure of human atrium to hypothermia results in prolongation of action potential together with a plateau level closer to the membrane potential. Thus, with the recovery of excitability, an ectopic response may be generated.[41] In addition, the presence of oscillations in the diseased fibrillating human atrium has been documented by microelectrode studies.[16,42] However, whether such experimental demonstrations of sustained arrhythmias of afterpotential have a clinical counterpart remains to be settled.

Another form of increased automatic activity is parasystole. This is a subsidiary automatic focus within the atrium or ventricle that fires at its own independent rate[43] and is immune from discharge by the normally conducted impulse by virtue of the presence around it of a zone of protection block. The independent firing of a parasystolic focus results in extrasystoles with varying coupling intervals. Moreover, parasystolic impulses are blocked if they fall within the refractory period of the normally conducted beats and may cause fusion beats when simultaneous depolarization of different areas of myocardium by normal and parasystolic impulses takes place.

It has been suggested[44] that the electrophysiologic basis of parasystole may be increased automaticity in the His-Purkinje system that results in partial depolarization and impaired propagation of an impulse. Thus increased automaticity in a segment of the conducting system may be the basis of protection (entrance) block seen in parasystole. More recently[12] it has been proposed that parasystolic impulses may be the result of cells with low resting membrane potentials and slow response activity. It is postulated that entrance or exit block can more easily occur in areas with properties of slow response, since conduction of a slow fiber action potential is more liable to block than a fast fiber.

ARRHYTHMIAS OF REENTRY

There are two types of reentrant mechanism involved in generation of ectopic impulses: circus reentry and focal reexcitation.

Circus Reentry

Circus reentry requires the establishment of a circular loop. This necessitates the presence of unidirectional block at some point on this loop, together with an alternate route of excitable tissue over which the impulse travels sufficiently slowly to excite, in a retrograde direction and in sequence, the area just distal to the block, then the block, and finally the tissue proximal to the block that by then has recovered its excitability. As such, factors that predispose to reentrant rhythms would include a short refractory period, slow conduction, and unidirectional block. Circus-type reentrant arrhythmias may involve reentry in the A-V junction, which is responsible for return extrasystoles, and reciprocating A-V nodal tachycardia. Others consist of microreentry in areas of local block, which may be responsible for certain premature beats and atrial or ventricular tachyarrhythmias. Reentrant circuits may also have a long track, such as in atrial flutter or Wolff-Parkinson-White (W-P-W) tachyarrhythmias.

Perhaps the most thoroughly studied example of reentrant activity is A-V reciprocal beating, first described by White half a century ago.[45] In this regard, physiologic evidence has been advanced by Moe and collaborators for the presence of a dual A-V conduction pathway.[46] More recently, Watanabe and Dreifus have demonstrated the functional dissociation of the A-V junctional tissue, due to a disparity of conduction velocity, and plotted the pathway of reciprocal beat down the left side and retrograde reentry into the right side of the A-V junction.[47,48] It should be noted that atrial and ventricular echoes are a special form of reentry, and their occurrence is facilitated by the extremely slow conduction velocity through the A-V junction. Thus conduction time required for completion of the reentrant circuit exceeds the refractory period of the myocardial fibers in the circuit. Another example of reentry involving the A-V junctional area is paroxysmal tachycardia associated with W-P-W syndrome.[49,50] In this condition there is anatomic dissociation of the dual pathway with the presence of an accessory bundle. Tachycardia is precipitated by an extrasystole that enters the circuit and is usually transmitted antegrade down the A-V node and retrograde up the accessory bundle. In the Lown-Ganong-Levine syndrome[49] there is a bypass tract that inserts into the distal end of the A-V node producing a short PR interval and a basis for reentrant tachycardia as in the W-P-W syndrome.[45] Moreover, the His electrogram and pacing studies in certain cases of supraventricular tachycardia indicate the presence of a bypass tract not reflected on the electrocardiogram.[51]

It appears that reentry may account for a number of tachycardias. This has been well demonstrated in atrial flutter.[52] Similarly, many cases of paroxysmal supraventricular tachycardia involve reentry through the A-V node.

An important type of circus reentry may develop in areas of local block with microreentry. It has been postulated that such reentrant arrhythmias may be generated within small segments of myocardium due to conduction delay and unidirectional block. This type of reentry has been the subject of much interest ever since 1928, when Schmitt and Erlanger demonstrated unidirectional block and reentry in a muscle strip.[53] Figure 7-4 is a diagrammatic illustration of this type of reentry. In panel A an impulse propagating down the Purkinje fibers encounters block in one branch leading to the ventricle and transmits down the alternate path. In panel B the impulse reenters the Purkinje fiber in a retrograde fashion and works its way through the blocked area, arriving at the proximal point after recovery of excitation, thus reentering the circuit. Another possible mechanism of reentry has been suggested by Hoffman.[44] In this mechanism, antegrade conduction is markedly slowed in the area of reentry, with complete retrograde block. Thus the impulse emerges from the area of slow conduction beyond the refractory period of peripheral tissue depolarized through alternate pathways, and reexcitation occurs. It is noted that in both hypotheses unidirectional block plays an important role. The major difficulty in accepting reentry as a basis of ectopic tachyarrhythmias has been the argument that in view of a long refractory period and fast conduction velocity in the ventricular conducting system and myocardium reentry pathways require an exceedingly long track. More recently this problem has been overcome by the demonstration that conduction velocities even slower than that of atrio-

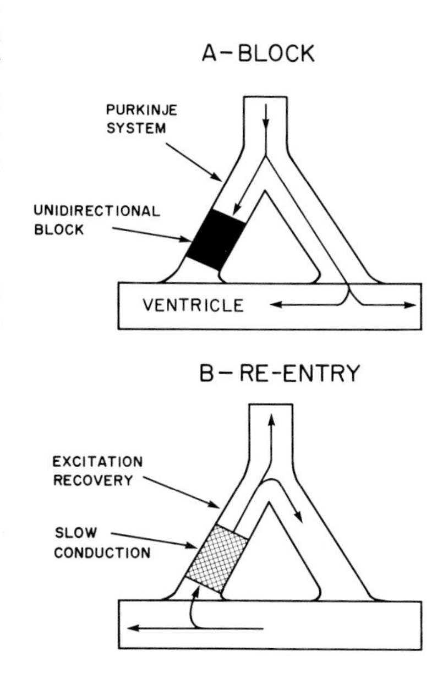

Fig. 7-4. Diagrammatic representation of the circus type of reentry. The drawing shows a peripheral segment of the Purkinje system and the adjoining ventricular cell. In panel A there is impaired conduction in one branch of the Purkinje system due to unidirectional block. The same impulse, however, conducts through an alternate normal branch to the ventricle. In panel B the impulse conducts from the ventricle through the area of unidirectional block in a retrograde course, arrives in the Purkinje system after recovery of excitation, and reenters the circuit. [Adapted from Hoffman BF: The possible mode of action of antiarrhythmic agent. In Briller S, Conn H Jr (eds): The Myocardial Cell. Philadelphia, University of Pennsylvania Press, 1966.]

ventricular node can be induced by depression of excitability in short segments of the ventricular conducting system and by interference of two impulses.[54,55] Similarly, in acute ischemia, Hope and collaborators[56] have found extremely slow conduction velocity leading to delay of activation in the epicardial zone of ischemic myocardium beyond the T wave of the standard electrocardiogram. In these studies ventricular tachycardia occurred when ischemic-zone epicardial activation delay extended into the T wave of the standard electrocardiogram. This marked delay of activation within the ischemic zone would argue for reentry rather than automaticity being the basis for arrhythmias of the early phase of acute ischemia. In addition, faster rates accelerated and slower rates impeded the development of ventricular tachycardia, further pointing away from automaticity as a cause of such arrhythmias.

The peripheral Purkinje–ventricular junction appears to provide an appropriate anatomic basis for unidirectional block. A Purkinje fiber is distributed among a number of ventricular fibers, and therefore slow conduction or block in a particular branch may set up a reentry loop by excitation of the ventricle through the normally conducting branches. Furthermore, the distribution of a terminal Purkinje fiber to a number of ventricular fibers results in convergence of impulses in the ventricular–Purkinje direction and divergence in the opposite,[57] causing a directional difference in conduction velocity. In addition, the disparity of the duration of action potential in the Purkinje–ventricular junction (vide infra) predisposes to unidirectional block. Finally, a short refractory period, a further predisposing factor to reentry, occurs during ischemia.[58] Thus recent studies appear to have moved reentry from the position of an attractive theory to that of an established phenomenon.

Focal Reexcitation

Focal reexcitation may result from boundary currents that arise when there is a significant difference between the membrane potentials of adjacent myocardial fibers. These depolarizing currents may cause excitation of the better polarized and therefore excitable fiber. Figure 7-5 demonstrates the theoretical basis of this type of impulse formation. At point *X* during repolarization, fiber *a* is completely repolarized, whereas fiber *b* is only partially repolarized. Consequently, a depolarizing boundary current will flow from *b* to *a*, causing reexcitation of this fiber. It should be noted that although the concept of boundary current is easily understandable, questions have been raised as to its validity. Thus it has been argued that if cardiac intercellular connections are of low internal resistance, a widely disparate action potential duration cannot exist in functionally connected cells.[59] Yet Han and Moe have produced evidence, through the determination of effective refractory period among a number of epicardial electrodes,[60] that disparity of refractoriness may exist within small segments of the myocardium. These studies have

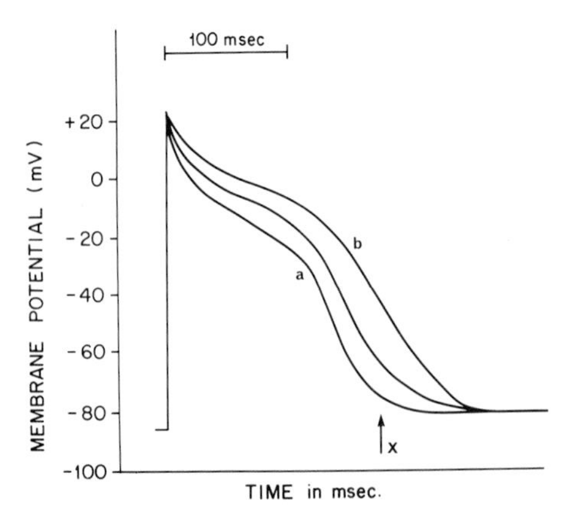

Fig. 7-5. Diagrammatic representation of the focal reexcitation type of reentry. The drawing illustrates transmembrane potentials of three adjacent myocardial fibers with different durations of action potential and refractory period. At the point in time designated *X* (arrow), fiber *a* is already repolarized to −80 mV, whereas fiber *b* has a less negative transmembrane potential. Depolarizing boundary currents flow from *b* to *a* resulting in reexcitation of this fiber and giving rise to closely coupled ectopic beats. (Adapted from Han J, Millet D, Chizzonitti B, et al.: Temporal dispersion of recovery of excitability in atrium and ventricle as a function of heart rate. Am Heart J 71:481, 1966.)

demonstrated that conditions that reduce ventricular fibrillation threshold, such as ischemia and stimulation of the left stellate ganglion, increase the dispersion of refractoriness. These results may indicate that increased irritability during myocardial ischemia may result from asynchrony of refractoriness within the ischemic segment. Our recent observations confirm the presence of increased dispersion of refractoriness within the ischemic segment during coronary ligation.[58] However, immediately following release of the coronary ligature, when there is a striking vulnerability to fibrillation,[61] no such increased dispersion within the ischemic segment was detected. On the other hand, in these studies there was a brief and abrupt additional shortening of the refractory period immediately upon reperfusion of the ischemic myocardium that resulted in a rise in the gradient of dispersion of refractoriness between the ischemic and nonischemic myocardium. Thus, in these studies, arrhythmias of acute ischemia appear to correlate with dispersion within the ischemic area, and those of reperfusion correlate with the gradient of dispersion across the boundary between the ischemic and normal myocardium.

Boundary current may also be the result of membrane potential difference between normal and depressed tissue in ischemic heart disease.

CONDUCTION AND ITS DISTURBANCES

Conduction velocity in the heart is largely determined by two factors, the rate of rise of the action potential (dv/dt) and the diameter of the conduction fiber.[11] The rate of rise of the action potential in turn is primarily a function of the resting membrane potential. Thus a deteriorated and poorly polarized fiber is a slow conductor. Similarly, when an extrasystole reaches a normal fiber before it is fully repolarized, slow conduction or block ensues. With regard to diameter, velocity of conduction is proportional to the square root of the width of the conducting fiber. Thus Purkinje fibers with a diameter of 30 μ have a high conduction velocity (1.5–3 m/sec), whereas atrioventricular fibers (diameter about 3 μ) possess a conduction velocity of 0.05 m/sec.[11]

THE ANATOMIC BASIS OF CONDUCTION

The impulse generated in the sinus node is transmitted along three internodal tracts to the A-V node, which is composed of three regions, AN, N, and NH. The AN and N regions are probably devoid of automaticity (vide supra), and therefore ectopic tachycardias of A-V nodal origin, of the type based on automaticity, probably arise in the NH region of the A-V node; hence the designation of junctional tachycardia for such arrhythmias is physiologically more correct. Distally, the A-V node joins the bundle of His, which divides into three fascicles—the right bundle and the anterior and posterior divisions of the left bundle.[62] The bundle branches terminate in the Purkinje network, which ends in the ventricular muscle. Each impulse triggers the next, which results in an action potential appropriate to its location. A point of electrophysiologic significance is the fact that there is a progressive prolongation of action potential duration through the A-V node, down the bundle branches to the Purkinje network. The maximum action potential duration occurs about 2 mm proximal to the termination of the Purkinje fibers in the muscle,[63] and then there is a progressive shortening of the action potential to the myocardial fiber. This area of the maximum action potential duration and refractoriness provides a "gate" that will not allow the passage of a closely coupled premature beat in either direction.[64]

With the exception of the AN and N regions of the A-V node, the conducting system from the sinus node through the Purkinje cells demonstrates automatic activity in a decreasing order. Thus sinus node has an automaticity of about 70 beats per minute, which decreases to 45 to 60 beats per minute in the A-V junction and 10 to 20 beats in the peripheral Purkinje cell. The clinical implication of this decreasing automaticity down the conduction pathway should be clear. Thus, should there be a failure of automaticity or a block in the propagation of impulses at any point down this pathway, the pacemaker immediately below the point of failure or block takes over the pacemaking function. Hence, in cases of atrioventricular block it is important to ascertain the level of the block, which determines whether there will be adequate heart rate to guard against impaired cerebral perfusion and Stokes-Adams attacks. Thus conduction block at the A-V node level results in a junctional escape rhythm with a reasonable rate (45–60 per minute), whereas a block at the level of bundle branches may result in extremely low idioventricular rate, inadequate circulation, Stokes-Adams syncope, or cardiac failure. There are a number of features that distinguish an A-V block at the level of the A-V node from that at the level of bundle branches. Thus the A-V nodal level of conduction block usually presents as a Wenckebach phenomenon (gradual PR prolongation); it has a narrow QRS complex, is often associated with inferior myocardial infarction, is frequently drug-induced, and responds favorably to atropine. A-V block at the level of bundle branches, on the other hand, has a fixed PR interval and a wide

QRS complex; it may be associated with anterior myocardial infarction, is unrelated to drugs, and does not respond to atropine.[65]

DISTURBANCES OF IMPULSE CONDUCTION

An important cause of the failure of impulse propagation is the presence of a refractory period ahead of the excitation front. This may occur when the fibers ahead of the propagating impulse have a longer refractory period. This disparity of action potential duration with progressive prolongation of refractory period is present as the impulse passes from the atrium through various regions of the A-V node to the bundle of His.[4] This disparity of refractoriness across the A-V node is probably the electrophysiologic basis for the protection of the ventricle against extremely fast atrial tachyarrhythmias (i.e., atrial fibrillation) or closely coupled atrial premature beats. Indeed, this physiologic mechanism provides a protective barrier that is the electrophysiologic basis for the relatively benign nature of supraventricular tachyarrhythmias. During atrial fibrillation many impulses entering the A-V node encounter fibers that are poorly polarized (by virtue of longer refractory periods) and thus have a slow upstroke and velocity of conduction. This form of altered conduction (that is, progressive impairment of conduction resulting from progressive decrease in upstroke and amplitude of the propagating impulse) is termed decremental conduction. Many impulses fail to emerge from the node through complete decrement and are said to be concealed, since they leave no record on the electrocardiogram. Although decremental conduction was first demonstrated in the A-V node,[66] it is by no means confined to it. Partial loss of membrane potential (whether through ischemia or high extracellular potassium) in other regions of the myocardium may, under certain circumstances, lead to progressive decrease in the upstroke of action potential of the slow response type and decremental conduction. In this regard, slow fiber response characteristics in the Purkinje fibers surviving myocardial infarction have been demonstrated in the dog.[37]

Decremental conduction is often unidirectional, since in many areas of the heart there is a specific directional disparity of the duration of refractoriness. Hence decremental conduction may set the stage for reentrant arrhythmias by virtue of providing two prerequisites of such arrhythmias, namely slow conduction and unidirectional block.

In the A-V node, particularly the N region, the fibers have (even under physiologic conditions) a lower membrane potential and a slower upstroke with slow fiber response characteristics (vide supra). Factors that are known to depress A-V conduction (vagal stimulation, digitalis glycosides,[67] and acetylcholine[68]) depress the upstroke even further and may lead to block. Thus decremental conduction in the A-V node may cause not only the normal A-V delay (PR interval) and concealed conduction,[69] but some of A-V conduction disturbances such as first-degree and second-degree block (Wenckebach) as well. Indeed, it has been demonstrated that the Wenckebach phenomenon is due to depression of conduction largely in the N region of the A-V node.[70]

Similarly, decremental conduction in partially depolarized Purkinje cells may occur due either to enhanced diastolic depolarization or to poor repolarization.[71] This may result in A-V block of the Mobitz type II, which has been shown to be the result of conduction disturbances in the more distal segment of the A-V conduction system.[70]

It is noted that increased automaticity in the His-Purkinje system may lead to partial depolarization of these cells at the time of excitation and result in poor conductivity, decremental conduction, and reentry. There is evidence that this mechanism may explain certain arrhythmias of digitalis glycoside toxicity.[72] In similar fashion, the bundle branch block that may follow marked bradycardia during carotid sinus massage may be the basis of partial depolarization due to prolongation of the automatic phase, resulting in conduction block.[73,74]

FIBRILLATION

The exact mechanism of fibrillation, atrial or ventricular, remains a subject for debate despite intensive effort by many investigators over the years. Two mechanisms have been proposed for the initiation of this arrhythmia: (1) repetitive impulse formation of unifocal or multifocal origin and (2) repetitive microreentry. Fibrillation constitutes a chaotic asynchronous fractionated activity of the heart that can be induced in diseased as well as normal myocardium.[75] Microelectrode studies have revealed almost complete electrical asynchrony and disorganization of the heart in ventricular fibrillation.[76] Extensive data on the subject have been excellently reviewed by Surawicz.[77]

Factors that increase the vulnerability to fibrillation include (1) a minimum critical mass of myocardium, (2) an initiating stimulus, and (3) a certain critical relation between impaired conduction velocity and shortened duration of refractory period. The initiating stimulus for fibrillation could be one of several mechanisms, i.e., automaticity, focal reexcitation, or circus reentry. Thus atrial cells from patients with atrial arrhythmias have been shown to have increased automatic activity.[78] In addition, automatic activity of the slow fiber response type has been demonstrated in the Purkinje system in experimental myocardial infarction.[33]

The role of temporal dispersion of refractoriness in the initiation of ventricular fibrillation has been emphasized by Han and Moe.[60] It has been argued that premature beats arising by focal reexcitation due to disparity of refractoriness tend to have shorter coupling intervals.[79] Hence the risk of fibrillation would be higher, since closely coupled beats fall in the vulnerable period.[22] Han and co-workers[60,80] have presented evidence that conditions that predispose to tachyarrhythmias such as ischemia and sympathetic stimulation result in increased temporal dispersion of refractoriness.

Other conditions that facilitate spontaneous ventricular fibrillation include prolongation of QT interval, presumably through increased dispersion of refractoriness. Thus congenital lengthening of the QT interval[81] frequently associated with deafness (Jervell–Lange-Nielson syndrome) is associated with ventricular tachyarrhythmias and sudden death. Other causes of prolongation of QT interval predisposing to ventricular tachyarrhythmias include hypothermia,[82] hypokalemia,[83] quinidine,[84] and phenothiazines.[85] Biochemical abnormalities may also be associated with ventricular fibrillation. Thus in hypothermia the onset of fibrillation has been related to loss of potassium and a gain of calcium by the heart.[86] Moreover, recent observations indicate that adrenal medullary secretions may also be involved in the predisposition to ventricular fibrillation following acute coronary ligation.[87]

ELECTROPHYSIOLOGIC BASIS OF ANTIARRHYTHMIC THERAPY

A sound electrophysiologic approach to the treatment of arrhythmias has been fraught with difficulties because of incomplete understanding of the electrophysiologic basis of many clinical arrhythmias as well as inadequate data on the effects of antiarrhythmic agents. Despite all this, however, a clear understanding of some of the basic mechanisms of cardiac arrhythmias has resulted in a more rational approach to the prevention and treatment of these disorders. Thus the knowledge that alterations of many environmental factors (i.e., electrolytes, pH, Pco_2, autonomic influences, and myocardial stretch) may result in changes of automatic activity and related arrhythmias has placed the prevention and treatment of such arrhythmias on a firm foundation through control of these factors.

ANTIARRHYTHMIC AGENTS

From the electrophysiologic point of view, antiarrhythmic agents may be classified into two groups.[88–90] Group I comprises quinidine and procaine amide. Group II consists of lidocaine and diphenylhydantoin. Propranolol is somewhat closer to Group I than to II. Despite this classification there are some features common to all of these antiarrhythmic agents. These features include:

1. Depression of phase-4 slow diastolic depolarization.[21] With respect to propranolol, this depressing effect on phase 4 is essentially a function of its antiadrenergic effect. This antiautomatic effect of antiarrhythmic agents can be expected to be effective in the treatment of arrhythmias of enhanced automaticity. Conversely, they would be expected to have an adverse effect on the whole spectrum of bradyarrhythmias. Group I agents not only reduce automaticity but also increase the level of threshold potential. Group II agents reduce the automaticity of the conducting system, probably through an increase in potassium permeability, but do not significantly alter the threshold potential. Group II agents do not cause ventricular tachyarrhythmias at toxic levels, which is a feature common to quinidine and procaine amide.

2. A second feature common to all antiarrhythmic agents is that they cause a shift of the membrane potential during repolarization, from which a response may be evoked to a more negative level. In other words, they prolong the effective refractory period relative to the action potential duration. This happens whether the duration of action potential is prolonged (as in Group I) or shortened (as in Group II). This effect occurs both in Purkinje and ventricular fibers. The effects of these two groups of antiarrhythmic agents on relative durations of effective refractory period and action potential differ; as seen in Fig. 7-6, those of Group I shift the membrane responsive curve (i.e., V_{max} at a given membrane potential) to the right, and hence decrease conduction velocity, while those of Group II cause a slight shift of the curve to the left. It is noted in Fig. 7-6 that for both groups of agents the earliest obtainable response is at a more negative membrane potential than the normal.

3. In simultaneous recordings of the Purkinje and ventricular action potentials, lidocaine, propranolol, and quinidine reduce the dispersion of refractoriness between the two types of fibers. Lidocaine, through this mechanism, can improve the conduction velocity of premature stimuli

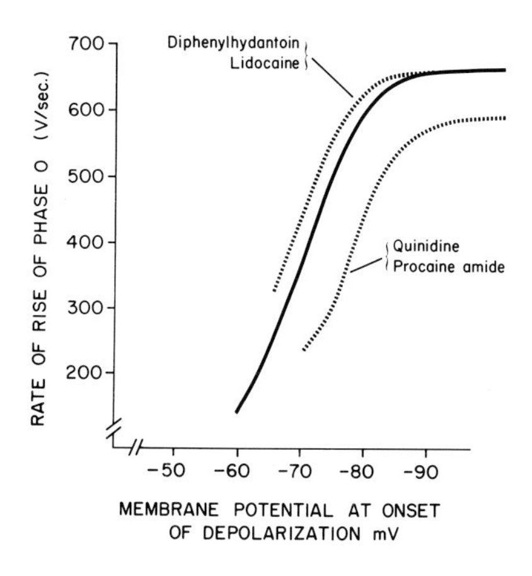

Fig. 7-6. Diagrammatic representation of membrane responsiveness curve of Purkinje cells. It is noted that normally there is a sinusoid curve (solid line) with much steeper rate of change of phase 0 (dv/dt) at larger magnitudes (more negative) of membrane potential at the onset of depolarization. It is noted that quinidine and procaine amide shift the curve to the right, but lidocaine and diphenylhydantoin cause a slight shift to the left. It is of interest that all agents shift the least negative membrane potential from which a response can be evoked to a more negative level. This is related to a differential alteration of action potential duration and effective refractory period by these agents (see text). (Adapted from Gettes LS.[21])

across the Purkinje–ventricular junction.[91] Thus lidocaine theoretically may improve conduction velocity and eliminate unidirectional block and thus break up a reentrant circuit. Since quinidine and procaine amide reduce conduction velocity, their effect on reentrant arrhythmias must clearly be on a different basis. It is possible that they may eliminate reentrant pathway by converting the area of unidirectional block to complete block. Recent work[92] indicates that procaine amide may indeed work in this manner. Thus it prolongs the coupling interval of premature beats until they are extinguished. However, under certain circumstances, through depression of conductivity, Group I agents may facilitate reentrant ectopic beats. Furthermore, quinidine at higher dose levels may cause dispersion of refractoriness, leading to focal reexcitation.[60]

4. Another feature common to all antiarrhythmic agents is that their effect is diminished by hypokalemia and their toxicity is enhanced by hyperkalemia.[93,94] Thus one cannot expect antiarrhythmic agents to work in the face of hypokalemia. Moreover, potassium should be considered an antiarrhythmic agent in its own right, since it possesses such electrophysiologic properties as the ability to decrease the rate of diastolic depolarization, to shift threshold potential to a less negative level, and to diminish the dispersion of refractoriness between Purkinje and ventricular fibers.[21] In addition, the potassium gradient is the most important determinant of the magnitude of resting membrane potential and, consequently, the rate of rise of phase 0 and conduction velocity.

Thus it has a dominant influence on both automaticity and conductivity in the concentration range encountered in the clinical setting.[95]

ANTIARRHYTHMIC AGENTS IN ISCHEMIC HEART DISEASE

The effects of antiarrhythmic agents in the early phase of ischemia may differ from those in the late phase. For example, Hope and collaborators[56] found that lidocaine, procaine amide, and propranolol (provided the heart rate was kept constant) were ineffective in impeding the development of ventricular tachyarrhythmias in acute myocardial ischemia. They postulated that the ineffectiveness was due to the fact that arrhythmias of early ischemia are reentrant in origin and, as demonstrated by Gamble and Cohn,[96] are less likely to be suppressed by these antiarrhythmic agents than are arrhythmias of automatic origin. However, an alternative explanation emerges from the more recent studies of the automatic activity of the fast and slow fibers. It appears that the commonly employed antiarrhythmic agents are relatively ineffective in suppressing the automaticity of the slow response type, as compared to the fast. Thus antiarrhythmic agents such as lidocaine and diphenylhydantoin suppress the automaticity of the Purkinje fibers at concentrations that have little effect on the sinus node.[12,97,98] Similarly, Wit and collaborators have demonstrated that a high concentration of quinidine does not abolish automaticity of the slow fiber type associated with partially depolarized Purkinje fibers.[12] Moreover, the same group has found that lidocaine has little antiautomatic effect on depressed Purkinje fibers surviving in areas of extensive infarction. In like manner, Parameswaran and associates have demonstrated that in diseased and poorly polarized human atrial muscle high concentrations of procaine amide fail to suppress automaticity.[99] It must be emphasized that the relevance of these electrophysiologic studies of antiarrhythmic agents to arrhythmias of clinical disease is at best conjectural. The studies in acute experimental ischemia cited could conceivably apply to the prehospital phase of myocardial infarction. Undoubtedly the efficacy of antiarrhythmic agents in arrhythmias of myocardial infarction in the hospital setting is supported by studies too numerous to require citation. However, some arrhythmias prove intractable to therapy, and it would be of interest to find out whether this is based on their origin in automaticity of slow fiber type and whether agents with specific effects on the calcium-dependent slow fiber depolarization[13,26] will prove more effective in such arrhythmias.

REFERENCES

1. Ling G, Gerard RW: The normal membrane potential of frog sartorius fibers. J Cell Comp Physiol 34:383, 1949
2. Draper MH, Weidmann S: Cardiac resting and action potentials recorded with an intracellular electrode. J Physiol 115:74, 1951
3. Woodbury LA, Hecht HH, Christopherson AR: Membrane resting and action potentials of single cardiac muscle fibers of the frog ventricle. Am J Physiol 164:307, 1951
4. Hoffman BF, Cranefield PF: Electrophysiology of the Heart. New York, McGraw-Hill, 1960
5. Cranefield PF, Wit AL, Hoffman BF: Genesis of cardiac arrhythmias. Circulation 47:190, 1973
6. Hodgkin AL: The ionic basis of nervous conduction (Nobel lecture). Science 145:1148, 1964
7. Huxley AF: Excitation and conduction in nerve: Quantitative analysis (Nobel lecture). Science 145:1154, 1964
8. Hodgkin AL, Katz B: The effect of sodium ion on the electrical activity of the grant axon of squid. J Physiol 108:37, 1949
9. Goldman DE: Potential impedence and rectification in membranes. J Gen Physiol 27:37, 1943
10. Cranefield PF, Wit AL, Hoffman BF: Conduction of the cardiac impulse. III. Characteristics of very slow conduction. J Gen Physiol 59:227, 1972
11. Trautwein W: Membrane currents in cardiac muscle fibers. Phys Rev 53:793, 1973
12. Wit AL, Rosen MR, Hoffman BF: Electrophysiology and pharmacology of cardiac arrhythmias. II. Relationship of normal and abnormal electrical activity of cardiac fibers to the genesis of arrhythmias. Am Heart J 88:515, 1974
13. Zipes DP, Mendez C: Action of manganese ions and tetrodotoxin on atrioventricular nodal transmembrane potentials in isolated rabbit hearts. Circ Res 32:447, 1973
14. Trautwein W: Generation and conduction of impulses in the heart as affected by drugs. Pharmacol Rev 15:277, 1963
15. Singer DH, Lazzara R, Hoffman BF: Interrelationships between automaticity and conduction in Purkinje fibers. Circ Res 21:537, 1967
16. Singer DH, Ten Eick RE: Electrophysiology of the heart and genesis of cardiac arrhythmias in cardiac and vascular diseases, in Conn HL, Horowitz O (eds): Cardiac and Vascular Diseases, vol I. Philadelphia, Lea & Febiger, 1971
17. Trautwein W, Dudel J: Aktionspotential und kontraktion des herzmuskels im sauerstofsmangel. Pfluegers Arch 266:324, 1958
18. Coraboeuf E, Boistel J: L'action des taux élevés de gaz carbonique sur le tissue cardiaque, etudiée à l'aide de microelectrodes intercellulaires. Compt Rend Soc Biol 147:654, 1953

19. Kassebaum DG, Van Dyke AR: Electrophysiological effects of isoproterenol on Purkinje fibers of the heart. Circ Res 19:940, 1966
20. Wallace AG, Daggett WM: Pacemaker activity during vagal escape rhythms. Circ Res 15:93, 1964
21. Gettes LS: The electrophysiologic effects of antiarrhythmic drugs. Am J Cardiol 28:526, 1971
22. Hoffman BF, Cranefield PF: Physiological basis of cardiac arrhythmias. Am J Med 37:670, 1964
23. Hoffman BF, Suckling EE: Effect of several cations on transmembrane potentials of cardiac muscle. Am J Physiol 186:317, 1956
24. Aronson RS, Cranefield PF: The electrical activity of canine Purkinje fibers in sodium-free, calcium rich solutions. J Gen Physiol 61:786, 1973
25. Brooks CMC, Lee HH: The Sinoatrial Pacemaker of the Heart. Springfield, Ill., Charles C. Thomas, 1972
26. Cranefield PF, Aronson RS, Wit AL: The effects of verapamil on the normal action potential and on a calcium-dependent action potential in canine cardiac Purkinje fibers. Circ Res 34:204, 1974
27. Vassalle M, Karis J, Hoffman BF: Toxic effects of ouabain on Purkinje fibers and ventricular muscle fibers. Am J Physiol 203:433, 1962
28. Cranefield PF, Aronson RS: The initiation of sustained rhythmic activity by single propagated action potentials in canine cardiac Purkinje fibers exposed to Na free solutions or to ouabain. Circ Res 34:477, 1974
29. Surawicz B: Use of chelating agent EDTA in digitalis intoxication and cardiac arrhythmias. Prog Cardiovasc Dis 2:432, 1959
30. Trautwein W, Kassebaum DG: On the mechanism of spontaneous impulse generation in the pacemaker of the heart. J Gen Physiol 45:317, 1961
31. Harris AS: Potassium and experimental coronary occlusion. Am Heart J 71:797, 1966
32. Griffiths J, Leung F: The sequential estimation of plasma catecholamines and whole blood histamine in myocardial infarction. Am Heart J 82:171, 1971
33. Solberg L, Ten Eick K, Singer D: Electrophysiological basis of arrhythmias in infarcted ventricles. Circulation 46:II–116, 1972
34. Scherlag BJ, Helfant RH, Haft JI, Damato AN: Electrophysiology underlying ventricular arrhythmias due to coronary ligation. Am J Physiol 219:1665, 1970
35. Scherlag BJ, El-Sherif N, Hope R, Lazzara R: Characterization and localization of ventricular arrhythmias resulting from myocardial ischemia and infarction. Circ Res 35:372, 1974
36. Hope R, Scherlag BJ, Valone T, et al: Effect of catecholamines on ventricular automaticity of normal and infarcted dog hearts in vivo. Circulation 49,50:III–146, 1974
37. Friedman PL, Stewart JR, Wit AL: Spontaneous and induced cardiac arrhythmias in subendocardial Purkinje fibers surviving extensive myocardial infarction in dogs. Circ Res 33:612, 1973

38. Lazarra R, El-Sherif N, Scherlag BJ: Electrophysiologic properties of canine Purkinje cells in one-day-old myocardial infarction. Circ Res 33:722, 1973
39. Matsuda K, Hoshi T, Kameyama S: Effects of aconitine on the cardiac membrane potential of the dog. Jap J Physiol 9:419, 1959
40. Brooks CMC, Hoffman BF, Suckling EE, Orias O: Excitability of the Heart. New York, Grune & Stratton, 1955
41. Sleator W, de Gubareff T: Transmembrane action potentials and contraction of human atrial muscle. Am J Physiol 206:1000, 1964
42. Trautwein W, Kassebaum MD, Nelson RM, Hecht HH: Electrophysiological study of human heart muscle. Circ Res 10:306, 1962
43. Pick A: Parasystole. Circulation 8:243, 1953
44. Hoffman BF: The electrophysiology of heart muscle and the genesis of arrhythmias, in Driefus LS, Likoff W (eds): Mechanisms and Therapy of Cardiac Arrhythmias. New York, Grune & Stratton, 1966
45. White PD: A study of atrioventricular rhythm following auricular flutter. Arch Intern Med 16:517, 1915
46. Moe GK, Preston JB, Burlington H: Physiologic evidence for a dual A-V transmission system. Circ Res 4:357, 1956
47. Watanabe Y, Dreifus LS: Newer concepts in the genesis of cardiac arrhythmias. Am Heart J 76:114, 1968
48. Watanabe Y, Dreifus LS: Inhomogeneous conduction in the A-V node. A model for reentry. Am Heart J 70:505, 1965
49. Durrer D, Schuilenburg RM, Wellens HJJ: Pre-excitation revisited. Am J Cardiol 25:690, 1970
50. Narula OS: Wolff-Parkinson-White syndrome: A review. Circulation 47:872, 1973
51. Spurrell RAJ, Krikler DM, Sowton E: Concealed bypasses of the atrioventricular node in patients with supraventricular tachycardia. Am J Cardiol 31:527, 1973
52. Rosenblueth A, Garcia Ramos J: Studies on flutter and fibrillation. II. The influence of artificial obstacles on experimental auricular flutter. Am Heart J 33:677, 1947
53. Schmitt FO, Erlanger J: Directional differences in conduction of the impulse through heart muscle and their possible relationships to extrasystoles and fibrillary contractions. Am J Physiol 87:326, 1928
54. Cranefield PF, Hoffman BF: Reentry: Slow conduction, summation and inhibition. Circulation 44:309, 1971
55. Cranefield PF, Hoffman BF: Conduction of the cardiac impulse. II. Summation and inhibition. Circ Res 28:220, 1971
56. Hope RR, Williams DO, El-Sherif N, et al: The efficacy of antiarrhythmic agents during acute myocardial ischemia and the role of heart rate. Circulation 50:507, 1974
57. Hoffman BF: Physiologic basis of disturbances of cardiac rhythm and conduction. Prog Cardiovasc Dis 2:319, 1959
58. Naimi S, Avitall B, Levine HJ: Abrupt shortening of the effective refractory

period following release of coronary ligation: Possible mechanism of reperfusion arrhythmias. Circulation 49,50:III–98, 1974
59. Moe GK, Mendez C: Physiologic basis of premature beats and sustained tachycardias. N Engl J Med 288:250, 1973
60. Han J, Moe GK: Non-uniform recovery of excitability in ventricular muscle. Circ Res 14:44, 1964
61. Battle WE, Naimi S, Avitall B, et al: Distinctive time course of ventricular vulnerability to fibrillation during and after release of coronary ligation. Am J Cardiol 34:42, 1974
62. Rosenbaum MB: The hemiblocks: Diagnostic criteria and clinical significance. Mod Concepts Cardiovasc Dis 39:141, 1970
63. Myerburg, RJ, Stewart JW, Hoffman BF: Electrophysiological properties of the canine peripheral A-V conduction system. Circ Res 26:361, 1970
64. Myerburg RJ: The gating mechanism in the distal atrioventricular conducting system. Circulation 43:955, 1971
65. Langendorf R, Pick A: Atrioventricular block, type II (Mobitz)—Its nature and clinical significance. Circulation 38:819, 1968
66. Cranefield PF, Klein HO, Hoffman BF: Conduction of cardiac impulse. I. Delay, block and one-way block in depressed Purkinje fibers. Circ Res 28:199, 1971
67. Watanabe Y, Dreifus LS: Electrophysiologic effects of digitalis on A-V transmission. Am J Physiol 211:1461, 1966
68. Cranefield PF, Hoffman BF, Paes de Carvalho A: The effect of acetylcholine on single fibers of the A-V node. Circ Res 7:19, 1959
69. Hoffman BF, Cranefield PF, Stuckey JH: Concealed conduction. Circ Res 9:194, 1961
70. Watanabe Y, Dreifus LS: Second degree atrioventricular block. Cardiovasc Res 1:150, 1967
71. Van Dam RT, Moore EN, Hoffman BF: Initiation and conduction of impulses in partially depolarized cardiac fibers. Am J Physiol 204:1133, 1963
72. Vassalle M, Greenspan K, Hoffman BF: Analysis of arrhythmias induced by ouabain in intact dogs. Circ Res 13:132, 1963
73. Hoffman BF, Cranefield PF: Physiological basis of cardiac arrhythmias (II). Mod Concepts Cardiovasc Dis 35:107, 1966
74. Wallace AG, Laszlo J: Mechanisms influencing conduction in a case of intermittent bundle branch block. Am Heart J 61:548, 1961
75. Moe GK, Abildskov JA, Han J: Factors responsible for the initiation and maintenance of ventricular fibrillation, in Surawicz B, Pellegrino ED (eds): Sudden Cardiac Death. New York, Grune & Stratton, 1964, pp 56–63
76. Hogancamp CE, Kardesch M, Danforth WH, Bing RJ: Transmembrane electrical potentials in ventricular tachycardia and fibrillation. Am Heart J 57:214, 1959
77. Surawicz B: Ventricular fibrillation. Am J Cardiol 28:268, 1971
78. Singer DH, Ten Eick RE: Aberrancy: Electrophysiologic aspects. Am J Cardiol 28:381, 1971

79. Han J: Mechanism of ventricular arrhythmias associated with myocardial infarction. Am J Cardiol 24:800, 1969
80. Han J, Garcia de Jalon PD, Moe GK: Fibrillation threshold of premature ventricular responses. Circ Res 18:18, 1966
81. Jervell A, Lange-Nielson F: Congenital deafmutism, functional heart disease, with prolongation of the Q-T interval and sudden death. Am Heart J 54:49, 1957
82. Covino BG, D'Amato HE: Mechanisms of ventricular fibrillation in hypothermia. Circ Res 10:148, 1962
83. Grumbach L, Howard JW, Merrill VI: Factors related to the initiation of ventricular fibrillation in the isolated heart; effect of calcium and potassium. Circ Res 2:452, 1954
84. Wegria R, Nickerson ND: Effect of paraverine, epinephrine and quinidine on fibrillation threshold of mammalian ventricles. J Pharmacol Exp Ther 75:50, 1942
85. Surawicz B, Lasseter K: Effect of drugs on the electrocardiogram. Prog Cardiovasc Dis 13:26, 1970
86. Covino BG, Hegnauer AH: Electrolytes and pH changes in relation to hypothermic ventricular fibrillation. Circ Res 3:575, 1955
87. Keliker GJ, Beasley A, Roberts J: Participation of the adrenal medulla in ventricular fibrillation after acute coronary artery occlusion. Am J Cardiol 35:148, 1975
88. Bigger JT Jr: Antiarrhythmic drugs in ischemic heart disease. Hospital Practice Nov. 1972, p 68
89. Bassett AL, Hoffman BF: Antiarrhythmic drugs; electrophysiological actions. Ann Rev Pharmacol 11:143, 1971
90. Hoffman BF, Bigger JT Jr: Antiarrhythmic drugs, in Di Palma JR (ed): Drill's Pharmacology in Medicine. New York, McGraw-Hill, 1970
91. Wittig JH, Harrison LA, Wallace AG: Effects of lidocaine on conduction and refractoriness in the distal Purkinje system. Circulation 43, 44:II–85, 1971
92. Giardina EG, Bigger JT Jr, Heissenbuttel R, Yu E: Antiarrhythmic action of procaine amide on ventricular arrhythmias. Circulation 43, 44:II–86, 1971
93. Watanabe Y, Dreifus LS, Likoff W: Electrophysiologic antagonism and synergism of potassium and antiarrhythmic agents. Am J Cardiol 12:702, 1963
94. Pamintuan JC, Dreifus LS, Watanabe Y: Comparative mechanisms of antiarrhythmic agents. Am J Cardiol 26:512, 1970
95. Fisch C: Relation of electrolyte disturbance to cardiac arrhythmias. Circulation 47:408, 1973
96. Gamble OW, Cohn K: Effect of propranolol, procaine amide and lidocaine on ventricular automaticity and reentry in experimental myocardial infarction. Circulation 46:498, 1972
97. Mandel WJ, Bigger JT Jr: Electrophysiologic effects of lidocaine on isolated canine and rabbit atrial tissue. J Pharmacol Exp Ther 178:81, 1971

98. Stramss HC, Bigger JT Jr, Bassett AL, Hoffman BF: Actions of diphenylhydantoin on the electrical properties of isolated rabbit and canine atria. Circ Res 23:463, 1968
99. Parameswaran R, Ten Eick RF, Singer DH: Effect of drugs on automaticity in diseased human atria. Circulation 44:II-84, 1971

Thomas W. Smith

8
Digitalis

Digitalis glycosides have occupied a prominent place in the management of congestive heart failure and certain cardiac rhythm disturbances for nearly 200 years. As Withering recognized in 1785, optimal use of digitalis requires considerable knowledge and skill on the part of the physician because of the unusually narrow therapeutic:toxic dose ratio of this class of drugs. Adequate appreciation of fundamental mechanisms of action and pharmacokinetics of these drugs is necessary to minimize the risk of toxicity and to provide maximum benefit to the patient.

SOURCES AND STRUCTURE-ACTIVITY RELATIONSHIPS

The term digitalis is used generally to refer to any of the cardioactive steroid glycoside compounds that have characteristic positive inotropic and electrophysiologic effects on the heart. Most of the drugs of this class currently in common use in the United States are steroid glycosides derived from the leaves of the common flowering plant known as foxglove or *Digitalis purpurea* (digitoxin, gitalin, digitalis leaf) or from *D. lanata* (digoxin, lanatoside c, deslanoside). An exception is ouabain, which is obtained from the seed of *Strophanthus gratus*.

All cardiac glycosides contain a steroid nucleus to which an α,β-unsaturated lactone ring is attached at the C-17 position. This portion of the molecule without attached sugars is called the genin (or aglycone) and is generally more transient and less potent in its myocardial effects than the parent glycoside. The structure of digoxin is shown in Fig. 8-1; digitoxin differs only in the absence of the C-12 hydroxyl group. Other

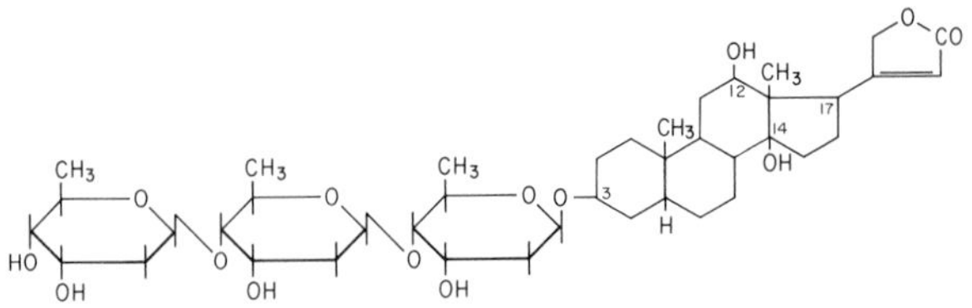

Fig. 8-1. Structure of digoxin (digoxigenin consists of the same steroid nucleus with a hydroxyl group at the C-3 position).

cardiac glycosides differ in substituent groups on the steroid nucleus as well as in the structure and number of sugars attached at the C-3 position. Molecular requirements for typical cardiac effects include the unsaturated lactone ring, the C-14 hydroxyl group, and an unusual cis fusion of the C and D rings of the steroid nucleus that is absent in endogenous steroids from mammalian species.

SUBCELLULAR MECHANISMS OF ACTION

Inotropy

Essentially every known area of myocardial cellular activity has been examined in the search for the subcellular mechanism of inotropic action of the cardiac glycosides.[1,2] Despite the mass of information in resulting publications, no detailed understanding of the mechanism of positive inotropy has emerged that has been subjected to rigorous experimental verification. One issue that has been settled, however, is that digitalis glycosides increase the force and velocity of contraction in the normal as well as the failing heart; thus mechanisms can be studied in intact or subcellular nonfailing myocardial preparations.

Several observations must be explained by any comprehensive model of cardiac glycoside-induced inotropy. These include the following: (1) Cardiac glycosides produce a positive inotropic effect on cardiac muscle but not on skeletal muscle.[3] (2) The magnitude of positive inotropy is dependent upon contraction frequency, lessening on either side of an optimum rate.[4] (3) Magnitude and rate of onset of positive inotropy is dependent upon concentrations of a number of cations, including K^+, Na^+, Ca^{++}, and Mg^{++}.[5]

Although cardiac glycosides in toxic doses can be shown to produce increased sympathetic activity in the heart,[6] numerous studies have failed to document a close correlation between myocardial catecholamine content and cardiac glycoside response.[7,8] Pronounced contractile responses persist in the presence of full β-adrenergic blocking doses of drugs such as propranolol.[9] Thus the major inotropic effects of cardiac glycosides are probably not mediated by catecholamine release or increased catecholamine sensitivity.

Direct cardiac glycoside actions on contractile proteins have also been studied.[10] Actin, myosin, and the troponin-tropomyosin system have all been postulated as possible sites of digitalis action, but there is no convincing evidence that these proteins play a primary role in mediating positive inotropy.

Primary effects of digitalis glycosides on myocardial energy metabolism have been sought,[2] with the conclusion that cardiac glycosides have no direct effect upon the intermediary metabolism of the heart that can be documented in isolated or well-defined enzyme systems. Those metabolic effects that have been described in isolated muscle preparations appear to be secondary to effects on transmembrane electrolyte movements.

Because of the vital role of calcium in cardiac excitation-contraction coupling and hence contractile state,[11] many investigators have sought to explain the positive inotropic effects of cardiac glycosides on the basis of increased availability of Ca^{++} to the contractile element at the time of excitation-contraction coupling.[2,12,13] The complexity of Ca^{++} compartmentalization in the myocardial cell[12] precludes any simple experimental approach to these problems, but at least two laboratories have observed cardiac glycoside-induced increases in a small labile pool of Ca^{++} that appears to be linked closely with the contractile state of the cell.[14,15] Although the adenylate cyclase system may be involved in changes in myocardial calcium handling brought about by positively inotropic agents, such as epinephrine and glucagon, cardiac glycosides do not appear to influence adenylate cyclase activity in the heart.[16]

A direct influence of cardiac glycosides on the uptake of Ca^{++} by cardiac sarcoplasmic reticulum ("relaxing system") has been sought with variable results. Lee and Klaus have recently concluded that cardiac glycosides have no consistent effect on the Ca^{++} transport mechanism of sarcoplasmic reticulum,[2] and most investigators would agree with this conclusion. The hypothesis that cardiac glycosides might act by increasing transmembrane Ca^{++} influx during the plateau phase of the action potential is interesting, but the limited evidence available provides no direct support for this concept.[2]

This lack of well-defined cardiac glycoside effects on other cellular functions has directed investigative interest to the highly specific ability of cardioactive steroids to inhibit transmembrane sodium and potassium movement by inhibition of the Mg^{++}- and ATP-dependent Na^{+}- and K^{+}-

activated transport enzyme complex known as $(Na^+ + K^+)$-ATPase. Schatzmann first noted the ability of cardiac glycosides to inhibit red-blood-cell monovalent cation transport,[17] and Skou subsequently described the membrane-bound enzyme system known as $(Na^+ + K^+)$-ATPase.[18] Repke systematically studied the relationship between cardiac activity and the ability of cardiac glycosides to inhibit $(Na^+ + K^+)$-ATPase.[19] On the basis of these observations he postulated that $(Na^+ + K^+)$-ATPase is a receptor for cardiac glycosides, and a number of other workers have extended and modified the concept that the $(Na^+ + K^+)$-ATPase system is involved in the mechanism of at least some cardiac glycoside actions. Although circumstantial, the weight of evidence favoring this link is substantial and includes the following:

1. $(Na^+ + K^+)$-ATPase is inhibited by cardioactive but not by inactive digitalis analogues, and a quantitative relationship exists between potency in $(Na^+ + K^+)$-ATPase inhibition and cardiac activity.[19]
2. Species cardiac sensitivity to digitalis glycosides is directly correlated with the ability of glycosides to inhibit myocardial $(Na^+ + K^+)$-ATPase from that species. The time course of $(Na^+ + K^+)$-ATPase inhibition parallels that of the inotropic effect in hearts of several species,[20] both in terms of onset and offset ofeffects.
3. A number of interventions, such as lowering of temperature, raising extracellular K^+ concentration, lowering Na^+ concentration, or lowering pH, all have a parallel tendency to reduce both the inotropic effect on isolated heart muscle and the inhibitory effect of cardiac glycosides on $(Na^+ + K^+)$-ATPase.[5]
4. Doses of cardiac glycosides producing a positive inotropic effect in the intact dog have been shown to cause inhibition of $(Na^+ + K^+)$-ATPase activity in myocardial enzyme preparations from the heart of the same experimental animal,[21,22]
5. Elevated extracellular K^+ concentrations have been shown to slow the uptake of cardiac glycosides by myocardial tissue, to slow the onset of inotropic effect,[23] and to slow the rate of binding of cardiac glycosides to myocardial $(Na^+ + K^+)$-ATPase.[24]
6. At least one group of compounds, the erythrophleum alkaloids, differ chemically from the cardiac glycosides, yet share both the positive inotropic effect and the ability to inhibit $(Na^+ + K^+)$-ATPase.[25]

Recognizing a specific interaction between cardiac glycosides and $(Na^+ + K^+)$-ATPase, it remains to correlate this interaction with the well-known effects of digitalis on the intact heart. As early as 1931 Calhoun and Harrison observed a marked decrease in myocardial K^+ concentration after toxic doses of cardiac glycosides.[26] Subsequent studies have documented decreases in intracellular K^+ and, in many cases, increases in intracellular Na^+ in response to toxic doses of cardiac glycosides. Results with smaller subtoxic but still inotropic doses have been more equivocal.[2] Some recent studies with highly sensitive isotopic techniques have provided evidence that positive inotropic responses are accompanied by a net loss of K^+ and a net uptake of Na^+, and by a net uptake of cellular Ca^{++}.[15] The appearance of toxic manifestations, such as ectopic beating and contracture, was accompanied by changes of still greater magnitude. Positive inotropic responses were not observed without evidence of some degree of inhibition of Na^+ transport. Langer has argued from these and other experiments that inhibition of active cellular Na^+ transport may result in enhancement of Ca^{++} uptake, in turn producing a positive inotropic response analogous to that which follows an increase in contraction frequency in the Bowditch staircase phenomenon. The mechanism of this effect might be enhanced exchange of intracellular Na^+ for extracellular Ca^{++}, as has been observed in the squid giant axon.[27] Such a mechanism would be consistent with the observed contraction and frequency dependence of cardiac glycoside-induced inotropy.[3,4] The Na^+ influx occurring with each action potential, together with decreased outward Na^+ pumping, would lead to the increased intracellular Na^+ concentration proposed to promote transmembrane Na^+–Ca^{++} exchange. This hypothesis has not been amenable to direct experimental testing in myocardial tissue.

Alternatively, it may be that the positive inotropic response, although dependent upon a specific interaction between cardiac glycosides and (Na^+ + K^+)-ATPase, does not directly depend upon inhibition of Na^+ and K^+ pumping by the enzyme. More direct interactions may occur between this enzyme system and the release of Ca^{++} during excitation-contraction coupling that could be modified by cardiac glycosides. Recent studies provide some evidence in support of the hypothesis that cardiac glycoside interaction with (Na^+ + K^+)-ATPase inhibits outward pumping of both Na^+ and Ca^{++} from the myocardial cell, resulting in an increase in the Ca^{++} pool available for excitation-contraction coupling.[28]

Cardiac Glycoside Uptake and Subcellular Localization

Availability of radioactively labeled cardiac glycosides has allowed detailed studies of uptake by intact heart muscle and of subcellular localization. A general conclusion is that more polar glycosides, such as ouabain, undergo principally specific binding to membrane receptor sites that have many of the properties of (Na^+ + K^+)-ATPase.[2] Less polar glycosides are bound nonspecifically to lipid membrane sites, and there is a very high degree of nonspecific binding of nonpolar glycosides such as digitoxin. Digoxin is intermediate both in polarity and in specificity of binding.[29]

The degree of specific binding to myocardial cell membrane components is dependent upon K^+ and Na^+ concentrations, with low extracellular K^+ and high Na^+ promoting binding. Metabolic energy appears to be necessary for specific binding to occur, and this may be related to the ATP requirement for binding of cardiac glycosides to $(Na^+ + K^+)$-ATPase. It has been observed that cardiac glycosides can inhibit monovalent cation transport only when present at the outer cell surface in the red blood cell[30] and the squid giant axon.[31] Cell fractionation procedures have shown that binding of cardiac glycosides to myocardium of sensitive species is mainly in the "microsomal" or membrane fraction.[29]

Electrophysiologic Effects

The transmembrane resting potential of cardiac cells is maintained by Na^+ and K^+ gradients (particularly the latter), which are in turn dependent upon the active Na^+–K^+ pump mechanism. Hence agents such as cardiac glycosides that inhibit the pump mechanism have marked effects on the electrophysiology of the intact heart as well as isolated muscle preparations. It is very likely that inhibition of $(Na^+ + K^+)$-ATPase underlies toxic effects on cardiac rhythm. Despite major advances in the understanding of cardiac electrophysiology, however, many unanswered questions remain concerning effects of digitalis on the electrical activity of the normal and diseased heart. Cells in various parts of the heart show differing sensitivities to digitalis, and both direct and neurally mediated effects are involved.

Within the specialized conduction system of the heart the refractory period is increased by digitalis, and conduction velocity is diminished. This tends to slow the ventricular response to atrial fibrillation and atrial flutter, or to prolong the PR interval in the presence of normal sinus rhythm. In atrial and ventricular myocardium, on the other hand, the refractory period tends to be shortened, and the more rapid recovery time results in a shortening of the QT interval of the electrocardiogram.

Detailed in vitro studies of the effects of digitalis on transmembrane action potentials of the mammalian ventricular Purkinje fiber have shown that at low digitalis concentrations the resting potential, action potential amplitude, and time course of depolarization and repolarization remain unchanged at the time when inotropic effects are first apparent. At higher cardiac glycoside concentrations and particularly at more rapid stimulation rates there is progressive loss of resting potential, and changes occur in the time course of depolarization and repolarization including decreased slope of the upstroke of the action potential, shortening of the plateau phase, and increased rate of spontaneous diastolic depolarization.[32] Because of uncertainty regarding the effective cardiac glycoside concentration in the conduction system of normally digitalized patients, it is difficult to tell which of these effects are part of the "therapeutic" effects of digitalis and which represent toxic effects. Recent studies of Rosen et al.,[33] in which isolated canine Purkinje fibers were perfused with the blood of intact donor dogs, showed that ouabain-induced changes in the donor dog's ECG were correlated with changes in the Purkinje fiber transmembrane potential. At the time of onset of early ouabain toxicity in the donor dog, Purkinje fiber recordings showed decreases in action potential amplitude, resting membrane potential, maximum velocity of the upstroke of the action potential, action potential duration, and plateau phase. Slowing of conduction was also apparent. Increased automaticity (increased slope of phase 4 spontaneous diastolic depolarization) was found to occur at the time of early toxicity and was more frequent when Purkinje fibers had been stretched and in the presence of hypokalemia, which correlated well with clinical observations of increased incidence of digitalis toxicity in the dilated, failing heart as well as in the presence of hypokalemia. There is as yet no clear answer to the question of whether ectopic beats and sustained tachyarrhythmias arising as a clinical manifestation of digitalis toxicity occur as a result of increased automaticity (i.e., increased spontaneous diastolic depolarization) or reentry. It may well be that both mechanisms are operative.

Neurally Mediated Effects

While vagal effects have long been known to be involved in the influence of digitalis on atrioventricular conduction, opinion has varied with respect to sympathetic effects of cardiac glycosides. Direct experimental evidence has been obtained from nerve recordings that cardiac glycosides can influence preganglionic cardiac sympathetic nerve activity in cats.[6] At doses in the toxic range, sympathetic nerve activity was substantially augmented, and high-intensity activity was temporally correlated with ventricular tachyarrhythmias including fatal ventricular fibrillation. Additional brainstem effects were manifest as ouabain-induced enhancement of traffic in vagus and phrenic nerves, the latter resulting in hyperventilation. Spinal cord section prevented these effects and produced an increase in the dose of ouabain required to produce ventricular arrhythmias.[34] Propranolol reduced ouabain-induced neural hyperactivity and usually converted ventricular arrhythmias to normal sinus rhythm. These studies support the concept that neural activation by cardiac glycosides may play a significant role in the development of cardiac rhythm disturbances and hyperventilation.

HEMODYNAMIC EFFECTS OF CARDIAC GLYCOSIDES

Myocardial Contractile Response

Only relatively recently has an accurate description of the effect of digitalis on the contractility of heart muscle been available. Wiggers and Stimson[35] showed in 1927 that digitalis increases the rate of development of intraventricular pressure at onset of systole, if heart rate and aortic pressure are maintained constant. A direct effect on cardiac muscle was demonstrated by the experiments of Cattell and Gold[36] in 1938 showing that ouabain increased the force of contraction of isolated cat papillary muscles. The inotropic action of digitalis was for many years thought to be confined to the decompensated heart. Further work has established that the inotropic action of digitalis is manifest in normal as well as failing heart muscle.[37] The effect of digitalis on the intact heart is shown in Fig. 8-2, in which ouabain shifts the ventricular function curve upward and to the left, so that at any given ventricular filling pressure more

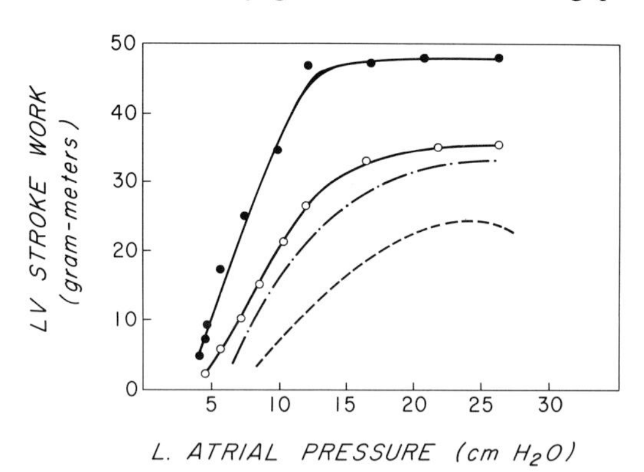

Fig. 8-2. Left ventricular function curves from a normal dog before (circles) and after (black dots) administration of ouabain. (Reproduced by permission from Cotten M deV, Stopp PE: Action of digitalis on the nonfailing heart of the dog. Am J Physiol 192:114, 1958). A hypothetical curve for a failing ventricle has been added (dashes), together with a representation of the response of the failing ventricle to ouabain (dashes with dots).

stroke work is generated in the digitalized preparation than in the control preparation.

Despite a demonstrable positive inotropic effect, the administration of cardiac glycosides results in no change or a slight decline in cardiac output in normal subjects.[38] Cardiac output is determined not only by cardiac contractile state but by ventricular filling pressure, peripheral arterial resistance, and heart rate. Digitalis significantly augments the contractile state of the normal myocardium in intact man, but reflex adjustments in the other determinants of cardiac output generally prevent any measurable increase in cardiac output.

Application of general principles discussed elsewhere in this text allows an understanding of the mechanisms by which digitalis benefits congestive heart failure. As disease processes (including ischemia, excess volume or pressure load, or intrinsic cardiac muscle defects) compromise contractility, compensatory mechanisms are brought into play. Increases in diastolic pressure enhance contractile force by the Frank-Starling mechanism. Elevated sympathetic tone promotes increasing systolic tension and contractile-element velocity. Ventricular hypertrophy provides more contractile elements. However, each of these compensatory mechanisms exacts a price. Pulmonary or peripheral edema occurs when end-diastolic ventricular pressures rise excessively, and tachycardia may be an undesirable effect of excess sympathetic tone. Increased myocardial oxygen consumption is the end result of all three compensatory mechanisms. As cardiac disease progresses and contractility diminishes further, the consequences of one of these compensatory mechanisms become dominant (e.g., pulmonary edema) or the compensatory mechanisms become insufficient to maintain cardiac output. Digitalis augments the diminished contractile state, resulting in less encroachment on compensatory mechanisms and a greater cardiac reserve. The ventricular function curve is shifted (Fig. 8-2) so that cardiac output is greater for any given left ventricular end-diastolic pressure. Clinical improvement is a consequence of diminished end-diastolic pressure (resulting in diminished pulmonary and systemic venous pressure) and increased cardiac output.

The inotropic action of digitalis increases progressively with increasing doses until toxic arrhythmias appear. Suppression of arrhythmias allows further increases in inotropy.[39] It is thus apparent that digitalization is not an all-or-none state. The task of the clinician is to determine the maximum dose consistent with an adequate margin of safety.

Although there is general agreement concerning the beneficial effects of digitalis glycosides in the treatment of the failing heart, there is less certainty concerning their usefulness in the absence of heart failure.

Digitalis is often given to patients with mitral stenosis regardless of their rhythm. It is clearly beneficial in slowing the ventricular rate in atrial fibrillation and thereby improving diastolic filling of the left ventricle through the narrowed valve orifice. In the presence of right heart failure it increases contractility and decreases end-diastolic pressure. However, in patients with mitral stenosis and normal sinus rhythm without right ventricular failure, cardiac glycosides do not have an appreciable beneficial effect on cardiac output, oxygen consumption, or severity of pulmonary hypertension.[40]

The therapeutic value of digitalis in the hypertrophied or dilated nonfailing heart is controversial. With the development of hypertrophy, but prior to the onset of overt failure, the work capacity of the myocardium at a given left ventricular end-diastolic pressure is reduced. Digitalis augments work capacity and reduces end-systolic volume and end-diastolic pressure. If cardiac output was not reduced, it does not increase in response to digitalis. However, the same stroke work and cardiac output can be delivered from a lower ventricular filling pressure, which should result in greater inotropic reserve.

These theoretical considerations are sometimes difficult to translate into clinical advice. The clinical value of digitalis in patients without heart failure has not been documented by long-term follow-up studies. Nevertheless, in the early stages of heart failure, cardiac output may be limited on exercise but relatively normal at rest. The clinical diagnosis of this intermediate state between compensation and overt failure is often difficult. It is likely that digitalis would improve exercise performance in such patients by increasing cardiac output on exercise and by preventing an undue rise in end-diastolic pressure.

Use of Digitalis in Ischemic Heart Disease

The chief determinants of myocardial oxygen consumption are ventricular wall tension and contraction velocity.[41] Increases in oxygen consumption with digitalization have been observed in the normal heart and are due to increased velocity of contraction and increased myocardial wall tension. In the failing heart, decreased oxygen consumption may occur due to a decrease in left ventricular end-diastolic pressure, resulting in a decrease in end-diastolic volume and thus (on the basis of the LaPlace relation) a decline in intramyocardial tension. Changes in oxygen consumption are always the net result of two opposing effects of digitalis—a potential reduction in wall tension with decreased ventricular size and an increase in contractility. Thus in the failing heart the net result depends upon the balance of these effects, and either a diminution of oxygen consumption or no change may be observed. In the well-compensated heart there is always an increase in oxygen consumption with digitalis administration.[41] These considerations are of clinical importance in deciding on the use of digitalis in patients with ischemic heart disease. Angina pectoris may improve with digitalization in patients with heart failure, but may also worsen in those who are well compensated.

There are still many unanswered questions concerning the role of digitalis therapy in the setting of acute myocardial infarction. It is felt by most clinicians that there is little to be gained by administration of the drug to patients who have uncomplicated infarction without cardiomegaly. There is little documentation of its value in cardiogenic shock. Indeed, rapid digitalization may, on occasion, be harmful on account of its vasoconstrictor properties. Digitalis is of use in the treatment of congestive heart failure complicating myocardial infarction, particularly when diuretics alone do not provide an adequate therapeutic response.

The clearest indication for digitalis following acute myocardial infarction is in the management of atrial fibrillation with a rapid ventricular response. Electrical cardioversion may be preferred in the treatment of other supraventricular tachyarrhythmias.[42] Although it is difficult to obtain unequivocal clinical evidence that patients with acute myocardial infarction are particularly prone to develop digitalis-induced arrhythmias,[43] animal experiments suggest that in experimental myocardial infarction the toxic dose of digitalis glycosides is reduced.[44,45] Inhomogeneity of digoxin uptake in the experimentally infarcted canine ventricle[46] suggests at least a partial explanation. Quantitative differences in digoxin concentrations in adjacent tissues, combined with local ischemia, would be expected to predispose to heterogeneity of recovery times and hence reentrant rhythm disturbances. Another issue in the use of digitalis in myocardial infarction relates to the observation by Maroko et al. that a number of inotropic agents including ouabain increase the severity and extent of ischemic injury in experimental coronary occlusion in the nonfailing heart.[47] Whether or not these observations weigh against the clinical use of digitalis under such circumstances is still uncertain.

Extracardiac Hemodynamic Effects

While the direct cardiac action of digitalis is of importance in understanding the hemodynamic effects of the drug, extracardiac effects also play a role. Digitalis glycosides constrict isolated arterial and venous segments,[48] and arteriolar and venous constriction has been demonstrated in intact experimental animals.[49,50] Elevation of total systemic arteriolar resistance has been demonstrated in conscious normal human subjects.[51] These effects appear to be mediated by both local action of digitalis on vascular smooth muscle and indirectly through the central nervous system. An increase in vascular resistance has been demonstrated in the isolated canine gracilis muscle vascular bed either by the direct addition of digitalis to the muscle perfusate or by injecting digitalis into the animal's systemic circulation. In the latter case, since the muscle is isolated from the circulation and only connected by a nerve, neural effects can be clearly demonstrated.[52] The neural effects are mediated through alpha receptor stimulation and appear to be the predominant mechanism by which the drug increases skeletal muscle vascular resistance.[53] These observations were extended in cross-circulation experiments in which the circulation of the central nervous system of a dog was isolated from the rest of the systemic circulation and perfused by a donor dog. Administration of digoxin either directly into the cerebral ventricles or into the circulation of the head isolated from the rest of the body produced vasoconstriction in both coronary and skeletal muscle beds, indicating that the central nervous system has an important role in mediating the vasoconstrictor effect of digoxin.[53]

In normal canine preparations digitalis produces generalized venoconstriction. This action is particularly marked in the hepatic veins and leads to pooling of blood in the portal venous system with diminished venous return.[50] Effects are probably less marked in man, but increased hepatic venous wedge pressures have been observed after administration of digitalis in patients without heart failure.[54] These direct and CNS-mediated peripheral actions contribute to the failure of digitalis to increase cardiac output in the normal circulation. Hepatic portal venous pooling tends to diminish venous return to the heart. General arteriolar constriction increases cardiac afterload. Both tend to counter the effect of augmented contractility of the myocardium. It is also likely that reflex circulatory readjustments mediated by baroreceptors and withdrawal of sympathetic tone play a major part.[55]

In congestive heart failure sympathetic augmentation of contractility is important in maintaining cardiac output. This increase in sympathetic nervous activity, producing systemic arteriolar and venous constriction,[51] serves to maintain blood pressure in the presence of diminished cardiac output and to redistribute this lowered output among regional circulations.[56] When digitalis is given to patients with heart failure, generalized vasodilation usually occurs,[51] instead of the vasoconstriction observed in the normal individual. This is probably related to increased cardiac output mediated by the inotropic action of the drug, resulting in reflex withdrawal of sympathetic vasoconstriction. This may account for the fact that venous pressure is often lowered prior to diuresis after the administration of digitalis.[51]

Caution must be exercised when digitalis glycosides are administered acutely, particularly in situations where transient increases in peripheral resistance would be deleterious. Increased mesenteric vascular resistance can also compromise splanchnic blood flow, subjecting the marginally perfused patient to increased risk of ischemic bowel necrosis.[57]

Digitalis has been shown to inhibit tubular reabsorption of sodium.[58] Direct infusion of ouabain into the renal artery inhibits renal (Na^+ + K^+)-ATPase and impairs both concentrating and diluting ability.[59] Relatively large doses are needed to demonstrate these effects, and it is unlikely that a direct renal action of digitalis plays an important part in the diuresis that occurs in the treatment of congestive heart failure.

Finally, a major hemodynamic effect of digitalis results from its ability to slow the ventricular response to supraventricular tachyarrhythmias. This is particularly important in patients with such lesions as mitral stenosis, where cardiac output is related to diastolic filling time of the left ventricle. Slowing of sinus tachycardia occurs in patients with congestive heart failure through withdrawal of enhanced sympathetic tone when the failure state is ameliorated by increased cardiac contractility and output.

PROPERTIES OF INDIVIDUAL PREPARATIONS

Among the numerous compounds with digitalis-like activity, only a few enjoy widespread clinical use. It has become axiomatic that the clinician need be intimately familiar with the properties of at most two or three of these preparations. Detailed knowledge of a single intermediate-acting agent such as digoxin is sufficient for most clinical needs. The agent of choice in the individual case depends on the speed of onset of action required, desired route of administration, and desired duration of action. Pure crystalline glycoside preparations have largely replaced powdered digitalis leaf because of better standardization of active drug content.

Major advances have occurred in recent years in the understanding of the pharmacokinetics of cardiac glycosides. Important knowledge gained by the use of radioactively labeled compounds continues to expand through the use of methods that allow measurement of

Table 8-1
Cardiac Glycoside Preparations

Agent	Gastrointestinal Absorption	Onset of Action*	Peak Effect	Average Half-Life†	Principal Metabolic Route (Excretory Pathway)	Average Digitalizing Dose		Usual Daily Oral Maintenance Dose‡
						Oral§	Intravenous¶	
Ouabain	Unreliable	5–10 min	30–120 min	21 hr	Renal; some gastrointestinal excretion	—	0.3–0.5 mg	—
Deslanoside	Unreliable	10–30 min	1–2 hr	33 hr	Renal	—	0.8 mg	—
Digoxin	55%–75%‖	15–30 min	1.5–5 hr	36 hr	Renal; some gastrointestinal excretion	1.25–1.5 mg	0.75–1.0 mg	0.25–0.5 mg
Digitoxin	90%–100%	25–120 min	4–12 hr	4–6 days	Hepatic;** renal excretion of metabolites	0.7–1.2 mg	1.0 mg	0.1 mg
Digitalis leaf	About 40%	—	—	4–6 days	Similar to digitoxin	0.8–1.2 g	—	0.1 g

Reproduced by permission from Smith TW, Haber E: Medical progress: Digitalis. N Engl J Med 289:945; 1010; 1063; 1125, 1973.

*For intravenous dose.

†For normal subjects (prolonged by renal impairment in case of digoxin, ouabain, and deslanoside, and probably by severe hepatic disease in case of digitoxin and digitalis leaf).

‡Average for adult patients without renal or hepatic impairment; will vary widely among individual patients and requires close medical supervision.

§Divided doses over 12–24 hr at intervals of 6–8 hr.

¶Given in increments for initial subcomplete digitalization, to be supplemented by further small increments as necessary.

‖For tablet form of administration (may be less in malabsorption syndromes and in formulations with poor bioavailability).

**Enterohepatic cycle exists.

nanogram (10^{-9} g) amounts of unlabeled cardiac glycosides. Table 8-1 summarizes data for the most commonly used preparations, listed in order of rapidity of onset and duration of action. It cannot be overemphasized that the values given represent averages and that variations in measurements such as half-life are substantial even in "normal" subjects.

Ouabain

This drug is the most polar and rapidly acting of the cardiac glycosides currently available in the United States for routine clinical use. As in the case of digoxin and digitoxin, its excretion from the body follows first-order kinetics, with a fixed proportion of the residual drug in the body being excreted each day. It is useful to think about drugs that are excreted in this exponential fashion in terms of half-lives. For ouabain, the plasma half-life in normal subjects is about 21 hr, similar to the half-lives of positive inotropic effect and ventricular-rate slowing in patients with atrial fibrillation.[60] This half-life is longer than some earlier estimates. The amount of ouabain in the body, and also the risk of toxicity, in a patient placed on any regular dosage schedule without a loading dose will continue to rise for four to five half-lives (4 to 5 days) until a plateau is reached. Impairment of renal function will prolong the half-life of ouabain and also the period during which accumulation will continue. The pharmacokinetics of ouabain are thus quite analogous to those of digoxin. Ouabain is poorly absorbed from the gastrointestinal tract and is not available for oral use.

Deslanoside (Desacetyllanatoside C, Cedilanid-D)

This drug is structurally identical to digoxin except for the presence of an added glucose residue attached to the terminal sugar. This alteration results in poor gastrointestinal absorption, and deslanoside cannot be recommended for oral use. Its half-life and renal excretion are essentially identical to those of digoxin. Although onset of action is slightly more rapid, it probably enjoys no substantial advantages over parenteral use of digoxin.

Digoxin

Because of flexibility of route of administration and intermediate duration of action, digoxin has become the glycoside most commonly used in hospitalized patients and, to a somewhat lesser extent, in office practice. It has recently become apparent that important differences in gastrointestinal absorption (bioavailability) of tablet preparations can exist as a result of differences in manufacturing procedures.[61] When an appropriate therapeutic response is not achieved one must consider the alternate possibilities that either the patient is digoxin-resistant or the digoxin is patient-resistant. Recent guidelines from the United States Food and Drug Administration should result in improved uniformity of bioavailability of oral digoxin preparations from various manufacturers.

Recent studies in normal human subjects indicate that the relative bioavailability of Lanoxin® (U.S.A.) tablets is about 55%, and that of the elixir about 65%, compared with slow intravenous infusion of digoxin taken as 100%.[62] Peak serum or plasma digoxin levels are reached about 1 hr after tablet ingestion in fasting subjects. Delayed gastric emptying for any reason will tend to delay this plasma peak, since little of the drug is absorbed before reaching the small bowel. When a rapid onset of effect is required, it is preferable to use increments of the drug intravenously. This route avoids some of the uncertainty regarding the time when cardiac effects will be maximal and reduces the likelihood of giving an additional dose before the prior increment has had its full effect. Intramuscular use is less desirable when a rapid effect is needed, because absorption may take as long as with oral administration. In addition, severe pain occurs at the intramuscular injection site.

Problems of absorption and bioavailability are even more complex in patients with malabsorption syndrome. Studies have shown that digoxin is poorly and erratically absorbed in these subjects, although maldigestion due to pancreatic insufficiency appears to have relatively little effect on digoxin absorption.[63] Cholestyramine binds digoxin in the gut lumen and decreases absorption when both agents are ingested concurrently.

Elimination of digoxin from the body occurs chiefly by glomerular filtration and excretion of the native molecule. The ratio of digoxin to creatinine renal clearance is close to unity in most clinical circumstances. This is consistent with lack of substantial tubular reabsorption or secretion of the drug and also with relatively minor binding of the drug by plasma proteins, estimated at about 23% protein binding at usual serum digoxin concentrations.[64] High rates of urine flow do not substantially accelerate digoxin excretion unless accompanied by increased glomerular filtration rate.

Digoxin is cleared from the body exponentially, with a half-life of about 36 hr in subjects with normal renal function. It is useful to remember that such a patient will excrete about 37% of the digoxin in his body every 24 hr, and a steady state is reached in maintenance therapy when the same amount is replaced each day. Patients begun on maintenance digoxin without a loading dose accumulate the drug until a plateau level has been reached after four to five half-lives, or about 1 week.[65] At this time, enough of the drug is presented to the kidney and ancillary excretory pathways to allow clearance of the amount given each day, and a final steady-state body concentration is reached that is independent of whether or not a loading dose has been given. Thus any compromise of renal function decreases the maintenance dose required to sustain a given body concentration.

Digoxin appears to have metabolic and excretory behavior in infants and children similar to that in adults. The drug crosses the placenta and establishes fetal umbilical-cord serum levels similar to those of a mother receiving digoxin.[66]

Digitoxin

This is the least polar and most slowly excreted of the cardiac glycosides in common use. Digitoxin constitutes the principal active agent in digitalis leaf, and the two may be considered together. Gastrointestinal absorption of digitoxin appears to be essentially complete, although no studies have been reported in patients with malabsorption syndrome or other gastrointestinal diseases.

Digitoxin binds avidly to human serum albumin, and about 97% of the plasma content of the drug is bound to albumin at clinically relevant concentrations.[64] Renal clearance of the unchanged drug is a minor excretory pathway, and extensive metabolism of digitoxin occurs, presumably in the liver, before excretion. An enterohepatic cycle exists for digitoxin and can be interrupted by resins such as cholestyramine that bind digitoxin in the gut lumen. A modest acceleration of the excretion of digitoxin has been documented in subjects given cholestyramine,[67] but the clinical efficacy of this approach in patients with digitoxin intoxication remains to be proved. Digoxin is not usually a quantitatively important metabolic product of digitoxin in man. It is of interest that drugs such as phenobarbital and phenylbutazone that are known to increase activity of hepatic microsomal enzyme systems have been shown to accelerate the metabolism of digitoxin in some patients.[68]

Half-times of digitoxin in plasma vary relatively little from patient to patient, usually remaining in the range of 4 to 6 days irrespective of renal function.[69]

CLINICAL USE

Indications

Congestive Heart Failure

Cardiac glycosides are of potential value in most patients with congestive heart failure due to ischemic, valvular, hypertensive and congenital heart disease, cardiomyopathies, and cor pulmonale. Improvement in impaired myocardial contractility will tend to increase cardiac output, promote diuresis, and reduce the filling pressure of the failing left or right ventricle (or both), with reduction of pulmonary vascular congestion and central venous pressure. The lack of demonstrable benefit in mitral stenosis with normal sinus rhythm (unless right ventricular failure has supervened) was discussed earlier. Little if any benefit may result in patients with constrictive pericarditis or pericardial tamponade. Idiopathic hypertrophic subaortic stenosis represents another process in which digitalis is often of little value and may actually be deleterious because it can increase left ventricular outlet obstruction by augmenting the contractility of the hypertrophied outflow tract segment. In the later stages of the process, in which congestive problems predominate over obstructive ones, digitalis glycosides may be beneficial, as in other cardiomyopathies. Cardiac glycosides may improve symptoms of angina pectoris when it coexists with cardiomegaly and congestive heart failure, as discussed previously. "Prophylactic" digitalization of the patient with diminished cardiac reserve about to undergo a major stress such as surgery is controversial. In the absence of cardiomegaly or other evidence of overt congestive heart failure, most clinicians prefer to withhold digitalis until a specific indication arises.

The availability of reliable pervenous cathether endocardial pacing methods has helped to resolve the problem of digitalis use in patients with marginal atrioventricular conduction or established atrioventricular block. Pacemaker implantation can now be carried out at minimal risk even in severely ill patients, and cardiac glycosides may then be given without fear of aggravating conduction problems.

Cardiac Rhythm Disturbances

Digitalis is of potential use in the management of four types of supraventricular tachyarrhythmias.

Paroxysmal supraventricular tachycardia, whether of atrial or atrioventricular junctional (nodal) origin, usually responds to digitalization when simpler measures such as carotid sinus pressure alone have failed. This vagal stimulatory maneuver should be repeated during the course of digitalization, since the combination of partial digitalization and carotid sinus pressure will often succeed when neither measure alone suffices. Maintenance digitalization usually abolishes or reduces the frequency of recurrent attacks. Use of digitalis in the setting of paroxysmal supraventricular tachycardia demands the exclusion of digitalis intoxication as a cause of the arrhythmia.

Atrial fibrillation with a rapid ventricular response is one of the most common indications for the use of digitalis. Both vagal and direct mechanisms result in increased blockade of impulses arriving at the atrioventricular junction, with slowing of the ventricular rate. Conversion to normal sinus rhythm may occur in the course of digitalization. Addition of propranolol may be useful in circumstances (e.g., untreated thyrotoxicosis) in which the ventricular rate is difficult to control without emergence of toxic symptoms and congestive heart failure is absent or minimal.

Atrial flutter, usually accompanied by 2:1 atrioventricular block in untreated cases, can often be managed with digitalis in doses sufficient to produce a degree of atrioventricular block resulting in a ventricular rate in the range of 70 to 100 per minute. This effect sometimes requires doses considerably in excess of the usual range. As in atrial fibrillation, when the arrhythmia is poorly tolerated by the patient it is often advisable to attempt direct-current cardioversion before administration of doses of digitalis that would render the procedure hazardous.

Finally, Wolff-Parkinson-White syndrome tachyarrhythmias may be terminated or prevented by digitalis in cases in which preferential effects on conduction or refractoriness in the normal or anomalous conduction pathways result in interruption of the reentrant circus movement. Quinidine or procainamide may be more effective in other cases, however, and an empirical approach is usually necessary.

Although ventricular premature beats are widely recognized as a manifestation of digitalis excess, they are also seen in conjunction with congestive heart failure and may disappear when failure is effectively treated with digitalis.

Dosage Schedules

Major emphasis in this section will be placed on digoxin. Careful and repeated observation of the patient remains a crucial aspect of optimal digitalis dosage, but recent advances in understanding of pharmacokinetics allow several useful generalizations to be made. There is usually no reason to use a loading dose far in excess of what the steady-state body content will be with usual maintenance doses. A patient with normal renal function who excretes 37% of the digoxin in his body each day, on a maintenance dose of 0.25 or 0.50 mg/day, will have a steady-state total-body content of about 0.67 or 1.35 mg, respectively. If an estimate of 75% absorption of the tablet form of digoxin is made, the estimates become about 0.5 and 1.0 mg. This amount could be given over a period of a day or so in several increments, or the same level of digitalization could be achieved over a period of about a week in a patient with normal renal function by administration of the daily maintenance dose without any loading dose. The latter procedure is often preferable in outpatient practice. It must be remembered, however, that severe renal impairment will prolong the half-life to a maximum of about 4.4 days and hence the period required to reach a steady-state plateau to a maximum of about 3 weeks. Lean body mass should be considered in selection of both loading and maintenance digoxin doses. In adult patients with cardiac disease, initial intravenous loading doses of about 0.5 to 0.75 mg per 100 lb of body weight, given in increments, are unlikely to produce toxicity and can be supplemented by further increments if clinically indicated.

The maintenance digoxin dose required to replace daily losses will vary from about 37% of the total-body content in patients with normal renal function to about 14% (average losses by nonrenal pathways) in patients who are essentially anephric. Between these extremes, digoxin excretion is linearly related to glomerular filtration rate or creatinine clearance (C_{Cr}). As Jelliffe suggests,[70] a useful approximation of daily percentage loss of digoxin is 14 + (C_{Cr} in ml/min)/5. Since accurate creatinine clearance values will often not be readily available, one can use the following estimate[70] based on a stable serum creatinine in milligrams per 100 ml, abbreviated as c:

$$C_{Cr}\,(\text{men}) = \frac{100}{c} - 12$$

$$C_{Cr}\,(\text{women}) = \frac{80}{c} - 7$$

These expressions can be combined so that

$$\%\ \text{daily loss (men)} = 11.6 + \frac{20}{c}$$

and

$$\%\ \text{daily loss (women)} = 12.6 + \frac{16}{c}$$

The daily maintenance digoxin dose is intended to replace daily losses after an appropriate loading dose, so that the above value for daily percentage loss multiplied by the loading dose that produced a satisfactory therapeutic response gives a useful first approximation to the proper daily maintenance dose.

Digitoxin half-times usually range within 20% of a mean value of about 4.8 days, and relatively little variation in the body pool among individual patients receiving a given maintenance dose of the drug would be expected.[71] The average steady-state digitoxin pool in a patient receiving 0.1 mg/day is about 0.8 mg, and as with digoxin there is usually no reason to give a loading dose substantially in excess of the expected steady-state body pool achieved with usual maintenance doses. A loading dose of about 10 to 12 μg/kg (0.45 to 0.55 mg/100 lb) allows for variation in body size. Gradual digitalization without a loading dose occurs with digitoxin, but the four to five half-lives required to reach the steady-state plateau will take 3 to 4 weeks.

The above estimates of loading and maintenance doses are average values intended only as first approximations, and they in no way diminish the need for further adjustments based on frequent and careful observation of the patient.

Evaluation of the Patient

A number of factors influence individual sensitivity to cardiac glycosides. These include serum electrolytes (potassium, magnesium, and calcium), adequacy of tissue oxygenation, acid–base balance, age, renal function, thyroid status, autonomic nervous system tone, other drugs concurrently received, and type and severity of the underlying heart disease. Low extracellular potassium levels result in more rapid myocardial uptake of digitalis, probably contributing to the tendency for digitalis toxicity to develop in hypokalemic patients. Adequate potassium replacement or use of potassium-sparing agents such as spironolactone or triamterene in patients receiving potent diuretics is of particular importance in digitalized patients. Magnesium deficiency also predisposes to digitalis intoxication, and increased magnesium losses accompanying diuretic therapy may require replacement. Calcium potentiates cardiac glycoside effects, necessitating caution when calcium is given parenterally to digitalized patients or when digitalis is given in the presence of hypercalcemia.

Hypoxemia and disturbances of acid–base balance decrease tolerance to cardiac glycosides and contribute to the problems encountered in digitalized patients with chronic lung disease and cor pulmonale. The effects of age, per se, are not well defined. An age-correlated factor tending to increase sensitivity to a given maintenance digoxin dose is diminished renal function, resulting in greater accumulation of the drug.

Thyroid disease affects digitalis sensitivity by two mechanisms. Hyperthyroidism increases, and hypothyroidism decreases, the rates of ouabain and digoxin excretion. In addition, the thyrotoxic heart appears to be less sensitive to any given dose or serum level of digitalis, at least with respect to control of the ventricular response to atrial flutter or fibrillation. Autonomic nervous system tone and drugs concurrently received by the patient, particularly those with direct or indirect autonomic effects or antiarrhythmic properties, can have marked effects. Diphenylhydantoin (like potassium ion) can sometimes abolish or prevent digitalis-induced rhythm disturbances and allow further inotropic effects beyond the ceiling initially imposed by the arrhythmia.

Type and severity of underlying heart disease have important influences on the response of the heart to cardiac glycosides. Focal myocardial ischemia produces electrophysiologic abnormalities similar in many respects to those occurring in the presence of toxic concentrations of digitalis. It is therefore not surprising that increased digitalis sensitivity may be encountered in some patients with active myocardial ischemia or acute myocardial infarction. Stretching of Purkinje fibers has been shown experimentally to potentiate digitalis toxic rhythm disturbances, and a clinical analogy may exist in patients with advanced cardiomegaly.

The influence of underlying heart disease on response to digitalis is illustrated in patients ingesting large doses of digoxin with suicidal intent. In young subjects without preexisting heart disease, atrioventricular block or sinoatrial exit block tends to develop and can usually be managed successfully with atropine. In contrast, older patients with coronary disease often develop multifocal ventricular premature beats, ventricular tachycardia, and quite possibly ventricular fibrillation in response to massive digitalis ingestion.[72]

It is important to recognize that optimal digitalization does not necessarily represent the largest dose that can be tolerated without emergence of overt toxicity. The safety margin for cardiac glycosides is small at best, and the availability of potent oral diuretics usually obviates the necessity of balancing the patient at the edge of toxicity. Electrocardiographic ST-segment and T-wave changes of "digitalis effect" are limited indicators of state of digitalization, and slowing of sinus tachycardia is not a reliable index of adequacy of digitalis dosage. In the presence of atrial flutter or fibrillation, control of the ventricular response provides a straightforward end point. Failure of atropine or exercise to increase the ventricular response has been used as a further indication of "full digitalization" in such patients, but overly vigorous pursuit of this objective may result in very slow resting heart rates or other evidence of impending or overt toxicity.

When congestive heart failure is the indication for use of digitalis, it should be remembered that positive inotropy is a graded response that is appreciable at doses well short of "full digitalization." Carotid sinus pressure can provide useful bedside clues to impending digitalis toxicity. Rhythm disorders such as second-degree atrioventricular block, accelerated atrioventricular junctional rhythm, and ventricular premature beats or bigeminy may emerge in response to carotid sinus stimulation before they occur spontaneously.[73]

Serum Cardiac Glycoside Concentrations

Increasing availability of serum or plasma cardiac glycoside concentration measurements offers the clinician another means of estimating body stores of these drugs. Since rhythm disturbances due to digitalis are dose-related, and serum levels rise linearly with increasing dosage, at least a statistical correlation between clinical state and blood level would be expected. With digoxin, the ratio of serum to myocardial concentration is relatively constant after completion of uptake and distribution of drug, so that serum concentration would be expected to reflect myocardial concentration. Finally, the ($Na^+ + K^+$)-ATPase previously discussed as a probable mediator of at least some digitalis actions is located in the cell membrane in close proximity to the extracellular compartment, which would tend to enhance the translation of plasma concentration to myocardial effect.

Several assay methods are in current use, including double-isotope dilution derivative, red-cell ^{86}Rb uptake inhibition, ($Na^+ + K^+$)-ATPase inhibition, ATPase enzymatic displacement, gas chromatography, and radioimmunoassay.[74] The literature reporting clinical experience with these techniques reflects general agreement that mean digoxin and digitoxin levels are significantly higher in patients with clinical evidence of toxicity than in nontoxic patients. Nevertheless, there is overlap in levels of toxic and nontoxic patients owing to the multiple factors influencing individual response, and serum-level data must be interpreted in the overall clinical context. One must be sure that the laboratory carrying out the determination has adequate quality control and is aware of the potential technical pitfalls of the particular method in use.

In general, measurements of serum levels tend to be useful

whenever an unanticipated response to digitalis is encountered, such as apparent toxicity or failure to obtain an adequate therapeutic response with usual doses. Concentration measurements have also proved useful in the evaluation of patients unable to give an adequate history and in complex situations such as after cardiac surgery. Problems raised by uncertain bioavailability or suspected malabsorption can be usefully evaluated with the aid of cardiac glycoside serum concentration determinations.

DIGITALIS TOXICITY

Digitalis toxicity is one of the most prevalent adverse drug reactions encountered in clinical practice. The incidence in hospitalized patients is reported to vary between 8% and 35%, with estimates near 20% being encountered most commonly.[75]

Mechanisms of Digitalis Intoxication

Manifestations of digitalis intoxication include gastrointestinal and central nervous system symptoms and cardiac rhythm disturbances. Anorexia, nausea, and vomiting appear to be mediated by chemoreceptors located in the area postrema of the medulla rather than by direct irritant effect of the drug on the gastrointestinal tract.[76]

Digitalis-induced disturbances of cardiac impulse formation and conduction can be explained in terms of alterations of refractory period, impulse transmission, and automaticity of cardiac tissues. Alterations in centrally mediated sympathetic activity and changes in vagal tone may also be of importance, as discussed earlier.

Sinus Node and Atrium

Digitalis-induced slowing of the sinus rate in patients without congestive heart failure is mediated chiefly by vagal effects on the sino-atrial node. The atrium is more sensitive than the sinus node to direct effects of digitalis. At higher doses depression in sinus node automaticity can be shown experimentally.[32] At toxic dose levels a combination of vagal and direct effects on the sinus node contributes to sinus bradycardia as well as occasional cases of sino-atrial arrest. Such bradycardias predispose to the emergence of junctional or ventricular escape rhythms.

Digitalis shortens the refractory period of the atrial myocardium, accounting for the increased atrial rate sometimes observed with digitalization of patients with atrial flutter, since the circus movement of depolarization is able to proceed at a more rapid rate.[32]

Atrioventricular Node

The effective refractory period of the atrioventricular node is prolonged by digitalis. As in the case of the sinus node, this is in part related to increased vagal activity and in part to direct action on nodal fibers. The vagal component appears to predominate in subjects without intrinsic disease of the cardiac conduction system.[72] Decreased amplitude and upstroke velocity of the action potential from cells within the node itself and from nodal His fibers has been observed in response to cardiac glycosides.[77]

The therapeutic effect of digitalis in slowing ventricular response to atrial flutter or fibrillation is in part dependent on the entry of atrial impulses into the atrioventricular node with failure to reach the His-Purkinje system due to decremental conduction within the node. When second- or third-degree atrioventricular block occurs as a result of digitalis intoxication, however, the principal mechanism is failure of propagation within the atrioventricular node.[78]

His-Purkinje System

Digitalis-induced increases in the automaticity of the His-Purkinje system are related to enhanced spontaneous diastolic (phase 4) depolarization. This may lead to the appearance of new pacemakers, which are manifest clinically by premature junctional or ventricular beats or by accelerated junctional or ventricular rhythms. The nonuniform effects of digitalis on ventricular and Purkinje fibers[79] and simultaneous enhancement of automaticity, depression of conduction velocity, and local block predispose to arrhythmias based on reentry mechanisms that may progress to ventricular tachycardia and fibrillation.

Clinical Manifestations of Digitalis Toxicity

Anorexia is often, but by no means always, an early manifestation of digitalis intoxication. Nausea and vomiting follow as consequences of digitalis overdose.[76] In clinical situations it is often difficult to attribute these symptoms to digitalis, since they may also be caused by cardiac failure or by associated illness.

Nervous system effects are well recognized but are largely documented only through case reports. They include headache, fatigue, malaise, neuralgic pain, disorientation, confusion, delirium, and seizures. Visual symptoms include scotomata, flickering, halos, and changes in color perception. As with gastrointestinal symptoms, it is often difficult to determine whether such symptoms are a consequence of digitalis excess, fluid or electrolyte disturbances, or associated illnesses. Allergic skin lesions are rare, but have been reported.[80] Gynecomastia occasionally occurs.[81]

Cardiac toxicity can take the form of essentially every known rhythm disturbance. Common arrhythmias include atrioventricular junctional escape rhythms, ventricular bigeminy or trigeminy, nonparoxysmal atrioventricular junctional tachycardia, unifocal or multifocal ectopic ventricular beats, and ventricular tachycardia. Atrioventricular junctional exit block, paroxysmal atrial tachycardia with atrioventricular block, sinus arrest, and Mobitz type 1 (Wenckebach) second-degree atrioventricular block also occur. Ther are no unequivocal electrocardiographic features that distinguish digitalis toxic rhythms from those caused by underlying cardiac disease, although rhythms combining features of increased automaticity of ectopic pacemakers with impaired conduction, such as paroxysmal atrial tachycardia with atrioventricular dissociation and accelerated atrioventricular junctional pacemaker, are strongly suggestive of digitalis toxicity. The cause of the arrhythmia may at times be clarified by demonstrating a reversion to normal rhythm upon withholding the drug.

Massive digitalis overdose, either suicidal or accidental, is not uncommon. Patients without underlying heart disease tend to tolerate large doses relatively well.[72] The principal manifestations are sinus bradycardia, first-, second-, or third-degree atrioventricular block, or sino-atrial exit block. Atropine alone is often successful in reversing these manifestations, but it is not invariably effective. Ventricular pacing with a pervenous endocardial catheter electrode is generally successful if atropine fails, although ventricular standstill unresponsive to pacing has been reported.[72,82] Patients with preexisting heart disease tend to be more difficult to manage, in that ectopic ventricular arrhythmias are often the initial manifestation of digitalis intoxication. At very high digoxin doses and serum concentrations, hyperkalemia, sometimes refractory to treatment, has been observed.[72,83] Such elevation of serum potassium is presumably a consequence of inhibition of (Na^+ + K^+)-ATPase throughout the body with consequent impairment of cell membrane monovalent cation transport.

Treatment of Digitalis Intoxication

Early recognition that a rhythm disturbance is related to digitalis excess is the key to successful management. Some common manifestations, such as occasional ectopic beats, marked first-degree atrioventricular block, or atrial fibrillation with a slow ventricular response, require only temporary withdrawal of the drug, electrocardiographic monitoring if indicated until the arrhythmia has subsided, and adjustment of dosage to prevent recurrence. Rhythm disturbances that impair cardiac output because of too rapid or too slow rates, or those that threaten to progress to ventricular fibrillation, require more active intervention. Ventricular tachycardia due to digitalis intoxication requires immediate vigorous treatment. Sinus bradycardia, sino-atrial arrest, and second- or third-degree atrioventricular block may be treated effectively with atropine as indicated previously. Occasionally electrical pacing will be required. It is recommended that nonparoxysmal atrioventricular junctional rhythms with rates greater than 90 or with exit block be treated actively.[84] Atrioventricular junctional escape rhythms may be simply monitored if the rate is satisfactory.

Diphenylhydantoin and lidocaine are useful drugs in the treatment of digitalis toxic ectopic arrhythmias. They have little adverse effect on sino-atrial rate, atrial conduction, atrioventricular conduction, or conduction in the His-Purkinje system.[85,86] Diphenylhydantoin may actually improve sino-atrial block and atrioventricular conduction under some circumstances.[87] A recommended regimen for diphenylhydantoin[84] is 100 mg slowly infused intravenously every 5 min until onset of toxicity or control of the arrhythmia, followed by a maintenance oral dose of 400–600 mg/day if the arrhythmia is controlled. Lidocaine is given intravenously in 100-mg bolus doses every 3 to 5 min followed by a continuous infusion of 15–50 μg/kg of body weight per minute as required to maintain control of the arrhythmia.

Therapy with potassium is recommended for active ectopic arrhythmias when hypokalemia is present, but must be used with great caution because of the risks associated with hyperkalemia. It is contraindicated in the presence of conduction disturbances, because elevations of plasma potassium concentration tend to impair atrioventricular conduction.[88]

Propranolol has been useful in the treatment of some digitalis toxic arrhythmias. Through its beta adrenergic blocking effects it causes a decrease in automaticity. By virtue of direct myocardial effects it shortens the refractory period of atrial and ventricular muscle and Purkinje fibers and slows the rate of depolarization and conduction velocity.[89] Potentially undesirable effects include depression of atrioventricular conduction and of sino-atrial and atrioventricular junctional pacemakers with asystole or marked bradycardia[90] and depression of myocardial contractility.[91]

Quinidine and procaine amide carry the risk of cardiac toxicity including depression of the sino-atrial node and of atrioventricular and

His-Purkinje conduction, as well as the potential for eliciting ventricular arrhythmias. They also tend to decrease myocardial contractility. Other agents are usually preferable for use in digitalis intoxication.

Although countershock should generally be avoided in the presence of digitalis intoxication because of the severe arrhythmias that may ensue, it occasionally must be used because all other methods have failed in the face of a life-threatening rhythm disturbance. The risk is decreased when lower energy levels are employed.[92,93]

There has been recent investigative interest in the possible use of digoxin-specific antibodies and Fab fragments in the management of advanced digoxin toxicity.[94,95]

Lastly, patient education efforts deserve strong emphasis, since most digitalized patients will remain on the drug for the rest of their lives.

REFERENCES

1. Glynn IM: The action of cardiac glycosides on ion movement. Pharmacol Rev 16:381, 1964
2. Lee KS, Klaus W: The subcellular basis for the mechanism of inotropic action of cardiac glycosides. Pharmacol Rev 23:193, 1971
3. Moran NC: The effects of cardiac glycosides on mechanical properties of heart muscle, in Marks BM, Weissler AM (eds): Basic and Clinical Pharmacology of Digitalis. Springfield, Ill., Charles C. Thomas, 1972, p 94
4. Koch-Weser J, Blinks JR: Analyis of the relation of the positive inotropic action of cardiac glycosides to the frequency of contraction of heart muscle. J Pharmacol Exp Ther 136:305, 1962
5. Farah A: The effect of the ionic milieu on the response of cardiac muscle to cardiac glycosides, in Fisch C, Surawicz B (eds): Digitalis. New York, Grune & Stratton, 1969, p 55
6. Gillis RA, Raines A, Sohn YJ, et al: Neuroexcitatory effects of digitalis and their role in the development of cardiac arrhythmias. J Pharmacol Exp Ther 183:154, 1972
7. Spann J, Sonnenblick EH, Cooper T, et al: Studies on digitalis. XIV. Influence of cardiac norepinephrine stores on the response of isolated heart muscle to digitalis. Circ Res 19:326, 1966
8. Boyajy LD, Nash CB: Influence of reserpine of arrhythmias, inotropic effects and myocardial potassium balance induced by digitalis materials. J Pharmacol Exp Ther 148:193, 1965
9. Koch-Weser J: Beta-receptor blockade and myocardial effects of cardiac glycosides. Circ Res 28:109, 1971
10. Katz AM: Contractile proteins of the heart. Physiol Rev 50:63, 1970

11. Bassingthwaighte JB, Reuter H: Calcium movements and excitation-contraction coupling in cardiac cells, in DeMello WC (ed): Electrical Phenomena in the Heart. New York, Academic, 1972, p 354
12. Langer GA: Ion fluxes in cardiac excitation and contraction and their relation to myocardial contractility. Physiol Rev 48:708, 1968
13. Entman M, Bressler R, Schwartz A: Proposed mechanism for the positive inotropic effect of digitalis, in Marks BH, Weissler AM (eds): Basic and Clinical Pharmacology of Digitalis. Springfield, Ill., Charles C. Thomas, 1972, p 144
14. Bailey LE, Harvey SC: Effect of ouabain on cardiac Ca^{45} kinetics measured by indicator dilution. Am J Physiol 216:123, 1969
15. Langer GA, Sarena SD: Effects of strophanthidin upon contraction and ionic exchange in rabbit ventricular myocardium: Relation to control of active state. J Mol Cell Cardiol 1:69, 1970
16. Entman ML, Cook JW Jr, Bressler R: The influence of ouabain and alpha angelica lactone on calcium metabolism of dog cardiac microsomes. J Clin Invest 48:229, 1969
17. Schatzmann HJ: Hertzglykoside als Hemmstaffe für den activen Kalium und Natriumtransport durch die Erythrocytenmembran. Helv Physiol Pharmacol Acta 11:346, 1953
18. Skou JC: Enzymatic basis for active transport of Na^+ and K^+ across cell membrane. Physiol Rev 45:596, 1965
19. Repke K: Metabolism of cardiac glycosides, in Wilbrandt W (ed): Proceedings of the First International Pharmacology Meeting, Stockholm, 1961, vol 3, New Aspects of Cardiac Glycosides. Oxford, Pergamon Press, 1963, p 47
20. Akera T, Baskin SI, Tobin T, Brody TM: Ouabain: Temporal relationship between the inotropic effect and the in vitro binding to, and dissociation from, (Na^+ + K^+)-activated ATPase. Naunyn Schmiedebergs Arch Pharmacol 227:151, 1973
21. Besch HR Jr, Allen JC, Glick G, Schwartz A: Correlation between the inotropic action of ouabain and its effects on subcellular enzyme systems from canine myocardium. J Pharmacol Exp Ther 171:1, 1970
22. Akera T, Larson FS, Brody TM: Correlation of cardiac sodium and potassium activated adenosine triphosphatase activity with ouabain-induced inotropic stimulation. J Pharmacol Exp Ther 173:145, 1970
23. Prindle KH Jr, Skelton CL, Epstein SE: Influence of extracellular potassium concentrations on myocardial uptake and inotropic effect of tritiated digoxin. Circ Res 28:337, 1971
24. Schwartz A, Matsui H, Laughter AH: Tritiated digoxin binding to (Na^+ + K^+)-activated adenosine triphosphatase: Possible allosteric site. Science 160:323, 1968
25. Bonting SL, Hawkins NM, Canady MR: Studies of sodium-potassium-activated adenosine-triphosphatase. VII. Inhibition by erythrophleum alkaloids. Biochem Pharmacol 13:13, 1964
26. Calhoun KA, Harrison TR: Studies on congestive heart failure. IX. The

effect of digitalis on the potassium content of the cardiac muscle of dogs. J Clin Invest 10:139, 1931
27. Baker PF, Blaustein MP, Hodgkin AL, Steinhardt RA: The influence of calcium on sodium efflux in squid axons. J Physiol 200:431, 1969
28. Schön R, Schönfeld W, Menke K-H, Repke KRH: Mechanism and role of Na^{+}/Ca^{++} competition in (NaK)-ATPase. Acta Biol Med Germ 29:643, 1972
29. Marks BH: Factors that affect the accumulation of digitalis glycosides by the heart, in Marks BH, Weissler AM (eds): Basic and Clinical Pharmacology of Digitalis. Springfield, Ill., Charles C. Thomas, 1972, p 69
30. Hoffman JF: The red cell membrane and the transport of sodium and potassium. Am J Med 41:66, 1966
31. Caldwell PC, Keynes RD: The effect of ouabain on the efflux of sodium from a squid giant axon. J Physiol 148:89, 1959
32. Hoffman BF, Singer DH: Effects of digitalis on electrical activity of cardiac fibers. Prog Cardiovasc Dis 7:226, 1964
33. Rosen MR, Gelband H, Hoffman BF: Correlation between effects of ouabain on the canine electrocardiogram and transmembrane potentials of isolated Purkinje fibers. Circulation 47:65, 1973
34. Levitt B, Cagin NA, Somberg J, et al: Alteration of the effects and distribution of ouabain by spinal cord transection in the cat. J Pharmacol Exp Ther 185:24, 1973
35. Wiggers CJ, Stimson B: Studies on cardiodynamic action of drugs. III. Mechanism of cardiac stimulation by digitalis and g-strophanthidin. J Pharmacol Exp Ther 30:251, 1927
36. Cattell M, Gold H: Influence of cardiac glycosides on force of contraction of mammalian cardiac muscle. J Pharmacol Exp Ther 62:116, 1938
37. Braunwald E, Bloodwell RD, Goldberg LI, Morrow AG: Studies on digitalis. IV. Observations in man on the effects of digitalis preparations on the contractility of the nonfailing heart and on total vascular resistance. J Clin Invest 40:52, 1961
38. Burwell CS, Neighbors DeW, Regen EM: The effect of digitalis upon the output of the heart in normal man. J Clin Invest 5:125, 1927
39. Williams JF Jr, Klocke FJ, Braunwald E: Studies of digitalis. XIII. Comparison of the effects of potassium on the inotropic and arrhythmia producing actions of ouabain. J Clin Invest 45:346, 1965
40. Beiser GC, Epstein SE, Stampler M, et al: Studies on digitalis. XVII. Effects of ouabain on the hemodynamic response to exercise in patients with mitral stenosis in normal sinus rhythm. N Engl J Med 278:131, 1968
41. Sonnenblick EH, Ross J, Braunwald E: Oxygen consumption of the heart. Am J Cardiol 22:328, 1968
42. Selzer A: The use of digitalis in acute myocardial infarction. Prog Cardiovasc Dis 10:518, 1968
43. Lown, B, Klein MD, Barr I, et al: Sensitivity to digitalis drugs in acute myocardial infarction. Am J Cardiol 30:388, 1972
44. Morris JJ, Taft CV, Whalen RE, McIntosh HD: Digitalis and experimental myocardial infarction. Am Heart J 77:342, 1969

45. Kumar R, Hood WB Jr, Joison J, et al: Experimental myocardial infarction. VI. Efficacy and toxicity of digitalis in acute and healing phase in intact conscious dogs. J Clin Invest 49:358, 1970
46. Beller GA, Smith TW, Hood WB Jr: Altered distribution of tritiated digoxin in the infarcted canine left ventricle. Circulation 46:572, 1972
47. Maroko PR, Kjekshus JK, Sobel BE, et al: Factors influencing infarct size following experimental coronary artery occlusions. Circulation 43:67, 1971
48. Leonard E: Alteration of contractile response of artery strips by a potassium-free solution, cardiac glycosides, and changes in stimulation frequency. Am J Physiol 189:185, 1957
49. Ross J Jr, Waldhausen JA, Braunwald E: Studies on digitalis. I. Direct effects on peripheral vascular resistance. J Clin Invest 39:930, 1960
50. Ross J Jr, Braunwald E, Waldhausen JA: Studies on digitalis. II. Extracardiac effects on venous return and on the capacity of the peripheral vascular bed. J Clin Invest 39:937, 1960.
51. Mason DT, Braunwald E: Studies on digitalis. X. Effects of ouabain on forearm vascular resistance and venous tone in normal subjects and in patients in heart failure. J Clin Invest 43:532, 1964
52. Stark JJ, Sanders CA, Powell WJ Jr: Neurally mediated and direct effects of acetyl strophanthidin on canine skeletal muscle and vascular resistance. Circ Res 30:274, 1972
53. Garan H, Smith TW, Powell WJ Jr: Mechanism of vasoconstrictor effect of digoxin in coronary and skeletal muscle. Fed Proc 32:718, 1973
54. Baschieri L, Ricci PD, Mazzuoli GF: Studi su la portata epatica nell 'uomo: Modificazioni del flusso epatico da digitale. Cuore Circ 41:103, 1957
55. Daggett WM, Weisfeldt ML: Influence on the sympathetic nervous system on the response of the normal heart to digitalis. Am J Cardiol 16:394, 1965
56. Zelis R, Mason DT: Compensatory mechanisms in congestive heart failure: The role of the peripheral resistance vessels. N Engl J Med 282:962, 1970
57. Shanbour LL, Jacobson ED: Digitalis and the mesenteric circulation. Am J Dig Dis 17:826, 1972
58. Strickler JC, Kessler RH: Direct renal action of some digitalis steroids. J Clin Invest 40:311, 1961
59. Torretti J, Hendler E, Weinstein E: Functional significance of Na-K-ATPase in the kidney: Effects of ouabain inhibition. Am J Physiol 222:1398, 1972
60. Selden R, Smith TW: Ouabain pharmacokinetics in dog and man: Determination by radioimmunoassay. Circulation 45:1176, 1972
61. Lindenbaum J, Mellow MH, Blackstone MO, Butler VP Jr: Variation in biological availability of digoxin from four preparations. N Engl J Med 285:1344, 1971
62. Greenblatt DJ, Duhme DW, Koch-Weser J, Smith TW: Evaluation of digoxin bioavailability in single dose studies. N Engl J Med 289:651, 1973
63. Heizer WD, Smith TW, Goldfinger SE: Absorption of digoxin in patients with malabsorption syndrome. N Engl J Med 285:257, 1971
64. Lukas DS, DeMartino AG: Binding of digitoxin and some related cardenolides to human plasma proteins. J Clin Invest 48:1041, 1969
65. Marcus FI, Burkhalter L, Cuccia C, et al: Administration of tritiated

digoxin with and without a loading dose: A metabolic study. Circulation 34:865, 1966
66. Rogers MC, Willerson JT, Goldblatt A, Smith TW: Serum digoxin concentrations in the human fetus, neonate, and infant. N Engl J Med 287:1010, 1972
67. Caldwell JH, Bush CA, Greenberger NJ: Interruption of the enterohepatic circulation of digitoxin by cholestyramine. II. Effect on metabolic disposition of tritium labeled digitoxin and cardiac systolic intervals in man. J Clin Invest 50:2638, 1971
68. Soloman HM, Abrams WB: Interactions between digitoxin and other drugs in man. Am Heart J 83:277, 1972
69. Lukas DS: Some aspects of the distribution and disposition of digitoxin in man. Ann NY Acad Sci 179:338, 1971
70. Jelliffe RW: Factors to consider in planning digoxin therapy. J Chronic Dis 24:407, 1971
71. Lukas DS: Of toads and flowers. Circulation 46:1, 1972
72. Smith TW, Willerson JT: Suicidal and accidental digoxin ingestion. Circulation 44:29, 1971
73. Lown B, Levine SA: The carotid sinus: Clinical value of its stimulation. Circulation 23:766, 1961
74. Smith TW, Haber E: The current status of cardiac glycoside assay techniques, in Yu PN, Goodwin JF (eds): Progress in Cardiology 2. Philadelphia, Lea & Febiger, 1973, p 49
75. Beller GA, Smith TW, Abelmann WH, et al: Digitalis intoxication: A prospective clinical study with serum level correlations. N Engl J Med 284:989, 1971
76. Borison HL, Wang SC: Physiology and pharmacology of vomiting. Pharm Rev 5:193, 1953
77. Watanabe Y, Dreifus LS: Interactions of lanatoside C and potassium on A-V conduction in rabbits. Circ Res 27:931, 1970
78. Watanabe Y, Dreifus LS: Electrophysiologic effects of digitalis on A-V transmission. Am J Physiol 211:1461, 1966
79. Moe GK, Mendez R: The action of the several glycosides on conduction velocity and ventricular excitability in the dog heart. Circulation 4:729, 1951
80. Brauner GJ, Green MH: Digitalis allergy: Digoxin-induced vasculitis. Cutis 10:441, 1972
81. LeWinn EB: Gynecomastia during digitalis therapy: Report of 8 additional cases with liver function studies. N Engl J Med 248:316, 1953
82. Asplund J, Edhag O, Mogensen L, et al: Four cases of massive digitalis poisoning. Acta Med Scand 189:293, 1971
83. Citrin D, Stevenson IH, O'Malley K: Massive digoxin overdose: Observations on hyperkalemia and plasma digoxin levels. Scot Med J 17:275, 1972
84. Bigger JT Jr, Strauss HC: Digitalis toxicity: Drug interactions promoting toxicity and the management of toxicity. Semin Drug Treatment 2:147, 1972
85. Bigger JT Jr, Mandel WJ: Electrophysiological actions of lidocaine on canine Purkinje and ventricular muscle fibers. J Clin Invest 49:63, 1970

86. Bigger JT Jr, Bassett AL, Hoffman BF: Electrophysiological effects of diphenylhydantoin on canine Purkinje fibers. Circ Res 22:221, 1968
87. Helfant RH, Scherlag BJ, Damato AN: The electrophysiological properties of diphenylhydantoin sodium as compared to procaine amide in the normal and digitalis intoxicated heart. Circulation 36:108, 1967
88. Fisch C, Knoebel SB, Feigenbaum H, Greenspan K: Potassium and the monophasic action potential, ECG, conduction and arrhythmias. Prog Cardiovasc Dis 8:387, 1966
89. Davis LD, Temte JV: Effects of propranolol on the transmembrane potentials of ventricular muscle and Purkinje fibers of the dog. Circ Res 22:661, 1968
90. Stephen SA: Unwanted effects of propranolol. Am J Cardiol 18:463, 1966
91. Parmley WW, Braunwald E: Comparative myocardial depressant and antiarrhythmic properties of d-propranolol, dl-propranolol and quinidine. J Pharmacol Exp Ther 158:11, 1967
92. Lown B, Kleiger R, Williams J: Cardioversion and digitalis drugs: Changed threshold to electric shock in digitalized animals. Circ Res 17:519, 1965
93. Ten Eick RE, Wyte RS, Ross SM, Hoffman BF: Postcountershock arrhythmias in untreated and digitalized dogs. Circ Res 21:375, 1967
94. Schmidt DH, Butler VP Jr: Reversal of digoxin toxicity with specific antibodies. J Clin Invest 50:1738, 1971
95. Curd JG, Smith TW, Jaton JC, Haber E: The isolation of digoxin-specific antibody and its use in reversal of the effects of digoxin. Proc Natl Acad Sci 68:2401, 1971
96. Smith TW, Haber E: Medical progress: Digitalis. N Engl J Med 289:945, 1010, 1063, 1125, 1973

Herbert J. Levine

9
Congestive Heart Failure

Congestive heart failure generally has been considered a clinical syndrome characterized by pulmonary and/or systemic venous congestion associated with a suboptimal cardiac output. In recent years, however, it has become clear that a broad definition of this sort may in some circumstances be misleading. The very expression *congestive* heart failure, so common in the physician's vernacular, has in itself fostered constraints to our full understanding of this syndrome. For example, while the word congestive ensures that venous congestion is present, we recognize that severe heart failure may exist with normal or even low atrial and ventricular filling pressures. Similarly, pulmonary and systemic venous congestion may be severe in some patients who have perfectly normal hearts. Much of this confusion can be avoided by making a clear distinction between myocardial failure on the one hand and congestive failure on the other. While it is appreciated that failure of heart muscle (myocardial failure) is generally accompanied by venous congestion, each may exist without the other.

Consider, for example, the postoperative patient with normal myocardial function who develops congestive failure as a consequence of acute volume overload.. In this situation pulmonary venous and systemic venous pressures are high and the ventricles are distended and noncompliant, but stroke volume and stroke work are increased and the "pump" functions efficiently. A similar situation is encountered in the patient with acute tubular necrosis or acute glomerulonephritis, where hypervolemia is the result of oliguria. While myocardial failure may be present in certain chronic high-output states such as systemic arteriovenous fistulae, anemia, cirrhosis, beriberi, thyrotoxicosis, and some forms of bone disease (i.e., polyostotic fibrous dysplasia and Paget's disease), these conditions may produce congestive failure with little impairment of myocardial contractility. Inflow obstructiuon to the ventricle as a result of mitral or tricuspid stenosis, constrictive pericarditis, or merely a ventricle rendered stiff by virtue of excessive hypertrophy, scarring, or an infiltrative process, may also produce congestion of the lungs or the systemic circulation without significant impairment of the contractile process. Additionally, diastolic filling may be impaired because of sustained rapid heart action, and thus venous congestion may evolve without myocardial failure.

On the other hand, true impairment of the contractile process may be present without associated congestive failure. Examples of this phenomenon would include myxedema and hypothermia, or on occasions following the administration of pharmacologic depressants such as barbiturates, quinidine, or procaine amide. In such circumstances, myographic studies of heart muscle function might yield results entirely similar to those observed in diseased, failing myocardium; yet congestion may be absent in the greater or lesser circulations. Clinicians also encounter this phenomenon in some patients with mitral insufficiency or ischemic cardiomyopathy where the heart muscle is severely diseased and hypokinetic yet ventricular filling pressures are normal. It should be evident, therefore, that physicians should make a strenuous effort to distinguish myocardial failure from congestive failure, since therapeutic efforts to improve one may worsen the other.

ETIOLOGY OF CONGESTIVE HEART FAILURE

The major causes of congestive heart failure are outlined in Table 9-1. For the most part these can be considered to be the result of three major processes: (1) conditions that produce myocardial failure, (2) diseases that produce inflow obstructiuon to the ventricle, and (3) conditions that produce hypervolemia and a high output state. Several of the headings listed in this table are purposely broad and in themselves include a wide variety of diseases. For example, nonischemic congestive cardiomyopathy would include chronic myocardial failure associated with the postpartum state, neuromuscular syndromes such as Friedreich's ataxia and muscular dystrophy, alcoholic cardiomyopathy, and many others. In some instances a single disease may produce congestive failure by more than one mechanism. Thus congestive failure due to coronary heart disease may present as a large hypokinetic ventricle or may mimic constrictive pericarditis, producing pulmonary congestion by virtue of a small noncompliant ventricle. Similarly, pulmonary conges-

Table 9-1
Causes of Congestive Heart Failure

- Conditions that produce myocardial failure
  - Pressure loads
    - Aortic stenosis
    - Pulmonic stenosis
    - Systemic hypertension
    - Pulmonary hypertension
    - Coarctation of aorta; supravalvular aortic stenosis
  - Volume loads
    - Aortic insufficiency
    - Mitral insufficiency
    - Atrial and ventricular septal defects; patent ductus arteriosus
  - Loss of normal myocardium
    - Coronary heart disease
    - Myocardial infarction
    - Myocarditis (bacterial, viral, chemical, immunologic, etc.)
    - Nonischemic congestive cardiomyopathy
- Diseases that produce inflow obstruction to the ventricle
  - Mitral stenosis; tricuspid stenosis
  - Constrictive pericarditis
  - Atrial myxoma
  - Cor triatriatum; mediastinal collagenosis; pulmonary venoocclusive disease
  - Diseases that reduce compliance of the ventricle
    - Hypertrophic cardiomyopathy
    - Infiltrative diseases, i.e., amyloid
    - Endomyocardial fibroelastosis
    - Severe hypertrophy from any cause
    - Ischemic cardiomyopathy (stiff heart syndrome)
- Hypervolemia with high cardiac output
  - Acute
    - Volume overload
    - Acute renal failure
    - Obstructive uropathy
  - Chronic
    - Arteriovenous fistula
    - Beriberi
    - Anemia
    - Cirrhosis
    - Hyperplastic bone disease (i.e., Paget's disease)
- Other causes
  - Thyrotoxicosis
  - Persistent rapid heart action
  - Severe cardiac slowing
  - High altitude
  - Hyperkinetic heart syndrome (rare)
  - Injured pulmonary endothelium (i.e., smoke inhalation)
  - Neurogenic pulmonary edema
  - Hypoalbuminemia

tion in a patient with obstructive hypertrophic cardiomyopathy is often the consequence of inflow obstruction from a thick, stiff ventricle, chronic pressure overload, and mitral insufficiency due to abnormal function of the anterior mitral valve leaflet. While it is generally rewarding to establish a true etiologic diagnosis in congestive heart failure, it may be more beneficial to the patient in terms of selecting appropriate therapy to understand the mechanical cause of heart failure than the disease that has caused it. A patient with coronary heart disease may have congestive failure associated with generalized hypokinesis of the ventricle, while another may have a strong ventricle but be suffering from severe mitral insufficiency due to papillary muscle dysfunction.

PATHOPHYSIOLOGY OF MYOCARDIAL FAILURE

The most conspicuous abnormality in the performance of the failing heart is that the muscle fibers contract with less than normal vigor. The mechanical expression of this defect, in turn, depends upon the loading conditions imposed upon the myocardium. To detect and analyze the defects of the failing myocardium, three different methods of analysis are employed: (1) an analysis of the pumping function of the heart, (2) an analysis of myocardial performance in terms of basic muscle mechanics, and (3) an analysis of overall and regional geometry of the ventricle using ventriculography. Each of these methods provides information that may not be detected by the other two, and an examination by all three methods is often necessary to adequately detect abnormal performance.

Pump Function

The time-honored method for examining the performance of the intact heart is by the construction of ventricular function or Starling curves. This is accomplished by relating some function of ventricular performance to an index of the end-diastolic dimensions of the ventricle. The former is represented as a measure of systolic performance (i.e., stroke volume, stroke work, stroke power, peak systolic stress, etc.), the latter by end-diastolic pressure, volume, fiber length, or stress, etc. In this manner a curve is constructed that characterizes the performance of the ventricle (see Fig. 9-1). Thus at a given end-diastolic size the ventricle is capable of a finite measurable performance. If the contractile state of the ventricle is increased or decreased, the curve will be displaced upward or downward, respectively.

Coordinates of systolic performance and diastolic dimensions,

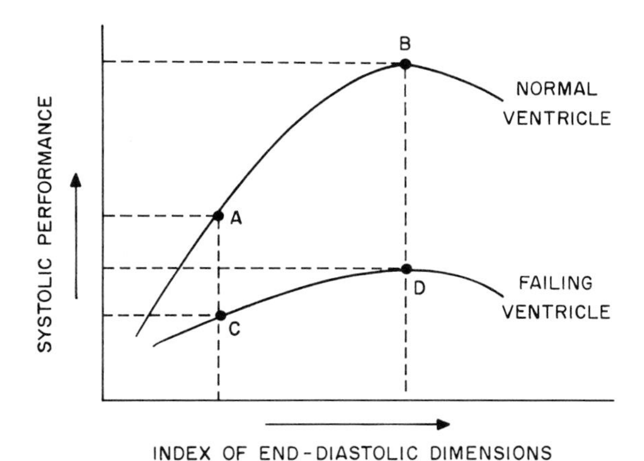

Fig. 9-1. Ventricular function curves. A measure of the pumping function of the heart (i.e., stroke work, stroke volume, etc.) is plotted against an index of the end-diastolic dimensions of the ventricle (end-diastolic pressure, volume, stress, etc.). At each level of relative end-diastolic dimensions the performance of the failing ventricle is less than normal. Point *A* represents the coordinates of a normal heart functioning at a normal ventricular filling pressure. Point *B* describes the consequences of an acute volume overload upon the normal ventricle, where a high-output state is achieved at the expense of venous congestion. Point *C* identifies the low-output state of a failing ventricle with normal filling pressures, while point *D* represents partial compensation of subnormal performance in the failing heart achieved by high filling pressures and venous congestion. That portion of the curve to the right of peak performance represents the descending limb of the ventricular function curve.

therefore, enable one to identify the two general mechanisms by which the heart is able to alter its activity. One of these is dependent wholly upon changes in fiber length and is called by some heterometric autoregulation. This is Starling's law of the heart and is graphically displayed as a single ventricular function curve. The ultrastructual basis for this phenomenon lies in the sarcomere length–tension relationship and is described in detail in Chapter 2. The other mechanism by which the heart regulates its performance (homeometric autoregulation) is independent of dimensional changes and is referred to as change in contractile state. When a sustained depression of the contractile state occurs, myocardial failure exists that generally manifests itself as a low output state. In most instances the failing heart attempts to compensate for this low output state by exploiting the reserve of the Frank-Starling mechanism. To accomplish this the central blood volume (that is, the volume of blood in the heart and lungs) is increased, primarily by renal conservation of sodium and water (see Chapter 10) and by important changes in adrenergically mediated alterations in venous and arterial tone (vide infra). By this means the distending pressure within the ventricle is increased, and performance of the heart is augmented by moving up along the depressed ventricular function curve (Fig. 9-1). In this manner cardiac output is increased somewhat, but at the expense of cardiac dilatation and pulmonary congestion.

For years there has been considerable debate as to whether the failing ventricle ever functions on the descending limb of a depressed ventricular function curve; that is, does cardiac performance diminish when the ventricle is distended beyond a critical filling pressure? The reasons for questioning this view stem from observations made on sarcomere length–tension relationships during diastolic loading of the ventricle. From the sliding-filament model of the myocardial sarcomere (see Chapter 2) it would be predicted that sarcomeres stretched to lengths greater than 2.2 μ would generate less than maximal forces, and this descending limb of the sarcomere length–tension curve would thus correspond to the descending limb of the ventricular function curve. Recently, however, evidence has accumulated to suggest that a descending limb of the sarcomere length–tension relationship is never encountered in the intact heart. Monroe et al.[1] were unable to observe overstretched sarcomeres at filling pressures as high as 60 mm Hg, and in studies of both acute and chronic overload in the intact dog heart Ross and associates[2] also found sarcomere lengths not to exceed optimal values. Ross has suggested that a descending limb of the ventricular function curve in the absence of overstretched sarcomeres may be explained by slippage of myocardial fibers between one another at high filling pressures, such that the internal surface area of the ventricle is increased without a corresponding increase in the length of a given fiber and its sarcomeres. In the chronic situation the role of such plastic deformation of the ventricle in the recovery following corrective cardiac surgery has not been adequately studied, but is likely to be of signal importance.

The construction of ventricular function curves has long proved to be a relatively simple and accurate method for assessing changes in either the contractile state or the Frank-Starling mechanism within a given heart. However, when one attempts to apply this technique to the diseased human heart the results are much less satisfactory. Part of this problem lies in the difficulty in accurately measuring regurgitant flows and asynchrony of contraction in some patients with heart disease, but a

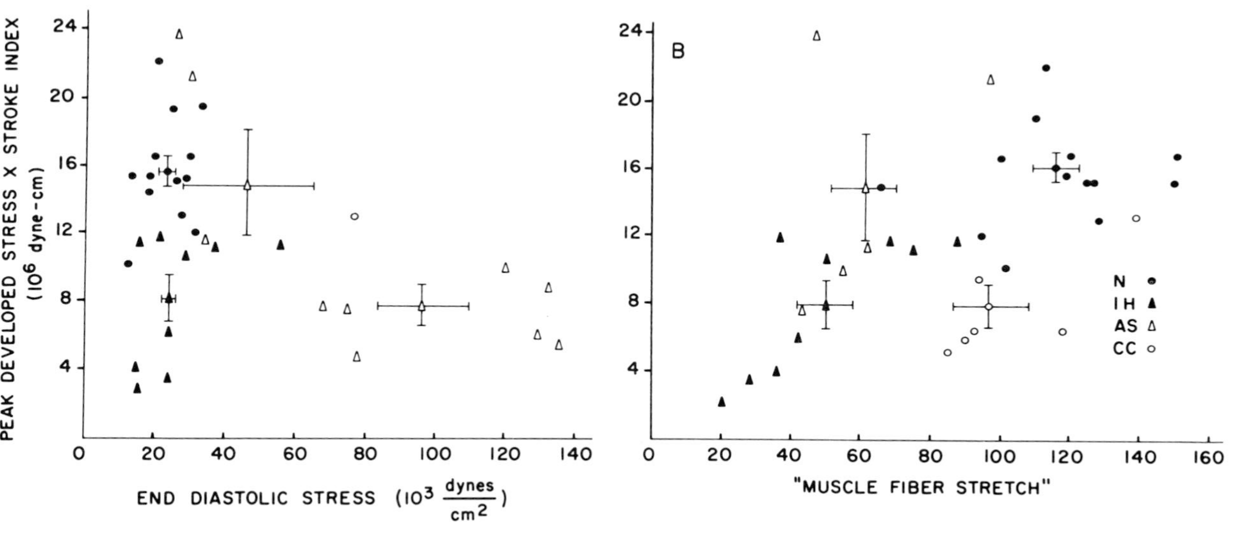

Fig. 9-2. Comparison of ventricular function plots using end-diastolic stress (left panel) and "muscle fiber stretch" (right panel) as indices of muscle strain. In both instances work is plotted on the ordinate as the product of peak developed systolic stress and stroke index. Values (including means ± SEM) are shown for 13 normals (N), 9 subjects with inappropriate hypertrophy (IH), 5 with aortic stenosis (AS), and 6 with congestive cardiomyopathy (CC). The analysis in the left panel shows end-diastolic stress to be normal in the IH group and high in the CC subjects. Similar directional changes in fiber strain are implied. In the right panel, estimates of end-diastolic strain, calculated as the product of end-diastolic stress and compliance ("muscle fiber stretch"), indicate that at end-diastole the fibers of the IH subjects are stretched less than normal, while those of the CC group are within the normal range (see text). (Adapted by permission from Gaasch WH, Battle WE, Oboler AA, et al: Left ventricular stress and compliance in man: With special reference to normalized ventricular function curves. Circulation 45:746, 1972.)

more important limitation of the Starling curve analysis concerns the problem of estimating the degree to which the myocardial fibers are stretched at end-diastole in diseased ventricles. Whether one chooses to use end-diastolic pressure, volume, or stress to gauge fiber stretch at end-diastole, none of these measurements will indicate relative stretch or strain of the fibers if the compliance of the ventricle is unknown. Ideally we would like to know what mean sarcomere length is at end-diastole. While end-diastolic pressure alone might prove reasonably representative of sarcomere length in a homogeneous population of normal ventricles, such is not the case in ventricles that are hypertrophied, scarred, or limited by pericardial constriction or an infiltrative process. Thus if the ventricular function curve analysis is to be applied to the diseased human heart, the estimation of initial fiber stretch (or sarcomere length) must consider not only the stress placed on the fibers at end-diastole but the compliance of the muscle at that moment as well. Such an approach has recently been proposed by Gaasch et al.[3] Figure 9-2 compares four groups of patients studied by cardiac catheterization and ventriculography in whom systolic performance was related to end-diastolic stress on the one hand and to end-diastolic muscle fiber stretch on the other. The latter expression, an index of muscle strain, was derived as the product of end-diastolic stress and compliance. This analysis illustrates that the relative position of a diseased heart on a ventricular function plot is influenced in a major way by a consideration of the diastolic properties of the ventricle. Once ventricular compliance can be measured accurately and with relative ease in the diagnostic catheterization laboratory, our ability to assess the performance of the diseased human ventricle will be greatly enhanced.

Muscle Mechanics

The mechanical performance of the failing ventricle can be analyzed with greater precision if one examines the heart in terms of its muscle function rather than pumping function. This approach has its origin in studies of isolated muscle experiments. In a series of epoch-making studies of skeletal muscle, Hill[4] postulated that the conversion of chemical to mechanical energy in muscle occurred in a portion of the muscle (the contractile element) that functionally existed both in series and in parallel with passive elastic elements. This concept gave rise to the functional analogue models of muscle depicted in Fig. 9-3. Abbott and Mommaerts,[5] Sonnenblick,[6] and Brady[7] extended these studies to cardiac muscle and demonstrated that here, too, an inverse relationship

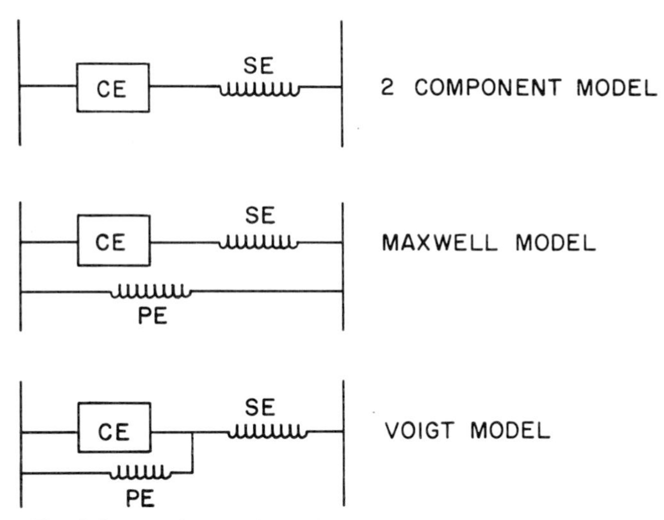

Fig. 9-3. Analog models of muscle contraction. CE represents the contractile element, which is freely extensible at rest in the Maxwell and Voigt models, but not so in the two-component model. The PE and SE indicate the parallel- and series-elastic components, respectively, each represented as passive nonlinear springs. Upon stimulation, the CE is capable of shortening and bearing force.

between force and velocity existed in series and in parallel with passive nonlinear springs. While the diseased ventricle may exhibit abnormalities of the contractile element (CE), the series elastic (SE) component, or the parallel elastic (PE) component, the major functional defect lies within the CE.

The 1967 Spann and associates[8] studied the mechanical behavior of papillary muscles from hypertrophied and failing right ventricles of cats subjected to pumonary artery constriction. When the length–tension relationships of these muscles were compared to normal cat papillary muscle, the active tension per unit cross-sectional area developed at any given muscle length was markedly diminished in the failing muscle (Fig. 9-4). Further insight into this abnormality is provided by an analysis of the force–velocity relationships of failing heart muscle. Figure 9–5 demonstrates that the maximum velocity of these failing muscles is markedly reduced, as well as the maximum isometric force generated in a single contraction. Thus when the muscle is faced with a near-zero load the maximum rate of shortening of the CE is diminished, and conversely when shortening of the muscle is prevented the maximum load that can

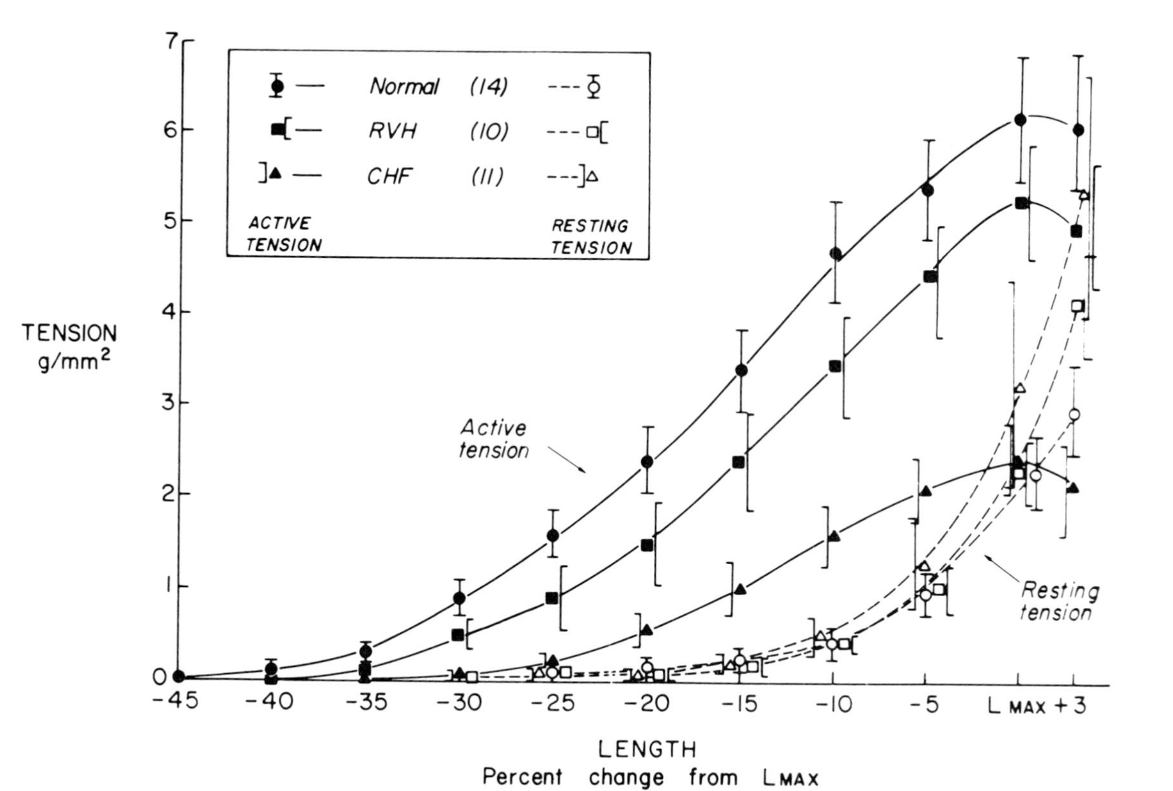

Fig. 9-4. Length–tension relationships of cat papillary muscles from normal (circles), hypertrophied (squares), and failing (triangles) right ventricles. Open symbols represent resting tension, and solid symbols indicate active tension. Values shown are averages for the group ±SEM. (Reproduced by permission from the American Heart Association, Inc. and Spann JF Jr, Buccino RA, Sonnenblick EH, Braunwald E: Contractile state of cardiac muscle obtained from cats with experimentally produced ventricular hypertrophy and heart failure. Circ Res 21:341, 1967.)

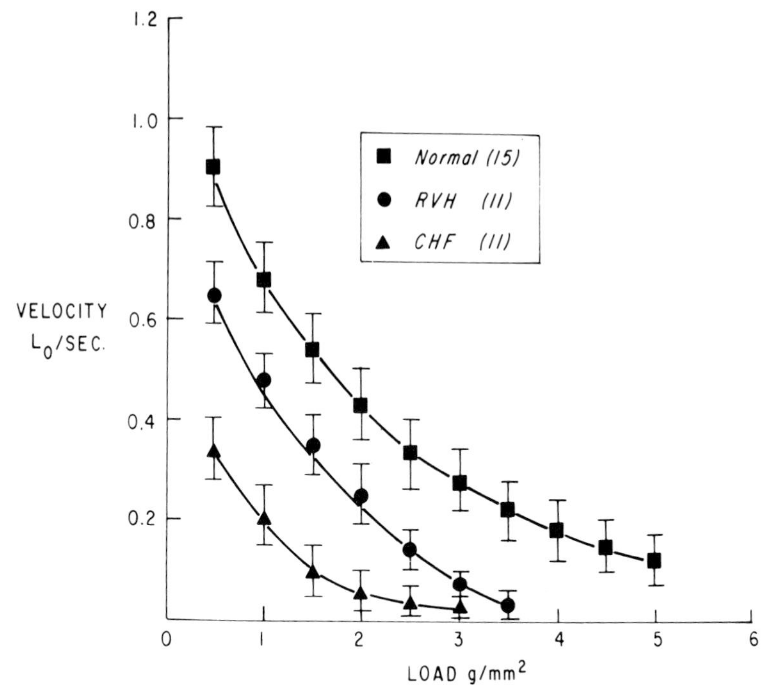

Fig. 9-5. Force–velocity relations of normal, hypertrophied (RVH), and failing (CHF) right ventricular cat papillary muscles. Values given represent averages for each group ± SEM. Velocity has been normalized as lengths per second (see text). (Reproduced by permission from the American Heart Association, Inc. and Spann JF Jr, Buccino RA, Sonnenblick EH, Braunwald E: Contractile state of cardiac muscle obtained from cats with experimentally produced ventricular hypertrophy and heart failure. Circ Res 21:341, 1967.)

be borne by the CE is reduced. In these analyses it must be remembered that the mechanical expression of the capabilities of the CE is influenced by the fact that it is shortening in series with a nonlinear spring. The true activity of the CE, therefore, would be best approximated if the SE were rigid. This has given rise to the concept of the *active state* of muscle, which, having dimensions of both intensity and duration, reflects the true characteristics of the CE.* With this in mind, let us reexamine the mechanical abnormalities of the failing muscle. It is generally believed

*While the intensity of the active state of muscle may be considered the force borne by the CE when it is neither lengthening nor shortening, this definition has been expanded to a mechanical measure of the chemical processes that generate force and shortening.[9]

that the fundamental defect of failing heart muscle is an impaired rate of energy release by the muscle and that the mechanical equivalent of the defect is a reduced maximum velocity of shortening. If this is so, why is maximum isometric force reduced as well as maximum unloaded velocity? When failing muscle contracts isometrically, a decrease in the velocity of shortening of the CE is expressed as a decrease in the rate of force development (dF/dt). With an unchanged duration of the active state, this decrease in dF/dt results in a fall in maximum force developed. In the experiments reported by Spann et al.[8] the time to peak force of an isometric contraction was not altered in both hypertrophied and failing heart muscle. Thus the duration of the active state of muscle was not shortened, but the intensity of the active state (that is, the mechanical product of energy liberation by the muscle) was depressed. It should be noted that in other studies of experimental hypertrophy in the rat heart, Bing et al.[10] found a significant prolongation of the duration of contraction, providing a partial compensatory mechanism for maintaining maximum isometric force in the face of depressed CE velocity.

Of the two passive elements in heart muscle, the series elastic component generally appears to play no important role in the development of myocardial failure. Parmley et al.[11] found no alterations in series elasticity in either hypertrophied or failing heart muscle. However, if one considers the ventricle as a whole, some cases of ventricular aneurysm might be viewed as a functional defect of the SE. Forwand and colleagues[12] found the overall functional series elastic component of dogs with experimental ventricular aneurysm to be more compliant than the normal ventricle. In this manner the shortening capabilities of the CE are wasted in stretching the overly compliant SE to a given load, and less shortening is left over to move blood out of the ventricle. Thus even in the presence of a normal CE, the stage is set for a low output state and its hemodynamic consequences. Alterations in the parallel elastic component, on the other hand, play a most important role in determining both systolic performance of the ventricle and the circulatory manifestations of congestion in heart failure. These abnormalities are considered in greater detail below in the discussion of the diastolic properties of the failing heart.

During the past 10 years considerable effort has been made to extend these observations of muscle mechanics in isolated muscle to the intact ventricle. In 1964 techniques were first developed to study force–velocity relationships in the intact heart, and an inverse relationship between CE force and velocity was demonstrated that permitted estimates of maximum isometric force and unloaded velocity.[13,14] Where each heart served as its own control, an analysis of measured and extrapolated values of CE force and velocity in the intact heart has proved to be a sensitive means of detecting changes in contractile state.[15] Attempts to utilize these methods to accurately assess the contractile state of the diseased human ventricle, however, have met with serious theoretical and practical difficulties. Uncertainty exists as to which of the muscle models shown in Fig. 9-3 most closely approximates the intact ventricle. Even were this known, the altered compliance of the PE in the diseased ventricle makes it difficult to estimate the extension of the CE by a given preload. Since CE force and velocity are influenced by CE length, calculation of these parameters becomes uncertain. In addition, while methods have been developed for estimating the series elasticity of the intact dog heart,[12] no such techniques have been reported for use in human subjects. Despite these and other shortcomings for studying force–velocity relationships in the diagnostic catheterization laboratory, some useful information has been obtained from human studies.

The peak systolic force generated by large failing hearts is uniformly increased by virtue of a large chamber radius. However, since the ventricular muscle mass is greater than normal in these hearts, a greater amount of muscle is available to bear this load, and the peak systolic force per unit cross-sectional area of muscle (wall stress) is generally normal or greater than normal in patients with chronic myocardial failure.[3,16] This observation, then, is at variance with the findings reported above in studies of isolated heart muscle from animals with experimental myocardial failure. It is likely, however, that this discrepancy does not represent a qualitative difference in the contractile defect of the human failing heart, but rather that the phenomenon observed in the intact organism has been altered by compensatory mechanisms that change the loading conditions imposed upon the ventricle and perhaps the duration of the active state as well. It is well known that the body will go to great lengths to maintain the arterial perfusion pressure within a relatively narrow range. With gradual impairment of the contractile process of heart muscle, the body "chooses" to allow this defect to become manifest as a low output state rather than as hypotension. Force development, therefore, is not impaired and indeed may actually be increased as the ventricle is forced to contract against a constant or even increasing load.

While measurements of CE force or stress can be made with reasonable accuracy in the diseased ventricle, estimation of CE velocity (V_{CE}) are more tenuous. On the basis of studies of isolated heart muscle[6,17] it has been assumed that estimates of V_{CE} made at low afterloads provide a measure of the contractile state of the intact ventricle. Since peak V_{CE} occurs early during contraction when the isovolumic pressure is still quite low, it is considered relatively uninfluenced by afterloading conditions and is therefore a useful measure of the contractile state. By making serial measurements of V_{CE} throughout the isovolumic period,* an inverse curve relating V_{CE} and isovolumic pressure may be constructed from which V_{max}, the theoretical unloaded V_{CE}, can be obtained by extrapolation. Since these measurements of V_{CE} are easily normalized and can be obtained from pressure measurements alone, it has been tempting to utilize measured and extrapolated values of isovolumic V_{CE} to assess the contractile state in the intact heart. Unfortunately there are serious limitations to this method, particularly in the diseased heart. First, as mentioned above, the series elasticity of the human heart cannot be measured and must be assumed constant from one heart to another.. Second, conditions that preclude a truly isovolumic contraction period, such as mitral regurgitation or ventricular asynchrony, invalidate this method for calculating isovolumic V_{CE}. Furthermore, considerable disagreement exists as to whether total pressure or developed pressure should be used in the calculation of isovolumic V_{CE}.[17–19] While this is of relatively little practical importance in the normal ventricle where end-diastolic pressure is small compared to isovolumic systolic pressure, this consideration is of great importance where the end-diastolic pressure is elevated. Furthermore, altered compliance of the diseased ventricle makes it difficult to translate a given preload into a predictable fiber stretch. With recent evidence that V_{CE} is partially dependent upon CE length, the use of isovolumic V_{CE} indices for estimation of the contractile state of the human heart has been criticized.[20,21]

Despite these objections, calculations of isovolumic V_{CE} indices for the detection of chronic myocardial failure have been reported to be of some, albeit limited, clinical value. Levine et al.[22] observed depressed values of V_{max}, peak V_{CE}, and V_{CE} at maximum dP/dt in patients with aortic stenosis who exhibited evidence of myocardial failure on the basis of conventional hemodynamics. More recent studies by Hugenholtz et al.[23] and Mirsky et al.[24] suggest that depressed values for peak V_{CE} in the presence of normal conventional hemodynamics are a useful prognostic sign and identify the former as a sensitive indicator of abnormal myocardial performance. Mason and associates[25] and Falsetti et al.[26] have also reported that isovolumic V_{CE} measurements are useful in assessing the contractile state in man. Others have utilized measurements of muscle velocity and stress during the ejection period to

*Since isovolumic V_{CE} equals series elastic velocity, $V_{CE} = (dF/dt)/kF$, where k equals the modulus of series elasticity. By simplification, $V_{CE} = (dP/dt)/kP$.

characterize cardiac function. Using cineangiographic and pressure recordings, Gault et al.[27] determined V_{CE} at the point of maximum wall stress, thus avoiding the use of an assumed series elasticity.* While a true coordinate of CE stress and velocity is obtained by this technique, this relationship is highly dependent upon afterload and is thus of limited value in assessing contractile state. Karliner et al.[28] found that the relatively simple measurement of mean fiber shortening rate proved useful in separating patients with myocardial failure from those with normal ventricular function. There is, however, no convincing evidence to date to indicate that velocity measurements, whether isovolumic or during ejection, reliably characterize the contractile state of the diseased human heart. In a recent review of isovolumic and ejection phase indices of myocardial performance in man, Peterson et al.[19] found isovolumic indices to be unreliable in assessing myocardial function in diseased hearts and concluded that ejection phase indices (i.e., mean fiber shortening rate and mean normalized systolic ejection rate) were clearly to be preferred.

Ventricular Geometry

The third method employed in the analysis of myocardial function consists of an examination of the overall and regional contraction patterns of the ventricle using ventriculography. Regurgitant flows may be estimated by comparing total ventricular output to forward cardiac output measured by other techniques. The ejection fraction (stroke volume/end-diastolic volume) provides an extremely useful measure of the ability of the ventricle to shorten its fibers against a given load. Localized areas of abbreviated shortening (hypokinesis), no shortening (alkinesis), or paradoxical systolic motion (dyskinesis) may be recognized and evaluated quantitatively.

Ventriculography also provides information regarding chamber geometry that is useful in understanding mechanisms of heart failure in certain patients with valvular heart disease. Measurements of the major and minor axes of diseased ventricles not unexpectedly indicate that a predictable chamber geometry is associated with chronic pressure overload and chronic volume overload. Thus the major/minor semiaxis ratios of the ventricle are greater in patients with high-pressure pumps and closer to unity in patients with high-flow ventricles. Since an efficient high-pressure pump would be expected to have a large stroke and a short bore (high force/unit area), the cylindrical shapes observed in aortic stenosis and IHSS are not accidental. Conversely, a short stroke and

*Since the series elastic component is at zero velocity at peak stress, V_{CE} at this moment equals fiber shortening rate.

large bore provide an efficient pump for the delivery of high flows at low pressures. Thus the more spherical shape of the ventricle in subjects with chronic mitral regurgitation represents an adaptive mechanism that might not be shared by the ventricle inflicted with acute subvalvular mitral regurgitation.

These observations go far in explaining, for example, the disastrous consequences of even modest mitral regurgitation in the patient with chronic aortic stenosis. In this instance the ventricle has for years molded itself into the optimum geometry for generating a high intraventricular pressure. Should, as the disease progresses, mitral valvular incompetence develop, the ventricle is suddenly faced with an obligatory flow load for which it is ill-prepared. Indeed, there is no way for the ventricle to efficiently deal with the dual task of generating sufficient pressure to maintain forward flow across the stenotic aortic valve *and* deliver a regurgitant volume into the left atrium. This hydraulic dilemma may be partially responsible for the rapid downhill course observed in patients with aortic stenosis once congestive heart failure has developed. In this instance the imminent threat to the patient may be less readily understood by an examination of muscle function or even overall pump function than by examination of hydraulics and ventricular geometry.

Consequences of Cardiac Dilatation

The force generated by the heart during systole can be represented by two variables: the intraventricular pressure and the radius of the ventricular chamber. Assuming a simplified, spherical geometry of the left ventricle, the force per unit length within the wall of the ventricle (F) is defined by the law of Laplace as $F = rP/2$, where r equals the radius and P the ventricular pressure at a given moment of systole. The total force acting tangentially at the equator of this sphere may be derived by multiplying this value by $2\pi r$, yielding $\pi r^2 P$. Thus the total force generated by a spherical ventricle varies directly with the systolic pressure and square of the chamber radius. While the geometry of the mammalian ventricle is considerably more complex than that of a thin-walled sphere, formulas have been derived for the calculation of wall stress that take into account ellipsoidal chamber shape and thick-walled theory. For a detailed discussion of these calculations, the reader is referred to other sources.[29]

The simplified spherical model, however, remains useful in demonstrating the mechanical consequences of dilatation in the failing heart. Let us examine the time course of wall force during systole in a normal

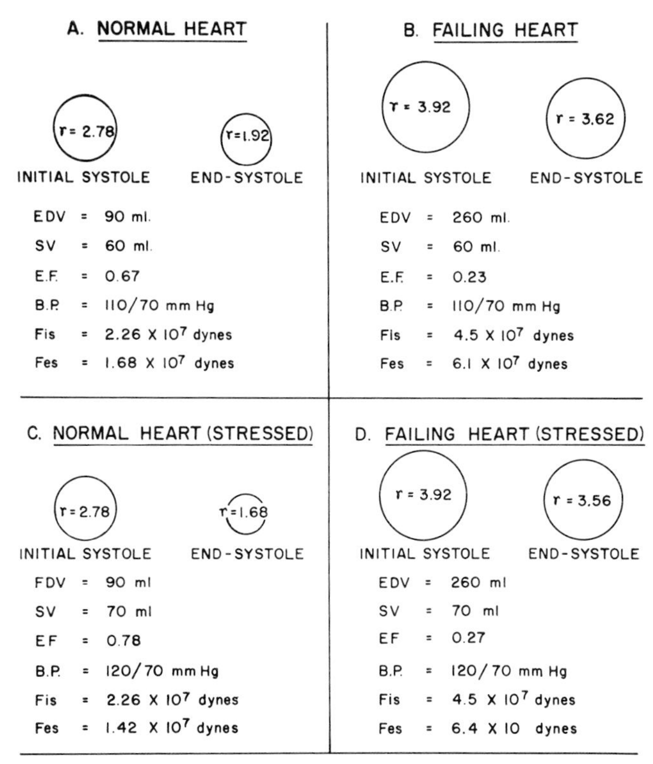

Fig. 9-6. Initial systolic and end-systolic wall forces for simplified spherical ventricles of normal and failing hearts. EDV = end-diastolic volume; SV = stroke volume; E.F. = ejection fraction; B.P. = arterial blood pressure; Fis = initial systolic total wall force; Fes = end-systolic total wall force. Fis and Fes are calculated from ventricular pressures and radii for a normal ventricle (panel A) and failing ventricle (panel B). The latter differs from the normal example only in that the EDV is larger and the EF is smaller. Note that not only are both Fis and Fes larger in the failing heart, but wall force falls during systole in the normal example, while it rises in the failing ventricle. In panels C and D, both ventricles are called upon to increase their stroke volume by 10 ml and systolic pressure by 10 mm Hg without a change in EDV. Fis remains the same in both ventricles, but Fes is decreased in the normal heart despite increased pressure, but is further increased in the failing heart (see text). (Adapted by permission from Levine HJ, Wagman RJ: Energetics of the human heart. Am J Cardiol 9:372, 1962.)

ventricle. To simplify matters we will assume that peak systolic pressure is achieved at the end of the ejection period. Thus if we know both the ventricular pressure and the dimensions of the ventricle at the beginning and end of ejection, we may calculate wall force at both these moments during systole. In Fig. 9-6A it will be seen that for the simplified normal ventricle, wall force is less at the end of ejection despite the fact that ventricular pressure has risen from 70 to 100 mm Hg. Since the ejection fraction (stroke volume/end-diastolic volume) in this ventricle is quite normal, r^2 falls more than P rises, and wall force declines during the course of ventricular emptying. In the case of the failing ventricle (Fig. 9-6B) the time course of wall force is quite different. While arterial pressure and stroke volume are the same as in the normal example, initial chamber volume is greatly increased and the ejection fraction reduced. Thus not only are all wall forces increased, but wall force at the end of ejection exceeds that at the onset of ejection. The implications of this abnormality become even more convincing when these ventricles are called upon to increase their performance. Let us consider that under stress both ventricles respond by an equal increase in stroke volume and pulse pressure without a change in their end-diastolic volume. In the case of the normal ventricle (Fig. 9-6C) wall force is unchanged at the onset of ejection, but falls even further at end-systole by virtue of more complete emptying. The failing ventricle, on the other hand, is faced with an even greater load at end-systole than in the control state (Fig. 9-6D). Therefore it would appear that the normal ventricle is in a good position to increase its work performance, since wall force is falling throughout systole. The failing ventricle, however, is encountering a progressively higher load throughout systole and finds it particularly difficult to increase stroke volume and pressure without a further increase in end-diastolic volume. This inefficiency is illustrated in yet another way. Since the energy consumed by the heart is determined in large part by force development, the increased work performed by the stressed normal heart in Fig. 9-6C may be accomplished with a fall in energy requirements and thus a real increase in mechanical efficiency (work performed/energy consumed). Indeed, increased mechanical efficiency has been observed in the exercising normal heart, while an unchanged or decreased efficiency is characteristically found in the stressed failing ventricle.[30]

While this analysis of a simplified spherical ventricle involves liberal and perhaps unjustified assumptions regarding ventricular dynamics and geometry, measurements of systolic wall force or stress in both normal subjects and those with myocardial disease support these conclusions.[16,27,31] At the same time it is important to emphasize that not all consequences of dilatation are harmful. It is perhaps apparent that a given amount of sarcomere shortening will result in a greater stroke volume if the initial volume of the heart is larger. That is to say that the sarcomeres need to shorten less to eject a given stroke volume when the end-diastolic volume is increased. For example, in terms of ventricular dimensions and total systolic force, the athlete's heart and the failing ventricle may appear to be almost identical at the onset of ejection, since both would be found to have a high systolic force at this moment. The former, however, has a normal ejection fraction and ejects a large stroke volume against a normally falling force–time plot.

DIASTOLIC PROPERTIES OF THE FAILING HEART

While the manifestations of myocardial failure are most clearly evident from an analysis of the systolic performance of heart muscle, it is generally the diastolic properties of the failing ventricle that are directly responsible for the clinical features and hemodynamic abnormalities of congestive failure. These diastolic properties, in turn, can be appreciated

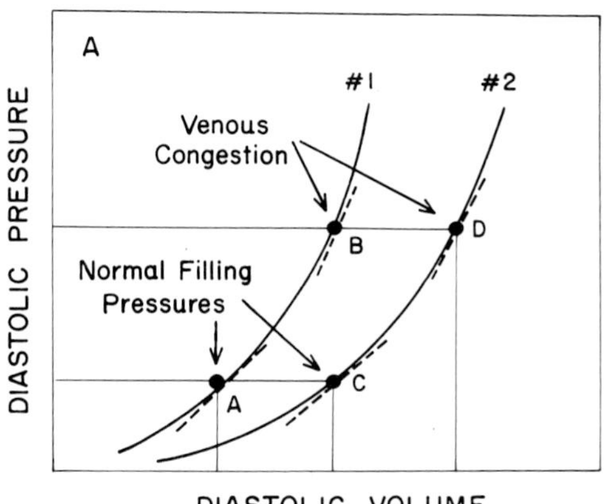

Fig. 9-7. Diastolic pressure–volume relationships of the ventricle. The slope of the pressure–volume curve (dP/dV), representing an index of the stiffness of the ventricle, increases at higher diastolic filling pressures. Thus dP/dV at point *B* is greater than at point *A*. If curves 1 and 2 represent two normal ventricles differing only in size, dP/dV will be less in the larger ventricle (i.e., point *C*) than in the smaller chamber (point *A*) (see text).

from an examination of the pressure–volume relationships of the relaxed ventricle (Fig. 9-7). Inspection of this curve reveals that, like most biologic tissue, the relaxed ventricle demonstrates nonlinear stress–strain characteristics, and except at the extremes of the pressure–volume curve this relationship generally has been found to be exponential.[3,32,33] Since the slope of this curve (dP/dV) is an index of the stiffness or elasticity of the ventricle, it is clear that diastolic stiffness is a function of the operative pressure; that is, the greater the diastolic pressure the stiffer a ventricle becomes. Thus in Fig. 9-7 dP/dV is greater at point B than it is at point A on the same pressure–volume curve. It is also implicit that ventricular size itself will influence this analysis. If, for example, curve 1 represents that of a normal child's left ventricle, the pressure–volume curve of a proportional normal adult ventricle would be expected to be shifted to the right (curve 2) as the heart grows in size, without invoking a change in the elasticity of the heart muscle itself. At point C, therefore, dP/dV would be less than at point A, while the stress–strain characteristics of a unit of muscle remain quite normal. Thus a comparison of points on two pressure–volume curves must consider not only identification of the operative pressure or stress but normalization for ventricular volume as well. The third factor that should be considered in estimating compliance of heart muscle is the ventricular mass–volume ratio.[34] By so doing one may distinguish the ventricle rendered stiff by hypertrophy alone from the ventricle whose low compliance is due to scarring, stretched fibers, or other factors that diminish the distensibility of a unit mass of myocardium.

In studies of the intact heart the reciprocal of the stiffness index (dV/dP) is commonly used as an index of distensibility or compliance of the ventricle, and normalization of this index by ventricular volume (dV/VdP) assists one in comparing hearts of different size.[35] It should be emphasized, however, that an accurate assessment of the diastolic properties of a ventricle requires measurements of pressure–volume coordinates over the entire operative range of diastolic filling pressures. Diamond and Forrester[36] have estimated the slope of the linear relationship of $\Delta P/\Delta V$ versus P (passive elastic modulus) in man by relating the ratio of $\Delta P/\Delta V$ to mean filling pressure. In these studies ΔP was derived as the pressure difference from initial to end-diastole and ΔV was the stroke volume. A similar approach has been used by McCullagh et al.[37] using the ratio $\Delta D/\mathrm{D}P$, where $\Delta\mathrm{D}$ represents the change in left ventricular diameter corresponding to a measured change in diastolic filling pressure.

In all of these analyses it should be remembered that the stress–

strain relationship of relaxed heart muscle exhibits hysteresis and may be influenced by alterations in ventricular geometry and by viscous elements that could alter the stress–strain relationships during rapid stretching of heart muscle. It is generally considered, however, that the major determinants of the diastolic pressure–volume relationship are elastic and are little influenced by changes in inotropic state.[35,38]

In the human subject with left ventricular failure, pulmonary congestion is invariably associated with a noncompliant left ventricle, although the pathogenesis of the latter may vary greatly. In the case of the normal heart subjected to acute volume overloading, the ventricle is rendered noncompliant by abrupt movement upward along its diastolic pressure–volume curve (point *D*, Fig. 9-7). The patient with constrictive pericarditis or amyloid cardiomyopathy may have equally high pulmonary venous and left ventricular filling pressures with near-normal ventricular dimensions, but in these instances the low compliance is due to factors other than the stretching of heart muscle.

Following experimental myocardial infarction, an early brief increase in ventricular compliance has been detected that quickly disappears.[39] Three to five days following experimental myocardial infarction, ventricular compliance is clearly reduced from the normal values,[40] and some patients with chronic coronary heart disease exhibit a "stiff heart syndrome" with severe pulmonary congestion and radiographically normal heart size.[41] In patients with angina pectoris induced by atrial pacing, the diastolic pressure–volume curve has been found to be displaced upward only to return to normal levels when the ischemic episode has resolved.[42] The effects of chronic volume overloading upon left ventricular dimensions and compliance have been studied by McCullagh and associates.[373] Following a large aortocaval shunt, the diastolic pressure–volume curve was observed to be gradually shifted to the right, but at the same time the slope of the diastolic pressure–diameter curve rose. The former was attributed to ventricular dilatation and hypertrophy and to the phenomenon of creep (time-dependent deformation in response to constant stress). The latter was considered to reflect alteration in the properties of the ventricular myocardium itself, and it is noteworthy that the observed decrease in compliance was in part reversible following closure of the shunt. In this regard, Gault et al.[43] have shown that patients with normal myocardial function undergoing surgical correction of free aortic regurgitation demonstrate improvement in their diastolic pressure–volume relationships postoperatively, while those with depressed contractile state have a fixed abnormality in ventricular compliance that persists after operation.

BIOCHEMICAL ABNORMALITIES IN CONGESTIVE FAILURE

Oxidative Phosphorylation

While it is generally agreed that substrates for oxidative metabolism are present in normal or near-normal quantities in the failing myocardium, there has been considerable debate as to whether energy production by mitochondria is impaired. Oxidative phosphorylation has been found normal in experimental heart failure by some[44–46] and decreased by others.[47,48] Chidsey et al.[49] found no abnormality of oxidative phosphorylation in mitochondria from failing human hearts, and while Lindenmayer and associates[50] observed lower rates of respiratory activity in failing human hearts than in nonfailing hypertrophied myocardium, activity of the failing hearts was at or slightly below the normal values. Surprisingly, the oxygen consumption of a unit mass of pressure-hypertrophied nonfailing myocardium is increased above control levels,[51] and it has been suggested by Cooper et al.[52] that this increase in nonphosphorylating oxygen consumption is related to increased calcium pumping by mitochondria isolated from pressure-hypertrophied myocardium. In failing heart muscle calcium uptake by mitochondria has been reported to be decreased.[53] It is of interest that a commensurate degree of hypertrophy induced by volume overload is not associated with this abnormality of oxygen metabolism.[54] Thus in volume overload hypertrophy the oxygen consumed by a unit weight of myocardium developing a given isometric load is quite normal. It would appear, therefore, that oxidative phosphorylation is little affected in chronic myocardial failure, while other energy-consuming mitochondrial functions (i.e., calcium transport) may be impaired.

Calcium Transport

Recent studies suggest that a defect in calcium transport by the sarcoplasmic reticulum may play an important role in the biochemical lesion in myocardial failure. Calcium accumulation by microsomes that contain fragments of sarcoplasmic reticulum operates by two kinetically dissimilar processes. Calcium binding is a rapid process capable of sequestering only small amounts of the ion. Calcium uptake is a slower process that has the potential of removing much larger amounts of calcium from the contractile sites. Calcium binding by sarcoplasmic reticulum has been found to be reduced in the hereditary cardiomyopathic hamster, while calcium uptake has been observed to be normal.[55] In other studies, Gertz et al.[56] reported calcium uptake to be decreased in spontaneously failing heart-lung preparations. Suko et al.[57] observed reduced calcium uptake and normal calcium binding in chronically failing calf hearts. On the basis of extensive studies of calcium-binding membranes in myocardial failure, Schwartz and associates[58] conclude that the earliest change in congestive failure is a decrease in the rate of calcium release by sarcoplasmic reticulum. The rate of calcium binding is next affected, and finally the total calcium uptake is reduced. Whether these abnormalities in calcium transport by the sarcoplasmic reticulum represent a primary biochemical defect responsible for secondary impaired contractility or whether impaired calcium pumping is merely the expression of a more fundamental biochemical defect remains to be shown.

Contractile Proteins

In 1967 Bárány reported a correlation between myosin ATPase activity and the speed of muscle shortening among skeletal muscles from different animal species.[59] Since the major mechanical abnormality of failing myocardium is a decrease in the unloaded contractile element velocity, V_{max}, reports of depressed contractile ATPase from failing myocardium have stimulated considerable interest. Alpert and Gordon[60] found a decrease in myofibrillar ATPase activity in failing human hearts studied at postmortem. This finding has been confirmed by others in experimental myocardial failure[61,62] and in humans with heart failure,[63] although Nebel and Bing[64] observed some increase in ATPase activity in the failing human heart. In light of these observations, it is tempting to infer that depressed contractile element velocities in the failing heart are the direct consequence of a primary abnormality of the contractile proteins, manifest as a reduced myosin ATPase. While this possibility remains, it is disturbing that digitalis glycosides, which increase CE velocities in the failing heart, do not evoke parallel increases in myosin ATPase.[63] Furthermore, Katz has suggested that both depressed myosin ATPase activity and V_{max} may represent a compensatory mechanism of the failing heart permitting the slower muscle to be more efficient in the development and maintenance of tension.[65]

Modulatory Proteins

In addition to the contractile proteins themselves, the modulatory proteins troponin and tropomyosin have been considered as primary culprits in the lesion of myocardial failure. Katz and Hecht[66] recently suggested that intracellular pH changes may influence the contractile process by means of altering the affinity of troponin for calcium. Intracellular acidosis, as a consequence of myocardial ischemia, would reduce calcium binding by troponin, which is the initial step in the reversal of the inhibitory effect of troponin upon the contractile proteins. Schwartz and associates[58] report preliminary studies suggesting that the major defect in chronically ischemic dog hearts resides in the troponin–tropomyosin complex rather than in the contractile proteins.

ADRENERGIC MECHANISMS IN CONGESTIVE HEART FAILURE

In the normal subject the sympathetic nervous system plays the dominant role in the autoregulation of the heart and circulation. By virtue of its almost instantaneous and profound effects upon heart rate, myocardial contractility, and both arterial and venous tone, sympathetic activity largely determines the response of the circulatory system to physical and emotional stress, and thus provides a measure of the cardiac reserve. One of the first abnormalities of the autonomic nervous system to be demonstrated in congestive heart failure was augmented sympathetic activity during physical exercise.[67] Further evidence for increased sympathetic activity was provided by the fact that the 24-hr excretion of norepinephrine was increased in patients with heart failure.[68] These observations are in sharp contrast to the finding that myocardial stores of norepinephrine are depleted both in failing human myocardium[68] and in experimentally induced myocardial failure.[69,70] Interestingly, Monroe et al.[71] have shown norepinephrine efflux from the myocardium when the ventricle generates high systolic pressures, but not when heart rate is increased or when very high diastolic filling pressures are sustained by the ventricle.

In an effort to clarify the mechanism of myocardial norepinephrine depletion in heart failure, studies of norepinephrine synthesis, turnover rates, binding, and degradation have been performed. Pool et al.[72] demonstrated reduced tyrosine hydroxylase activity in failing ventricles and thus identified a defect in the rate-limiting enzyme system necessary for the synthesis of norepinephrine. In addition, the uptake of labeled exogenous norepinephrine by failing heart muscle is reduced.[69] On the other hand, turnover rates for norepinephrine are the same in failing and normal myocardium. Krakoff et al.[73] studied the activity of monoamine oxidase and catechol *o*-methyltransferase (the two enzymes responsible for the breakdown of norepinephrine) in cats with experimental right ventricular failure and concluded that neither activity could account for the striking reduction in myocardial norepinephrine stores.

Since it is well known that exogenous norepinephrine will agument the contractile state of failing heart muscle, it is reasonable to question whether depletion of myocardial norepinephrine stores may in itself be responsible for the depressed contractility of the failing heart. That this is not the case has been convincingly demonstrated by studies showing that depletion of norepinephrine stores following cardiac denervation to levels even less than that observed in the failing heart is associated with normal contractile function.[74] Thus it must be concluded that the depletion of myocardial catechols that occurs in myocardial failure is not in itself responsible for depression of the intrinsic contractile state. However, this depletion interferes with the augmentation of contractility mediated through the adrenergic system.

When the failing heart is challenged with the administration of exogenous norepinephrine or other beta adrenergic agonists, the mechanical response, in terms of augmented performance, is actually greater than normal—a response consonant with Cannon's law of denervation. Thus while myocardial norepinephrine stores are low the failing heart is supersensitive to beta adrenergic amines, and is not only accustomed to increased sympathetic activity but is acutely aware of deprivation of this support. It is well known that propranolol, reserpine, and guanethidine may cause abrupt worsening of cardiac function in patients with even latent congestive heart failure, while subjects with normal cardiac function generally tolerate large doses of these drugs without adverse circulatory effects.

The elevated plasma levels and urinary excretion of norepinephrine observed in congestive heart failure appear to be due, at least in part, to augmented sympathetic activity in the peripheral vascular bed[75] and perhaps to an adrenal medullary origin as well. The increased peripheral vascular resistance characteristically found in low-output failure, in turn, is due only partially to augmented sympathetic activity. It has been shown that increased vascular stiffness in congestive failure, related to a higher vascular sodium content, contributes to the reduced ability of the resistance vessels to dilate.[76]

Defects in parasympathetic activity have also been identified in congestive heart failure. Following beta adrenergic blockade, baroreceptor-induced slowing of heart rate has been found to be attenuated; similarly, the heart rate response to atropine has been found to be less than normal in patients with congestive failure.[77] The sympathetic–parasympathetic interactions in the heart are extremely complex, and the reader is referred to other sources for further information on this important subject.[78,79]

CLINICAL MANIFESTATIONS OF HEART FAILURE

The clinical manifestations of heart failure are for the most part the result of low or suboptimal cardiac output and of venous congestion. In the early or latent state, fatigue and/or dyspnea are noticed only during periods of physical stress. As the ability of the myocardial fibers to shorten adequately is diminished, stroke volume falls and pulse pressure is diminished. In an effort to compensate for this decrease in cardiac output, the heart rate is increased at each level of activity, culminating in a resting sinus tachycardia. Regional circulations best able to withstand low flow rates suffer disproportionately. Thus skin blood flow and renal blood flow are depressed, while cerebral and coronary flow are maintained. Since the most efficient mechanism for the elimination of body heat is radiation from the skin, dissipation of heat is impaired and compensation is achieved by vaporization (sweating). The skin, therefore, is generally cool and moist, except in patients who have high-output congestive failure. In advanced heart failure this defect in the elimination of heat may produce low-grade fever that will disappear following effective therapy.

Decreased renal blood flow initiates mechanisms for the renal conservation of sodium and water (see Chapter 10). This expansion of the intravascular blood volume, together with adrenergically mediated increases in arterial and venous tone, elevates venous pressure of the systemic and/or pulmonary circulations. The jugular veins become distended and the liver engorged, and dependent edema often develops. Hepatic congestion not infrequently gives rise to anorexia, abdominal fullness, or frank pain. However, if systemic venous congestion is gradual, severe congestive hepatomegaly may develop with little effect upon conventional liver function tests, except for increased retention of Bromsulphalein. With acute right heart failure, on the other hand, or with anoxia resulting from circulatory collapse, bilirubinemia and extraordinary elevations of serum transaminases may be observed. Chronic severe systemic venous congestion, as seen in constrictive pericarditis or tricuspid valve disease, is often associated with conspicuous ascites, not infrequently without dependent edema.

With left heart failure, left atrial and ventricular filling pressures rise to partially compensate (via the Frank-Starling mechanism) for the depressed contractile state of the failing myocardium. Pulmonary capillary pressures increase, resulting in engorged pulmonary lymphatics and perivascular edema. With acute severe pulmonary venous hypertension, frank transudation occurs when the plasma oncotic pressure (25–30 mm Hg) is exceeded and pulmonary edema results. Alveolar edema, however, is not a prerequisite for dyspnea due to pulmonary venous congestion. Fishman[80] has shown that interstitial edema precedes alveolar edema, and the former alone may evoke tachypnea. Interestingly, this increase in interstitial edema stimulates an increase in pulmonary lymphatic flow, which in turn may become a limiting factor in relieving pulmonary edema. With chronic pulmonary congestion, expansion of the pulmonary lymphatic system may enable patients to tolerate levels of pulmonary venous hypertension that would ordinarily evoke pulmonary edema.[81] With the development of pulmonary venous congestion, lung compliance falls, the work of breathing is increased, and vital capacity is reduced. In more advanced cases, arterial hypoxemia may be present, generally with some respiratory alkalosis. In some instances, particularly in patients with mitral valve disease, pulmonary congestion may present as bronchospasm simulating primary endobronchial disease. Orthopnea is especially characteristic of pulmonary venous congestion. However, since total lung volumes and vital capacity are reduced in all persons in the supine position, some patients with severe primary lung disease may complain of orthopnea. Paroxysmal nocturnal dyspnea is almost diagnostic of congestive heart failure, but must be distinguished from the early morning clearing of pooled secretions experienced by some patients with chronic lung disease. Pleural effusion, particularly on the right side, is common in patients with congestive failure. Recumbent cough is also a very useful sign of left heart failure. In rare cases of rheumatic mitral regurgitation a giant left atrium may be responsible for a constant irritative cough, hoarseness from recurrent laryngeal nerve palsy, or even dysphagia.

Most of the cardiac signs of congestive heart failure are more the manifestation of a noncompliant ventricle than the result of a weakened, hypokinetic ventricle. One of the most useful of these signs is the ventricular gallop (S_3). While this is a normal finding in some children and young adults, the presence of an S_3 in persons over the age of 35 years is generally a reliable sign of congestive failure. Frequently this elusive sound can be brought out by the simple stress of a few coughs, isometric handgrip, elevation of the legs, or several situps.

As the diastolic pressures within the ventricle rise, the compliance of the ventricle falls (vide supra) and the normal respiratory variations in filling pressure evoke lesser changes in cardiac dimensions than when the ventricle is functioning at low filling pressures. Consequently the normal sinus arrhythmia disappears and fixed or near-fixed splitting of the second heart sound may be observed, particularly when biventricular failure is present. The characteristic "square wave" response of the failing heart to the Valsalva maneuver may be observed,[82,83] since the reduction in effective filling pressure during the strain phase of the maneuver evokes a smaller change in ventricular volume at high diastolic pressures and thus the heterometric response of the heart is attenuated. Since the failing ventricle is likely to function on the flattened portion of the Starling curve, changes in stroke volume due to changes in filling pressure are lessened even further.

Some patients with chronic congestive heart failure may have profound cachexia suggesting disseminated neoplasm. A number of different phenomena may be responsible for this clinical picture. These include (1) decreased absorption by the gut, (2) increased protein loss from the body, (3) increased metabolic demands, and (4) anorexia, often iatrogenic. In those with long-standing systemic venous congestion, malabsorption syndromes may contribute to cardiac cachexia.[94] A protein-losing enteropathy associated with lymphangiectasia of intestinal villae has been reported in some cases of constrictive pericarditis,[85] and several instances of nephrotic syndrome have been observed[86,87] that further deplete the body of protein. Hypermetabolic states of sufficient intensity to suggest thyrotoxicosis have been reported in patients with large failing hearts.[88,89] In these subjects the source of this "oxygen trap" has been identified as the myocardium itself.[90] In one such instance, for example, the myocardial oxygen consumption was found to be over 90 ml/min and accounted for 27% of the total-body oxygen consumption. The wasting process in chronic congestive failure may be understood more readily when it is appreciated that a myocardial oxygen consumption of this magnitude requires more than 600 calories daily merely to sustain cardiac activity.

In the terminal phase of congestive failure, cardiac output falls to dangerously low levels. Oxygen extraction by the body widens, forcing the tissues to function at Po_2's that evoke anaerobic metabolism in those organ systems that have this option. Lactic acid is produced at a rate in excess of the body's ability to utilize this substrate, and acidosis ensures. The bone marrow responds with an outpouring of reticulocytes and nucleated red blood cells[91] in an agonal effort to improve the oxygen-carrying capacity of the blood. Renal blood flow falls; prerenal azotemia develops as peripheral vasoconstriction maintains blood pressure to the end.

With the exception of the chest x-ray, conventional laboratory tests are of little specific value in the diagnosis of congestive failure. Systemic venous pressure can generally be assessed quite adequately at the bedside. While invasive measurements of central venous or pulmonary pressures are often useful in acutely ill patients, examination of the jugular venous pulse and a careful look at the chest x-ray generally provide sufficient information to estimate venous pressures. The radiographic changes of pulmonary venous congestion are primarily those of (1) dilatation of the superior pulmonary veins (the antler appearance), (2) increased interstitial density of central lung markings (pulmonary clouding), (3) thickened septums and lymphatics (Kerley *B* lines), (4) alveolar edema, (5) dilated superior vena cava and azygos vein, (6) free pleural fluid, and (7) interlobar fluid (pseudotumor). Each of these findings may be present with or without pulmonary rales.

The timed vital capacity is an extremely useful technique in evaluating a patient complaining of dyspnea or in following the course of patients with known cardiac disease. For example, a perfectly normal vital capacity may alert the physician to search for other explanations for dyspnea, i.e., neurocirculatory asthenia or the hyperventilation syndrome. In a patient suspected of having congestive failure, determination of the vital capacity before and after a diuretic trial may be diagnostically helpful. Measurement of the circulation time is rarely indicated, but occasionally may be helpful in identifying a high-output state in patients with congestive failure.

THERAPY

The treatment of heart failure is aimed at each of the following, singly or in combination: (1) improvement of myocardial function, (2) relief of adverse loads imposed upon the failing heart, (3) a decrease in pulmonary and/or systemic venous pressure, (4) removal of accumulated sodium and water, and (5) an increase in oxygen availability to the tissues.[92]

When the basis for heart failure is myocardial failure, the most important lasting form of therapy is inotropic support of the heart, generally with digitalis glycosides. The effect of these agents upon both the normal and failing heart is such as to shift the ventricular function curve upward and to the left (see Fig. 9-1). In the failing heart this is manifest as an increase in cardiac output associated with renal losses of sodium and water and a fall in ventricular filling pressures. With this improvement in contractility, ventricular dimensions are decreased and the net effect upon myocardial energetics is generally a fall in myocardial oxygen requirement. Short-term inotropic support may be accomplished by the intravenous administration of beta adrenergic agonists such as isoproterenol, dopamine, epinephrine, etc., or other inotropic agents such as glucagon. Beneficial effects have also been reported in chronic congestive heart failure with oral therapy using L-Dopa as the source of beta adrenergic stimulation.

All too frequently inattention to the relief of a flow or pressure load upon the heart frustrates the success of treatment of congestive heart failure. The patient with hypertensive heart disease who develops congestive failure will profit immediately by a reduction in ventricular afterload. With a lowering of systemic arterial pressure stroke volume will rise (via the inverse force–velocity relationship), and the associated reduction in ventricular size that follows systolic unloading enables the mechanical efficiency of the ventricle to rise. A reduction in systolic pressure is particularly beneficial to the patient with mitral insufficiency, since the magnitude of the regurgitant leak is governed largely by the systolic gradient across the mitral valve. Thus a 50% reduction in mean systolic pressure will reduce mitral insufficiency by 22%, since turbulent flow across an orifice is proportional to the square root of the pressure gradient. Recognition and correction of a high-output state due to anemia, hypoxia, or a hypermetabolic state will also reduce the demand upon an already compromised failing heart. For example, it would be predicted that a fever of 3°C would produce a rise in total-body oxygen consumption of approximately 30%. In the patient with critical mitral stenosis, a commensurate increase in cardiac output may be of signal importance in the precipitation of pulmonary edema, particularly if this rise in cardiac output is associated with tachycardia and shortening of the diastolic filling period. Indeed, in the patient with mitral stenosis the left atrial and pulmonary venous pressures will be determined almost wholly by cardiac output and the duration of diastole.

A decrease in systemic and pulmonary venous pressures may be achieved temporarily by altering the tone of the capacitance vessels or by sequestering blood in the extremities or viscera, but finally and more permanently by the removal of excess sodium and water from the body. Central venous pressures are reduced immediately by the upright position or by venous tourniquets. Morphine has a profound lowering effect on pulmonary venous pressure, largely by virtue of splanchnic pooling of venous blood. However, it is the thiazide diuretics and furosemide and ethacrynic acid that form the mainstay of diuretic therapy in congestive heart failure. Each warrants replacement of renal potassium losses, particularly in the digitalized patient, where hypokalemia may produce serious digitalis intoxication arrhythmias. This complication may often be prevented by combining the above diuretics with potassium-retaining diuretics such as spironolactone or triamterene. In refractory cases of congestive failure, utilization of diuretics that work on different portions of the renal apparatus, i.e., furosemide and spironolactone, may produce particularly gratifying results.

In recent years serious and refractory cases of congestive heart failure have been treated successfully with vasodilating drugs. The basis for this therapy is that reduction in peripheral vascular resistance permits the failing heart to increase its meager stroke volume, and coincident with arterial unloading this improved ventricular emptying of the ventricle reduces ventricular filling and venous pressures. This has been achieved using phentolamine, nitroprusside, nitroglycerin, chlorpromazine, and other drugs that reduce systemic vascular resistance. This form of therapy is not without danger, however, and careful monitoring of pulmonary or ventricular filling pressures in addition to central arterial pressure is required.

Finally, care should be taken to assure that extracardiac factors contributing to the clinical syndrome of heart failure are dealt with appropriately. Thus anemia, hypoxia, renal failure, obstructive uropathy, myocardial depressant drugs, inadequate salt restriction, etc., may all thwart successful treatment of the patient with congestive heart failure. At all times, too, the physician should be alert to the role of cardiac surgery in the correction of major hemodynamic abnormalities that are responsible for heart failure.

REFERENCES

1. Monroe RG, Gamble WJ, LaFarge CG, et al: Left ventricular performance at high end-diastolic pressures in isolated perfused dog hearts. Circ Res 26:85, 1970
2. Ross J Jr, Sonnenblick EH, Taylor RR, et al: Diastolic geometry and sarcomere lengths in the chronically dilated canine left ventricle. Circ Res 28:49, 1971
3. Gaasch WH, Battle WE, Oboler AA, et al: Left ventricular stress and compliance in man: With special reference to normalized ventricular function curves. Circulation 45:746, 1972
4. Hill AV: The heat of shortening and the dynamic constant of muscle. Proc R Soc Lond [Biol] 126:136, 1938
5. Abbott BC, Mommaerts WFHM: A study of inotropic mechanisms in the papillary muscle preparation. J Gen Physiol 42:533, 1959
6. Sonnenblick EH: Implications of muscle mechanics in the heart. Fed Proc 21:975, 1962

7. Brady AJ: Time and displacement dependence of cardiac contractility: Problems in defining the active state and force–velocity relations. Fed Proc 24:975, 1965
8. Spann JF Jr, Buccino RA, Sonnenblick EH, Braunwald E: Contractile state of cardiac muscle obtained from cats with experimentally produced ventricular hypertrophy and heart failure. Circ res 21:341, 1967
9. Podolsky R: Mechanochemical basis of muscular contraction. Fed Proc 21:964, 1962
10. Bing OHL, Matsushita S, Fanburg BL, Levine HJ: Mechanical properties of rat cardiac muscle during experimental hypertrophy. Circ Res 28:234, 1971
11. Parmley WW, Spann JF Jr, Taylor RR, Sonnenblick EH: The series elasticity of cardiac muscle in hyperthyroidism, ventricular hypertrophy and heart failure. Proc Soc Exp Biol Med 127:606, 1968
12. Forwand SA, McIntyre KM, Lipana JG, Levine HJ: Active stiffness of the intact canine left ventricle. Circ Res 19:970, 1966
13. Levine HJ, Britman NA: Force–velocity relations in the intact dog heart. J Clin Invest 43:1383, 1964
14. Fry DL, Griggs DM Jr, Greenfield JC Jr: Myocardial mechanics: Tension–velocity–length relationship of heart muscle. Circ Res 14:73, 1964
15. Covell JW, Ross J Jr, Sonnenblick EH, Braunwald E: Comparison of the force–velocity relation and the ventricular function curve as measures of the contractile state of the intact heart. Circ Res 19:364, 1966
16. Hood WP Jr, Rackley CE, Rolett EL: Wall stress in the normal and hypertrophied human left ventricle. Am J Cardiol 22:550, 1968
17. Brutsaert DL, Sonnenblick EH: Cardiac muscle mechanics in the evaluation of myocardial contractility and pump function: Problems, concepts and directions. Prog Cardiovasc Dis 16:337, 1973
18. Mirsky I, Pasternac A, Ellison RC, Hugenholtz PG: Clinical application of force–velocity parameters and the concept of a normalized velocity, in Mirsky I, Ghista DN, Sandler H (eds): Cardiac Mechanics: Physiological, Clinical and Mathematical Considerations. New York, John Wiley & Sons, 1974
19. Peterson KL, Skloven D, Ludbrook P, et al: Comparison of isovolumic and ejection phase indices of myocardial performance in man. Circulation 49:1088, 1974
20. Noble MIM, Bowen TE, Hefner LL: Force–velocity relationship of cat cardiac muscle, studied by isotonic and quick release method technics. Circ Res 24:821, 1969
21. Pollack GH: Maximum velocity as an index of contractility in cardiac muscle: A critical evaluation. Circ Res 26:111, 1970
22. Levine HJ, McIntyre KM, Lipana JG, Bing OHL: Force–velocity relations in failing and non-failing hearts of subjects with aortic stenosis. Am J Med Sci 259:79, 1970
23. Hugenholtz PG, Ellison RC, Urschel CW, et al: Myocardial force–velocity relationships in clinical heart disease. Circulation 41:191, 1970
24. Mirsky I, Ellison RC, Hugenholtz PG: Assessment of myocardial

contractility in children and young adults, from ventricular pressure recordings. Am J Cardiol 27:359, 1971
25. Mason DT, Spann JF Jr, Zelis R: Quantitation of the contractile state of the intact human heart. Am J Cardiol 26:248, 1970
26. Falsetti HL, Mates RE, Greene DG, Bunnell IL: V_{max} as an index of contractile state in man. Circulation 43:467, 1971
27. Gault JH, Ross J Jr, Braunwald E: Contractile state of the left ventricle in man: Instantaneous tension–velocity–length relations in patients with and without disease of the left ventricular myocardium. Circ Res 22:451, 1968
28. Karliner JS, Gault JH, Eckberg D, et al: Mean velocity of fiber shortening: A simplified measure of left ventricular myocardial contractility. Circulation 44:323, 1971
29. Mirsky I: Review of various theories for the evaluation of left ventricular wall stresses, in Mirsky I, Ghista DN, Sandler H (eds): Cardiac Mechanics: Physiological, Clinical and Mathematical Considerations. New York, John Wiley & Sons, 1974
30. Levine HJ, Messer JV, Neill WA, Gorlin R: The effect of exercise on cardiac performance in human subjects with congestive heart failure. Am Heart J 66:731, 1963
31. Wilcken DEL: Load, work and velocity of muscle shortening of the left ventricle in normal and abnormal human hearts. J Clin Invest 44:1295, 1965
32. Noble MIM, Milne EN, Goerke RJ, et al: Left ventricular filling and diastolic pressure–volume relations in the conscious dog. Circ Res 24:269, 1969
33. Diamond G, Forrester JS, Hargis J, et al: Diastolic pressure–volume relationship in the canine left ventricle. Circ Res 29:267, 1971
34. Mirsky I, Parmley WW: Evaluation of passive elastic stiffness for the left ventricle and isolated heart muscle, in Mirsky I, Ghista DN, Sandler H (eds): Cardiac Mechanics: Physiological, Clinical and Mathematical Considerations. New York, John Wiley & Sons, 1974
35. Levine HJ: Compliance of the left ventricle. Circulation 46:423, 1972
36. Diamond G, Forrester JS: Effect of coronary artery disease and acute myocardial infarction in left ventricular compliance in man. Circulation 45:11, 1972
37. McCullagh WH, Covell JW, Ross J Jr: Left ventricular dilatation and diastolic compliance changes during chronic volume overloading. Circulation 45:943, 1972
38. Covell JW, Ross J Jr: Nature and significant of alterations in myocardial compliance. Am J Cardiol 32:449, 1973
39. Forrester JS, Diamond G, Parmley WW, Swan HJC: Early increase in left ventricular compliance after myocardial infarction. J Clin Invest 51:598, 1972
40. Hood WB Jr, Bianco JA, Kumar R, Whiting RB: Experimental myocardial infarction. IV. Reduction of left ventricular compliance in the healing phase. J Clin Invest 49:1316, 1970

41. Dodek A, Kassebaum DG, Bristow JD: Pulmonary edema in coronary artery disease without cardiomegaly. N Engl J Med 286:1347, 1972
42. Barry WH, Brooker JZ, Alderman EL, Harrison DC: Changes in diastolic stiffness and tone of the left ventricle during angina pectoris. Circulation 49:255, 1974
43. Gault JH, Covell JW, Braunwald E, Ross J Jr: Left ventricular performance following correction of free aortic regurgitation. Circulation 42:773, 1970
44. Olson RE: Myocardial metabolism in congestive heart failure. J Chronic Dis 9:442, 1959
45. Plaut GWE, Gertler MM: Oxidative phosphorylation studies in normal and experimentally produced congestive heart failure in guinea pigs: A comparison. Ann NY Acad Sci 72:515, 1959
46. Sobel BE, Spann JF Jr, Pool PE, et al: Normal oxidative phosphorylation in mitochondria from the failing heart. Circ Res 21:355, 1967
47. Schwartz A, Lee KS: Study of heart mitochondria and glycolytic metabolism in experimentally induced cardiac failure. Circ Res 10:321, 1962
48. Argus MF, Arcos JC, Sardesai VM, Overby JL: Oxidative rates and phosphorylation in sarcomeres from experimentally-induced failing rat heart. Proc Soc Exp Biol Med 117:380, 1964
49. Chidsey CA, Weinbach EC, Pool PE, Morrow AG: Biochemical studies of energy production in the failing human heart. J Clin Invest 45:40, 1966
50. Lindenmayer GE, Sordahl LA, Harigaya S, et al: Some biochemical studies on subcellular systems isolated from fresh recipient human cardiac tissue obtained during transplantation. Am J Cardiol 27:277, 1971
51. Gunning JF, Coleman HN: Myocardial oxygen consumption during experimental hypertrophy and congestive heart failure. J Mol Cell Cardiol 5:25, 1973
52. Cooper G, Satava RM, Harrison CE, Coleman HN: Mechanism for the abnormal energetics of pressure-induced hypertrophy of cat myocardium. Circ Res 33:213, 1973
53. Sordahl LA, McCollum WB, Wood WG, et al: Mitochondria and sarcoplasmic reticulum function in cardiac hypertrophy and failure. Am J Physiol 224:497, 1973
54. Cooper G, Puga FJ, Zujko KJ, et al: Normal myocardial function and energetics in volume overload hypertrophy in the cat. Circ Res 32:140, 1973
55. Sulakhe PV, Dhalla NS: Excitation-contraction coupling in the heart. VII. Calcium accumulation in subcellular particles in congestive heart failure. J Clin Invest 50:1019, 1971
56. Gertz EW, Hess ML, Lain RF, Briggs FN: Activity of the vesicular calcium pump in the spontaneously failing heart lung preparation. Circ Res 20:477, 1967
57. Suko J, Vogel JHK, Chidsey CA: Intracellular calcium and myocardial contractility. III. Reduced calcium uptake and ATPase of the sarcoplasmic reticular fraction prepared from chronically failing calf hearts. Circ Res 27:235, 1970
58. Schwartz A, Sordahl LA, Entman ML, et al: Abnormal biochemistry in myocardial failure. Am J Cardiol 32:407, 1973

59. Bárány M: ATPase activity of myosin correlated with speed of muscle shortening. J Gen Physiol 50:197, 1967
60. Alpert NR, Gordon MS: Myofibrillar adenosine triphosphatase activity in congestive heart failure. Am J Physiol 202:940, 1962
61. Chandler BM, Sonnenblick EH, Spann JF Jr, Pool PE: Association of depressed myofibrillar adenosine triphosphatase and reduced contractility in experimental heart failure. Circ Res 21:717, 1967
62. Berson G, Swynghedauw B: Cardiac myofibrillar ATPase and electrophoretic pattern in experimental heart failure produced by a two-step mechanical overloading in the rat. Cardiovasc Res 7:464, 1973
63. Gordon MS, Brown AL: Myofibrillar adenosine triphosphatase activity of human heart tissue in congestive failure: Effects of ouabain and calcium. Circ Res 18:534, 1966
64. Nebel ML, Bing RJ: Contractile proteins of normal and failing human heart. Arch Intern Med 111:190, 1963
65. Katz AM: Contractile proteins in normal and failing myocardium. Hospital Practice 7:67, 1972
66. Katz AM, Hecht HE: The early pump failure of the ischemic heart. Am J Med 47:497, 1969
67. Chidsey CA, Harrison DC, Braunwald E: Augmentation of plasma norepinephrine response to exercise in patients with congestive heart failure. N Engl J Med 267:650, 1962
68. Chidsey CA, Braunwald E, Morrow AG: Catecholamine excretion and cardiac stores of norepinephrine in congestive heart failure. Am J Med 39:442, 1965
69. Spann JF Jr, Chidsey CA, Pool PE, Braunwald E: Mechanism of norepinephrine depletion in experimental heart failure produced by aortic constriction in the guinea pig. Circ Res 17:312, 1965
70. Chidsey CA, Kaiser GA, Sonnenblick EH, et al: Cardiac norepinephrine stores in experimental heart failure in the dog. J Clin Invest 43:2386, 1964
71. Monroe RG, LaFarge CG, Gamble WJ, et al: Norepinephrine release and left ventricular pressure in the isolated heart. Circ Res 19:774, 1966
72. Pool PE, Covell JW, Levitt M, et al: Reduction of cardiac tyrosine hydroxylase activity in experimental congestive heart failure. Circ Res 20:349, 1967
73. Krakoff LR, Buccino RA, Spann JF Jr, DeChamplain J: Cardiac catechol o-methyltransferase and monoamine oxidase activity in congestive heart failure. Am J Physiol 215:549, 1968
74. Spann JR Jr, Sonnenblick EH, Cooper T, et al: Cardiac norepinephrine stores and the contractile state of heart muscle. Circ Res 19:317, 1966
75. Kramer RS, Mason DT, Braunwald E: Augmented sympathetic neurotransmitter activity in the peripheral vascular bed of patients with congestive heart failure and cardiac norepinephrine depletion. Circulation 38:629, 1968
76. Zelis R, Mason DT: Compensatory mechanisms in congestive heart failure: The role of the peripheral resistance vessels. N Engl J Med 282:962, 1970
77. Eckberg DL, Drabinsky M, Braunwald E: Defective cardiac parasympathetic control in patients with heart disease. N Engl J Med 285:877, 1971

78. Levy MN: Sympathetic–parasympathetic interactions in the heart. Circ Res 24:437, 1971
79. Rutenberg HL, Spann JF Jr: Alterations in cardiac sympathetic neurotransmitter activity in congestive heart failure. Am J Cardiol 32:472, 1973
80. Fishman AP: Pulmonary edema: The water exchanging function of the lung. Circulation 46:390, 1972
81. Uhley HN, Leeds SE, Sampson JJ, Friedman M: Role of pulmonary lymphatics in chronic pulmonary edema. Circ Res 11:966, 1962
82. Sharpey-Shafer EP: Effects of Valsalva maneuver on the normal and failing circulation. Br Med J 1:693, 1955
83. Gorlin R, Knowles JH, Story CF: The Valsalva maneuver as a test of cardiac function: Pathologic physiology and clinical significance. Am J Med 22:197, 1957
84. Pittman JG, Cohen P: Cardiac Cachexia. New York, Grune & Stratton, 1965
85. Davidson JD, Waldman TA, Goodman DS, Gordon RS Jr: Protein-losing gastroenteropathy in congestive heart failure. Lancet 1:899, 1961
86. Burack WR, Pryce J, Goodwyn JF: Reversible nephrotic syndrome associated with congestive heart failure. Circulation 18:562, 1958
87. Pastor BH, Cahn M: Reversible nephrotic syndrome resulting from constrictive pericarditis. N Engl J Med 262:872, 1960
88. Boas EP, Shapiro S: Diastolic hypertension with increased basal metabolism. JAMA 84:1558, 1925
89. Smith JA, Levine SA: Aortic stenosis with elevated metabolic rate simulating hyperthyroidism. Arch Intern Med 80:265, 1947
90. Levine HJ, Wagman RJ: Energetics of the human heart. Am J Cardiol 9:372, 1962
91. Frumin AM, Mendel TH, Mintz SS, et al: Nucleated red blood cells in congestive heart failure. Circulation 20:367, 1950
92. Levine HJ: The treatment of congestive failure. Med Clin North Am 46:1261, 1962

Zalman S. Agus
Martin Goldberg

10

Renal Function in Congestive Heart Failure

One of the hallmarks of congestive heart failure is the inability of the kidneys to normally regulate the excretion of sodium, resulting in a positive sodium balance and ultimately the accumulation of edema fluid. This abnormal pattern of renal tubular sodium reabsorption may stem from a variety of factors, including changes in the filtered load of sodium, hyperaldosteronism, and accelerated tubular sodium reabsorption at various sites in the nephron. The primary stimulus or initiating event responsible for these changes is generally felt to be a decrease in the effective arterial plasma volume, so that there is no longer adequate organ perfusion. This concept of the effective circulating arterial volume will be used throughout this chapter and should not be confused with the actual plasma volume, which may be increasing at a time when patients are functionally hypovolemic.

It is important for several reasons to understand the pathophysiology of these changes in renal function. For example the success of a diuretic regimen is largely dependent upon its site of action within the tubule and its interaction with the intrarenal sites of accelerated tubular sodium reabsorption. In addition, certain clinically significant fluid and electrolyte disturbances, such as hypokalemia and hyponatremia, are closely related to abnormal patterns of tubular sodium reabsorption. The purpose of this chapter is first to review normal sodium and water metabolism, and then discuss the pathophysiology of renal function in congestive heart failure. We will then outline the use of modern diuretics and consider the pathophysiology and treatment of fluid and electrolyte disturbances often associated with the management of congestive heart failure.

NORMAL PHYSIOLOGY OF BODY FLUID REGULATION

Body Fluid Compartments

Approximately 60% of the adult body is composed of water, the exact figure depending upon the fat content and therefore varying with age, sex, and body habitus. The total-body water of an average 70-kg man is thus approximately 42 liters. This fluid, a solution of organic and inorganic solutes, is divided into two main compartments: intracellular fluid (ICF), comprising two-thirds of the total-body water, and extracellular fluid (ECF), which comprises the other one-third. The extracellular fluid in turn is composed of plasma (one-fourth of ECF and therefore one-twelfth of total-body water and 5% of body weight) and interstitial fluid (three-fourths of ECF and 15% of body weight).

Regulation of ECF Volume

Sodium is present mainly in the extracellular fluid, where it and its accompanying anions (principally chloride and bicarbonate) constitute the major osmotic forces in this fluid space. As solute concentration of body fluids (osmolality) is fixed within narrow limits by water metabolism, as will be discussed below, net gain or loss of sodium from the body will normally obligate a similar gain or loss of water from the ECF compartment. Thus the regulation of external sodium balance is the single most important factor determining the volume of this fluid compartment. There is normally no significant extrarenal sodium loss from the body, so that sodium balance is accomplished by matching sodium excretion to the dietary sodium intake. This is illustrated in Fig. 10-1: A subject ingesting a diet containing 100 mEq of sodium is abruptly placed on a low-sodium (10 mEq) diet. Sodium excretion falls slowly, reaching the intake level after 3 or 4 days. During this period of negative sodium balance, a weight loss of 2–3 kg is incurred, representing 2–3 liters of ECF. When the dietary sodium intake is changed abruptly to a high level, the opposite sequence of events occurs. There is a transient period of positive sodium balance resulting in progressive expansion of ECF until urine sodium reaches the level of sodium intake and a new steady state is achieved. Two important concepts are illustrated by this example. First, changes in urinary sodium excretion prevent marked shifts in ECF volume, but do not prevent small oscillations in response to changes in

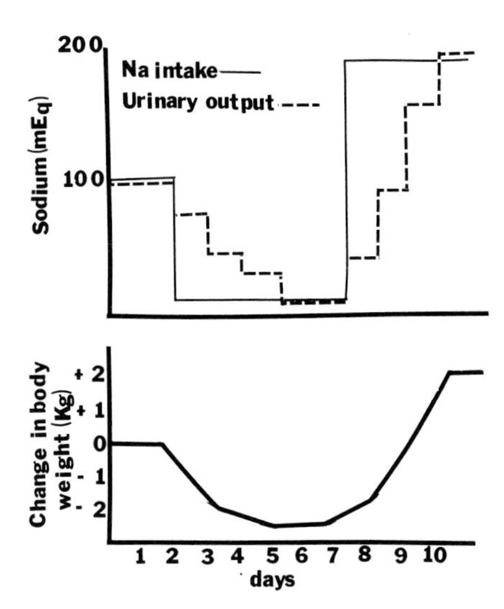

Fig. 10-1. Normal response to variations in oral sodium intake.

sodium intake. Second, this example demonstrates why there is no such thing as a normal fixed value for sodium excretion. Values for urinary sodium must always be interpreted in relation to sodium intake. The factors that mediate the renal response to changes in ECF volume include glomerular filtration rate, aldosterone, and a variety of hemodynamic, neural, and hormonal factors that act in concert to alter proximal and distal tubular sodium reabsorption. Before reviewing these factors individually, it is useful to review the overall pattern of renal sodium handling and the transport characteristics of the various segments of the nephron.

Renal Sites and Mechanisms of Sodium Transport

Renal Circulation and Glomerular Filtration

Under normal conditions one-fifth of the cardiac output (approximately 1 liter/min) reaches the renal arteries, representing the renal blood flow. At the hilar region the renal artery divides into two major sets of branches, each supplying one-half of the kidney. These primary branches give off interlobar arteries that subdivide at the junction of the cortex and outer medulla to form arcuate arteries from which arise the interlobular arteries that pass toward the capsular surface of the kidney. Throughout the renal cortex afferent arterioles arise from the interlobular arteries that pass toward the capsular surface of the kidney. The course of the postglomerular blood depends upon the type of nephron that is supplied. There are two distinguishable types of nephrons. The juxtamedullary nephrons, comprising 20% of the nephrons, are located at the corticomedullary junction and are characterized by a long loop of Henle; the other type of nephron, possessing a short loop of Henle, is found throughout all but the inner cortex. The efferent glomerular arteriole of the cortical nephron leads to the peritubular capillary system surrounding the convoluted tubules and most of the loop of Henle. In contrast, the efferent arterioles of the juxtamedullary nephron may enter the medulla as large capillaries arranged in bundles (vasa rectae) coursing toward the medullary papilla where they loop and ascend back toward the cortex to join the renal venous system.

The perfusion pressure in the renal arteries is equivalent to the mean aortic blood pressure (approximately 95 mm Hg), while the peritubular capillary pressure is on the order of 15 mm Hg. The major pressure drop is between the glomerular and the peritubular capillaries (i.e., the efferent arteriole). The second largest resistance site is at the level of the afferent arteriole. Thus intraglomerular pressure (and consequently glomerular filtration rate) may be increased by increased efferent arteriolar tone, decreased afferent arteriolar tone, or a combination of these changes. Approximately 10% of the renal blood flow or 20% of the renal plasma flow (normally 100 ml/min) is ultrafiltered at the glomerulus into the proximal tubule, as shown in Fig. 10-2. The primary force at the

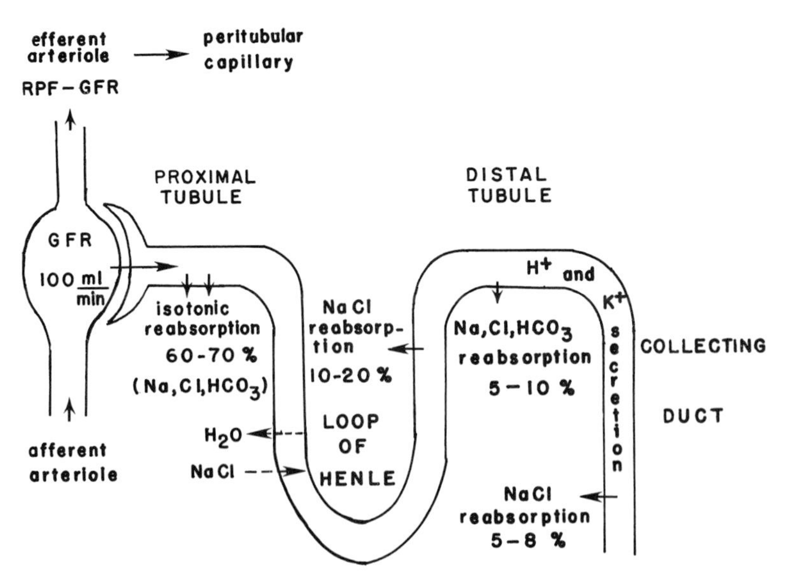

Fig. 10-2. Functional anatomy of a single nephron unit.

glomerulus responsible for filtration is the glomerular capillary hydrostatic pressure, opposed by the pressure within Bowman's space and the oncotic pressure of the plasma proteins, which are nonfilterable except for a very small fraction.[1-4]

Proximal Tubule

Approximately two-thirds of the glomerular filtrate is reabsorbed isotonically in the proximal tubule. Sodium reabsorption is an active process (i.e., it can occur against an electrochemical gradient). The majority of the sodium ions are reabsorbed with chloride, which is thought to follow sodium passively. The remainder of the sodium ions are reabsorbed with bicarbonate via a process involving hydrogen-ion secretion. Water passively follows electrolyte reabsorption in order to maintain isotonicity.

The pathways through which salt and water are reabsorbed across the proximal tubular epithelium have not been completely defined. One currently popular model is diagrammed in Fig. 10-3. Each proximal tubular cell has luminal (apical) and peritubular (basal) cell membranes. Adjacent cells are connected apically by tight junctions. The tight junctions and the intracellular space thus constitute an extracellular compartment within the tubular epithelium. The process of sodium reabsorption begins with passive diffusion of sodium ions into the proximal tubular cells down an electrochemical gradient. (The intratubular sodium concentration is approximately 140 mEq/liter, while inside the cell it is only 10–20 mEq/liter and the electrical potential difference is −60 mV.) Sodium is then actively pumped via an energy-requiring process into the intercellular spaces and the interspaces at the basal infoldings of the tubular cell. This accumulation of sodium salts creates an area of hyperosmolarity that abstracts water either from the tubule across the tight apical junction or via the cell compartment. As the fluid accumulates in the interspace, hydrostatic pressure builds up, and this pressure gradient forces fluid to the basal parts of the compartment and then into the interstitium. The final step in the reabsorptive process is the movement across the capillary wall from the interstitium. The uptake of interstitial fluid by the peritubular capillary is governed by Starling forces, i.e., mainly the net balance of hydrostatic and colloid osmotic (oncotic) pressure differences across the capillary wall.[5,6]

Under ordinary conditions the balance of these forces is such as to draw water into the capillary, dragging sodium and other dissolved solutes passively with it. The factors influencing uptake at the capillary are listed in Table 10-1. The hydrostatic pressure gradient is the difference between the intracapillary pressure and the interstitial

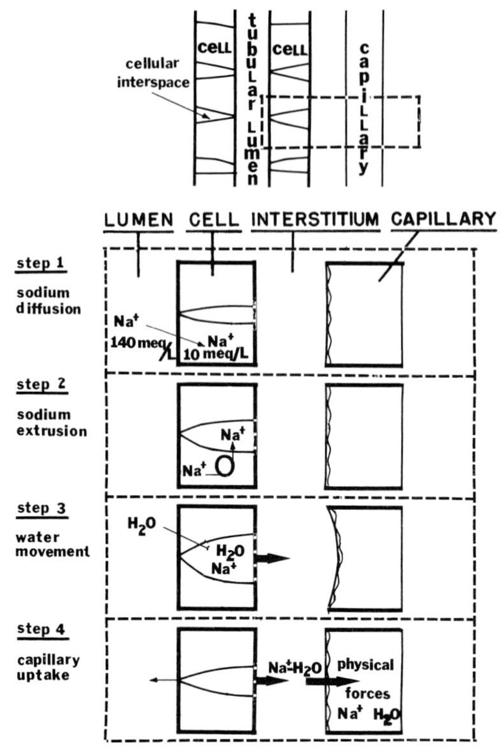

Fig. 10-3. Diagrammatic representation of mechanism of fluid and electrolyte reabsorption in the proximal tubule. Upper figure depicts anatomic components of the system. The tubular cell has a luminal and peritubular cell membrane. Adjacent cells are joined luminally by tight junctions. Located between tubule cells is an extracellular interspace. The area bounded by the dotted line is enlarged in the bottom of the figure. The reabsorptive process begins with diffusion of sodium down a chemical gradient into the cell (step 1) from which it is actively pumped into the extracellular interspace (step 2) creating an area of local hypertonicity. Water follows along the osmotic gradient expanding these interspaces. Fluid and sodium movement out of the interspace toward the capillary is accomplished by elevation of the hydrostatic pressure within these compartments (step 3). Finally, fluid and sodium are taken up at the capillary, the result of net balance of hydrostatic and oncotic pressure across the capillary wall (step 4). There is a small backleak of fluid and sodium across the tight junction into the tubule due to the hydrostatic pressure gradient.

Table 10-1
Peritubular Capillary Uptake*

Determinants of net capillary oncotic pressure
1. Plasma protein concentration
2. Filtration fraction
3. Interstitial fluid protein concentration
Determinants of net capillary hydrostatic pressure
1. Systemic arterial pressure
2. Renal vascular resistance
3. Interstitial hydrostatic pressure

*Capillary Uptake = f (oncotic pressure—hydrostatic pressure).

pressure. The capillary hydrostatic pressure is determined by the perfusion pressure and the renal vascular resistance. The oncotic pressure gradient is largely a function of the protein concentration in the postglomerular capillary circulation, as the interstitial fluid protein concentration is very small. Two factors determine the protein concentration within the capillary: the plasma protein concentration and the filtration fraction (GFR/RPF). Thus the greater the proportion of renal plasma flow filtered at the glomerulus (i.e., the higher the filtration fraction), the more concentrated will be the protein leaving the glomerulus in the efferent arteriole. The resultant elevated protein concentration in the peritubular capillary will increase the capillary uptake of water and dissolved solutes from the interstitium. This concept of the influence of filtration fraction upon proximal tubular sodium reabsorption is potentially a very important one and will be used in subsequent discussions of glomerulotubular balance.

In summary, two-thirds of the glomerular filtrate is reabsorbed in the proximal tubule by a process involving active sodium transport at the tubular cellular level and certain forces at the capillary level. Thus the amount of sodium reabsorbed in the proximal tubule can be influenced by factors affecting either the active energy-requiring step or the peritubular capillary uptake. Inhibition of the active cellular component of sodium transport would lead to an accumulation of sodium within the cell, which would markedly diminish diffusion from the proximal tubule, thereby decreasing net sodium transport. This may be the mechanism of action of certain agents such as parathyroid hormone and acetazolamide that have been shown to inhibit proximal tubular sodium transport without measurable hemodynamic changes. At the level of the peritubular capillary, sodium transport may be inhibited without altering the active cellular component. Thus increases in perfusion pressure or renal

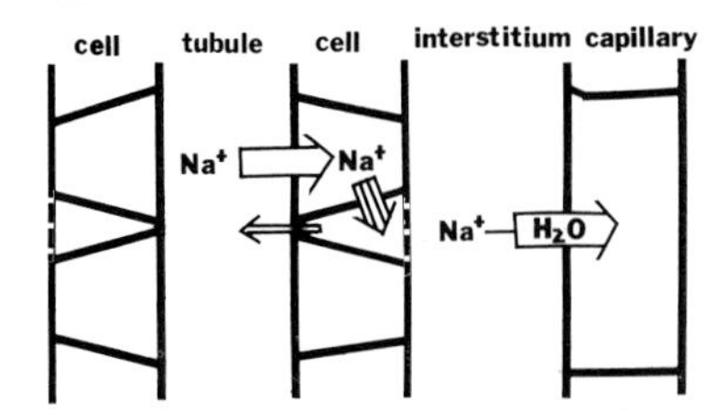

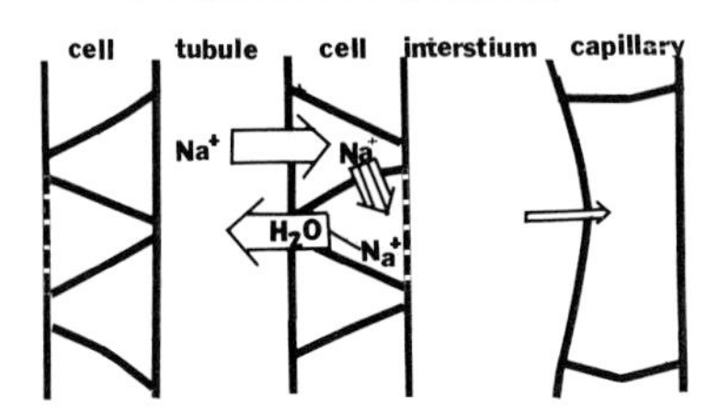

Fig. 10-4. Diagrammatic representation of enhanced backleak and inhibition of net fluid and sodium reabsorption in the proximal tubule with alteration of Starling forces at the peritubular capillary.

vasodilatation may increase capillary hydrostatic pressure and diminish uptake of sodium and water. Decreased capillary uptake can also be produced by a decrease in peritubular capillary protein concentration. The decreased peritubular capillary uptake is reflected in an increase in pressure in the intercellular spaces (Fig. 10-4) and consequently a backleak of sodium and water into the tubule, effectively decreasing net sodium and water reabsorption for the tubule.

Loop of Henle

Beyond the proximal tubule sodium and water reabsorption are not necessarily parallel, because of differing permeabilities of the various segments. It is this property that allows the process of concentration and dilution of urine to take place. As fluid enters the descending limb of the loop of Henle it equilibrates with the hypertonic interstitium by a process involving abstraction of water and possibly entry of solute. In the ascending limb of the loop, 20%–25% of the filtered load of sodium and chloride is reabsorbed. Whether or not there is active transport in the thin limb of the loop of Henle remains controversial, and alternative concepts have been proposed.[7] As this segment is virtually impermeable to water, a hypotonic fluid is generated, so that by the beginning of the distal tubule the urine contains a sodium concentration of 40–50 mEq/liter and an osmolality of 100–120 mOsm/kg. The sodium and chloride extracted from the urine are extruded into the interstitium surrounding the loop. Through the process of countercurrent multiplication, an area of very high solute concentration is produced in the medulla, such that the concentration gradient from cortex to medulla goes from 300 to 1200 mOsm/kg. Recent evidence has established that in contrast to other parts of the nephron there is a positive transtubular potential difference across this segment, and rather than sodium, chloride appears to be actively transported. This process is energy-dependent and is inhibited by certain diuretics (e.g., ethacrynic acid and furosemide.[8,9] The factors that control sodium reabsorption in this segment have not been defined. It is known that the ascending limb of the loop of Henle appears to have a remarkably high salt reabsorptive capacity and generally will tend to reabsorb the major portion of sodium and chloride ions delivered to it out of the proximal tubule. Thus, as will be discussed later, inhibition of sodium reabsorption in the proximal tubule alone will not increase urinary sodium excretion to a large extent, since the extra load will be absorbed in the loop. Also, chloride is virtually the only anion reabsorbed in the loop of Henle, in contrast to the proximal and distal tubules, where bicarbonate is reabsorbed as well.

Distal Tubule

Sodium reabsorption in the distal convoluted tubule amounts to 10%–15% of the filtered load and proceeds against large sodium concentration gradients. Thus a ratio of tubular fluid to plasma concentration of 0.3 for sodium can be maintained despite a large electrical gradient with a lumen-to-peritubular-fluid transepithelial potential difference of −35 to −60 mV. This indicates that compared to proximal tubules the distal tubule either has a stronger pump or the distal tubular epithelium is characterized by different permeability characteristics. A major factor controlling sodium reabsorption in the distal tubule is aldosterone. The permeability of the terminal portion of the distal tubule to water is dependent upon antidiuretic hormone (ADH). In the absence of ADH, fluid becomes more dilute in its passage through the distal tubule, reaching a minimum of 40–50 mOsm/liter at the end of the distal tubule.

Collecting Duct

In the collecting duct another 3%–4% of the filtered load of sodium is reabsorbed, and the sodium concentration can be reduced to extremely low values. As in the distal tubule the water permeability of the collecting duct and the osmolal concentration of the final urine are dependent upon ADH. Factors controlling sodium reabsorption in this segment have yet to be definitively established, but some recent evidence suggests the possibility of an important role for this segment in controlling the final sodium excretion.[10]

Regulation of Sodium Excretion

The sodium excreted in the urine is the difference between that filtered at the glomerulus and that reabsorbed throughout the nephron. Factors that conceivably could alter sodium excretion then would include those that alter the filtered load (glomerular filtration rate) and the factors that alter tubular reabsorption (hemodynamic, physical, and hormonal). In this section we will discuss these various factors individually and then review the renal response to changes in ECF volume.

Glomerular Filtration Rate (GFR)

The first factor to be considered as a potential potent regulator of sodium excretion is GFR. For example (Table 10-2), if the normal GFR is 100 ml/min (or 14.4 liters/day) and normal sodium concentration is 140 mEq/liter, then 20,160 mEq of sodium would be filtered per day. If 160 mEq are excreted (assuming dietary intake is 160 mEq/day), then 20,000 mEq of sodium are reabsorbed by the tubules per day. A spontaneous increase in GFR of 10% unaccompanied by a concomitant change in tubular reabsorption would produce an increase in sodium excretion of 2,016 mEq/day. This amount is equivalent to the total amount of sodium in the ECF space. Although spontaneous changes in GFR of this magnitude are common, this is not associated with such massive natriuresis because of a phenomenon known as glomerulotubular balance. This means a nearly constant fraction of the filtered load of sodium is reabsorbed by the tubules, so that raising GFR by 10% will only increase excretion by 10%, i.e., 16 mEq/day. It is now clear that changes in GFR per se cannot be a very significant factor in the regulation of sodium excretion. While many theories have been advanced to explain the

Table 10-2
Glomerulotubular Balance

		10% Increase in GFR	
	Control	Constant Net Reabsorption	Constant Fractional Reabsorption
Glomerular filtration rate	100 ml/min	110	110
Plasma sodium concentration	140 mEq/liter	140	140
Filtered load of sodium	20,160 mEq/day	22,176	22,176
Tubular sodium reabsorption	20,000 mEq/day	20,000	22,000
% filtered load reabsorbed	99.2	91	99.2
Urinary sodium excretion	160 mEq/day	2,176	176
% filtered load excreted	0.8	9.0	0.8

phenomenon of glomerulotubular balance, recent micropuncture studies have clearly shown that this is a property of the proximal tubule and its peritubular environment. A recent hypothesis suggests that this balance is achieved by changes in filtration fraction and the resultant peritubular protein concentration.[11,12] Thus at a constant renal plasma flow of approximately 500 ml/min, a change in GFR from 100 to 110 ml/min would change the filtration fraction from 0.20 to 0.22 and increase proportionately the protein concentration in the peritubular capillary. As discussed earlier in the chapter, this increased capillary oncotic pressure would lead to an increase in sodium reabsorption in the proximal tubule. The net effect is to maintain fairly constant fractional proximal tubular sodium reabsorption with moderate fluctuations in GFR as long as they are associated with reciprocal changes in the filtration fraction. Thus glomerulotubular balance prevents major changes in sodium excretion with small fluctuations in GFR. With large decreases in GFR, however (i.e., 30%–40% or greater), it should be noted that sodium excretion will be appreciably decreased (30%–40%) despite glomerulotubular balance. On the other hand, comparably large increases in GFR will be blunted by the ability of the distal nephron to reabsorb increased loads.

Hemodynamic and Physical Forces

It is now clear that a major factor controlling proximal tubular sodium reabsorption is the peritubular environment. It should be apparent that a large variety of physiologic changes may alter sodium reabsorption through this mechanism (Table 10-1). For example, proximal tubular sodium reabsorption will be enhanced by an elevated peritubular capillary protein concentration produced by either increased plasma protein concentration or increased filtration fraction. Additionally, decreasing the peritubular capillary hydrostatic pressure either by a decrease in systemic arterial pressure or an increase in renal vascular resistance proximal to the capillary will enhance proximal tubular sodium reabsorption. Conversely, reversing these forces (vasodilatation and increased perfusion pressure) will lead to a decrease in proximal tubular sodium reabsorption.[5,6,11–16]

Hormones. Another factor of prime importance in the regulation of sodium reabsorption is aldosterone, a hormone secreted by the adrenal cortex that stimulates sodium reabsorption in conjunction with secretion of hydrogen and potassium ions in the distal tubule. It is currently thought that the major stimulus to aldosterone secretion is a decrease in extracellular fluid volume via the stimulation of renin secretion (Fig. 10-5). Renin is an enzyme synthesized in specialized cells located between the afferent arterioles and the early distal tubule (the

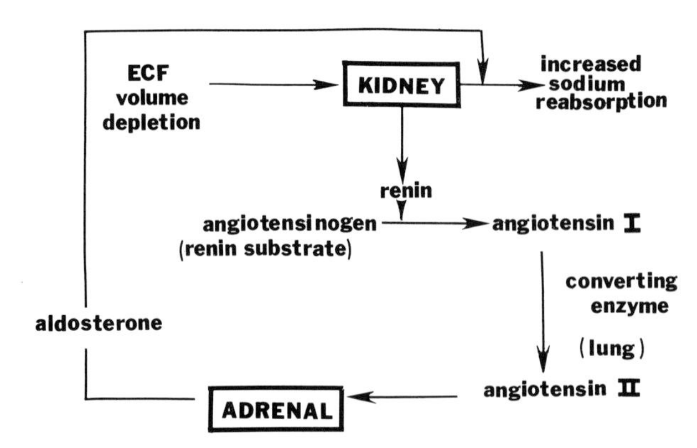

Fig. 10-5. Schematic summary of activation of the renin–angiotensin–aldosterone system by extracellular fluid volume contraction.

juxtaglomerular cells). Renin released into the circulation in response to extracellular fluid volume depletion acts upon circulating renin substrate (angiotensinogen), an α_2-globulin, to convert it to angiotensin I, a decapeptide. In the pulmonary circulation angiotensin I is converted to angiotensin II by converting enzyme, and it is this octapeptide that stimulates aldosterone secretion by the zona glomerulosa of the adrenal gland. Aldosterone then acts upon the distal tubule to stimulate sodium reabsorption and thereby expand the ECF volume. The mechanism by which volume depletion stimulates renin secretion remains unclear. Two current theories include pressure receptors in the afferent arteriole and/or a mechanism sensing sodium delivery to the macula densa.

It should be noted that there are two other factors affecting aldosterone production and secretion, ACTH and serum potassium. ACTH regulation is not a major factor in acute changes in aldosterone secretion, but rather appears to determine the baseline level of adrenal steroidal biosynthesis and secretion thereby affecting the magnitude of changes in aldosterone in response to other stimuli. Changes in serum potassium can alter aldosterone secretion, and in nephrectomized individuals there is a linear relationship between serum potassium and aldosterone levels.[17]

The importance of aldosterone in the regulation of renal sodium excretion in response to changes in extracellular fluid volume is undisputed. It is clear, however, that it cannot be the only factor and indeed may not be the most important. The reasons for this are: (1) With dietary sodium restriction, urinary sodium excretion begins to fall prior to measurable increases in aldosterone secretory rate. (2) Acute saline loading in experimental animals is associated with a decrease in proximal tubule sodium reabsorption, which is uninfluenced by administration of mineralocorticoids.[18] (3) Administration of spironolactone to many patients with edema and sodium retention often does not produce a natriuresis. (4) Normal man "escapes" from the sodium-retaining effects of aldosterone after several days. It has been recognized for many years that patients with primary aldosteronism do not develop edema. This escape phenomenon can be reproduced experimentally by giving daily mineralocorticoids to normal subjects. For the first several days subjects respond with salt retention and go into positive sodium balance with a weight gain of 2–5 lb. Between 4 and 6 days, however, sodium excretion rises to match intake, and a new steady state of ECF volume is reached. Even in the face of elevated mineralocorticoid levels this escape phenomenon is triggered by volume expansion, but how it is mediated in the kidney is unclear. As changes in GFR (because of glomerulotubular balance discussed above) and aldosterone are not responsible for this escape, other controlling factors must be invoked. Initially attempts were made to demonstrate the existence of a new hormone that inhibited sodium reabsorption in the proximal tubule, but these proved unsuccessful.[19,20] Subsequently much evidence has supported the concept that peritubular forces can adequately explain changes in proximal tubular transport associated with changes in extracellular fluid volume. More recently, however, interest in a natriuretic hormone has been reawakened. Two groups of investigators have reported the presence of a natriuretic factor in the blood of Doca-"escaped" dogs and patients with primary hyperaldosteronism or on a high salt intake.[21,22] Both of these groups have proposed that this natriuretic hormone has its major effects in the distal nephron, but estimates of molecular weight vary from 1000–3000[21] to 70,000.[22] A third group has found an inhibitor of sodium transport in the serum and urine of uremic patients that is absent when uremia is associated with a sodium-retaining state.[23,24] The source of this potentially important hormone is unknown, but recently a natriuretic factor has been isolated from the posterior lobe of the pituitary gland.[25]

Redistribution. It has been recognized for a long time that the kidney is composed of two major types of nephrons (the superficial outer cortical nephrons that possess short loops of Henle and the juxtamedullary nephrons that possess long loops of Henle), and there may be functional differences in these two groups of nephrons.[26] Studies using the ^{85}Kr washout technique have shown that some sodium-retaining states, including experimental heart failure, are characterized by reduction of outer cortical blood flow, while inner cortical and outer medullary flow is increased.[27] These alterations persist for several weeks, during which period plasma volume and total-body sodium are known to increase, and it has been proposed that in congestive heart failure the decreased sodium excretion is related to the redistribution of blood flow within the kidney. A superficial redistribution of blood flow has been found in human subjects on high-salt diets,[28] and micropuncture studies in rats have demonstrated a disproportionately greater superficial filtration rate in rats on a high salt intake as compared to rats on a low salt intake.[29,30] Two possible mechanisms have been postulated to explain this relationship. First, the redistribution may have resulted in a greater proportion of filtrate entering the juxtamedullary nephrons, which by virtue of their longer loops of Henle possess a more avid reabsorptive capacity for sodium. Second, it has been suggested that enhanced medullary blood flow may accelerate removal of sodium from the medullary interstitium, resulting in decreased interstitial sodium concentration. This decrease in concentration would then promote net sodium concentration in the loop of Henle by reducing backdiffusion from interstitium to tubular lumen. However, others have suggested that an increase in medullary blood flow decreases sodium reabsorption in the ascending limb of the loop of Henle.[31,32] More recent studies in the rat have shown that there is no change in either total renal plasma flow or intracortical distribution of plasma flow with wide variations in dietary sodium intake.[33] A similar absence of redistribution was found in a different model, that of chronic intravenous salt loading in the rat.[34] Thus while intrarenal redistribution of blood flow may occur in association with some alterations of sodium balance, it seems unlikely that this mechanism alone accounts for major alterations in sodium excretion.

Nervous System

The renal blood flow under resting conditions is near maximal, and it seems unlikely that the sympathetic nervous system plays a major role in controlling the renal circulation under ordinary conditions.[35] However, under adverse conditions, particularly severe hypovolemia, the renal vascular bed may be subject to vasoconstrictor influences, either through its sympathetic innervation or via circulating catecholamines.

The direct effects of the sympathetic nervous system upon renal tubular function remain undefined, but there is a body of data suggesting an effect of the adrenergic nervous system upon urinary sodium excretion. Thus sympathetic blockade in normal man diminishes the ability to conserve sodium in response to sodium deprivation.[36] Additionally, clearance studies in hypophysectomized dogs during water diuresis were interpreted as showing enhancement of proximal tubular sodium reabsorption by alpha adrenergic stimulation and inhibition by beta adrenergic stimulation.[37] Two mechanisms were suggested as mediators of these effects—alterations in the renal cortical adenyl cyclase systems (cyclic AMP is an inhibitor of proximal transport,[38] and urinary CAMP excretion is increased by adrenergic stimulation) and redistribution of renal blood flow. Subsequent investigators, however, using a similar protocol in hydropenic animals, were unable to detect any alteration in proximal tubular sodium reabsorption with either alpha or beta adrenergic stimulatiuons.[39] Thus the precise role of the sympathetic nervous system remains unclear, but it seems likely that it may play an important role in influencing renal tubular transport through alterations in the renal vasculature.

Intrarenal Site of Alterations in Sodium Transport with Changes in ECF Volume

The first micropuncture observation in this area was the demonstration that proximal tubular sodium reabsorption was inhibited with acute saline administration in the dog.[18] Subsequent studies have shown that the most likely mechanisms for these changes are alterations in hemodynamic and physical forces at the peritubular capillary.[12–16] Acute volume contraction is associated with enhanced proximal reabsorption, which can also be related to alterations in Starling forces at the capillary level.[40,41] The role of the more distal components of the nephron in response to acute changes in ECF volume remains somewhat controversial. In acute volume contraction, delivery of sodium out of the proximal tubule is diminished and can account for most of the decrement in sodium excretion. With acute volume expansion, however, the responses of the distal nephron to the increased load from the proximal tubule may be very important. At present the data are conflicting,[42] but it seems likely that an adjustment in distal sodium reabsorption contributes to the increased sodium excretion associated with increases in ECF volume.

While the intensive study of the renal tubular response to acute volume expansion has led to tremendous increases in our knowledge of renal physiology, it must be recognized that normal sodium homeostasis is not necessarily an analogous situation. Thus in day-to-day changes in sodium intake, the renal response tends to be sluggish, and chronic changes in extracellular fluid volume (or, more appropriately, effective circulating arterial volume) such as occur in oral salt loading or deprivation, congestive heart failure, nephrotic syndrome, or primary aldosteronism may produce changes in sodium excretion through mechanisms unrelated to those involved in the response to acute volume changes.

Studies of chronic changes of ECF volume have to date produced conflicting results.[44,47] Recent micropuncture studies suggest that the final controlling event in sodium excretion (and perhaps the most critical one) may occur in the collecting duct.[10,48]

Studies of this problem in man are necessarily limited. Clearance studies in normal subjects and patients with hyperaldosteronism using the excretion of phosphate as an index of changes in proximal sodium reabsorption also suggest, however, that the distal nephron is the primary site of adjustment to variation in dietary sodium intake.[49]

In summary, there are many potential factors involved in the regulation of sodium excretion in response to alterations in extracellular fluid volume. Glomerular filtration rate is probably not critical because of the phenomenon of glomerulotubular balance. The role of the adrenergic nervous system and redistribution of renal blood flow and/or glomerular filtration rate has not been definitely established, and there is experimental evidence both for and against these concepts. At present the factors that appear to be most influential include hemodynamic and physical forces, particularly in the proximal tubule, and hormonal influences (aldosterone and natriuretic hormone) that are probably operative in the distal nephron (loop of Henle, distal tubule, and collecting duct). In addition to understanding the mechanisms involved, it is important to be able to define the intrarenal site of adjustments in sodium excretion. It seems clear that acute or transient changes in extracellular fluid volume produce alterations in proximal reabsorption via changes in GFR and Starling forces operating at the peritubular capillary. The distal nephron acting upon the changes in sodium delivery to it from the proximal nephron modulates final urinary excretion. The influence of the distal nephron then will be less significant in states of acute volume contraction where delivery to it is low than in conditions of acute or transient ECF expansion where there is increased delivery. In contrast, sustained changes in ECF volume, particularly when a new steady state is reached, appear to be associated with minor changes in proximal tubular reabsorption. In these states the distal nephron (possibly via hormonal influences) appears to play the critical role in regulating urinary sodium excretion in response to alterations in ECF volume.

Regulation of Body Fluid Concentration

As discussed previously the total-body water is divided into two compartments: intracellular fluid, which contains two-thirds of TBW, and extracellular fluid, containing the remaining one-third. The solute composition differs in these two compartments, with sodium and its anions predominating in ECF, and potassium, magnesium, and phosphate in the ICF. Despite these differences, however, because water is able to move freely across cell membranes, the total solute concentration is the same throughout the total-body water. Normal osmolality is 280 to 300 mOsm/kg water and is finely regulated within a given individual. Just as sodium content is regulated by the balance between intake and urinary excretion, osmotic (and sodium) concentrations are determined by the balance between electrolyte-free water intake and output. Water is obligatorily lost by insensible means (skin and lungs), whereas electrolyte-free water losses can be regulated by renal excretion.

Urinary Concentration

Antidiuretic hormone is formed in hypothalamic supraoptic and paraventricular nuclei and then is transported to the posterior lobe of the pituitary gland. When a normal individual is deprived of water, the solute concentration (or osmolality) of the body fluids rises because of continued insensible loss. Osmoreceptors in the hypothalamus sense this change (probably via decrease in cellular volume), and this stimulates the release of preformed ADH. This is a very sensitive system, and changes in osmolality of 1%–3% are sufficient to markedly alter ADH release.[50] ADH acting via stimulation of adenyl cyclase and increased production of cyclic 3′,5′-AMP from ATP increases the permeability of the distal tubule and collecting duct epithelium to water, allowing concentration of the urine. In a normal individual, after 12–15 hr of dehydration urine osmolality should exceed 900 mOsm/kg water and in most individuals will reach 1000–1200. It should be pointed out that although change in osmolality is the primary stimulus to ADH release, a variety of other stimuli such as a fall in ECF volume have been found to influence ADH secretion. These are listed in Table 10-3. These nonosmotic stimuli are potent and usually can override a simultaneous osmotic stimulus. In addition, inhibition of ADH secretion can be produced by alcohol ingestion and procedures that expand ECF volume or increase distention of central venous reservoirs, specifically the left atrium.

Once ADH is released the final urinary concentration is dependent upon the circulating levels of ADH and a complex series of physiologic processes within the kidney. These are summarized in Fig. 10-6. Isotonic fluid leaving the proximal tubule enters the loop of Henle. The ascending limb of Henle's loop is impermeable to water but actively reabsorbs sodium and chloride, resulting in dilution of the fluid remaining in the tubule. The extracted sodium in extruded into the interstitium, and through the process of countercurrent multiplication an area of very high

Table 10-3
Nonosmotic Factors Affecting ADH Release

Stimulatory	Inhibitory
1. ECF volume contraction via left atrial and other baroreptors	1. ECF volume expansion via left atrial and other baroreptors
2. Pain	2. Alcohol
3. Anxiety	3. Paroxysmal tachycardia (probably via left atrial receptors)
4. Anesthesia	4. Alpha adrenergic drugs or stimuli
5. CNS depressant drugs (Demerol, morphine, barbiturates)	5. Other drugs; e.g. diphenyl hydantoin, lithium carbonate
6. Cholinergic drugs	
7. Beta adrenergic drugs or stimuli	
8. Other drugs; e.g. Clofibrate, Vincristine	

sodium concentration is produced in the medulla. This increase in interstitial fluid concentration is maintained by the countercurrent flow of blood in the vasa recta, so that an osmotic gradient is established in the interstitium from the cortex (isotonic to plasma) to the papillary tip where the osmolality may reach 1000–1200 mOsm/kg water. Tubular fluid reaches the distal tubule with an osmolality of 75-100 mOsm/kg

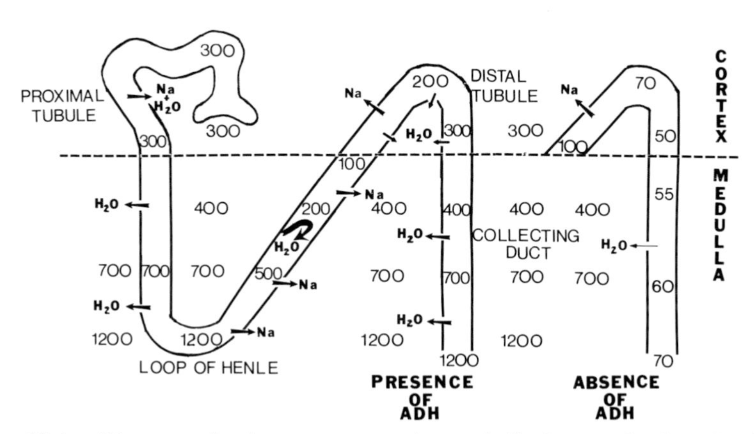

Fig. 10-6. Diagram of urinary concentrating and diluting mechanism. In the presence of ADH, distal tubule and collecting-duct epithelium are permeable to water, allowing concentration of final urine as tubular fluid passes through the hyperosmotic renal medulla. In the absence of ADH, dilute urine formed at the medullary diluting site (ascending limb of Henle's loop) and the cortical diluting site (distal tubule) passes through the collecting duct subject only to a small amount of non-ADH-dependent backdiffusion.

water, and in the presence of ADH water is reabsorbed, so that by the end of the distal tubule or early in the collecting duct the osmolality inside the tubule is equilibrated with the interstitium at 280–300 mOsm/kg water. The isotonic fluid then enters the collecting duct, and in its passage more water is extracted as the fluid equilibrates with the interstitium and the final urine osmolality of 1000–1200 mOsm/kg water is reached in man.

Urinary Dilution

With ingestion of an oral water load plasma osmolality is diluted and cellular volume increases. The osmoreceptors inhibit ADH release, and urine osmolality falls as urine volume rises. The water diuresis persists until the water load has been excreted. Normally 80%–85% of an administered water load is excreted in less than 5 hr. In the kidney sodium is reabsorbed without water in the ascending limbs of the loop of Henle (diluting site one), allowing hypotonic fluid to reach the early distal tubule (Fig. 10-6). In the absence of ADH additional dilution is accomplished in the distal tubule by reabsorption of sodium and chloride (diluting site two), so that by the end of the distal tubule the osmolality is 40–50 mOsm/kg water. In the collecting duct, even in the absence of ADH, the tremendous osmotic gradient between tubular fluid and interstitium is powerful enough to draw some water across.[51] This ADH-independent back diffusion of water is inversely related to the flow through the collecting duct. Under normal circumstances urine osmolality is increased by this mechanism from 40–50 to 50–70 mOsm/kg water. When tubular flow is slowed, however, as occurs in volume-contracted states because of accelerated reabsorption of fluid in more proximal sites, the collecting duct fluid osmolality can be increased to greater than 300 mOsm/kg via this backdiffusion mechanism alone. Thus the capacity to excrete a volume of dilute urine sufficient to prevent dilution of body fluids with a water load depends primarily on two factors. First, there must be suppression of ADH to allow urinary dilution. Second, there must be adequate delivery of sodium and water out of the proximal tubule to the diluting sites, so that there will be sufficient volume of dilute urine to excrete. (A urine output of 10 ml/hr will not prevent hypo-osmolality when water intake is greater than 10 ml/hr, no matter how dilute the urine.) An abnormality in either of these components could result in water retention and hyponatremia when water intake is increased beyond insensible losses. The retained water will distribute throughout the total-body water producing cellular edema which probably is responsible for the cerebral manifestations of hyponatremia. It should also be noted that hyponatremia reflects positive water balance and not sodium depletion.

The serum sodium level is a concentration term indicating how much water relative to sodium is present in body fluids and bears little relationship by itself to the state of body balance of sodium.

CONGESTIVE HEART FAILURE

Models of Experimental Congestive heart Failure

Experimentally induced low- or high-output failure has been studied in the dog and rat to evaluate the pathophysiology of chronic sodium retention and edema. The models most widely used are thoracic vena caval constriction (low output) and aorta-to-vena-cava (A-V) fistula (high output). The questions asked in these studies are twofold: Where in the nephron is sodium reabsorption enhanced, and what role do various factors such as filtration rate, filtration fraction (peritubular oncotic pressure), redistribution of glomerular filtrate, renal nervous system, and hormonal factors play in the production of the antinatriuresis characteristic of these models?

Site of Enhancement of Sodium Reabsorption

As discussed in the previous section, the distal nephron appears to play a major role in the response to chronic volume expansion, while acute or transient changes in extracellular fluid volume are associated with alterations in proximal tubular function as well. The same pattern of response appears to hold true for animals with experimentally induced congestive heart failure. Thus in the dog with acute contriction of the thoracic vena cava, impaired natriuresis was felt to be mediated by increased proximal tubular sodium reabsorption.[52] Similarly, enhanced proximal sodium reabsorption was found in a micropuncture study of rats the first several days after creation of an A-V fistula.[53] When these preparations were studied in the chronic state, however, proximal tubular sodium reabsorption both before and after saline loading was not different than that in normal animals in chronic A-V dogs, chronic caval dogs, or chronic A-V rats.[46,47,53,54] The chronic salt retention as well as the inability of these preparations to excrete a saline load appears therefore to be caused by an enhanced rate of reabsorption along the distal nephron (loop of Henle, distal tubule, and/or collecting duct).

Role of Various Factors in Antinatriuresis of Experimental Heart Failure

Reductions in glomerular filtration rate cause a diminution in urinary sodium excretion. However, because of the phenomenon of glomerulotubular balance (discussed above) minor changes in GFR cause only small changes in sodium excretion, and other factors are required to explain fully the antinatriuresis of congestive heart failure. It is important to bear in mind, however, that when markedly reduced, GFR can be a very important determinant of urinary sodium excretion.

The role of aldosterone in the antinatriuresis of experimental congestive heart failure as a determinant of sodium reabsorption in the distal nephron seems clearly established. However, the lack of the normal escape phenomenon (described above) and the minimal effects of aldosterone antagonists in patients with congestive failure suggest that other factors are also important in chronic salt retention.

Physical forces, including perfusion pressure, filtration fraction, and tubular capillary oncotic pressure, appear to exert a major role in the regulation of sodium reabsorption in the proximal tubule (see previous section). Accordingly, these factors are most likely involved in the earliest stages of sodium retention and edema formation in acute preparations when transient changes in systemic and renal hemodynamics are most pronounced. Thus the early increase in proximal sodium reabsorption seen in acute caval and A-V preparations would be analagous to the changes seen with acute reduction in renal hemodynamics produced by clamping the renal artery or acute extracellular fluid volume contraction. The role of physical factors in the distal nephron remains undefined, but an improvement in sodium excretion has been demonstrated in saline-loaded caval dogs with restoration of renal perfusion pressure.[47,55] Thus it is possible that capillary pressure may be an important determinant of distal nephron sodium reabsorption in a manner similar to that previously described for the proximal tubule (see above). If this is correct, then the difference between acute and chronic preparations would lie in the sensitivity of the proximal and distal nephrons to varying changes in systemic and renal hemodynamics.

Redistribution of renal blood flow and/or glomerular filtrate has been suggested as a possible mechanism for sodium retention in congestive heart failure, and by using the krypton washout technique evidence has been presented suggesting redistribution of renal blood flow from superficial to deeper nephrons.[56] Whether or not redistribution of blood flow can cause alterations in sodium excretion, however, has yet to be demonstrated. Recently (as described above) several groups of investigators utilizing micropuncture studies in caval and A-V fistula rats and dogs have been unable to find any evidence of redistribution of glomerular filtrate, and thus this mechanism does not appear to be necessary for the development of chronic sodium retention.

An important role for renal nerve stimulation in the antinatriuresis of congestive heart failure was suggested in 1959 when it was shown that the intrarenal infusion of Dibenzyline produced a natriuresis in dogs with experimentally induced failure.[57] It was subsequently found that autonomic ganglionic blockade would produce a natriuresis in cortisone-treated adrenalectomized dogs with chronic thoracic caval constriction.[36] Additionally, the markedly reduced renal response to intravenous saline loading in acute caval dogs was improved but not restored to normal by acute renal denervation.[58] Thus these studies suggest that the renal nerves constitute a potentially important efferent pathway for at least part of the antinatriuresis and blunted response to saline loading characteristic of this experiment model. In direct contrast, more recently it has been found that the antinatriuresis of acute thoracic caval constriction is unaltered by renal denervation.[59] Thus, as discussed above, the importance of the renal nervous system and the role it plays in the regulation of renal sodium excretion have yet to be clearly defined.

A role for a "natriuretic hormone" in congestive failure remains within the realm of speculation. Because of the multiplicity of factors involved in the experimental models of congestive heart failure it has not been possible to demonstrate its importance by exclusion. Most of the data suggesting the presence of the hormone have been obtained in volume-expanded models and are summarized above. If the hormone is important, analogy from these data would suggest that it plays a regulatory role in the distal nephron, which appears to be the site of major importance in the enhanced sodium reabsorption of congestive heart failure.

Physiology of Sodium Retention and Edema in Congestive Heart Failure

Based on the experimental observations summarized above, it is possible to construct a hypothetical series of physiologic events that may explain the various stages of fluid retention seen in clinical congestive failure. The initial event is a fall in cardiac output relative to the metabolic needs of the various tissues and organs. This leads to decreased effective filling of the systemic arterial tree. The decreased circulating volume is accentuated by transudation of fluid from the intravascular space to the interstitial compartment because of increased venous pressure. The arterial hypovolemia and resultant hypoperfusion of the kidney stimulate renal retention of sodium and water in an attempt to repair the functional volume deficit and increase cardiac output by increasing venous return. This compensatory mechanism may produce a new steady state whereby arterial volume is restored to near normal, but in the presence of an elevated venous pressure and an expanded interstitial fluid space this expansion is detected initially as occult weight gain and later as obvious edema, and finally it may progress to generalized anasarca. In the ensuing discussion of the pathogenesis of cardiac edema, we have arbitrarily divided the progression of events into three phases. These appear to be relatively distinct in terms of the profiles of sodium reabsorption in the nephron, but in reality there is probably much overlap between these various stages. Nevertheless, this type of conceptual analysis facilitates a rational approach to therapeutic management.

Mild Congestive Heart Failure

Following early mild decline in cardiac performance, there is a fall in renal blood flow. This decrease in renal blood flow is not precipitous and is within the autoregulatory range of the kidney, so that glomerular filtration rate remains stable. This decrease in renal blood flow out of proportion to changes in GFR may be mediated via increased renal adrenergic activity with renal vasoconstriction. The fall in blood flow in conjunction with an unchanged GFR results in an increase in filtration fraction. This increase in filtration fraction in early congestive failure has been noted clinically by a number of investigators and has been reproduced experimentally in dogs with surgically created valvular lesions of the heart.[60] As discussed previously, an increase in filtration fraction leads to a rise in postglomerular peritubular capillary protein concentration. This increased capillary oncotic pressure is then associated with an increase in peritubular uptake of water and sodium, which is ultimately reflected in an increase in net reabsorption of sodium and water in the proximal tubule (primarily by decreased backdiffusion as discussed earlier). (This enhancement of proximal sodium reabsorption by increased filtration fraction should not be confused with G-T balance, where the change in filtration fraction is proportional to the change in GFR.) At this stage sodium and water retention occur, but are probably mild and manifest as occult weight gain (up to 8 lb of fluid can be retained before edema is clinically apparent). It is possible that additional physiologic mechanisms such as aldosterone may also be present to influence distal sodium reabsorption, but systematic observations in this very early stage are lacking.

Moderate Congestive Heart Failure

As the heart progressively decompensates, other mechanisms are now brought into play in response to the more severe effective arterial hypovolemia. Thus renin hypersecretion leads to an increase in aldosterone secretion, which directly stimulates reabsorption of sodium and water in the distal tubule. The circulating aldosterone level may also be elevated by impaired degradation in a congested liver. However, the importance of aldosterone in the increased reabsorption of sodium varies from patient to patient. Thus aldosterone secretion is not invariably elevated in all patients with congestive failure, and in addition, blockade of its renal action with spironolactone often produces little change in urinary sodium excretion.

Experimental observations suggest that in this stage the proximal tubule now plays a relatively minor role in contributing to further sodium retention, possibly because the initial hemodynamic stimuli to the kidney are less intense because of compensatory mechanisms. The major renal mechanism in edema formation at this stage is enhanced distal reabsorption of sodium. The mechanisms responsible for increased sodium reabsorption in the nephron distal to the proximal tubule are unclear, but the possibility of a decrease in the release of a natriuretic substance besides aldosterone must be considered.

There is some evidence to suggest that in patients with heart failure there may be redistribution of blood flow within the kidney. This redistribution theoretically shifts blood preferentially to nephrons located deeper in the juxtamedullary portions of the kidney. These nephrons with longer loops of Henle may have a greater capacity for sodium transport, so that the net effect is increased sodium reabsorption in the loop of Henle. Although redistribution has been shown to occur in some forms of experimental congestive heart failure, its role in the pathogenesis of salt retention and edema in man remains to be established.

Thus in this stage of congestive heart failure the distal nephron (possibly via changes in aldosterone, natriuretic hormone, or as yet undefined hemodynamic alterations) is the major site of enhanced sodium reabsorption. It should be emphasized that we are describing a sustained steady-state condition. Any fluctuation in cardiac performance would be reflected in alterations in proximal tubular reabsorption superimposed on this steady-state condition.

Advanced Congestive Heart Failure

In very severe congestive heart failure the reduction in cardiac output is so severe that intense and sustained pathologic hemodynamic stimuli to the kidney result. This causes additional persistent and progressive enhancement of proximal tubular sodium reabsorption, so that the entire nephron is responsible for the elimination of sodium from the urine. Furthermore, GFR and filtered load of sodium are sharply reduced. Now all of these intensive efforts by the kidney are unsuccessful in restoring effective arterial blood volume, preventing the development of a new steady state. At this stage of the disease these pathophysiologic abnormalities result in several fluid and electrolyte abnormalities such as hyponatremia and refractoriness to diuretic therapy.

Treatment and the Use of Diuretics

A diuretic is any agent capable of producing an increase in urine flow, and thus in the broadest sense the term includes not only specific drugs that inhibit renal tubular reabsorption but also water, sodium salts, and other volume-expanding agents. It has become customary, however, to restrict the use of the term to drugs that act on the kidney to inhibit tubular reabsorption of sodium and water and thereby produce a net loss of sodium from the body fluids.

Under ordinary circumstances diuretics will result essentially in isotonic loss of sodium and water, either through decreased tubular reabsorption of water secondary to the diminished sodium reabsorption (as in the proximal tubule) or through a physiologic adjustment of water balance. There are significant exceptions to this generality, however, and hyponatremia is not unusual as a complication of diuretic therapy (it will be discussed further below). Although the mechanisms of action of most diuretics are unknown, recent studies have now localized the site of action within the nephron.[61] Since it is frequently necessary to use more than one diuretic, knowledge of the site of action allows one to make a rational choice of drugs that will have an additive effect. In the following discussion, therefore, individual diuretics will be classified according to their sites of action within the nephron, as diagrammed in Fig. 10-7.

Diuretics Acting in the Proximal Tubule

Acetazolamide is a potent inhibitor of carbonic anhydrase, the enzyme responsible for H^+ secretion and therefore bicarbonate reabsorption in the proximal tubule. Acetazolamide also inhibits proximal tubular NaCl reabsorption, probably as a consequence of the inhibition of sodium bicarbonate reabsorption. Although producing a marked inhibition of sodium reabsorption proximally, acetazolamide rarely produces a natriuresis of greater than 3%–4% of filtered sodium because of the ability of more distal sites of the nephron to reabsorb a large portion of the proximally rejected sodium. The increased delivery of sodium to the distal nephron combined with inhibition of H^+ secretion markedly enhances secretion of potassium. Thus the pattern of electrolyte excretion produced is one of increased sodium, potassium and bicarbonate producing a hypokalemic hyperchloremic metabolic acidosis. This side effect is usually self-limited, since acidosis prevents acetazolamide from producing its usual effects. Because acetazolamide

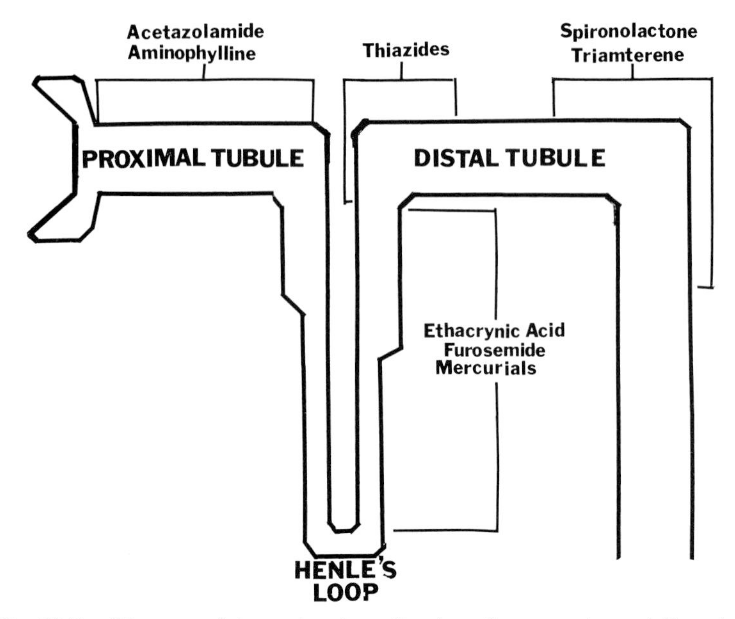

Fig. 10-7. Diagram of the major sites of action of commonly used diuretics.

directly inhibits proximal sodium reabsorption, it is most useful in congestive failure when excessive proximal reabsorption produces hyponatremia and refractoriness to more distally acting agents, as will be discussed below.

The most widely used of the xanthine diuretics, aminophylline, produces renal vasodilatation, secondarily increasing GFR; and it may also have a direct inhibitory effect upon sodium reabsorption in the proximal tubule. Thus, as with acetazolamide, it can be used both to potentiate the action of more distally acting diuretics and in the teatment of hyponatremia when associated with enhanced proximal tubular reabsorption.

Diuretics Acting in the Ascending Limb of Henle's Loop

Furosemide and ethacrynic acid are the most potent diuretics currently in use, producing a natriuresis of up to 25% of the filtered load of sodium. Both have a primary site of action upon active chloride reabsorption in the ascending limb producing a major sodium and chloride diuresis. Furosemide, a weak carbonic anhydrase inhibitor, also has a minor action in the proximal tubule and theoretically therefore may be more useful than ethacrynic acid in the management of a patient with congestive failure and hyponatremia.[62] As with most diuretics that increase sodium delivery to the distal K^+ secretory sites, marked kaliuresis is characteristic, and hypokalemia is commonly seen. Both drugs are effective orally, with a rapid onset of action of 20 to 30 min orally and 5 min when given intravenously. Activity peaks rapidly and is usually gone within 6 to 8 hr. One danger of the intravenous administration of ethacrynic acid that appears to be less common with furosemide is the sudden or delayed appearance of deafness, which may be irreversible.[63,64] The major complication with the use of these agents, however, is their remarkable potency, which may produce ECF volume depletion. Consequently, diuretic therapy in congestive heart failure should begin with the less potent thiazides.

Until the introduction of ethacrynic acid and furosemide, the organomercurials were the most potent diuretics available. In addition to oral ineffectiveness, they are ineffective when given during metabolic alkalosis. There is good evidence for a site of action within the loop of Henle, but in addition they must also have an action in the distal nephron at a K^+ secretory site, because they are unique among the potent diuretics in being relatively nonkaliuretic. The major toxicity of the mercurials is due to accumulation of mercuric ion, which may produce nephrotoxicity.[65]

Diuretics Acting in the Distal Tuble

All of the benzothiadiazine derivatives have a similar site of action, and except for minor differences in onset and duration of action they are interchangeable. They produce a sodium chloride diuresis of 5%–8% of filtered load and are well tolerated orally, so that they are the best initial diuretics to try in a patient with generalized edema. While the thiazides are sulfonamide derivatives and have varying degrees of carbonic anhydrase inhibitory activity, this effect is minimal at usual doses, and there is little evidence of action in the proximal tubule. Rather, their major site of action is in the distal nephron at the cortical diluting site. By increasing delivery of sodium to H^+ and K^+ secretory sites, thiazides are frequently associated with hypokalemic alkalosis. This can be controlled with either potassium chloride supplementation or the addition of a potassium-sparing diuretic. Other side effects include hyperuricemia (discussed below) and aggravation of hyperglycemia in the overt diabetic or prediabetic, an action probably mediated via hypokalemia.

Spironolactone is a competitive inhibitor of aldosterone at the distal tubular site where H^+ and K^+ are secreted in conjunction with reabsorption of sodium. Used alone, even in states of severe secondary hyperaldosteronism, it is usually a weak natriuretic agent (1%–2% of filtered sodium) because of the enhanced sodium reabsorption at more proximal sites in these patients. It is most useful when used in conjunction with a more potent natriuretic agent. A number of adverse reactions to spironolactone have been reported, including painful gynecomastia, hirsutism, skin rash, and gastrointestinal irritation. The most common problem, however, is related to its mechanism of action. Since it blocks potassium and hydrogen-ion excretion, hyperkalemia and metabolic acidosis may occur, particularly in patients with a decreased GFR and in diabetic patients. Lethal hyperkalemia has been reported when combined with potassium chloride therapy.

Triamterene and amiloride have the same effects of urinary electrolyte excretion as spironolactone, but they do not require the presence of aldosterone and thus are more predictable in their effects. Toxicity is unusual, but a rise in BUN unrelated to hypovolemia has been reported. The most serious problem is hyperkalemia, which may develop suddenly at any time during therapy. With all of these agents serum potassium should be determined daily until the desired diuretic dosage is reached and then weekly or semimonthly even if serum potassium appears to be stable.

Disturbances in Body Fluid Composition and Complications of Diuretic Therapy

Several electrolyte and acid-base abnormalities may arise in the course of a patient with congestive heart failure, as listed in Table 10-4. These abnormalities may present simply as a consequence of the pathophysiology of the disease state or may arise as complications of diuretic therapy. In either case their presence or potential development should be taken into account in the management of these patients.

Table 10-4
Disturbances in Body Fluid Composition and Complications of Diuretic Therapy

1. Hyponatremia
2. Hypokalemia
3. Metabolic alkalosis
4. Hyperkalemia
5. Metabolic acidosis
6. Hyperuricemia
7. Diuretic refractoriness

Hyponatremia

The concentration of sodium in the extracellular fluid is determined principally by the relationship between the intake and excretion of water, rather than by the relationship between the intake and output of salt. As hyponatremia is due to an excess of water for the amount of sodium present, it can coexist with either sodium depletion or sodium excess. Hyponatremia and hypo-osmolality are synonymous terms, with two exceptions. First, the addition of a nondiffusible osmotically active substance such as glucose or mannitol will increase the osmolality of the serum and ECF, but in doing so will draw water out of the cells, lowering the serum sodium concentration and producing both hyponatremia and hyperosmolality. In the case of hyperlipemia, there is a different situation. The large lipid molecules increase the solid content of a unit of plasma and thus diminish the water content in that unit just by occupying space. The concentration of sodium in 1 ml of plasma is decreased (because 1 ml of plasma is now really only 0.9 ml plasma and 0.1 ml lipid), whereas the concentration of sodium in 1 ml of plasma water is normal, as is the effective osmotic pressure. This is termed pseudohyponatremia. Thus in the absence of hyperglycemia and hyperlipemia, a low serum sodium means dilution of sodium by water and an inability to excrete ingested water sufficiently rapidly to maintain a normal serum osmolality.

As previously discussed in the section on water homeostasis, the ability to adequately excrete a water load is dependent upon suppression of ADH and appropriate reabsorption of sodium and water throughout the nephron. Dysfunction may occur at any point in the system, and there are thus four major pathophysiologic mechanisms of hyponatremia (Table 10-5):

1. ADH may be secreted in response to a nonosmotic stimulus, such as hypovolemia, which will prevent the excretion of dilute urine even in the face of hypo-osmolality (nonosmotic causes of elevated ADH levels other than hypovolemia are causes of the syndrome of inappropriate secretion of ADH).

2. A decrease in glomerular filtration rate and/or an increase in proximal tubule reabsorption in response to a decrease in effective circulating blood volume will increase the proportion of filtrate reabsorbed isotonically. This results in decreased delivery of fluid to the loop of Henle and the distal tubule where dilution takes place; the volume of solute-free water that can be excreted is thereby diminished.

3. In order to produce dilute urine, the diluting segments of the kidney must be functional. Thus administration of diuretics that inhibit sodium reabsorption in the loop of Henle (such as ethacrynic acid and

Table 10-5
Pathophysiologic Mechanism of Hyponatremia

1. Abnormal release of ADH
   - Nonosmotic stimuli
   - ADH-secreting tumors
   - Endocrinopathies (hypoadrenalism, hypopituitarism, myxedema)
   - Pulmonary tuberculosis
   - Cerebral disease
   - Acute intermittent porphyria
   - Thiazide diuretics and hypokalemia (?)
2. Decreased delivery of glomerular filtrate to diluting sites
   - Decreased glomerular filtration rate
   - ECF volume depletion:
     - Salt-losing nephritis
     - Adrenal insufficiency
     - Diuretic administration
     - Gastrointestinal losses
   - Decreased effective circulating plasma volume (edematous states)
     - Congestive heart failure
     - Nephrotic syndrome
     - Cirrhosis of the liver
3. Dysfunction of renal diluting sites (loop of Henle and distal tubule)
   - Diuretics
   - Hypercalcemia
   - Hypokalemia (?)
   - Tubular disease (interstitial nephritis)
4. Abnormal permeability of distal nephron to water
   - Glucocortoid deficiency (?)
5. Reset osmostat

furosemide) will decrease the ability of the kidney to dilute the urine and excrete solute-free water.

4. Certain diseases may alter the water permeability of the distal nephron.

As the serum sodium concentration does not reflect sodium balance, hyponatremia may be associated with either excessive, normal, or decreased total-body sodium. Hyponatremia with a normal body sodium content often reflects inappropriate ADH secretion, while in the other two circumstances there is a decrease in circulating arterial volume. Therapy is quite different in these situations, and the first step in evaluating any patient with hyponatremia is to estimate his total-body sodium (Table 10-6). Sodium balance determines the volume of ECF. The presence of generalized edema therefore indicates a significant in-

Table 10-6
Approach to Hyponatremia

1. Verify that hyponatremia represents hypo-osmolality in absence of
   a. hyperlipidemia
   b. hyperglycemia
   c. laboratory error
2. Define state of total body sodium (TB_{Na})
   a. Low TB_{NA} means salt depletion with actual hypovolemia
      Rx = replacement
   b. High TB_{Na} with edema in absence of renal disease implies "effective" hypovolemia
      Rx = water restriction and improvement of renal perfusion and delivery to diluting sites
   c. Normal TB_{Na} and normal renal function implies water excess due to inappropriately elevated ADH
      RX = (1) If acute and symptomatic, elevation of serum osmolality indicated to reduce cerebral edema. (2) If chronic and minimally symptomatic, water restriction to less than intake is preferred.

crease in ECF volume and excludes a diagnosis of sodium depletion—even in the presence of severe hyponatremia. If a patient has significant sodium depletion, then he will manifest the signs of hypovolemia (hypotension and oliguria) without generalized edema.

The treated patient with congestive heart failure may have either excess or deficient total-body sodium. The most common reason for development of sodium depletion in a patient with heart disease is excessive renal sodium loss with diuretic therapy, particularly while on a low-sodium diet. The fall in ECF volume provides a nonosmotic stimulus to ADH release, and if water is ingested hypotonicity will develop. Proper therapy in the patient who shows the signs of ECF volume depletion *without edema* is administration of sodium. This can be given isotonically, as correction of hypovolemia will allow osmotic inhibition of ADH release and correction of the hypo-osmolality. Hypertonic saline should only be used when water intoxication has produced signs of cerebral edema.

When hyponatremia develops in an edematous patient, total-body sodium is high. Usually the appearance of hyponatremia in congestive failure is a late finding and represents a symptom of a marked decrease in effective arterial blood volume and renal hypoperfusion. Despite the obvious excess of total-body sodium and water manifest as edema, the kidney "sees" effective extracellular fluid depletion and responds to this with a low level of sodium excretion. The combination of decreased cardiac output, elevated venous pressure, and fluid transudation from capillaries is responsible for a decreased effective circulating blood volume. This in turn results in a decrease in plasma flow and increase in filtration fraction, and a corresponding increase in proximal tubular sodium reabsorption. As congestive failure worsens, glomerular filtration rate beings to fall. The combination of these two events results in a markedly decreased delivery of filtrate out of the proximal tubule to the diluting site. As discussed previously, there is also intensified sodium reabsorption at the distal sites. Despite this, however, the net effect is a decrease in the amount of solute-free water delivered to the distal tubule and collecting duct. The bulk of the filtrate that is reabsorbed has been reabsorbed isotonically at more proximal sites. This decrease in delivery results in slowed flow through the collecting duct, and as previously discussed, even in the absence of ADH there is some backdiffusion of water across the collecting duct due to the tremendous osmotic gradient. The slower the flow through the collecting duct, the more time there is for this equilibration to take place; so even if ADH secretion were totally suppressed excess water would not be excreted normally, primarily because intense proximal tubular reabsorption markedly diminishes delivery of sodium and water to the portion of the nephron where dilution takes place. To make matters worse, however, as the effective hypovolemia progresses with increasing severity of the congestive heart failure, ADH is also released in response to this hypovolemic stimulus.

Correction of hyponatremia in an edematous patient may be very difficult. The primary goals are to improve the effective circulating plasma volume and to establish a balance between intake and excretion of electrolyte-free water. Mild to moderate hyponatremia may be effectively improved by restricting water intake to reduce the intake of electrolyte-free water to the level of its excretion. Recognizing that there is an insensible water loss of approximately 500–800 ml/day, prevention of progressive hyponatremia can usually be accomplished by restricting water intake to 1000–1500 ml/day. The second step is the use of agents such as acetazolamide and aminophylline that decrease proximal tubular reabsorption and thereby increase delivery to the diluting segments of the nephron.

The use of hypertonic saline to raise serum sodium is certainly contraindicated in patients who suffer from a surfeit of sodium and water. The administration of hypertonic solutions results in a sudden expansion of ECF volume and may thereby precipitate pulmonary edema—in addition to increasing the body sodium content of patients who are unable to excrete it. In severe hyponatremia with cerebral symptoms, acute elevation of plasma osmolality and subsequent decrease in intracellular volume may be required. Theoretically this may be accomplished by the use of hypertonic saline, but a preferable therapy is hypertonic mannitol—which in addition to elevating plasma osmolality will inhibit water reabsorption and promote the excretion of solute-free water. Once symptoms have subsided and sodium concentration approaches 118 to 120 mEq/liter, more conservative measures such as water restriction should suffice.

It should also be recognized that the usual treatment of congestive heart failure can aggravate the existing potential for hyponatremia. For example, a low-sodium diet without restrictions on water intake effectively increases the intake of electrolyte-free water. Diuretics aggravate hyponatremia by at least three mechanisms. First, their continued use in the absence of edema may cause actual sodium depletion, as discussed above. Second, administration of loop diuretics (ethacrynic acid and furosemide), by inhibiting the major diluting sites, may decrease the ability to excrete electrolyte-free water. This is less likely with furosemide, which may have a more pronounced proximal tubular effect than ethacrynic acid that would act to enhance free water excretion.[61] Finally, there is suggestive evidence that chronic potassium depletion may alter ADH release. In this situation it is thought that cells may become depleted of osmotically active solute (potassium and its anions are the bulk of intracellular solute). Thus the cells would regulate cellular volume at a lower total solute concentration than normal. ADH will then be released at a lower osmolality, and chronic hypotonicity will result. This syndrome has been described most frequently with thiazide administration and is corrected with repletion of total-body potassium stores.[66–68]

Hypokalemia and Metabolic Alkalosis

Any diuretic that inhibits sodium reabsorption proximal to the potassium-hydrogen-ion secretory site in the distal tubule is capable of producing hypokalemia and alkalosis. This occurs for two reasons. First, edema-forming states are characterized by hypovolemia and secondary aldosteronism, as discussed previously. The elevated levels of aldosterone accelerate sodium reabsorption distally concomitantly with hydrogen and potassium secretion. This is not usually manifest clinically as hypokalemia and alkalosis, as avid sodium retention in the proximal tubule and loop of Henle prevents sodium from reaching this site in the tubule. When more proximally acting diuretics are superimposed, however, sodium and chloride are delivered to this site and hypokalemia and alkalosis become apparent. This results in a preferential loss of chloride in the urine simultaneously with accelerated generation of new bicarbonate. Since the urinary electrolyte pattern contains more chloride relative to sodium than is normally present in the ECF, a state of relative chloride depletion results. In the sense that diuretics have produced a contraction of the ECF space by the removal of a hyperchloric fluid, bicarbonate concentration rises and a "contraction" alkalosis may be said to have occurred acutely. This acute alkalosis is maintained by effective hypovolemia, the low chloride state and accelerated proximal bicarbonate reabsorption, and distal bicarbonate generation. The low-chloride state perpetuates this proximally and distally; as with continued sodium reabsorption, in order to maintain electrical neutrality, cations such as H^+ and K^+ distally must be secreted when there is insufficient chloride to go with the sodium. Thus two points should be clear. First, the more potent the diuretic, the more severe will be chloride depletion and attendent hypokalemia and alkalosis. Second, most of the alkalinizing and potassium-depleting processes outlined above can be reversed or ameliorated by the administration of chloride salts, particularly potassium chloride. The provision of chloride will promptly result in excretion of the excessive bicarbonate in association with decreased excretion of hydrogen and potassium. Potassium depletion associated with metabolic alkalosis should not be treated with the more palatable alkalinizing potassium salts (bicarbonate, gluconate, acetate); this would not only perpetuate the alkalosis but make repair of the potassium deficits more difficult. Other chloride salts available include the acidifying chloride salts (ammonium chloride and arginine hydrochloride). Ammonium chloride will suffice if oral therapy is feasible, but it may be toxic when given intravenously. If severe derangements are present or if oral therapy is not possible, arginine hydrochloride may be used safely and effectively as a provider of both chloride and hydrogen ions.

Another method of preventing hypokalemia and alkalosis is with the simultaneous administration of a potassium-sparing diuretic such as spironolactone or triamterene. While spironolactone is a true competitive inhibitor of aldosterone, the latter drugs interfere directly with distal sodium transport and K^+ and H^+ secretion (by unknown mechanisms) and are effective even after adrenalectomy. As a group these drugs are minimally effective as diuretics, but they are useful as adjunctive agents to combat hypokalemia and alkalosis. Used alone, as discussed above, they may produce hyperkalemia and hyperchloremic acidosis. They should rarely be used in conjunction with potassium supplements, as under these circumstances several instances of hyperkalemia have been reported. Triamterene is felt by many to depress GFR, which although reversible is undesirable, particularly in patients in whom GFR may already be compromised. Spironolactone is therefore to be preferred, and in addition it is less likely to produce dangerous hyperkalemia, as hyperkalemia itself induces a further rise in aldosterone secretion that would tend to dilute the spironolactone effect (competitive inhibition).

Hyperkalemia

Hyperkalemia is unusual in patients with congestive heart failure and is usually iatrogenic in origin. The most common situation in which it is seen is in association with the use of potassium-sparing diuretics (triamterene, amiloride, or spironolactone) in conjunction with either potassium supplements or low-sodium diets. Low-sodium diets, particularly those with salt substitutes, are usually high in potassium content, as the sodium substitute used most often is potassium. Recently it has become clear that insulin-dependent diabetics have a defect in cellular and renal handling of potassium, rendering them peculiarly susceptible to potassium-sparing agents; and the use of these drugs in these patients is probably contraindicated.[69–71] Management of true hyperkalemia (pseudohyperkalemia refers to a falsely elevated serum potassium produced by either hemolysis during blood collection or release of potassium from platelets during clotting) begins with recognition and elimination of the causes. Significant hyperkalemia (greater than 6.5 mEq/liter) can be treated with a combination of sorbitol and the ion-exchange resin kayexalate.

Metabolic Acidosis

Metabolic acidosis is unusual in uncomplicated congestive failure unless renal function is significantly impaired. Acetazolamide characteristically produces a hypokalemic metabolic acidosis by inhibiting carbonic anhydrase, producing primarily a sodium and potassium bicarbonate diuresis. All of the diuretics that inhibit distal cation exchange can produce systemic acidosis usually associated with hyperkalemia, as discussed above.

Hyperuricemia

Hyperuricemia in congestive heart failure unexplained by the concurrence of primary gout or renal failure is most commonly a complication of diuretic therapy. Two mechanisms have been invoked to explain diuretic-induced hyperuricemia. First, excessive diuresis, by producing hypovolemia, enhances proximal tubular sodium reabsorption and simultaneously stimulates reabsorption of other constituents of proximal tubular fluid that are reabsorbed concomitantly with sodium. The mechanism for this parallelism, while incompletely understood, may be related to enhanced uptake at the peritubular capillary level secondary to changes in Starling forces. Additionally, diuretics may also produce renal urate retention by interfering with uric acid secretion, as thiazides, ethacrynic acid, and furosemide are all weak organic acids secreted by the probenecid-sensitive pathway in the proximal tubule.[72–74] Thus hyperuricemia may ensue with the use of any of the commonly administered diuretics, and in susceptible individuals it may be associated with acute gouty attacks. The question of whether or not chronic persistent hyperuricemia contributes to renal damage remains controversial,[75,76] and the use of allopurinol in this situation remains open to question.

Diuretic Management of Congestive Heart Failure and Refractoriness to Therapy

Edema and interstitial fluid accumulation require two factors—a primary stimulus and renal response with salt retention and production of a positive sodium balance. Thus the initial mode of therapy for congestive heart failure must always be an attempt to improve cardiac output—correction of undue demands upon the heart (hypertension, thyrotoxicosis, anemia, valvular lesions, etc.) and improvement of cardiac performance with digitalization. Once this has been done, secondary modes of therapy are aimed at preventing renal sodium retention. This is done in two ways. In early mild congestive heart failure, modestly reducing sodium intake may prevent significant sodium retention, and this can be accomplished by omitting the salt that is usually added to a normal diet. If this simple measure is ineffective then it is unlikely that further dietary salt restriction alone will prevent further sodium retention. In patients with moderate to severe congestive failure, daily sodium balance will remain positive unless sodium intake is almost completely restricted, thus producing a diet that is unpalatable and an unnecessary burden upon the patient. In these patients diuretic therapy in conjunction with moderate restriction (50–60 mEq sodium) is a much simpler approach. It should be recognized that except in severe end-stage patients with congestive heart failure the primary renal lesion is not decreased filtration of sodium but rather enhanced tubular reabsorption that is therefore amenable to pharmacologic intervention.

The abnormal sodium retention in congestive heart failure is based upon a progression of disturbances, and the diuretic regimen must change as the physiologic disturbances change. Therapy is aimed at producing sodium balance, with the least harmful regimen beginning with the thiazides, progressing to aldosterone antagonists, and then moving proximally with loop blockers and finally proximal tubular agents as diuretic refractoriness and hyponatremia become manifest. If possible, it is preferable to administer diuretics on an alternate-day regimen to allow equilibration of plasma volume with edema fluid, thereby minimizing diuretic-induced hypovolemia.

There are three major reasons for failure to respond to any given diuretic agent: (1) The development of acid–base disorders that nullify the action of a specific drug. Thus metabolic alkalosis and hypochloremia render organic mercurials ineffective, and acidosis markedly impairs the action of carbonic anhydrase inhibitors. The newer agents such as ethacrynic acid and furosemide and to a lesser extent thiazides are equally effective in acidosis and alkalosis. (2) The presence of organic renal disease that markedly impairs renal function and therefore reduces responsiveness to diuretics. Most diuretics are ineffective when the glomerular filtration rate drops below 20%–25% of normal, but ethacrynic acid and furosemide have been shown to be effective in patients with renal disease whose creatinine clearance may be less than 10 ml/min. Of course, under these conditions intravenous administration of these drugs in maximal dosage is usually required, and occasional reports of nerve deafness from these agents has been reported with massive doses in these circumstances. (3) Progression of the primary edema-producing disease to an advanced stage that results in certain physiologic changes in the kidney that markedly alter diuretic responsiveness. Thus as the hypovolemia progresses GFR falls and the intensity of the stimuli to tubular sodium reabsorption increases. Aldosterone secretion increases and sodium reabsorption within the proximal tubule and loop of Henle increases, as previously discussed. Under these conditions combinations of diuretics will be most effective. First, spironolactone should be added, and if the patient remains refractory because of reduced delivery out of the proximal tubule, acetazolamide should be added. In extreme congestive failure, however, there will remain a group of patients who may remain refractory even with attempted pharmacologic blockade of sodium reabsorption within the entire nephron. The extreme hypovolemia responsible for this situation obviously cannot be combated by plasma volume expanders (albumin, plasma, and dextran, for example), as would be useful in other states associated with inadequate filling of the arterial tree, such as the nephrotic syndrome and cirrhosis. In this type of patient one can sometimes improve the hemodynamic status and correct electrolyte abnormalities through the short-term use of peritoneal dialysis.[77] The resultant improvement may occasionally restore diuretic responsiveness or improve the patient's condition sufficiently to allow more definitive treatment of a correctable cardiac lesion.

REFERENCES

1. Deen WM, Robertson CR, Brenner BM: A model of glomerular ultrafiltration in the rat. Am J Physiol 223:1178, 1972
2. Brenner BM, Troy JL, Daugherty TM, Deen WM: Dynamics of glomerular ultrafiltration in the rat. II. Plasma-flow dependence of GFR. Am J Physiol 223:1184, 1972
3. Robertson CR, Deen WM, Troy JL, Brenner BM: Dynamics of glomerular ultrafiltration in the rat. III. Hemodynamics and autoregulation. Am J Physiol 223:1191, 1972
4. Deen WM, Troy JL, Robertson CR, Brenner BM: Dynamics of glomerular ultrafiltration in the rat. IV. Determination of the ultrafiltration coefficient. J Clin Invest 52:1500, 1973
5. Giebisch G: Coupled ion and fluid transport in the kidney. N Engl J Med 287: 913, 1972
6. Lewy JE, Windhager EE: Peritubular control of proximal tubular fluid reabsorption in the rat kidney. Am J Physiol 214:943, 1968
7. Kokko JP, Rector FC Jr: Counter current multiplication system without active transport in inner medulla. Kidney Int 2:214, 1972
8. Rocha AS, Kokko JP: Sodium chloride and water transport in the medullary thick ascending limb of Henle. J Clin Invest 52:612, 1973
9. Burg MB, Green N: Function of the thick ascending limb of Henle's loop. Am J Physiol 224:659, 1973
10. Stein JH, Reineck HJ: Regulation of the excretion of sodium and other electrolytes by the collecting duct. Kidney Int 6:1, 1974
11. Brenner BM, Troy JL: Postglomerular vascular protein concentration: Evidence for a causal role in governing fluid reabsorption and glomerulotubular balance by the renal proximal tubule. J Clin Invest 50:336, 1971
12. Brenner BM, Troy JL, Daugherty TM, MacInnes RM: Quantitative importance of changes in postglomerular colloid osmotic pressure in mediating glomerulotubular balance in the rat. J Clin Invest 52:190, 1973
13. Earley LE, Friedler RM: The effects of combined renal vasodilatation and pressor agents on renal hemodynamics and the tubular absorption of sodium. J Clin Invest 45:542, 1966
14. Earley LE, Martino JA, Friedler RM: Factors affecting sodium reabsorption by the proximal tubule as determined during blockade of distal sodium reabsorption. J Clin Invest 45:1668, 1966
15. Martino JA, Earley LE: Demonstration of a role of physical factors as determinants of the natriuretic response to volume expansion. J Clin Invest 46:1963, 1967
16. Hayslett JP: Effect of changes in hydrostatic pressure in peritubular capillaries on the permeability of the proximal tubule. J Clin Invest 52:1314, 1973
17. Walker WG, Cooke CR: Plasma aldosterone regulation in anephric man. Kidney Int 3:1, 1973
18. Dirks JH, Cirksena WJ, Berliner RW: The effect of saline infusion on so-

dium reabsorption by the proximal tubule of the dog. J Clin Invest 44:1160, 1965

19. Rector FC, Martinez-Maldonado M, Kurtzman NA: Demonstration of a hormonal inhibitor of proximal tubular reabsorption during expansion of extracellular volume with isotonic saline. J Clin Invest 47:761, 1968
20. Wright FS, Brenner BM, Bennett CM: Failure to demonstrate a hormonal inhibitor of proximal sodium reabsorption. J Clin Invest 48:1107, 1969
21. Buckalew VM Jr, Lancaster CD Jr: The association of a humoral sodium transport inhibitory activity with renal escape from chronic mineralocortocoid administration in the dog. Clin Sci 42:69, 1972
22. Sealey JE, Kirshman JD, Laragh JH: Natriuretic activity in plasma and urine of salt loaded man and sheep. J Clin Invest 48:2210, 1969
23. Bourgoignie JJ, Hwang KH, Espinel C, et al: A natriuretic factor in the serum of patients with chronic uremia. J Clin Invest 51:1514, 1972
24. Bourgoignie JJ, Hwang KH, Ipakchi E, Bricker NS: The presence of a natriuretic factor in urine of patients with chronic uremia. J Clin Invest 53:1559, 1974
25. Gitelman HJ, Blythe WB: Isolation of a natriuretic factor of the posterior pituitary, in Villarreal H (ed): Proceedings of the Fifth International Congress of Nephrology. 1972, p 91
26. Jamison RL: Intrarenal heterogeneity. Am J Med 54:281, 1973
27. Sparks HV, Kopald HH, Corriere S, et al: Intrarenal distribution of blood flow with chronic congestive heart failure. Am J Physiol 223:840, 1972
28. Hollenberg NK, Epstein M, Guttman RD, et al: Effect of sodium balance on intrarenal distribution of blood flow in normal man. J Appl Physiol 28:312, 1970
29. Horster M, Thurau K: Micropuncture studies of the filtration rate of single superficial and juxtamedullary glomeruli in the rat kidney. Pfluegers Arch 301:162, 1968
30. de Rouffignac C, Bonvalet JP: Etude chez le rat des variation du debit individuel de filtration glomerulaire des nephrons superficiels et profond en fonction de l'apport sole. Pfluegers Arch 317:141, 1970
31. Earley LE, Friedler RM: Changes in renal blood flow and possibly the intrarenal distribution of blood during natriuresis accompanying saline loading in the dog. J Clin Invest 44:929, 1965
32. Earley LE, Friedler RM: Studies on the mechanism of natriuresis accompanying renal blood flow and its role in the renal response to extracellular volume expansion. J Clin Invest 44:1857, 1965
33. Blantz RC, Wallin JD, Rector FC Jr, Seldin WD: Effect of variation in dietary NaCl intake on the intrarenal distribution of plasma flow in the rat. J Clin Invest 51:2790, 1973
34. Daugharty TM, Ueki IF, Nicholas DP, Brenner BM: Renal response to chronic intravenous salt loading in the rat. J Clin Invest 52:21, 1973
35. McGiff JC: The renal vascular response to hemmorrhage. J Pharmacol Exp Ther 145:181, 1964
36. Gill JR Jr, Bartter FC: Adrenergic nervous system in sodium metabolism.

II. Effects of guanethidine on the renal response to sodium deprivation in normal man. N Engl J Med 275:1466, 1966
37. Gill JR Jr, Casper AGT: Depression of proximal tubular sodium reabsorption in the dog in response to renal beta adrenergic stimulation by isoproterenol. J Clin Invest 50:112, 1971
38. Agus ZS, Puschett JB, Senesky D, Goldberg M: Mode of action of parathyroid hormone and cyclic adenosine 3′,5′-monophosphate on renal tubular phosphate reabsorption in the dog. J Clin Invest 50:617, 1971
39. Blendis LM, Auld RB, Alexander EA, Levinsky NG: Effect of renal beta and alpha adrenergic stimulation of proximal sodium reabsorption in dogs. Clin Sci 43:569, 1972
40. Dirks JH, Cirksena WJ, Berliner RW: Micropuncture study of the effect of various diuretics on sodium reabsorption by the proximal tubules of the dog. J Clin Invest 45:1875, 1966
41. Brenner BM, Berliner RW: The relationship between extracellular volume and fluid reabsorption in the rat nephron. Am J Physiol 217:6, 1969
42. Knox FG, Schneider EG, Willis LR, et al: Effect of volume expansion on sodium excretion in the presence and absence of increased delivery from superficial proximal tubule. J Clin Invest 52:1642, 1973
43. Knox FG, Schneider EG, Dresser TP, Lynch RE: Natriuretic effect of increased proximal delivery in dogs with salt retention. Am J Physiol 219:904, 1970
44. Willis LR, Schneider EG, Lynch RE, Knox FG: Effect of chronic alteration of sodium balance on reabsorption by proximal tubule of the dog. Am J Physiol 223:34, 1972
45. Schneider EG, Dresser TP, Lynch RE, Knox FG: Sodium reabsorption by proximal tubule of dogs with experimental heart failure. Am J Physiol 220:952, 1971
46. Levy M: Effects of acute volume expansion and altered hemodynamics on renal tubular function in chronic caval dogs. J Clin Invest 51:922, 1972
47. Auld RB, Alexander EA, Levinsky NG: Proximal tubular function in dogs with thoracic caval constriction. J Clin Invest 50:2150, 1971
48. Sonnenberg H: Proximal and distal tubular function in salt-deprived and in salt-loaded deoxycorticosterone acetate-escaped rats. J Clin Invest 52:263, 1973
49. Rastegar A, Agus ZS, Connor TB, Goldberg M: Renal handling of calcium and phosphate during mineralocorticoid "escape" in man. Kidney Int 2:279, 1972
50. Verney EB: The antidiuretic hormone and the factors which determine its release. Proc R Soc Lond [Biol] 135:25, 1974
51. Berliner RW, Davidson DG: Production of hypertonic urine in absence of pituitary antidiuretic hormone. J Clin Invest 36:1416, 1959
52. Cirksena WJ, Dirks JH, Berliner RW: Effect of thoracic cava obstruction on response of proximal tubule sodium reabsorption to saline infusion. J Clin Invest 45:179, 1966
53. Stumpe KO, Reinelt B, Ressel C, et al: Urinary sodium excretion and

proximal tubule reabsorption in rats with high-output failure. Nephron 12:261, 1974

54. Schneider EG, Dresser TP, Lynch RE: Sodium reabsorption by proximal tubule of dogs with experimental heart failure. Am J Physiol 220:952, 1971.
55. Friedler RM, Belleau LJ, Martino JA, Earley LE: Hemodynamically induced natriuresis in the presence of sodium retention resulting from constriction of the thoracic inferior vena cava. J Lab Clin Med 69:565, 1967
56. Sparks HV: Intrarenal distribution of blood flow with chronic congestive heart failure. Am J Physiol 223:840, 1972
57. Barger AC, Muldowney FP, Liebowitz MR: Role of the kidney in the pathogenesis of congestive heart failure. Circulation 20:273, 1959
58. Azer M, Gannon R, Kaloyanides GJ: Effect of renal denervation on the antinatriuresis of caval constriction. Am J Physiol 222:611, 1972
59. Schrier RW, Humphreys MH: Factors involved in the antinatriuretic effects of acute constriction of the thoracic and abdominal inferior vena cava. Circ Res 29:479, 1971
60. Barger AC, Rudolph AM, Yates EF: Sodium excretion and renal hemodynamics in normal dogs, dogs with mild valvular lesions of the heart, and dogs in congestive heart failure. Am J Physiol 183:595, 1965
61. Goldberg M: The renal physiology of diuretics, in Orloff J, Berliner RW (eds): Handbook of Physiology, Section 8, Renal Physiology. American Physiological Society, 1973
62. Schrier RW, Lehman C, Zacherle B, Earley LE: Effect of furosemide on free water excretion in edematous patients with hyponatremia. Kidney Int 3:30, 1973
63. Schreider WJ, Becker EL: Acute transient hearing loss after ethacrynic acid therapy. Arch Intern Med 117:716, 1966
64. Pillay VKG, Schwartz FD, Aimi K, Kark RM: Transient and permanent deafness following treatment with ethacrynic acid in renal failure. Ann Intern Med 70:1095, 1969
65. Freeman RB, Maher JF, Schreiner GE, Mostofi FK: Renal tubular necrosis due to nephrotoxicity of organic mercurial diuretics. Ann Intern Med 57:34, 1962
66. Maffly RH, Edelman ES: The role of sodium potassium and water in the hypo-osmotic states of heart failure. Prog Cardiovasc Dis 4:88, 1961
67. Laragh JH: The effect of potassium chloride on hyponatremia. J Clin Invest 33:807, 1954
68. Fishman MP, Vorherr H, Kleeman CR, Telfer N: Diuretic-induced hyponatremia. Ann Intern Med 75:853, 1971
69. Walker BR, Capuzzi DM, Alexander F, et al: Hyperkalemia after triamterene in diabetic patients. Clin Pharmacol Ther 13:643, 1972
70. Manning RT, Behrie FC: Use of spironolactone in renal edema effectiveness and association with hyperkalemia. JAMA 176:97, 1961
71. McNay JL, Oran E: Possible predisposition of diabetic patients to hyperkalemia following administration of the potassium-retaining diuretic, Amiloride (MK 870). Metabolism 19:58, 1970

72. Demartini FE, Wheaton EA, Healey LA, Laragh JH: Effect of chlorothiazide on renal excretion of uric acid. Am J Med 32:572, 1962
73. Cannon PJ, Heineman HO, Stason WB, Laragh JH: Ethacrynic acid: Effectiveness and mode of diuretic action in man. Circulation 31:5, 1965
74. Muth RG: Diuretic properties of furosemide in renal disease. Ann Intern Med 69:249, 1968
75. Cannon PJ, Stason WB, Demartini FE, et al: Hyperuricemia in primary and renal hypertension. N Engl J Med 275:457, 1966
76. Rastegar A, Thier SO: The physiologic approach to hyperuricemia. N Engl J Med 286:470, 1972
77. Cairns KB, Porter KB, Kloster FE, et al: Clinical and hemodynamic results of peritoneal dialysis for severe cardiac failure. Am Heart J 76:227, 1968

Rolf M. Gunnar
Henry S. Loeb
Sarah Ann Johnson

11
Cardiogenic Shock

The clinical syndrome of shock is due to inadequate tissue perfusion, and its clinical manifestations are those of poor perfusion of vital organs. The patient is obtunded because of inadequate cerebral perfusion and is oliguric because of inadequate renal perfusion; he appears pale, cold, and somewhat cyanotic because of poor perfusion of the skin and mucous membranes. Cardiogenic shock is this syndrome of poor perfusion caused by depression of cardiac function. The form of cardiogenic shock that arouses most interest at present is that associated with acute myocardial infarction. However, the cardiogenic shock that occasionally follows cardiovascular surgery is somewhat more stable, and many of the early physiologic studies were of this syndrome. Cardiogenic components do appear in other forms of shock, such as septic shock and hemorrhagic shock, and in the terminal phases of these illnesses may play a major role in preventing recovery.

Shock developing in the course of acute myocardial infarction is not entirely cardiogenic. Hypovolemia and peripheral vasomotor changes may be major factors in preventing the heart from maintaining adequate perfusion and perfusion pressure. In this chapter we will discuss the effects of pump failure, as well as the reflex peripheral vascular changes resulting from myocardial infarction, as they affect the ability of the cardiovascular system to adjust to evolving myocardial damage.

Supported in part by NIH Grant HL-15040 from the National Heart and Lung Institute, Bethesda, Maryland.

EXPERIMENTAL MODELS OF CARDIOGENIC SHOCK

It has not been easy to develop an animal model for study of cardiogenic shock, particularly the shock associated with myocardial infarction. Early studies in Wiggers's laboratory demonstrated the difficulty of producing cardiogenic shock by ligation of the coronary vessels. Very few animals survived ligation of the left main coronary artery. Even though ventricular function was depressed with ligation of the anterior descending branch, only a few animals became hypotensive. Most either recovered their arterial pressure or developed sudden ventricular fibrillation and died.[1-3] Direct injuries of myocardium using formalin,[4] cautery,[5] trauma,[6] and hypothermal injury[7] all failed to produce the syndrome of cardiogenic shock. In 1951 Agress et al. were finally able to produce a syndrome of protracted hypotension and shock in a dog by injecting microspheres into the coronary circulation.[8,9] Initially these microspheres were injected into the aortic root, but there was considerable peripheral delivery of the microspheres particularly to the cerebral circulation.[10] More recently the microspheres have been injected directly into the coronary circulation. This technique produced an animal that exhibited shock for many hours. Agress concluded from his model that to induce shock similar to that seen in patients with myocardial infarction there needed to be reflex inhibition of the normal peripheral vascular responses to the drop in cardiac output.[11]

In 1961 Lluch et al. reported on a model of cardiogenic shock using selective injection of mercury into a coronary artery in the unanesthetized dog.[12] This led to rather diffuse lesions throughout the distribution of the artery injected. With sufficient myocardial damage the animal became hypotensive and developed shock that lasted for several hours. Left ventricular end-diastolic pressure in these animals was elevated as the cardiac output decreased, indicating deterioration in myocardial function.

More recently Murphy et al. have used a method of localized electrocoagulation to cause coronary thrombosis in the two branches of the left coronary artery; in this model the shock syndrome was produced.[13] Feola et al. used a model of repetitive coronary ligation for myocardial infarction and, in addition, transected the spinal cord and vagus nerve in order to isolate the cardiac from the reflex autonomic changes. They produced an animal model with cardiogenic shock that is stable over several hours, making it possible to study the effects of various interventions.[14]

SHOCK IN MYOCARDIAL INFARCTION: PERIPHERAL VASCULAR RESPONSE

Patients with myocardial infarction have a depression of myocardial function dependent to a great extent on the mass of ventricular muscle damaged. If sufficient myocardium is destroyed, the cardiac output will decline despite an elevation in left ventricular end-diastolic pressure. If the myocardial damage is only moderate and the systemic vascular resistance increases sufficiently, mean arterial pressure will not fall despite the decrease in cardiac output. This is the usual situation in patients with acute myocardial infarction who have a slight decrease in cardiac output, an elevation in systemic vascular resistance, and normal arterial pressure. Significant arterial hypotension and the shock syndrome will develop if the expected compensatory response of the vascular resistance to a drop in cardiac output fails to occur, or if the decrease in cardiac output is so profound that the shock syndrome develops despite intense vasoconstriction. The first hemodynamic studies done in the human with shock and myocardial infarction were reported by Gilbert et al.[15] and were quickly followed by studies by Freis et al.[16] These studies confirmed the decrease in cardiac output that was expected in myocardial infarction, but more than half of the patients studied had normal systemic vascular resistance despite the drop in cardiac output.

Agress, on the basis of his experimental model, thought the lack of peripheral vascular response noted in clinical studies was due to inhibition of sympathetic tone by a reflex arising in the small coronary vessels or in the ischemic tissue surrounding the small vessels.[17] The afferent arc of this reflex appeared to lie in the dorsal sympathetic chain, since it was necessary to interrupt these nerves in order to abolish the reflex. By measuring action potentials in the efferent greater splanchnic nerves, these workers were able to demonstrate a decrease in sympathetic outflow in their animal preparation after microsphere injection.[11] These findings suggested that following experimental myocardial infarction produced by injection of plastic microspheres a reflex with its afferent pathway at least partly in the sympathetic chain causes central inhibition of sympathetic nervous system activity.

An additional mechanism for inhibition of compensatory vasoconstriction may be the activation of the baroreceptors in the left ventricle, as proposed by Daly and Verney[18] in 1967. The existence of these receptors was confirmed by Aviado et al.[19] in 1951. Ross et al. demonstrated that these receptors were activated by stretching the myocardium either in systole or diastole and that stretching the ventricular myocardium caused a decrease in peripheral vascular resistance.[20] They further proposed that these receptors are continuously activated under normal circumstances, and they demonstrated that sudden loss of activation by a cessation of left ventricular filling caused an immediate rise in systemic vascular resistance. On the basis of this type of evidence, Constantine related the inhibition of peripheral vasoconstriction in patients with acute myocardial infarction and a decrease in cardiac output to systolic bulging of the ischemic myocardium. Systolic expansion of the infarct results in activation of the myocardial stretch receptors and consequent inhibition of sympathetic outflow.[21] Constantine, however, concluded that the afferent limb of this reflex was in the vagus nerve and felt that he could abolish the reflex by vagotomy. More recent work by Kezdi et al. would tend to confirm the vagal pathway for the afferent arc of this reflex.[22] On the other hand, Brown, using the cat as the experimental animal, has demonstrated that there is an increase in the afferent discharge in cardiac sympathetic nerve fibers after coronary occlusion. He proposed that there is a reflex arising in the ischemic myocardium that is transmitted centrally along the sympathetic nerves.[23] Since he could not produce the same effect with anoxia or hypercapnia, he postulated that ischemia itself was the stimulus for activation of this reflex. He also felt that the receptors were the same as the stretch receptors of the myocardium. Hanley et al., using a dog model and looking at the isolated limb for evidence of change in arteriolar resistance, demonstrated that both an increase in ventricular volume and an increase in severity of ischemia augmented the vasodilator response.[24] They further demonstrated a difference between vasodilatation in the paw and the rest of the hind lmb and suggested that reflex responses to the stimulus of ischemia or stretch were not uniform throughout all vascular beds. This latter concept is very helpful in explaining differences in vasoconstriction in various organs and demonstrates why vasoconstriction may be apparent clinically even though hemodynamic measurements as calculated for the entire body reveal normal systemic vascular resistance.

From these experimental observations we can postulate that the vasomotor center in the brain receives conflicting signals during severe myocardial ischemia and infarction. The baroreceptors in the left ventricle are activated and call for a decrease in sympathetic outflow and thus vasodilatation. At the same time the baroreceptors in the aortic arch and carotid sinus sense a decrease in the rate of arterial pressure rise and a decreased pulse pressure and signal for an increase in sympathetic tone and thus vasoconstriction. The response of the vasomotor center is a result of the integration of these, and possibly other, opposing afferent impulses. The lack of uniformity and response of various vascular beds can also be explained by these complex signals reaching the vasomotor center. Therefore it is not surprising that patients with acute myocardial infarction exhibit such a wide range of peripheral vascular response to their illness.[25]

In a study of 50 patients with acute myocardial infarction and shock, we found the systemic vascular resistance to be elevated above 20 mm Hg/liter/min in 1 of 9 survivors and in 18 of 41 nonsurvivors. In the group as a whole only 38% had resistance values above the level of 20 mm Hg/liter/min. Thirty-two percent had values below 16 mm Hg/liter/min, and 12% had calculated values below 12 mm Hg/liter/min. More recent studies by Hughes et al. have shown that in patients with acute myocardial infarction there is an abnormal response of the peripheral vasculature to tilting,[26] again an indication of some inhibition of normal vasoconstrictor responses.

The significance of these peripheral vascular changes is not entirely in the mild hypotension that may develop, but more importantly in the effect of hypotension on extension of the area of infarction. Since there are numerous areas of partial occlusion of the coronary vessels in most patients who undergo myocardial infarction, and since the ischemic area around the myocardial infarct has coronary vessels that are maximally vasodilated, much of the viable remaining myocardium appears to be supplied by vessels that are dependent upon pressure for their flow. To allow these areas to remain at decreased flow levels can only enhance myocardial damage and change ischemic myocardium to necrotic myocardium. The recent work of Maroko et al. demonstrates very well that agents that reduce pressure or increase myocardial oxygen needs increase the area of myocardial injury.[27] The difficulty, of course, is to increase coronary perfusion pressures optimally, so that improved coronary flow occurs without increasing the work of the heart to such an extent that the increase in oxygen demands of the myocardium exceeds improvement in coronary blood flow.

PUMP FAILURE

The common denominator of all forms of cardiogenic shock, including that associated with acute myocardial infarction, is depression of myocardial function. It is not difficult to visualize such a depression in patients or experimental animals with acute myocardial infarction where a definable area of myocardium becomes noncontractile. As Swan et al. have pointed out, noncontractile myocardium with normal diastolic compliance would retain that compliance throughout systole and therefore would bulge inordinately with each contraction of the remaining functioning myocardium.[28] Looking at the ventricle as a single unit, this bulging could have a serious effect on ventricular function, much like the effect of mitral insufficiency. Shifting blood into the dyssynergic infarcted area until wall tension was great enough to elevate ventricular cavity pressure to systemic arterial levels would dissipate the work of the contracting ventricle. While this apparently does apply at the onset of infarction, the infarcted area becomes stiff rather quickly and no longer bulges excessively with each ventricular contraction. To maintain an adequate stroke volume, however, the remaining myocardium must increase its diastolic fiber length, and if this moves the diastolic pressure–volume curve to its steep portion there will be a substantial elevation of ventricular end-diastolic pressure. Thus if a sufficient area of the wall of the left ventricle becomes infarcted and stiff the remaining myocardium will need to operate near peak diastolic fiber length in order to maximally utilize the Frank-Starling mechanism to maintain stroke output. According to this model of the injured ventricle, the remaining myocardium would continue to support cardiac output by increasing its function in response to elevated filling pressures induced partially by the noncompliant infarcted portion of the ventricle. The cardiac output would thus be maintained in association with an elevated filling pressure. As more and more of the ventricular mass is involved in the infarct, cardiac output would be sustained only by an increase in heart rate. If enough ventricle were destroyed, limitation in stroke volume would be so great that, despite increases in heart rate, cardiac output would fall to levels incompatible with circulatory integrity.

Measurements of cardiac output before treatment in our group of 50 patients with acute myocardial infarction and shock revealed that 10 of the 18 patients who recovered from the shock syndrome had a cardiac index above 2.0 liters/min/m^2, while only 9 of the 32 who died in shock had a cardiac index above this level. Thus cardiac index less than 2.0 liters/min/m^2 in this syndrome appears to have a poor prognosis. However, estimation of prognosis based on the measurements of cardiac output necessarily includes the assumption that at the time cardiac output is measured filling pressure is high and arterial pressure is low.

CLINICAL ASSESSMENT OF VENTRICULAR FUNCTION

With the development of sophisticated mechanical and surgical methods for treating the complications of acute myocardial infarction, it became necessary to define more accurately ventricular dysfunction in patients with myocardial infarction and the shock syndrome in particular. In order to do this, more detailed information above left ventricular filling pressure was needed, particularly since measurements of central venous pressure were shown to be poor reflections of left ventricular filling pressure. Hamosh and Cohn began by measuring left ventricular pressure directly in 40 patients with acute myocardial infarction.[29] They used a fluid-filled catheter passed retrograde across the aortic valve without advantage of fluoroscopic visualization. Twelve of their 40 patients had evidence of congestive heart failure, and 14 manifested the shock syndrome. Left ventricular end-diastolic pressure, although elevated in most patients without shock or cardiac failure, was highest in the patients who manifested congestive heart failure and was elevated above 14 mm Hg in all but 2 of the patients with shock. Also, the patients with shock had the most markedly reduced cardiac output.

Because of the importance of knowing left ventricular pressure, we examined the accuracy of various indirect means of measuring filling pressure and correlated these measurements with direct measurement of left ventricular pressure. We had noted, as had Hamosh and Cohn, that in many patients with acute myocardial infarction a considerable increase in left ventricular diastolic pressure occurs during atrial contraction (Fig. 11-1). This end-diastolic pressure elevation is particularly important in the assessment of ventricular function, because it sets the point on the ventricular function curve for ventricular contraction. Simultaneous measurement of mean pulmonary capillary wedge pressure as measured through a small Swan-Ganz catheter recorded values that corresponded quite well to the left ventricular diastolic pressure prior to atrial contraction, but did not give the exact information needed for accurate assessment of ventricular function. However, since mean wedge pressure is an accurate means of assessing mean pulmonary venous pressure, it is useful in predicting the development of pulmonary edema. Pulmonary artery pressure at end-diastole also correlated well with mean or pre-A-wave left ventricular diastolic pressure, particularly in those patients with pulmonary vascular resistance below 2 mm Hg/liter/min. The methods of indirect assessment of left ventricular filling pressure correlated best with mean left ventricular diastolic pressure. The greatest discrepancies between these

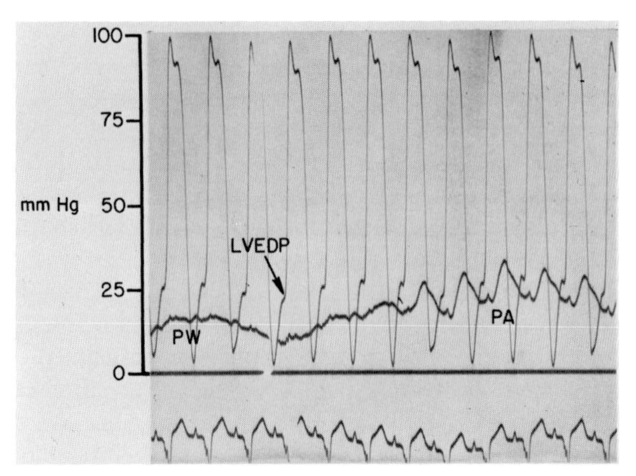

Fig. 11-1. Simultaneous left ventricular and pulmonary artery wedge pressures in a patient with acute myocardial infarction. The wedge pressure was measured through a Swan-Ganz catheter with the catheter tip balloon expanded, and the pulmonary artery pressure was measured as the balloon was allowed to deflate. The discrepancy between pulmonary artery wedge and left ventricular end-diastolic pressure can be seen.

measurements and actual end-diastolic pressures were in the patients with the highest end-diastolic pressures.[30] With these discrepancies in mind, we can review the reports of filling pressures in patients with shock and acute myocardial infarction.

In our own series of patients with shock and acute myocardial infarction, diastolic pressure was measured at end-diastole and found to be between 24 and 32 mm Hg with a mean of 20.8 mm Hg. In the 14 shock patients in the series of Hamosh and Cohn, end-diastolic pressure was between 8 and 34 mm Hg with a mean of 21.1 mm Hg. Russell et al. measured the pulmonary capillary wedge pressure in patients with myocardial infarction and demonstrated the need to increase the left ventricular filling pressure above 20 mm Hg before concluding that the damaged ventricle could not sustain an adequate cardiac output.[31] These findings are all in keeping with the model of Swan et al. described above. The elevated diastolic pressure may represent a change in compliance that may be an advantage to the remaining viable myocardium. It is appealing to think that the elevated end-diastolic pressure in the left ventricle may not represent failure, but rather a change in compliance. However, these elevated filling pressures leave very little leeway between the pressure needed for adequate stretch of the remaining normal myocardial fibers and the level of left atrial pressure that will precipitate pulmonary edema. Therefore caution must be exercised in treating patients with acute myocardial infarction and in altering left ventricular filling pressure in either direction by adjusting intravascular volume. Sudden massive diuresis imposed on the patient for treatment of very mild pulmonary congestion may decrease left ventricular filling pressure well below the level needed for maintenance of an adequate cardiac output, thus changing a patient from relatively uncomplicated myocardial infarction to the shock state. On the other hand, volume loading, if not appropriately monitored, can precipitate pulmonary edema as the noncompliant left ventricle quickly elevates its diastolic pressure in response to a volume load.

In order to define several points on the ventricular function curve upon which the heart is operating, Russell et al. gave rapid infusions of low-molecular-weight dextran and measured left ventricular filling pressure and cardiac output after each 100–200 ml injected.[31] They studied 19 patients within the first 72 hr after acute myocardial infarction, but the left ventricular filling pressure exceeded 15 mm Hg prior to the volume infusion in only 7 patients. They divided their patients into two groups. Fourteen patients had an increase in stroke index exceeding 0.30 ml/mm Hg increase in left ventricular filling pressure, with a mean increase of 1.06 ± 0.56 ml/mm Hg. Five patients had lesser increases in stroke index. the mean increase in stroke index for this group was 0.20 ± 0.07 ml/mm Hg. All 5 patients in this latter group were thought to have depressed ventricular function curves and were in clinical congestive failure. Four of the 5 died either in the hospital or within 2 months following discharge. Russell et al. also carried out this type of investigation in 22 patients who were in shock following acute myocardial infarction. In 12 of these patients they were able to assess left ventricular function by rapid intravenous infusion of low-molecular-weight dextran.[32] The survivors each increased their cardiac index during dextran infusion, with a mean increase for the group of 241 ml/min/m^2 per millimeter increase in left ventricular filling pressure. The nonsurvivors had an increase in cardiac index of only 78 ml/min/m^2 per millimeter increase in left ventricular filling pressure. This seems to indicate that the nonsurvivors of the group were operating on a flatter or poorer ventricular function curve than the survivors. The differentiation, however, was not a clear separation of the two groups, since 1 of the nonsurvivors almost tripled his cardiac index while doubling his left ventricular filling pressure during volume infusion.

Our own experience with 24 patients who had uncomplicated acute myocardial infarction and were studied during plasma volume expansion has also indicated that patients fall into two groups.[33] Eight patients increased their cardiac output by more than 20%, with an average increase of 46%, while 16 patients increased their cardiac index by less than 20% and had an average change of +1%. Since there were no significant differences between the two groups in the hemodynamic measurements prior to volume expansion, the separation could only be made on the basis of response to plasma volume expansion. At 6 months follow-up, 6 of the 8 patients in the group that responded well to volume loading were found and were all alive. On the other hand, 7 of the 16 patients that showed no positive response in cardiac output to volume expansion were dead at 6 months follow-up. The therapeutic implications of these studies appeared to be that left ventricular filling pressure must be assessed and elevated to levels in the range of 20 to 24 mm Hg before assuming that myocardial damage is so great that persistence of low cardiac output and pressure as seen in shock is cardiogenic and certainly before prescribing any form of mechanical circulatory assistance to sustain the patient.

THERAPY OF SHOCK IN MYOCARDIAL INFARCTION

A patient in shock with acute myocardial infarction presents the clinical syndrome because of a decrease in cardiac output with or without a fall in arterial pressure. After analgesics have been given for pain and adequate ventilation and oxygenation have been established, a sequence of therapeutic endeavors can be undertaken with the realization that any prolongation of the shock state increases the extent of myocardial damage. If the shock state is due preponderantly to loss of myocardial muscle mass, the immediate prognosis will be very poor unless some mechanical or surgical interventions are capable of increasing the mechanical efficiency of the heart.

PLASMA VOLUME EXPANSION

If the central venous pressure of a patient in shock with acute myocardial infarction is below 10 cm of water, there is a reasonable likelihood that inadequate filling pressure may play a significant role in the shock state. Although we feel that a measure of left ventricular filling pressure should be made in all patients in shock with acute myocardial infarction, it is fair to state that we have seldom found patients with elevated central venous pressures who were volume-depleted in the face of acute myocardial infarction and shock. However, it is possible to encounter patients in shock with very high left ventricular filling pressures in the presence of low central venous pressure. The single static measurement of 10 cm of water only identifies those patients in shock who should be given a trial of plasma volume expansion before moving to more dangerous forms of treatment. If one fails to expand the intravascular volume of the patient with this syndrome who is volume-depleted, he may allow a patient with good long-term outlook to die needlessly. On the other hand, if one volume-expands the patient with severe myocardial damage and precipitates pulmonary edema it is unlikely that he will have done harm to a patient who had any significant long-term viability.

We usually rely upon low-molecular-weight dextran for plasma volume expansion, although crystalloid solutions and solutions of reconstituted albumin can also be used. In using crystalloid solutions, however, one must give the infusion rapidly enough to expand the intravascular volume and elevate ventricular filling pressures before the fluid leaves the intravascular compartment. It is now an accepted practice in most coronary care units to measure left ventricular filling pressure during such plasma volume expansion. If indirect means that measure pre-A-wave filling pressure are used, then the left ventricular pressure should be increased to approximately 20 mm Hg before discontinuing volume expansion. If left ventricular pressures are measured directly, left ventricular end-diastolic pressure should be brought to approximately 24 mm Hg. If central venous pressure is used as a monitor, all of its inaccuracies must be recognized, and methods of administration must be precise, as described previously.[34] During the period of volume expansion it may be necessary to maintain arterial pressure using one of the inotropic agents described below if the hypotension is immediately life-threatening.

NOREPINEPHRINE

Norepinephrine remains the agent of choice for treatment of patients with shock and acute myocardial infarction. It has a significant inotropic effect at low dosage levels and in addition is useful in maintaining arterial pressure at levels adequate for coronary artery filling. It should be used in amounts just sufficient to elevate mean arterial pressure to levels of between 75 and 80 mm Hg. Even in previously hypertensive patients this is usually a level adequate for renal perfusion. If oliguria persists despite maintenance of the arterial pressure at these levels for over 1 hr and the patient is known to be hypertensive, it is logical to elevate the pressure to slightly higher levels to see if this can overcome the oliguria.

In studying 33 patients with acute myocardial infarction and shock during norepinephrine infusion, we found an increase in cardiac output averaging 18%, with an increase in arterial pressure averaging 43%.[35] Unlike the normal subject given norepinephrine, the patient in shock with acute myocardial infarction tends to increase his cardiac output during norepinephrine infusion until the arterial pressures reach levels that activate the carotid and aortic baroreceptor and cause reflex inhibition of cardiac contractility. This is consistent with animal experiments showing that at very small doses norepinephrine will increase cardiac output before significantly elevating arterial pressure and causing vasoconstriction.[36] At these small dosage levels the major effect of the agent is through activation of the beta adrenergic receptors of the heart.

ISOPROTERENOL

Isoproterenol is a synthetic sympathomimetic amine that activates the beta receptors in the heart and in the blood vessels of skeletal muscle. Therefore it has a very active inotropic and chronotropic effect on the heart, and it dilates peripheral vessels. The result is a marked increase in cardiac output and fall in peripheral vascular resistance. It has been our experience in treating patients in shock with this agent that it is seldom possible to increase the arterial pressure to adequate levels unless the underlying defect is, in large part, bradycardia.[37] Many patients will deteriorate abruptly when given isoproterenol because the drug increases myocardial oxygen requirements at a time when the perfusion pressure of the coronary vessels is being decreased. Mueller et al., using production of lactate as the index of ischemia, have found that isoproterenol increases anaerobic metabolism. They studied 6 patients during isoproterenol infusion and found lactate production to increase in all as arterial pressure fell. On the other hand, norepinephrine tended to decrease lactate production because it elevated arterial pressure levels. However, norepinephrine was not very effective in increasing cardiac output, and mechanical means were used to support the circulation.[38]

DOPAMINE

Dopamine is a sympathomimetic amine that is a natural precursor of norepinephrine. It stimulates the alpha and beta receptors, although the major beta receptor activity is found in the heart alone.[39] In addition to these receptors, dopamine activates receptors in the splanchnic and renal circulation, causing vasodilatation of these vascular beds.[40] This agent has been used to treat various shock states.

We have reported on the effects of this agent in 62 patients who were in shock. Five of these patients were in shock due to acute myocardial infarction.[41] In comparing dopamine to norepinephrine and isoproterenol in patients with various forms of shock, dopamine was found to increase output more effectively than norepinephrine, but not to the same extent as isoproterenol. Dopamine increased arterial pressure particularly at higher dosage levels as the alpha receptors were activated more effectively. This is in contradistinction to isoproterenol, which did not increase arterial pressure. However, dopamine did not elevate arterial pressure as effectively as norepinephrine. In using dopamine it was interesting to note that patients who had initially high systemic vascular resistance showed a decrease in vascular resistance with this agent as cardiac output increased and sympathetic discharge decreased. On the other hand, patients who had very low systemic vascular resistance initially tended to elevate their vascular resistance as dopamine infusion continued. Dopamine appeared to be an intermediate agent between isoproterenol and norepinephrine. It has chronotropic effects similar to isoproterenol, but no evidence of the direct vasodilator effect on the skeletal muscles that limits the usefulness of isoproterenol. It is perhaps not as useful as norepinephrine in the profoundly hypotensive patient, but had less of the marked vasoconstrictor activity that limits the inotropic usefulness of norepinephrine.

DOBUTAMINE

Tuttle et al. have manipulated the dopamine formula in an attempt to preserve the inotropic effects of the sympathomimetic amine while decreasing the chronotropic and peripheral vascular effects. Dobutamine was selected for trial.[42] This agent has been shown in the experimental animal to increase cardiac output and the force of myocardial contraction, with little change in heart rate and no evidence of significant change in peripheral vascular resistance. We have given this agent to 36 patients and found a significant increase in systolic pressure, cardiac index, and stroke index associated with a stable diastolic pressure, some fall in left ventricular end-diastolic pressure, and systemic vascular resistance, and a modest increase in heart rate.[43] This agent has been shown in the experimental animal with coronary ligation to decrease infarct size.[44] It is not available for general use and is just reaching the stage of therapeutic trial. It differs from dopamine mainly in having less

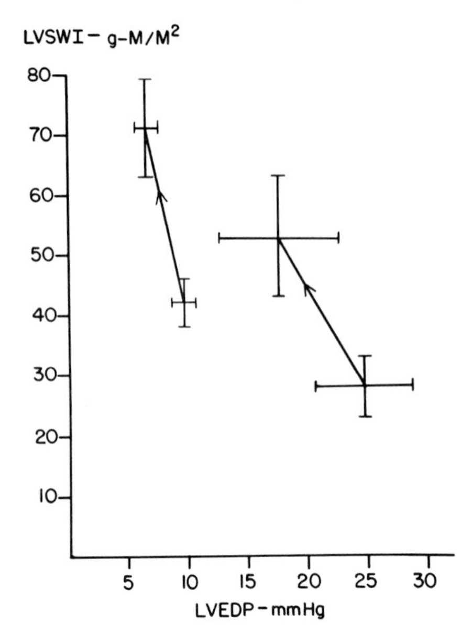

Fig. 11-2. Improvement in ventricular function during dobutamine infusion. Left ventricular stroke work index (LVSWI) increases as left ventricular end-diastolic pressure (LVEDP) falls. The bars represent the standard error of the mean and intersect at the mean. The arrow indicates the direction of change in values measured during dobutamine infusion as compared to measurements made before the infusion. The group on the left are measurements made in patients with normal ventricles, and the group on the right are measurements in patients with ventricular dysfunction.

chronotropic effects and apparent lack of significant peripheral alpha or beta effect (Fig. 11-2).

VASODILATOR THERAPY

Patients in shock with acute myocardial infarction and low intraarterial pressure are in general not candidates for vasodilator therapy, since this would merely decrease the filling pressure of the coronary vessels and increase myocardial ischemia. However, patients with severe cardiac failure following acute myocardial infarction may decrease their cardiac output enough to manifest the shock syndrome without a decrease in arterial pressure. Cuff pressure may be quite reduced because of the small pulse pressure. The discrepancy between intraarterial pressure as measured directly and cuff pressure can be as great as 164 mm Hg.[45,46] If arterial pressure is in the normal range and cardiac output is markedly reduced, and if in addition left ventricular filling pressure has been elevated to maximum extent, one therapeutic maneuver that may be helpful to the patient would be to decrease the impedance to left ventricular emptying. This may be particularly useful in patients who have associated mitral insufficiency.[47]

Majid et al. used phentolamine in treating severe heart failure in 12 patients with ischemic heart disease. Four had acute myocardial infarction.[48] With drug infusion rates between 1 and 2 mg/min, they lowered systemic arterial pressure by 20 to 30 mm Hg, increased cardiac output, and reduced the manifestations of pulmonary congestion. Kelly et al. treated 11 patients who had hypertension and acute left ventricular failure following acute myocardial infarction and noted similar increases in cardiac output and decreases in pulmonary congestion as arterial pressure fell.[49] These investigators did not use this agent in normotensive or hypotensive patients. Gold et al. used nitroglycerin in the treatment of patients with congestive failure following acute myocardial infarction. They did not treat the hypotensive patients and cautioned that this form of treatment should be accompanied by accurate measurement of arterial pressure as well as left ventricular filling pressures.[50] Franciosa et al. demonstrated improved left ventricular function during nitroprusside infusion in patients with acute myocardial infarction.[51] Chatterjee et al. used nitroprusside infusion in the treatment of patients who were hypotensive and who had severe pump failure and severe mitral insufficiency.[47] They were able to demonstrate a marked increase in cardiac output and a decrease in the evidence of mitral insufficiency in these patients. As the mitral insufficiency decreased, arterial pressure rose. However, it must be the very unusual shock patient with severe mitral insufficiency who responds dramatically to nitroprusside with enough of a decrease in regurgitant volume to allow elevation of arterial pressure. In general it is the patient who is slightly hypertensive in association with the shock or pump failure syndrome who will benefit from these forms of left ventricular unloading.

CORTICOSTEROIDS

Although corticosteroids have been advocated for their hemodynamic effect in treatment of patients with cardiogenic shock, it has been our experience that these agents produce little hemodynamic effect. If vasodilator therapy is indicated this should be achieved with the use of the more potent vasoactive agents described above.[52] On the other hand, corticosteroids may have an effect in preserving integrity of cell membranes, improving cellular metabolism, and stabilizing lysosomes. If these effects are significant in patients with acute myocardial infarction in shock, they probably will not be manifested by acute hemodynamic change, but rather by a change in overall mortality. Libby et al., in studies in the experimental animal with acute myocardial infarction, have shown that steriods as well as hyaluronidase do limit the area of ischemic necrosis in experimentally induced myocardial infarction.[53] For this reason steroids may be beneficial in the treatment of patients with cardiogenic shock, but this conclusion will have to wait results of appropriate clinical trials.

DIGITALIS AND DIURETICS

There has been undue controversy over the use of digitalis in patients with acute myocardial infarction and particularly in those with the shock syndrome. Digitalis is an inotropic agent that increases contractility in patients with acute myocardial infarction. Patients with uncomplicated myocardial infarction do show evidence of an increase in contractility when given digitalis. A decrease in ventricular diastolic pressure accompanies the rise in V_{max} within the first hour after ouabain, but there is little evidence of change in cardiac output, arterial pressure, heart rate, or stroke work.[54] This would be expected in the patients whose circulation was not appreciably compromised by the myocardial infarct. In the shock syndrome the beneficial effect of digitalis has been difficult to demonstrate, although there have been a large number of studies done on the use of this agent. Our own trials would indicate that digitalis does have a beneficial effect in the occasional patient in shock with acute myocardial infarction, particularly if there is evidence of congestive heart failure and cardiac dilatation.[55] Digitalis has a negative chronotropic effect and depresses A-V conduction and is therefore an extremely useful agent in the treatment of atrial fibrillation associated with acute myocardial infarction. Digitalis is also a treacherous agent, since the incidence of serious arrhythmias appears to be increased in the digitalized patient with acute myocardial infarction. Because of the propensity to precipitate arrhythmias, we have used approximately two-thirds of the normal digitalizing dose.

Furosemide and ethacrynic acid have been advocated for the treatment of pulmonary edema associated with pump failure and acute myocardial infarction. It has been suggested that the immediate effects of furosemide in reducing left ventricular filling pressure are due to a sudden increase in venous capacitance rather than to any direct effect on the renal tubules, since a fall in filling pressure precedes the onset of diuresis by several minutes.[56] It has also been noted that the use of furosemide in large doses may precipitate power failure by reducing left ventricular filling pressure at a time when very high filling pressures are needed to maintain adequate cardiac output.[57]

GLUCAGON

Glucagon appears to have limited usefulness in the treatment of the shock syndrome of acute myocardial infarction, but should be kept in the armamentarium of drugs that have an inotorpic effect. It has been shown to increase the cardiac output, heart rate, and rate of left ventricular pressure development.[58] Glucagon was initially used by Linhart et al. in patients with low-output states following cardiac surgery for heart valve replacement.[59] Murtagh et al.,[60] as well as Diamond et al.,[61] treated patients with acute myocardial infarction in shock with glucagon and noticed an inotropic response. However, patients with chronic congestive heart failure tend to have a lesser response to this agent.[62] Glucagon acts by activating the adenyl cyclase system, which mobilizes the high-energy phosphate stores to produce cyclic AMP. Its effectiveness can be enhanced by addition of aminophylline, which blocks phosphodiesterase, the enzyme responsible for breakdown of cyclic AMP.[63,64]

Glucagon given in combination with theophylline has been shown to cause maximal activation of the adenyl cyclase system in isolated myocardial preparations.[65] It has the advantage of activation of this system without the need for the beta receptors, since it is just as effective after beta adrenergic receptor blockade. It also has little chronotropic effect and can be used in the presence of digitalis excess. The major disadvantage of this agent is that it causes nausea on a dose-related basis. This limits the amount of drug that can be given. It also seems to be ineffective in long-term congestive heart failure where the glucagon receptor mechanism appears to be disrupted.[62] Recently there have been reports of accumulation of microaggregates in the pulmonary capillaries after prolonged use of intravenous glucagon. This effect has been the subject of a drug warning. In the desparately ill patient, however, it is possible to give a single intravenous bolus of 4 to 6 mg of glucagon and to observe the effects over the next 20 to 30 min. During this time the drug will be at its peak effectiveness. The occasional patient who responds favorably to this agent can be maintained for several hours with an infusion of 4 to 16 mg of glucagon per hour.

MECHANICAL CIRCULATORY ASSISTANCE: COUNTERPULSATION

Although various forms of circulatory support for patients with acute myocardial infarction have been proposed and occasionally tried, it was not until the development of counterpulsation devices that effective circulatory support became available clinically. Counterpulsation involves the withdrawal of blood from the central aorta during left ventricular systole and reinfusion of blood during diastole. This can be done without actual removal of blood by inserting an inflatable balloon into the dorsal aorta just distal to the left subclavian artery. By inflating the balloon during diastole, blood is forced out of the aorta, and arterial diastolic pressure rises. The balloon is rapidly deflated at the end of diastole, thus reducing the intraaortic volume at the onset of systole. This process allows a drop in central aortic pressure as ventricular ejection occurs and reduces the work requirements of the left ventricle. Thus diastolic pressure is increased while systolic peak pressure is decreased. Cardiac output can be enhanced by this means. In addition, by delivering a high pressure during diastole when the coronary circulation is filling, one may increase coronary blood flow. Scheidt et al. have recently reviewed the cooperative study of 87 patients with cardiogenic shock treated with intraaortic balloon counterpulsation.[66] Their definition of the shock state suggested that an 85% mortality would be expected. In those patients in whom physiologic measurements could be made, the balloon counterpulsation did indeed decrease peak systolic pressure, increase diastolic pressure, and increase cardiac output. There was little change in mean arterial pressure. However, urine flow increased, and myocardial lactate metabolism studies showed an increase in lactate extraction indicating an increase in myocardial aerobic metabolism. Thirty-five of the 87 patients survived discontinuation of circulatory assistance, but only 15 of these lived to leave the hospital. This study demonstrated the feasibility of using this device to support the circulation with relatively few unavoidable adverse side effects from the device itself. It also demonstrated that the investigators had selected a group of patients who had a large area of permanently damaged myocardium. No matter how the patient was supported, the eventual outcome was probably predetermined by the volume of myocardium destroyed. There was not enough functioning myocardium left to allow the heart to be an effective pump for long-term survival without support.

Our own experience in a small group of patients suggests that the intraaortic balloon pump does allow the next step to be taken in treatment of cardiogenic shock associated with acute myocardial infarction. That step is the evaluation of the patient for surgical treatment of the mechanical defect. We have used intraaortic balloon counterpulsation in support of 7 patients in shock following acute myocardial infarction (Table 11-1). These patients are unique in that 5 are long-term survivors and 3 are now in functional class II. In analyzing these good long-term results, two groups of patients should be discussed. Four

Table 11-1
Clinical and Laboratory Data on Seven Patients in Shock with Acute Myocardial Infarction Treated with Intraaortic Balloon Counterpulsation

Name	Age/Sex		EKG	Arrhythmia	Angiography	Surgery	Results
C.D.	57	M	LBBB	VT	100% LAD 80% MAR Apical aneurysm	Bypass LAD Aneurysmnectomy	IABP 56 hr Func. class II
E.M.	46	M	LBBB ALMI	AVD	95% LAD Apical aneurysm	Bypass to LAD Aneurysmnectomy	IABP 48 hr Func. class I
M.R.	49	M	RBBB ALMI	VT	90% LAD 100% MAR Ant + apical aneurysm	Bypass CIRC Aneurysmnectomy	Expired 11 days P.O. Rec. VT
J.S.	57	M	Ant MI RBBB	AVD Asyst	No cath	None	Expired 12 hr after IABP
J.O.	51	M	Lat MI LBBB	AVD Asyst	100% LAD 70% CIRC Apical aneurysm	None	IABP 98 hr Func. class II–III
J.K.	38	F	ALMI	None	Apical aneurysm	None	IABP 72 hr Func. class II
M.S.	56	M	ALMI	AVD Asyst VT	100% LAD 50% MAR Ant + apical aneurysm	Bypass MAR Aneurysmnectomy	IABP 72 hr Func. class II–III

All except patient J.K. had transvenous pacemaker inserted; 3 patients had left bundle branch block (LBBB); 1 patient had right bundle branch block (RBBB); 4 patients had anteroseptal and lateral myocardial infarction (ALMI); 3 of the patients experienced ventricular tachycardia (VT); 4 patients had AV dissociation (AVD).

Angiographic studies refer to percentage occlusion of the left anterior descending coronary artery (LAD) and marginal branch (MAR) of the circumflex artery (CIRC). Patient J.K. had normal coronary arteries.

New York Heart Association functional classification at 3 month follow-up is noted in the results column.

patients had paradoxical motion of the large infarcted area simulating aneurysm formation, and 3 had been in shock for a prolonged period prior to balloon insertion. All 4 had obstruction of the left anterior descending coronary artery near its origin. Use of the intraaortic balloon allowed angiographic study of these patients, definition of their lesions, and resection of the aneurysmal area as well as revascularization of viable areas of myocardium. Three patients survived surgery and were able to discontinue balloon support postoperatively within 56 hr. Two of these patients are now functional class II, while the third patient is still symptomatic with dyspnea on moderate exertion. At another institution we saw 1 additional patient with a similar aneurysmal acute infarct causing shock that was relieved by resection of the aneurysm. Recognition of this type of patient requires mechanical circulatory support to stabilize him for angiographic study and surgical correction. It is our opinion that patients with acute myocardial infarction should not be allowed to die of pump failure without exact anatomic and physiologic definition of the lesion to determine if surgery is feasible and would be helpful.

The other unique patient is a 38-year-old woman with massive anterolateral infarction. This patient recovered after intraaortic balloon support but without surgery. She is now functional class II, but still has occasional episodes of angina. Angiographic studies revealed normal coronary arteries and a large noncontracting area in the anterolateral wall of the left ventricle.

CARDIOGENIC COMPONENT TO SEPTIC SHOCK

Although most of this chapter is directed toward shock associated with acute myocardial infarction, there is a cardiogenic component to septic shock, and there may be a cardiac component associated with hemorrhagic and traumatic shock. Glenn and Lefer proposed the production of a myocardial depressant factor in shock states and think that this material originates in the splanchnic circulation, perhaps in the pancrease.[67] Patients whom we have studied with septic shock do have evidence of left ventricular dysfunction in that they do not respond adequately to elevation of ventricular filling pressures. The hearts are operating on a flat or abnormally depressed left ventricular function curve.[52] Our experience in older individuals on general medical wards indicates that the outline of treatment and the agents used do not differ much from the agents used in shock associated with acute myocardial infarction, except that in general volume expansion can be carried out with less precaution and with a better relationship of the central venous pressure to the left ventricular filling pressure. Isoproterenol has been found in our experience to be an ineffective agent, and norepinephrine and dopamine appear to be better agents in inducing diuresis. We would suspect, however, that in a younger group of patients without diffuse vascular disease, such as patients seen on obstetrical wards, isoproterenol might be a much more effective agent. Mechanical circulatory assistance has not been attempted in these patients, since increasing cardiac output can usually be accomplished with drugs alone. The high mortality is ordinarily associated with the underlying disease rather than with the inability to correct the hemodynamic disorder. The interest in septic shock now is to identify the mediators of the hemodynamic change and perhaps block the release or the effect of these mediators. In addition to the myocardial depressant factor described by Lefer, Nies et al. have shown activation of the kinin system with endotoxin.[68] We have demonstrated that the complement system is frequently activated through the alternate pathway with activation of C3 and C5.[69] These latter components of complement are associated with release of vasoactive substances classed together as anaphylatoxin. It may be that chronic release of these substances under continued endotoxin stimulation produces some of the myocardial depression seen in gram-negative sepsis. There has also been a suggestion that the damage is partly mediated by the release of lysosomal enzymes into the circulation, although the presence of these enzymes may be more a mark of cell destruction than the cause of myocardial depression.[70]

In summary, cardiogenic shock is a syndrome of failure of the heart as a pump to maintain adequate flow and pressure to vital organs. It can best be treated if one knows the physiologic parameters as therapy is designed. A measure of left ventricular filling pressure and measure of intraarterial pressure are absolute prerequisites to appropriate therapy. A measure or an index of cardiac output is helpful, but not essential. If these measurements are available, filling pressure should be elevated to levels of 20 to 24 mm Hg, at which point the maximum effect of the Starling mechanism in adjusting cardiac output will have been reached. If arterial pressure is not depressed, one can increase cardiac output by decreasing impedance to forward flow using an agent such as nitroprusside or phentolamine. If, on the other hand, arterial pressure is reduced, an agent such as norepinephrine, dopamine, or dobutamine can be used to enhance contractility and thus increase cardiac output at the same filling pressure. When these agents do not promptly correct the hemodynamic derangement and the patient does not quickly reduce his need for catecholamine infusion, mechanical circulatory assistance should be undertaken with the hope of stabilizing the patient for angiographic studies to determine if there is a surgically correctable lesion.

ACKNOWLEDGMENT

We wish to thank Patrick J. Scanlon, M.D., for providing data that are presented in Table 11-1.

REFERENCES

1. Orias O: The dynamic changes in the ventricles following ligation of the ramus descendens anterior. Am J Physiol 100:629, 1932
2. Tennant R, Wiggers CJ: The effect of coronary occlusion on myocardial contraction. Am J Physiol 112:351, 1935
3. Wiggers CJ: The functional consequences of coronary occlusion. Ann Intern Med 23:158, 1945
4. Lohmann A: Uber die funktion der brückesfasern an stelle der grossen venen die führung der herztätigkeit beim säugetiere zu übernehmen. Arch Gesamte Physiol 123:628, 1908
5. Cushing EH, Feil H, Stanton EJ, Wartman WB: Infarction of the cardiac auricles (atria). Clinical, pathological and experimental studies. Br Heart J 4:17, 1942
6. Bright EF, Beck CS: Non-penetrating wounds of the heart (A clinical and experimental study). Am Heart J 10:293, 1935
7. Taylor CB, Davis CB Jr, Vawter GF, Hass GM: Controlled myocardial injury produced by a hypothermal method. Circulation 3:239, 1951
8. Agress CM, Rosenburg MJ, Binder MJ, et al: Blood volume changes in protracted shock resulting from experimental myocardial infarction. Am J Physiol 166:603, 1951
9. Agress CM, Rosenberg MJ, Jacobs HI, et al.: Protracted shock in the closed-chest dog following coronary embolization with graded microspheres. Am J Physiol 170:536, 1952
10. Guzman SV, Swenson E, Mitchell R: Mechanism of cardiogenic shock. Circ Res 10:746, 1962
11. Agress CM, Binder MJ: Cardiogenic shock. Am Heart J 54:458, 1957
12. Lluch S, Moguilevsky HC, Pietra G, et al: A reproducible model of cardiogenic shock in the dog. Circulation 39:205, 1969
13. Murphy SD, Charrette EJP, Lynn RB: Experimental coronary artery thrombosis for production of cardiogenic shock. Can J Surg 13:189, 1970
14. Feola M, Limet R, Glick G: Synergistic effects of phenylephrine and counterpulsation in canine cardiogenic shock. Am J Physiol 224:1044, 1973
15. Gilbert RP, Aldrich SL, Anderson L: Cardiac output in acute myocardial infarction. J Clin Invest 30:640, 1951 (abstract)

16. Freis ED, Schnaper HW, Johnson RL, Schreiner GE: Hemodynamic alterations in acute myocardial infarction. I. Cardiac output, mean arterial pressure, total peripheral resistance, "central" and total blood volumes, venous pressure and average circulation time. J Clin Invest 31:131, 1952
17. Agress CM, Glassner HF, Binder MJ, Fields J: Hemodynamic measurements in experimental coronary shock. J Appl Physiol 10:469, 1957
18. Daly IDB, Verney EB: The localization of receptors involved in the reflex regulation of the heart rate. J Physiol 62:330, 1967
19. Aviado DM Jr, Li, TH, Lalow W, et al.: Respiratory and circulatory reflexes from the perfused heart and pulmonary circulation of the dog. Am J Physiol 165:261, 1951
20. Ross J Jr, Frahm CJ, Braunwald E: The influence of intracardiac baroreceptors on venous return, systemic vascular volume and peripheral resistance. J Clin Invest 40:563, 1961
21. Constantin L: Extracardiac factors contributing to hypotension during coronary occlusion. Am J Cardiol 11:205, 1963
22. Kezdi P, Misra SN, Kordenat RK, et al: The role of vagal afferents in acute myocardial infarction. Am J Cardiol 26:642, 1970 (abstract)
23. Brown AM: Excitation of afferent cardiac sympathetic nerve fibers during myocardial ischaemia. J Physiol 190:35, 1967
24. Hanley HG, Costin JC, Skinner NS: Differential reflex adjustments in cutaneous and muscle vascular beds during experimental coronary artery occlusion. Am J Cardiol 27:513, 1971
25. Kuhn L: The treatment of cardiogenic shock. Am Heart J 74:578, 725, 1967
26. Hughes JL, Amsterdam EA, Mason DT, et al.: Abnormal peripheral vascular dynamics in patients with acute myocardial infarction: Diminished reflex arteriolar constriction. Clin Res 19:321, 1971
27. Maroko PR, Kjekshus JK, Sobel BE, et al.: Factors influencing infarct size following experimental coronary artery occlusion. Circulation 43:67, 1971
28. Swan HJC, Forrester JS, Diamond G, et al.: Hemodynamic spectrum of myocardial infarction and cardiogenic shock: A conceptual model. Circulation 45:1097, 1972
29. Hamosh P, Cohn JN: Left ventricular function in acute myocardial infarction. J Clin Invest 50:523, 1971
30. Rahimtoola SH, Loeb HS, Ehsani A, et al.: Relationship of pulmonary artery to left ventricular diastolic pressures in acute myocardial infarction. Circulation 46:283, 1972
31. Russell RO Jr, Rackley CE, Pombo J, et al.: Effects of increasing left ventricular filling pressure in patients with acute myocardial infarction J Clin Invest 49:1539, 1970
32. Ratshin RA, Rackley CE, Russell RO Jr: Hemodynamic evaluation of left ventricular function in shock complicating myocardial infarction. Circulation 45:127, 1972
33. Loeb HS, Rahimtoola SH, Rosen KM, et al.: Assessment of ventricular function after acute myocardial infarction by plasma volume expansion. Circulation 47:720, 1973

34. Gunnar RM, Loeb HS: Use of drugs in cardiogenic shock due to acute myocardial infarction. Circulation 45:1111, 1972
35. Gunnar RM, Loeb HS, Pietras RJ, Tobin JR Jr: The hemodynamic effects of myocardial infarction and results of therapy. Med Clin North Am 54:235, 1970
36. Laks M, Callis G, Swan HJC: Hemodynamic effects of low doses of norepinephrine in the conscious dog. Am J Physiol 220:171, 1971
37. Gunnar RM, Loeb HS, Pietras PJ, Tobin JR Jr: Ineffectiveness of isoproterenol in the treatment of shock due to acute myocardial infarction. JAMA 202:1124, 1967
38. Mueller H, Ayers SM, Giannelli S, et al.: Effect of isoproterenol, *l*-norepinephrine, and intraaortic counterpulsation on hemodynamics and myocardial metabolism in shock following acute myocardial infarction. Circulation 45:335, 1972
39. McNay JL, Goldbert LI: Hemodynamic effects of dopamine in the dog before and after alpha adrenergic blockade. Circ Res 18:I–110, 1964
40. McNay JL, McDonald RH Jr, Goldberg LI: Direct renal vasodilation produced by dopamine in the dog. Circ Res 16:510, 1965
41. Loeb HS, Winslow EBJ, Rahimtoola SH, et al.: Acute hemodynamic effects of dopamine in patients with shock. Circulation 44:163, 1972
42. Tuttle RR, Mills J: Dobutamine: Development of a new catacholamine to selectively increase cardiac contractility. (to be published)
43. Gunnar RM, Loeb HS, Klodnycky M, et al.: Hemodynamic effects of dobutamine in man. Circulation 48:IV–132, 1973
44. Tuttle RR, Pollock GD, Todd G, Tust R: Dobutamine: Containment of myocardial infarction size by a new inotropic agent. Circulation 48:IV–132, 1973
45. Cohn JN: Blood pressure measurement in shock. JAMA 199:972, 1967
46. Gunnar RM, Loeb HS, Pietras RJ, Tobin JR Jr: Hemodynamic measurements in coronary care unit. Prog Cardiovasc Dis 11:29, 1968
47. Chatteerjee K, Parmley WW, Ganz W, et al.: Hemodynamic and metabolic responses to vasodilator therapy in acute myocardial infarction. Circulation 48:IV–152, 1973
48. Majid PA, Sharma B, Taylor SH: Phentolamine for vasodilator treatment of severe heart failure. Lancet 2:719, 1971
49. Kelly DT, Delgado CE, Taylor DR, et al.: Use of phentolamine in acute myocardial infarction associated with hypertension and left ventricular failure. Circulation 47:729, 1973
50. Gold HK, Leinbach RC, Sanders CA: Use of sublingual nitroglycerine in congestive failure following acute myocardial infarction. Circulation 46:839, 1972
51. Franciosa JA, Limas CJ, Guiha NH, et al.: Improved left ventricular function during nitroprusside infusion in acute myocardial infarction. Lancet 1:650, 1972
52. Winslow EJ, Loeb HS, Rahimtoola SH, et al.: Hemodynamic studies and results of therapy in 50 patients with bacteremic shock. Am J Med 54:421, 1973

53. Libby P, Maroko PR, Bloor CM, et al.: Reduction of experimental myocardial infarct size by corticosteroid administration. J Clin Invest 52:599, 1973
54. Rahimtoola SH, Sinno MZ, Chuquimia R, et al.: Effects of ouabain on impaired left ventricular function in acute myocardial infarction. N Engl J Med 287:527, 1972
55. Gunnar RM, Pietras RJ, Stavrakos C, et al.: The physiologic basis for treatment of shock associated with myocardial infarction. Med Clin North Am 51:69, 1967
56. Dikshit K, Vyden JK, Forrester JS, et al.: Renal and extrarenal hemodynamic effects of furosemide in congestive heart failure after acute myocardial infarction. N Engl J Med 288:1087, 1973
57. Kiely J, Kelly DT, Taylor DR, Pitt B: The role of furosemide in the treatment of left ventricular dysfunction associated with acute myocardial infarction. Circulation 48:581, 1973
58. Parmley WW, Glick G, Sonnenblick EH: Cardiovascular effects of glucagon in man. N Engl J Med 279:12, 1968
59. Linhart JW, Barold SS, Cohen LS, et al.: Cardiovascular effects of glucagon in man. Am J Cardiol 22:706, 1968
60. Murtagh JG, Binnion PF, Lal S, et al.: Hemodynamic effects of glucagon. Br Heart J 32:307, 1970
61. Diamond G, Forrester J, Danzig R, et al.: Hemodynamic effects of glucagon during acute myocardial infarction with left ventricular failure in man. Br Heart J 33:290, 1971
62. Goldstein RE, Skelton CL, Levey GS, et al.: Effects of chronic heart failure on the capacity of glucagon to enhance contractility and adenyl cyclase activity of human papillary muscles. Circulation 44:638, 1971
63. Epstein SE, Levey GS, Skelton CL: Adenyl cyclase and cyclic AMP: Biochemical links in the regulation of myocardial contractility. Circulation 43:437, 1971
64. Parmley WW, Sonnenblick EH: Glucagon: A new agent in cardiac therapy. Am J Cardiol 27:298, 1971
65. Marcus ML, Skelton CL, Prindle KH Jr, Epstein SE: Influence of theophylline on the inotropic effects of glucagon. Circulation 42:III–181, 1970
66. Scheidt S, Wilner G, Mueller H, et al.: Intraaortic balloon counterpulsation in cardiogenic shock. Report of a cooperative clinical trial. N Engl J Med 288:979, 1973
67. Glenn TM, Lefer AM: Significance of splanchnic proteases in the production of a toxic factor in hemorrhagic shock. Circ Res 29:338, 1971
68. Nies AS, Forsyth RP, Williams HE, Melmon KL: Contribution of kinins to endotoxin shock in inanesthetized rhesus monkeys. Circ Res 22:155, 1968
69. Robinson JA, Gunnar RM, Klodnycky M: Endotoxin prekallikrin complement in systemic vascular resistance (SVR): Sequential measurements in man. Clin Res 22:38, 1974
70. Miller RL, Reichgott MJ, Melmon KL: Biochemical mechanisms of generation of bradykinin by endotoxin. J Infect Dis 128:s144, 1973

Michael V. Herman
Richard Gorlin

12
Pathophysiology of Ischemic Heart Disease

Unlike other diseases of the heart, angina pectoris and myocardial ischemia had at one time been understood solely in anatomic terms, not at all in physiologic terms. It is only recently that the segmental nature of the disease has been appreciated and that techniques permitting appreciation of local regional events in the heart have been used. As a result there has been found a precise pattern of events that relate directly to local coronary arterial obstruction. This chapter intends to describe the interrelations of these anatomic physiologic phenomena and the clinical picture arising therefrom.

NORMAL CORONARY CIRCULATION

Anatomy

The right and left coronary arteries arise from the aortic root and subdivide into branches that follow an epicardial course over the cardiac surface (Figs. 12-1 and 12-2). The intramural blood supply,[1] particularly of the thick-walled left ventricle, occurs through two separate types of vessels (Fig. 12-3). The A vessels arise from the epicardial arteries at a shallow angle and branch within and supply the outer half of the myocardium. The inner portion of the myocardial wall is supplied by the B vessels, which arise at a right angle and plunge directly through the myocardial substance to ramify in the subendocardial plexus. Thus it can be seen that the B vessel by its very anatomic course is susceptible to

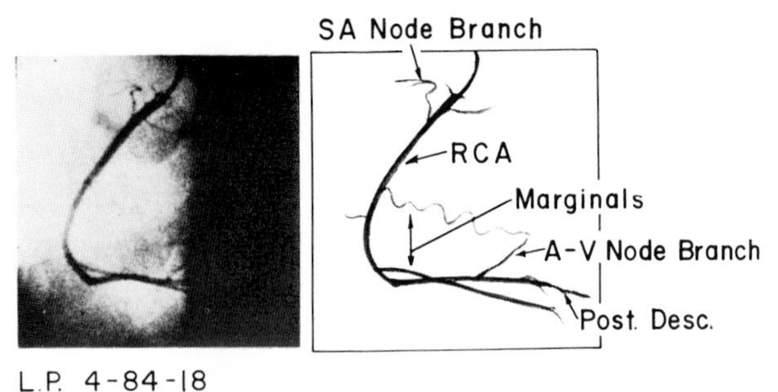

Fig. 12-1. Selective angiogram and schematic drawing of normal right coronary artery. The major epicardial branches are identified.

changes in balance between intramyocardial compressive forces and coronary perfusion pressure.

Determinants of Myocardial Oxygen Consumption

The energy requirement of the human heart in relation to its size is the greatest of any tissue in the body. Furthermore, with applied cardiovascular stress these energy requirements may double or triple. The source of this energy is primarily through oxidative pathways, and the human heart has a relatively limited capacity to generate high-energy phosphate from glycolysis. This means that the nature of the oxygen delivery system is critical to sustained cardiac activity. The oxygen delivery system is essentially a high-extraction low-flow system. While most organs remove about 25% of the delivered arterial oxygen, the heart ex-

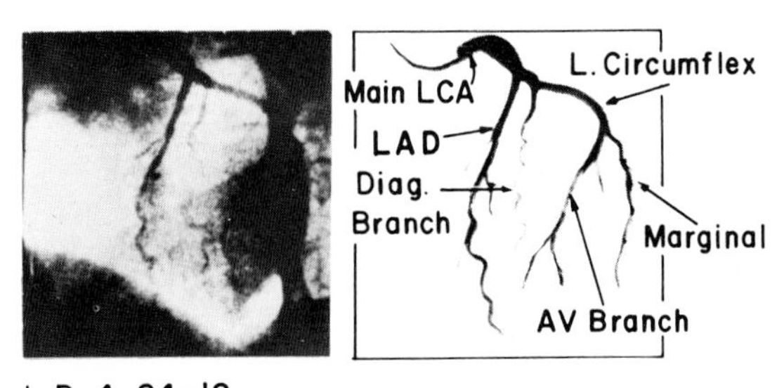

Fig. 12-2. Selective angiogram and schematic drawing of left coronary artery. The major epicardial branches in the right anterior oblique view are identified.

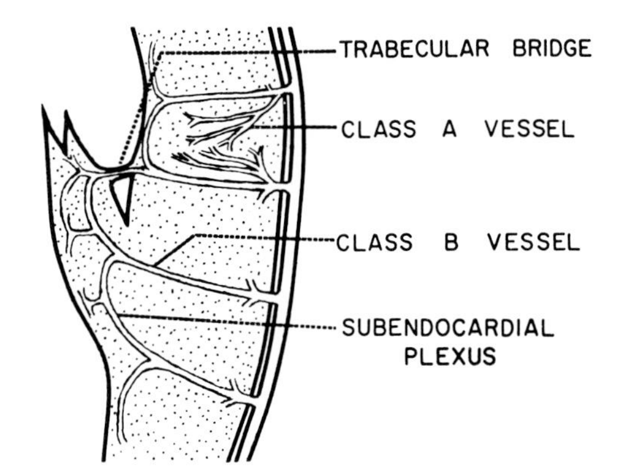

Fig. 12-3. Intramyocardial vascular supply. The A and B vessels are shown in cross section. See text for discussion and reference.

tracts approximately 75% of delivered oxygen (Fig. 12-4). Since this arteriovenous oxygen difference across the heart is near maximum, coronary blood flow becomes the key factor to varying the oxygen supply.

Myocardial oxygen consumption is dependent on the mechanical activity of the heart at any given moment.[2–4] The component factors that determine this mechanical activity are wall force (dependent on ventricular volume and pressure), heart rate, and contraction velocity of the heart muscle. Thus increases and decreases in all of these factors will profoundly influence the need for or create a deficiency in coronary flow.

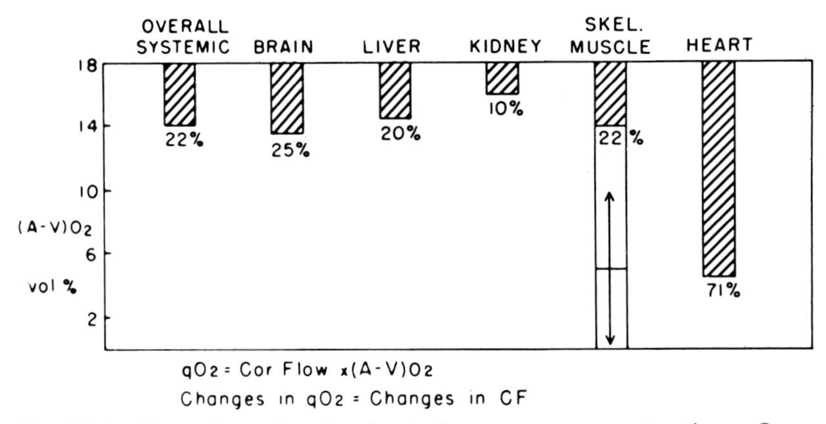

Fig. 12-4. Systemic and regional arteriovenous oxygen extraction. qO_2 = oxygen consumption per minute. Exercising skeletal muscle extracts and increasing but varying amount of O_2 during effort (arrow). Cardiac muscle extraction remains relatively constant and exceeds that of other organs.

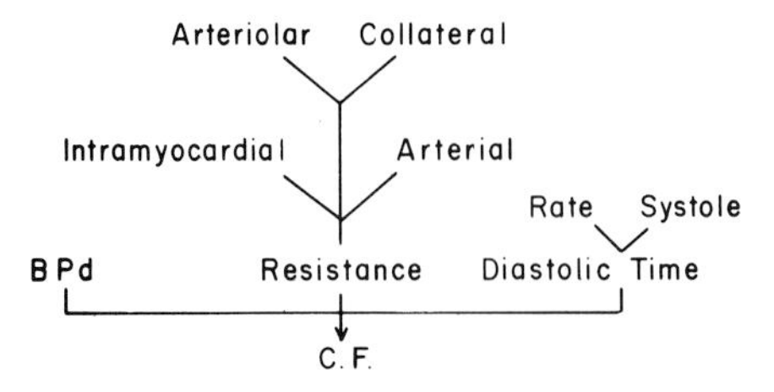

Fig. 12-5. Hydraulic factors affecting coronary flow (C.F.). Diastolic blood pressure (BPd) and time act in concert with resistance factors to deliver coronary flow. There are multiple points of resistance to blood flow.

Mechanical Factors Affecting Delivery of Coronary Blood Flow

Coronary blood flow, as in any other vascular bed, is a direct function of blood pressure and an inverse function of the degree of small vessel resistance (Fig. 12-5). Blood pressure is a primary variable, but because coronary flow occurs mainly in diastole, the diastolic level of pressure and the duration of diastole (heart rate effect) become important modifying factors. Similarly, intramyocardial resistance will add to or subtract from resistance forces intrinsic to the small vessels per se. Small-vessel resistance is a function of a variety of neurogenic and chemical effects on its smooth-muscle cells. Small changes in the diameter of resistance vessels result in profound changes in blood flow for any given pressure.

Other Factors Influencing Coronary Blood Flow

Metabolic Factors

The coronary system is extremely responsive to increased requirements for oxygen. Myocardial oxygen extraction remains fairly constant, while coronary blood flow increases to meet increased demands. Local tissue hypoxia is the most important stimulus to increase coronary flow.[5,6] The exact means by which hypoxia initiates this feedback mechanism has not been defined. It has been suggested, however, that the accumulation of a breakdown product of the adenine nucleotides, some of which are potent vasodilators, serves as the instigating mechanism.[5] Such a hypoxic servomechanism might be operative under the following circumstances: (1) inadequacy of blood flow due to coronary stenosis or to excessive muscular hypertrophy outstripping the oxygen delivery potential or (2) reduction in arterial oxygen content. The same control system could augment flow in response to increased mechanical activity of the heart under normal conditions.

Neurohumoral Factors

There is evidence that the coronary bed contains beta adrenergic receptors that result in dilatation (and also constriction), as well as alpha adrenergic receptors that affect constriction alone. Circulating norepinephrine is a primary coronary vasoconstrictor that results in reduction in flow and a decrease in coronary venous oxygen saturation. On the other hand, epinephrine has a mild vasodilating action. In addition to the effects of circulating products of sympathetic activity, the sympathetic nervous system directly influences arteriolar tone of the coronary bed. Bradykinin can produce vasodilatation. Angiotensin and vasopressin (Pitressin) are potent coronary vasoconstrictors, but whether they are

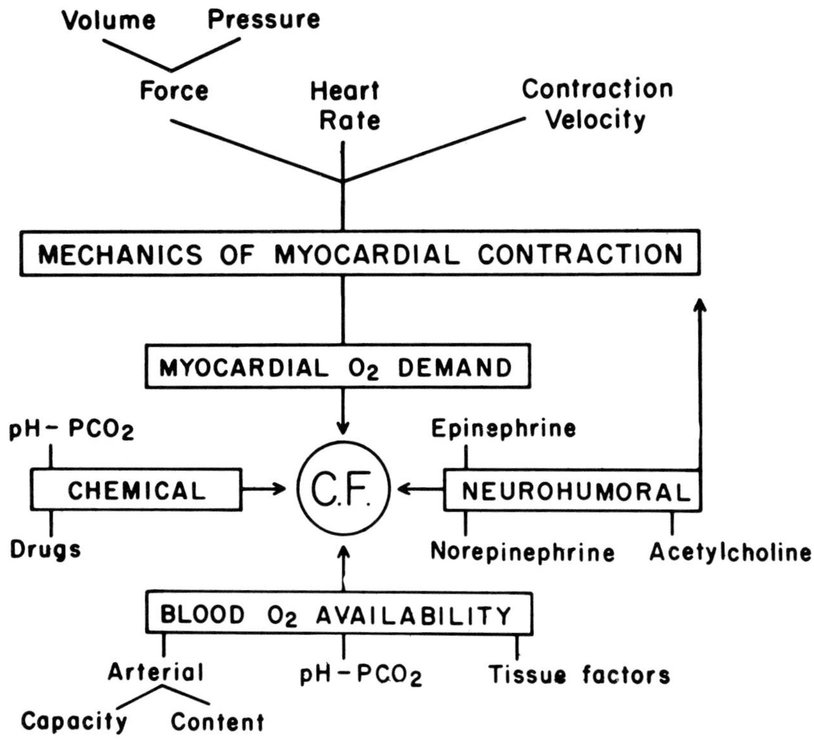

Fig. 12-6. Factors regulating coronary flow (C.F.). These factors can be subdivided into those affecting myocardial O_2 requirements (above), those affecting O_2 availability per unit flow (below), and those acting directly on the arterioles. Neurohumeral factors can affect coronary flow not only through primary vasomotion but also through altered O_2 demands. Likewise pH can affect both arterial resistance and O_2 availability.

released into the circulation under physiologic circumstances is unknown.

Overall Regulation of Coronary Blood Flow

Exclusive of the factors that are mechanically responsible for the actual delivery of coronary blood flow (Fig. 12-5), a schema of factors known to influence coronary flow is presented in Fig. 12-6. These factors can be subdivided into those affecting myocardial oxygen requirements and oxygen availability per unit flow and those acting directly on the arteriole, namely chemical and neurohumoral. Besides their action through primary vasomotion, the neurohumoral factors can also affect coronary flow through altered oxygen demands. Likewise, pH can affect both arteriolar resistance and oxygen availability.

Basal coronary flow will depend in part on the available oxygen per liter of blood flow. Therefore arterial anoxemia or anemia will result in increased coronary flow to keep venous Po_2 and corresponding tissue Po_2 constant. The factors that regulate myocardial oxygen consumption through the mechanical activity of contraction have already been discussed.

Summary

The coronary circulation may be characterized as follows:

1. Coronary flow to the left ventricle occurs principally during diastole rather than throughout the heart cycle as in the right ventricle.
2. Coronary flow is sensitive to small changes in oxygen requirements and thus to changes in myocardial activity.
3. Oxygen extraction is high and flow is relatively low, but there is a large reserve capacity for flow (up to five times resting values); this preserves the high and fairly fixed oxygen extraction characteristics of the system under a wide variety of circumstances.
4. Many factors affect both myocardial oxygen demand and oxygen delivery; they may affect each in opposite directions.

CORONARY ATHEROSCLEROSIS

Basic Pathology

The characteristic lesion (Fig. 12-7) in coronary atherosclerosis is the atheroma, an intimal plaque composed of lipid substances that may become fibrotic or calcified. As the atherosclerotic process progresses

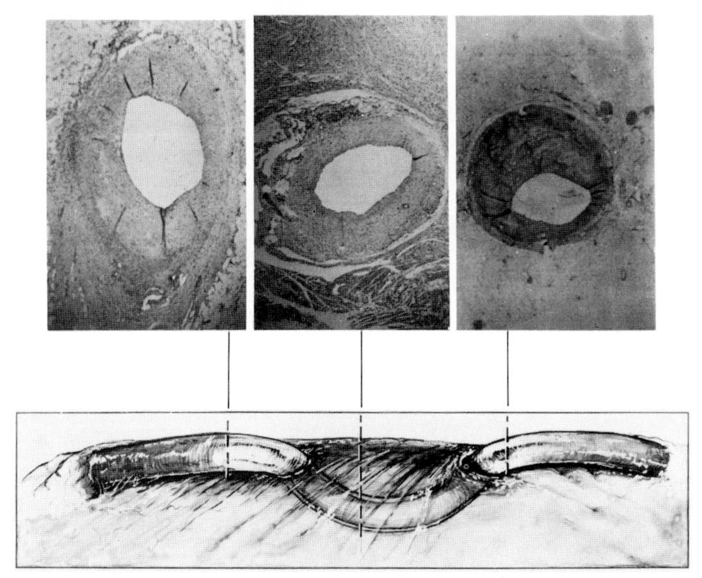

Fig. 12-7. Representations of epicardial (left figure), intramyocardial (center), and distal epicardial (right figure) cross sections of a coronary artery. The site of major atherosclerosis is found in the epicardial portions of the coronary arteries (left figure). The intramyocardial vessels show little or no atherosclerosis.

vessels become deformed, narrowed, and occluded. Certain cardinal characteristics about coronary atherosclerosis must be stressed:[7–8]

1. The epicardial vessels are the site of major atherosclerosis; the intramyocardial vessels show little or no atherosclerosis.

2. Major atherosclerotic lesions are generally found only in the proximal 4 cm of each major coronary artery, although exceptions do occur (Figs. 12-8 and 12-9).

3. Coronary atherosclerosis is a heterogeneous disease in all respects: distribution and severity of obstruction and extent and uniformity of collateral compensation exhibit major and unpredictable variation.

Effects of Coronary Atherosclerosis on Coronary Blood Flow

There is important quantitative significance to the degree of obstruction caused by an atherosclerotic lesion. Figure 12-10 illustrates a discrete atheroma in a coronary artery with several degrees of luminal

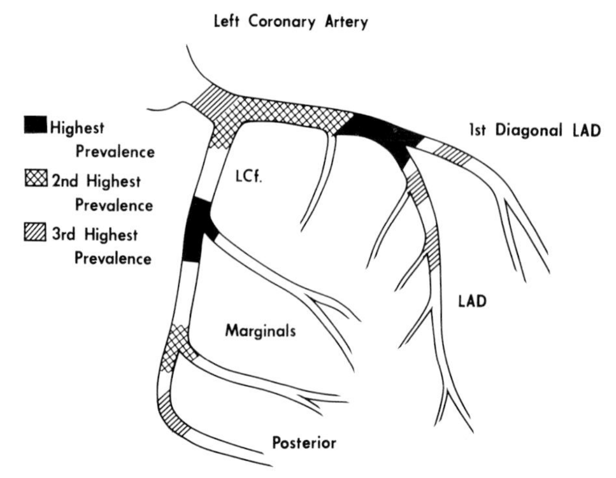

Fig. 12-8. Zones of predilection of the atherosclerotic lesion; left coronary artery. The zones of highest prevalence, second highest prevalence, and third highest prevalence are indicated. Note the occurrence of atherosclerotic disease in the proximal portions of the vessel and at the sites of bifurcations.

narrowing. The lesions producing 25%–50% obstruction probably have little effect in restricting distal coronary flow to the myocardium under stress. It is generally held that a restriction increased blood flow will occur only when the lesion causes greater than 80% luminal narrowing.[9] If, however, the myocardium is hypertrophied or has a chronically increased workload, even the lesser degrees of stenosis can become flow-limiting.

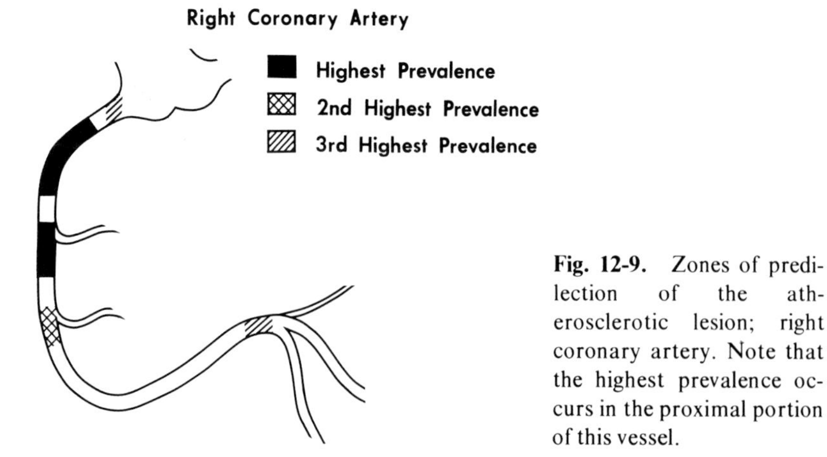

Fig. 12-9. Zones of predilection of the atherosclerotic lesion; right coronary artery. Note that the highest prevalence occurs in the proximal portion of this vessel.

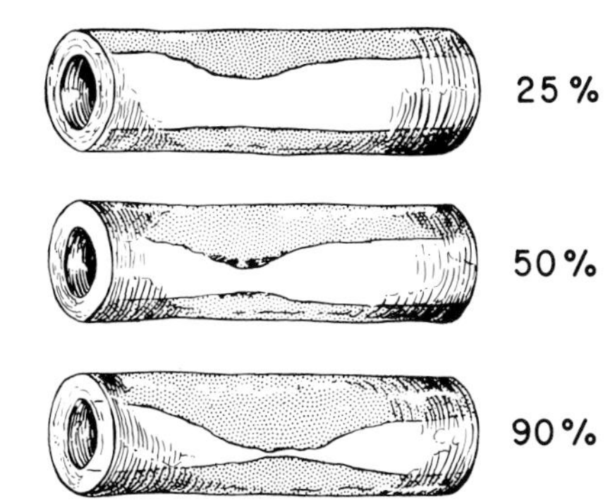

Fig. 12-10. Quantitative significance of an atherosclerotic lesion. Stenotic narrowings of 25%, 50%, and 90% are represented. Only lesions of approximately 80% or greater luminal narrowing restrict blood flow to the distal vessel and myocardium.

Two heterogeneities in blood flow distribution have resulted from coronary atherosclerosis: deep versus superficial myocardial blood flow differences and topographic or regional blood flow differences.

Deep versus Superficial Blood Flow

Owing to the greater compression of the deeper vessels by the contracting heart, an obstruction that progressively reduces perfusion pressure in distal coronary arteries will reduce blood supply to the deep layers before the superficial ones are affected.[10] This, of course, is recognized clinically in ST-segment depression in the electrocardiogram recorded during an attack of angina pectoris. In some cases ischemia of the deeper layers may even occur without significant coronary obstruction.[11]

Regional Differences in Blood Flow

It has been demonstrated in man with coronary atherosclerosis that myocardial flow is regionally nonuniform.[12] The degree of nonuniformity is dependent not only on the degree of local arterial obstruction but also on the extent of compensating collateral vessels. The region of low flow is supplied almost invariably at a greatly reduced arterial perfusion pressure. It is probable that local blood supply is maintained through constant generation of local metabolites that affect vasodilatation.

CHARACTERISTICS OF MYOCARDIAL ISCHEMIA

Definition

Myocardial ischemia exists when energy supply is inadequate for the energy demand of the system.[13] This energy imbalance results in certain biochemical, mechanical, and electrical derangements in the myocardium that may be used to evaluate the human subject with coronary atherosclerosis.

Biochemistry of Myocardial Ischemia

In normal subjects the first response to an increased oxygen demand is an increase in coronary flow. However, in some patients with significant coronary disease flow rates may not increase adequately for the demands of induced stress, and a second adaptive mechanism is necessary; that is, increased extraction of oxygen.[13] If this fails to yield sufficient oxygen the metabolism of the myocardial cell no longer can

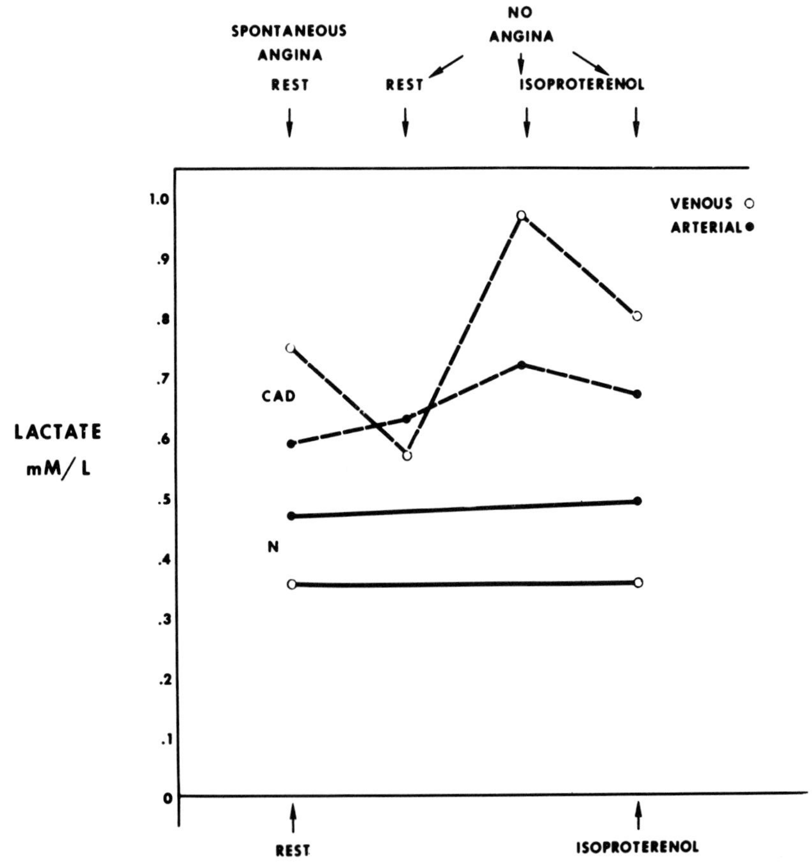

Fig. 12-11. Myocardial lactate metabolism in a normal subject (N) and a subject with coronary artery disease (CAD) under various conditions. See text for explanation.

remain completely aerobic. The cells therefore increase glycolysis in excess of oxidation in order to derive small amounts of high-energy phosphate from glucose or glycogen for the maintenance of cellular activity. This source of cellular energy is limited and cannot support all myocardial cell function for long periods. Lactic acid is produced as the end point of cytoplasmic glycolysis.

This phenomenon of myocardial lactate production has been used as an important tool with which to detect myocardial ischemia in man. This is accomplished by sampling blood simultaneously from an artery and from the coronary sinus. The normal heart is completely aerobic and can extract and actually utilize lactate as a fuel. A change to lactate production with venous levels higher than arterial lactate levels usually indicates inadequacy of oxygenation or ischemia.[14] Figure 12-11 demonstrates this clinical application of myocardial lactate metabolism, with lactate concentration given on the ordinate and conditions of the study on the abscissa. The lower portion of the panel shows a patient with a normal heart, with continuing extraction of lactate both at rest and during the

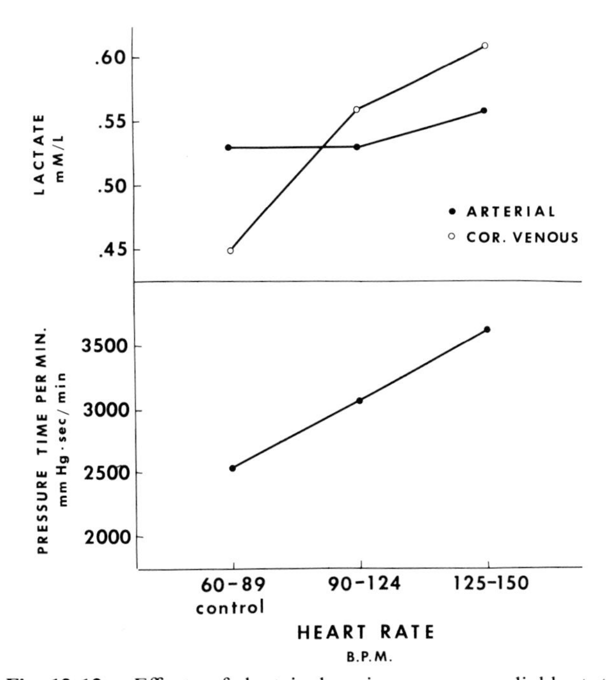

Fig. 12-12. Effects of electrical pacing on myocardial lactate metabolism. A progressive increase in heart rate results in an increase in pressure time per minute and lactate production, with higher heart rates and mechanical stress to the heart.

catecholamine infusion (isoproterenol). The upper panel shows the typical patterns seen in a patient with established coronary artery disease. Extraction of lactate at rest is converted to production under conditions of an anginal attack and during catecholamine-induced augmentation of cardiac activity. Figure 12-12 exhibits the typical results seen with pacing-induced tachycardia at two different heart rates. These points represent pooled patient data. Pressure time per minute is plotted against the myocardial lactate A-V difference. Lactate reversal occurred in subjects with coronary artery disease. Production on the average increased the greater the heart rate or mechanical stress to the heart.

Myocardial Ischemia and Ventricular Function

Cineventriculographic and coronary arteriographic studies in patients have revealed some form of abnormal resting ventricular wall motion in about 50% of patients who present with significant coronary artery lesions.[15] These abnormalities of wall motion are known as left ventricular asynergy (Fig. 12-13) and can be grouped as four distinct abnormalities: (1) akinesis, or local total failure of a portion of the wall to contract; (2) dyskinesis, or paradoxical systolic expansion of a portion of the ventricular wall; (3) asyneresis, or localized hypokinetic contraction; and (4) asynchrony, or a disturbed temporal sequence of contraction.[15] Similar abnormalities can be seen in the myocardium during acute

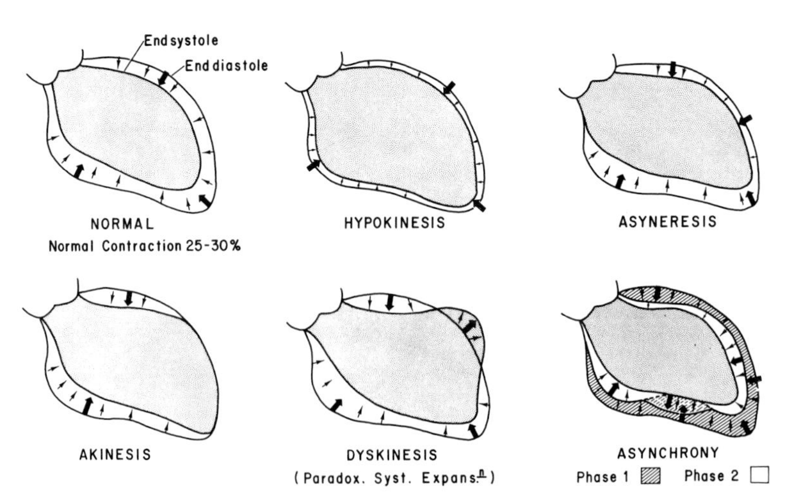

Fig. 12-13. Localized and generalized abnormalities in cardiac contraction. Motion from end-diastole to end-systole is illustrated by arrows. See text for definition and descriptions. Hypokinesis is a generalized reduction in normal degree of contraction.

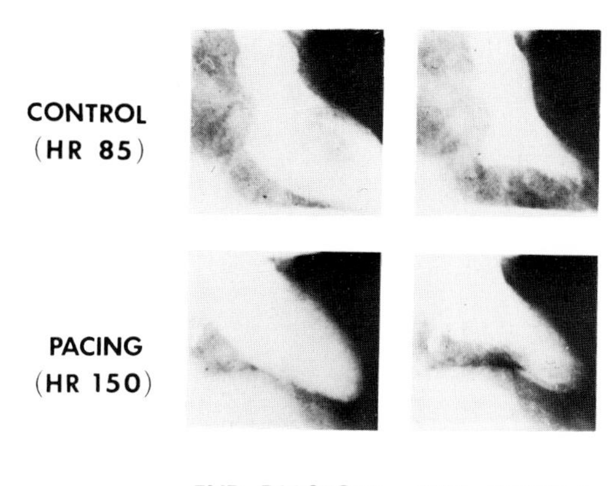

Fig. 12-14. Extensive asynergy induced by atrial pacing. Cineventriculogram at rest (control) and during atrial pacing at 150 beats per minute. Note the development of anterior akinesis during the pacing state (lower panel).

ischemia.[16] This has been shown in isolated muscle preparations studied during acute ischemia,[17] and similarly in man during the same stresses that are known to produce biochemical defects.[18] Figure 12-14 shows cine sequences from patients with known coronary heart disease but with adequate patterns of left ventricular contraction in the resting ventriculogram. Repeat ventriculogram during pacing tachycardia at a rate of 150 beats per minute showed the development of gross akinesis of the anterior apical portion of the left ventricle. This contraction abnormality developed in the distribution of a significant left anterior descending coronary artery lesion. In other patients the acute development of mitral insufficiency has been best demonstrated during acute induced ischemia, suggesting ischemia of the papillary muscle, with resulting dysfunction of the mitral valve apparatus.[19]

Often there is a close correlation with hemodynamic abnormalities, since the hemodynamic function of the entire ventricle can be affected by poorly contracting segments. Overall performance represents the final resultant of poorly contracting and normally contracting segments. Even though basal left ventricular function is unrelated to the degree of coronary artery disease, function during angina is dependent on the diffuseness of the arterial disease and extent of induced ischemia. Evidence of "muscle failure" may first be demonstrated by a decrease in the highly sensitive indices of myocardial contractility recorded during the isovolumic phase of systole. The summation of contraction of normal and ischemic myocardium may then result in a diminished ejection fraction, even before the traditional parameters of elevated and diastolic pressure and reduced cardiac output ("pump failure") are observed.

The dysfunction of the left ventricle can often be suspected on clinical grounds. The particular findings include the recognition of abnormal systolic ventricular bulges (asynergy) and abnormalities of diastolic filling (third and fourth heart sounds).[20]

Myocardial Ischemia and Electrical Factors

Alterations in the electrocardiogram, particularly the ST segment, have been used as indicators of myocardial ischemia. This is classically seen in the postexercise electrocardiogram where significant ST-segment depression has been positively equated with underlying myocardial ischemia. The electrocardiogram during other types of induced stress, such as catecholamine infusion[21] or pacing tachycardia,[22] has also been used to identify the presence of ischemia. Figure 12-15 shows the electrocardiographic and myocardial lactate responses to an isoproterenol infusion in a patient with an isolated ostial stenosis of the left coronary artery. During stress the electrocardiograms showed a 3-mm ST-segment depression associated with gross lactate production by the myocardium.

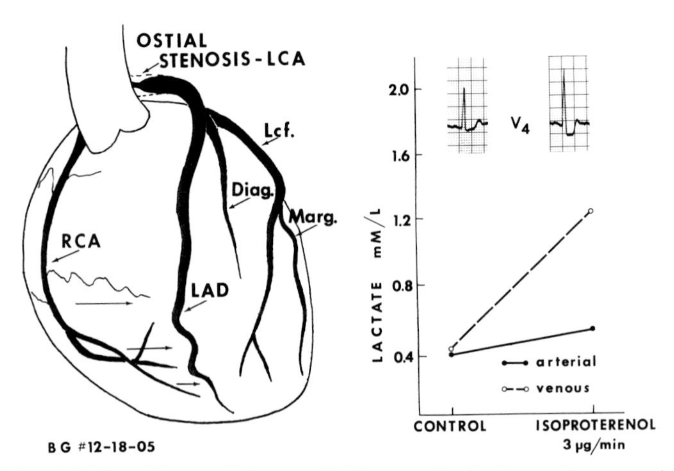

Fig. 12-15. Myocardial metabolism and electrocardiogram in ostial stenosis of the left coronary artery. During a control state, slight production of lactate and a normal cardiogram are noted. During isoproterenol infusion, gross production of lactate and a 3-mm ST-segment depression on the electrocardiogram are noted.

CLINICAL PATTERNS OF ANGINA PECTORIS

In general, it has not been possible to identify pain patterns corresponding to specific types of coronary lesions, nor to correlate clinical patterns of angina pectoris with the presence of stenosis as opposed to complete occlusion. Neither has there been a consistent pain pattern that could be correlated with the presence or absence of a collateral vessel. There has been no correlation of the site of ischemia in the myocardium with the somatic site to which pain has been referred. It has been shown, however, that a sudden change in tempo of the pain of angina pectoris or the frequency of pain may herald an impending myocardial infarction. Such a change usually predicates the immediate need for careful investigation. Certain associated factors should be considered in such a patient. Any hyperkinetic state (e.g., hyperthyroidism or anemia) or a ventricular hypertensive state (e.g., hypertension, aortic stenosis) can induce angina readily by increasing the workload and taxing the oxygen delivery. Factors affecting oxygen-carrying capacity (e.g., acute anemia or hypoxia) or increased viscosity of blood (e.g., polycythemia or macroglobulinemia) can also increase the tempo of angina.

SPECIFIC ANGINAL SYNDROMES

Effort Angina

When the work of the heart is increased, so is its oxygen consumption. It has been shown that a hemodynamic threshold for angina symptoms exists in most patients with significant coronary artery disease. This hemodynamic threshold may be expressed as the product of the heart rate and the systolic arterial pressure with the onset of anginal pain.[23] This threshold has been shown to be reproducible in individual patients, regardless of the type of stress. Thus there is a point at which myocardial energy demands as indicated by the hemodynamic threshold exceed myocardial oxygen delivery, resulting in ischemia.

There is another group of patients who have an unpredictable pain response to a given degree of effort. In these patients pain varies from day to day, but the patient can be made more or less prone to pain by such ancillary factors as eating, cold, or emotional tension. One occasionally encounters another small group of patients who have a "second wind" phenomenon in which the pain appears and then subsides on persistent effort (or after a painful effort, if exertion is repeated, the pain does not recur). The mechanism for this is unclear, but may be related to delayed onset of vasodilatation of the arterioles or collateral vessels in response to metabolic factors such as the arterial lactatic acidemia of persistent exercise.

Emotional Angina

Similar pain responses are observed from the autonomic discharge provoked by excitement, fear, anger, or anxiety. A careful search for changes in heart rate and blood pressure may indicate the importance of the sympathetic nervous system discharge under these states. Such stresses as mental arithmetic have been associated with angina, at which time heart rate and blood pressure have increased.

Spontaneous Angina

During episodes of angina decubitus, or pain that occurs at rest without apparent external provocation, there often is evidence of sympathetic overactivity manifested by a rise in heart rate and blood pressure. Under such circumstances local coronary arteriolar tone may be an important determinant of ischemia. Laboratory studies of patients with spontaneous angina have indicated that such patients have more severe disease and show shifts toward myocardial lactate production and rises in left ventricular systolic and end-diastolic pressure during these attacks.

The initial appearance, or sudden reappearance, of unprovoked angina should arouse suspicion of the preinfarctional state until proven otherwise.

Thermal Angina

It has long been known that cold reduces exercise tolerance in angina patients. It has been proposed that cold results in a reflex vasoconstriction of the peripheral and possibly the coronary circulation. The peripheral constriction results in hypertension and increased impedance to left ventricular ejection and results in the ischemic state.

Nocturnal Angina

Two types of nocturnal anginal patterns are recognized. In one group angina rarely occurs except at night and may be related to a drop in perfusion pressure across a coronary artery stenosis. More commonly, however, nocturnal angina indicates severe, extensive coronary disease. It has been suggested that such pain is preceded by increased dream activity.[24] Therefore, sympathetic discharge, similar to that seen in excitement, may mediate the increased oxygen requirement of the myocardium. It seems also possible that nocturnal angina may be the equivalent of paroxysmal nocturnal dyspnea and acute left ventricular failure.

Prandial Angina

The mechanism of pain production associated with eating, as well as the use of cigarettes, coffee, and drugs, is poorly understood. This type of angina usually is the sign of advanced coronary disease. Variables that may be important in the production of prandial angina include the sympathetic nervous system response, conditioned reflexes, and concomitant gastrointestinal disease associated with the summation of stimuli. No significant difference in the incidence of gastrointestinal or gallbladder disease in those patients who have prandial angina as opposed to those without prandial angina has been found.

Angina Associated with Tachycardia or Arrhythmia

Tachycardia causes a reduction in coronary filling time during diastole, while simultaneously increasing myocardial oxygen requirements. Occasionally the patient may not be aware of the presence of tachycardia or arrhythmia, and such episodes may be called spontaneous angina.

THERAPEUTIC AGENTS AND ANGINA PECTORIS

Nitroglycerin and Related Compounds

Although nitroglycerin is the most widely used antianginal preparation, its mechanism of action in relieving the pain of myocardial ischemia is still uncertain. Two theories, not mutually exclusive, have received widespread acceptance: relief of angina may be due either to increased coronary blood flow or to a reduction in the work of the heart.

In the first instance nitroglycerin is thought to increase perfusion in ischemic regions by altering regional blood flow. Its site of action may be both large and small coronary arteries, as well as collateral vessels.

In the second instance, indirect actions of nitroglycerin are considered most important: smooth-muscle relaxation in the peripheral systemic vessels (arteries, veins, or both) results in a reduction of intraventricular volume, pressure, and most importantly, intramyocardial wall tension. As a result, oxygen demands of the heart are reduced and do not exceed oxygen delivery through the diseased coronary system. These effects have been demonstrated both at rest and during stress such as exercise.[25-27]

The mode of action of longer acting nitrate preparations is similar to that of nitroglycerin, but the therapeutic advantages of these preparations, as compared to nitroglycerin, are still largely unproven.

Beta Adrenergic Blocking Agents

The cardiac excitatory effects (increased heart rate and increased myocardial contractility) produced by sympathomimetic amines are mediated mainly by the beta adrenergic receptors. Blockade of these receptors will thereby result in a decrease in each of these factors and a decrease in the energy requirements of the heart.[28,29] Propranolol is the most widely used of these agents and has been shown to reduce both myocardial oxygen requirements and coronary blood flow by causing a decrease in heart rate and myocardial contractile state.[30] The drug's clinical effectiveness in relieving angina has been demonstrated by an improved exercise tolerance in patients with effort-induced angina. In such patients the postexercise electrocardiogram demonstrates less evidence of ischemia. Laboratory observations have demonstrated that for the same total-body workload during supine leg exercise, propranolol attenuates the rise in myocardial oxygen consumption in direct proportion to the reduced cardiac effort.[30]

Double-blind clinical trials in the treatment of angina with propranolol have shown a significant improvement with reduced severity and frequency of attacks of angina pectoris and reduction in nitroglycerin requirement in about 80% of patients with significant coronary artery disease. Adverse side effects include bradycardia, heart block, and the development of heart failure and bronchospasm. In clinical practice these adverse effects are usually minimal and seldom contraindicate the use of the drug.[30]

Other Agents

Digitalization of anginal patients with ventricular dilatation or heart failure may lead to a reduction in ischemic pain. Digitalization causes a reduction in heart size and heart rate and developed tension in such individuals. This may effectively counteract the increase in myocardial oxygen consumption caused by the drug's enhancement of contractile state. Since so many patients with angina pectoris do not have cardiomegaly or congestive failure, and because of the interplay of the above-mentioned factors, it is not surprising that variable responses in exercise tolerance are observed following digitalization of patients with coronary artery disease and anginal complaints. In practice, digitalization is rarely effective.

Stimulation of the carotid sinus nerves has proved useful in the treatment of incapacitating angina in selected patients.[31] Carotid sinus nerve stimulation reduces myocardial oxygen requirements by a reflex decrease in sympathetic activity. A fall in heart rate and systemic blood pressure occurs, reducing the anginal threshold accordingly.

Phlebotomy of patients during anginal episodes (provoked by atrial pacing) was shown to be of benefit in ameliorating pain.[32] When the blood was reinfused, angina returned. Reduction in left ventricular volume (and stroke work) was the postulated mode of action.

SUMMARY

The blood vessels that supply the heart are variably afflicted by both the location and the degree of obstruction caused by coronary atherosclerosis. This means that the disease will affect the heart muscle in a patchy and unpredictable fashion. Furthermore, not only are the changes that take place in cardiac muscle nonuniform, but also there will be variability in the type and degree of stress that can bring out the inadequacy of blood supply to the given affected segments. As a result of the abnormalities in blood supply to the working heart muscle, there occur certain distortions in the chemistry of the heart muscle cell. These distortions in chemistry affect the way in which the electrical impulse of the heart is propagated and the muscle cell activated and restored. These chemical disturbances may interfere with proper contraction of the cardiac muscle. The first type of disturbance, namely the biochemical change, can be detected by studying abnormalities in the blood that leaves the vein draining the heart muscle proper. Abnormalities in the substances found in this vein have given an exquisitely sensitive means of detecting lack of adequate blood supply to the heart muscle. Electrical disturbances are recognized by study of the electrocardiogram under a variety of conditions. Mechanical disturbances in contraction have been observed by simple palpation and by auscultation of abnormal heart sounds, as well as by complex techniques such as cineventriculography to study the contraction of the heart muscle. Cineventriculography may exhibit disturbances in the way various segments of the heart contract when these are starved for blood supply. A combination of these various techniques permits an accurate appraisal of the state of the cardiac muscle in relation to its blood supply. The integration of all this information permits objective and accurate planning of therapy.

REFERENCES

1. Estes EH Jr, Entman ML, Dixon HB II, Hackel DB: The vascular supply of the left ventricular wall. Am Heart J 71:58, 1966
2. Sarnoff SJ, Braunwald E, Welch GH Jr, et al: Hemodynamic determinants of oxygen consumption of the heart with special reference to the tension-time index. Am J Physiol 192:148, 1958
3. Rolett EL, Yurchak PM, Hood WB Jr, Gorlin R: Pressure-volume correlates of left ventricular oxygen consumption in the hypervolemic dog. Circ Res 17:499, 1965
4. Sonnenblick EH, Ross J Jr, Covell JW, et al: Velocity of contraction as a determinant of myocardial oxygen consumption. Am J Physiol 209:919, 1965
5. Gregg DE: Coronary Circulation in Health and Disease. Philadelphia, Lea & Febiger, 1950
6. Berne RM: Regulation of coronary blood flow. Physiol Rev 44:1, 1964
7. Blumgart HL, Schlesinger MJ, Davis D: Studies on the relation of the clinical manifestations of angina pectoris, coronary thrombosis and myocardial infarction to the pathological findings. Am Heart J 19:1, 1940
8. Zoll PH, Wessler S, Blumgart HL: Angina pectoris, a clinical and pathologic correlation. Am J Med 11:331, 1951
9. Smith SC Jr, Gorlin R, Herman MV, et al: Myocardial blood flow in man: Effects of coronary collateral circulation and coronary artery bypass surgery. J Clin Invest 51:2556, 1972
10. Griggs DM Jr, Nakamura Y: Effects of coronary constriction on myocardial distribution of iodoantipyrine-I^{131}. Am J Physiol 215:1082, 1968
11. Buckberg GD, Fixler DE, Archie JP, Hoffman JIE: Experimental subendocardial ischemia in dogs with normal coronary arteries. Circ Res 30:67, 1972
12. Sullivan JM, Taylor WJ, Elliott WC, Gorlin R: Regional myocardial blood flow. J Clin Invest 46:1402, 1967
13. Gorlin R: Pathophysiology of cardiac pain. Circulation 32:138, 1965
14. Herman MV, Elliott WC, Gorlin R: Electrocardiographic, anatomic, and metabolic study of zonal myocardial ischemia in coronary heart disease. Circulation 35:834, 1967
15. Herman MV, Heinle RA, Klein MD, Gorlin R: Localized disorders in myocardial contraction. N Engl J Med 277:222, 1967
16. Tennant R, Wiggers CJ: Effect of coronary occlusion on myocardial contraction. Am J Physiol 112:351, 1935

17. Tyberg JV, Parmley WW, Sonnenblick EH: In-vitro studies of myocardial asynchrony and regional hypoxia. Circ Res 25:569, 1969
18. Pasternac A, Gorlin R, Sonnenblick EH, et al: Abnormalities of ventricular motion induced by atrial pacing in coronary artery disease. Circulation 45:1195, 1972
19. Brody W, Criley JM: Intermittent severe mitral regurgitation. N Engl J Med 283:673, 1970
20. Cohn PF, Vokonas PS, Williams RA, et al: Diastolic heart sounds and filling waves in coronary artery disease. Circulation 44:196, 1971
21. Kemp HG, Most AS, Gorlin R: Correlation between electrocardiographic and metabolic evidence of myocardial ischemia in man. Circulation [Suppl 6] 38:113, 1968
22. Parker JO, Chiong MA, West RO, Case RB: Sequential alterations in myocardial lactate metabolism, ST segments, and left ventricular function during angina induced by atrial pacing. Circulation 40:113, 1969
23. Robinson BF: Relation of heart rate and systolic blood pressure to the onset of pain in angina pectoris. Circulation 35:1073, 1967
24. Nowlin JB, Troyer WG Jr, Collins WS, et al: Association of nocturnal angina pectoris with dreaming. Ann Intern Med 63:1040, 1965
25. Brachfeld N, Bozer J, Gorlin R: Action of nitroglycerin on the coronary circulation in normal and mild cardiac subjects. Circulation 19:697, 1959
26. Bernstein L, Friesinger GC, Lichtlen PR, Ross RS: The effect of nitroglycerin on coronary circulation in man and dog: Myocardial blood flow measured with xenon^{133}. Circulation 33:107, 1966
27. Najmi M, Griggs DJ, Kasparian H, Novack P: Effects of nitroglycerin on hemodynamics during rest and exercise in patients with coronary insufficiency. Circulation 35:46, 1967
28. Hamer J, Grandjean T, Malendez L, Sowton E: Effect of propranolol (inderal) in angina pectoris: Preliminary report. Br Med J 2:720, 1964
29. Hamer J, Sowton E: Effects of propranolol on exercise tolerance in angina pectoris. Am J Cardiol 18:354, 1966
30. Wolfson S, Gorlin R: Cardiovascular pharmacology of propranolol in man. Circulation 40:501, 1969
31. Epstein SE, Beiser GD, Goldstein RE, et al: Treatment of angina pectoris by electrical stimulation of the carotid sinus nerves: Results in 17 patients with severe angina. N Engl J Med 280:971, 1969
32. Parker JO, Case RB, Khaja F, et al: The influence of changes in blood volume on angina pectoris: A study of the effect of phlebotomy. Circulation 41:593, 1970

Charles E. Rackley
William P. Hood, Jr.

13
Aortic Valve Disease

Disease of the aortic valve imposes mechanical burdens on the left ventricle that alter its function and stimulate various adaptive mechanisms over a period of time. Primary abnormalities of the aortic valvular apparatus are stenosis or incompetence or combinations of the two. With the exception of acute disruption of the aortic valve apparatus, which can produce sudden severe aortic incompetence, lesions that produce regurgitation or stenosis of the valve are chronic in nature. Patients generally tolerate disturbance of the aortic valve function for lengthy periods of time before development of clinical symptoms.[1-4] During this period compensatory mechanisms are operative and maintain the mechanical performance of the left ventricle adequately at rest and during times of stress.

An appreciation of the overload imposed on the left ventricle by these disturbances of aortic valve function and the compensatory mechanical adaptations of the ventricle can enhance our understanding of the clinical manifestations. Therefore the nature of the physiologic disturbances imposed on the left ventricle, the methods of quantitating these overloads, the compensatory mechanisms, and the clinical and laboratory findings will be reviewed.

Supported by the Myocardial Infarction Research Unit Program, Department of Health, Education and Welfare, Contract PH43-67-1441 and by the Cardiovascular Research and Training Center, Department of Health, Education and Welfare, Program Project Grant HE11,310.

PHYSIOLOGIC ASPECTS OF PUMP PERFORMANCE OF THE LEFT VENTRICLE

The pump function of the left ventricle can be more clearly understood if the mechanisms of contraction and ejection and the loads imposed on the myocardial fibers are appreciated. A useful approach to this analysis is the delineation of the pressure–volume relationships of the left ventricle under normal circumstances and alterations that follow mechanical disturbances in the aortic valve.

The law of Laplace is a mathematical principle that permits an estimate of wall tension in a chamber with an infinitely thin wall. The mathematical expression of this principle is $P = T(1/R_1 + 1/R_2)$, where P = pressure, T = tension, and R_1 and R_2 = radii of curvature.[5] Therefore pressure within the chamber is proportional to the tension divided by the radius of curvature, and conversely tension is proportional to the pressure in the chamber times the chamber radius. Then tension described in mechanical terms is the force tending to disrupt the membrane wall as determined by the pressure within the chamber and the radius of the chamber. Early physiologic experiments employed the term tension interchangeably with pressure.[6,7] However, tension as properly defined represents the product of chamber pressure and radius.

Two important considerations must be emphasized in applying the law of Laplace to any estimation of forces acting within the wall of the left ventricle. First, an appropriate geometric model must be chosen. In recent experimental and clinical studies a frequently selected geometric reference figure for the intact left ventricle is the prolate ellipsoid.[8,9] Common examples of a prolate ellipsoid are footballs and eggs; each of these three-dimensional figures has two minor or cross-sectional axes that are similar and a major or long axis.

Second, the left ventricle must be recognized as a chamber with a finite wall thickness. If measurements of pressure can be obtained within the ventricular cavity and the radius measured, endocardial tension can be calculated; but this term has limitations, since consideration is not given to the thickness of the ventricular wall.[10] Another engineering term that is more appropriate for the analysis of mechanical ventricular function, and in particular force distribution across myocardial fibers, is stress.[10–13] The basic Laplace relationship can be modified to take into account wall thickness: pressure/h is proportional to stress/R. Therefore, stress is proportional to $P \times R/h$. Stress forces are exerted in several planes within or on the ventricular wall: circumferential or equatorial, longitudinal or meridional, and radial or perpendicular to the endocardium.[13] Figure 13-1 shows diagrammatically circumferential and

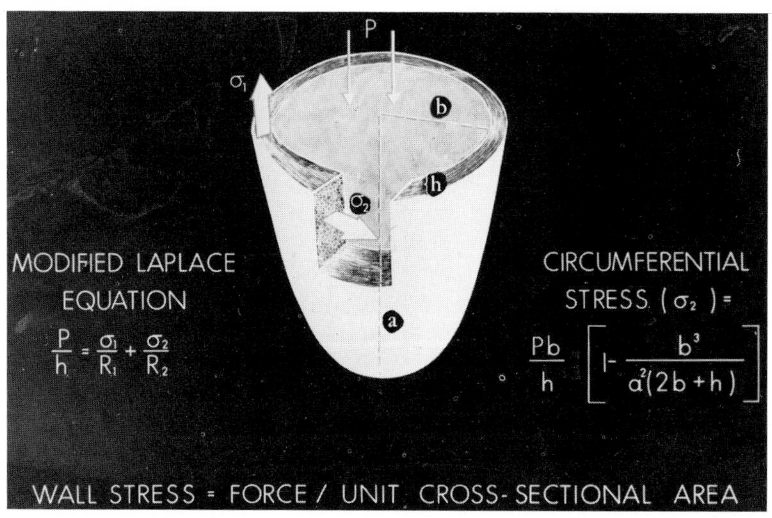

Fig. 13-1. A diagrammatic cross section of the left ventricle to show wall forces. The modified Laplace equation is given for the calculation of stress as force per unit of cross-sectional wall area. P = left ventricular pressure; h = left ventricular wall thickness; a = major semiaxis; b = minor semiaxis; σ_1 = longitudinal wall stress; σ_2 = circumferential wall stress; R_1 and R_2 = major and minor radii of curvature.

longitudinal stress in a model of the left ventricle. In the passive sense stress is the force tending to rupture the ventricular wall. In its active sense stress is the force within the wall that generates pressure within the chamber.

The basic Laplace relationships and wall stress are important in understanding the overload imposed on the normally functioning left ventricular myocardium by pathologic conditions of the aortic valve.[14] Since the pressure in the root of the aorta is higher than that in the left ventricle in diastole, incompetence of the aortic valve will allow regurgitation during the diastolic phase of the cardiac cycle. In addition, the left ventricle will receive the normal diastolic volume from the pulmonary veins and left atrium. A volume overload on the ventricle will be produced, and in order to accommodate this volume overload the ventricle must increase its diastolic dimensions (radius). Obstruction of flow across the aortic valve created by stenosis of the valve will result in a higher pressure in the left ventricle than in the aorta during the systolic phase of ejection, but no increase in chamber radius. The stress in the ventricular wall will be proportional to the elevation of the systolic chamber pressure. Therefore a volume overload on the left ventricle will

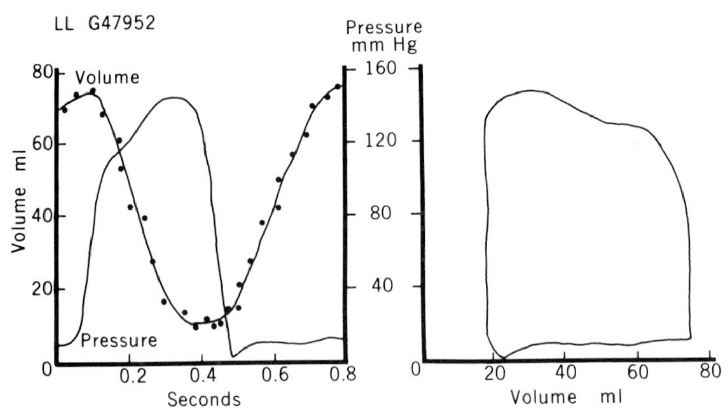

Fig. 13-2. Volume and pressure changes obtained from a patient with a normal left ventricle are displayed on the left, timed with respect to onset of the QRS complex in the electrocardiogram. On the right, the pressure and volume are related to construct a pressure–volume loop. Reproduced by permission from Grossman W (ed.): Cardiac Catheterization and Angiography. Philadelphia, Lea & Febiger, 1974, Chap. 17, p 176.

increase the radius of curvature (R), and a pressure overload will increase the pressure (P) for the calculations of wall stress.

Since changes in chamber pressure, ventricular dimensions, and wall thickness occur both in the systolic and diastolic phases of the cardiac cycle, it will be useful to review the normal pressure–volume relationships and wall forces in the course of the single contraction of the left ventricle. Figure 13-2 illustrates normal left ventricular pressure and volume curves from the ventricle of a patient without heart disease. For timing purposes these measurements are related to the onset of the QRS complex of the electrocardiogram for the cardiac cycle. Throughout diastole there is a comparatively low pressure within the left ventricle. Following atrial contraction, ventricular filling is complete. After depolarization of the ventricular myocardium, there is a rapid rise in the ventricular pressure curve with closure of the mitral valve. The phase of rapid pressure development without change in chamber volume is the isovolumic phase of contraction. After the aortic valve has opened and blood is ejected into the systemic circulation, the pressure curve reaches its peak during the systolic ejection phase. The ventricular volume has remained unchanged during the isovolumic phase, but following opening of the aortic valve there is rapid decline in chamber volume, which reaches its low point at end-systole. This represents the end-systolic volume of the left ventricle. Thereafter the chamber pressure falls rapidly, the aortic valve closes, and the isovolumic relaxation phase ensues. There is no change in ventricular volume until the pressure falls below that of the left atrium. With the opening of the mitral valve, rapid diastolic filling occurs, with a rapid increase in chamber volume associated with very small changes in pressure. These pressure–volume changes then represent the four phases of a cardiac cycle: isovolumic contraction, systolic ejection, isovolumic relaxation, and diastolic filling. The pressure and volume changes throughout the cardiac cycle have been combined in Fig. 13-2 to display a pressure–volume diagram. At end-diastole the chamber pressure is low and the chamber volume is maximum. Following ventricular depolarization there is a rapid elevation in pressure without change in volume, which is reflected in the ascending (right-hand) portion of the diagram. Aortic valve opening is followed by reduction in volume but continued elevation in pressure, as represented by the superior aspect of the diagram, the ejection phase. Aortic valve closure results in rapid fall of pressure without volume change, as reflected in the descending (left-hand) portion of the diagram, the isovolumic relaxation phase. Mitral valve opening is then associated with rapid increase in volume with a very small increase in pressure—the diastolic filling phase. The pressure–volume loop is the mechanical expression of work done by the left ventricle.[15] Total work is the area within and beneath the loop. Work done on the ventricle during diastolic filling is the area beneath the loop, and the net systolic work is then the area within the loop.

Measurements of left ventricular chamber pressure and left ventricular volume provide two components in the modified Laplace equation. In order to calculate wall stress, an additional measurement of left ventricular wall thickness is required. This can be obtained from angiocardiograms of the left ventricle by measuring a segment of the left ventricular wall along the left cardiac border. Figure 13-3 shows a left parent in consideration of the Laplace expression modified to include stress and wall thickness changes throughout a cardiac cycle in a patient without heart disease. All measurements were related to the onset of the QRS cycle of the electrocardiogram. At end-diastole both the ventricular pressure and wall stress are low, and it is apparent that wall thickness is thin. With the onset of the isovolumic contraction phase there is a rapid increase in wall stress that slightly precedes that of left ventricular pressure. Once the ventricular pressure exceeds the aortic pressure, ventricular ejection occurs and the volume of the left ventricle is proportionately reduced. During the early systolic ejection phase, wall

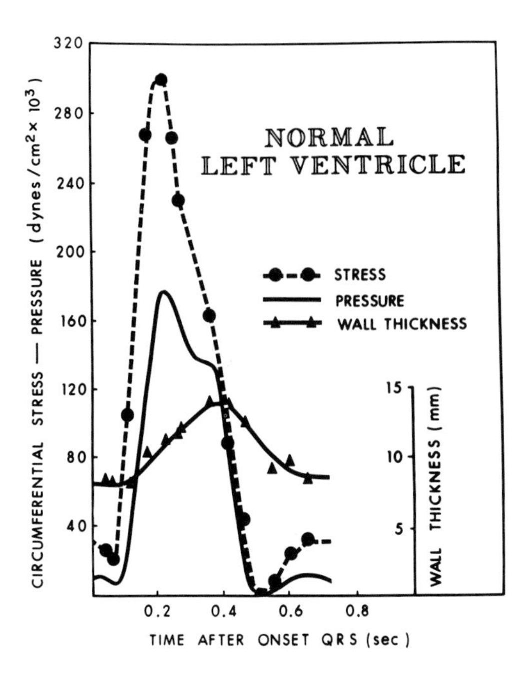

Fig. 13-3. Left ventricular pressure recording and wall stress measurements throughout the cardiac cycle are described in a patient with a normal left ventricle. Changes in left ventricular wall thickness are also represented. Reproduced by permission from Brest AN (ed.): Diagnostic Methods in Cardiology, Philadelphia, Davis, 1975, pp 287, 289, 290.

stress and pressure continue to rise. Wall stress peaks early during the systolic ejection phase along with the left ventricular pressure, but wall stress begins to fall earlier than pressure. The explanation becomes apparent in consideration of the Laplace expression modified to include wall thickness. Stress is directly proportional not only to the chamber pressure but also to the radius or ventricular dimension and is indirectly proportional to the wall thickness. Therefore, even though the pressure remains elevated during the latter phase of systole, there occurs a reduction in left ventricular volume and hence a reduction in the chamber radius, while there is an increase in the wall thickness. This concomitant reduction in chamber volume and increase in wall thickness reduce the stress within the ventricular wall even though chamber pressure is being maintained at its near-maximum level during late systole. The stress curve rather than the pressure curve represents the force resisting shortening, or afterload, throughout systole.

These considerations of the Laplace expression of wall stress and pressure–volume relationships of the left ventricle can be useful not only in delineating normal mechanics of left ventricular function but also in appreciating the extent and severity of altered ventricular contraction following disruption of normal aortic valve function.

PATHOPHYSIOLOGY OF AORTIC STENOSIS

Pathology

Aortic stenosis results from an inability of the aortic valve to open adequately during the systolic ejection phase of the cardiac cycle. Alterations in valve function can be the result of congenital abnormalities of the aortic valve. Normally the aortic valve is tricuspid, but bicuspid valves may be present at birth. Bicuspid valves have a propensity to lose their flexibility during systolic opening and also to calcify during the course of the patient's life. Significant obstruction to the aortic flow can occur in childhood, and a bicuspid valve is the most frequent cause in this setting.[16] Rheumatic disease also affects the aortic valve, and the subsequent scar formation can restrict the opening of the valve, initially by fusion of the commissures with gradual loss of tissue flexibility and ultimate calcification. Elderly patients may also develop a form of aortic stenosis, but the calcium present in these valves may be localized in the sinuses of Valsalva, while the commissures are not as severely damaged as in rheumatic aortic stenosis. Pathologic studies in recent years have suggested that most instances of isolated aortic stenosis can be attributed to a bicuspid valve at birth, whereas aortic valve disease associated with mitral valve disease is rheumatic in origin.[17,18]

Physiologic Overload

Aortic stenosis, regardless of etiology, inflicts a pressure overload on the left ventricular myocardium, as the ventricle has to generate higher pressures to eject the same volume of blood per beat across a stenotic aortic valve.

In general, Poiseuille's law determines flow through the vascular system, where ordinarily flow rate is relatively low and pressure losses are frictional. In the case of orifices, however, the flow rate is high and significant pressure energy is converted to velocity, so that Poiseuille's law does not apply. One must then employ Toricelli's orifice equations where flow $F = A \times V \times C_C$, where A = area of orifice, V = velocity of flow, and C_C = coefficient of orifice contraction.[19]

Velocity may be calculated from the equation $V = C_V \sqrt{2g} \times \Delta P$, where g = acceleration due to gravity (980 cm/sec^2), ΔP = pressure differential across the orifice, and C_V = coefficient of velocity. These equations for flow and velocity may be combined, resulting in the equation for area: $A = F/(44.5 \sqrt{\Delta P} \times C_V \times C_C)$.[19]

Gorlin has applied this expression to the calculation of aortic valve area from measurements of aortic valve flow and the square root of the pressure gradient during the period of ejection.[19,20] The normal size of the aortic valve varies from 2.6 cm^2 to 3 cm^2, and the critical aortic valve area for development of significant symptoms is less than 0.8 cm^2.[21]

Compensatory Mechanisms

As the cross-sectional area of the aortic valve is reduced and the ventricular systolic pressure is commensurately elevated, the pressure load influences the myocardial fibers in several ways. The ventricular chamber undergoes concentric hypertrophy characterized by normal end-diastolic and end-systolic volumes, a normal ejection fraction, and a

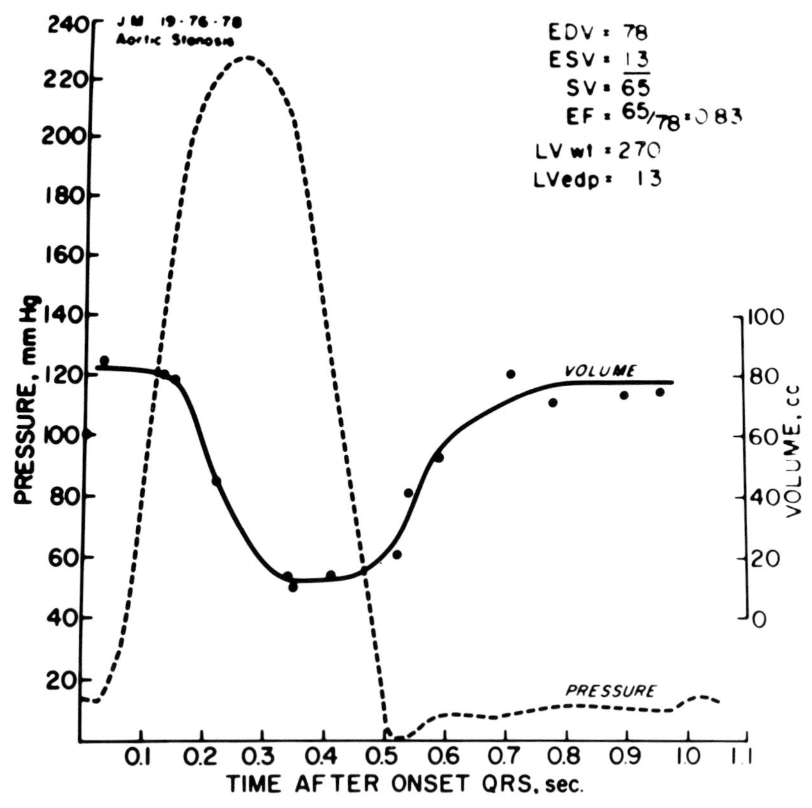

Fig. 13–4. Left ventricular pressure and volume measurements in a patient with compensated aortic stenosis. End-diastolic volume is normal, ejection fraction is slightly higher than normal, and left ventricular mass is significantly increased. These findings indicate concentric hypertrophy of the ventricle in response to the pressure overload. Reproduced by permission from Brest AN (ed.): Diagnostic Methods in Cardiology, Philadelphia, Davis, 1975, pp 287, 289, 290.

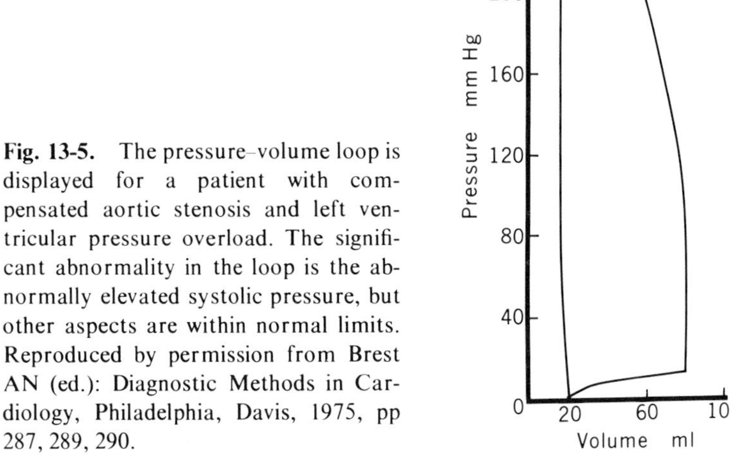

Fig. 13-5. The pressure–volume loop is displayed for a patient with compensated aortic stenosis and left ventricular pressure overload. The significant abnormality in the loop is the abnormally elevated systolic pressure, but other aspects are within normal limits. Reproduced by permission from Brest AN (ed.): Diagnostic Methods in Cardiology, Philadelphia, Davis, 1975, pp 287, 289, 290.

marked increase in wall thickness and left ventricular mass.[22–24] If the increase in wall thickness is proportional to the increase in pressure, the wall stress values are maintained within normal limits.[11,25] Figure 13-4 displays the pressure and volume measurements in a patient with compensated aortic stenosis. Chamber volumes are normal, and ejection fraction is slightly higher than normal at 0.83. Left ventricular mass is significantly increased. This represents an example of concentric hypertrophy with normal volumes and preservation of myocardial function. In Fig. 13-5 the pressure–volume diagram for the same patient with compensated aortic stenosis is displayed. The increase in ventricular work is that contributed by the elevation in left ventricular systolic pressure, since the chamber volume is normal.

In Fig. 13-6 the measurements of left ventricular pressure, wall thickness, and calculated wall stress are displayed for a patient with compensated aortic stenosis. It should be noted that the diastolic wall thickness is significantly increased. During the systolic ejection phase wall thickness further increases, so that there is a rapid fall in systolic wall stress. Peak systolic wall stress is normal in this patient with compensated pressure overload, even though the chamber pressure itself is quite elevated. This mechanism of concentric hypertrophy and maintenance of myocardial contractility is expressed by a significant increase in wall thickness during the systolic ejection phase, resulting in normalized wall stress values.

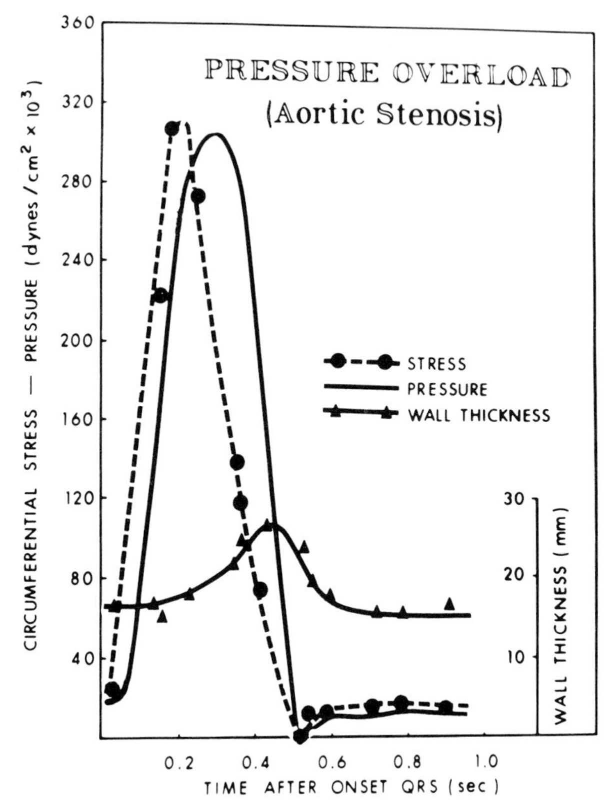

Fig. 13-6. Left ventricular pressure and circumferential wall stress measurements for a cardiac cycle in a patient with aortic stenosis and pressure overload. The diastolic wall thickness is abnormally increased in this patient, and further thickening during the late phase of systole results in a rapid decline of circumferential wall stress.

Deterioration

Although the mechanism of ventricular hypertrophy can continue for protracted periods, gradual deterioration in the contractile state will eventually result in changes in overall mechanical performance. Although the ejection fraction may be normal, some experimental and clinical studies suggest that the contractile state of the myocardium with pressure-induced hypertrophy may be impaired.[26,27] With further deterioration in the contractile state of the myocardium, systolic emptying is less effective; that is, ejection fraction falls and end-diastolic and end-sys-

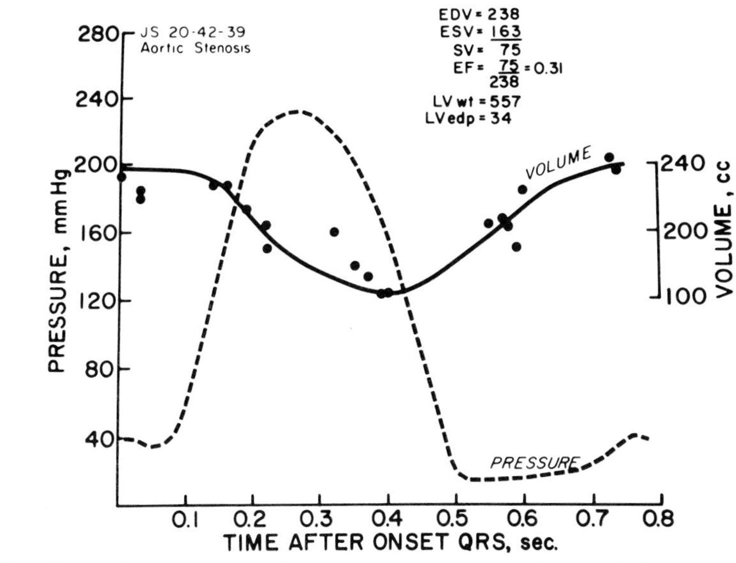

Fig. 13-7. Pressure and volume curves from a patient with decompensated aortic stenosis. Left ventricular stroke volume is normal, but end-diastolic and end-systolic volumes are abnormally increased. Hence ejection fraction is significantly reduced. The left ventricular end-diastolic pressure is elevated, and left ventricular mass is markedly increased.

tolic volumes will gradually increase.[28] Gradual increase in end-diastolic volume will be reflected in further increase in diastolic pressure. In Fig. 13-7 the pressure and volume curves from a patient with decompensated aortic stenosis are shown. The end-diastolic pressure is abnormally elevated to 34 mm Hg, while the systolic pressure is still quite high due to the aortic stenosis. The volume curve indicates that increases in end-diastolic volume and end-systolic volume have occurred, with preservation of the normal stroke volume. However, the ejection fraction is abnormally reduced to 0.31, and there has been a marked increase in left ventricular mass at 557 g. Therefore, in association with clinical symptoms of heart failure, this patient demonstrated chamber dilatation and a fall in ejection fraction despite compensatory hypertrophy. In Fig. 13-8 the pressure and volume curves of this patient are combined to construct a pressure–volume loop. The increased pressure–volume work in this case is similar to that seen in the patient with compensated aortic stenosis (Fig. 13-5), but the pressure–volume loop has been shifted to the right by the dilatation. In Fig. 13-9, pressure, stress, and tension curves for the patient with decompensated aortic stenosis are illustrated. Al-

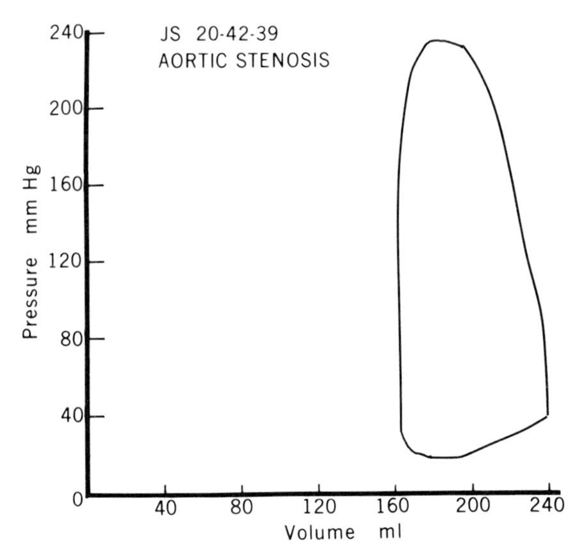

Fig. 13-8. A left ventricular pressure–volume loop is constructed in a patient with decompensated aortic stenosis. Although the contour and area under the pressure–volume loop are similar to those in a patient with compensated aortic stenosis, the left ventricular stroke volume is maintained by significant increases in end-diastolic and end-systolic volumes. Also, the left ventricular end-diastolic pressure is quite abnormal.

though the height of the pressure curve is similar to that of the patient with compensated aortic stenosis because of chamber dilatation and inadequate hypertrophy, both the stress and tension curves are abnormally high. Furthermore, the stress curve remains elevated throughout the systolic phase and does not exhibit the normal rapid fall during early ejection. This patient has, therefore, demonstrated that secondary dilatation can be imposed on the long-standing pressure overload of severe aortic stenosis. At this stage further increases in wall thickness may be insufficient to maintain wall stress within normal limits.[11]

PATHOPHYSIOLOGY OF AORTIC VALVE INCOMPETENCE

Pathology

Disturbances of the aortic valve apparatus that result in incompetence can be both acute and chronic.[29] An acute situation may result from bacterial endocarditis that erodes one of the valve leaflets,

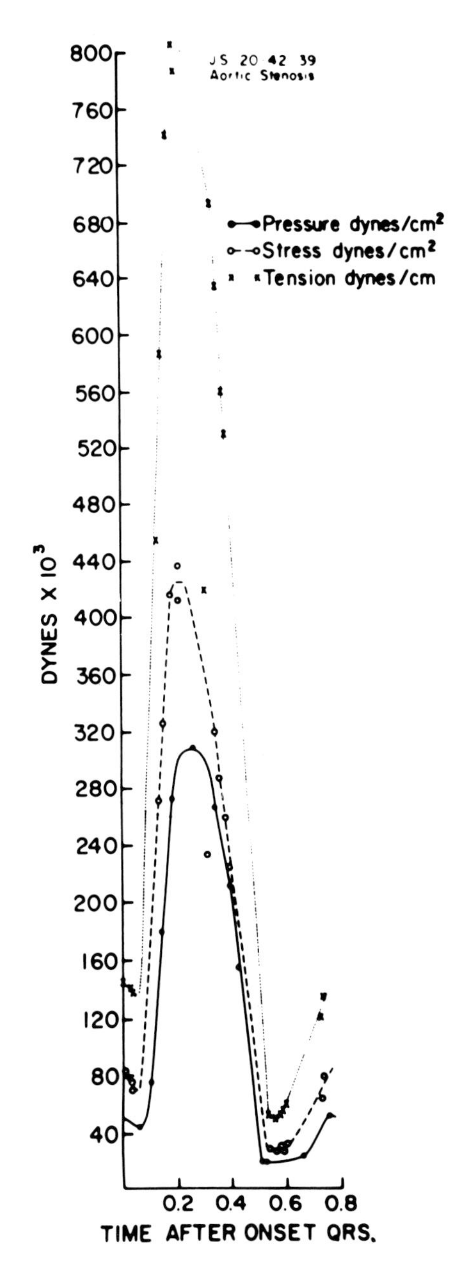

Fig. 13-9. Left ventricular pressure, stress, and tension curves for the patient with decompensated aortic stenosis are illustrated. Both peak systolic wall stress and wall tension are abnormally elevated in this patient.

acute rheumatic fever producing eversion of the aortic cusps, and dissection of the ascending aorta that distorts the valve ring. However, the more frequent causes of aortic regurgitation are chronic in nature, due to scarring by a rheumatic process and to deficiencies in the integrity of the valve caused by various diseases in the aorta and leaflets. In addition to rheumatic fever, long-standing hypertension and arteriosclerosis can both cause incompetence of the aortic valve. Similarly, connective tissue diseases of the aorta such as Marfan's syndrome and other situations that weaken the valve ring can result in aortic incompetence.

Physiologic Overload

In both acute and chronic aortic regurgitation the overload imposed on the left ventricle is that of volume. During diastole the high pressure in the aorta and systemic arterial system results in filling of the left ventricle in varying degrees, in addition to the normal filling from the left atrium. In acute aortic regurgitation, although the imposed regurgitant volume may be considerable, the left ventricle does not have sufficient time to dilate in order to accommodate this added volume. This lack of distensibility produces an inordinate increase in left ventricular filling pressure, which is reflected backward as severe left atrial and pulmonary venous hypertension. In chronic aortic regurgitation (the process being gradual) the regurgitant volume increases over many years, and the left ventricle does have adequate time to dilate appropriately. Nevertheless, this long-standing regurgitation can become excessive as the compensatory mechanisms of the ventricle begin to fail.

Compensatory Mechanisms

The increased filling of the left ventricle is reflected primarily in an increase in the end-diastolic volume.[23,24] This increase in end-diastolic volume is also associated with an increase in total left ventricular stroke volume ejected with each beat. Early in the course of the disease the effective stroke volume (that is, the total left ventricular stroke volume minus regurgitant volume) remains normal. Although end-diastolic volume is significantly elevated, the ejection fraction tends to remain within normal limits.[24] The increase in chamber radius associated with normal chamber pressures and wall thickness will then produce an initial increase in peak systolic wall stress. This elevated peak systolic wall stress serves as a stimulus for an increase in the mass or weight of the left ventricle. Subsequently the increase in wall thickness will serve to maintain the peak systolic wall stress within normal limits.[11,25] The pressure–volume work of the left ventricle in aortic incompetence is increased due to the large end-diastolic volume that develops.[15] Systolic and diastolic pressures may remain normal during the early course of significant aortic regurgitation, but the total pressure–volume work may be massively increased due to the large changes in left ventricular stroke volume.

The mechanisms responsible for chamber dilatation in aortic regurgitation are several.[24] There may be fiber slippage, which would result in a larger chamber with a thinner ventricular wall. As the chamber radius enlarges, however, the hypertrophy mechanism is also stimulated. Replication of sarcomeres in series could then lengthen the myocardial fibers. Chronic volume overload characteristically produces eccentric hypertrophy characterized by an increased wall thickness proportional to the increased chamber radius.[22] In this manner wall stress tends to be maintained within the normal range.[11,25]

Deterioration

The dilatation and hypertrophy mechanisms maintain the large stroke volume and end-diastolic volume in patients with aortic regurgitation until impairment of the contractile state of the myocardium develops. Deterioration of muscle performance is reflected primarily by an increase in end-systolic volume disproportionate to the increase in end-diastolic volume. Thus ejection fraction and eventually effective left ventricular stroke volume decline.[28] As further dilatation of the ventricle occurs, the degree of hypertrophy may not be sufficient to maintain wall stress within normal limits. With impairment of the hypertrophy mechanism, often the compliant properties of the diastolic ventricle change, and the increased end-diastolic pressure results in pulmonary venous hypertension.

Figure 13-10 concerns a patient with decompensated aortic regurgitation. Although this patient was symptomatic from his left ventricular failure, the left ventricular end-diastolic pressure is only minimally elevated. However, the regurgitation is associated with an end-diastolic volume of 430 ml, with a left ventricular stroke volume of 101 ml and a regurgitant volume of 34 ml. Forward stroke volume is maintained at 67 ml despite an ejection fraction of 0.23. This patient's pressure–volume diagram (Figure 13-11) shows essentially normal pressure–volume work, but the loop is shifted markedly to the right.

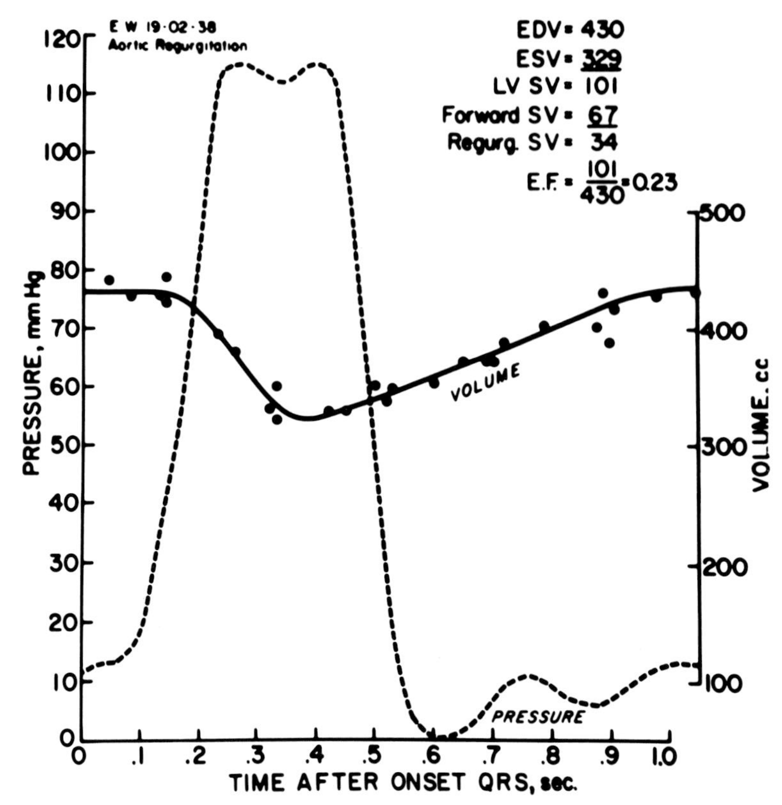

Fig. 13-10. Left ventricular pressure and volume curves are displayed in a patient with aortic regurgitation in the decompensated state. End-diastolic and end-systolic volumes are extremely elevated. The reduced ejection fraction indicates that in order to maintain forward stroke volume the dilatation mechanism has been exhausted.

Figure 13-12 illustrates the left ventricular pressure, wall stress, and wall thickness curves in this same patient with decompensated volume overload from long-standing aortic regurgitation. Of particular note is the extreme elevation of wall stress, which remains elevated during the entire phase of systolic ejection. Although the diastolic wall thickness is increased, there is relatively little additional thickening during the systolic phase. This reduced wall thickness during systole coupled with the large end-systolic volume contributes to the abnormal peak wall stress and the prolonged elevation of stress during the ejection phase.

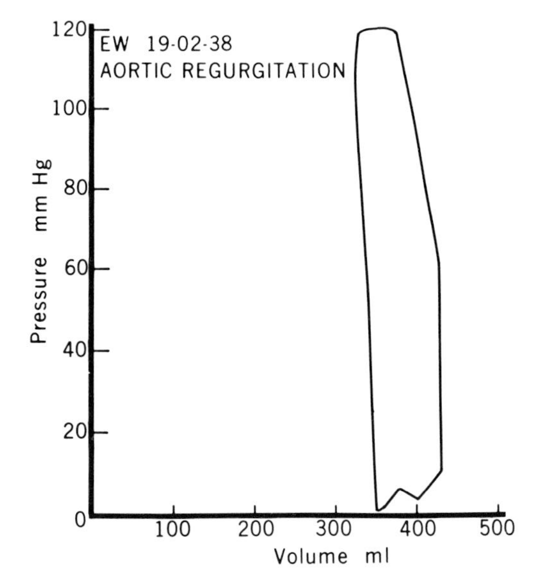

Fig. 13-11. A pressure–volume loop from a patient with decompensated aortic regurgitation. Abnormal features of the loop are extreme displacement to the right along the volume scale, loss of the isovolumic relaxation phase, and elevation in left ventricular end-diastolic pressure.

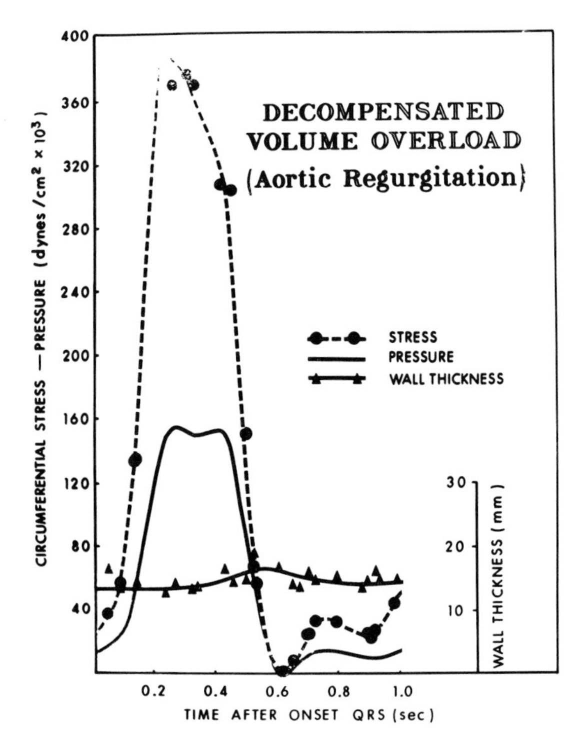

Fig. 13-12. Left ventricular pressure and wall stress are displayed in a patient with decompensated volume overload from aortic regurgitation. There is diminished wall thickening during ejection, and wall stress remains elevated during the systolic phase.

COMBINED AORTIC STENOSIS AND INCOMPETENCE

Pathology

Quite frequently, disturbances in the aortic valve apparatus result in both stenosis and incompetence. The diseases that produce the individual lesions may also be the common mechanism underlying the combined lesion. Particularly, congenital bicuspid valve disease, rheumatic aortic stenosis, and calcific aortic stenosis often present combined aortic stenosis and incompetence. Pathologic examination of the valve reveals stiffness, scarring of the commissures, and calcium, which results in a thickened valve orifice that is both stenotic and incompetent.

Physiologic Overload

The reduced orifice size impairs left ventricular ejection, which results in generation of excessive left ventricular pressure and exposure of the ventricular myocardium to a systolic pressure overload. The fixed orifice also produces aortic regurgitation, which increases the volumes of the ventricle, both at end-diastole and end-systole, and places a volume overload on the myocardium. Therefore the combined lesions of stenosis and incompetence produce pressure and volume overload in a manner similar to the isolated valve lesions.

Effects of Overload

In combined stenosis and incompetence of the valve, both increased chamber pressure and enlarged chamber dimensions contribute to increased wall tension values. Early in the course of the disease when wall thickness is normal, wall stress may be increased. The hypertrophy mechanism of the myocardial fibers is stimulated to increase wall thickness and left ventricular mass.[23] The pressure–volume work, as reflected in a very large pressure–volume loop, is increased by both the elevation of systolic pressure and the increased left ventricular stroke volume.[15]

Compensatory Mechanisms

The initial compensatory mechanisms are those of dilatation and hypertrophy of the myocardium resulting from the increase in peak systolic wall stress. The increase in left ventricular mass and dimensions and appropriate elevation in left ventricular systolic pressure will maintain a normal effective stroke volume for a period of time. The ejection fraction tends also to remain normal.

Deterioration

Pressure and volume overloads with combined aortic stenosis and incompetence may be tolerated until the contractile state of the muscle becomes impaired. This results in an increase in the end-systolic volume, a fall in ejection fraction, and further increase in end-diastolic volume. Because of the impaired hypertrophy mechanism, wall forces will then be increased in a manner similar to those found in decompensated aortic stenosis or decompensated aortic regurgitation.[11] As these changes occur the compliant properties of the ventricle are also reduced, and the end-diastolic pressure becomes abnormally elevated due to the failure of the contracting muscle to empty in systole and to the diminished compliance during diastolic filling.[30] As a result, left atrial and pulmonary venous hypertension develop and contribute to the clinical symptoms of pulmonary congestion.

CLINICAL SYMPTOMS

Aortic Stenosis

The most frequent symptoms encountered by patients with aortic stenosis are those of exertional chest pain, syncope, and heart failure.[31] These symptoms often become apparent when compensatory mechanisms are exhausted and deterioration of mechanical performance of the ventricle has occurred. Ischemic-type chest pain is probably due to areas of the subendocardium being inadequately perfused in the face of a tremendous increase in muscle mass, even though coronary arterial lesions may be absent. However, as patients (particularly males) approach the middle age range, coronary disease may also contribute to the ischemic pain.

Loss of consciousness, particularly with exertion, is another common symptom in patients with aortic stenosis. A small fixed orifice size limits the increase in cardiac output required during periods of exertion. The associated peripheral vasodilatation with exercise further contributes to an overall reduction in cerebral blood flow, which renders the patient unconscious.

Clinical congestive heart failure occurs when pulmonary venous hypertension has resulted from deterioration of ventricular function and loss of compensatory mechanisms. By this time the ventricle has begun to dilate despite extreme concentric hypertrophy. Impaired ventricular contractility as well as decreased compliance result in pulmonary venous hypertension.

Aortic Incompetence

Patients with aortic incompetence may develop symptoms similar to those in aortic stenosis, such as exertional chest pain, syncope, and arrhythmias.[31] However, the most common clinical manifestation is that of left ventricular heart failure. The heart failure results from the failure of compensatory mechanisms and deterioration in the contractile state of the myocardium. The increase in both end-diastolic and end-systolic volumes and the accompanying decrease in the compliance of the ventricle result in elevation of the end-diastolic pressure of the ventricle.

Aortic Stenosis and Incompetence

Patients with mixed aortic disease may present with dominant hemodynamic abnormalities and clinical manifestations of either lesion or a combination of the two. Angina, syncope, and heart failure are the usual symptoms.

PHYSICAL FINDINGS

Aortic Stenosis

Patients with aortic stenosis usually demonstrate characteristic findings in examination of the peripheral circulation as well as of the heart.[31] The arterial pulse is diminished in amplitude and has a marked delay in upstroke, which can be appreciated in the carotid or other peripheral arteries. The pulse pressure may be narrow, and the systolic blood pressure may be reduced. Examination of the carotid arteries often reveals a palpable shudder due to transmitted systolic vibrations.

Palpation of the precordium frequently reveals a systolic thrill at the right upper sternal edge, which reflects turbulent flow through the stenotic valve and may reveal systolic outward movement over the aortic outflow tract associated with poststenotic dilatation of the aorta. Unless the ventricle has become dilated the apical impulse on inspection may not be displaced laterally. The concentric hypertrophy characteristically produces a sustained or heaving apical thrust, and a prominent presystolic A wave, representing the palpable equivalent of an atrial gallop, may be present.

Auscultation may reveal abnormalities in the heart sounds as well as a characteristic diamond-shaped murmur. The first sound at the apex is usually unremarkable, but is frequently preceded by an atrial gallop sound and may be followed by a prominent ejection click. The aortic component of the second sound is diminished in proportion to the extent of the stiffness and calcification of the valve, and in advanced cases it is completely absent. A systolic crescendo–decrescendo murmur is heard over the aortic outflow tract, with radiation into the carotids and along the left sternal border to the apex. Sometimes, especially in patients with emphysema, the murmur can be heard only at the apex. Thus murmur has its onset after a detectable delay following the first sound and characteristically terminates before the second aortic sound. It is typically harsh and low-pitched, but occasionally is musical or even cooing in quality. A ventricular diastolic gallop is not heard unless the patient also has volume overload or has developed left ventricular failure.

Aortic Incompetence

The circulatory and ventricular dynamics in aortic regurgitation reflect the increased stroke volume ejected into the aorta, which during the diastolic phase regurgitates across the incompetent aortic valve, and also associated peripheral vasodilatation.[31] The peripheral pulse pressure characteristically increases and manifests itself as capillary pulsation that can sometimes be observed in the fundi and fingernails. The carotid pulse is hyperdynamic and may be bisferious in contour. The wide pulse pressure is related to a significant increase in systolic blood pressure and a lowered diastolic pressure, which may be audible down to extremely low levels and in some instances even to zero. There may be visible bobbing of the patient's head synchronous with ejection of a large stroke volume into the aortic arch and carotid vessels.

Examination of the heart reveals the findings of significant volume overload of the left ventricle. The apical impulse is displaced laterally and inferiorly and is frequently visible. On palpation, the impulse is hyperdynamic. If the aortic incompetence is associated with dilatation of the aortic root, systolic pulsation may also be visible or palpable along the right sternal border. On auscultation the apical first sound is usually preserved, but in instances of acute severe aortic regurgitation it may be absent due to premature closure of the mitral valve by the regurgitant flow. The murmur of aortic incompetence is typically a high-pitched early diastolic decrescendo blowing murmur heard best along the left sternal border, in the aortic area, and occasionally at the apex. The murmur is heard best with the patient in the sitting position and with the breath held at full expiration. A diastolic rumble at the apex may be due to striking of the anterior mitral leaflet by the directed aortic regurgitant jet during diastolic inflow, or it may be due to late diastolic mitral regurgitation if left ventricular diastolic pressure is quite high.[32,33] A systolic ejection murmur, accompanied by a thrill, may be present at the aortic area and apex in the absence of organic aortic stenosis, due to high flow. A ventricular diastolic gallop (S_3) may be audible, reflecting volume overload with preserved ventricular function or, late in the disease, with impaired myocardial function.

Aortic Stenosis and Incompetence

In combined aortic valvular stenosis and incompetence, the peripheral findings often reflect both abnormalities. The pulse pressure as well as the blood pressure will reflect the dominant lesion, either stenosis or regurgitation. Examination of the heart frequently demonstrates significant displacement of the apical impulse inferiorly and laterally due to dilatation from volume overload. However, minimal aortic incompetence may accompany the aortic stenosis and produce only slight chamber enlargement with a normally placed apical impulse. Precordial movements along the aortic outflow tract would again suggest dilatation of the aorta due to post-stenotic dilatation or dilatation due to a basic defect in the aorta that produces secondary incompetence of the valve.

On auscultation the first sound at the apex is usually preserved and the diamond-shaped murmur of aortic stenosis is detectable. The murmur of aortic incompetence is heard best along the left sternal border and over the outflow tract. The aortic second sound may be diminished or absent in combined lesions. If tachycardia accompanies left ventricular failure, the systolic and diastolic murmurs of the aortic valve may sound continuous. However, it is very important to separate "to-and-fro" murmurs from continuous murmurs suggesting a left-to-right shunt.[34]

ELECTROCARDIOGRAPHIC AND RADIOGRAPHIC FINDINGS

In aortic valve disease with hypertrophy and dilatation, the electrocardiogram often reflects only the hypertrophy. Characteristic findings are increased QRS voltage and ST–T wave changes in the precordial leads.[31,35] Although attempts have been made to distinguish hypertrophy due to pressure loading from that due to volume overloading by ECG changes, generally these have not been successful.[36]

The chest x-ray in aortic stenosis often reveals no cardiac enlargement, and therefore the degree of concentric hypertrophy is not readily detected.[37] Calcification in the aortic valve and post-stenotic dilatation of the ascending aorta may be apparent. In aortic incompetence the dilatation and hypertrophy produce significant increases in left ventricle and overall cardiac size. The increase in cardiothoracic ratio reflects primarily dilatation of the ventricle, and this cannot be usefully analyzed to estimate the extent of hypertrophy in either stenosis or insufficiency.

CARDIAC CATHETERIZATION

Aortic Stenosis

The diagnosis of aortic stenosis is made at cardiac catheterization by demonstrating a systolic pressure gradient between the left ventricle and the aorta.[38] This gradient can be recorded either by advancing a catheter retrograde to the left ventricle and withdrawing it across the aortic valve or by means of simultaneous recordings from two catheters, one in the aorta and one in the left ventricle. Cardiac output can then be measured by the Fick principle or the thermodilution technique. From the measurement of pressure gradient and flow across the valve, aortic valve area can be calculated using the formula described earlier.[19,20] Quantitative angiocardiography can also be performed to obtain chamber dimensions and wall thickness measurements.[24] Table 13-1 contrasts hemodynamic and quantitative angiographic findings in a patient with compensated aortic stenosis and another patient with decompensated aortic stenosis. Both patients exhibited pressure gradients across the aortic valve in systole. In the decompensated state there was an increase in the left ventricular end-diastolic pressure, dilatation of the ventricle with increases in end-diastolic and end-systolic volumes, a significant reduction in ejection fraction, a large increase in left ventricular mass, and an increase in the peak systolic wall stress. The pressure–volume work was similar in the compensated and decompensated conditions.

Aortic Regurgitation

At cardiac catheterization the diagnosis of aortic insufficiency is made by an injection of contrast material into the aortic root, with filming of the ventricle to detect regurgitation of the contrast material

Table 13-1
Aortic Stenosis

Measurement	Compensated	Decompensated
LVEDP mmHg	13	34
EDV ml	78	238
ESV ml	13	163
LVSV ml	65	75
FSV ml	56	59
Regurg V ml	9	16
EF	0.83	0.31
LV Mass gms	270	557
ED Wall Stress dynes/cm^{2}X10^3	18	-
Peak Sys Wall Stress dynes/cm^{2}X10^3	304	426
LVSW gm-m/beat	209	204

Measurements of various left ventricular parameters in representative patients with compensated aortic stenosis and decompensated aortic stenosis. LVEDP = left ventricular end-diastolic pressure; EDV = end-diastolic volume; ESV = end-systolic volume; LVSV = left ventricular stroke volume; FSV = forward stroke volume; Regurg V = regurgitant volume; EF = ejection fraction; LVSW = left ventricular stroke work.

into the ventricle.[39] Quantitative angiography permits quantitation of the regurgitant flow and of ventricular function.[24] In Table 13-2 data from a patient with compensated aortic regurgitation and another with decompensated aortic regurgitation are compared. There are only slight differences in left ventricular end-diastolic pressures and end-diastolic

Table 13-2
Aortic Regurgitation

Measurement	Compensated	Decompensated
LVEDP mmHg	5	13
EDV ml	400	430
ESV ml	159	329
LVSV ml	241	101
FSV ml	69	67
Regurg V ml	172	34
EF	0.60	0.23
LV Mass	366	561
ED Wall Stress dynes/cm^{2}X10^3	18	47
Peak Sys Wall Stress dynes/cm^{2}X10^3	364	386
LVSW gm-m/beat	222	155

Measurements of left ventricular pressure, chamber volume, ejection fraction, mass, wall forces, and stroke work are shown in a patient with compensated aortic regurgitation and in another with decompensated aortic regurgitation.

volumes. However, in the decompensated patient end-systolic volume is much higher and ejection fraction much lower. Left ventricular mass is increased in both circumstances, but more so in the decompensated state. Left ventricular stroke work is higher than normal even in the patient with decompensation.

Table 13-3
Aortic Stenosis and Regurgitation

Measurement	Compensated	Decompensated
LVEDP mmHg	8	36
EDV ml	436	491
ESV ml	219	403
LVSV ml	217	88
FSV ml	97	36
Regurg V ml	120	52
EF	0.50	0.18
LV Mass gms	474	541
ED Wall Stress dynes/cm^{2}X10^3	31	-
Peak Sys Wall Stress dynes/cm^{2}X10^3	439	542
LVSW gm-m/beat	387	153

Measurements of left ventricular end-diastolic pressure, chamber volume, ejection fraction, mass, wall forces, and stroke work are compared in representative patients with compensated aortic stenosis and regurgitation versus the decompensated state.

Aortic Stenosis and Regurgitation

Combined aortic stenosis and regurgitation is confirmed by demonstration of a pressure gradient across the aortic valve in systole and regurgitation of contrast material across the valve from the aortic root in diastole. Pressure, volume, mass, wall force, and work values for a patient with compensated aortic stenosis and another patient with decompensated aortic stenosis and regurgitation are shown in Table 13-3.

Left ventricular end-diastolic pressure is severely elevated in the decompensated state. End-diastolic volumes and masses are similar, but the forward stroke volume is significantly reduced in the decompensated condition. Similarly, the ejection fraction is markedly reduced. The left ventricular stroke work is larger than normal in both examples, particularly in the compensated state.

REFERENCES

1. Ross J Jr, Braunwald E: Aortic stenosis. Circulation [Suppl V] 37:61, 1968
2. Frank S, Johnson A, Ross J Jr: Natural history of valvular aortic stenosis. Br Heart J 35:41, 1973
3. Spagnuolo M, Kloth H, Taranta A, et al: Natural history of rheumatic aortic regurgitation: Criteria predictive of death, congestive heart failure, and angina in young patients. Circulation 44:368, 1971
4. Goldschlager N, Pfeifer J, Cohn K, et al: The natural history of aortic regurgitation: A clinical and hemodynamic study. Am J Med 54:577, 1973
5. Laplace PS: Théorie de l'action capillaire, in: Traité de Mécanique Céleste, Supplement au livre X. Paris, Courcien, 1805
6. Starling EH: The Linacre Lecture on the Law of the Heart (Cambridge, 1915). London, Longmans, Green and Co., 1918
7. Sarnoff SJ, Braunwald E, Welch GH Jr, et al: Hemodynamic determinants of oxygen consumption of the heart with special references to the tension-time index. Am J Physiol 192:148, 1958
8. Koushanpour E, Collings WD: Validation and dynamic applications of an ellipsoidal model of the left ventricle. J Appl Physiol 21:1655, 1966
9. Sandler H, Ghista DN: Mechanical and dynamic implications of dimensional measurements of the left ventricle. Fed Proc 28: 1344, 1969
10. Sandler H, Dodge HT: Left ventricular tension and stress in man. Circ Res 13:91, 1963
11. Hood WP Jr, Rackley CE, Rolett E: Wall stress in the normal and hypertrophied human left ventricle. Am J Cardiol 22:550, 1968
12. Hood WP Jr, Thomson WJ, Rackley CE, Rolett E: Comparison of calculations of left ventricular wall stress from thin-walled and thick-walled ellipsoidal models. Circ Res 24:575, 1969
13. Wong AYK, Rautaharju PM: Stress distribution within the left ventricular wall approximated as a thick ellipsoidal shell. Am Heart J 75:649, 1968
14. Linzbach AJ: Heart failure from the point of view of quantitative anatomy. Am J Cardiol 5:370, 1960
15. Bunnell IL, Grant C, Greene DG: Left ventricular function derived from the pressure volume diagram. Am J Med 39:881, 1965
16. Ellis FH, Kirklin JW: Congenital valvular aortic stenosis: Anatomic findings and surgical technique. J Thorac Cardiovasc Surg 43:199, 1962

17. Roberts WC: Anatomically isolated aortic valvular disease: The case against its being a rheumatic etiology. Am J Med 49:151, 1970
18. Roberts WC: The structure of the aortic valve in clinically isolated aortic stenosis. Circulation 42:91, 1970
19. Gorlin R, Gorlin SG: Hydraulic formulae for calculation of the area of the stenotic mitral valve, other cardiac valves, and central circulatory shunts. Am Heart J 41:1, 1951
20. Gorlin R: Shunt flows and valve areas, in Zimmerman HA (ed): Intravascular Catheterization (ed 2). Springfield, Ill., Charles C. Thomas, 1966, p 545
21. Wood P: Aortic stenosis. Am J Cardiol 1:553, 1958
22. Grant C, Greene DG, Bunnell IL: Left ventricular enlargement and hypertrophy: A clinical and angiocardiographic study. Am J Med 39:895, 1965
23. Kennedy JW, Twiss RD, Blackmon JR, Dodge HT: Quantitative angiocardiography. III: Relationships of left ventricular pressure, volume, and mass in aortic valve disease. Circulation 38:838, 1968
24. Rackley CE, Hood WP Jr: Quantitative angiographic evaluation and pathophysiologic mechanisms in valvular heart disease. Prog Cardiovasc Dis 15:427, 1973
25. Hood WP Jr: Dynamics of hypertrophy in the left ventricular wall of man, in Alpert NR (ed): Cardiac Hypertrophy. New York, Academic, 1971, p 445
26. Spann JF Jr, Buccino RA, Sonnenblick EH, Braunwald E: Contractile state of cardiac muscle obtained from cats with experimentally produced ventricular hypertrophy and heart failure. Circ Res 21:341, 1967
27. Simon H, Krayenbuehl HP, Rutishauser W, Preter OB: The contractile state of the hypertrophied left ventricular myocardium in aortic stenosis. Am Heart J 79:587, 1970
28. Dodge HT, Baxley WA: Left ventricular volume and mass and their significance in heart disease. Am J Cardiol 23:528, 1969
29. Cobbs BW: Clinical recognition and medical management of rheumatic heart disease and other acquired valvular disease, in Hurst JW, Logue RB (eds): The Heart, Arteries and Veins (ed 2). New York, McGraw-Hill, 1970, p 773
30. Rackley CE, Hood WP Jr, Rolett EL, Young DT: Left ventricular end-diastolic pressure in chronic heart disease. Am J Med 48:310, 1970
31. Reichek N, Shelburne JC, Perloff JK: Clinical aspects of rheumatic valvular disease. Prog Cardiovasc Dis 15:491, 1973
32. Wong M: Diastolic mitral regurgitation. Hemodynamic and angiographic correlation. Br Heart J 31:468, 1969
33. Fortuin NJ, Craige E: On the mechanism of the Austin-Flint murmur. Circulation 45:558, 1972
34. Craige E, Millward DK: Diastolic and continuous murmurs. Prog Cardiovasc Dis 14:38, 1971
35. Singer DH, Perloff JK: Electrocardiogram of free aortic insufficiency. Circulation 26:786, 1962

36. Cabrera EC, Monroy JR: Systolic and diastolic loading of the heart. Am Heart J 43:661, 1952
37. Klatte EC, Tampas JP, Campbell JA, Lurie PR: The roentgenographic manifestations of aortic stenosis and aortic valvular insufficiency. Am J Roentgenol Radium Ther Nucl Med 88:57, 1962
38. Gorlin R, McMillan IKR, Medd WE, et al: Dynamics of the circulation in aortic valvular disease. Am J Med 18:855, 1955
39. Baron MG: Angiocardiographic evaluation of valvular insufficiency. Circulation 43:599, 1971

Thomas J. Ryan

14
Mitral Valve Disease

Abnormalities of the mitral valve have long been recognized as a significant cause of heart disease. These abnormalities arise from a wide variety of etiologic factors, and they result in clinical symptoms, physical findings, and electrocardiographic and roentgenographic changes that are quite characteristic. The resultant clinical picture of mitral valve disease is attributable to specific disturbances in circulatory dynamics that set it apart from other valvular disorders and warrant discussion as a separate entity. The symptoms are frequently disabling, and the physical findings, while often pathognomonic, are usually subtle. Using the simple tools of careful bedside examination the skillful clinician can precisely diagnose the type and extent of mitral valve disease. While this might explain the clinician's fascination for the disease, it should be recognized that it was the application of scientific measurement provided by cardiac catheterization that led to our current understanding of the disorder. A voluminous literature has arisen over the last quarter century describing the pathophysiology of mitral valve disease, and it continues to burgeon as new noninvasive techniques are applied to the study of this disease model. As a result of this continued investigation, it has become increasingly apparent that the mitral valve is more than an orifice placed between the left atrium and ventricle. The entire mitral valve apparatus is a complex conduit and is comprised of the anulus fibrosis, anterior and posterior leaflets, first-, second-, and third-order chordae tendineae numbering 120–124 in all, and two papillary muscles with a terminal arcuate microcirculation. Abnormalities of this apparatus can be divided into two functional types: (1) those causing obstruction (mitral stenosis) and (2) those causing incompetence (mitral regurgitation). Each in turn results in a primary circulatory abnormality that differs sharply from that of the other. Mitral stenosis represents a systolic overload to the left atrium and creates pressure work for the right side of the heart as the major disturbance in circulatory dynamics. Mitral regurgitation, on the other hand, represents a diastolic overload to the left ventricle, with the principal circulatory abnormality being flow work for that chamber. Viewed in this light it is understandable that the two conditions should differ in natural history, clinical symptoms, physical findings, and hemodynamic measurements.

MITRAL STENOSIS

Etiology

The known causes of the clinical and hemodynamic picture of circulatory blockade at the mitral valve level are: (1) rheumatic mitral stenosis,[1] (2) congenital mitral stenosis,[2] (3) left atrial tumor,[3] (4) cor triatriatum,[4] and (5) nonrheumatic massive calcification of the mitral anulus.[5] Since mitral stenosis is almost always due to rheumatic endocarditis, the following discussion will relate primarily to this entity, with only a limited commentary on the other listed causes, which are all extremely rare and are of importance mostly from the standpoint of differential diagnosis. A specific etiologic relationship between streptococcal infections and rheumatic fever has been well established, with evidence that the group A streptococcus has antigens that cross-react with the structural glycoprotein of heart valves.[6]

Natural History

The course of mitral stenosis from its inception during acute carditis to the point of earliest symptoms has been called the latent period by Wood, and in his series it averaged 19 years.[1] While the mitral leaflets are involved with the initial endocarditis, they are incapable of producing actual narrowing and stenosis at this stage. There is a temporary malfunction of the mitral valve apparatus resulting in clinical signs of mitral regurgitation, but these do not indicate permanent damage. Bland and Jones[7] found that it required more than a decade to develop definite clinical evidence of mitral stenosis in their study of 1,000 patients followed for 20 years. The progression of symptoms, once they appear, from onset to the point of total disability was estimated by Wood to span 7 years. In reporting the natural history of 271 patients with mitral stenosis, Olesen[8] showed a 70% mortality over an 11-year period that increased to 83% at the end of 18 years. White and his colleagues[9] followed a group of 250 patients with pure mitral stenosis for 20 years and pointed out that while 79% were dead at the end of 20 years, 25% of those patients who were asymptomatic at the onset remained so 20 years later. In a more recent review of the natural history of mitral stenosis, Selzer[10] has pointed out that while symptoms usually begin in the fourth and fifth decades there is evidence that the course of mitral stenosis in areas with a lower standard of living is considerably accelerated and progresses rapidly to serious disability in early life. Serial catheterization studies have shown progressive narrowing of the valve orifice in some patients at a rather rapid rate, while in others it remains relatively stable. In the progressive form there is clinical worsening, while in the nonprogressive form symptoms develop less frequently and are usually related to such complications as intercurrent bronchial infection and the onset of atrial fibrillation. This would seem to hold true in the postvalvotomy population as well, since restenosis is found in some (progressive) and not in others (nonprogressive). Selzer has also called attention to the pathogenesis of the later stages of mitral stenosis, disclaiming "smoldering" chronic rheumatic activity as the explanation for the progression of the disease in the fourth and fifth decades of life. Referring to the work of Edwards,[11] he suggests that the early scarring process of rheumatic valvulitis produces changes in the mitral valve that are capable of perpetuating themselves later on in a nonspecific manner. This would be similar to the pathogenesis of calcific aortic stenosis arising from a congenital bicuspid valve, wherein it is now felt that abnormal flow patterns traumatize the valve, leading to thickening, fibrosis, and calcification of the cusps.[12] This is a particularly attractive hypothesis when one considers the three distinct types of mitral stenosis seen pathologically:[13] (1) commissural type (fusion of commissures only), (2) cuspal type (thickened, fibrous leaflets that eventually calcify), and (3) chordal type (chordae are fused, thickened, and shortened and principally interfere with valve mobility). It is suggested that in certain instances the acute rheumatic process results in commissural fusion only, resulting in minimal trauma with no significant blockade to the flow of blood from atrium to ventricle. Such involvement would result in mild mitral stenosis that could persist for life. In other instances more severe involvement of the commissures could be envisioned, with the resultant cuspal and chordal involvement altering flow patterns through the mitral orifice. This would set the stage for progressive valvular deformity on a traumatic basis with inexorable progression of symptoms. The natural history of mitral stenosis seems quite consistent with this concept of a nonspecific pathogenesis. It would account for a lifetime of nonprogressive mild mitral stenosis in certain individuals and the late rapid progression in others.

Clinical Profile

Symptoms

The classic description of mitral stenosis continues to be Wood's treatise of the subject derived from his personal experience with more than 350 cases up to 1954.[1,14,15] The reader should be aware that not all of these patients had their findings substantiated by either surgery or a cardiac catheterization and that 50 patients were thought to have predominant mitral regurgitation. Although considerable sophistication has been added to hemodynamic studies over the past two decades, the clinical picture and hemodynamic profile offered by Wood has, in large measure, been well substantiated in most other reported large series. Our own experience with a slightly more select group of 320 patients with hemodynamically significant mitral stenosis, defined as a mitral valve area of 1.3 cm^2 or less, differs only in an analysis of left ventricular function. Such discrepancy is readily understood when one considers that he had no available method for directly measuring left ventricular pressure or assessing mitral regurgitation by angiographic techniques.

The predilection for pure mitral stenosis to develop in females is well established (M:F sex ratio is 1:4) and poorly understood. It is equally curious that this ratio reaches a 2:1 balance as pure mitral regurgitation dominates the picture. Wood obtained a history of the active rheumatic state in 68% of his patients, but it is traditionally taught that there is a negative past history of acute rheumatic fever in approximately one-half of all patients with rheumatic heart disease.[16] While this implies that subclinical acute rheumatic fever is rather common, it may also suggest that a substantial amount of adult valvular disease is of nonrheumatic origin. We continue to be equally concerned about the false positive past history of acute rheumatic fever. It is not rare in our experience to encounter an adult with a significant heart murmur who has grown up with imposed restrictions and certain penalties relating to employability and insurability because the diagnosis of rheumatic fever was incorrectly made at an early age. Such individuals are frequently found to have a mild congenital abnormality—a floppy mitral valve or idiopathic hypertrophic suboartic stenosis. The presence of a murmur, intensified by fever associated with a nonspecific illness in childhood, perhaps occurring in association with a pharyngitis, too often results in the unfortunate label of rheumatic fever. In obtaining a history of prior rheumatic fever one should not accept such statements as "I was told I had rheumatic fever as a child." It is important to obtain a convincing description of the acute illness, which most often includes a recollection of acute arthritis or, less commonly, chorea.

Dyspnea. Dyspnea is the cardinal symptom of mitral stenosis, and this has been attributed to increased rigidity of the lungs caused by changes in the interstitial tissue. The most important consequence of mitral stenosis is left atrial hypertension, which elevates pressures in the pulmonary vascular system. To prevent the transudation of fluid from the pulmonary capillaries under high hydrostatic pressure, the lymphatic system increases its capacity to drain excess fluid. Since mitral stenosis develops gradually over a period of time, the pressure increments in the pulmonary capillaries are small and permit adaptive processes to take place. Translated into a patient's history, it is important to appreciate the gradual onset and patient adaptation to this dyspnea. Oftentimes subjective awareness of shortness of breath is quite obscure and even denied by the patient with mitral stenosis. He has developed a life style that spares him this unpleasant feeling, and one must probe beyond the traditional query of how many stairs are climbed. In our present-day world where automation has sharply reduced effort requirements for most individuals it is especially feasible for the housewife to function quite successfully in her role without developing significant symptoms. Climbing stairs and carrying wash are becoming less a daily chore, and one frequently gets the first clue that exertional dyspnea is present by inquiring about casual recreational activity. The abandonment of a once-pleasurable pastime such as bicycle riding, swimming, or walking on a sandy beach is often the first indication of these pulmonary changes. For the same reason, frank pulmonary edema occurs much less frequently in mitral stenosis than in other forms of heart failure where sudden elevation of left ventricular end-diastolic pressure results in equally elevated levels of left atrial pressure. In this latter instance there is insufficient time for the secondary and adaptive mechanisms to develop and work effectively in preventing the transudation of fluid from the pulmonary capillaries to the aveoli. It has been suggested that the perivascular cuffing that results from chronically elevated intravascular pressure results in a capillary–alveolar interstitial barrier that prevents this transudation of fluid. Recognizing that pressure is a function of flow × resistance, it would seem evident that the level of cardiac output would be another major determinant in the production of pulmonary edema. This is suggested in Wood's finding that normal sinus rhythm was twice as frequent as atrial fibrillation in those individuals with mitral stenosis who gave a history of acute pulmonary edema. Furthermore, the average age of patients with pulmonary edema was 32 years, compared with 37 for the series as a whole. This suggests that time-dependent mechanisms develop to protect against pulmonary edema.

Hemoptysis. Hemoptysis occurs in approximately 35% of patients with mitral stenosis and is important as an early symptom. It is usually a striking event for the patient, and its persistence invariably leads to medical attention. In approximately one-third of the patients there is a history of recurrent attacks of bronchitis. Since the bronchial veins drain into the pulmonary veins, edema of the bronchial mucosa secondary to elevated pulmonary venous pressure is believed to be responsible for the severity of the symptoms if not the susceptibility to infection.

Systemic embolism. The incidence of systemic embolism has been estimated to range from 9% to 21% in patients suffering from rheumatic mitral valve disease, and the embolism is cerebral in 60% of the cases.[15,17,18] Szekely[19] reported 89 embolic episodes among his 754 patients with an overall incidence of 1.5% per patient-year. For patients in atrial fibrillation the incidence was 5% per patient-year, compared to 0.7% per patient-year for patients in sinus rhythm. Of the 14 patients who experienced embolic recurrence, 40% experienced recurrence within 1 month and 66% within 1 year. He found the incidence of recurrent embolic phenomena to be 3.4% per patient-year among 23 patients who received anticoagulant therapy, compared to 9.6% per patient-year among 46 patients who survived the first embolic episode and who were not placed on anticoagulants. In a review of 839 cases, Coulshed[20] concluded that (1) systemic embolism is as common with mitral regurgitation as with mitral stenosis, (2) atrial fibrillation with clot in the left atrium is the main cause, and (3) the size of the atrial appendage has no influence on the occurrence of embolic phenomena, and appendagectomy has no influence on postoperative emboli. Only in patients with atrial fibrillation and an operable valve is there thought to be any reduction in the number of emboli after surgery. We currently anticoagulate patients with atrial fibrillation and mitral stenosis who are essentially asymptomatic and thus not considered surgical candidates. Any patient having a systemic embolus in association with mitral stenosis is promptly anticoagulated, and cardiac catheterization is planned for the immediate future, unless a severe neurologic deficit persists. If the valve area is found to be 1.5 cm^2 or less, the patient is recommended for surgery. Postoperative anticoagulation has been carried out only in those patients receiving a mitral valve prosthesis.

Other symptoms. Other symptoms attendant with mitral stenosis that are less common include subacute bacterial endocarditis, anginal-like chest pain, dry cough, and hoarseness. It is virtually axiomatic that bacterial endocarditis rarely complicates established cases of pure mitral stenosis. Wood encountered only 1 instance in over 500 patients, and we have seen but 2 examples in our 320 patients. It is quite true that bacterial endocarditis occurs more frequently when mitral regurgitation is the dominant lesion. Wood found a 10% incidence of chest pain indistinguishable from that encountered in occlusive coronary disease in his population that was dominantly female with an average age of 36 years. He attributed this to functional impairment of coronary blood flow rather than occlusive coronary disease and found it twice as common in cases with extreme stenosis and high pulmonary vascular resistance. For the past 4 years we have been performing selective coronary cineangiographic studies at the time of cardiac catheterization in our patients over 35 years of age who have mitral valve disease. In over 100 such studies we have found the incidence of significant obstructive coronary disease to be approximately 10%. Usually the coronary lesion was an unsuspected finding in a patient who had no complaints of chest pain. Patients who complained of typical angina pectoris and had physical findings consistent with mitral stenosis almost invariably had significant coronary narrowing. A few patients who complained of rather severe dyspnea but had physical findings suggesting only mild mitral stenosis were found to have significant coronary lesions and segmental abnormalities of ventricular contraction typically found in patients with prior myocardial infarctions. Curiously enough, these few patients did not complain of chest pain. For our group as a whole, complaints of chest pain were usually atypical for coronary artery disease and presented more often in patients with pulmonary hypertension. We thus feel that patients with mitral stenosis who complain of typical angina pectoris are very likely to have significant coronary artery disease.

Physical Signs

Physical findings in mitral stenosis are extensively covered in standard textbooks of cardiology,[2,15,16] and the emphasis here will be to comment on those features of the physical examination that have most commonly influenced our diagnosis. I have found it personally a great help to categorize physical signs into three broad categories in attempting to arrive at a diagnosis with clinical certainty: (1) those findings I consider virtually essential and whose presence is required, (2) those findings that virtually exclude the diagnosis and whose absence I search for, and (3) those findings that are merely helpful, but whose presence or absence is nonessential.

Inspection. The malar flush ascribed to mitral stenosis is present in many patients with rheumatic valvular disease, but is not specific for mitral stenosis. It is more often associated with class III or IV symptoms of whatever cause. The patient with mitral stenosis is usually slender, and we still consider it something of a rarity to find severe stenosis in the obese patient. The jugular venous pulse is helpful in assessing the presence or absence of congestive failure, but is nonspecific for mitral stenosis. A monophasic pulsation bespeaks atrial fibrillation, and it is well to remember that it is the negative shadow of the skin deflection that coincides with the Y trough that calls attention to the V wave. The absence of this negative trough strongly suggests tricuspid stenosis or elevated right ventricular end-diastolic pressure. The precordium is usually quiet in patients with mitral stenosis, but a gentle heave along the left sternal border is frequently appreciated in patients with significant pulmonary hypertension.

Palpation. The carotid pulse is never bounding in patients with pure mitral stenosis, and the presence of this finding indicates some associated problem such as aortic regurgitation, anemia, or thyrotoxicosis. The carotid upstroke is normal, although the amplitude is characteristically small. The precordium is nonheaving, and the PMI must be located in the fifth intercostal space to be consistent with mitral stenosis. It is frequently displaced laterally as far as the anterior axillary line due to displacement from right ventricular enlargement. The PMI should not be displaced downward to the sixth intercostal space unless there is associated left ventricular enlargement caused by other conditions. In mitral stenosis it is tapping and nonsustained, which means it should fall away by midsytole. A right ventricular heave is appreciated by feeling a significant lift across the palm of the hands when the heel of the hand is placed along the manubrium sternum and the fingers extended horizontally toward the anterior axillary line. The impulse may extend up to but not beyond the metacarpophalangeal joints. Any lift of the distal phalanges in this position is strongly suggestive of associated left ventricular enlargement. A diastolic thrill is virtually pathognomonic of mitral stenosis when it is felt at the apex.

Auscultation. On auscultation the hallmarks of mitral stenosis include: (1) An accentuated pulmonic closure with persistence of very narrow physiologic splitting. (2) An accentuated and single S_1. (3) An opening snap appreciated as a third heart sound occurring 0.06–0.12 sec after the second sound at the left sternal edge or at the apex. While there is nothing "snapping" about this sound, it differs considerably from other third heart sounds. Instead of appreciating a dull thud typical of the ventricular diastolic gallop or S_3, this sound, although quite faint, is sharp in quality and appreciated as a sound coming toward the ear. It is occasionally extremely loud and is often appreciated in the aortic area when one first places the stethoscope on the precordium. (4) The diastolic rumble is a low-pitched noise following the opening snap that with rare exceptions extends to the first sound in significant mitral stenosis. It usually requires the extremely light application of the bell to be appreciated, and it is well to remember the aphorism of Sir William Osler: "The murmur of mitral stenosis may be concealed under a quarter of a dollar." Since it is often quite localized it should be judged absent only after provocative maneuvers have been carried out. This usually requires nothing more than turning the patient to the left lateral decubitus position, eliciting a cough, or performing several situps. In addition to these specific auscultatory events, there is an overall cadence that is heard in mitral stenosis. In patients with normal sinus rhythm the dominant beat of this cadence is the presystolic accentuation of the diastolic rumble that leads to the characteristically loud first sound. With atrial fibrillation the presystolic accentuation disappears, but not totally. There is a striking accentuation of the irregularly occurring first heart sound, with the louder sound occurring after the shorter R-R intervals. This specific cadence is particularly helpful in differentiating mitral stenosis from other clinical conditions that mimic it, due to the presence of accentuated pulmonic second sound, a loud first sound associated with early third sound, and diastolic rumbles. Conditions most commonly confused with mitral stenosis on auscultation are thyrotoxicosis, atrial septal defect with large left-to-right shunt, and occasionally cardiomyopathy. None of these conditions has the cadence of mitral stenosis. A reduplicated first heart sound virtually excludes significant mitral stenosis, but in relying on this finding one must be careful not to confuse an early ejection click with splitting of the first sound.

Concomitant findings of congestive heart failure indicate that the mitral stenosis is hemodynamically significant. The absence of failure, on the other hand, does not mean significant stenosis is excluded. The presence of a right ventricular heave is perhaps the most reliable sign of pulmonary hypertension, and this does not occur in mild stenosis. The time interval between the second sound and the opening snap referred to as the 2-OS time best reflects the level of left atrial pressure. Since this time interval corresponds to isovolumic relaxation occurring between the closure of the aortic valve and the opening of the mitral valve, it will vary inversely with the height of left atrial pressure. This in turn correlates well with the degree of mitral stenosis. Left ventricular pressure will fall below an elevated left atrial pressure sooner in time after aortic valve closure than it will when the left atrial pressure is low. In general, the loudness of the murmur of mitral stenosis relates more to the level of cardiac output than it does to the degree of mitral stenosis. While the presence of symptoms usually indicates hemodynamically significant mitral stenosis, the absence of symptoms does not permit one to conclude that the degree of stenosis is insignificant.

Electrocardiogram

A broad notched P wave, often most visible in Lead II, has been labeled "P-mitrale" because of its frequent association with mitral stenosis and sinus rhythm. The first peak represents right atrial depolarization and the second left atrial depolarization. It is preferable, however, to rely on stricter criteria for left atrial enlargement. Morris and co-workers[21] have proposed an analysis of the P terminal force in V_1 as the most exact means of separating normal subjects from patients with left-sided valvular lesions. They found that the algebraic product of duration (seconds) and amplitude (millimeters) of the V_1 terminal force ranged from +0.01 to −0.03 in normals and from −0.03 to −0.30 in those with left-sided disease. Kasser[22] found that left atrial volume was a more important determinant of abnormal atrial forces on the electrocardiogram than left atrial pressure. He likewise found that the V_1 terminal force accurately diagnosed increased left atrial size in 90% of patients with mitral stenosis. Criteria for right ventricular hypertrophy vary widely, but when convincingly present they indicate significant associated pulmonary hypertension.

In assessing the severity of mitral stenosis we have found the most helpful electrocardiographic finding to be the frontal-plane mean QRS axis. Rightward displacement +60° or greater correlates well with a mitral valve area 1.3 cm^2 or less. In an earlier study of a group of patients with pure mitral stenosis,[23] 89% of those with a frontal-plane axis +60° or greater had a mitral valve area less than 1.3 cm^2. Conversely, 82% of those patients with an axis of less than +60° had valve areas 1.3 cm^2 or larger. We have applied this rule of thumb to over 400 patients with mitral stenosis and continue to find a 90% confidence rate in predicting mitral valve area, provided the patient does not have associated valvular disease or significant hypertension.

X-Ray

A "mitral configuration" to the cardiac silhouette on chest x-ray features a wide-waisted heart with a diminutive aortic knob and a distinct pulmonary vascular pattern. The legendary "straight left heart border" is nonspecific and is found in cardiac silhouettes of many normal subjects. An enlarged left atrium makes its first appearance in the PA projection as a convex bulge immediately below the main pulmonary artery trunk, or the so-called pulmonary conus. As left atrial hypertrophy occurs, there is an increased density to the radiographic shadow that is recognized as a "double density" within the center of the cardiac silhouette immediately below the bifurcation of the trachea. Continued enlargement results in a convex bulge on the right heart border. As right ventricular dilatation occurs, an additional bulge is seen along the left border in the PA projection immediately below the left atrial shadow. In the lateral projection, left atrial enlargement is seen as a discrete bulge rather high up on the posterior margin of the cardiac silhouette. A more gentle bulge lower down on the posterior margin is indicative of left ventricular enlargement and should not be present in cases of mitral stenosis. Thus the typical x-ray appearance of severe mitral stenosis may show five separate convex bulges along the left heart border that are, from the top down: (1) a small aortic knob, (2) a prominent pulmonary conus, (3) an enlarged atrium, (4) a dilated right ventricle, and (5) the normal left ventricle.

In addition to the dilated main pulmonary artery segments, the vascular pattern seen in mitral stenosis is described as showing preferentially increased blood flow to the upper zones of the lung fields.[24] Simon[25] has called attention to the lower lobe venoconstriction that occurs when hydrostatic pressure reaches a critical point. In upright man this occurs earlier in the lower lobe veins, resulting in a disparate dilatation of the pulmonary venous markings in the upper lung fields. Using tomographic techniques, Dulfano[26] found this preferential lower zone constriction in supine individuals as well.

Hemodynamics

Technical Considerations

It is important to recognize that much of the hemodynamic information relating to left atrial pressure is derived indirectly from right heart catheterization and measurement of the pulmonary capillary wedge pressure as first described by Hellems.[27] Until the development of the newer balloon occluding catheters, this required wedging a cardiac catheter in the smallest pulmonary branch that size would permit and the demonstration of 95% saturated hemoglobin upon withdrawal of a blood sample from this position. Such a wedge measurement has been validated as reflecting left atrial pressure, particularly since the introduction of the transseptal technique of left atrial catheterization[28] that has permitted ready comparison of the two measurements. Discrepancies between them, however, have been reported to exist at higher levels of left atrial pressure.[29] This was based on a retrospective analysis of catheterization data, and the authors found no significant difference between the average wedge and left atrial pressures until the wedge pressure exceeded 25 mm Hg. Having investigated the accuracy of these two measurements in a prospective fashion, we find that meticulous attention to detail in pressure recording results in virtually identical measurements at all levels of pressure. While the transseptal catheter has the advantage of providing a more distinct waveform and obviates the need for correcting pulse wave transmission delay, there is a recognized morbidity and mortality to its use.[30]

It is of more than historic interest to review the physiologic method for calculation of the cross-sectional area of the mitral valve based on the application of the standard hydrokinetic formula for orifices as proposed by Gorlin and Gorlin.[31] After two decades of use the Gorlin formula is still considered an accurate index of the degree of mitral stenosis and remains the cornerstone for most clinical pathologic studies. It is based on the following formulas for fixed orifices:

$$F = C_c A \cdot V \tag{1}$$

With a fixed orifice A, changes in flow F are associated with proportionate changes in velocity V. C_c is the coefficient of orifice contraction.

$$V = C_c \sqrt{2gh} \tag{2}$$

Changes in velocity V through the orifice are proportional to the square root of the pressure in height h of a given liquid standing above the orifice in question; g is acceleration due to gravity and in the CGS system equals 980 cm/sec^2. C_v is the coefficient of velocity. Combining equations (1) and (2) and solving for A:

$$A = F/C \cdot 44.5 \ \sqrt{\Delta P} \tag{3}$$

F equals flow rate through the orifice; C equals discharge coefficient (empirical constant); $44.5 = \sqrt{2g}$ or $\sqrt{1960}$; ΔP equal pressure gradient across the orifice.

The coefficients C_c and C_v were incorporated into one coefficient C, which basically contains the conversion factor from millimeters of mercury to centimeters of water, 1.36. The square root of 1.36, being 1.17, establishes the basic constant $C = 1$. In comparing the application of this formula to estimates of mitral valve orifice with data obtained from autopsy measurements and those obtained at the time of surgery there was a resultant overestimation of the severity of stenosis by the formula. Accordingly the empiric constant was changed by a factor of 0.7, resulting in the commonly used constant 31. The basic limitation in the early use of this formula was the assumption of a normal mean left ventricular diastolic pressure of 5 mm Hg. Another limitation was the need to estimate diastolic filling period from the peripheral arterial pressure pulse. It is measured from the dicrotic notch of the brachial arterial tracing to the upstroke of the next arterial pressure pulse. Although these limitations no longer exist with the routine use of directly measured left ventricular pressure, correlative measurements with necropsy specimens are not always in agreement.[32] It is our experience and that of others that the best correlation occurs when the empiric constant is somewhere between 31 and 44.5. The formula used in our laboratory is expressed as follows:

$$\mathrm{MVA} = \frac{\mathrm{MVF}}{38\sqrt{\mathrm{MVG}}}$$

where MVA = mitral valve area in cm^2, MVF = mitral valve flow in cc/min = [cardiac output (cc/min)]/[diastolic filling period (sec) × heart rate], MVG = mitral valve gradient, and $38 = C\sqrt{2g} = 0.86\sqrt{1960}$.

Considerations of Pulmonary Vascular Resistance

Mechanical obstruction of blood flow arising from mitral stenosis results in a rise in pressure proximal to the mitral valve and accounts for the primary hemodynamic abnormalities measured at cardiac catheterization. The orifice size of the normal adult mitral valve is approximately 5 cm^2. A normal cardiac output 5.0 to 6.0 liters/min is associated with a normal left atrial pressure that ranges from 6 to 10 mm Hg. When orifice size is reduced to ~1.5 cm^2, left atrial pressure rises above these normal values, and it promptly reaches levels of 20–25 mm Hg when the narrowing reaches the critical level of 1.0 cm^2. As Dexter[33] has pointed out, all patients are thoroughly symptomatic with this degree of valve narrowing, since the rise of left atrial pressure is accompanied by a similar rise in pressure in the lungs and in the right ventricle. Considering the pulmonary capillaries as the locus minoris resistentiae, any rise of pressure above the osmotic pressure of plasma (25 mm Hg) should result in a transudation of fluid from capillaries to aveoli. Based on the early and still classic hemodynamic studies of mitral stenosis performed in Dexter's laboratory,[34–36] the concept of pulmonary hypertension serving as a protective mechanism preventing pulmonary edema evolved. In mitral stenosis, when pulmonary edema impends as a chronic state, pulmonary vascular changes appear that produce an obstruction to blood flow through the lung and effect a reduction in cardiac output. There are thus two stenoses—the primary blockade of the mitral valve and the secondary blockade at the pulmonary arterioles. It has been reasoned that this secondary vascular obstruction, by causing a further reduction in blood flow, allows the mitral valve to become progressively smaller, with the pressure in the left atrium remaining at or near 25 mm

Hg. At the same time, pressure in the pulmonary artery rises inordinately to force blood through to the left side of the heart.

The concept that left atrial hypertension triggers a reflex pulmonary vasoconstriction has been controversial from the outset. Gorlin[37] claimed that the degree of anatomic mitral stenosis affected the degree of pulmonary capillary pressure rise in exponential fashion. Araujo[38] disclaimed any obligatory relationship. Two subsequent studies offer rather convincing evidence that the development of high pulmonary vascular resistance does not develop as a protective mechanism in severe mitral stenosis of critical severity. In serial studies over a 3-year period Selzer[39] demonstrated a fourfold to 10-fold increase in pulmonary vascular resistance that did not coincide with a significant increase in mitral valvular obstruction or concomitant rise in left atrial pressure. He found a consistant reduction in pulmonary vascular resistance following surgical correction of mitral stenosis and admitted that left atrial hypertension plays some role in sustaining high pulmonary vascular resistance. He did not consider it the sole stimulus, however, since elevated pulmonary resistance fails to develop in many persons with high left atrial pressure and advanced mitral stenosis. Hugenholtz et al.[23] studied a group of asymptomatic patients (classes I and II) with mitral stenosis and found critical valve narrowing (1.0 cm^2 or less) with elevated left atrial pressures and reduced cardiac output in the absence of increased pulmonary vascular resistance. In contrast, their symptomatic group (class III) had significantly elevated pulmonary vascular resistance. It was concluded that if any relationship between mitral valve area and pulmonary vascular resistance exists it will be biased by the relative proportion of symptomatic (class III) to asymptomatic (classes I and II) patients present in any given series.

An adaptation of the Poiseuille equation is utilized to quantify the degree of pulmonary vascular resistance:

$$\begin{array}{l}\text{Resistance (R)}\\ \text{(Wood units)}\end{array} = \frac{\text{mean PA pressure} - \text{mean LA pressure (mm Hg)}}{\text{pulmonary blood flow (C.O. liters/min)}}$$

This expression of resistance in Wood units may be converted to fundamental units of force by converting pressures in mm Hg to $dynes/cm^2$ and flow to cm^3/sec. A conversion factor of 80, derived from the specific gravity of mercury (13.59) and the acceleration due to gravity (981.2), is used to express resistance in $dynes\text{-}sec\text{-}cm^{-5}$.

$$\text{R (dynes-sec-cm}^{-5}\text{)} = \frac{\text{mean PA pressure} - \text{mean LA pressure (mm Hg)} \times 80}{\text{C.O. (liters/min)}}$$

Table 14-1
Severe Pulmonary Hypertension and Mitral Stenosis:
Clinical Profile of 22 Patients

	Female	Male	Mean
Age (years)	50.2 (17)	60.6 (5)	52.5 years
AF	7	4	57.6 years
NSR	10	1	42.7 years
Valve Ca^{++}	68% (15/22)		
Syst. emboli	23% (5/22)		
Mortality	29% (5/17)		

AF = atrial fibrillation; NSR = normal sinus rhythm; Valve Ca^{++} = calcification of the mitral valve; Syst. emboli = systemic emboli.

While pulmonary hypertension and elevated pulmonary vascular resistance are common findings in mitral stenosis, neither value ordinarily reaches the extreme degree often seen in congenital heart disease with pulmonary vascular obstruction. It is still the unusual case of mitral stenosis that demonstrates a pulmonary artery pressure equal to systemic pressure. The overall incidence of pulmonary vascular resistance greater than 10 units in the Wood series was 8%. Such extreme and inappropriate pulmonary hypertension in certain cases of mitral stenosis must in part be due to an abnormality of the reflex vasoconstriction that has been amply demonstrated to occur in response to left atrial hypertension. The infusion of hexamethonium has been shown to markedly decrease pulmonary hypertension and thus supports the concept of a reflex mechanism.[40] The marked reduction in pulmonary pressure following mitral valve surgery further suggests the reactive nature of the pulmonary hypertension.[41] In our own series of 320 patients with severe mitral stenosis the incidence of extreme pulmonary hypertension defined as (1) PA systolic pressure 80% of aortic systolic pressure, (2) PA mean pressure 70% of aortic mean pressure, and (3) pulmonary vascular resistance greater than 9 Wood units (700 dynes-sec-cm^{-5}) was 7%. The clinical profiles of the 22 patients who met these criteria are shown in Table 14-1. It can be seen that this level of pulmonary hypertension was not limited to the younger patients and those with normal sinus rhythm; half the group had atrial fibrillation and a mean age of 57 years. They tended to be the sicker patients and had an overall high mortality. In only one of the group was a pulmonary embolus documented to have occurred. Applying linear regression analysis to the relationship of mean pulmonary to mean left atrial pressure in 200 patients with pure mitral stenosis, Walston[42] found 36 patients with increased pulmonary pressure greater than two standard errors from the regression line. Hemody-

Table 14-2
Severe Pulmonary Hypertension and Mitral Stenosis: Hemodynamic Profile

22 Patients (Boston)		36 Patients (Duke)
119/64 85	Ao mm Hg	—
110/47 70	PA mm HG	61
30	PC mm Hg	27
22	MDG mm Hg	20
3.04	C.O. liters/min	3.2
1154	PAR dynes-sec-cm^{-5}	—
14.4	PAR Wood units	10.9

Ao = aortic pressure; PA = pulmonary artery pressure; PC = pulmonary capillary wedge; MDG = mitral diastolic gradient; C.O. = cardiac output; PAR = pulmonary arteriolar resistance; Boston = author's series; Duke = Walston's series.[42]

namically they were quite comparable to our group (Table 14-2). However, evidence of associated pulmonary emboli or severe obstructive lung disease was present in 15 (40%) of these patients, and it was suggested that the extreme pulmonary hypertension was secondary to these complications. We found no evidence of significant obstructive lung disease in our 22 patients who had a mean value of 73% for the FEV 1 and an MBC of 72%. Only 2 patients in the group had 1-sec vital capacities below 65%. Furthermore, those patients who underwent successful mitral valve surgery and who were studied postoperatively all showed a marked reduction in the pulmonary artery pressure. We continue to feel that the problem of extreme pulmonary hypertension in mitral stenosis is limited to less than 10% of patients and is more often the result of an excessive vasoreactivity than it is a sequela to chronic lung disease or pulmonary thromboembolism.

Cardiac Output

It had been thought that mildly symptomatic patients with mitral stenosis differed little if any from normal[43] and that severely symptomatic patients had decreased cardiac output.[44] When a group of 26 patients, minimally symptomatic (classes I and II), with mitral stenosis were compared with 18 symptomatic patients (class III), 17 of the mildly symptomatic group had valve areas below 1.5 cm^2, similar to all the patients in class III.[23] The mean cardiac index for the mildly symptomatic group was 2.81 liters/min/m^2, significantly lower than the normal value of 3.6 liters/min/m^2 and almost identical to the mean value for the symptomatic group, 2.82 liters/min/m^2. Since the symptomatic group had elevated pulmonary vascular resistance and the mildly symptomatic

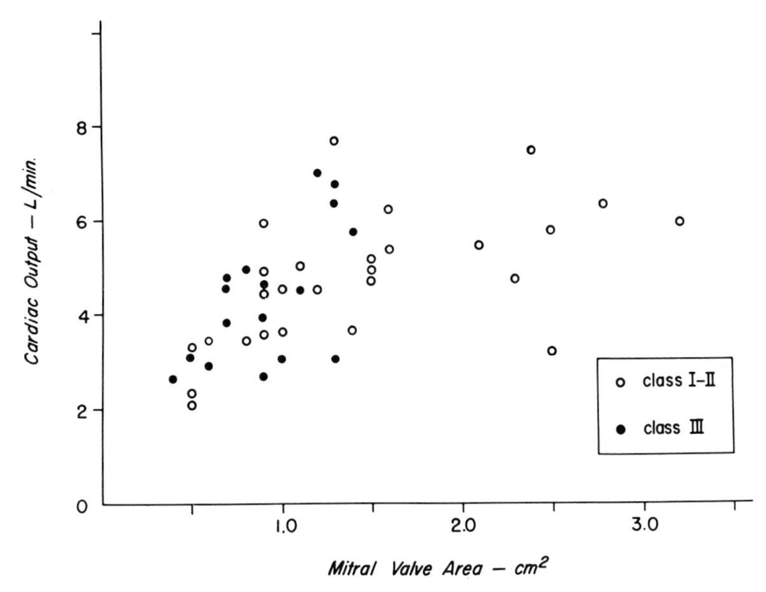

Fig. 14-1. A direct relationship between mitral valve area and cardiac output was obtained from a group of 43 patients with mitral stenosis who were mildly symptomatic (open circles) as well as severely symptomatic (black dots).

group did not, these observations indicate that a decrease in cardiac output observed in mitral stenosis can occur before the onset of pulmonary vascular changes. Carman has reported similar findings.[45] The exact mechanism by which the cardiac output is lowered early in the course of mitral stenosis may be the result of many factors,[46] but it is quite possibly related to baroreceptors or stretch receptors that are sensitive to pressure or wall tension changes in the left atrium and pulmonary veins. Volume receptors have been demonstrated in the left atrial wall that influence urinary flow.[47] Such a reflex system could conceivably affect circulatory volume. It is our feeling that low cardiac output remains the hallmark of significant mitral stenosis regardless of the severity of symptoms and is primarily dependent upon the degree of mitral valve narrowing. This is seen in the linear relationship that exists when one plots cardiac output against mitral valve area (Fig. 14-1).

Atrial Fibrillation

While it is appreciated that atrial fibrillation is associated with a decreased cardiac output, that does not of itself explain the lowered cardiac output observed early in the course of patients with mitral stenosis. Table 14-3 presents some of the hemodynamic consequences of

Table 14-3
Atrial Fibrillation in Mitral Stenosis

	Mitral Stenosis				
	NSR	AF	Mean	Units	Normals
C.I.	2.6	2.3	2.4 ± .11	liters/min/m^2	3.1 ± 0.18
EDV/M^2	74.7	67.3	69.9 ± 3.2	ml/m^2	66.6 ± 2.7
ESV/M^2	30.1	32.2	31.5 ± 1.6	ml/m^2	24.5 ± 1.2
EF	59.1	54	55.8 ± 1.4	%	63.5 ± 1.8
LVEDP	6.7	8.3	7.8 ± .5	mm Hg	7.5 ± 0.5
Age	44.1	54.2	50.5 ± 4	years	33.7 ± 3.4

The influence of atrial fibrillation versus regular sinus rhythm on hemodynamics and volumetric measurements of patients with mitral stenosis. The group as a whole is also compared to a group of patients without known heart disease. C.I. = cardiac index; EDV = end-diastolic volume; ESV = end-systolic volume; EF = ejection fraction; LVEDP = left ventricular end-diastolic pressure.

atrial fibrillation occurring in association with mitral stenosis. Measurements of flow, volume, and pressure are shown for normal subjects and patients with mitral stenosis in sinus rhythm and in atrial fibrillation. It can be seen that the parameters of cardiac index, ventricular volume, and ejection fraction differ from normal in patients with mitral stenosis irrespective of the cardiac rhythm. The development of atrial fibrillation exaggerates these abnormalities, but the magnitude of difference between the atrial fibrillation group and sinus rhythm group is small. The age difference between those two groups is significant and supports the well-known fact that the incidence of atrial fibrillation is related to the age of the patient more than to any other factor. It is generally agreed that atrial fibrillation is precipitated by left atrial enlargement and eventually perpetuated by disintegration of the architecture of the atrial muscle.[48]

Myocardial Factor

An important consideration in the understanding of cardiac performance in mitral stenosis is an appreciation of the influence of the disease on the left ventricle. A myocardial factor has been suggested by Harvey[49] and Fleming[50] to account for circulatory dysfunction that is not explained by the mechanical difficulties of stenosis. It was hypothesized that ventricular dysfunction could be the result of intrinsic myocardial damage arising from the acute rheumatic process. With the demonstration that cardiac output in severe mitral stenosis did not return to normal in many patients following successful valvotomy,[51,52] this concept attained new significance. Feigenbaum[53] observed abnormal rises in ventricular diastolic pressure for a given increase in diastolic volume and implied that mitral stenosis reduced left ventricular compliance. His data assumed that left ventricular volume was normal or decreased in patients with mitral stenosis. Kirch,[54] as early as 1929, pointed out that the posterior wall or inflow tract of the left ventricle was markedly shortened in nearly all hearts with mitral stenosis. Grant[55] substantiated these findings in his classic article on the architectonics of the heart and concluded that this shortening was due to selective atrophy of the myocardium of the posterior wall of the left ventricle. He found the circumference and the anteroposterior diameter of the mitral ring to be greatly increased in mitral stenosis. This was due to shortening of the posterior wall and not to stretching of the ring or dilatation of the ventricle. He attributed these changes to the thickening of the valve leaflets and fibrosis of the chordae tendineae that converted the valves into a rigid cylinder of scar tissue that served as an internal cast to immobilize the posterior wall of the left ventricle. Much as with skeletal muscle, he hypothesized that focal atrophy followed such immobilization. He found no evidence of a diffuse fibrosis that could be related to earlier inflammatory changes.

Heller and Carleton[56] carried out an angiographic evaluation of 25 patients with pure mitral stenosis and found slightly elevated values for left ventricular end-diastolic volumes and substantially increased end-systolic volumes in these patients compared to normal subjects. Ejection fractions were consequently reduced from normal. Applying the technique of segmental ventricular cineangiographic analysis, they found qualitative changes in the contraction pattern of the postero-basal area of the left ventricle. These qualitative changes would tend to support the hypothesis of Grant. Our own analysis of ventricular function in over 40 patients with pure mitral stenosis utilizing single-plane cineangiographic techniques is substantially in agreement with the findings of these workers and is presented in Table 14-4. Our mean value for end-diastolic volume among patients with mitral stenosis was increased over normal subjects, but was not as elevated as the value reported by Heller and Carleton (84.6 ml/m^2). Wall tension was increased as compared to normal subjects, and it related to the increased short-diameter measurement characteristic of these patients. Mean velocity of circumferential fiber shortening was slightly reduced. While these measurements imply altered ventricular function, the degree of dysfunction is actually quite small. The principal abnormality is found in the ratio of the long diameter L to the short diameter D. The foreshortened long diameter L pointed out more than 40 years ago by Kirch gives the spade-shape appearance to the ventriculogram of patients with mitral stenosis. This is in marked contrast to the football appearance of a normal ventriculogram.

Table 14-4
Left Ventricular Function in Mitral Stenosis

	Mitral Stenosis		Normals	*p* Value
EDV	69.9 ± 3.2	ml/m^2	66.6 ± 2.7	NS
ESV	31.5 ± 1.6	ml/m^2	24.5 ± 1.2	0.001
EF	55.8 ± 1.4	%	63.5 ± 1.8	0.001
T (mean)	287 ± 13		221 ± 11	0.01
V_{CF}	1.14 ± 0.05		1.4 ± 0.05	0.01
L/D	1.55 ± 0.03		1.74 ± 0.03	0.001

EDV = end-diastolic volume; ESV = end-systolic volume; EF = ejection fraction; T = wall tension; V_{CF} = mean velocity of circumferential fiber shortening; L/D = ratio between long diameter and short diameter.

Unlike Heller and Carleton, we did not find consistent segmental abnormalities of contraction in the postero-basal segment. We attributed the occasional irregularities of opacification that occurred in this region to varying degrees of entrapment of dye in and around the scarred mitral apparatus. Clear-cut abnormalities of ventricular contraction were difficult to ascertain, and when the ventriculograms were viewed in the opposite oblique no abnormalities of posterior wall motion were seen. It is quite likely that exaggerated rotational changes of the long axis create an illusion of postero-basal hypokinesis. We believe that segmental abnormalities of ventricular contraction are the result of associated coronary artery disease when seen in patients with mitral stenosis.

Echocardiogram

Since Edler and Herz first demonstrated that ultra-high-frequency sound waves could be reflected from cardiac structures to provide useful information,[57] echocardiography has become a well-established noninvasive means of studying cardiac motion. This has had particularly successful application in the study of the mitral valve, with the demonstration that a distinctive mitral echo truly originated from the anterior mitral leaflet.[58] With the patient positioned in a slight degree of left obliquity and a transducer located in the third or fourth left intercostal space about 1 to 3 cm lateral to the left sternal edge, a characteristic waveform of mitral valve motion can usually be readily obtained. There are a number of classic articles and standard texts to which the reader is referred for a detailed analysis of this waveform and its correlation to the various phases of the cardiac cycle.[59–62] In mitral stenosis the characteristic abnormality seen on the echocardiogram is the delay in the

Table 14-5
Mitral Valve Area by Ultrasound:
50 Patients with Mitral Stenosis

E-F Slope	MVA (cm^2)
1–15 mm/sec	0.4–0.8
15–25 mm/sec	0.8–1.6
25–35 mm/sec	1.6–2.4

A correlation of E-F slope velocity determined by ultrasound with measurements of mitral valve area determined by cardiac catheterization in 50 patients with mitral stenosis.

downslope from the E point (representing the point of maximum opening of the valve) to the point F (the end of the rapid-filling phase). The E-to-F slope represents the movement of the anterior leaflet toward closure during and immediately after the rapid inflow of blood into the ventricle. In the case of the stenotic mitral valve there is impairment of rapid inflow, and a significant delay is seen in the normal floating motion of the valve leaflets toward each other in this phase of the cardiac cycle. A measure of this slope, therefore, correlates well with orifice size.[63] Normal subjects have been found to have a velocity (E-to-F slope) of 85 to 170 mm/sec. In cases of mild mitral stenosis the velocity ranges from 25 to 40 mm/sec, and in moderate stenosis this is reduced to 10–25 mm/sec. Tight mitral stenosis is indicated by a slope of 10 mm/sec or less. Our own experience with this technique of estimating the severity of mitral stenosis is summarized in Table 14-5. This shows the correlation between mitral valve area determined at cardiac catheterization and the E-to-F slope in 50 patients with varying degrees of mitral stenosis. The typical echocardiogram of a patient with mitral stenosis is seen in Fig. 14-2. We have found echocardiography to be particularly useful in assessing the results of mitral valvuloplasty by comparing E-to-F slopes preoperatively and postoperatively (Fig. 14-3).

It is important to remember that the E-to-F slope is also reduced in states of low compliance and elevated end-diastolic pressure. Therefore a decreased E-to-F slope is not diagnostic of mitral stenosis. We consider the echocardiographic findings that establish the diagnosis of mitral stenosis to be (1) a diminished C-E amplitude (diminished excursion), (2) a decreased diastolic slope E to F, (3) an absence of the A wave of the anterior leaflet motion of the mitral valve in patients with regular sinus rhythm, and (4) concordant anterior movement of the posterior leaflet with the anterior leaflet.

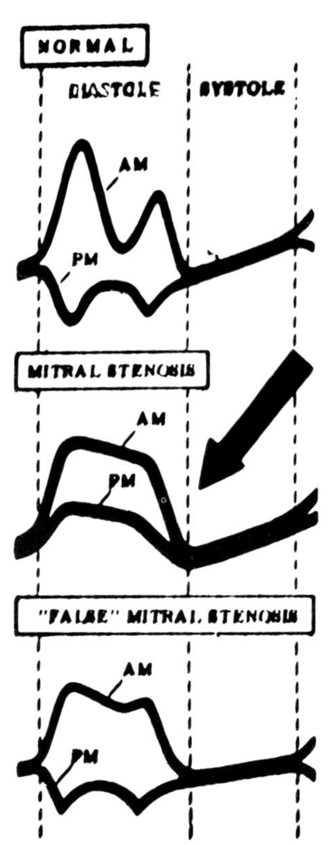

Anterior leaflet mitral valve

Fig. 14-2. A delayed E-F slope in a patient with mitral stenosis is shown in the right-hand panel. Schematized line drawings representing anterior and posterior leaflet motion are seen in the left-hand panel. Upper: normal leaflet motion. Middle: concordant anterior movement of both leaflets with a delayed E-F slope in mitral stenosis. Note the absence of a secondary peak (A wave). Lower: Delayed E-F slope, but posterior leaflet maintains normal posterior movement, excluding mitral stenosis.

Conclusions

While the clinical picture, physical findings, and natural history of mitral stenosis remain immutable observations, our appreciation of the pathophysiology involved continues to change with the development of newer techniques and a broader understanding of the basic physiologic processes. To propose a unifying concept that would embrace and explain all the findings that have been developed over the past 25 years would be hazardous at best and quite likely incorrect. If one should exist it will most likely center about the disease process on the mitral apparatus itself. This would explain the observed relationships between the

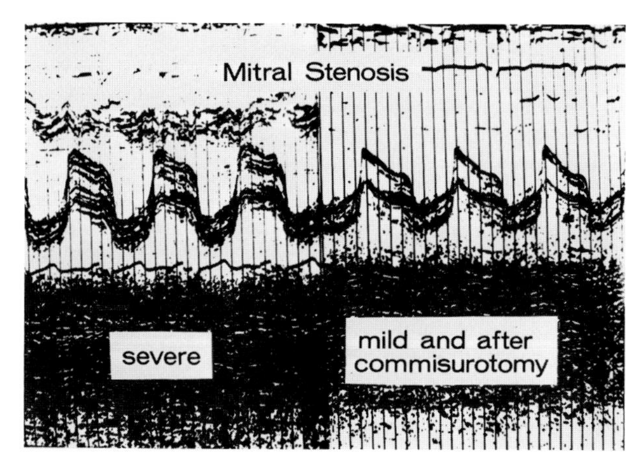

Fig. 14-3. Preoperative (left panel) and postoperative (right panel) mitral valve echo showing improved E-F slope velocity in a patient who underwent successful valvotomy.

degree of stenosis and cardiac output as well as the limited ventricular dysfunction that appears to exist. All else would follow from left atrial hypertension via known reflex pathways. The primary challenge for the next quarter century will be to alter the natural history by improved surgical techniques and to eradicate the basic disease by manipulation of such fundamental processes as the immune mechanism.

OTHER CAUSES OF MITRAL BLOCKADE

Left Atrial Tumors

Myxomas constitute 35%–50% of all primary cardiac tumors,[64] and over 75% of these arise from the left atrial septum to which they are attached by a pedicle close to the foramen ovale. Goodwin[3] constructed a distinct clinical picture from an analysis of 45 patients with left atrial myxoma pooled from the experience of members of the British Cardiac Society. He found that the clinical course of these patients was characterized by three types of findings: (1) obstructive, (2) embolic, and (3) constitutional. While the physical findings closely resemble those of pure mitral stenosis, a distinguishing feature is the delay in the third heart sound (tumor plop) when compared to the opening snap of mitral stenosis.[65] Since herniation of the tumor into the left ventricle occurs after atrial emptying, it comes to rest against the ventricular wall during the rapid ventricular filling phase. This occurs at a point in time after the

0 point of the apex cardiogram and is too delayed for even the latest opening snap. A late opening snap is found in association with minimal or mild mitral stenosis. Left atrial myxoma should come to mind when one is faced with a patient who has apparent mitral stenosis and severe symptoms with advanced failure and pulmonary hypertension and who demonstrates a paradoxically prolonged opening snap. It should be remembered that atrial fibrillation is present in approximately 13% of patients with myxoma. It is also rare to see massive cardiomegaly and extreme left atrial enlargement in this condition.

Helpful clues to the diagnosis of left atrial myxoma can be obtained by phonocardiography and apex cardiography.[66] These noninvasive techniques are particularly helpful in alerting a catheterization team to the likelihood of left atrial myxoma, thus assuring that appropriate angiographic studies will be carried out at the time of catheterization. The echocardiogram is perhaps the single most useful tool in arriving at the diagnosis of left atrial myxoma by the finding of a characteristic cluster of echoes immediately beneath the anterior leaflet of the mitral valve. Martinez[67] has stressed the importance of a systematic echocardiographic examination when searching for a left atrial myxoma and has established criteria by which other diagnostic possibilities, including artifacts, may be eliminated.

Calcification of the Mitral Anulus

Calcification of the anulus fibrosis of the mitral valve is commonly found in older people at autopsy. Although this can be the result of a rheumatic, rheumatoid, or other type of endocarditis, histologic evidence of previous inflammation is usually absent.[68] It is generally regarded as an end stage of a degenerative process and most frequently involves the posterior segment of the anulus. Extensions of the calcification into adjacent myocardium and interventricular septum are well recognized and explain its common association with the picture of complete heart block due to an interruption of the special conducting tissue. A holosystolic apical murmur consistent with mitral regurgitation is frequently found in association with calcification of the anulus, and apical diastolic murmurs attributed to turbulent blood flow over the jutting margins of the calcified anulus have been described.[69] Korn[5] has reported 14 cases of calcification of the entire anulus that resulted in anatomic deformity of the ostium of the mitral valve when examined at postmortem. In 5 of the 14 cases it was thought to represent the primary problem resulting in the patient's death. There was no evidence of healed rheumatic heart disease in any of these patients. All were female, and they ranged in age from 56 years to 88 years (mean 75 years). A significant degree of mitral stenosis was present in 9, secondary to protrusion of the main calcified ledge into the cardiac lumen. Gross extension of the calcification into the base of the anterior leaflet was demonstrated, but it was never seen to involve the posterior leaflet. It was postulated that mitral stenosis occurred either as the direct result of protrusion of the calcification into the cardiac lumen or by extension into the anterior mitral leaflet where it produced an immobile rigid shelf. Since all of these patients had grades II–IV systolic murmurs, it was evident that there was associated mitral regurgitation. The loss of the normal sphincteric action of the anulus in its calcified state results in an orifice gap too large to be bridged by the leaflets. The rigid shelf of calcium was also seen to elevate the base of the leaflet, which in turn stretched the chordae and resulted in the leaflet margin being curved downward. Such a curvature would prevent normal apposition of the leaflet margins in systole. There was also a high incidence of visible calcification of the tracheobranchial tree and the costal cartilages in these patients.

Cor Triatriatum

This is an exotic disorder of the heart resulting from the persistence of a septum in the left atrium that can give rise to many of the clinical features of mitral stenosis with pulmonary vascular congestion. In his review of the literature up to 1965, McGuire[4] found only 10 adults with this anomaly. The lesion is amenable to surgical correction in the second and third decades of life, but there have been instances in which cardiac surgery directed at the mitral valve was unsuccessful because of a lack of awareness of this rare cause of pulmonary venous obstruction. Embryologically it results from a failure of the common pulmonary vein to be incorporated into the left atrium in a normal manner. In the presence of an abnormal septum or membrane separating the pulmonary venous orifices from the mitral valve, a common pulmonary vein dilates to become a sac that is referred to as a third atrium.[70] Dyspnea, orthopnea, and pronounced hemoptysis are the leading symptoms. Signs of significant pulmonary hypertension are evident by a prominent parasternal impulse and marked accentuation of the pulmonic closure sound. While diastolic murmurs at the apex are usually present, no opening snap is present, and this serves to distinguish it from mitral stenosis. The electrocardiogram usually shows unequivocal right ventricular hypertrophy, but there is no evidence of left atrial enlargement. In this setting the absence of "P-mitrale" should be helpful in differentiating the lesion from mitral stenosis. At cardiac catheterization the finding of an elevated pulmonary capillary wedge pressure in conjunction with a normal left atrial pressure indicates obstruction at the pulmonary venous level. Angiography has been successful in demonstrating the lesion.

MITRAL REGURGITATION

An interesting historical perspective to our current concepts of mitral regurgitation has been presented by Perloff.[71] Although nineteenth-century clinicians such as Hope and Flint considered the condition a form of organic heart disease, mitral regurgitation was viewed as a functional disorder by Sir James Mackenzie and Sir Thomas Lewis in the early days of the present century. These eminent teachers questioned the pathologic significance of an apical systolic murmur and virtually denied mitral insufficiency, describing only 7 instances in a series of 1,846 autopsies, and rendered the opinion that the diagnosis of mitral regurgitation without stenosis is never justified since there are no physical signs by which it can be recognized.[72] It was not until P. D. White and S. A. Levine in the second quarter of this century emphasized the significance of the apical systolic murmur that the diagnosis of organic mitral regurgitation as a clinical entity was reestablished. With the development of techniques permitting catheterization of the left and right heart and the advent of surgical correction for mitral stenosis, a keen interest and new understanding of the significance of mitral regurgitation developed. Unlike mitral stenosis, where the cardinal circulatory abnormality is delayed inflow into the left ventricle, mitral regurgitation is best characterized as an abnormality of unloading the left ventricle during systole. Braunwald[73] and others[74] have studied the myocardial mechanics produced by this valvular lesion and found the decline in tension during systole to be the primary adaptive mechanism of the left ventricle in this situation. This allows the contractile energy of the heart to be expended in shortening rather than in developing tension, and as a consequence no significant increase in myocardial oxygen consumption ensues. These considerations contribute significantly to our understanding of both the natural history and clinical findings of patients with pure mitral regurgitation.

Natural History

Wood[15] found mitral regurgitation to be the dominant valvular lesion more often than aortic stenosis or aortic regurgitation, but not as often as aortic disease as a whole. In his experience patients with mitral regurgitation usually had a more malignant form of acute rheumatic fever. He viewed the condition as existing from the time of active carditis with no latent interval. This general view is concluded from other large series[71,75,76] wherein a greater percentage of patients with mitral regurgitation have a prior history of rheumatic fever, as compared to patients with mitral stenosis. It is possible the early occurrence and ease of recognition of the systolic murmur of mitral insufficiency, in contrast to the easily missed and later developing diastolic murmur of stenosis, are responsible for a large number of these patients being diagnosed as having rheumatic fever in the acute stage. While it is generally felt that these patients are symptom-free for a longer period of time than patients with mitral stenosis, there is some disagreement about the clinical course once symptoms have developed. Wood observed a more progressive downhill course that averaged 5.3 years. Ellis,[75] on the other hand, underscored the long and often benign course of patients with severe mitral regurgitation.

Considering the time interval from the first knowledge of heart disease to Ellis's last personal observation of 42 patients with severe insufficiency, 6 were known to have had the disease from 49 to 66 years. Of the remaining 36 patients, 15 had recognized severe mitral insufficiency for more than 20 years, and all were managed by medical therapy. Fifteen of the 28 patients in whom heart disease had first been recognized in the asymptomatic phase were observed to be in class I from 22 to 65 years. There are multiple etiologies of organic incompetence in addition to acute rheumatic fever and bacterial endocarditis. These include ruputure of chordae tendineae, rupture of papillary muscle secondary to trauma or infarction, calcification of the anulus fibrosis, and congenital malformations that include ostium primum defects, isolated anomalous chordal insertion, and myxomatous degeneration of the mitral valve. A clinical spectrum of severe mitral regurgitation includes, at one end, patients with acute mitral regurgitation resulting from spontaneous rupture of a chordae tendineae due to subacute bacterial endocarditis or rupture of a head of a papillary muscle due to infarction, and, at the other end, the group with long-standing mitral regurgitation due to rheumatic heart disease. The former have a thick-walled, small left atrium with marked left atrial hypertension, and the latter present with markedly dilated left atria with low atrial pressures. These groups may be equally disabled, but their clinical examinations may differ significantly, as discussed below.

Clinical Profile

Following systolic regurgitation of blood into the left atrium as the result of mitral incompetence, there is prompt and massive early ventricular filling in diastole. As a consequence, a short diastole does not prevent proper ventricular filling, so these patients are less embarrassed by tachycardia. In comparison to patients with mitral stenosis they have less exertional dyspnea and half the incidence of pulmonary edema, hemoptysis, and angina. The physical signs of pure mitral regurgitation include a characteristic peripheral pulse that is rapidly rising, sharp but poorly sustained. It is almost water-hammer in quality, but lacks the amplitude of the pulse of aortic regurgitation. This is the result of a rise in the peak velocity of left ventricular ejection, which is more rapid and occurs early in mitral regurgitation.[73] The precordium reveals a hyperdynamic apex with the PMI displaced downward and to the left. There is often a pseudo right ventricular heave that is appreciated as a sustained lift of the precordium just to the left of the sternal edge. It is difficult to distinguish from a true right ventricular heave due to pulmonary hypertension. In the situation of mitral regurgitation where pulmonary hypertension is less common, it is the result of an anterior displacement of the heart resulting from atrial expansion during ventricular systole.

On auscultation the most striking finding is the presence of a holosystolic murmur usually well heard along the left sternal border but usually maximum at the apex with frequent radiation to the axilla. The murmur begins with the first sound and extends to or through the second sound, since with an incompetent mitral valve left ventricular pressure exceeds left atrial pressure from the onset of isometric contraction to isometric relaxation. While this murmur may have varying configurations, sometimes with midsystolic accentuation that will create the impression of an ejection murmur, it has several features that distinguish it from the murmur of aortic stenosis that may be equally intense at the apex. The latter murmur invariably begins after the first sound and ends before the second. Mitral regurgitation will show no intensification of the murmur in the postpremature beat when compared to control beats, while the converse is true of aortic stenosis. Both murmurs will diminish in intensity during a Valsalva maneuver, provided severe congestive heart failure is not present. The murmur of idopathic hypertrophic subaortic stenosis will fail to diminish and even intensify in response to this maneuver. The infusion of pressor amines results in an amplification and lengthening of the murmur of mitral regurgitation,[71] while the murmur of subaortic stenosis usually diminishes. Amyl nitrite, on the other hand, results in a diminution in the intensity of the regurgitant murmur despite the reflex tachycardia and the transient increase in cardiac output that results from its use. It is well to remember that the murmur of mitral regurgitation can radiate anteriorly toward the base of the heart and simulate aortic stenosis. It is suggested that this radiation results from incompetence of the posterior mitral leaflet, which tends to direct the regurgitant stream forward and medially against the atrial septum. With anterior leaflet incompetence the stream is directly posterolaterally and radiates to the axilla and back.[77] The holosystolic murmur of mitral regurgitation is often followed by an early diastolic to middiastolic rumble, and in severe cases this often creates the impression of a to-and-fro murmur at the apex.

Careful attention to the intensity of the first heart sound is quite helpful in identifying mitral regurgitation. Perloff[71] found this to be a single sound in 26 of 33 patients, and it was of normal intensity in 25. The second heart sound at the base almost invariably demonstrates some degree of splitting. In 14 of 33 patients it was considered to be widely split by Perloff.[71] In the presence of mitral regurgitation the left ventricle is unable to sustain intraventricular pressure, and left ventricular pressure falls off abruptly as systole proceeds. Therefore aortic valve closure occurs prematurely and the first component of the second heart sound occurs earlier than normal. This explains the wide splitting of the second heart sound with maintenance of the normal respiratory variation. A prominent ventricular diastolic gallop is usually heard in association with mitral regurgitation. In Perloff's series it occurred an average of 0.14 sec after S_2 and was present in 26 of the 33 patients studied. It is attributed to the rapid filling wave causing abrupt stretching of the ventricular muscle. It diminished in intensity or totally disappeared during the inhalation of amyl nitrite. Fourth heart sounds or atrial gallops are extremely uncommon in established cases of mitral regurgitation. When present they are usually a consequence of associated pulmonary hypertension. The S_4 is more apt to be present when there is resistance to ventricular filling associated with increase force of atrial contraction.[78] The presence of an atrial gallop is considered the hallmark of recent-onset mitral regurgitation and persists until the left atrium loses its normal compliance. In the average case this will take 9 to 18 months.[73] An opening snap has been described in cases of pure mitral regurgitation[79] and was found in 4 of the 33 patients analyzed in Perloff's study. It is found in those situations where the anterior leaflet is normal or slightly enlarged with the regurgitation occurring exclusively via the posterior mitral leaflet. Thus it is seen in those patients where the systolic murmur radiates anteriorly toward the base of the heart.

The x-ray appearance of mitral regurgitation is characterized by left atrial enlargement and is frequently difficult to distinguish from cases of mitral stenosis. However, the largest atria result from mitral regurgitation. The syndrome of giant left atrium usually bespeaks massive mitral regurgitation. There is frequently evidence of left ventricular enlargement, and the aortic knob tends to be larger than in cases of pure mitral stenosis. Calcification of the mitral valve is rarely seen in pure mitral regurgitation.

The electrocardiogram differs from that of pure mitral stenosis by the frequent association of left ventricular hypertrophy, which is rarely present in cases of mitral stenosis. The frontal-plane axis is usually less rightward in mitral regurgitation, but the incidence of atrial fibrillation tends to be higher than in cases with mitral stenosis. It is felt that atrial fibrillation occurs at an earlier age in patients with pure mitral regurgitation. It is well to remember that left ventricular hypertropy is not present in one-third of the cases. When present it is rarely as striking as the concentric hypertropy seen in patients with aortic valvular disease.

Mitral regurgitation cannot be diagnosed echocardiographically, but it can be suspected in the presence of a large left atrial dimension with no evidence of a delayed E-to-F slope. Generally speaking, there is an enhanced excursion with the C-E amplitude increased over normal, as well as an exaggerated excursion of the posterior left ventricular wall greater than 1.5 cm. These findings indicate a volume overload and are not specific for mitral regurgitation.

MITRAL VALVE PROLAPSE: CLICK SYNDROME

Intense interest has been focused in recent years on the clinical findings of nonejection clicks and late systolic murmurs ascribed to mitral valve origin. Although initially thought to be of extracardiac origin, Humphries and McKusick[80] used the term auscultatory-electrocardiographic syndrome to describe patients with midsystolic clicks, late systolic murmur, and associated T-wave abnormalities that they attributed to residual pericarditis. Reid,[81] in South Africa, recorded phonocardiograms with clicks in 8 patients, some with and others without murmurs, and concluded that the sounds arose in the mitral valve and probably represented a "chordal snap." Both the clicks and the late systolic murmurs were subsequently proven to arise from the mitral valve by intracardiac phonocardiographic analyses.[82] Barlow[83] demonstrated that the late systolic murmur behaved like the murmur of mitral regurgitation, decreasing in intensity with amyl nitrite inhalation and increasing with phenylephrine infusion. He also demonstrated that this syndrome was associated with protrusion of the posterior mitral cusp into the left atrium during the middle third of systole and that mitral reflux, usually minimal in volume, occurred with the murmur. These findings, coupled with a distinctive electrocardiographic pattern, constitute a specific syndrome, appropriately bearing his name. Anatomic features have been documented at the time of surgery and postmortem examination and include voluminous and scalloped leaflets of the mitral valve as well as myxomatous degeneration of the valve substance with absence of inflammatory changes. Read[84] coined the expression floppy valve syndrome in describing 9 patients found to have myxomatous valvular transformation. Since the histologic appearance of the valve leaflets in these cases was identical to that described primarily as an incidental lesion in Marfan's syndrome, he considered it a possible forme fruste of Marfan's syndrome. Of the 5 females who presented with symptomatic mitral insufficiency in this series, 2 had had bacterial endocarditis.

Patients with mitral valve prolapse/click syndrome have a variety of symptoms that include chest pain, lightheadedness, dyspnea, fatigue, and palpitations. In one of the larger series of well-studied patients, Jeresaty[85] found chest pain as the chief complaint in 61 of 100 patients. He attributed this to ischemia of the posterior papillary muscle resulting from excessive stretching and suggested this may explain the electrocardiographic changes of inverted T waves in Leads II, III, and AVF as well. Barlow[86] has suggested that according to Laplace's law the chordae attached to a billowing leaflet are under increased tension and such forceful traction on the papillary muscle might occlude or thrombose its vascular supply. Jeresaty found dyspnea and fatigue the presenting symptom in 60% of his patients, while 46% were noted to complain of palpitations. He also called attention to psychiatric manifestations that were present in 15% of his group.

A snapping mid-late-systolic extra sound, loudest at the apex or just inside the apex, is considered the auscultatory hallmark of this syndrome. In some patients it has a scratchy quality, and multiple clicks may create the erroneous impression of a friction rub. The click is often followed by a late systolic murmur, but not invariably so. In other patients only a pansystolic murmur is audible. Jeresaty demonstrated mitral valve prolapse angiographically in 12 patients in whom neither click nor murmur was heard in various positions. It is important to realize that the midsystolic click may not be present in the supine position but emerges as the patient assumes the left lateral decubitus or upright positions. The click occurs earlier during inspiration, in the straining phase of the Valsalva maneuver, and following inhalation of amyl nitrite. It moves toward the second heart sound following vasoactive agents and squatting. The murmur, on the other hand, increases in intensity after the infusion of vasoactive drugs and decreases in response to amyl nitrite inhalation or in the straining phase of the Valsalva maneuver. Changes in left ventricular volume provide the most logical explanation for the observed variations in the timing of the click and the systolic murmur. Reduction of the systolic size of the ventricle causes the chordae to become excessively long relative to the long axis of the chamber and allows the leaflets to billow into the left atrium before tensing can occur. This would favor an earlier and more pronounced prolapse. The erect position, the straining phase of the Valsalva maneuver, and amyl nitrite inhalation are associated with a smaller left ventricular volume and an enhanced left ventricular emptying. The reverse occurs in the supine position or after infusion of vasoactive substances such as phenylephrine.

Ventricular irritability is considered a frequent complication of this syndrome, and is estimated to occur in 25% of cases. Sudden death has been reported.[85] The exercise stress test frequently unmasks ventricular irritability in these patients, and the appearance of ventricular irritability postexercise is an indication to undertake antiarrhythmic therapy preferably with propranolol. The resting electrocardiogram frequently shows inverted T waves in Leads II, III, and AVF with and without minimal ST depression. Jeresaty found these changes in 42 of 100 patients, with 24 of them showing similar changes in the left precordial leads. Normal coronary arteriograms were demonstrated in 26 patients of this series, 16 of whom demonstrated ST-T-wave abnormalities in Leads II, III, and AVF. The diagnosis of an acute coronary episode is frequently made in these individuals because of the association of chest pain. It is therefore important to listen carefully for a midsystolic click in both the supine and upright positions in patients with these commonly encountered electrocardiographic findings.

The angiographic pattern of mitral valve prolapse has been well

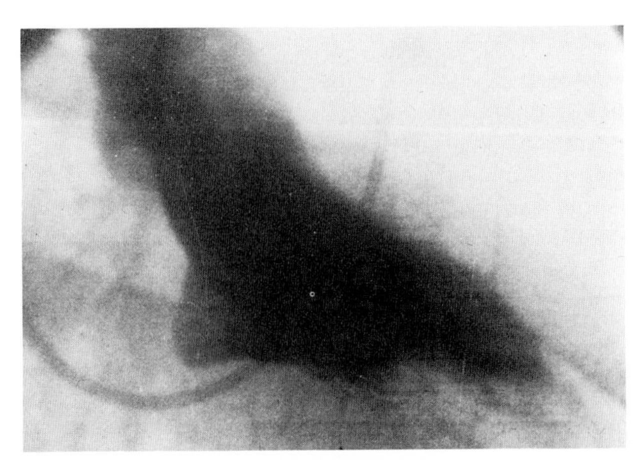

Fig. 14-4. A normal left ventriculogram. A cine frame viewed in the RAO projection of end-systole in a normal heart.

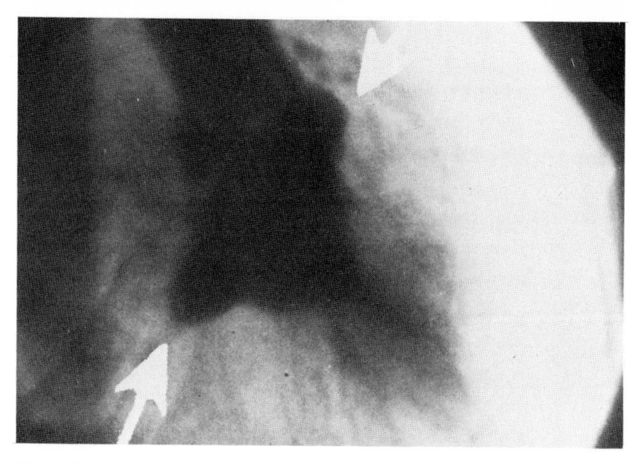

Fig. 14-5. Anterior and posterior leaflet prolapse (upper and lower arrows, respectively) give a doughnut appearance to the mitral valve ring when viewed in the RAO projection. Note the excessive emptying of the left ventricle at end-systole when compared to Fig. 14-4.

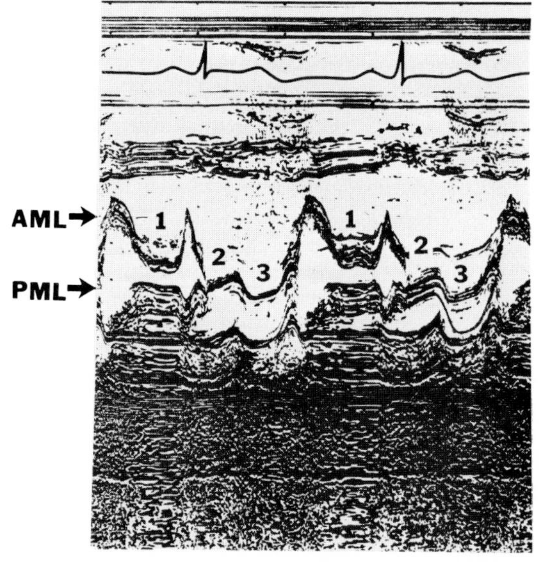

Fig. 14-6. Echocardiogram of a patient with mitral valve prolapse/click syndrome. AML: anterior mitral leaflet. PML: Posterior mitral leaflet. (1) Middiastole: AML and PML "floating" toward each other at end of rapid filling phase. (2) Early systole: AML and PML unite. (3) Late systole: AML and PML widely separate as PML billows into the left atrium, away from the transducer, creating a negative deflection in late systole.

described[87] and correlated with the known morphology of the human mitral valve.[88] In its most obvious form, with prolapse of both anterior and posterior leaflets, the prolapsed mitral valve resembles a doughnut viewed in the right anterior oblique projection. This is illustrated in Fig. 14-4 and 14-5. The anterior leaflet tends to prolapse anterosuperiorly and appears as a hump in the upper anterior wall of the left ventricle located immediately beneath the aortic valve. The posterior leaflet extends posteroinferiorly and is more difficult to ascertain. Since the posterior leaflet is divided into three scallops, a middle scallop and two small commissural scallops, mitral valve prolapse of the posterior leaflets should be limited to those cases where a serrated or scalloped margin is demonstrated. Jeresaty has observed excessive emptying of the left ventricular cavity associated with prominence of the papillary muscles in most patients with prolapse of both leaflets. In his experience, isolated prolapse of the posterior leaflet is rarely complicated by mitral regurgitation, whereas prolapse of the anterior leaflet is usually associated with the late systolic or pansystolic murmur and significant mitral regurgitation. Electrocardiographic abnormalities are more common when the prolapse involves both leaflets.

The echocardiogram may be helpful in establishing the diagnosis of mitral valve prolapse. A separation of the two mitral leaflets during systole, usually a late systolic posterior movement of the posterior mitral leaflet, is characteristically seen. This is illustrated in Fig. 14-6.

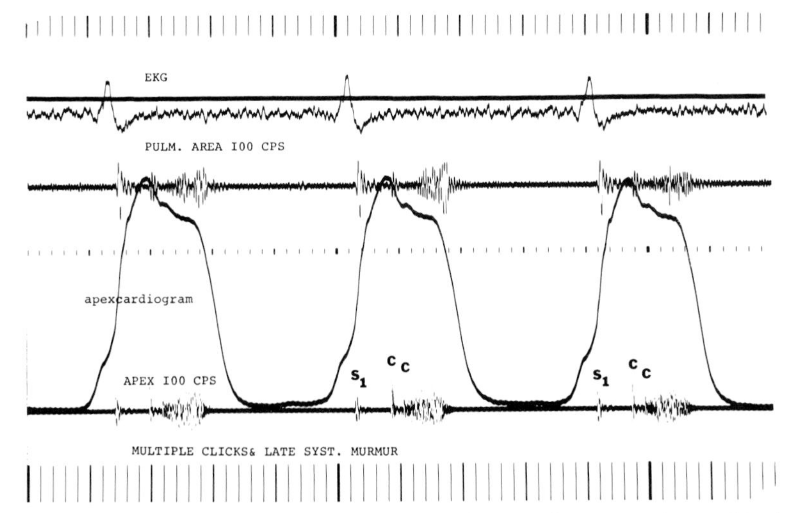

Fig. 14-7. Phonocardiogram and apex cardiogram from a patient with the mitral valve prolapse/click syndrome. A late systolic murmur is recorded at both the pulmonic area and apex. S_1 = mitral first sound, of normal intensity; C = nonejection clicks appearing in midsystole.

Considering the differential diagnosis of mitral regurgitation, the following findings would favor mitral valve prolapse secondary to myxomatous degeneration: (1) a midsystolic to late systolic murmur or the late accentuation of a pansystolic murmur, (2) the presence of a midsystolic click, (3) the absence of accentuation of the first sound and of an opening snap, (4) multiple premature ventricular contractions with T-wave inversions in Leads II, III, and AVF, (5) severe symptomatology with normal to slightly enlarged cardiac silhouette, and (6) the absence of mitral valve calcification. The auscultatory findings of this syndrome are illustrated in Fig. 14-7.

REFERENCES

1. Wood P: An appreciation of mitral stenosis. Part I. Clinical features. Br Med J 1:1051, 1954
2. Friedberg CK: Congenital mitral stenosis, in: Diseases of the Heart. Philadelphia, W. B. Saunders, 1966, p 1294
3. Goodwin JF: Diagnosis of left atrial myxoma. Lancet 1:464, 1963
4. McGuire LB, Nolan TB, Reeve R, Dammann JF: Cor triatriatum as a problem of adult heart disease. Circulation 31:263, 1965
5. Korn D, DeSanctis RW, Sell S: Massive calcification of the mitral anulus. A clinicopathological study of fourteen cases, N Engl J Med 267:900, 1962
6. Goldstein I, Halpern B, Robert L: Immunological relationship between streptococcus polysaccharide and the structural glycoprotein of heart valve. Nature 213:44, 1967
7. Bland EF, Jones TD: Rheumatic fever and rheumatic heart disease. A twenty year report of 1000 patients followed since childhood. Circulation 4:836, 1951
8. Olesen KH: The natural history of 271 patients with mitral stenosis under medical treatment. Br Heart J 24:349, 1962
9. Rowe JC, Bland EF, Sprague HB, White PD: The course of mitral stenosis without surgery: Ten- and twenty-year perspectives. Ann Intern Med 52:741, 1960
10. Selzer A, Cohn KE: Natural history of mitral stenosis: A review. Circulation 45:878, 1972
11. Edwards JE: The congenital bicuspid aortic valve. Circulation 23:485, 1961
12. Roberts WC: The structure of the aortic valve in clinically isolated aortic stenosis. Circulation 42:91, 1970
13. Rusted IE, Scheifley CH, Edwards JE: Studies of the mitral valve: II. Certain anatomic features of the mitral valve and associated structures in mitral stenosis. Circulation 14:398, 1956
14. Wood P: An appreciation of mitral stenosis. II. Investigation and results. Br Med J 1:1113, 1954
15. Wood P: Diseases of the Heart and Circulation (ed 3). Philadelphia, J. P. Lippincott, 1968

16. Cobbs RW: Mitral valve disease, in Hurst JW, Logue RB, Schlant RC, Wenger NK (eds): The Heart (ed 3). New York, McGraw-Hill, 1974, p 840
17. Ellis LB, Harken DE: Arterial embolization in relation to mitral valvuloplasty. Am Heart J 62:611, 1961
18. Casella L, Abelmann WH, Ellis LB: Patients with mitral stenosis and systemic emboli. Arch Intern Med 114:773, 1964
19. Szekely P: Systemic embolism and anticoagulant prophylaxis in rheumatic heart disease. Br Med J 1:1209, 1964
20. Coulshed N, Epstein EJ, McKendrick CS, et al: Systemic embolism in mitral valve disease. Br Heart J 32:26, 1970
21. Morris JJ Jr, Estes H Jr, Whalen RE, et al: P-wave analysis in valvular heart disease. Circulation 29:242, 1964
22. Kasser I, Kennedy JW: The relationship of increased left atrial volume and pressure to abnormal P waves on the electrocardiogram. Circulation 39:339, 1969
23. Hugenholtz PG, Ryan TJ, Stein SW, Abelmann WH: The spectrum of pure mitral stenosis. Am J Cardiol 10:773, 1962
24. Doyle AE, Goodwin OF, Harrison CV, Steiner RE: Pulmonary vascular patterns in pulmonary hypertension. Br Heart J 19:353, 1957
25. Simon M: The pulmonary veins in mitral stenosis. J Fac Radiologists 9:25, 1958
26. Dulfano MJ, Adler H: Tomographic evaluation of hemodynamic changes in mitral stenosis. Circulation 23:177, 1961
27. Hellems HR, Haynes FW, Dexter L, Kinney TD: Pulmonary capillary pressure in animals estimated by venous and arterial catheterization. Am J Physiol 155:98, 1948
28. Ross J Jr: Transeptal left heart catheterization. A new method of left atrial puncture. Ann Surg 149:395, 1959
29. Walston A, Kendal ME: Comparison of pulmonary wedge and left atrial pressure in man. Am Heart J 86:159, 1973
30. Braunwald E: Transseptal left heart catheterization. Circulation 37:III–74, 1968
31. Gorlin R, Gorlin SG: Hydraulic formula for calculation of the area of the stenotic mitral valve, other cardiac valves and central circulatory shunts. Am Heart J 41:1, 1951
32. Richter HS: Mitral valve area: Measurement soon after catheterization. Circulation 28:451, 1963
33. Dexter L: Physiologic changes in mitral stenosis. N Engl J Med 254:829, 1956
34. Dexter L, Dow JW, Haynes FW, et al: Studies of the pulmonary circulation in man at rest. Normal variations and the interrelationship between increased pulmonary blood flow, elevated pulmonary arterial pressure and high pulmonary "capillary" pressure. J Clin Invest 29:602, 1950
35. Gorlin R, Haynes FW, Goodale WT, et al: Studies of the circulatory dynamics in mitral stenosis. II. Altered dynamics at rest. Am Heart J 41:30, 1951

36. Gorlin R, Sawyer CG, Haynes FW, et al: Effects of exercise on circulation dynamics in mitral stenosis III. Am Heart J 41:192, 1951
37. Gorlin R, Lewis BM, Haynes FW, et al: Factors regulating pulmonary "capillary" pressure in mitral stenosis IV. Am Heart J 41:834, 1951
38. Araujo J, Lukas DS: Interrelationships among pulmonary "capillary" pressure, blood flow and valve size in mitral stenosis. The limited regulatory effects of the pulmonary vascular resistance. J Clin Invest 31:1082, 1952
39. Selzer A, Malmborg RO: Some factors influencing changes in pulmonary vascular resistance in mitral valvular disease. Am J Med 32:532, 1962
40. Davies LG, Goodwin JF, VanLeuven BD: The nature of pulmonary hypertension in mitral stenosis. Br Heart J 16:440, 1954
41. Braunwald E, Braunwald NS, Ross J Jr, Morrow AG: Effects of mitral valve replacement on the pulmonary vascular dynamics of patients with pulmonary hypertension. N Engl J Med 273:509, 1965
42. Walston A, Peter RH, Morris JJ, et al: Clinical implications of pulmonary hypertension in mitral stenosis. Am J Cardiol 32:650, 1973
43. Wadi G, Werko L, Eliasch H, et al: The hemodynamic basis of the symptoms and signs in mitral valvular disease. Q J Med 21:361, 1952
44. Ball JD, Kopelman H, Witham AC: Circulatory changes in mitral stenosis at rest and on exercise. Br Heart J 14:363, 1952
45. Carman GH, Lange RL: Variant hemodynamic patterns in mitral stenosis. Circulation 24:712, 1961
46. Sarnoff SJ, Mitchell JH: The regulation of the performance of the heart. Am J Med 30:747, 1961
47. Henry JP, Gauer OH, Reeves JL: Evidence of the atrial location of receptors influencing urine flow. Circ Res 4:85, 1956
48. Bailey GWH, Braniff BA, Hancock EW, Cohn KE: Relation of left atrial pathology to atrial fibrillation in mitral valve disease. Ann Intern Med 69:13, 1968
49. Harvey RM, Ferrer MI, Samet P, et al: Mechanical and myocardial factors in rheumatic heart disease with mitral stenosis. Circulation 11:531, 1955
50. Fleming HA, Wood P: The myocardial factor in mitral valve disease. Br Heart J 21:117, 1959
51. Donald KW, Bishop JW, Wade OL, Wormald PN: Cardiorespiratory function two years after mitral valvotomy. Clin Sci 16:325, 1957
52. Feigenbaum H, Linback RE, Nasser WK: Hemodynamic studies before and after instrument mitral commissurotomy. Circulation 38:261, 1968
53. Feigenbaum H, Campbell RW, Wunsch CM, Steinmetz EF: Evaluation of the left ventricle in patients with mitral stenosis. Circulation 34:462, 1966
54. Kirch E: Alterations in size and shape of individual regions of heart in valvular disease. Verk Deutsch Kong Inn Med 41:324, 1929
55. Grant RP: Architectonics of the heart. Am Heart J 46:405, 1953
56. Heller SJ, Carleton RA: Abnormal left ventricular contraction in patients with mitral stenosis. Circulation 42:1099, 1970
57. Elder I, Herz CH: The use of the ultrasonic reflectoscope for the continuous

recording of movement of heart walls. Kungl Fysiogr Sallsk Lund Forhandle 24:40, 1954
58. Edler I, Gustofson S, Karefors T, Christenson B: Ultrasound cardiography. Acta Med Scand 170:67, 1961
59. Edler I: Ultrasoundcardiography in mitral valve stenosis. Am J Cardiol 19:18, 1967
60. Feigenbaum H: Echocardiography. Philadelphia, Lea & Febiger, 1972
61. Feigenbaum H: Clinical application of echocardiography. Prog Cardiovasc Dis 14:531, 1972
62. Pohost GM, Dinsmore RE, Rubenstein JJ, et al: The echocardiogram of the anterior leaflet of the mitral valve: A correlation with hemodynamic and cineroentgenographic studies in dogs. Clin Res 22:296A, 1974
63. Joyner CR, Reid JM, Bond JP: Reflected ultrasound in the assessment of mitral valve disease. Circulation 27:503, 1963
64. Wenger NK: Tumors of the heart, in Hurst, JW, Logue RB, (eds): The Heart (ed 2). New York, McGraw-Hill, 1970, p. 1278
65. Case Records of the Massachusetts General Hospital (Case 37-1972) N Engl J Med 287:555, 1972
66. Craige E, Algary WP: Left atrial myxoma. Arch Intern Med 129:470, 1972
67. Martinez EC, Giler TD, Burch GE: Echocardiographic diagnosis of left atrial myxoma. Am J Cardiol 33:281, 1974
68. Simon MA, Liu SF: Calcification of the mitral valve anulus and its relation to functional valvular disturbance. Am Heart J 48:497, 1954
69. Clawson BJ, Bell ET, Hartzell TB: Valvular disease of the heart with special reference to pathogenesis of old valvular defects. Am J Pathol 2:193, 1926
70. Jordan JD, McNamara DG, Marcontelli J, Rosenberg HS: Cor triatriatum: An anatomic and physiologic study. Cardiovasc Res Center Bull 1:79, 1963
71. Perloff JK, Harvey WP: Auscultatory and phonocardiographic manifestations of pure mitral regurgitation. Prog Cardiovasc Dis 5:172, 1962
72. Cabot RC: Facts on the Heart. Philadelphia, W. B. Saunders, 1926.
73. Braunwald E: Mitral regurgitation. Physiologic and surgical considerations. N Engl J Med 281:425, 1969
74. Eckberg DL, Gault JH, Bouchard RL, et al: Mechanics of left ventricular contraction in chronic severe mitral regurgitation. Circulation 47:1252, 1973
75. Ellis LB, Ramirez A: The clinical course of patients with severe "rheumatic" mitral insufficiency. Am Heart J 78:406, 1969
76. Bentivoglio L, Uricchio J, Goldberg H: Clinical and hemodynamic features of advanced rheumatic mitral regurgitation. Am J Med 31:372, 1961
77. Edwards JE, Burchell HB: Endocardial and internal lesions (jet impact) as possible sites of origin of murmurs. Circulation 18:946, 1958
78. Leatham A: Auscultation of the heart. Lancet 2:703, 1958
79. Nixon P, Wooler G, Radigan L: The opening snap in mitral incompetence. Br Heart J 22:395, 1960
80. Humphries JO, McKusick VA: The differentiation of organic and "innocent" systolic murmur. Prog Cardiovasc Dis 5:152, 1962
81. Reid JVO: Midsystolic clicks. S Afr Med J 35:353, 1961

82. Leon DF, Leonard JJ, Kroetz FW, et al: Late systolic murmurs, clicks and whoops arising from the mitral valve. Am Heart J 72:325, 1966
83. Barlow JB, Bosman CK, Pocock WA, Marchand P: Late systolic murmurs and non-ejection ("mid-late") systolic clicks. Br Heart J 30:203, 1968
84. Read RC, Thal AP, Wendt VE: Symptomatic valvular myxomatous transformation (the floppy valve syndrome): A possible forme fruste of the Marfan syndrome. Circulation 32:897, 1965
85. Jeresaty RM: Mitral valve prolapse—Click syndrome. Prog Cardiovasc Dis 15:623, 1973
86. Pocock WA, Barlow JB: Etiology and electrocardiographic features of the billowing posterior mitral leaflet syndrome: Analysis of a further 130 patients with a late systolic murmur or nonejection systolic click. Am J Med 51:731, 1971
87. Jeresaty RM: Ballooning of the mitral valve leaflet: Angiographic study of 24 patients. Radiology 100:45, 1971
88. Ranganathan N, Lam JHC, Wigte ED, Silver MD: Morpology of the human mitral valve. II. The valve leaflets. Circulation 41:459, 1970

Aram V. Chobanian

15
Pathophysiology of Systemic Hypertension

Multiple complex interacting factors are involved in the regulation of blood pressure. Perturbations in these control mechanisms may lead ultimately to sustained elevations of arterial pressure. Such increases in pressure are exceedingly common and represent a major cause of disability and death in Western society.

MEASUREMENT OF BLOOD PRESSURE

Direct measurement of blood pressure with an intraarterial needle connected to a pressure transducer is the most accurate technique, although the method is rarely employed clinically because of practical considerations. The systolic pressure is the maximum pressure reached at the height of systole, and the diastolic pressure the minimum pressure between the pulse waves. The pulse pressure is the difference between systolic and diastolic pressure and represents the pressure that expands the aorta. The mean blood pressure, which is the integrated mean of the systolic and diastolic pressures, generally approximates a calculated value of the diastolic plus one-third the pulse pressure.

In the indirect measurement of blood pressure using a sphygmomanometer, the pressure is first increased in the cuff to greater than systolic pressure. As the pressure is reduced slowly in the cuff, successions of sounds, the so-called Korotkoff sounds, are heard by auscultation over the artery. The first clear sounds appear at the level of systolic pressure. The second phase is characterized by a softening or even a disappearance of the sounds (auscultatory gap), followed in phase 3 by an increase in intensity. In phase 4 the sounds suddenly become muffled, and in phase 5 they disappear. In most individuals the diastolic pressure more closely approximates the phase-4 rather than the phase-5 reading and should be recorded as such. When an appreciable discrepancy is present between phase-4 and phase-5 readings, both should be recorded (e.g., 130/80/75).

A proper cuff should be wide enough to allow adequate transmission of pressure to the center of the arm and long enough to permit firm application to the arm. For most adults a cuff 12–13 × 27–30 cm is adequate. However, in markedly obese individuals or in small children, larger or smaller cuffs will be required to prevent significant errors in measurement.

The pressure should be determined with the arm at the level of the heart. Measurement of pressures in the lower extremity may also be of value in patients with suspected aortic obstructive disease, but the measurement generally should be performed with a wide cuff. Pressures in the leg are normally about 20/10 mm Hg greater than in the arm. The blood pressure should be measured in both arms to rule out obstructive vascular diseases such as coarctation or dissecting aneurysms of the aorta. The pressure should also be evaluated in the upright as well as the recumbent or sitting positions. Orthostatic hypotension may occur in certain hypertensive states (e.g., pheochromocytoma, primary aldosteronism, renovascular hypertension) or may develop following antihypertensive therapy.

REGULATION OF BLOOD PRESSURE

Blood pressure is a function of cardiac output and peripheral vascular resistance. Mean pressure (mm Hg) = cardiac output (ml/sec) × peripheral resistance (dynes-cm^{-5}-sec). Compensatory mechanisms normally prevent large changes in blood pressure despite major swings in either cardiac output or peripheral resistance.

Baroreceptors

A major regulatory mechanism of blood pressure involves pressor receptors in the arterial system. A number of such receptors, including those in the thoracic aorta, cerebral arteries, atria, and ventricular walls, have been identified.[1] The most important are located at the origin of the internal carotid artery and at the aortic arch. Sensory nerve endings located in the arterial adventitia are stimulated to increase their rate of discharge with increase in distending pressure of the artery. The impulses are transmitted to vasomotor centers in the medulla oblongata. Sympathetic vasoconstrictor outflow is inhibited, and the vagus nerve is stimulated, with resultant decreases in peripheral resistance and heart rate.

An abnormality in baroreceptor activity has been postulated as a possible cause of hypertension. Deinnervation of the carotid sinus in animals is associated with hypertension, but in contrast to the case with human hypertensive diseases, marked tachycardia also ensues.[3] Reduced baroreceptor activity has been observed in experimental animals with renal hypertension. A diminished response in heart rate to acute infusions of pressor agents in patients with long-standing hypertension has also been reported.[4] However, it is unclear whether such observed abnormalities in baroreceptor activity are an effect rather than a cause of the hypertension. Reduction in blood pressure has been produced in patients with severe hypertension by chronic carotid sinus stimulation,[5] but the method is not of practical value in the general management of the disease.

Central Mechanisms

Higher centers of the brain and the vasomotor center in the medulla play an important though poorly understood role in blood pressure regulation. Elevation and reduction of blood pressure have been observed in animals following stimulation of a number of areas in the medulla, hypothalamus, thalamus, and cerebral cortex.[6] Both sustained hypertension and lowering of blood pressure have been produced in subhuman primates by influencing behavior with operant conditioning.[7]

Sympathetic Nervous System

The sympathetic nervous system exerts the major control over the peripheral vasculature. Vascular smooth muscle is innervated by sympathetic nerve fibers, which contain norepinephrine. With sympathetic nerve discharge, the norepinephrine that is stored in granules is released by exocytosis, and a potent although brief vasoconstrictor effect is produced through stimulation of the so-called alpha receptor. The adrenal gland releases both norepinephrine and epinephrine in response to emotional stresses, exercise, hypoglycemia, and hypoxemia. Epinephrine can produce vasodilatation (beta receptor stimulation) in certain vascular beds (i.e., skeletal muscle and heart) as well as vasoconstriction in others. Under normal conditions beta receptor activity does not appear important in peripheral circulatory control. Both epinephrine and norepinephrine have a powerful ionotropic effect on the heart (beta effect).

The exact nature of the alpha and beta receptors has not been fully characterized. Recently a membranous cellular fraction isolated from myocardial cells has been shown to exhibit some features of the beta receptor.[8] The membranes appear to bind norepinephrine, epinephrine, and other beta stimulators, and this binding is inhibited by beta blockers.

Norepinephrine and epinephrine are synthesized according to the pathway illustrated in Fig. 15-1. The major rate-limiting step involves the mitochondrial enzyme tyrosine hydroxylase. Experimental inhibition of the enzyme in man with the drug alpha-methyltyrosine is associated with a significant decrease in catecholamine synthesis.[9] The *N*-methylating enzyme involved in the conversion of norepinephrine to epinephrine is confined to the adrenal gland and to adrenal rest tissues. This finding has been utilized for the localization of pheochromocytomas since extraadrenal chromaffin tumors secrete predominantly norepinephrine, while tumors of adrenal origin produce both norepinephrine and epinephrine.

The enzyme dopamine-β-hydroxylase is present in the granules of sympathetic nerve terminals and is released with norepinephrine upon nerve stimulation. Recent investigations have suggested that changes in the serum levels of this enzyme may reflect sympathetic nervous activity.[10] Serum dopamine-β-hydroxylase activity appears to be reduced in patients with familial dysautonomia and secondary orthostatic hypotension.[11] Its levels in hypertensive states have not been clearly delineated as yet.

Inactivation of the norepinephrine liberated from nerve endings occurs primarily as a result of reuptake by the sympathetic nerves. In addition, catecholamines are metabolized in the liver (catechol-*o*-methyl transferase) or in the nerve terminals (monoamine oxidase). The major metabolites and the pathways for their reactions are illustrated in Fig. 15-2.

The contribution of the sympathetic nervous system and of catecholamines to the etiology of hypertension remains unclear, although different lines of evidence suggest an important role (see section on essential hypertension). Vascular hyperreactivity[12] and increased plasma norepinephrine concentrations[13] have been reported in hypertensive patients. Furthermore, many of the pharmacologic agents found to be of value in the treatment of hypertension influence the sympathetic nervous system.

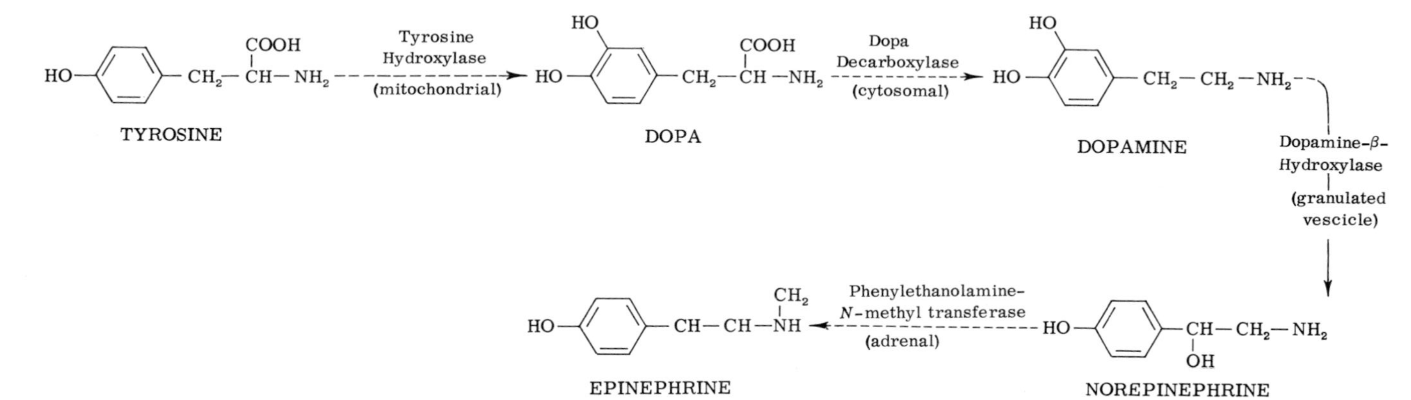

Fig. 15-1. The synthesis of norepinephrine and epinephrine from tyrosine.

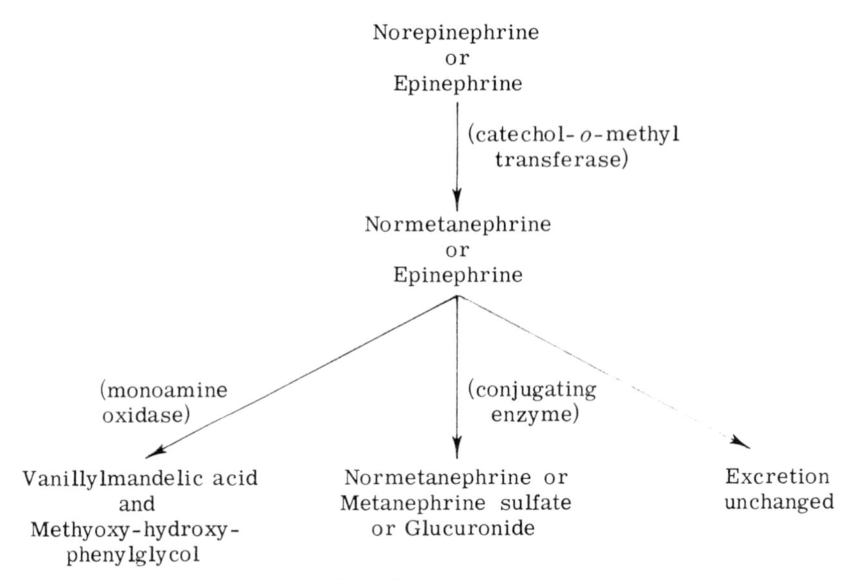

Fig. 15-2. Metabolism of norepinephrine and epinephrine.

Renin–Angiotensin System

The renin–angiotensin system is involved in the regulation of blood pressure and sodium balance. It also appears to play an important role in the pathogenesis of certain forms of hypertension.

Renin is a proteolytic enzyme synthesized in the juxtaglomerular cells of the media of afferent arterioles of the kidney. The enzyme catalyzes the conversion of a polypeptide substrate to the decapeptide angiotensin I, which in turn is converted to the octapeptide angiotensin II by an enzyme located primarily in the pulmonary circulation (Fig. 15-3).[14] Almost all angiotensin I is converted to angiotensin II with a single passage through the lung. Angiotensin II can, in turn, be converted to a heptapeptide, angiotensin III, which may be more potent than angiotensin II in stimulating aldosterone secretion.[15] Angiotensin II, on the other hand, has a more pronounced effect on the vasculature than angiotensin III and is the most potent pressor agent known. In addition, angiotensin II also appears to have a direct intrarenal effect on sodium reabsorption[16] and may increase the release of the prostaglandin PGE by the renal medulla.[17]

Renin release from the kidney is controlled by a number of factors. Reductions in afferent arteriolar intravascular pressure at the site of the juxtaglomerular cell, systemic hypotension, decreased intravascular volume, and reductions in serum sodium or body sodium balance all tend

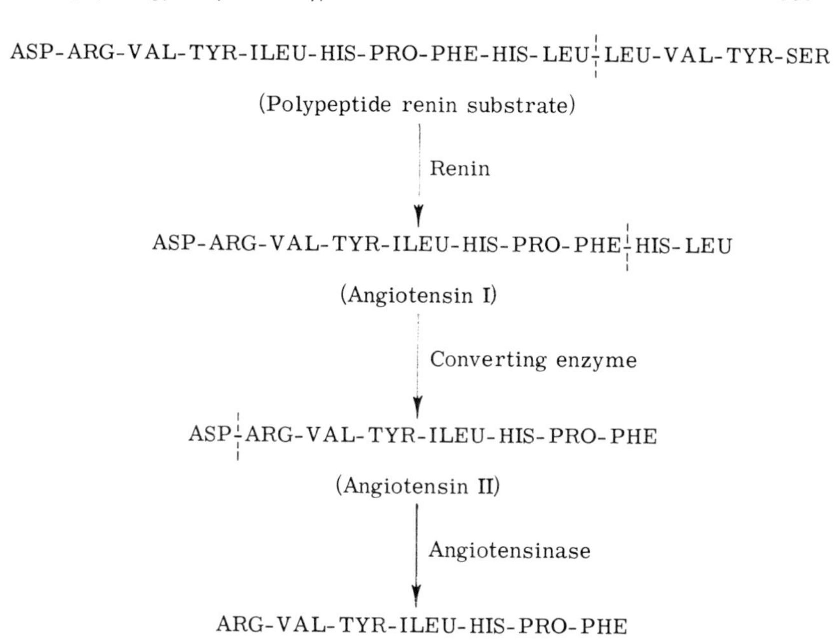

Fig. 15-3. Peptide structure and pathway of angiotensin formation.

to increase plasma renin activity.[18] Increased sympathetic nervous activity and catecholamines also enhance renin release by way of a beta adrenergic effect, and beta adrenergic blockade will significantly depress renin secretion.[19,20] The response in renin secretion is generally rapid and can be demonstrated within a minute or two in man.[21,21a] Angiotensin itself exerts a negative feedback control over renin secretion.[22]

The exact role of the renin–angiotensin system in hypertension is uncertain. In experimental animals the development of hypertension following renal artery clamping is associated with increased plasma renin activity, although renin activity may decrease toward normal with time despite persistence of hypertension. Conflicting results had been reported regarding the effects of angiotensin antibodies or competitive inhibitors of angiotensin on experimental renovascular hypertension.[23] However, recent studies utilizing the competitive antagonist for angiotensin II (1-sar-8-ala-angiotensin), saralasin, have indicated that blood pressure may be normalized by such therapy in animals with hypertension induced by unilateral renal arterial clamping but not in animals who have had unilateral renal artery clamping and contralateral nephrectomy.[24] The lack of response in the latter group has been attributed to sodium retention

Table 15-1
Causes of Hyperreninemia

Without Hypertension	With Hypertension
	High Renin Essential Hypertension
Sodium depletion	Primary hyperreninemia
Pregnancy	JG cell tumor
Addison's disease	Secondary hyperreninemia
Congestive heart failure	Accelerated hypertension
Nephrotic syndrome	Renovascular hypertension
Hepatic cirrhosis with ascites	Oral contraceptive drugs
Idiopathic edema	Diuretic therapy
JG cell hyperplasia	Congestive heart failure

and volume expansion, which might prevent any hypotensive effect of renin inhibition. Angiotensin may also influence blood pressure by enhancing sympathetic activity. The hormone may act on the area postrema of the brain medulla and increase the outflow of sympathetic nerve activity from the brain.[24a] In addition, angiotensin may increase responsiveness to endogenous catecholamines.

Increased levels of plasma renin activity and angiotensin may be observed in a number of clinical states, both with or without hypertension (Table 15-1). Only a relatively small fraction (~ 10%) of the hypertensive population exhibits increased renin activity. Most of the high-renin group have evidence of accelerated or malignant hypertension and marked renal arteriolar disease. In some of the patients, the increase in renin activity is on the basis of renal artery stenosis and renovascular hypertension.

Abnormally low renin activity is present in approximately 40 percent to 50 percent of black and 10 percent to 15 percent of white hypertensive subjects.[25,26] This decrease in activity has been attributed to an increase in mineralocorticoids; however, to date, increased levels of mineralocorticoid hormones have been demonstrated in only a small fraction of the low-renin group. A further discussion of the renin–angiotensin system in hypertension is provided in subsequent sections on essential and renal hypertension.

Aldosterone

Aldosterone is a potent mineralocorticoid hormone secreted by the adrenal cortex. It regulates sodium and potassium balance in the body by promoting reabsorption of sodium and secretion of potassium in the distal convoluted tubule of the kidney. Aldosterone also has been postulated to have a direct effect on blood vessels in increasing their sensitivity to vasoactive agents, although the evidence for this hypothesis is limited.

Aldosterone release is controlled primarily by three factors: the renin–angiotensin system, plasma concentrations of ACTH, and serum potassium.[27] Under normal conditions the renin–angiotensin system is the most important of these. Reductions in circulating blood volume will increase renin and angiotensin, which in turn stimulate aldosterone release and thereby act to correct the volume deficit. ACTH is thought to have a permissive role in regulating aldosterone release. It is probably responsible for the increase in aldosterone secretion observed during stress and may be important in anephric patients who lack the normal angiotensin system for control. Increased plasma potassium can stimulate the adrenal cortex directly to secrete aldosterone. This mechanism may be important in patients with renal failure and may act to prevent dangerous levels of hyperkalemia.

Increased secretion of aldosterone is observed in a number of clinical states, both with and without hypertension. In hypertensive subjects aldosterone excess may be due to tumors of the adrenal cortex. Secondary forms of hyperaldosteronism are observed in hypertensive patients with increased angiotensin levels (Table 15-1), as with malignant or accelerated hypertension, renin-producing tumors of the kidney, diuretic therapy, or oral contraceptive drug therapy (see subsequent sections).

Vasodilators

A number of vasodilator substances appear to be involved in blood pressure regulation.

Prostaglandins

Prostaglandins are unsaturated cyclical fatty acids with widespread biologic activities. Their presence in most tissues and their reported ability to modify the activation of cyclic AMP suggest that the prostaglandins may have multiple important regulatory roles.

The prostaglandins of the A and E series (Fig. 15-4) are potent vasodilators. Their major site of synthesis may be in the interstitial cells of the renal medulla, although synthesis may also occur in small amounts in many other areas. Indomethacin and aspirin are commonly used drugs that can inhibit prostaglandin synthesis.

Compounds of the E series are rapidly inactivated by the pulmonary vascular bed, and their effect in tissues such as the kidney may be localized to near their site of release. PGA compounds withstand

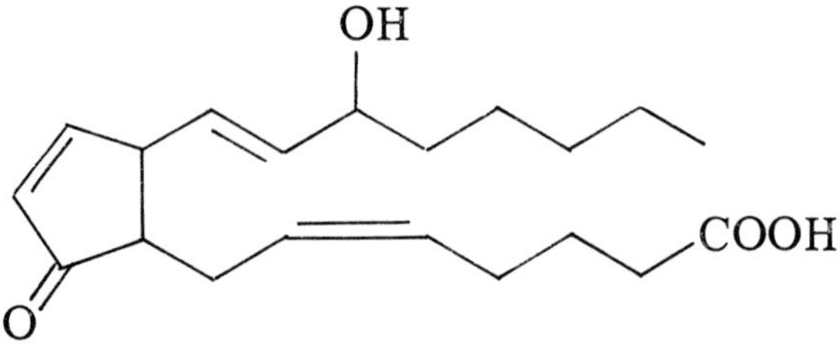

Prostaglandin A_2

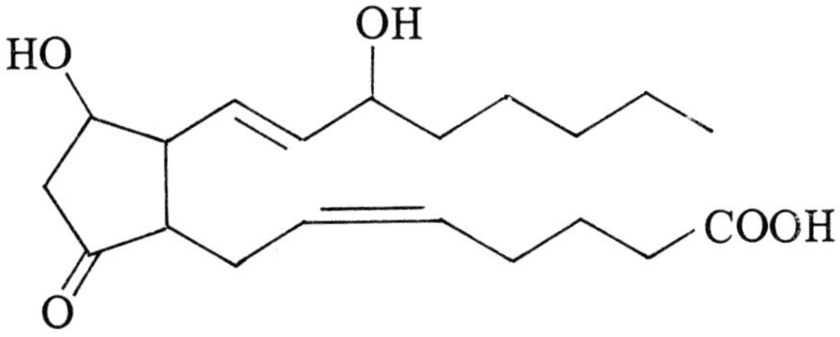

Prostaglandin E_2

Fig. 15-4. Chemical structure of prostaglandins A_2 and E_2.

pulmonary inactivation and can be isolated from the circulating blood. Both PGA and PGE have a direct effect on arteriolar smooth muscle, producing vasodilatation that is probably independent of the sympathetic nervous system. Studies in hypertensive man have demonstrated significant decreases in peripheral vascular resistance and blood pressure along with compensatory rises in cardiac output and heart rate following infusions of PGA.[28]

Prostaglandins appear to be involved in the intrarenal regulation of blood flow and in sodium homeostasis.[29] PGA and PGE infusions will increase renal cortical blood flow and induce a sodium diuresis, presumably by inhibiting proximal tubular reabsorption of sodium. Plasma levels of PGA in man have been reported to increase significantly with dietary sodium restriction.[30] PGA may also stimulate directly the release of both renin and aldosterone.[31,32] The release of prostaglandins from the kidney, in turn, may be stimulated by angiotensin, norepinephrine, and bradykinin and by increased sympathetic nervous activity.[33,34]

Little information is currently available concerning the role of prostaglandins in the pathogenesis of hypertension. The antihypertensive function of the kidney and the development of renoprival hypertension suggest a physiologic function of the prostaglandins in blood pressure

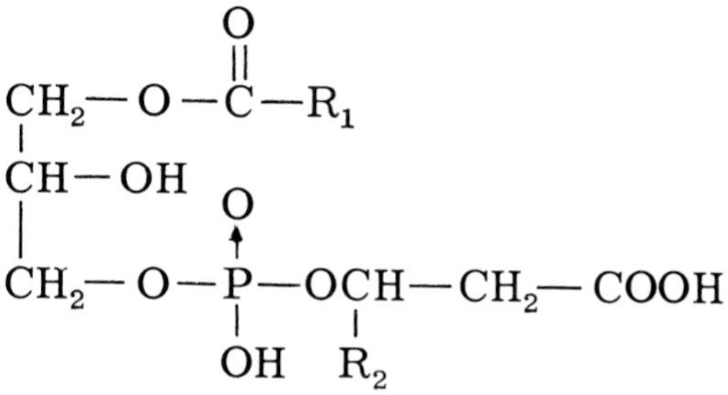

$$
\begin{array}{l}
CH_2-O-\overset{\overset{\displaystyle O}{\|}}{C}-R_1 \\
\;| \\
CH-OH \\
\;| \\
CH_2-O-\underset{\underset{\displaystyle OH}{|}}{\overset{\overset{\displaystyle O}{\uparrow}}{P}}-O\underset{\underset{\displaystyle R_2}{|}}{C}H-CH_2-COOH
\end{array}
$$

Fig. 15-5. Chemical structure of the vasodepressor lysophospholipid.

regulation. A further discussion of this topic is provided in the section on essential hypertension.

Other Renal Medullary Lipids

A neutral lipid isolated from extracts of renal medulla has been shown to exhibit a potent antihypertensive effect in animals with renoprival and renal hypertension.[35,36] The material is apparently synthesized in the interstitial cells of the renal medulla. Its exact structure and function remain to be identified, but this substance appears to be a major factor responsible for the antihypertensive function of the kidney.

A renal phospholipid similar in structure to phosphatidyl serine may also exhibit antihypertensive properties.[37] The compound is apparently converted in the plasma to a lysophospholipid that is a potent inhibitor of renin. This lysophospholipid (Fig. 15-5) has been shown to reduce the blood pressure of renal hypertensive animals, but not of normotensive animals.

Bradykinin and the Kallikrein–Kinin System

Bradykinin and related kinins are relatively small polypeptides that can act as potent vasodilators. They appear to exert a direct action on arterioles and venules that is independent of an effect on alpha and beta receptors. These kinins are inactivated by the pulmonary circulation and therefore have a brief half-life in the blood. While the physiologic functions of bradykinin are uncertain, its excessive release has been reported with endotoxin shock, carcinoid syndrome, dumping syndrome, and angioneurotic edema.[38–41] Increased bradykinin levels may be present in certain patients with a familial form of orthostatic hypotension in which bradykinin inactivation may be faulty.[42] Bradykinin also appears to have a potent natriuretic effect and may play a role in sodium regulation.[43]

Bradykinin and other kinins are formed by the action of the enzyme kallikrein on an α_2-globulin precursor, as illustrated in Fig. 15-6. The nonapeptide bradykinin is inactivated by kininases that facilitate

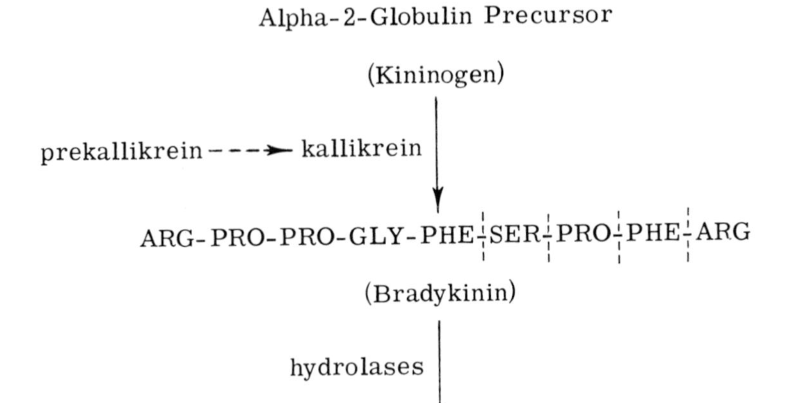

Fig. 15-6. Pathway of formation and degradation of bradykinin (and other kinins).

hydrolysis of multiple peptide bonds yielding smaller inactive peptides. An interesting relationship between the renin–angiotensin and kinin systems has been elucidated. Specific inhibitors of bradykininase also inhibit the conversion of angiotensin I to angiotensin II.[44]

The enzyme kallikrein is found in the kidney and is excreted in the urine. Recent investigations have demonstrated abnormalities in urinary kallikrein excretion in hypertension. Kallikrein excretion appears to be significantly increased in mineralocorticoid hypertension of both man and animals and reduced in patients with idiopathic hypertension.[45] Kallikrein excretion may also be increased by dietary sodium restriction.[46] While the significance of these findings is uncertain, the results suggest a role of the kallikrein–kinin system in blood pressure regulation and sodium homeostasis.

Intracellular Cations

Evidence is accumulating to indicate that the effects of many, if not all, vasoactive substances may be mediated ultimately by changes in intracellular calcium of vascular smooth muscle.[47] An increase in intracellular calcium may induce contraction, and a decrease may induce relaxation, of the smooth muscle. These effects appear to be mediated by its influence on the contractile protein actomyosin.[48] Calcium appears to permit the activation of the ATPase of actomyosin and thus cause muscle contraction. ATP itself is a potent vasodilator, as are ADP and AMP.

Other Vasodilator Substances

Acetylcholine will produce vasodilatation in certain vascular beds such as the pulmonary circulation, but probably does not play an important role in normal circulatory regulation. Stimulation of parasympathetic fibers innervating glandular organs may produce vasodilatation, perhaps by causing release of bradykinin. Histamine, which is released as a result of tissue injury, will initiate intense local vasodilatation. Serotonin also may induce vasodilatation in some vascular beds (e.g., coronary circulation), although vasoconstriction may be expected in other areas (e.g., pulmonary vascular bed).

Sodium and Blood Pressure

Numerous lines of evidence suggest that sodium plays an important role in the etiology of hypertension. Epidemiologic studies have demonstrated increased incidence of hypertension in populations with relatively high dietary sodium intake.[49] Conversely, hypertension is rare in populations with markedly restricted dietary sodium. In natives of northern New Guinea, who normally ingest less than 10 mEq of sodium per day, hypertension is almost nonexistent.[50] Severe sodium restriction and diuretic drugs are well-known antihypertensive maneuvers in man.

Inbred strains of hypertensive rats have been developed that are either highly sensitive or resistant to dietary sodium. The sensitive animals will develop severe hypertension with high sodium intake.[51] The sodium-sensitive mice have increased secretion of the mineralocorticoid hormone 18-hydroxydesoxycorticosterone.[52] In addition, mineralocorticoid drugs will produce hypertension in experimental animals only in the presence of adequate dietary sodium.

Increased sodium, calcium, and water contents of blood vessels from experimental animals have been demonstrated in a number of studies.[53–55] It is uncertain, however, whether these changes are causally related or are a consequence of the hypertension. Increases in arteriolar calcium and sodium could conceivably influence intracellular calcium uptake and contractility of vascular smooth muscle.

The exact relationship between sodium and hypertension in man remains elusive. Total-body sodium and extracellular fluid volume are not elevated in hypertensive subjects,[56] except in the presence of increased aldosterone, as with primary aldosteronism, accelerated hypertension, or renovascular hypertension.[57] Plasma volume actually appears to be depressed in uncomplicated hypertension, the degree of reduction correlating with the level of diastolic pressure.[58] Abnormalities in the renal handling of sodium have been observed in hypertension, with hyperexcretion of sodium in response to saline loads[59] and a paradoxical natriuresis with angiotensin administration.[60]

SYSTEMIC HYPERTENSION IN MAN

What Is Hypertension?

The levels of blood pressure in any given population represent a continuum. The definition of abnormality must take into consideration the age of an individual as well as the pressure level above which a detrimental effect can be expected. Epidemiologic studies and actuarial data from life insurance statistics have shown a progressive increase in the incidence of ischemic heart disease with systolic blood pressures above 120–130 mm Hg and diastolic pressures exceeding 80 mm Hg.[61,62] The World Health Organization has defined abnormality to include any systolic pressure $\gtrsim$ 160 mm Hg and diastolic pressure $\gtrsim$ 95 mm Hg. The latter values should be revised downward in young individuals in whom lower readings would be clearly abnormal. An unusually high (20%–30%) incidence of hypertension has recently been demonstrated in black populations in the United States.[63] The incidence in whites in this country may approximate 10%–15%.

Lability of Blood Pressure

Blood pressures normally can vary during any given day by as much as 40–50 mm Hg. The lowest levels are recorded during sleep and early morning, and the highest in the afternoon and evening. Labile hypertension may be observed in some individuals, particularly under stressful situations. Such patients may exhibit symptoms and signs consistent with hyperactivity of the beta adrenergic system as increased heart rate and cardiac output[64] and increased renin release in response to certain stresses.[65] Labile hypertension may be the forerunner of fixed hypertension, although many such patients may remain labile throughout life.

Hemodynamics of Hypertension

The sequence of hemodynamic alterations occurring in hypertensive patients has not been clearly established. An increased peripheral resistance in conjunction with a relatively normal or slightly reduced cardiac output is a characteristic feature of long-standing hypertension. Recent studies suggest that an increase in cardiac output is an initial hemodynamic abnormality in at least some patients with idiopathic hypertension.[65,66] Changes in peripheral resistance observed later could develop secondary to prolonged elevations of pressure. Support of this hypothesis is offered by the observation that reduction in cardiac output in such hypertensive patients by beta adrenergic blockade may normalize blood pressure.[67] Elevation of cardiac output has also been reported in renovascular hypertension.[68]

An early increase in cardiac output has been reported in the spontaneous hypertensive rat, an animal with genetically determined hypertension that has many of the characteristic features of the human disease.[69] With increasing duration of hypertension in these animals, peripheral resistance increases and the output appears to return to normal.

Left ventricular hypertrophy is a frequent occurrence with prolonged hypertension and is associated with a reduced prognosis.[70] Overt left ventricular failure may develop with severe or prolonged hypertension and represents the cause of death in approximately 40% of untreated patients.

CLASSIFICATION OF HYPERTENSION

Systolic Hypertension

Systolic hypertension without appreciable diastolic pressure increase may be observed in clinical states characterized by either decreased distensibility or obstruction of the aorta or by increased stroke output of the left ventricle (Table 15-2).

Systolic hypertension occurs most commonly in elderly individuals. Increase in aortic calcification and degeneration of aortic elastin are common with increasing age. These changes lead to reduced distensibility and capacitance of the aorta. Systolic hypertension by itself is an important risk factor in the development of either ischemic heart disease or strokes.[61] It may be difficult to treat, particularly when present in elderly patients who often require potent antihypertensive drugs but cannot tolerate marked blood pressure reductions because of coexisting atherosclerotic disease of the coronary and cerebral circulations.

With coarctation and other obstructive disease of the aorta the blood pressure is reduced in the lower extremities, and the pulses are both decreased in amplitude and delayed in their appearance. Pressures in the upper part of the body and the left ventricle will be markedly increased, and a prominent collateral circulation from above the level of obstruction may be apparent. The rib notching produced by dilated intercostal arteries in aortic coarctation reflects the latter change. The markedly increased pressures may lead to left ventricular failure,

Table 15-2
Etiologic Classification of Systolic Hypertension

Decreased aortic distensibility
Arteriosclerosis
Coarctation
Increased stroke output of left ventricle
Hyperthyroidism
AV fistula
Anemia
Paget's disease
Complete heart block
Aortic regurgitation
Left-to-right shunts

cerebrovascular accidents, or dissecting aneurysms of the aorta. If diastolic hypertension becomes prominent with coarctation, involvement of the renal circulation and superimposed renovascular hypertension should be suspected.

Management of systolic hypertension associated with increased stroke volume is best achieved by treatment of the underlying cause of the elevated cardiac output. However, in selected individuals therapy utilizing beta adrenergic blockade may be useful (see previous section).

Diastolic Hypertension

Idiopathic or Essential Hypertension

The diagnosis of essential hypertension is generally made by exclusion when the diagnostic workup does not reveal evidence of a known secondary cause (Table 15-3). More than 90% of hypertensive patients fall into this category. Approximately 70% of these exhibit a positive family history of the disease. A familial aggregation of blood pressure may be apparent,[71] even in children,[72] suggesting that the hypertension may manifest itself much earlier than previously suspected.

Current evidence would suggest that idiopathic or essential hypertension does not represent a homogeneous group. Multiple etiologies of the hypertension are to be expected as we learn more about the disease. Behavioral factors and abnormalities in sympathetic nervous function and catecholamine metabolism may be involved in its pathogenesis. While stress is exceedingly difficult to quantify, individuals such as air-traffic controllers who are subjected to chronic stress have an increased

Table 15-3
Etiologic Classification of Diastolic Hypertension

- Idiopathic or essential hypertension
- Renal hypertension
  - Parenchymal
    - Pyelonephritis
    - Glomerulonephritis
    - Diabetic nephropathy
    - Connective-tissue diseases
    - Renal tumors (juxtaglomerular cell tumor, hypernephroma, Wilms' tumor)
    - Polycystic kidneys
    - Other (amyloidosis, gouty nephritis, hematomas)
  - Obstructive—hydronephrosis
  - Renovascular
    - Renal artery atherosclerosis
    - Fibrous stenosis of renal arteries
    - Thrombotic or embolic occlusion (tumors, inflammation, pseudoxanthoma elasticum)
    - Other diseases involving renal artery
  - Renoprival
    - Renal failure
    - Anephric state
- Adrenal hypertension
  - Mineralocorticoid
    - Primary aldosteronism
    - Idiopathic aldosteronism
    - 18-hydroxy-DOC hypertension
    - DOC hypertension
    - Hydroxylation deficiency syndromes
    - ? 16 beta-hydroxydehydroepiandrosterone excess
  - Pheochromocytoma
  - Cushing's disease
  - Adrenogenital syndrome
- Coarctation of the aorta
- Toxemia of pregnancy
- Neurogenic hypertension
  - Increased intracranial pressure (brain tumors, hematomas)
  - Neuroblastomas
  - Neuropathies (polyneuritis, porphyria, lead poisoning, tabes)
  - Spinal cord transection
  - Encephalitis
  - Bulbar poliomyelitis
  - Diencephalic syndrome

Table 15-3 *cont.*

Drug-induced
- Oral contraceptive drugs
- Monoamine oxidase inhibitors plus tyramine
- Sympathomimetic drugs (amphetamines, cold remedies, etc.)

Other
- Hypercalcemia
- Licorice ingestion
- Myxedema

incidence of hypertension.[73] Of interest is the fact that persistent hypertension may be induced in animals by operant conditioning techniques. Some patients with idiopathic hypertension exhibit evidence of a hyperactive sympathetic nervous system, with increased sweating, tachycardia, markedly labile blood pressures, and elevated cardiac output.[66] A significant increase in plasma catecholamines has been reported in essential hypertension,[13] as has an increase in urinary norepinephrine in response to stress.[74] The plasma turnover of norepinephrine may be abnormal.[75] Increase in sympathetic vascular tone and vascular hyperreactivity to pressor agents and to various stresses may also occur.[12] While these multiple lines of evidence all point to abnormal activity of the autonomic nervous system in essential hypertension, the extent to which these changes are primary in nature or are secondary to the hypertension remains a major unresolved question.

Abnormalities in the renin–angiotensin system also have been identified in essential hypertension. Persistently reduced plasma renin activity is present in approximately one-fifth of the group.[25,26] Whether or not the low renin activity is related to mineralocorticoid excess has been the subject of considerable speculation. Many of these patients will become normotensive when treated with large doses of the aldosterone antagonist spironolactone or with the inhibitor of steroidogenesis aminoglutethimide.[76,77] Plasma aldosterone and progesterone concentrations may be increased to a mild degree in some patients with idiopathic hypertension, and the metabolic clearance of aldosterone by the liver may be reduced.[78] Primary aldosteronism appears to have been ruled out as a factor in most of the low-renin group, but some other unidentified mineralocorticoid hormone, or an abnormality in mineralocorticoid metabolism as yet unidentified, may ultimately be involved. In this regard, it has recently been reported that the urinary excretion of 16 β-hydroxy-dehydroepiandrosterone is increased in patients with low-renin hypertension, but the significance of this observation is as yet uncertain.[78]

Recent data have created interest in the possible involvement of vasodepressor factors in essential hypertension. Plasma prostaglandins of the A series may be abnormally reduced in some subjects.[79] In addition, urinary kallikrein excretion may also be decreased. In patients with renovascular hypertension, increased prostaglandin levels have been observed in renal venous blood from ischemic kidneys.[80]

Renal Hypertension

Renal diseases represent the most frequent secondary cause of hypertension. The renal abnormality may involve either the parenchyma, the collecting system, or the circulation.

Renal parenchymal diseases. Chronic pyelonephritis is probably the most common cause of renal parenchymal disease. The mechanism for the hypertension is uncertain. Its severity may be unrelated to the extent of the pyelonephritis, except with renal failure. Abnormalities in the renin–angiotensin system are variable. Studies in experimental animals have suggested that pyelonephritis may develop more frequently in the presence of hypertension, so that the coexistence of the two diseases may not always signify an etiologic role of pyelonephritis.

Both acute and chronic glomerulonephritis are associated with hypertension. With chronic nephritis the hypertension does not tend to be severe unless azotemia supervenes.

Renal tumors as well as space-occupying lesions of any type may be associated with blood pressure elevations. Juxtaglomerular cell tumors that produce renin in huge amounts have recently been described.[81] The tumors tend to be small and difficult to visualize by standard radiographic procedures. However, they can produce very severe hypertension, hyperreninemia, hypokalemia, and other findings of secondary aldosteronism. Wilms' tumors may also produce renin-like material on rare occasions and therefore may induce a similar syndrome.

Urinary tract obstruction with hydronephrosis is also a cause of hypertension. Increased renin secretion has been described in such patients, with a decrease in renin and blood pressure levels when the hydronephrosis is relieved.[82]

Renovascular hypertension. Renovascular hypertension is involved in approximately 1% of the hypertensive population and may represent the most frequent surgically curable cause of the disease. Its recognition is complicated by the fact that renovascular lesions are relatively common in normotensive as well as hypertensive patients, and coexistence does not necessarily imply a cause–effect relationship. The diagnosis cannot be made on clinical grounds alone, since few clinical features other than the presence of an upper abdominal bruit distinguish the disease from idiopathic hypertension.[83] Intravenous pyelography is a useful screening procedure and will indicate an abnormality in approximately 80% to 85% of patients. Radioisotope renography may also be of value in screening.

Renal angiography and renin assays are both needed to confirm the diagnosis of renovascular hypertension. The renal arterial abnormality is generally one of two types—renal artery atherosclerosis or fibrous stenosis. The atherosclerotic lesions typically occur in older individuals, particularly males, and involve the proximal one-third of the renal artery. The fibrous lesions are generally present in younger patients, usually females, and involve the distal half of one or both renal arteries. Fibrous stenosis can be of many types, but the most common form is associated with marked intimal or medial fibrosis, elastic tissue degeneration, and aneurysmal dilatation of the vessel. Other medium-sized vessels, as well as the renal artery, may be involved.

The measurement of renin or angiotensin activity in renal venous blood is the most specific and sensitive assay for determining the functional significance of renal artery stenosis. The renin activity in renal venous blood from an ischemic kidney will usually exceed that from the normal contralateral kidney by more than 50%. Agents that stimulate renin release such as diuretics, vasodilators, and upright posture will typically widen the resting difference in renin concentration between the two renal veins and may improve the sensitivity of the procedure. Drugs that inhibit renin activity such as propranolol may narrow differences. A new diagnostic approach which appears to have considerable value in the diagnosis of renovascular or other angiotensin-dependent forms of hypertension, involves the use of the angiotensin antagonist 1-sar-8-ala-angiotensin II (saralasin). A significant blood pressure reduction in response to short-term infusions of saralasin would suggest that angiotensin is involved in the maintenance of the hypertension.[84] Use of an angiotensin-converting enzyme inhibitor may also prove to have merit in the identification of patients with angiotensin-induced hypertension.[84a]

The mode of therapy of patients with renovascular hypertension generally depends on the ease and success of the therapy in controlling blood pressure.[85] Even with complete diagnostic assessment, revascularization procedures may produce a cure of the hypertension in only 70% to 80% of the total group, and the risk of renovascular surgery is not insignificant. A conservative approach utilizing medical therapy is warranted with mild or moderate hypertension, which is easily managed with medications. Beta adrenergic blockade can reduce renin activity markedly and may be particularly useful in the medical management of the disease.

Table 15-4
Hypertension, Hypokalemia, and Mineralocorticoid Excess

With Reduced Plasma Renin Activity	With Increased Plasma Renin Activity
Primary aldosteronism	Diuretic drugs
Idiopathic aldosteronism	Oral contraceptive drugs
DOC hypertension	Accelerated hypertension
18-OH-DOC hypertension	Renovascular hypertension
17α-hydroxylase deficiency	JG cell tumor
11β-hydroxylase deficiency	Congestive heart failure
? 16-beta-hydroxydehydroepiandrosterone	
Licorice ingestion	

Primary Aldosteronism

Hypertension may occur secondary to aldosterone-secreting adenomas of the adrenal cortex.[86] The incidence of the disease is uncertain, but it may approximate 1% of the hypertensive population. The most characteristic feature of primary aldosteronism is potassium depletion and hypokalemia. The major symptoms and signs of muscle weakness, polyuria, metabolic alkalosis, tetany, cardiac arrhythmias, and carbohydrate intolerance are related to the potassium wasting. Mild sodium retention and hypernatremia are common, but peripheral edema is rare.

Hypokalemia is present in a number of hypertensive states and is common following therapy with most diuretics. Measurement of plasma renin activity may be very useful as a screening procedure for distinguishing between primary states of mineralocorticoid excess and secondary forms of aldosteronism (Table 15-4). The renin activity is typically increased in secondary aldosteronism, but is abnormally reduced or normal in primary states. A therapeutic trial of the aldosterone antagonist spironolactone is also of diagnostic value. The drug in doses of 200 to 400 mg/day will normalize blood pressure and serum potassium levels by 4 to 6 weeks in almost all patients with primary aldosteronism.[87] Plasma and/or urinary aldosterone measurements are needed to confirm the diagnosis.

Approximately one-fourth of patients with the syndrome of primary aldosteronism will have either multiple tumors or bilateral hyperplasia of the adrenal cortex (idiopathic aldosteronism). In contrast to the patients with solitary nodules, most of whom become normotensive postoperatively, the patients with idiopathic aldosteronism respond relatively poorly to adrenal surgery despite an excellent antihypertensive response to spironolactone therapy. Assay of adrenal venous aldosterone concentration and adrenal venography is useful for distinguishing between the primary and idiopathic forms and should be performed preoperatively whenever possible. Surgery is clearly contraindicated in the idiopathic group, but may also be unnecessary in certain patients with solitary nodules. The tumors are nonmalignant in nature, and the blood pressure typically will respond well to spironolactone or even to other diuretics.[88] If medical therapy is elected, then careful attention to potassium balance as well as to blood pressure is required.

Other Conditions of Mineralocorticoid Hypertension

18-hydroxy-DOC hypertension. This syndrome characterized by hypertension, low plasma renin activity, and low aldosterone secretion in association with increased secretion of 18-hydroxy-DOC has recently been described.[89] 18-hydroxy-DOC is a relatively weak mineralocorticoid hormone, but may possibly be responsible for the hypertension in these individuals. Hypokalemia is a variable finding in this syndrome.

17 α-hydroxylase deficiency hypertension. This genetic defect of 17 α-hydroxylase activity in the adrenal may produce hypertension, hypokalemia, and reduced plasma renin concentration.[90] Aldosterone secretion is low, but increased production of deoxycorticosterone (DOC) occurs. In addition, because of the defect in 17 α-hydroxylation, there is a reduction in the synthesis of the sex hormone precursors 17 α-hydroxypregnenolone and hydroxyprogesterone, and gonadal insufficiency occurs. Treatment of such patients with dexamethasone will correct the hypertension and hypokalemia.

11 β-hydroxylase deficiency hypertension. This condition results from a defect in adrenal 11 β-hydroxylase activity that interferes with the conversion of 11 β-desoxycortisol to cortisol and of DOC to corticosterone. The increased levels of DOC appear to be responsible for the hypertension.

DOC hypertension. In addition to the above metabolic disorders, increased levels of DOC in the plasma of a significant percentage of patients with idiopathic hypertension have been reported recently.[91] The nature and incidence of this syndrome has not been clarified.

Licorice-induced hypertension. Ingestion of licorice (50 to 100 g per day) may produce a reversible form of hypertension that is associated with hypokalemia.[92] The effects are apparently related to the presence in licorice of glycyrrhizinic acid, which produces a mineralocorticoid effect.

Pheochromocytoma

Pheochromocytoma is a tumor of chromaffin tissue that produces hypertension as a result of excess secretion of catecholamines. The tumors originate typically from the adrenal gland, but may in 5% to 10% of patients occur at other sites, such as the organ of Zuckerkandl, sympathetic nerve ganglia, bladder, or thorax. Pheochromocytomas may be malignant in a small percentage of cases.

The most common symptoms and signs are headaches, excessive sweating, paroxysmal hypertension, orthostatic hypotension and tachycardia, fever, and carbohydrate intolerance. Sustained hypertension is present in approximately one-half of patients. Pheochromocytomas may be familial in nature and may be associated with medullary thyroid carcinomas, parathyroid adenomas, neuromas, and adrenal cortical hyperplasia, the so-called multiple endocrine adenomatosis Type II. The inheritance appears to be autosomal dominant in nature. Dysplasia of the neural crest cells may be present.[93] Multiple pheochromocytomas are common in this syndrome. Neurocutaneous disorders, cerebellar hemangioblastomas, polycystic disease, coarctation of the aorta, and Turner's syndrome also may occur with somewhat increased frequency in patients with pheochromocytoma.

The diagnosis of pheochromocytoma is made by biochemical means. Assay of urinary metabolites of norepinephrine (see Fig. 15-2), particularly urinary metanephrines and vanillylmandelic acid, should uncover almost all patients. Normotensive patients with a history of paroxysmal hypertension may also require a provocative test utilizing intravenous administration of histamine (0.025–0.050 mg) or glucagon (1.0 mg) in conjunction with urinary measurements. Localization of the tumor can be achieved by angiographic techniques, although invasive procedures of this type are hazardous. Regional venous catheterization with assay of venous blood for catecholamines is also useful for localization. In addition, since the conversion of norepinephrine to epinephrine occurs only in the adrenal medulla or in tissues of adrenal origin, assay of the urine for individual norepinephrine and epinephrine derivatives may be of value. A nonadrenal site for the tumor may be expected if norepinephrine derivatives represent more than 80% of the total.

Toxemia of Pregnancy

Toxemia of pregnancy is characterized by hypertension, albuminuria, and edema developing generally during the latter half of pregnancy. Blood pressure normally does not change appreciably with pregnancy in normotensive individuals or in patients with idiopathic hypertension. Blood pressure levels tend to decrease during the first trimester, return to normal during the second trimester, and remain unchanged or be slightly increased in the final trimester. The factors leading to the development of toxemia and hypertension are unknown. The marked fluid retention present in this condition has suggested that the renin–angiotensin–aldosterone system may be involved in the pathogenesis of the disease. However, recent studies have indicated that the marked increases in renin and aldosterone activity normally observed with pregnancy are actually inhibited with the development of toxemia and preeclampsia.

The treatment of toxemia is currently a highly controversial topic. Previous approaches utilizing dietary sodium reduction and diuretic therapy have been questioned as a result of recent evidence suggesting that the effective circulating volume may be reduced in such individuals and further reductions might prove injurious.

The development of eclampsia in patients with toxemia constitutes a hypertensive crisis that demands acute therapy with parenteral antihypertensive drugs. Vasodilators such as hydralazine and potent diuretics such as furosemide are particularly useful agents for managing this serious complication.

Oral Contraceptive Drugs and Hypertension

The most important reversible cause of hypertension in women currently appears to be treatment with oral contraceptive drugs.[94] Hypertension may develop in 1 percent to 5 percent of subjects treated over prolonged periods. The hypertension may take weeks or months either to develop or to recede following discontinuation of the drug. The mechanism for the hypertension has not been clearly delineated, but may involve abnormalities in the renin–angiotensin–aldosterone system and resultant volume expansion. Increases in plasma renin activity and aldosterone secretion have been noted in approximately one-half of patients and increase in renin substrate in virtually all.[95] Both estrogenic and progestational agents are capable of producing these biochemical changes and hypertension although the estrogens appear to be more important. The majority of patients developing hypertension may have a positive family history of the disease, suggesting that a genetic predisposition to hypertension may influence the response to the drugs. Oral contraceptive drugs should be avoided by hypertensive patients, subjects with a strong family history of the disease, and individuals with a prior history of toxemia during pregnancy.

ADVERSE EFFECTS OF SYSTEMIC HYPERTENSION

Hypertension is a major cause of death and disability, and if untreated it may take 20 years off the average life expectancy.

Hypertension and Atherosclerosis

Acceleration of the atherosclerotic process occurs in most hypertensive patients. Hypertension is the most important risk factor predisposing to ischemic heart disease or cerebrovascular disease. Increased risk for ischemic heart disease is observed even in very mild or labile hypertension. The etiologic basis for this relationship has not been delineated. Atherosclerotic disease is difficult to produce in experimental animals with hypertension alone, although increased atherogenesis from hypertension does occur readily when it is combined with hyperlipidemia. Hypertension may somehow damage the arterial endothelial and subendothelial layers, leading to increased rate of entry of plasma lipoproteins and other plasma constituents into the vessel wall. Subsequent changes including smooth-muscle cell migration and proliferation, lipid deposition, increased lysosomal activity, and vascular degeneration could result. Recently an influence of renin or angiotensin on the atherosclerotic process has been suggested by the report that hypertensive patients with low plasma renin activity have a reduced incidence of myocardial infarcts and strokes.[96] An influence of angiotensin on increasing vascular permeability by widening junctional gaps between endothelial cells has also been suggested.[97] However, other studies have not confirmed a protective effect of low plasma renin activity.[98]

Hypertension and Arteriolar Sclerosis

Arteriolar sclerosis involving multiple vascular beds is a characteristic feature of systemic hypertension. Its severity generally reflects the height and duration of hypertension. Assessment of retinal arteriolar disease is the major feature of the Keith-Wagener classification of the severity of hypertension. The impairment in renal function occurring as a consequence of hypertension may be attributed predominantly to nephrosclerosis. Severe blood pressure elevation from any source may also lead to fibrinoid degeneration of arterioles and the syndrome of malignant hypertension (see below).

Complications of Hypertension

Ischemic heart disease represents the most important cause of death in the hypertensive patient. Angina pectoris and myocardial infarction are exceedingly common in patients with long-standing hypertension, and approximately one-half of all patients who develop a myocardial infarct have an antecedent history of hypertension. Control of hypertension from an early age should reduce the incidence of coronary heart disease, although the effects of antihypertensive therapy on the process following prolonged periods of hypertension are uncertain.[99] Reduction of blood pressure in hypertensive patients decreases the work load of the heart and tends to improve the adequacy of myocardial perfusion relative to demand. As a result, antihypertensive therapy rarely will induce coronary insufficiency or myocardial infarction unless vasodilators such as hydralazine are used.

Congestive heart failure can occur as a result of the excessive work load placed upon the heart by elevated pressure. This complication, which is a major cause of death in untreated patients, can be prevented almost entirely or reversed by effective lowering of blood pressure.[99,100] In those patients in whom cardiac compensation is not restored with treatment, complicating ischemic heart disease generally will be the cause of continuing heart failure.

Cerebral vascular disease and resultant strokes are usually related to hypertension. Most patients who develop either cerebral thrombosis or hemorrhage have a history of blood pressure elevation. With cerebral hemorrhages the elevation tends to be marked. A marked decrease in incidence of both cerebral thrombosis and hemorrhage has been observed in hypertensive patients treated effectively with medication.[99,100]

Renal insufficiency results primarily from nephrosclerosis, although renal artery atherosclerosis and other complicating renal diseases may also play a role. Severe renal failure is unusual on the basis of hypertension alone, unless the malignant phase of the disease is present. Treatment of the hypertension will help prevent or delay the deterioration of renal function, even in patients with marked azotemia.[101]

Dissecting aneurysms of the aorta occur most commonly in hypertensive patients, particularly those with marked pressure elevations. Antihypertensive therapy has been shown to be of value in both the prevention of the disease and the management of patients who develop the problem. *Aortic aneurysms* with rupture also occur with increased frequency in hypertensive subjects.

Hypertensive retinopathy of severe form with retinal hemorrhages, exudates, or papilledema may lead to impairment of vision. These changes can generally be prevented or reversed by treatment.

Malignant hypertension is characterized by marked blood pressure elevation (diastolic pressure usually > 130) and papilledema. Necrotizing arteriolitis that is somehow related to the high level of pressure will develop in the kidneys and other organs as well. Marked increases in renin activity and secondary aldosteronism develop as a result of the renal vascular disease. Renal failure and death will occur in less than a year in untreated patients with malignant hypertension.

Hypertensive encephalopathy is a serious complication manifested by markedly elevated blood pressure levels, headaches, and agitation, and at times by convulsions and coma. Cerebrospinal fluid pressure is elevated in most of the patients. Hypertensive encephalopathy is a medical emergency that can lead to death if untreated.

ANTIHYPERTENSIVE THERAPY

The overwhelming benefits of antihypertensive therapy are incontrovertible.[99,100] A detailed approach to therapy is not within the scope of this paper, but is available elsewhere.[102] Other than for patients with severe hypertension where an aggressive program may be necessary, treatment initially should utilize mild agents. Any stepwise increase in dosage and addition of new drugs should be gradual. A combination of mild agents has proved to be preferable to large doses of a single drug. Following long-term control of blood pressure, a reduction in the number of dosage of medications is to be expected, although complete cessation of medications almost always will lead to recurrence of

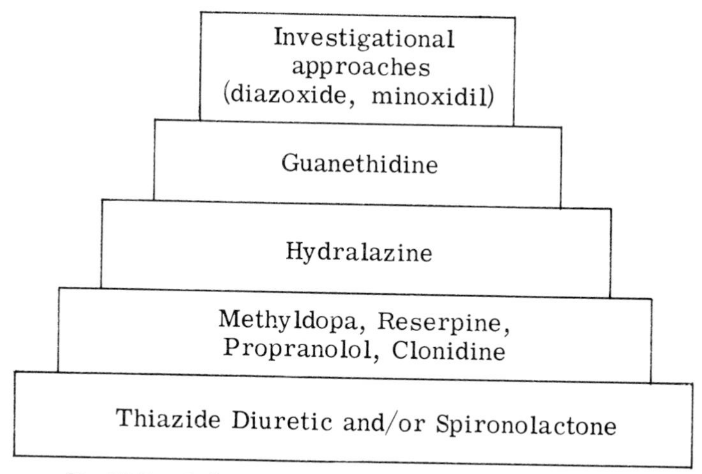

Fig. 15-7. A diagrammatic representation of the stepwise progression of drug therapy.

hypertension. The major factor responsible for therapeutic success involves patient adherence to the treatment program.

The individual drugs used most commonly at the present time can be divided into three major categories: diuretics (e.g., thiazides, spironolactone, quinethazones), compounds that influence the sympathetic nervous activity (e.g., reserpine, methyldopa, clonidine, guanethidine), and vasodilators (e.g., hydralazine, diazoxide). In addition, experimental approaches currently being studied utilizing beta adrenergic blockers (e.g., propranolol) to reduce cardiac output or renin activity also will undoubtedly prove of value in the future. Our step-care approach with these drugs is illustrated in Fig. 15-7. The numbers and varieties of antihypertensive drugs provide an adequate armamentarium for treating almost all hypertensive patients.

REFERENCES

1. Heymans C, Neil E: Reflexogenic Areas of the Cardiovascular System. Boston, Little, Brown, 1958
2. McCubbin JW, Green JH, Page IH: Baroreceptor function in chronic renal hypertension. Circ Res 4:205, 1956
3. Kezdi P: in: Baroreceptors and Hypertension. Proceedings of an International Symposium. London, Pergamon Press, 1967
4. Gribbin B, Pickering TG, Sleight P, Peto R: Effect of age and high blood pressure on baroreflex sensitivity in man. Circ Res 29:424, 1972
5. Tuckman J, Reich T, Lyon AF, et al: Electrical stimulation of the sinus nerve in hypertensive patients. Hypertension 16:23, 1967
6. Monnier M: Neurohumoral regulation of blood circulation, in Monnier M (ed): Function of the Nervous System. Amsterdam, Elsevier, 1968, p 314
7. Benson H, Herd JA, Moise WH, Kelleher RT: The behavioral induction of arterial hypertension and its reversal. Am J Physiol 217:30, 1969
8. Lefkowitz RJ: Characterization and purification of the cardiac beta-adrenergic receptor, in Mehlman M (ed): The Role of Membranes in Metabolic Regulation. New York, Academic, 1972, p 261
9. Sjoerdsma A: Pheochromocytoma: Current concepts of diagnosis and treatment. Ann Intern Med 65:1302, 1966
10. Weinshilboum RM, Thoa NB, Johnson DG, et al: Proportional release of norepinephrine and dopamine-β-hydroxylase from sympathetic nerves. Science 174:1349, 1971
11. Weinshilboum RM, Axelrod J: Reduced plasma dopamine-β-hydroxylase activity in familial dysautonomia. N Engl J Med 285:938, 1971
12. Mendlowitz M: Vascular reactivity in systemic arterial hypertension. Am Heart J 85:252, 1973

13. Engelman K, Portnoy B, Sjoerdsma A: Plasma catecholamine concentrations in patients with hypertension. Circ Res 37:I-141. 1970
14. Skeggs LT Jr, Lentz KE, Kahn JR, Shumay NP: The synthesis of tetradecapaptide renin substrate. J Exp Med 108:283, 1958
15. Goodfriend TL, Peach MJ: Angiotension III: (Des-aspartic-acid) angiotension II. Circ Res 36:I-38, 1975
16. Laragh JH, Cannon PJ, Bentzel CJ, et al: Angiotensin II, norepinephrine, and renal transport of electrolytes and water in normal man and in cirrhosis with ascites. J Clin Invest 42:1179, 1963
17. McGiff JC, Crowshaw K, Terragno NA, Lonigro AJ: Release of a prostaglandin-like substance into renal venous blood in response to angiotensin II. Circ Res 27:I-121, 1970
18. McCubbin JW, Bunag RD, Kaneko Y, Vander AJ: Control of renin release, in Page IH, McCubbin JW (eds): Renal Hypertension. Chicago, Year Book Medical Publishers, 1969, p 100
19. Vander AJ: Effect of catecholamines and renal nerves on renin secretion in anesthetized dogs. Am J Physiol 209:659, 1965
20. Assaykeen TA, Ganong WF: The sympathetic nervous system and renin secretion, in Martini L, Ganong WF (eds): Frontiers in Neuroendocrinology. New York, Oxford University Press, 1971
21. Oparil S, Vassaux C, Sanders CA, Haber E: Role of renin in acute postural homeostasis. Circulation 41:89, 1970
21a. Chobanian AV, Lille RD, Tercyak A, Blevins P: The metabolic and hemodynamic effects of prolonged bed rest in normal subjects. Circulation 49:551, 1974
22. Genest J, de Champlain J, Veyrat R, et al: Role of the renin-angiotensin system in various physiological and pathological states. Hypertension 13:95, 1965
23. MacDonald GJ, Louis WJ, Renzini V, et al: Renal-clip hypertension in rabbits immunized against angiotensin II. Circ Res 27:197, 1970
24. Gavras H, Brunner HR, Thurston H, Laragh JH: Reciprocation of renin dependency with sodium volume dependency in renal hypertension. Science 188:1316, 1975.
24a. Ferrario CM, McCubbin JW: Neurogenic factors in hypertension. Hosp. Prac. Dec. 1974, 71
25. Helmer OM: Renin activity in blood from patients with hypertension. Can Med Assoc J 90:221, 1964
26. Channick BJ, Adlin EV, Marks AP: Suppressed plasma renin activity in hypertension. Arch Intern Med 123:131, 1969
27. Ganong WF, Biglieri EG, Mulrow PJ: Mechanisms regulating adrenocortical secretion of aldosterone and glucocorticoids. Recent Prog Horm Res 22:381, 1966
28. Lee JB, McGiff JC, Kannegiesser H, et al: Prostaglandin A_1: Antihypertensive and renal effects. Ann Intern Med 74:703, 1971
29. Lee JB: Renal homeostasis and the hypertensive state. A unifying

hypothesis, in Ramwell PW (ed): The Prostaglandins. New York, Plenum, 1967, p 133

30. Zusman RM, Spector D, Caldwell BV, et al: The effect of chronic sodium loading and sodium restriction on plasma prostaglandin A, E, and F concentrations in normal humans. J Clin Invest 52:1093, 1973
31. Werning CW, Vetter P, Weidmann HV, et al: Effect of prostaglandin E_1 on renin in the dog. Am J Physiol 220:852, 1971
32. Saruta T, Kaplan NM: Adrenocortical steroidogenesis: The effects of prostaglandins. J Clin Invest 51:2239, 1972
33. McGiff JC, Crowshaw K, Terrogno NA, et al: Differential effect of noradrenaline and renal nerve stimulation on vascular resistance in the dog kidney and the release of a prostaglandin E-like substance. Clin Sci 42:223, 1972
34. McGiff JC, Terrogno NA, Malik KV, Lonigro AJ: Release of a prostaglandin E-like substance from canine kidney by bradykinin. Circ Res 31:36, 1972
35. Muirhead EE, Brown GB, Bermain GS, Leach BE: The renal medulla as an antihypertensive organ. J Lab Clin Med 76:641, 1970
36. Muirhead EE, Brooks B, Pitcock JA, Stephenson P: Renomedullary antihypertensive function in accelerated (malignant) hypertension. J Clin Invest 51:181, 1972
37. Smeby RR, Sens S, Bumpus FM: A naturally occurring renin inhibitor. Circ Res 21:II-129, 1967
38. Nies AS, Forsyth RP, Williams HE, Melmon KL: Contributions of kinins to endotoxic shock in unanesthetized rhesus monkeys. Circ Res 22:155, 1968
39. Oates JA, Pettinger WA, Doctor RB: Evidence for the release of bradykinin in carcinoid syndrome. J Clin Invest 45:173, 1966
40. Wong PY, Talamo RC, Babior BM, et al: Kallikrein-bradykinin system in post-gastrectomy early dumping (PED) syndrome. Clin Res 21:529, 1973
41. Lewis GP: Kinins in inflammation and tissue injury, in Erdos EG (ed): Handbook of Experimental Pharmacology, vol 25. Berlin, Springer-Verlag, 1972, p 516
42. Streeten DHP, Kerr LP, Kerr CB, et al: Hyperbradykininism: A new orthostatic syndrome. Lancet 2:1048, 1974
43. Willis LR, Ludens JH, Hook JB, Williamson HE: Mechanism of natriuretic action of bradykinin. Am J Physiol 217:1, 1969
44. Greene LJ, Camarga ACM, Kreiger EM, et al: Inhibition of the conversion of angiotensin I to II and potentiation of bradykinin by small peptides present in bothrops jararaca venom. Circ Res 30,31:II-62, 1972
45. Margolius HS, Geller RG, de Jong W, et al: Altered kallikrein excretion in rats with hypertension. Circ Res 30,31:II-125, 1972
46. Margolius HS, Horwitz D, Pisano JJ, Keiser HR: Urinary kallikrein and human hypertension: Effects of salt and mineralocorticoids. Clin Res 21:699, 1973
47. Holloway EF, Sitrin MD, Bohr DF: Calcium dependence of vascular

smooth muscle from normotensive and hypertensive rats, in Genest J, Koiw E (eds): Hypertension 1972. Berlin, Springer-Verlag, 1972, p 400

48. Shibata N, Rosenthal J, Hollander W: The role of cations, ATP and vasoactive substances on the activity of contractile proteins of arteries, in Genest J, Koiw E (eds): Hypertension 1972. Berlin, Springer-Verlag, 1972, p 184
49. Dahl LK, Love RA: Etiological role of sodium chloride intake in essential hypertension in humans. JAMA 164:397, 1957
50. Maddocks I: Blood pressure in Melanesians. Med J Aust 1:1123, 1967
51. Dahl LK, Heine M, Tassinari L: Role of genetic factors in susceptibility to experimental hypertension due to chronic excess salt ingestion. Nature 194:480, 1962
52. Rapp JP, Dahl LK: Adrenal steroidogenesis in rats bred for susceptibility and resistance to the hypertensive effect of salt. Endocrinology 88:52, 1971
53. Tobian L, Janacek J, Tomboulian A, Ferreira D: Sodium and potassium in the walls of arterioles in experimental renal hypertension. J Clin Invest 40:1922, 1961
54. Tobian L, Chesley G: Calcium content of arteriolar walls in normotensive and hypertensive rats. Proc Soc Exp Biol Med 121:340, 1966
55. Hollander W, Kramsch DM, Farmelant M, Madoff IM: Arterial wall metabolism in experimental hypertension from coarctation of the aorta of short duration. J Clin Invest 47:1221, 1968
56. Hollander W, Chobanian AV, Burrows BA: Body fluid and electrolyte composition in arterial hypertension. I. Studies in essential, renal and malignant hypertension. J Clin Invest 40:408, 1961
57. Chobanian AV, Burrows BA, Hollander W: Body fluids and electrolyte composition in arterial hypertension. II. Studies in mineralocorticoid hypertension. J Clin Invest 40:416, 1961
58. Tarazi RC, Dustan HP, Frohlich ED: Relation of plasma to interstitial volume in essential hypertension. Circulation 30:357, 1969
59. Hollander W, Judson WE: Electrolyte and water excretion in arterial hypertension. I. Studies in non-medically treated subjects with essential hypertension. J Clin Invest 36:1460, 1957
60. Dustan H, Nijensohn C, Corcoran AC: Natriuretic-diuretic effect of angiotensin in essential hypertension. J Clin Invest 34:931, 1955
61. Kannell WB, Kagan A, Revotskie N, Stokes J: Factors of risk in the development of coronary heart disease—Six year follow-up experience. Ann Intern Med 55:33, 1961
62. Society of Actuaries: Build and Blood Pressure Study. Chicago, Society of Actuaries, 1959
63. Wilber JA, Millward D, Baldwin A, et al: Atlanta community high blood pressure program methods of community hypertension screening. Circ Res 31:II-101, 1972
64. Eich RH, Peters RJ, Cuddy RP, et al: The hemodynamics in labile hypertension. Am Heart J 63:188, 1962
65. Kuchel O, Cuche JL, Hamet P, et al: The relationship between the

adrenergic nervous system and renin in labile hyperkinetic hypertension, in Genest J, Koiw E (eds): Hypertension 1972. Berlin, Springer-Verlag, 1972, p 118

66. Frohlich ED, Tarazi RC, Dustan HP: Hyperdynamic beta adrenergic circulatory states. Arch Intern Med 123:1, 1969
67. Frohlich ED: Beta adrenergic blockade in the circulatory regulation of hyperkinetic states. Am J Cardiol 27:195, 1971
68. Frohlich ED, Ulrych M, Tarazi RC, et al: A hemodynamic comparison of essential and renovascular hypertension. Circulation 35:289, 1967
69. Pfeffer MA, Frohlich ED: Hemodynamic and myocardial function in young and old normotensive and spontaneously hypertensive rats. Circ Res 31:I-23, 1973
70. Kannel WB, Castelli WP, McNamara PM, et al: Role of blood pressure in the development of congestive heart failure. N Engl J Med 287:781, 1972
71. Miall WE, Oldham PD: The heredity factor in arterial blood pressure. Br Med J 1:75, 1963
72. Zinner SH, Levy PS, Kass EM: Familial aggregation of blood pressure in childhood. N Engl J Med 284:401, 1971
73. Cobb S, Rose RM: Hypertension, peptic ulcer, and diabetes in air traffic controllers. JAMA 224:489, 1973
74. Nestel PJ: Blood pressure and catecholamine excretion after mental stress in labile hypertension. Lancet 1:692, 1969
75. Gitlow SE, Mendlowitz M, Wilk EK, et al: Plasma clearance of *dl*-H^3-norepinephrine in normal human subjects and patients with essential hypertension. J Clin Invest 43:2009, 1964
76. Crane MG, Harris JJ, Johns VJ: Hyporeninemic hypertension. Am J Med 52:457, 1972
77. Woods JW, Liddle GW, Stant EG, et al: Effect of an adrenal inhibitor in hypertensive patients with suppressed renin. Arch Intern Med 123:366, 1969
78. Genest J, Nowaczynski WJ, Kuchel O, Sasaki C: New evidences of disturbances of mineralocorticoid activity in benign uncomplicated essential hypertension. Trans Am Clin Climatol Assoc 83:134, 1972

78a. Sennett JA, Brown RD, Island DP, et al: Evidence for a new mineralocorticoid in patients with low-renin essential hypertension. Circ Res 36:I-2, 1975.

79. Zusman R, Spector D, Caldwell B, et al: Prostaglandin A concentrations in plasma of normal and hypertensive humans. J Clin Invest 52:93a, 1973
80. Edward WG Jr, Strong CG, Hunt JC: A vasodepressor lipid resembling prostaglandin E_2 (PGE_2) in the renal venous blood of hypertensive patients. J Lab Clin Med 74:389, 1969
81. Conn JW, Cohen EL, Lucas CP, et al: Primary reninism. Hypertension, hyperreninemia due to renin-producing juxtaglomerular cell tumors. Arch Intern Med 130:682, 1972
82. Belman AB, Krobb KA, Simon NM: Renal-pressor hypertension secondary to unilateral hydronephrosis. N Engl J Med 278:1133, 1968

83. Simon N, Franklin SS, Bleifer KH, Maxwell MH: Clinical characteristics of renovascular hypertension. JAMA 220:1209, 1972
84. Streeten DHP, Anderson GH, Freiberg JM, Dalakos TG: Use of an angiotensin II antagonist (Saralasin-P113) in the recognition of angiotensinogenic hypertension. New Eng J Med 292:657, 1975
84a. Gavras H, Brunner HR, Laragh JH, et al: The use of an angiotensin converting enzyme inhibitors to identify and treat vasoconstrictor and volume factors in hypertensive patients. New Eng J Med 291:817, 1974
85. Dustan HP, Meaney TF, Page IH: Conservative treatment of renovascular hypertension, in Gross F (ed): Antihypertensive Therapy: Principles and Practice. Berlin, Springer-Verlag, 1966, p 544
86. Conn JW: The evolution of primary aldosteronism, 1954–1967. Harvey Lect 62:257, 1967
87. Melby JC: Primary aldosteronism (symposium on endocrinology). Practitioner 200:519, 1968
88. Dustan HP, Tarazi RC, Bravo EL, Dart RA: Plasma and extracellular fluid volume in hypertension. Circ Res 32:I-73, 1973
89. Melby JC, Dale SL, Wilson TE: 18-hydroxy-deoxycorticosterone in human hypertension. Circ Res 28,29:II-143, 1971
90. Biglieri EG, Herron MA, Brust N: 17 α-hydroxylation deficiency in man. J Clin Invest 45:1946, 1966
91. Brown JJ, Ferriss JB, Fraser R, et al: Apparently isolated excess deoxycorticosterone in hypertension. A variant of the mineralocroticoid excess syndrome. Lancet 2:243, 1972
92. Koster M, David GK: Reversible severe hypertension due to licorice ingestion. N Engl J Med 278:1381, 1968
93. Keiser HR, Beaven MA, Doppman J, et al. Sipple's syndrome: medullary thyroid carcinoma, pheochromocytoma, and parathyroid disease. Ann Int Med 78:561, 1973
94. Lille RD, Gould J, Viera M, Chobanian AV: Management of hypertension in an inner city area. Clin Res 21:539, 1973
95. Laragh JH, Sealey JE, Ledingham JGG, Newton MA: Oral contraceptives, renin, aldosterone, and high blood pressure. JAMA 201:918, 1967
96. Brunner HR, Laragh JH, Baer L, et al: Essential hypertension: Renin and aldosterone, heart attack and stroke. N Engl J Med 286:441, 1972
97. Robertson AL, Khairallah PA: Effects of angiotensin II and some analogues on vascular permeability in the rabbit. Circ Res 31:923, 1972
98. Mroczek WJ, Finnerty FA, Catt KJ: Lack of association between plasma-renin and history of heart-attack or stroke in patients with essential hypertension. Lancet 1:464, 1973
99. Veterans Administration Cooperative Study Group on Antihypertensive Agents: Effects of treatment on morbidity in hypertension. Results in patients with diastolic blood pressures averaging 115 through 129 mm Hg. JAMA 202:1028, 1967
100. Veterans Administration Cooperative Study Group on Morbidity in

Hypertension: Results in patients with diastolic blood pressures averaging 90 through 114 mm Hg. JAMA 213:1143, 1970
101. Woods JW, Blythe WB: Management of malignant hypertension complicated by renal insufficiency. N Engl J Med 277:57, 1967
102. Chobanian AV, Lanzoni V: Current concepts of the drug therapy of hypertension. Medical Counterpoint 3:18, 1971

Abraham M. Rudolph
Michael A. Heymann

16

Neonatal Circulation and Pathophysiology of Shunts

During the perinatal period dramatic changes occur in the circulation associated with the transfer of the function of gas exchange from the placenta to the lungs. A brief review of the fetal circulation is in order as an introduction to the adaptations immediately after birth.

FETAL CIRCULATION

The course of the circulation in the fetal heart and great vessels is shown in Fig. 16-1.[1,2] Blood returns to the heart through the vena cavae. Inferior vena caval return consists of the venous return from the lower body of the fetus and an admixture of umbilical venous return; the umbilical venous return either passes through the hepatic circulation and then through the hepatic veins into the inferior vena cava or bypasses the liver through the ductus venosus. A little more than half the inferior vena caval blood enters the right atrium, while the remainder passes directly through the foramen ovale and enters the left atrium.[3] Essentially all superior vena caval and coronary sinus return drains through the right atrium across the tricuspid valve and into the right ventricle. The right ventricle ejects about two-thirds of the total cardiac output (combined right and left ventricular outputs), whereas the left ventricle ejects only

Supported by Program Project Grant HL-06285 from the National Heart and Lung Institute. M.A.H. was recipient of Research Career Development Award HD-35398 from the National Institute of Child Health and Human Development.

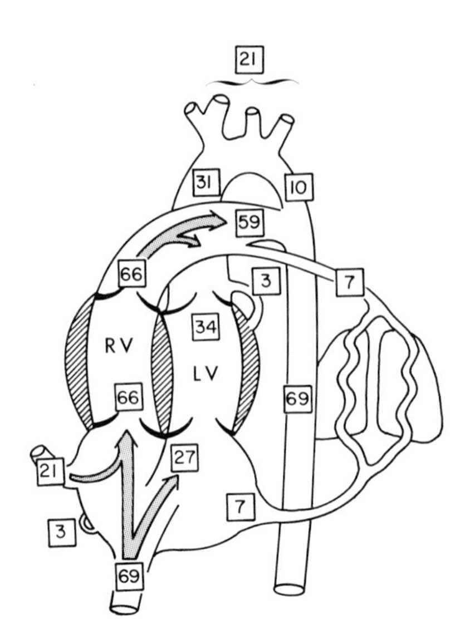

Fig. 16-1. The proportions of combined left and right ventricular outputs (C.V.O.) that flow through the major vessels and shunts in the fetal heart are shown in the squares. RV = right ventricle; LV = left ventricle.

about one-third of the cardiac output.[3] The blood ejected by the right ventricle into the pulmonary trunk is largely directed away from the lungs through the very large ductus arteriosus present in the fetus, and only an average of 7%–8% of total cardiac output reaches the lungs.[4] The ductus arteriosus therefore carries about 59% of the cardiac output to the descending aorta. The placenta has a relatively low resistance as compared to other segments of the fetal circulation, and it receives about 40% of the total cardiac output.[4]

The presence of the large ductus arteriosus also results in an equalization of pressures in the aorta and pulmonary trunk, and left and right ventricular systolic pressures are therefore essentially equal.[5] Oxygen uptake in the fetus occurs in the placenta, and umbilical venous blood therefore has a Po_2 of about 28–32 torr. The inferior vena caval blood, which receives umbilical venous return, therefore has a higher Po_2 than superior vena caval blood. Since the inferior vena cava provides the major venous return to the left atrium and left ventricle, blood in the ascending aorta has a Po_2 of about 23–26 torr. Blood entering the right ventricle consists of a mixture of superior vena caval and inferior vena caval return and has a Po_2 of about 17–20 torr. The descending aortic blood consists largely of blood that has passed through the ductus arteriosus from the pulmonary trunk with a small contribution across the aortic arch and has a Po_2 of about 20–23 torr, which is lower than that in the ascending aorta.[5]

CHANGES IN THE CIRCULATION AT BIRTH

The two major events that occur at birth are elimination of the umbilical placental circulation and ventilation of the lungs with air. Constriction or clamping of the umbilical vessels removes the relatively low resistance placental circulation and thus effectively increases the vascular resistance on the systemic side of the circulation. It also results in a marked decrease of total blood returning to the heart via the inferior vena cava, since 40% of fetal cardiac output is distributed to the placenta and is returned to the heart through the inferior vena cava. Almost simultaneously with removal of the placental circulation the lungs are ventilated with air. This results in a rapid and marked fall in pulmonary vascular resistance.

Changes in Pulmonary Circulation

In the fetus, pulmonary vascular resistance is high. Although it has been suggested that this may be due to the fact that the lungs are not expanded,[6] more recent studies have indicated that the main cause of the high resistance to flow in fetal lambs is related to intense constriction of the pulmonary vascular smooth muscle.[7–10] This muscle in fetal pulmonary vessels is very sensitive to small changes in the level of oxygen to which the vessels are exposed (Fig. 16-2).[11] In utero they are exposed to the Po_2 of the blood perfusing them, which is about 17–19 torr. An increase in Po_2 produces pulmonary vasodilatation. Ventilation of the lungs with air has been shown to produce a marked increase in pulmonary blood flow. Although this effect is partly related to simple physical expansion of the lungs with gas, the most striking change is related to a fall in pulmonary vascular resistance associated with an increase in alveolar oxygen concentration. The main resistance to flow in the pulmonary circulation is at the precapillary level in vessels of arteriolar size (30–200 μ diameter). These vessels have a very thick medial smooth-muscle layer. It has been difficult to understand why an increase in alveolar oxygen level, which would mainly influence Po_2 in pulmonary capillaries, could decrease vascular resistance at the arteriolar level. It has been shown, however, that oxygen diffuses from alveoli surrounding the arterioles directly through the wall into the lumen.[12] Thus a change in oxygen produced by ventilation with air may directly affect the smooth muscle in these vessels. It is estimated that pulmonary flow increases about 8- to 10-fold in the first 10 min after birth associated with the release of the pulmonary vasoconstriction.

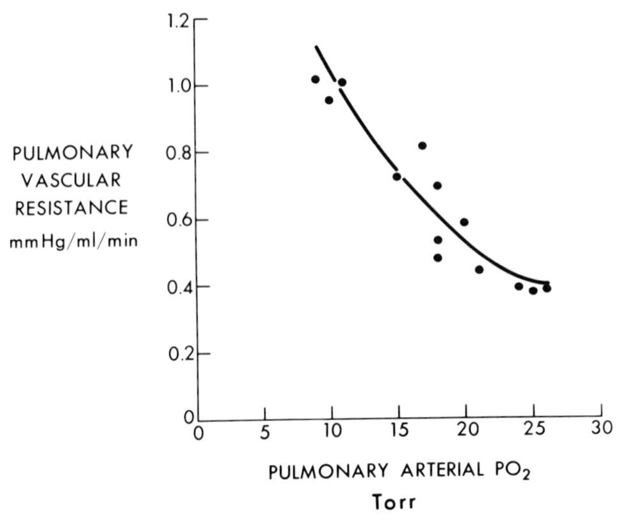

Fig. 16-2. The relationship of calculated pulmonary vascular resistance to pulmonary arterial Po_2 in a fetal lamb.

Closure of the Ductus Arteriosus

The ductus arteriosus in the fetus has a diameter similar to that of the descending aorta. A large amount of smooth muscle is present in its wall. Whereas the smooth muscle of the pulmonary arterioles relaxes when Po_2 is increased, that in the wall of the ductus arteriosus constricts.[13] In the mature fetal lamb constriction first begins to occur when the Po_2 rises above 40 torr, and intense constriction occurs above levels of 100–150 torr. In the immature fetal lamb the ductus arteriosus is less sensitive to increases in Po_2, a higher level of Po_2 is required before constriction develops, and very high Po_2 levels are necessary to produce marked constriction.[14] As seen in Fig. 16-3, the less mature the animal the less striking is the constrictor effect of oxygen.

In the mature infant the ductus arteriosus is usually functionally closed by constriction of its muscle within 12–15 hr after birth. Final closure is accomplished by thrombosis and fibrosis of the wall over a period of 1 to 2 weeks. This results in a separation of the pulmonary and systemic circulations in a normal infant. The lowering of pulmonary vascular resistance after birth permits the pulmonary circulation to be perfused at a lower pressure than the systemic circulation, in which the resistance is much higher after birth.

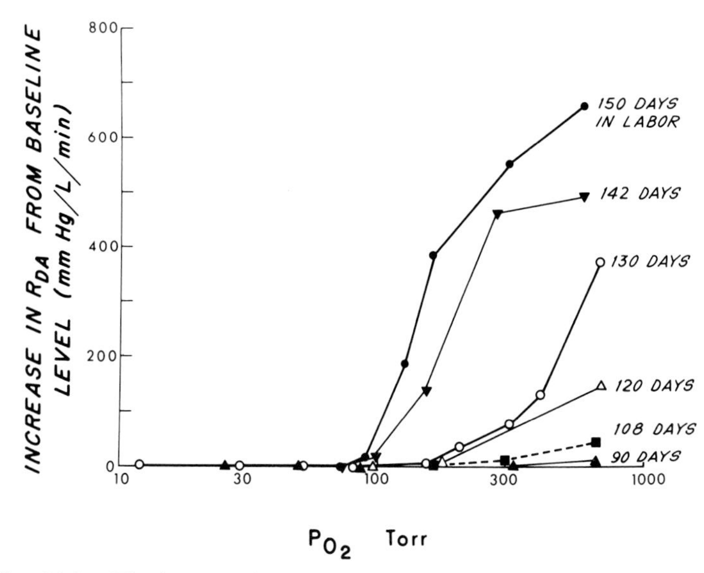

Fig. 16-3. The increase from resting level at Po_2 15–35 in calculated ductus arteriosus resistances (R_{DA}) produced by raising the Po_2 of the perfusing solution of the isolated ductus arteriosus obtained from fetal lambs of different gestational ages.

Closure of the Foramen Ovale

The foramen ovale is a slit-like opening in the lower portion of the atrial septum.[5] The upper margin of the foramen ovale (crista dividens) projects into the right atrium, whereas the lower margin is a thin valve-like structure that projects into the left atrium. Streaming of blood from the inferior vena cava directly into the left atrium tends to hold the valvelike flap open in the fetus. When the umbilical circulation is removed, inferior vena caval return and pressure are reduced. Simultaneously the increase in pulmonary blood flow results in a marked increase of venous return to the left atrium and an increase in pressure in the left atrium. These two events result in deflection of the valve of the foramen ovale against the upper margin, thus effectively closing the opening.

Postnatal Changes in Pulmonary Circulation

The fall in pulmonary vascular resistance immediately after birth is predominantly associated with pulmonary vasodilatation due to the increase in alveolar Po_2, but the vessels still retain a thick muscle layer

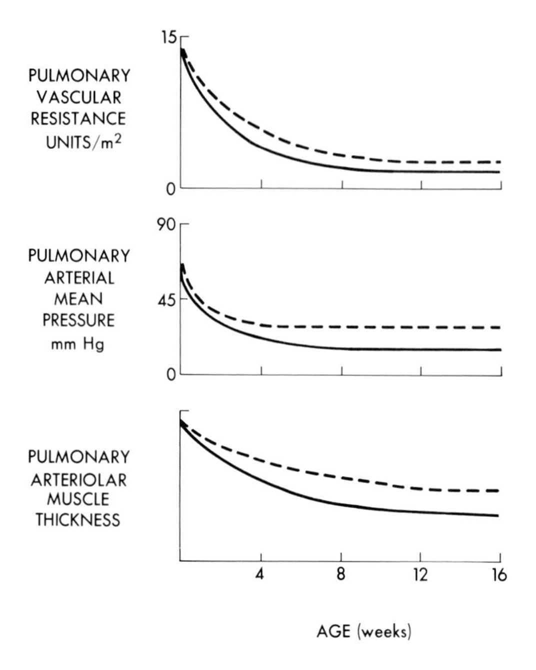

Fig. 16-4. Diagrammatic representation of the changes in calculated pulmonary vascular resistances, pulmonary arterial pressure, and pulmonary arteriolar muscle thickness in the first 16 weeks after birth (— normal; --- altitude).

that has developed during fetal life. Pulmonary arterial pressure is still considerably higher than that noted in older children or adults. Within 6–10 weeks after birth the amount of pulmonary vascular smooth muscle progressively decreases, and the vessels assume the morphologic characteristics of adult pulmonary vessels, with large lumens, thin walls, and a minimal amount of smooth muscle (Fig. 16-4).[15,16] Associated with this disappearance of muscle, pulmonary arterial pressures gradually fall to

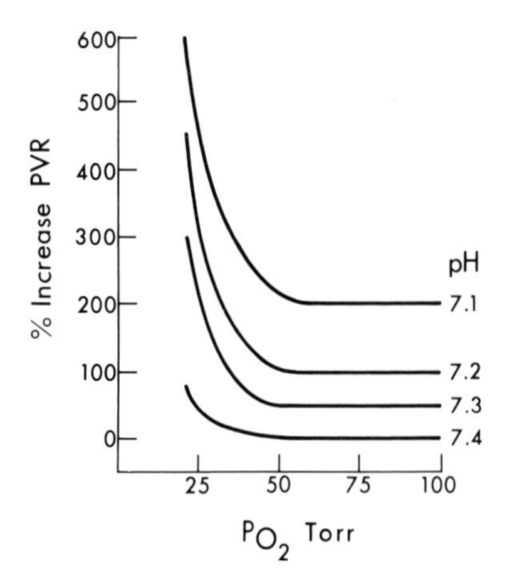

Fig. 16-5. The percentage increases in calculated pulmonary vascular resistance (PVR) in newborn calves are shown for changes in systemic arterial Po_2 and pH alone and in combination.

adult levels (systolic 20–25 mm Hg, diastolic 8–12 mm Hg, mean 12–15 mm Hg).[17,18] While the vessels still retain their smooth-muscle layer, they are very responsive to changes of Po_2 and pH.[19] The pattern of response is shown in Fig. 16-5. In those circumstances in which alveolar Po_2 is decreased, pulmonary vascular resistance does not fall normally after birth, and the regression of smooth muscle is delayed or even prevented. This phenomenon is noted in infants born at altitude, where the partial pressure of oxygen is reduced, as diagrammatically demonstrated in Fig. 16-4.[20–22] It is also noted in infants with ventilatory problems in whom alveolar Po_2 may be reduced.[23]

FACTORS INFLUENCING SHUNTING

The main factors determining the magnitude and direction of shunts in the circulation are: (1) the size of the communication between the chambers and vessels involved, (2) the pressure differences between the

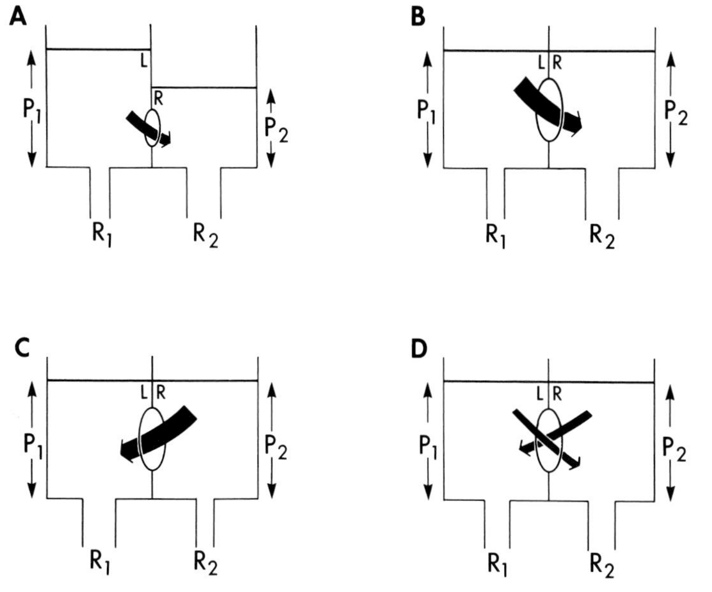

Fig. 16-6. Diagrammatic representation of the factors controlling shunting between two chambers. (Arrows indicate direction of shunt.) P_1 = pressure in left-sided chamber, L = left-sided chamber, R_1 = outflow resistance of left-sided chamber, P_2 = pressure in right-sided chamber, R = right-sided chamber, R_2 = outflow resistance of right-sided chamber.

chambers and vessels involved, and (3) the outflow resistances of the chambers. These three factors are frequently interrelated. In Fig. 16-6A the shunt through the small communication shown will be largely determined by the pressure difference, $P_1 - P_2$. The outflow resistances, R_1 and R_2, do not significantly alter the magnitude of the shunt except by influencing the height of the respective pressures. A large communication (Fig. 16-6B–D) will result in an equalization of pressures between the two chambers, so that P_1 equals P_2. Under these circumstances the shunting will be predominantly determined by the respective outflow resistances, R_1 and R_2. In Fig. 16-6B, where R_2 is lower than R_1, shunting will occur through the defect from left to right resulting in a greater flow out of the right side as compared to the left. In Fig. 16-6C, where R_2 is greater than R_1, shunting will occur from right to left, resulting in a decreased flow out of the right side as compared to the left. In Fig. 16-6C, where R_1 and R_2 are equal, either there is no shunt or both small left-to-right and right-to-left shunting may occur.

APPLICATION OF THESE PRINCIPLES TO THE CIRCULATION

Fetal and Neonatal Circulation

In the normal fetus the large ductus arteriosus produces an equalization of pressures in the aorta and pulmonary trunk and in the left and right ventricles. Pulmonary vascular resistance is very high, but systemic vascular resistance is low due to the presence of the low-resistance placental vascular bed. Blood is thus shunted from the pulmonary to the systemic circulation through the ductal communication. After birth the systemic vascular resistance is increased by elimination of the placental bed, and concomitantly the pulmonary vascular resistance is reduced. While the ductus arteriosus remains open, blood may be shunted from the aorta to the pulmonary artery. As the ductus arteriosus constricts, this shunting will progressively decrease.

Shunting in Congenital Cardiac Disease

In cardiac lesions in which there is an abnormal communication between the systemic and pulmonary circulations, blood is frequently shunted between them. It has been customary to designate shunts from the systemic to the pulmonary circulation as left-to-right shunts and shunts from the pulmonary to the systemic circulation as right-to-left shunts. Thus in a left-to-right shunt oxygenated blood will be recirculated through the lungs without entering the peripheral arterial circulation, whereas in right-to-left shunting systemic venous blood will recirculate to the systemic arteries without being oxygenated. In some complicated lesions shunts may occur in both directions. The application of these principles may be readily appreciated by considering an individual with a ventricular septal defect. In Fig. 16-6 the left ventricle may be represented by L and the right ventricle by R; P_1 and P_2 represent left and right ventricular pressures, respectively, while R_1 and R_2 indicate the outflow resistances of the left and right ventricles, respectively. In the normal child or adult the left ventricular systolic pressure is about four to five times that in the right ventricle, and the left ventricular outflow resistance as represented by systemic vascular resistance is about 20 times that of the pulmonary vascular resistance, which is the outflow resistance of the right ventricle. If a small ventricular septal defect is present, as in Fig. 16-6A, a small left-to-right shunt from the left to the right ventricle will occur. If a large ventricular septal defect is present, as in Fig. 16-6B–D, the pressures between the left and right ventricles will equalize. If the pulmonary vascular resistance is less than the systemic resistance, as in Fig. 16-6B, a left-to-right shunt will occur; whereas if right ventricular outflow resistance is greater, either due to severe pulmonic stenosis or a high pulmonary vascular resistance, right-to-left shunting will occur through the defect as in Fig. 16-6C. If left and right ventricular outflow resistances are equal, insignificant shunting will occur in either direction. In the absence of aortic or pulmonic stenosis, the relationship between pulmonary vascular resistance and systemic vascular resistance will determine the magnitude and direction of shunting. Since systemic vascular resistance is usually quite high and does not change very much, alterations in pulmonary vascular resistance will play the major role in the regulation of the shunt through the defect. On the basis of these considerations it has been possible to delineate two distinct types of left-to-right shunting patterns in patients with congenital heart lesions:[18] (1) The first is dependent shunting, in which shunting is determined by the ratio of the pulmonary to the systemic vascular resistance. This type of shunting occurs in many cardiac defects in which there are large communications between the aorta and the pulmonary artery (patent ductus arteriosus, aortopulmonary fenestration, truncus arteriosus communis), the left and right ventricles (ventricular septal defects, single ventricle, double-outlet right ventricle), or the left and right atria (atrial septal defect, single atrium).[2] The second type is an obligatory shunt, in which a communication exists between a high-pressure and a low-pressure chamber or vessel, such as in a peripheral arteriovenous fistula, a direct communication between the left

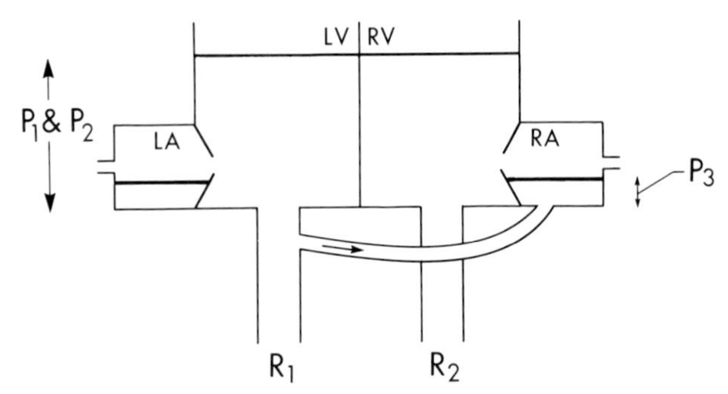

Fig. 16-7. Diagrammatic representation of factors controlling obligatory shunting in a communication between high- and low-pressure chambers. (Arrow indicates direction of shunt.) LV = left ventricle, RV = right ventricle, LA = left atrium, RA = right atrium, P_1 and P_2 = pressures in left ventricle and right ventricle, P_3 = pressure in left atrium and right atrium, R_1 = left ventricular outflow resistance, R_2 = right ventricular outflow resistance.

ventricle and right atrium, or a communication between the aorta and the right atrium (sinus of Valsalva fistula). In this latter group of defects the pulmonary vascular resistance is not important in determining the magnitude of the shunt (Fig. 16-7).

Dependent Shunts

Aortopulmonary Communications

When a large communication with a diameter equal to or larger than the aortic valve diameter exists between the aorta and the pulmonary artery (Fig. 16-8), aortic and pulmonary arterial systolic and diastolic pressures will be equal. The respective flows through the pulmonary and systemic circulations will therefore be determined by the relationship between systemic and pulmonary vascular resistances. As pulmonary vascular resistance falls after birth, a progressively increasing left-to-right shunt through the defect will occur, and pulmonary blood flow will increase. This results in an increase in venous return to the left atrium, with an increase in left atrial and left ventricular end-diastolic volume. In the presence of this left-to-right shunt the left ventricular stroke volume will increase in an attempt to maintain an adequate systemic blood flow. If pulmonary venous return is markedly increased the left ventricle may not be capable of ejecting the large addi-

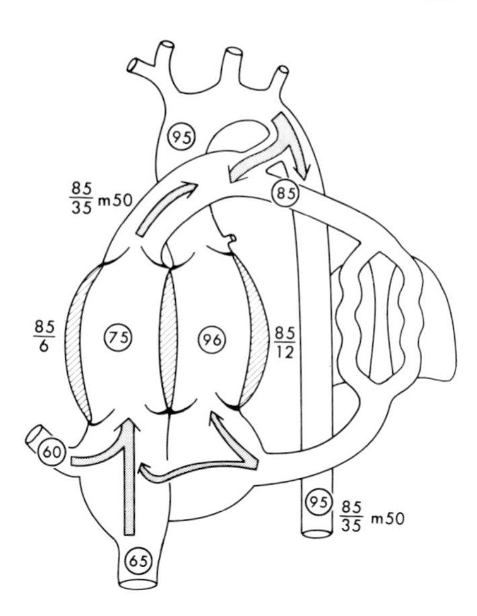

Fig. 16-8. Diagram of heart and great vessels in an infant with a patent ductus arteriosus. Arrows indicate direction of flow, and representative saturations (in circles) and pressures are shown.

tional volume load placed on it, and marked elevation of left ventricular end-diastolic and left atrial pressures will occur, resulting in left-sided failure and pulmonary edema. The clinical manifestations of these changes are varied. There may be a progressive increase in precordial activity due to the large volume handled by the heart, and this will be predominantly left ventricular hyperactivity. A murmur that is variable in character may develop. Usually a systolic murmur is first evident, and with an increasing shunt the murmur extends into diastole and becomes louder. However, with very large communications large shunts may occur with soft or even absent systolic murmurs, since little turbulence is created in the presence of a large communication. The increased flow across the normal mitral valve frequently causes an audible mid-diastolic rumble due to relative mitral stenosis. On chest roentgenography overall heart size increases, pulmonary vascular markings progressively increase, the left atrium enlarges, and the ascending aorta may also enlarge. Since there is a large diastolic flow from the systemic to the pulmonary circulation, the systemic arterial as well as pulmonary arterial diastolic pressures are decreased and pulse pressure is widened. As left ventricular failure develops, the infant shows evidence of increasing respiratory distress associated with pulmonary edema. Later evidence of right-sided failure may occur and be manifested by hepatomegaly. The presence of a large communication allows pulmonary arterial pressures to equal systemic arterial pressures, and since pulmonary arterial pressures are elevated to systemic levels, a loud

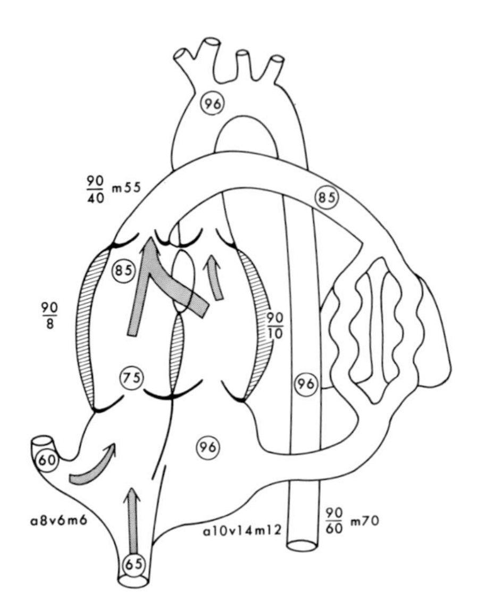

Fig. 16-9. Diagram of heart and great vessels in an infant with a ventricular septal defect. Arrows indicate direction of flow, and representative saturations (in circles) and pressures are shown (a = a wave, v = v wave, m = mean pressure).

pulmonic component of the second sound is heard. The pulmonary hypertension is not necessarily a reflection of pulmonary vascular resistance elevation, but could be due to the large communication. The magnitude of the shunt is dependent on the pulmonary-to-systemic vascular resistance ratio, and an actual pressure difference of large size between the two ventricles or aorta and pulmonary artery is not necessary for shunting to occur.

Ventricular Communications

In the presence of a large ventricular communication (Fig. 16-9)[24] equal to or greater than the aortic valve orifice, right and left ventricular systolic and diastolic pressures are equalized. Pulmonary and systemic arterial systolic pressures are also equal, but diastolic pressures may differ. The sequence of events following birth as pulmonary vascular resistance falls is similar to that described for aortopulmonary communications. An increased shunt develops, resulting in cardiac failure that is first predominantly left ventricular, and right ventricular failure occurs later. Cardiac activity involving both left and right ventricles increases, and a systolic murmur that becomes progressively louder and longer is heard. A mid-diastolic rumble is heard at the apex, and similar radiologic features develop, but the aorta is not dilated. Although aortic and pulmonary arterial systolic pressures are equal, systemic diastolic pressure is not decreased. Pulmonary arterial diastolic pressure is increased to a level similar to that in the aorta when pulmonary vascular

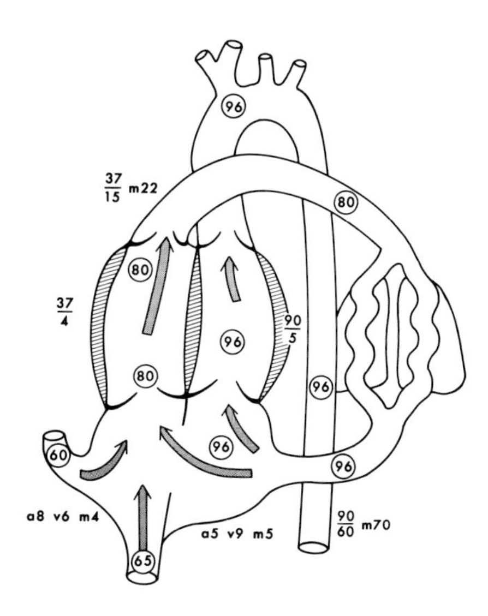

Fig. 16-10. Diagram of heart and great vessels in an infant with an atrial septal defect.

resistance is high, but tends to be reduced when pulmonary vascular resistance is decreased. The pulmonic component of the second sound is accentuated, and as with systemic–pulmonary arterial communications the increased pulmonary second sound reflects the presence of pulmonary hypertension and not the status of the pulmonary vascular resistance in these patients.

Atrial Septal Defects

A large atrial septal defect (Fig. 16-10)[25] results in equalization of pressures in the left and right atria. Since the atria empty through the atrioventricular valves into the respective ventricles, the volume of blood flowing into each ventricle will be related to at least two factors: (1) the compliance of the ventricles and (2) the diastolic pressures in each ventricle. At the time of birth the left and right ventricles have similar wall thicknesses, and it has been assumed that they have similar compliances (that is, that they show similar pressure–volume characteristics). It had generally been thought that because left and right ventricular wall thicknesses were similar at birth, insignificant shunting would occur through an atrial septal defect; however, cardiac catheterization studies in infants with atrial septal defects have shown significant left-to-right shunts soon after birth.[25] This is almost certainly related to the difference in end-systolic volumes and pressures in the two ventricles. In the presence of a large atrial septal defect the filling pressures of the two ventricles during diastole would be essentially equal, and if compliance characteristics were the same they would fill to the same extent.

However, it has been shown that the degree of emptying of a ventricle during systole is related to afterload or outflow resistance on that ventricle.[26] As pulmonary vascular resistance falls after birth, the right ventricular afterload will fall, and in the presence of an atrial septal defect the volume of blood ejected by the right ventricle will increase. This will result in a lower end-systolic volume and pressure in the right ventricle as compared with the left, with greater filling of the right ventricle in early diastole, and shunting from left to right atrium. Pulmonary arterial and right ventricular pressures will fall normally. The fall in right ventricular pressure is associated with a relative thinning of the right ventricular muscle, as compared to the left ventricular muscle, and a gradual increase in right ventricular compliance relative to the left. At the same filling pressure, therefore, there will also be considerably greater filling of the right ventricle, and blood will therefore preferentially flow from the left to the right atrium. Thus in patients with atrial septal defects the development of shunting is related to the changes in pulmonary vascular resistance, and they may also be considered to have dependent shunts. The clinical manifestation associated with these changes is an increase in right ventricular activity due to the larger volume ejected by the right ventricle. Loud murmurs do not originate at the site of the atrial septal defect because of the low pressure and relatively nonturbulent flow across the defect. The large right ventricular stroke volume creates a rapid flow across the normal pulmonary valve, with the production of a systolic ejection murmur. The increased flow across the tricuspid valve may also result in the production of a mid-diastolic rumble. The second sound in the pulmonic area becomes widely split and remains well split throughout the respiratory cycle. This possibly has been explained on the basis of the increase in right ventricular stroke volume throughout the respiratory cycle. Cardiomegaly involving predominantly the right ventricle and the main pulmonary artery is noted on chest roentgenography, and the pulmonary vascular markings are increased. The pulmonary arteries are not, however, as large as in patients with ventricular septal defects or aortopulmonary arterial communications, in whom pulmonary arterial hypertension is also present.

Obligatory Shunts

The presence of a communication between a high- and low-pressure chamber or vessel results in shunting independent of the pulmonary vascular resistance and changes in the pulmonary vascular resistance, or the relationship of the pulmonary to the systemic vascular resistance.

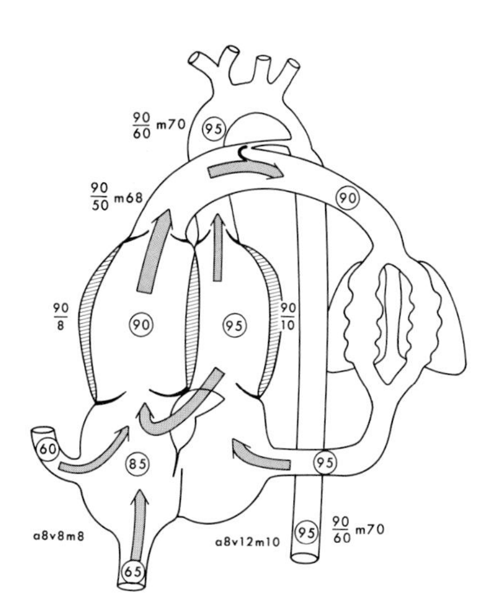

Fig. 16-11. Diagram of heart and great vessels in an infant with an endocardial cushion defect.

This may be readily appreciated in peripheral arteriovenous communications such as cerebral or hepatic arteriovenous fistulae. This type of shunt may also occur in certain intracardiac communications such as a fistula between the left ventricle and the right atrium. It also occurs commonly in patients with partial or complete atrioventricular canal defects (Fig. 16-11). In ostium primum defects or partial atrioventricular canal defects blood may be ejected from the left ventricle in systole through a cleft mitral valve and a low atrial defect directly into the right atrium. In atrioventricularis communis or complete atrioventricular canal defects blood may be ejected directly from the left ventricle through the common atrioventricular orifice into the right atrium. In these lesions the shunting is of the obligatory type. It is possible for both obligatory and dependent shunting to be present in the same patient with an atrioventricular canal defect, in that shunting from the left ventricle to the right atrium is obligatory, whereas shunting directly from the left ventricle to the right ventricle occurring across the ventricular septal defect will be dependent. The clinical manifestations of obligatory shunts are not different from dependent shunting and are related to the level at which the shunt occurs.

Circulatory Responses to Left-to-Right Shunts: Myocardial Responses

In the presence of an aortopulmonary or ventricular communication a left-to-right shunt removes a proportion of left ventricular output from the systemic circulation. In an attempt to maintain a normal systemic blood flow left ventricular output is increased, and in large left-to-right shunts it may increase threefold to fourfold. The work of the left ventricular myocardium is thus markedly increased. The sympathetic adrenal system is very important in stimulating the myocardium to respond to these increased demands. In the adult the myocardium is richly innervated with sympathetic nerves that release norepinephrine locally in response to stress. It has been shown that patients with cardiac failure have an increased catecholamine activity as reflected by elevated urinary excretion of metabolic by-products of catecholamine utilization.[27] The importance of the sympathetic adrenal system in adaptation to left-to-right shunting has also been demonstrated in animals with experimentally produced left-to-right shunts.[28,29] A shunt that may be well tolerated with an intact sympathetic adrenal system frequently produces severe cardiac failure after administration of beta adrenergic or ganglionic blocking agents. In many species sympathetic innervation to the myocardium is either not developed or incompletely developed at the time of birth.[30] The myocardium in the newborn infant may therefore be less able to respond to an equivalent stress than that of the adult; furthermore, the myocardium in the premature infant may be even more limited in its ability to respond to stress of a large left-to-right shunt. The responses of the myocardium in premature infants may be even further limited by structural differences, as compared to full-term or adult myocardium.[31] This is corroborated by the fact that the fetal heart is not capable of increasing its output in response to volume loading to the same extent as is the adult.[31,32]

The time course over which the left ventricular myocardium is exposed to volume overloading will determine the type and extent of response of the myocardium. If the load is presented rapidly, cardiac failure is very likely to occur. If presented more gradually, the myocardium will have the opportunity to develop hypertrophy and/or hyperplasia, so that it will be capable of maintaining a higher output without failure. Thus in infants with dependent shunts the rate of decline of the pulmonary vascular resistance, which will determine the development of left-to-right shunting, will also be crucial in determining the sequence of events in the myocardium. It is therefore possible that differences in the amount of smooth muscle in the pulmonary arterioles of premature infants as compared to mature infants may be one of the factors accounting for the onset of early failure in premature infants with large left-to-right shunts (Fig. 16-12). A lesser amount of smooth muscle in the pulmonary arterioles in the premature infant may permit pulmonary vascular resistance to reach adult levels more rapidly, thus resulting in earlier development of maximal dependent shunting.

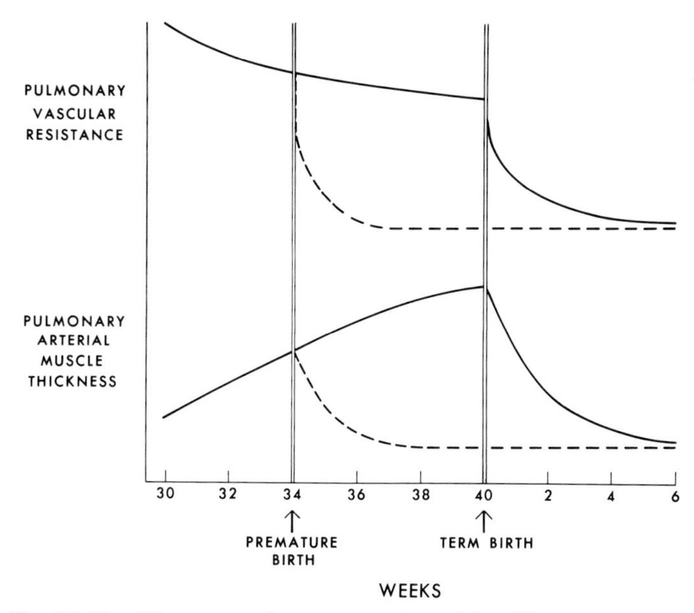

Fig. 16-12. Diagrammatic representation of the difference between prematurely born and full-term infants in the postnatal changes in calculated pulmonary vascular resistance and amount of smooth muscle in the pulmonary arterioles.

The increased work load on the ventricle is associated with an increase in the oxygen requirements of the myocardium. Oxygen supply is determined by coronary blood flow and oxygen content of arterial blood. Since most coronary blood flow to the left ventricle occurs during diastole, a lowering of aortic diastolic pressure, as occurs in aortopulmonary shunts, may interfere with adequate myocardial oxygenation. Also, when left ventricular end-diastolic pressure is markedly increased, coronary blood flow may be impaired.[33] An additional factor that may be important in infants with cardiac failure is a reduction in the duration of diastole due to marked increases in heart rate that occur with cardiac failure.[33] Also, oxygen content of arterial blood normally falls after birth in relationship to the physiologic decreases in hemoglobin levels. This may be an additional factor in preventing the achievement of adequate myocardial oxygenation in response to stress. Should anemia develop, as frequently occurs in premature infants, myocardial oxygen supply may be seriously reduced under conditions of stress. Anemia may not only directly interfere with myocardial oxygen supply but may place a further demand on the heart to increase its output requirements to supply adequate oxygen to the systemic circulation. Similarly, infection, which in normal infants is associated with an increase in left ventricular output, may place further demands on an already compromised heart and precipitate cardiac failure in infants with left-to-right shunts.

Metabolic Disturbances Associated with Cardiac Failure in Infants

Recently it has been appreciated that infants, and particularly premature infants with cardiac failure, are very prone to develop hypoglycemia and hypocalcemia.[34] The mechanisms responsible for these disturbances are not yet clearly understood. Hypoglycemia may be related to increased insulin secretion due to alpha adrenergic stimulation of pancreatic islet cells by the norepinephrine released in response to stress. It is also possible that prolonged adrenal stimulation, which produces glucose release into the circulation, may cause rapid depletion of glycogen stores, particularly in premature or dysmature infants. Hypocalcemia has been frequently observed, but no adequate explanation for its occurrence in premature infants has yet been discovered. Severe hypocalcemia or hypoglycemia may seriously interfere with myocardial function and thus with ventricular output.

Effects of Shunt Lesions on Pulmonary Circulation

In patients who have a large aortopulmonary or ventricular communication after birth, pulmonary arterial and right ventricular systolic pressures will be maintained at or near systemic levels due to equalization of pressure across the defect. Probably in part because of the persistent high pulmonary arterial pressure, the pulmonary arterioles do not undergo the normal maturational changes after birth. There is some degree of thinning of the smooth-muscle coat in the medial layer of the arterioles, but it does not regress as rapidly or to the same extent as normal (Fig. 16-13). In these infants, pulmonary vascular resistance falls rapidly immediately after birth, due to the release of vasoconstriction associated with the low P_{O_2} to which the vessels are exposed in utero. Following this there is a more gradual decline of pulmonary vascular resistance, as compared with the normal, so that it reaches its low level after only 3–4 months, in contrast to the normal 1–2 months. The pulmonary vascular resistance also does not reach the same level achieved in the normal infant, but remains about three to four times the normal level. In view of the time sequence of changes in pulmonary vascular resistance, the left-to-right shunt progressively increases, and the onset of cardiac failure may be delayed until 2–3 months after birth.

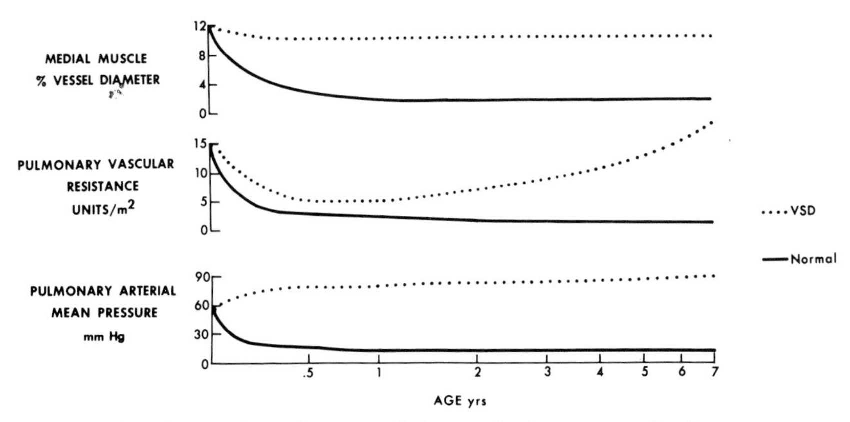

Fig. 16-13. Comparison of postnatal changes in the amount of pulmonary arteriolar medial muscle, calculated pulmonary vascular resistance, and pulmonary arterial mean pressure in a child with a ventricular septal defect and a normal child.

In infants with atrial septal defects there is usually a normal reduction of pulmonary arterial pressure and pulmonary vascular resistance and a normal regression of pulmonary vascular smooth muscle after birth.

Patients with obligatory shunts will have abnormally high pulmonary blood flows immediately after birth. There is considerable experimental evidence to indicate that when a high blood flow is presented to the lungs after birth the regression of pulmonary vascular smooth muscle is delayed, and there is thus a slower than normal reduction of pulmonary vascular resistance and pulmonary arterial pressure. This is also noted in infants with large obligatory shunts. In infants with aortopulmonary or ventricular communications, or with obligatory shunts, in whom the pulmonary vascular resistance is increased, initially the medial smooth-muscle layer is thickened, but the intima is normal. Later intimal thickening occurs, due at first to cellular proliferation. These changes progress and involve an increasing number of vessels and are later associated with thrombosis of small vessels, with complete or almost complete occlusion of the lumina. The thrombi become organized, and hyalinization and fibrosis of the vessels follow. A progressive increase in pulmonary vascular resistance occurs, so that the left-to-right shunt decreases and evidences of left ventricular failure diminish. As pulmonary vascular resistance rises above systemic levels, shunting from right to left occurs, resulting in cyanosis that progressively increases with further obliterative changes in the pulmonary vessels. When these changes are severe, exercise tolerance is reduced because severe hypoxemia develops, the patient may have syncope, and later right heart failure develops. In severe pulmonary vascular obstructive disease, thin-walled arteriovenous shunts may develop, and hemoptysis may occur if they rupture. The time course over which these changes occur is very variable. It is unusual to find marked intimal proliferative changes during the first year after birth, but after this there is an increasing involvement of the small pulmonary vessels. Some infants may have marked pulmonary vascular obstruction during the second year, and yet other patients may not develop severe changes until late adolescence or early adult life. We do not know all the factors that influence the rapidity with which these changes occur.

In patients with atrial septal defect, who have normal maturation of the muscular component of the pulmonary vessels, intimal proliferative changes rarely occur in infancy or childhood. There is, however, an increasing incidence of pulmonary vascular obstruction beyond the age of 20 years in these individuals. The time course of these changes is also very variable. Rarely, severe degrees of pulmonary vascular obstruction may occur by the age of 8–10 years; in adults with large atrial septal defects, marked obliterative changes may be delayed to the age of 40–50 years, and in some individuals they may not be apparent even at age 60 years.

PRINCIPLES OF TREATMENT

Based on the pathophysiology of left-to-right shunt lesions described above, certain important principles of management are apparent. During the infancy period the main clinical manifestations of these lesions are those due to cardiac failure. The infant with cardiac failure should be actively treated with digitalis glycosides and diuretics and, if necessary, restriction of sodium in the diet by providing a milk formula low in sodium content. If the cardiac failure cannot be controlled by these measures, however, the failure can be treated by reducing or eliminating the left-to-right shunt. In lesions such as patent ductus arteriosus, surgery can be performed with very low risk in infancy, and the defect can be corrected. In some centers, ventricular septal defects can be closed with little risk in infants, and the defect can be corrected. However, when the lesion is more complicated and cannot be corrected in infancy or when there is a high risk of corrective surgery in infancy, another approach can be used to alleviate cardiac failure when it is associated with lesions in which the left-to-right shunt is of the dependent type. The procedure of banding the pulmonary artery consists of constricting the main pulmonary artery by a broad ligature of Teflon or umbilical tape. This increases the total outflow resistance of the right ventricle and has the same effect as increasing pulmonary vascular resistance. The left-to-right shunt and pressure in the pulmonary arteries distal to the band are reduced, and the pulmonary vascular obliterative changes are prevented.

Banding of the pulmonary artery is not appropriate in treatment of infants with obligatory shunts. Since the shunt is from a very high to a low-pressure chamber, the level of pulmonary vascular resistance is not of great importance in determining the magnitude of the left-to-right shunt. Banding of the pulmonary artery will thus not significantly reduce the shunt and will also produce an increase in impedance on the right ventricle, increasing its pressure further; this may result in rapid onset of right ventricular failure. Thus in infants with lesions that produce obligatory shunts, if medical management is not successful a direct approach to treating the lesion surgically will have to be undertaken.

One of the main concerns in managing patients with systemic–pulmonary communications is the development of an increased pulmonary vascular resistance. There is now considerable experience indicating that when an elevated pulmonary vascular resistance is due to thickening of the medial muscle layer alone reduction of the pulmonary arterial pressure and blood flow will allow the resistance to drop to normal or near normal levels. However, when extensive intimal proliferation and luminal obstruction is present, closing the communication will not significantly influence the high resistance. In patients with large systemic pulmonary communications it is thus most important to determine whether pulmonary vascular obstructive changes are developing, so that severe organic changes can be prevented, either by treating the underlying cardiac lesion by surgery or, if this is not possible, by banding the pulmonary artery in the patient with dependent shunting. In those patients in whom there is a high risk of corrective surgery in early life, repeated cardiac catheterization may have to be performed at 9–12-months intervals in order to evaluate the state of the pulmonary circulation, because clinical evaluation may not be reliable.

REFERENCES

1. Dawes GS, Mott JC, Widdicombe JG: The foetal circulation in the lamb. J Physiol (Lond) 126:563, 1954
2. Rudolph AM, Heymann MA: The fetal circulation. Ann Rev Med 19:195, 1968

3. Heymann MA, Creasy RK, Rudolph AM: Quantitation of blood flow pattern in the foetal lamb in utero, Comline KS, Cross KW, Dawes GS, Nathanielsz PW (eds) In: Foetal and Neonatal Physiology. Proceedings of the Sir Joseph Barcroft Centenary Symposium. Cambridge, Cambridge University Press, 1973, p 129
4. Rudolph AM, Heymann MA: Circulatory changes during growth in the fetal lamb. Circ Res 26:289, 1970
5. Heymann MA, Rudolph AM: Effects of congenital heart disease on fetal and neonatal circulations. Prog Cardiovasc Dis 15:115, 1972
6. Reynolds SRM: Fetal and neonatal pulmonary vasculature in guinea pigs in relation to hemodynamic changes at birth. Am J Anat 98:97, 1956
7. Dawes GS, Mott JC: The vascular tone of the foetal lung. J Physiol (Lond) 164:465, 1962
8. Cook CD, Drinker PA, Jacobson NH, et al: Control of pulmonary blood flow in foetal and newly born lamb. J Physiol (Lond) 169:10, 1963
9. Lauer RM, Evans JA, Aoki H, Kittle CF: Factors controlling pulmonary vascular resistance in fetal lambs. J Pediatr 67:568, 1965
10. Cassin S, Dawes GS, Mott JC, et al: The vascular resistance of the foetal and newly ventilated lung of the lamb. J Physiol (Lond) 171:61, 1964
11. Rudolph AM, Heymann MA: Pulmonary circulation in fetal lambs. Pediatr Res 6:341, 1972
12. Staub NC: Site of action of hypoxia on the pulmonary vasculature. Fed Proc 22:453, 1963
13. Oberhansli-Weiss I, Heymann MA, Rudolph AM, Melmon KL: The pattern and mechanisms of response to oxygen of the ductus arteriosus and umbilical artery. Pediatr Res 6:693, 1972
14. McMurphy DM, Heymann MA, Rudolph AM, Melmon KL: Developmental changes in constriction of the ductus arteriosus: Responses to oxygen and vasoactive substances in the isolated ductus arteriosus of the fetal lamb. Pediatr Res 6:231, 1972
15. Naeye RL: Arterial changes during the perinatal period. Arch Pathol 71:121, 1961
16. Phillips CE Jr, DeWeese JA, Manning JA, Mahoney EB: Maturation of small pulmonary arteries in puppies. Circ Res 8:1268, 1960
17. Rudolph AM, Auld PAM, Golinko RJ, Paul MH: Pulmonary vascular adjustments in the neonatal period. Pediatrics 28:28, 1961
18. Rudolph AM: The changes in the circulation after birth: Their importance in congenital heart disease. Circulation 41:343, 1970
19. Rudolph AM, Yuan S: Response of the pulmonary vasculature to hypoxia and H+ ion concentration changes. J Clin Invest 45:399, 1966
20. Arias-Stella J, Saldana M: The muscular pulmonary arteries in people native to high altitude. Med Thorac 19:292, 1962
21. Sime F, Banchero N, Penaloza D, et al: Pulmonary hypertension in children born and living at high altitudes. Am J Cardiol 11:150, 1963
22. Penaloza D, Sime F, Banchero N, et al: Pulmonary hypertension in healthy men born and living at high altitudes. Am J Cardiol 11:150, 1963

23. Naeye RL, Letts HW: The effects of prolonged neonatal hypoxia on the pulmonary vascular bed and heart. Pediatrics 30:902, 1962
24. Hoffman JIE: Ventricular septal defect: Indications for therapy in infants. Pediatr Clin North Am 18:1091, 1971
25. Hoffman JIE, Rudolph AM, Danilowicz D: Left to right atrial shunts in infants. Am J Cardiol 30:868, 1972
26. Wilcken DEL, Charlier AA, Hoffman JIE, Guz A: Effects of alterations in aortic impedance on the performance of the ventricles. Circ Res 14:283, 1964
27. Chidsey CA, Braunwald E, Morrow AG: Catecholamine excretion and cardiac stores of norepinephrine in congestive heart failure. Am J Med 39:442, 1965
28. Rudolph AM, Mesel E, Levy JM: Epinephrine in the treatment of cardiac failure due to shunts. Circulation 28:3, 1963
29. Rudolph AM: The effects of postnatal circulatory adjustments in congenital heart disease. Pediatrics 36:763, 1965
30. Lipp JA, Rudolph AM: Sympathetic nerve development in the rat and guinea-pig heart. Biol Neonate 21:76, 1972
31. Friedman WF: The intrinsic physiologic properties of the developing heart, in Friedman WF, Lesch M, Sonnenblick EH (eds): Neonatal Heart Disease. New York, Grune & Stratton, 1973, p 21
32. Heymann MA, Rudolph AM: Effects of increasing preload on right ventricular output in fetal lambs in utero. Circulation 48:IV–37, 1973
33. Buckberg GB, Fixler DE, Archie JP, Hoffman JIE: Experimental subendocardial ischemia in dogs with normal coronary arteries. Circ Res 30:67, 1972
34. Benzing G III, Schubert W, Hug G, Kaplan S: Simultaneous hypoglycemia and acute congestive heart failure. Circulation 40:209, 1969

David H. Spodick M.D., D.Sc. (Hon.)

17
Pathophysiology of Disorders of the Pericardium

The pericardium normally facilitates cardiac action, limits responses to certain undue hemodynamic loads, and shields the heart from external trauma (Table 17-1). Disease may compromise these functions, and at the worst it may convert the pericardium from being the protector of the heart to its deadly enemy.

THE NORMAL PERICARDIUM

Pericardial anatomy, normal physiology, and their important interrelationships are considered in detail elsewhere.[1,2] Briefly, the pericardium comprises a thin semitransparent but tough fibrous jacket (*fibrosa*) externally clasping a delicate monocellular sac (*serosa*) that envelops the heart and juxtacardiac portions of the great vessels and contains 20 to 50 ml of serous fluid. Serosal mesothelial cells permit rapid transfer of ions and water and slower transport of large molecules between the blood and pericardial fluid. The *fibrosa* varies in thickness in inverse proportion to the underlying heart chamber walls and has remarkably specialized organization of its fibers in mechanical support of the atria, the aortic arch, and other structures. The fibrosa with the serosa applied to it (*parietal pericardium*) are rather sparsely supplied with lymphatics, whereas the portion of the serosa applied to the heart (*epicardium, visceral pericardium*) overlies rich epicardial lymphatic networks. *Elastic tissue,* varying in amount with age, permits limited active stretching of the parietal pericardium during rapid increases in the volume of the heart and/or pericardial fluid.

Table 17-1
Functions of the Pericardium

- Mechanical function: Promotion of cardiac efficiency, especially during stress
  - Limitation of excessive acute dilatation
    - Defense of the integrity of the Starling curve
      - Maintenance of output response to venous inflow loads
      - Protection against excessive acute ventriculoatrial regurgitation
        - Maintains ventricular function curves
        - Limits effect of increased left ventricular end-diastolic pressure
        - Favors equality of transmural end-diastolic pressure throughout ventricle, therefore uniform stretch of muscle fibers (preload)
      - Limits right ventricular stroke work during increased resistance to left ventricular outflow
    - Maintenance of output response to rate fluctuations
  - Maintenance of functionally optimum heart geometry
  - Provision of a closed chamber:
    - In which pressure changes aid the intracardiac flow of blood
    - Which applies compensated hydrostatic pressures on all external cardiac surfaces during changes in gravitational and inertial forces
  - ? mutually restrictive chamber favoring equality of output from right and left ventricles over several beats
  - Maintenance of normal ventricular compliance (volume–elasticity relationship)
  - Limits hypertrophy associated with chronic exercise
- Membrane function: Shielding the heart
  - Reduction of external friction due to heart movements
  - Barrier to inflammation from contiguous structures
  - Buttressing of thinner portions of the myocardium
    - Atria
    - Right ventricle
  - Feedback circulatory regulation; stimulation:
    - Neuroreceptors (via vagus)—lower heart rate and blood pressure
    - Mechanoreceptors—lower blood pressure and contract spleen
- Ligamentous function: Optimum functional position of the heart; limitation of undue cardiac displacement

Adapted by permission from Spodick DH: Chronic and Constrictive Pericarditis. New York, Grune & Stratton, 1964.

Although human beings and animals tolerate well the removal of the pericardium, experimental pericardiotomy and pericardiectomy have demonstrated the functions summarized in Table 17-1. In particular, ventricular pressure–circumference and pressure–volume curves (mainly determined by apparent myocardial compliance) show higher pressure levels for a given circumference or volume with an intact pericardium.[3] Tight pericardial scarring or excessive pericardial fluid exaggerate this normal effect, yielding higher pressure levels at any ventricular volume.[4] During experimental tamponade, induced hypervolemia increases cardiac output while raising the pericardial pressure–volume curve, giving higher intrapericardial pressures for equal increments of fluid added to the sac.[5]

THE DISEASED PERICARDIUM: GENERAL CONSIDERATIONS

Many experimental studies have been carried out in dogs. In the earlier work the effects of anesthesia and use of open-chest animals may have modified responses. Moreover, the dog lacks the pericardial "sleeves" found on the pulmonary veins of humans. Yet most animal results seem applicable to the pathologic physiology of human pericardial disease. Classic contributions are reviewed elsewhere.[1,2] Our current understanding owes much to the work of Isaacs,[4] Shabetai,[5] Fowler,[6] Guntheroth,[7] and Dock[8] and their associates.

Physiologically significant pericardial disorders occur as a result of particular effects of pericardial inflammation or trauma: intrapericardial accumulation of excessive fluid (liquid/gas) and constrictive pericardial scarring (adhesions/fibrosis). Within limits, these have no detectable effects unless they cause external compression (by fluid or scar) or entrapment by scarring of all or parts of the heart and juxtracardiac great vessels. Effusion and scarring may coexist, and less frequently either one may be localized. The present discussion will be confined to "pure" generalized forms of excessive fluid accumulation (cardiac tamponade) and constrictive scarring.

CONSTRICTIVE PERICARDITIS

Although there are very few laboratory models[4] of constrictive pericarditis, the physiology of cardiac constriction has been studied in many more patients than has that of cardiac tamponade, for which experimental models abound. Most hemodynamic studies have yielded uniform results in the typical[2] forms of constrictive pericarditis (Table 17-2) in which there is more or less dense adherence of the scarred pericardium over most of the cardiac surface. This scar markedly decreases the compliance of the physically coupled myocardium and pericardium (Fig. 17-1), resulting in an abrupt end to filling of both ventricles in early

Table 17-2
Hemodynamics in Constrictive Pericarditis

	Pressure (mm Hg)	
	Range	Modal Range
Right ventricular		
Systolic	15–50	25–35
Dip (nadir)	2–20	5–10
End-diastolic (plateau)	10–35	15–20
Pulse		15–20
Ratio EDP/SP		> 1.3
Right atrial		
Mean	6–28	10–20
a (or ac)	8–30	10–15
z		1.4 mm below a
c		7–15
x	4–15	5–10
v	6–28	8–15
y	2–12	3–8
Pulmonary artery		
Systolic	20–45	30–40
Diastolic	10–30	15–25
Mean	15–35	20–30
Pulmonary "capillary" mean	8–30	15–20
Stroke volume (resting)	20–95	30–60
Cardiac output (resting)	2–7.5	2.5–4.5
Arteriovenous oxygen difference		5–9 ml/100 ml

Adapted by permission from Spodick DH: Chronic and Constrictive Pericarditis. New York, Grune & Stratton, 1964.

diastole. This results in a brief characteristic early diastolic dip in ventricular pressure curves permitting torrential, but limited, rapid filling (Fig. 17-2). Filling is halted prematurely by the unyielding myocardial-pericardial wall, so that pressure then rises steeply (the "square-root" configuration) to a high mid-to-end-diastolic plateau during which there is virtual equilibration of ventricular pressures with atrial and central venous pressures. Systolic pressure may be high or normal, so that ventricular pulse pressure tends to be reduced. In the right ventricle the plateau often exceeds one-third of systolic pressure (Table 17-2). The steepness of the upstroke from the dip and the level of the plateau pressure reflect the degree of cardiac compression. Because of the tendency to equal involvement, the diastolic portions of the right and left

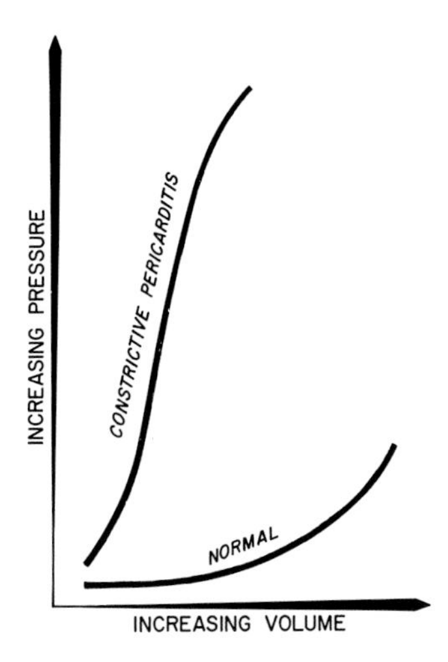

Fig. 17-1. Schema of ventricular (or pericardial) pressure–volume relationships. The brusque ascent of the curve during constrictive pericarditis is the result of the tightening and loss of elasticity of the scarred pericardium. By contrast, with a normal pericardium volume increments are first accommodated (flat portion), and cause later less-sharp pressure increments. (Reproduced by permission from Spodick DH: Chronic and Constrictive Pericarditis. New York, Grune & Stratton, 1964.)

ventricular pressure curves are superimposable (allowing 20 to 30 msec time difference).[2] During sinus rhythm, a brief presystolic ventricular pressure rise due to atrial contraction into the low-compliance ventricles probably contributes little to the strength or output of the ensuing systole. Stroke volume is limited, since little or no filling occurs during the plateau, so that ventricular systolic pressure and stroke volume tend to be independent of the duration of the preceding diastole (Fig. 17-2). Minute cardiac output depends directly on heart rate, since tachycardia

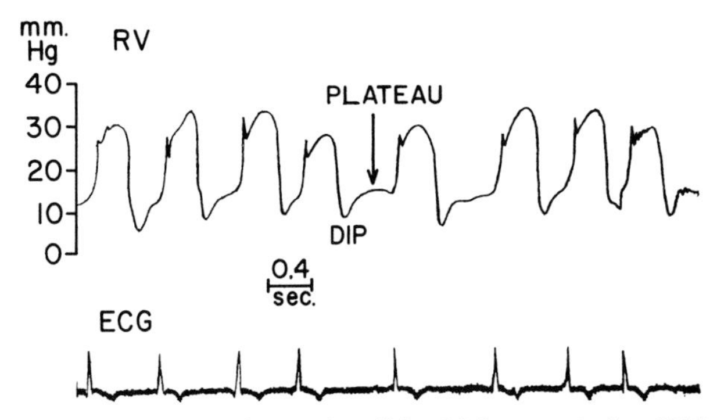

Fig. 17-2. Constrictive pericarditis: Right ventricular (RV) pressure in a patient with atrial fibrillation. Early diastolic pressure dip (rapid filling phase) followed by mid-to-late-diastolic plateau, during which there is little or no filling. Faster heart rates (shorter cycles) amputate the plateau phase at little or no expense to filling. (Reproduced by permission from Spodick DH: Chronic and Constrictive Pericarditis. New York, Grune & Stratton, 1964.)

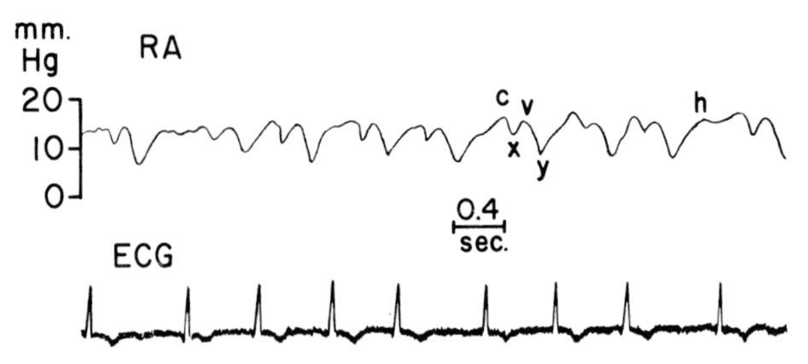

Fig. 17-3. Constrictive pericarditis: Right atrial (RA) pressure curve during atrial fibrillation (no a wave). M configuration owing to high-pressure c and v waves with x and very sharp y troughs; h shoulder in longer diastoles. (Reproduced by permission from Spodick DH: Chronic and Constrictive Pericarditis. New York, Grune & Stratton, 1964.)

(short of extreme rates) amputates only the nonfilling plateau phase ("diastasis") while multiplying the number of strokes and (rapid) filling periods.

Constriction may include variable myocardial impairment owing to previous myocarditis, involvement of muscle and coronary vessels by scar, myocardial atrophy (in chronic constriction), or unrelated heart disease. Burch[9] has suggested that the myocardium may perform useless work in systole by "spring-loading" the system, i.e., deforming the thickened pericardium, which undergoes an abrupt early diastolic elastic recoil. Following systole the combined myocardial-pericardial heart wall may thus create the dip by springing back like a rubber bulb, which might aid ventricular filling by a suction effect.[2]

Atrial and caval pressure curves (Fig. 17-3) show characteristic "W" or "M" contours dominated by high-pressure *a, c* and *v* waves with conspicuous *x* and especially *y,* descents. The *x* descent coincides with the onset of ventricular ejection. The *y* trough virtually coincides with the nadir of the ventricular dip; its steepness reflects the rate of rapid filling, and the subsequent rate of reascent reflects the brusque rise of the ventricular pressure from the nadir of the dip. Forward venoatrial flow[5] occurs during the *x* and *y* descents. Venoatrial and atrioventricular pressure *gradients* tend to be like those in normal hearts—almost unmeasurable or only a few millimeters of mercury. Hepatic and portal venous pressures tend to parallel the caval pressures. Pulmonary artery (PA) pressures tend to be at most moderately elevated, even with exercise—the diastolic pressure being proportionately greater and approximating the "plateauing" of end-diastolic pressures throughout the central circulation[2] (Table 17-2).

Respiratory Effects and Pulsus Paradoxus

Unlike cardiac tamponade, constriction abolishes or severely limits any increase in venoatrial or atrioventricular flow during inspiration. Increased inspiratory PA flow does occur, possibly because of decreased impedance to right ventricular ejection. With very severe constriction, inspiratory increase in absolute PA pressure occasionally occurs, and such cases may show decrease in flow and increase in pressure in the venae cava,[10] clinically seen as inspiratory swelling of the neck veins (Kussmaul's sign). By contrast, in most cases of constriction, respiration has little or no effect on caval flow or pressure. This probably explains the absence in most cases of noteworthy (i.e., over 12 mm Hg) pulsus paradoxus.[2] Indeed, some patients with definite pulsus paradoxus show a peculiar abrupt (within one beat) splitting of the second heart sound due to an immediate inspiratory fall in left ventricular ejection time.[11] This occurs before any possible transmission of right-sided inspiratory events.

Myocardial Mechanics

Constriction shifts the ventricular pressure–volume loops upward and to the left[12] in conjunction with end-diastolic volume (EDV) restriction, which may be mild (> 50 ml/m^2), moderate (25–50 ml/m^2), or severe (<25 ml/m^2).[13] Because of the reduced cavity radius, systolic wall stress is low and the ventricles are underloaded despite high apparent filling pressures. Lewis and Gotsman[13] showed that (1) stroke index, stroke work index, and left ventricular ejection rate (*measurements related to absolute fiber shortening*) are decreased linearly relative to the reduction in EDV; (2) ejection fraction, filling fraction, and circumferential fiber shortening (*measurements related to relative fiber shortening*) are normal; (3) peak LV dp/dt and mean velocity of circumferential fiber shortening (*velocity measurements*) are normal or slightly reduced. These changes are reflected in the *systolic time intervals:*[12–15] Electromechanical systole (EMS), left ventricular ejection time (LVET), and preejection period (PEP) all tend to be normal, although LVET may be short (for heart rate) at very low stroke volumes; LVET and EMS show little change with changing heart rate, e.g., during atrial fibrillation. Normal PEP and a near-normal ratio of PEP to LVET reflect near-normal ejection fraction and stand in marked contrast to patients with comparable diastolic ventricular restriction owing to cardiomyopathy.

CARDIAC TAMPONADE

Cardiac tamponade is defined as the decompensated phase of cardiac compression resulting from an unchecked rise in intrapericardial fluid pressure.[16] Relentless accumulation of inflammatory exudate, pus, blood, chyle, or gas first fills the pericardial recesses,[1] then stretches the parietal pericardium to its elastic limit, compressing the heart and decreasing apparent myocardial compliance. (Local venous obstruction with or without lymphatic obstruction increases the "massaging out" of fluid from the myocardial interstitial spaces through the epicardium.[7]) Depending on the *rate of accumulation,* compensating mechanisms fail, as quite small increments of fluid cause sharp rises in intrapericardial pressure. Thus 150 ml of blood from a cardiac wound or aortic dissection can cause lethal tamponade, while slow accumulation of liters of fluid may be well tolerated.[18] This is because the few elastic elements "give" quickly, but the fibrous tissue can yield only gradually, which probably accounts for the shape of the pericardial pressure–volume curve in acute experiments—gradual initial rise with a very steep later portion.

Elevated intrapericardial pressure primarily opposes diastolic expansion of the ventricles, hindering ventricular filling and causing rapid ascent of ventricular pressure early in diastole (Fig. 17-4). This results in early atrioventricular valve closure owing to premature obliteration of the A-V pressure gradient. Underfilling lowers ventricular stroke volume and minute cardiac output. The latter is reflected in a wide arteriovenous oxygen difference and decreased aortic pressure.

Auxiliary mechanisms further compromise cardiac output. Normally, during diastole intracavitary pressure ranges from zero to a few millimeters of mercury, and the external (epicardial) pressure upon the heart reflects slightly negative pleural pressure. This situation is radically altered in tamponade—the heart wall is clamped between greatly elevated pressures on its endocardial and epicardial aspects. (Indeed, in experimental tamponade the net transmural pressure tends to be negative, and the intrapericardial pressure can actually exceed even systolic pressure in the right ventricle.[10]) The coronary arteries and veins are compressed to some degree even in diastole. This effect combines with the low aortic pressure to diminish the coronary perfusion gradient;[18] thus coronary flow tends to be impeded and its phasic quality distorted. Tachycardia also reduces coronary flow. As a result, myocardial performance must be impaired. The stroke volume tends to be diminished not only by underfilling but also in association with shorter than normal fiber length at the premature termination of effective diastole.

Hindrance to atrial emptying is the corollary of obstructed ven-

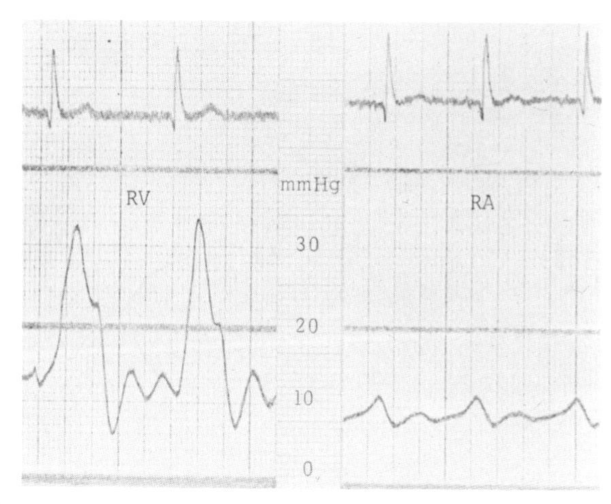

Fig. 17-4. Cardiac tamponade: Left: Right ventricular (RV) pressure showing rapid ascent from low early diastolic level. The diastolic oscillations may be artifacts of fluid-filled cathers. (Unlike constrictive pericarditis, the dip is not a common feature of tamponade.) Right: Right atrial (RA) pressure showing conspicuous x through and minimal y descent. Catheterization courtesy of Dr. H. J. Levine.

tricular filling. Mean left atrial, pulmonary venous, pulmonary arterial, right atrial, and vena caval pressures are thus elevated. When compensation is adequate each pressure increment is of similar magnitude, maintaining pressure differences across orifices (atrioventricular and venoatrial gradients) on both sides of the heart. The peripheral venous pressure usually exceeds 140 mm H_2O (11 mm Hg) and may rise beyond 300 mm.[1] Changes in the venous pressure level reflect the intrapericardial pressure and vary directly with it. Although pulmonary venous pressure is correspondingly high, pulmonary edema is quite rare in uncomplicated acute tamponade because of the reduction in right ventricular output (i.e., signs of left-heart "failure" cannot occur without right heart "success").

Compensation of Cardiac Compression by Fluid

"Cardiac tamponade" can be used as a generalization to describe *any* degree of cardiac compression; in the present context it is considered to be the phase of significant impairment of cardiac function. Most pericardial effusions, in fact, do not cause such significant cardiac compression, owing to a very slow rate of fluid production, efficient reabsorption by pericardial serosa and lymphatics, or compensatory mechanisms that either maintain normal resting cardiac output or stabilize pressure changes at some lower, but not critically diminished, output level.

The compensatory mechanisms provide a far-reaching delaying action that may maintain the circulation until the fibrosa can stretch. Unless they are overborne by progressive effusion, they are remarkably effective; arterial pressure, for example, changes only slightly until the limit of compensation is approached. The appearance of significant hypotension evokes release of catecholamines, producing an increased inotropic state, vasoconstriction, and tachycardia. *Increased inotropy* improves stroke volume at any end-diastolic pressure and decreases ventricular volume, increasing the effective pericardial space.[19] *Tachycardia* supports cardiac minute output in the presence of low stroke volume by multiplying the number of strokes per minute. *Arterial constriction* supports arterial pressure. *Venous hypertension* supports cardiac filling by maintaining the venoatrial gradients, and is itself the result of other circulatory adjustments. The latter include *elevated venous tone, redistribution of blood* into the venous system by arterial constriction, and above all *increased blood volume.* Indeed, expansion of blood volume is necessary for significant rises in venous pressure that can increase cardiac output.[16] (This may not occur soon enough in rapid tamponade, e.g., brisk intrapericardial hemorrhage.) Blood volume expansion is due primarily to renal sodium and water conservation; initially this follows venous hypertension alone, but later diminished renal plasma flow and glomerular filtration increase tubular sodium and water retention.[1] Finally, ventricular filling during tamponade should in theory be aided by diastolic suction, although knowledge of this mechanism is incomplete.[2]

The effects of various interventions reflect some of the compensatory and reserve factors in tamponade.[6] Beta adrenergic stimulation (isoproterenol) decreases peripheral resistance and decreases both end-systolic and end-diastolic volumes. End-diastolic ventricular, atrial, and pericardial pressures fall, while ejection fraction and stroke volume rise. Since alpha stimulation (norepinephrine) and digitalis do not do this, it is apparent that any improvement from tachycardia or increased inotropy may not be as effective as decreasing peripheral resistance. Finally, beneficial effects of expanded central blood volume are illustrated by the capacity of intravenous infusions to further increase central venous pressure and raise cardiac output.[6]

Decompensated Acute Cardiac Compression: Acute Cardiac Tamponade

Compensatory processes are defeated when, in the face of inexorably mounting ventricular diastolic pressure, stroke volume falls so much that tachycardia cannot maintain cardiac output and vasoconstriction is maximal; arterial pressure now drops sharply. Further effective rise in venous pressure is forestalled because the decline in arterial pressure has dropped the mean circulatory pressure, depleting the *vis a tergo* for cardiac filling. A vicious cycle is thus set up whereby decreasing cardiac output exacerbates defective cardiovascular function, which in turn paralyzes the mechanisms supporting cardiac output. In addition to feeding upon itself, this cycle is accelerated by steadily increasing intrapericardial pressure. As a result the process gains momentum, and pressure relationships plunge toward the liquidation of all gradients: circulatory collapse.

An instructive contrast to the usual situation in inflammatory pericardial effusions is found in cardiac tamponade from unchecked hemopericardium in many persons with cardiac wounds or intrapericardial rupture of aneurysms of the heart and great vessels. Compensatory mechanisms are preempted in such instances because acute blood loss and hemorrhagic shock have already depleted blood volume and induced considerable vasoconstriction.[16]

Circulatory Dynamics

Increased pericardial pressure restricts diastolic fiber length and increases ventricular end-diastolic pressure, limiting filling pressure and increasing atrial pressure. Reduced end-diastolic volume plus normal end-systolic volume produce a decreased ejection fraction.[6,20] With the decreased mean systolic ejection rate, this suggests decreased ventricular contractility. If the measured pressures are used, the left ventricular pressure–volume loop[20] is shorter, narrower, and shifted up and to the left. Yet when intrapericardial pressure is subtracted from atrial pressure to calculate the *effective filling pressure,* ventricular function curves are normal. Stroke volume and ventricular work are decreased, but MVO_2 does not fall commensurately, probably owing to tachycardia, which increases the relative duration of systole.[21,22] Tachycardia also shortens diastole, and the filling period is increasingly confined to early diastole by the high VEDP. Rapid filling should occur at this time, but in severe tamponade even the rate of early diastolic filling is slowed.[20] This is reflected in the atrial and venous pulses by the absent, greatly reduced, or even directionally reversed y descent (Fig. 17-4) and the usual (but not invariable) absence of an early diastolic dip in ventricular pressure. Vena cava flow shows the expected reciprocal relationship to the pressure pattern and has a normal inspiratory acceleration (which contributes a greater than normal increment in venous return).[5]

The coronary arteries and veins are undoubtedly clamped within the compressed heart walls, contributing an oligemic element of unknown importance that must tend to impair myocardial function.

Respiratory Effects

During critical tamponade the intrapericardial pressure does not show its normal parallel relationship to intrapleural pressure. Tandem inspiratory drops may occur in both, but the intrapericardial pressure decrease is damped or actually increased by the expansion of the right heart owing to increased venous return.[10] Pulmonary artery diameter, transmural pressure, and flow all increase in inspiration. Absolute PA pressure (PA pressure minus pericardial pressure) tends to fall, but may rise in severe tamponade.[5] These inspiratory effects are followed after two or three beats by increased pulmonary vein diameter and flow. However, during atrial systole regurgitant flow into the pulmonary veins increases.

PULSUS PARADOXUS

One of the most important effects of cardiac tamponade is pulsus paradoxus, not only because it reflects the cardiocirculatory disorder but also because it is easily appreciated at the bedside.[1,16] Kussmaul emphasized that the arterial pulse is "paradoxic" because it is palpably diminished or absent during inspiration, while cardiac movements continue unchanged. The small (5–10 mm Hg) normal inspiratory drop in arterial pressure, usually not palpable, has been accounted for in several ways. In acute cardiac tamponade there is superadded a much sharper decline in systemic arterial pressure owing to marked underfilling of the left ventricle during inspiration. Shabetai, Fowler, and associates[5,6,10] have demonstrated that during acute cardiac tamponade venous return to the right heart, transpericardial pressure, and pulmonary arterial pressure and flow all increase, while left ventricular stroke output declines and the systemic arterial pressure "paradoxically" falls. When inspiratory return to the right heart is prevented, transmural pericardial pressure does not rise during inspiration, and pulsus paradoxus is abolished. On the other hand, with respiration suspended, phasic simulation of increased venous return to the right heart results in corresponding diminution in left ventricular output and systemic arterial pressure. Thus it appears that the principal contribution to pulsus paradoxus in tamponade is (surprisingly, in view of the high pericardial pressure) inspiratory increase in caval blood flow. Moreover, increased right ventricular stroke volume and decreased left ventricular stroke volume in tamponade both occur at the onset of inspiration, i.e., before there could be any input from increased caval flow. This would appear to support Dornhorst's view that despite tamponade the right heart expands in inspiration, further raising the intrapericardial pressure, which would compress the left heart, reducing its input and consequently its output. On the other hand, subsequent increases in left ventricular stroke volume and aortic systolic pressure lag behind the right ventricular stroke volume changes by one or two beats.[23] Thus immediate manifestations of pulsus paradoxus reflect certain right-heart changes, plus later effects that are delayed during passage through the lung vessels.

REFERENCES

1. Spodick DH: Acute Pericarditis. New York, Grune & Stratton, 1959
2. Spodick DH: Chronic and Constrictive Pericarditis. New York, Grune & Stratton, 1964
3. Hefner LL, Coghlan CH, Jones WB, Reeves TJ: Distensibility of the dog left ventricle. Am J Physiol 201:97, 1961
4. Isaacs JP, Carter BN, Haller JA: Experimental pericarditis: The pathologic physiology of constrictive pericarditis. Bull Johns Hopkins Hosp 90:259, 1952
5. Shabetai R, Fowler NO, Guntheroth WG: The hemodynamics of cardiac tamponade and constrictive pericarditis. Am J Cardiol 26:480, 1970
6. Fowler NO, Holmes JC: Hemodynamic effects of isoproterenol and norepinephrine in acute cardiac tamponade. J Clin Invest 48:502, 1969
7. Guntheroth WG, Morgan BC, Mullins GL: Effect of respiration on venous return and stroke volume in cardiac tamponade: Mechanism of pulsus paradoxus. Circ Res 20:381, 1967
8. Dock W: Inspiratory traction on the pericardium: The cause of pulsus paradoxus in pericardial disease. Arch Intern Med 108:837, 1961
9. Burch GE, Giles TD: Theoretic considerations of the post-systolic "dip" of constrictive pericarditis. Am Heart J 86:569, 1973
10. Shabetai R, Fowler NO, Braunstein JR, Gueron M: Transmural ventricular

pressures and pulsus paradoxus in experimental cardiac tamponade. Dis Chest 39:557, 1961
11. Beck W, Schrire V, Vogelpoel L: Splitting of the second heard sound in constrictive pericarditis. Am Heart J 64:767, 1972
12. Vogel JHK, Horgan JA, Strahl CC: Ventricular function in chronic constrictive pericarditis: Observations on fiber shortening. Cardiol Digest 3:20–27 1971
13. Lewis BS, Gotsman MS: Left ventricular function in systole and diastole in constrictive pericarditis. Am Heart J 86:23, 1973
14. Kumar S, Spodick DH: Study of the mechanical events of the left ventricle by atraumatic techniques. Am Heart J 80:401, 1970
15. Kesteloot H, Denef B: Value of reference tracings in diagnosis and assessment of constrictive epi- and pericarditis. Br Heart J 32:675, 1970
16. Spodick DH: Acute cardiac tamponade: Pathologic physiology, diagnosis and management. Prog Cardiovasc Dis 10:64, 1967
17. Miller AJ, Pick R, Johnson PJ: The production of acute pericardial effusion. Am J Cardiol 28:463, 1971
18. Spodick DH: Electric alternation of the heart. Its relation to kinetics and physiology of the heart during cardiac tamponade. Am J Cardiol 10:155, 1962
19. Stein L, Shubin H, Weil MH: Recognition and management of pericardial tamponade. JAMA 225:503, 1973
20. Craig RJ, Whalen RE, Behar VS, McIntosh HD: Pressure and volume changes of the left ventricle in acute pericardial tamponade. Am J Cardiol 22:65, 1968
21. Binion JT, Morgan WJ, Welch GH, Sarnoff ST: Effect of sympathomimetic drugs in acute experimental cardiac tamponade. Circ Res 4:705, 1956
22. O'Rourke RA, Fischer DP, Escobar EE, et al. Effect of acute pericardial tamponade on coronary blood flow. Am J Physiol 212:549, 1967
23. Ruskin J, Bache RJ, Rembert JC, Greenfield JC: Pressure–flow studies in man: Effect of respiration on left ventricular stroke volume. Circulation 48:79, 1973

Bernard F. Schreiner, Jr.
Paul N. Yu

18

Pulmonary Circulation, Congestion, and Edema: Anatomic and Physiologic Considerations

This chapter deals in sequence with the functional anatomy of the pulmonary circulation, with emphasis on the relationships among the capillary bed, interstitial space, alveoli, and lymphatics. The dynamics of liquid exchange in the lung are summarized and regional variations noted. Evidence is presented suggesting that the Starling hypothesis of capillary tissue exchange may not be adequate to explain the observed sequence of events in the lung. Physiologic parameters of plumonary blood flow, pressure, resistances, and intravascular and extravascular volumes are discussed.

The response of the pulmonary circulation to stress in the form of exercise, hypoxemia, and high altitude is illustrated. A subsequent section deals with responses to pharmacologic interventions, including digitalis, adrenegic drugs, aminophylline, acetylcholine, and diuretics.

Finally, a section is devoted to pulmonary congestion and edema, where emphasis is placed upon hydrostatic forms, those caused by altered permeability such as O_2 poisoning, and those of unknown or mixed etiologies such as high altitude and narcotic overdosage.

FUNCTIONAL ANATOMY OF THE PULMONARY CIRCULATION

Airways

The pulmonary circulation communicates with the environment by means of the tracheobrachial tree—two main bronchi and branching bronchioles of smaller and smaller caliber. After approximately 18 divisions there are several hundred thousand respiratory bronchioles, and after six additional divisions the airway terminates in approximately 300 million alveoli[1] where gas exchange occurs with pulmonary capillaries.

Alveoli

The alveoli form from outpouchings of the alveolar ducts. Under conventional light microscopy the alveoli appear to be in intimate contact with pulmonary capillaries, separated from them only by a thin frequently indistinct membrane composed of alveolar epithelium and capillary endothelium. The observations of Weibel and Gomez[2,3] indicate that the human alveoli are approximately 250 μ in diameter and are juxtaposed to capillaries on a one-to-one basis. Studies in intact lungs and in isolated perfused preparations, however, clearly demonstrate that not all alveoli are ordinarily used for gas exchange, nor are all capillaries perfused in proportion to their ventilation.[4–8] Thus capillary blood volume varies directly with lung volume, alveolar size, and the degree of capillary filling.

During the past 10 years increasing use of electron microscopy has added important anatomic and physiologic dimensions to concepts of alveolar–capillary exchange. As emphasized by Weibel[2] and by Gil,[9] the alveolar walls are composed mainly of epithelial cells that have a central nucleus and that possess long thin cytoplasmic extensions. The latter over large areas of the alveoli surface are in intimate contact with similar cytoplasmic extensions of capillary endothelial cells and are separated from each other only by tightly fused basement membranes. The alveolar cells are lined by an ultrathin lipoprotein layer called pulmonary surfactant,[2,9] which is responsible for maintaining alveolar architectural integrity. In other words, it prevents alveolar collapse by decreasing surface tension at air–liquid interfaces. This material is believed to be synthesized from distinctive metabolically highly active type II pneumocytes.[10,11] In addition, these cells are capable of replacing alveolar epithelial cells following injury.[12]

A third type of cell found in the walls of alveoli is the alveolar macrophage. This cell is highly permeable and is capable of taking up bacteria, iron, and iron containing components such as hemosiderin.[13,14]

Pulmonary Circulation

Arteries

The pulmonary arteries are thinner than systemic arteries and contain relatively more elastic tissue. As they bifurcate they progressively decrease in caliber and lie adjacent to the bronchial tree. Near the respiratory bronchioles the arterial bifurcations encompass the airway.[15] These vessels constitute the small muscular branches, containing well-formed media, and probably are a major site of resistance change. However, this resistance function may also be mediated by the passive distention imposed by the adjacent airways.[16] The pulmonary arterioles arising from these vessels contain no contractile elements in their walls, except at their origins.[17]

Veins

The venules and veins corresponding to the arterioles and arteries have even less well organized smooth muscle and elastic fibers and have relatively little vasomotor activity.

Capillary Bed

The capillary bed forms a network of varying length, diameter, and number of vessels that are located between adjacent alveolar walls. Fishman has emphasized the concept that the capillaries are incorporated into the alveolar walls in an eccentric fashion.[18] On the thinner side the capillary, composed of a cell nucleus and its long cytoplasmic extensions, is separated from the alveolar cavity only by the fused basement membranes, the cytoplasma extensions of the alveolar epithelial cells, and a layer of surfactant. This is the picture given by conventional light microscopy that earlier gave rise to the concept of the lung as a two-compartment system involving a single cellular barrier that separated capillary from alveolus. Ultrastructure studies have now clearly indicated that the anatomy is much more complex. While gases diffuse through the alveolar–capillary barrier, substances such as water, electrolytes, and even protein molecules appear to traverse the capillary wall on the opposite side where the capillary endothelium is separated from an alveolus by the interstitial space (Fig. 18-1). The latter contains collagenous fibers, fibroblasts, and fibroblast extensions that are interposed between basement membranes of the endothelium and epithelium.[3] The interstitial space is continuous and is anchored by fibers extending from central structures such as bronchioles and pulmonary arteries to the peripheral pleura and interlobular septa. Although this provides support for the capillary bed, its main function is the exchange of lung water by facilitating water and protein transport from capillaries toward lymphatics in the interstitium and into more remote interstitial spaces.

Electron microscopy suggests that the pulmonary capillaries have a highly complicated structure, even though physiologic observations indicate that their walls function as semipermeable membranes. They are highly permeable to water, solutes of small ions (Na and Cl), and me-

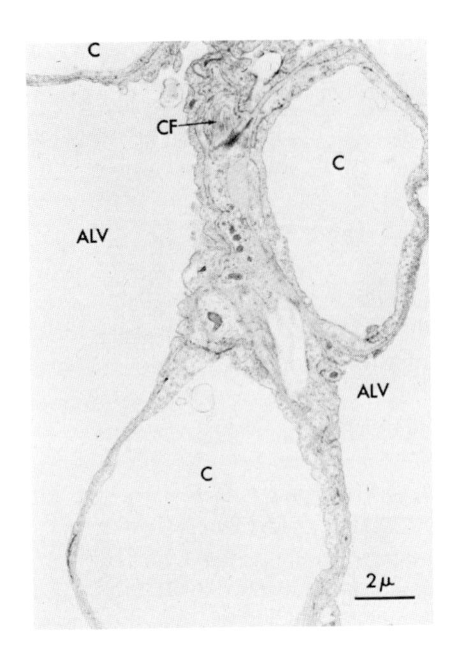

Fig. 18-1. Electron microphotograph of portions of three capillaries (C) in the septum between two alveoli (ALV). On the right side of the middle capillary the membrane is composed only of fused capillary endothelium and alveolar epithelium separated only by their fused basement membranes. The opposite side is thick and contains collagen fibers (CF) in the interstitial space between capillary endothelium and alveolar epithelium. (Reproduced by permission from Fishman AP: Pulmonary edema. Circulation 46:390, 1972, and by permission of the American Heart Association, Inc.)

tabolites such as urea and glucose.[19,20] Proteins, on the other hand, are restricted in movement but appear to cross the barrier by diffusion.[19–24] Lipid-insoluble compounds (< 90,000 d molecular weight) probably traverse the endothelium into the interstitial space by way of endothelial pores.[25] The mechanisms of transport of large molecules (> 90,000 d) are debatable. Some favor the concept of a large pore system, while others favor transport by pinocytosis.[3]

That the concept of pore permeability is dynamic has been emphasized by observations indicating that pore size may be influenced by hydrostatic pressure.[26,27] These concepts of fluid accumulation in the interstitial space secondary to increased capillary permeability on the one hand and increased hydrostatic pressure in the capillary bed on the other cannot be entirely separated either at the clinical or ultrastructural level. Moreover, once capillary permeability is increased and large molecules enter the interstitium the emigration of water into the interstitium is increased.

Interstitial Space

The function of the interstitial space is essential to normal pulmonary gas exchange and to the maintenance of the integrity of fluid balance within the lung. Gas exchange appears to be solely dependent upon the integrity of the alveolar–capillary membrane, including the

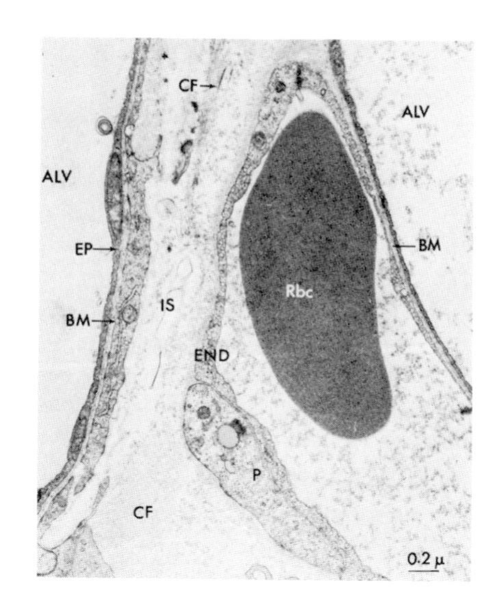

Fig. 18-2. Electron microphotograph of experimentally induced hemodynamic interstitial pulmonary edema. The interstitial space (IS) of the thick portion of the alveolar septum is distended by edema fluid. On the opposite, the part containing the fused basement membranes (BM) is unchanged. The lacy material in the alveoli, surrounding the red blood cell (RBC) and staining the collagen fibers (CF), is stroma-free hemoglobin used as a tracer. EP is alveolar epithelium. END is capillary endothelium. (Reproduced by permission from Fishman AP: Pulmonary edema. Circulation 46:390, 1972, and by permission of the American Heart Association, Inc.)

surfactant layer and balanced alveolar ventilation and capillary perfusion. In contrast, liquid exchange is dependent primarily upon interactions among capillaries, the interstitial space, and pulmonary lymphatics. Under normal conditions liquid preferentially moves from capillary to interstitial space and/or pulmonary lymphatics. When this mechanism becomes saturated, interstitial edema exists. This is manifested anatomically by widening of the interstitial space and separation of collagen fibers. Although these changes indicating pericapillary interstitial edema are demonstrable by ultrastructure (Fig. 18-2), they are not discernible by light microscopy.[28] As the process continues, eventually alveolar edema ensues, associated with destruction or inactivation of surfactant and deformation of the alveolar epithelium.[26] Thus intraalveolar edema must be considered a very late complication of interstitial congestion.

J Receptors

The adaptive responses of the mammalian organism are not dependent upon end-stage features of pulmonary edema, but become operative at relatively early stages of pulmonary interstitial edema. This response, characterized by increased ventilatory activity, has been attributed to stimulation of stretch receptors in the vicinity of collagen in the interstitial spaces that is transmitted centrally over the vagus nerve.[29,30] The stimulus appears to be an increase in pulmonary interstitial volume followed by the response of augmented ventilation that favors removal of that volume.

Extraalveolar Vessels

These vessels, larger than capillaries and perhaps as large as precapillary vessels up to 1 mm in diameter, probably participate in gas exchange,[16] leak fluid into the interstitial space,[31] and participate in modifying pulmonary vascular resistance.[32] The degree of fluid leak may be monitored by pericytes that are encased by the basement membranes of the endothelial cells.[33] The resistance function of these vessels appears to be determined by lung volume. As the lung expands these vessels are opened, but the ultimate caliber is determined not only by the lung volume but by the opposing forces of the elastic tissue in the vessel walls and by smooth-muscle tone. The increase in vascular resistance at low lung volumes appears to be dependent upon narrowing of these extraalveolar vessels.

On the other hand, the increase in vascular resistance encountered at large lung volumes is most likely dependent upon narrowing of pulmonary capillaries resulting from increasing tension in the walls of the alveoli.[34]

Pulmonary Lymphatics

The lungs have an elaborate lymphatic system. If one were to consider the lung as only a two-compartment system consisting of capillaries, alveoli, and their respective membranes it would be difficult to attribute functional significance to the extensive lymphatic capillary network. However, in the context of complexities of capillary–interstitial-space–alveolar arrangement the lymphatics undoubtedly have an important function. Primary lymphatics are seen in the perivascular connective tissue even as far as the alveolar walls.[35,36] Liquids and protein molecules enter the lymphatics from the interstitial space by means of subatmospheric pressure gradients.[26] As larger lymphatics are reached, centripetal movement toward the hilus is promoted by respiratory movement. Tobin has clearly demonstrated these channels, which accompany both pulmonary arteries and veins. Although this drainage is usually centripetal, it has been shown that subpleural lymphatics can provide collateral pathways to carry lymph around local sites of obstruction within pulmonary segments to other points of entry into the lymphatics leading toward the mediastinum.[37]

The total functional significance of the lymphatic system has not been delineated. It is obvious that the fluid and other plasma constituents that leave the pulmonary capillaries must ultimately be matched by an equal lymphatic drainage. There is experimental evidence, however, that indicates a lag time between interstitial fluid accumulation and increased lymphatic flow.[38] This may be explained in part by such factors as the relatively large capacity of the lymphatic system, preferential return of protein, and the limited lymphatic drainage related to the size of the right lymphatic duct and the thoracic duct.[39] These anatomic factors may explain why estimates of increase in lymph flow have been quite variable in response to experimentally provoked pulmonary edema. However, some general observations may be summarized as follows:[40–43] (1) an increase in pulmonary lymph flow accompanies pulmonary edema, but the increase may be delayed depending upon the duration of left atrial hypertension and may be of small magnitude if left atrial pressure is less than 25 mm Hg. (2) In acute experiments, lymph flow may increase 300%–400% when mean left atrial pressure exceeds 25 mm Hg. (3) In chronic experiments lymph flow does not increase when left atrial pressure is less than 25 mm Hg.

Whether or not systemic venous hypertension can influence lymphatic drainage to a clinically significant degree is debatable. In conditions in which left heart pressures and pulmonary capillary pressures are elevated, clinical pulmonary congestion may be aggravated by the concomitant presence of right ventricular failure and ensuing systemic venous hypertension.[39] However, in acute and chronic isolated right-heart decompensation this factor is unlikely to evoke clinical evidence of pulmonary edema.

Bronchial Circulation

The systemic arterial supply to the lungs constitutes only 1%–2% of the left ventricular output.[44,45] Venous drainage is mainly via the azygous system, with a small percentage entering the pulmonary veins.[45] Under physiologic circumstances bronchial veins form a latticework in the adventitia and submucosa of the bronchial tree.[46] Dog experiments by Pietra et al. have shown that histamine introduced by any route results in interstitial edema in the lung by uniquely causing leakage of fluid from these vessels.[47] One possible mechanism for this property is contraction of contractile filaments in the endothelium of these vessels.[18]

In many cardiac and pulmonary pathologic states components of the bronchial circulation enlarge to different degrees in response to various stimuli. Precapillary anastomoses, thought not to be present normally, becomes prominent communications between the systemic and pulmonary systems.[46]

Under normal conditions bronchial venous drainage is mostly returned to the right side of the heart.[16] In the presence of post-pulmonary capillary hypertension, of which mitral stenosis is a notable example, the bronchial venous drainage to the right heart is greatly enhanced.[46] This hemodynamic situation gives rise to bronchial vein varicosities and is a common cause for hemoptysis. On the other hand, in isolated right-heart failure, such as cor pulmonale secondary to pulmonary emphysema, bronchial venous flow is also increased, but flow is directed into the pulmonary veins and left atrium because of high right ventricular filling pressures.[16]

In the presence of inflammatory pulmonary parenchymal disease the bronchial arteries proliferate. Under these circumstances systemic pressures and a large-volume flow in the form of a left-ventricle–systemic-artery–pulmonary-vein shunt may be created that may burden both the pulmonary circulation and the left ventricle.

DYNAMICS OF LIQUID EXCHANGE IN THE LUNG

It has become evident in recent years that the movement of water and solutes across capillary walls cannot be determined solely by physical and physiochemical forces based upon the traditional Starling concept.[48–50] According to Starling, the rate of net liquid movement across the unit surface area of the capillary (J_v) is governed by factors as listed in the following formula:

$$Jv = Kf\,(P_c - P_{IP}) - (\pi_{PL} - \pi_{IP})$$

where Kf = capillary filtration coefficient, P_c = hydrostatic pressure in the capillaries, P_{IP} = hydrostatic pressure in the interstitial space, π_{PL} = oncotic pressure of plasma, and π_{IF} = oncotic pressure of the interstitial space.

Thus at equilibrium the theoretical capillary hydrostatic pressure may average +7 mm Hg and the capillary oncotic pressure −28 mm Hg with respect to the interstitium. The corresponding interstitial osmotic pressure, based upon analysis of protein in lung lymph, may average −9 mm Hg. At equilibrium the net capillary pressure of −21 mm Hg would be balanced by the combination of the interstitial oncotic pressure of −9 mm Hg and an interstitial fluid pressure of −12 mm Hg (total −21 mm Hg) (Fig. 18-3). The calculated value for interstitial fluid pressure has varied in experimental situations from −6 to −13 mm Hg.[50–53] It is unlikely that there is a uniform balance between capillary and interstitial forces either regionally in the lung or momentarily in any single portion of the lung. The net forces undoubtedly favor capillary filtration, so that there is a net fluid movement into the interstitium. For this reason the interstitial fluid pressure in Fig. 18-3 is given as −13 mm Hg. Thus there is

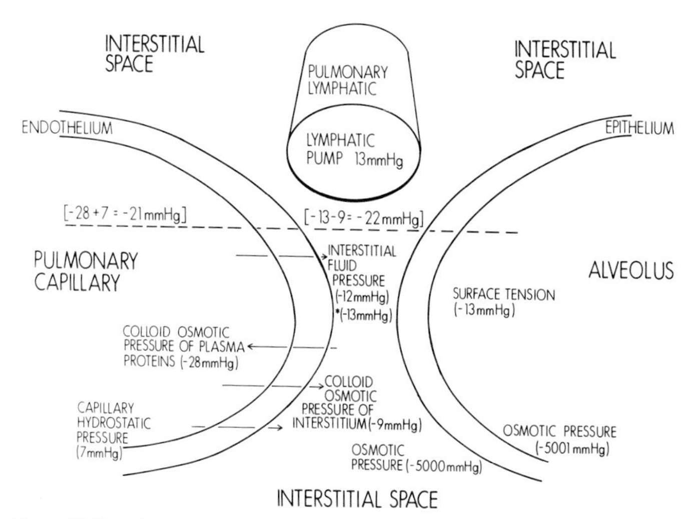

Fig. 18-3. A schematic representation of capillary, interstitial alveolar, and lymphatic pressures thought to be operative in fluid exchange within the lung. The numerical values are those given by Guyton.[49] See text for details.

a net imbalance of pressures between capillary and interstitium of 1 mm Hg favoring fluid movement into the interstitial space. Alveolar surfaces are wet, and there is a constant lymph drainage from the lungs.[54–56] Furthermore, pulmonary edema may occur at capillary pressures below 25 mm Hg because of changes in capillary permeability.

The mechanisms by which interstitial fluid pressures are maintained at a subatmospheric level are not clearly defined. It may include the pumping action of lymphatics, the respiratory motion of the lung, and its elastic recoil.[57,58] The negative interstitial fluid pressure tends to maintain the relative dryness of the alveolar spaces and to promote drainage of fluid from the pulmonary vascular tree. Once water accumulation exceeds the capacity of the interstitial space and the capacity of the lymphatics to remove it, interstitial fluid pressure becomes less negative or even positive with respect to atmosphere. Thus large increments of water are accumulated with only small changes in interstitial fluid pressure.[52,53] Under these conditions the interstitial fluid space therefore becomes more compliant.

Because of gravitational effects, interstitial fluid pressures are less negative at the bottom than at the top of the lung. In contrast, capillary hydrostatic pressures are greatest at the bottom of the lung and decrease in magnitude toward the apices. These factors further emphasize the regional inhomogeneity of intravascular and extravascular forces that affect fluid exchange. Using estimates of pulsatile flow pressures obtained in isolated perfused ventilated lungs with normal pulmonary artery, pulmonary capillary, and left atrial pressure, Lee has calculated fluid pressures for the normal human lung under equilibrium conditions.[59] The colloid osmotic capillary pressure was assumed to be −28 mm Hg, and the colloid osmotic pressure of the interstitial space was assumed to be −6 mm Hg. Under these circumstances the capillary pressure varied from 17 mm Hg at the bottom of the lung to 7 mm Hg at midthorax and −3.0 mm Hg at the apex. In contrast, the hydrostatic interstitial pressure varied from −5.0 mm Hg at the lung base to −15 mm Hg at midthorax and −25 mm Hg at the apex. Since colloid osmotic pressures in the capillary (−28 mm Hg) and in the interstitium (−6 mm Hg) were constant, the total fluid pressures varied from −11 mm Hg at the base to −21 mm Hg at midthorax and −31 mm Hg at apices. These hydraulic considerations serve to further emphasize the relatively small margin of safety at the lung bases. Thus if capillary pressure should increase significantly at the lung bases, either due to increased left atrial pressure or increased lung perfusion associated with exercise, the intravascular hydrostatic pressure would approach or exceed plasma oncotic pressure.

Surface tension forces at the alveolar membranes have also been elucidated to some extent.[49,60] As described by Guyton, an average interstitial fluid pressure of −13 mm Hg is matched by a pressure of −13 mm Hg caused by surface tension at the alveolar membrane pores (Fig. 18-3). As interstitial fluid pressure attempts to pull fluid from the alveolus, the fluid surface at the pore develops an equal opposing pressure.[49] Calculations suggest that the negative pressure required to pull air through the pore into the interstital space would be on the order of −100,000 mm Hg.[61] The osmotic pressures on either side of the alveolar membrane have been estimated to be in the range of 5,000 mm Hg because of experimental evidence indicating that alveolar epithelial cells are tightly joined to one another, so that the major route of transit of fluid molecules is through alveolar membranes rather than between alveolar cells.[49] There is a slightly greater osmotic pressure in the alveolus than on the interstitial side to account for continuous evaporation of water from the alveolar surface and its replenishment (Fig. 18-3). Since pressure differences are small, it is unlikely that elevation of surface tension forces plays a prominent role in the development of pulmonary edema.

The effects of alveolar forces on interstitial forces and pulmonary vascular hydrostatic pressures are difficult to quantitate. Intraalveolar pressure fluctuates from negative (during inspiration) to atmospheric (where air flow is nil) to positive (during voluntary expirations). The effect of these changing forces on capillary hydrostatic pressure varies depending upon lung volume and the size and location of the vessels.[14]

The studies of Meyer et al.[53] suggest that directional changes in alveolar pressure result in similar pressure changes in both capillary hydrostatic and interstitial tissue pressure.

The basic determination of whether interstitial fluid accumulates in excess, and whether this excess leads to intraalveolar edema as well, resides in the function of the pulmonary lymphatics. The interstitial fluid pressure is negative to a variable extent throughout the lung. Fluid is transferred into the lymphatics despite this negative interstitial fluid pressure. This is made possible because of pumping action derived from each lung inflation and the unidirectional centripetal function of the lymphatic valves. These valves in series could theoretically provide large lymphatic pumping pressures, which when added to that provided by respiratory movements could exceed 13 mm Hg[49] (Fig. 18-3).

In summary, the classic concept that the only significant driving force that determines fluid movement across the pulmonary capillary–alveolar interfaces is the algebraic sum of the pulmonary capillary–hydrostatic pressures and the plasma oncotic pressure requires further examination. It is a reasonable hypothesis that the net vectorial sum of Starling forces across pulmonary capillaries is positive, at least in the dependent portions of the lung. It is also reasonable to assume that the interstitial fluid pressure is negative and is an important factor that tends to maintain alveoli in a relatively dry state. Accumulations of interstitial fluid tend to decrease this negative pressure that favors capillary transudation of fluid and creates an interstitial milieu of increasing compliance, which further favors fluid migration into the interstitium. As long as the capacity of the interstitium can be increased and as long as pulmonary lymphatics are capable of channeling this fluid out of the free interstitial spaces, alveolar membrane integrity will be preserved and alveolar edema will not supervene. However, should capillary endothelium integrity be decreased by factors that increase permeability, the manifestations of a positive capillary-to-interstitial gradient would be enhanced and would occur at a lower capillary hydrostatic presence.

PHYSIOLOGIC PARAMETERS IN THE PULMONARY CIRCULATION

Pulmonary Blood Flow

In normal resting man the mean pulmonary blood flow, the output of right and left ventricles, and the pulmonary capillary flow are identical. However, it must be emphasized that these are time-averaged values under steady-state conditions and do not reflect acute transients in which flows are altered by factors such as phase of respiration, changes in lung volume and body posture, and changes in sympathetic tone or stimuli during anxiety or discomfort. Perhaps the most frequently neglected concept is that the pulmonary blood flow is very pulsatile and that it varies from 8–18 liters/min during systole to 1–3 liters/min during diastole.[62] At peak systolic flow the pulmonary artery pressure will exceed alveolar gas pressures, so that the majority of vessels will be perfused, while during diastole when the flow rate and pressure decrease some pulmonary capillaries may not be perfused. The decrease in perfusion will be greatest toward the lung apices in the upright posture and will extend downward as the pulmonary arterial pressure transiently falls below alveolar pressure. Segmental lung perfusion will vary with the time course of the cardiac cycle and with respect to gravity. Perfusion becomes more uniform during systole, during exercise, and especially in the supine position, whereas it is most inhomogeneous in the erect position at rest. In human investigations these important differences tend to be minimized or ignored because of the technique of cardiac catheterization and the fact that most studies have been made in the supine position.

These factors should be considered in analyzing not only determinations of pulmonary blood flow but the distribution of intravascular pulmonary pressure.

Fick Method

A century ago Fick proposed that pulmonary blood flow could be determined from the oxygen uptake divided by the arteriovenous oxygen difference.[63] The relationship is as follows:

$$\text{cardiac output (liters/min)} = \frac{\text{oxygen consumption (ml/min)}}{\text{arteriovenous oxygen difference (ml/liter)}}$$

Its validity depends upon both analytic and biologic factors. The former include the precision of measurement of oxygen consumption and of arterial and mixed venous O_2 content of blood. These factors are quite well standardized. On the other hand, the biologic factors involved are numerous, frequently complex, and dependent upon the existence of a steady state. These may be summarized as follows: (1) A steady state exists only if the gas exchange ratio is relatively constant. (2) It is further reflected in constancy of heart rate, right and left heart pressures, and patient composure as evaluated clinically. (3) It can be defined only if the measured O_2 consumption is reasonably constant. (4) The steady-state analysis strictly applies only when blood and expired air samples are collected not only simultaneously with respect to time but with respect to *volume*. That is, the volume of sampling should reflect the volume flow of both blood and mixed expired air.[64–66]

Indicator Dilution Method

In this method an indicator that is confined to the circulation between injection and sampling sites and that traverses a mixing chamber (right or left ventricle) is detected downstream from the point of injection and its time-concentration distribution plotted. The cardiac output is calculated from the following relationship:

$$\text{cardiac output (liters/min)} = \frac{I \times 60}{AC \times t}$$

where I = amount of indicator injected (mg), AC = average concentration of the dye during its primary circulation (mg/liter), and t = duration of the primary circulation curve (sec). Effects of recirculation on the primary curve are removed by plotting the time-concentration of the downslope as an exponential function of concentration as originally proposed by Stewart and Hamilton et al.[67,68] The exponential extrapolation to the baseline allows completion of the primary curve, with the exclusion of the recirculation portion. It is fortuitous that the downslope of concentration is linear when plotted as an exponential. However, the validity of the technique is largely based upon this finding. The theoretical and mathematical validity of the method has been extensively documented.[69–72] The reader is referred to the excellent discussions of the method by Zierler.[71,72]

A variety of indicators have been employed. They include: (1) physiologically inert dyes such as indocyanine green, Evans blue (T-1824), methylene blue, indigo carmine, and Bromsulphalein; (2) radiopharmaceuticals such as ^{125}I- or ^{131}I-human serum albumin (RISA) or radioactive krypton; (3) substances that alter current flow at a positively charged platinum electrode or change a potential at a potentiometric electrode, such as hydrogen gas and ascorbic acid; (4) substances that can be detected because of thermal changes that they induce in blood, most commonly cool isotonic saline. Of these indicators the inert dyes (indocyanine green and Bromsulphalein), radioiodinated human serum albumin, and normal saline (thermodilution) are used most frequently. Specific advantages of the dyes and of thermodilution are that the dilution curves can be recorded repetitively with the use of on-line recording equipment and computerized calculation of cardiac output and mean transit time. RISA curves may be recorded from external counters positioned over the heart, as well as by arterial sampling using sodium iodide crystals and count-rate computers. The latter technique is particularly advantageous when used in conjunction with the beta-emitter tritiated water to determine simultaneously pulmonary intravascular and extravascular volumes. However, radioactive techniques have limitations imposed by the quantity of ionizing radiation that may be safely administered to the human subject, even when the thyroid uptake is blocked by prior administration of sodium iodide.

Comparison of Fick Method with Indicator Technique

Historically the Fick method employing analysis of time-averaged collections of mixed expired air and arterial and mixed venous oxygen contents was used as the reference standard for cardiac output determinations. In a representative study reported by Cournand, the average cardiac index in basal supine subjects was 3.12 liters/min/m² (SD ± 0.40) with a corresponding oxygen consumption of 138 ml/min/m² (SD ± 14).[73] According to Reeves and associates the corresponding arteriovenous oxygen difference averages 38.4 ml/liter (SD ± 6.4).[74]

Analysis of cardiac indices determined nearly simultaneously by the Fick and the indicator dilution methods indicated that the two methods failed to agree in 25% of instances, but that there was no systematic error.[75] In addition, these authors summarized data collected from normal individuals in whom cardiac output was determined by either the Fick or the indicator dilution method. Prior to 1961 they found 45 reports with 510 measurements divided approximately evenly between the two techniques. The most frequent range of mean values for each method at rest was between 3.2 and 3.8 liters/min/m².[75] The mean value for cardiac index noted in 15 normal patients studied in our laboratory was 3.37 ml/min/m² (SE ± 0.12).[76] Similar values have been reported from a number of other laboratories.

Regional Pulmonary Blood Flow

Although time-averaged values for pulmonary blood flow have been generally employed, the early studies of DuBois and co-workers clearly demonstrated its pulsatile character with instantaneous values varying from 15 liters/min during early systole to less than 3 liters/min during diastole.[62] Appreciation of this concept is especially important when considering regional blood flow and derived measurements such as pulmonary vascular resistance (where resistance = pressure/flow). Regional blood flow differences within the lung have been clearly delineated by a variety of techniques, including angiocardiography of the pulmonary vasculature,[77–79] differential bronchospirometry,[80] and external detection of inhaled or injected radiopharmaceuticals using external radioisotope scanning.[6,81–83]

In the normal upright subject, blood flow increases progressively from apex to base. Thus high ventilation–perfusion relationships at the apex progressively decrease until the ratio is less than one toward the inferior borders of the lungs.[8,81,82] It is generally accepted that these differences in regional lung perfusion are related to gravity. During exertion in the upright position blood flow increases proportionally more in the apical than in the basal regions, resulting in a more even distribution of blood flow and hence ventilation–perfusion ratios.

In contrast to normal subjects, in upright patients with pulmonary hypertension secondary to left-heart disease such as mitral stenosis the pulmonary blood flow is more evenly distributed, with relative hyperemia of the upper lobes.[6,8,83] With progression of the disease, inversion of the normal flow pattern occurs, so that apical blood flow exceeds basal flow.

West and co-workers studied the waterfall concept of the distribution of blood flow in isolated upright dog lung preparations.[84] Increases in pulmonary artery or pulmonary venous pressure resulted in more even distribution of blood flow, while increases in alveolar pressure caused underperfusion of the apices of the lungs. The lung was divided into three zones. In zone 1 (apex) alveolar pressure exceeds both pulmonary artery and pulmonary venous pressure, so that flow is nil, presumably due to the direct effect of alveolar pressure upon the capillaries. In zone 2 (midlung) the flow is determined by the magnitude of the pulmonary artery pressure, which is greater in the lower part of the zone due to gravity, and the alveolar pressure, which remains constant. At the bottom of the lung in zone 3 flow is determined by the pressure difference between pulmonary artery and pulmonary vein. Flow is greatest in this zone because the pulmonary artery pressure is greatest and because vessels are large in caliber due to higher transmural pressures.[7,84]

The external estimation of inspired air distribution using xenon combined with estimations of perfusion distributions using O_2 and O_2-labeled carbon dioxide led to the quantitative expression of ratio of alveolar ventilation (VA) to perfusion (Q) in apical to base regions of the upright normal lung. The VA/Q is high toward the apices but falls progressively toward the base to reach quite low values. Although gas tensions in regional capillaries and alveoli would obviously vary, the overall efficiency of the lung to take up oxygen and eliminate carbon dioxide remains unimpaired. During the breathing of ambient air near sea level, arterial–alveolar carbon dioxide gradients of 1–3 mm Hg have been noted.

In 1949 Rahn and Riley and co-workers independently devised a conceptual model of a homogeneous ideal lung.[85,86] Although subsequent observations revealed inhomogeneity as noted above, the model served as a reference to which these complexities could be compared. Typically, inhomogeneity was referred to its effect on alveolar–arterial oxygen and carbon dioxide gradients. Using maneuvers that altered the inspired oxygen concentration, the alveolar–arterial oxygen gradient ($AaDO_2$) could be characterized as having three components: (1) that related to diffusion characteristics at the alveolar–capillary interface, especially evident at low inspired oxygen tensions; (2) that related to ventilation–perfusion inequalities at normal ambient inspired oxygen tensions; (3) that related to true venous admixture at high inspired oxygen tension. The third category would detect anatomic shunting of blood from the right to the left side of the circulation through the lungs by means of bronchial-vein–pulmonary-vein anastomoses and from thebesian veins draining into the left side of the heart. Calculation by Rahn and Farhi suggested that the $AaDO_2$ would vary from 7 to 10 mm Hg at alveolar O_2 tension of 50–100 mm Hg to approximately 10 mm Hg at alveolar O_2 tension of 100–110 mm Hg and to 35–40 mm Hg at alveolar O_2 tension at 1 atm.[87]

One of the most important aspects of this analysis concerned the relationship between alveolar ventilation and capillary perfusion at ambient inspired air. It was conceived that alveoli perfused in excess of their ventilation ($VA/Q < 0.8$) contributed to venous admixture ($AaDO_2$), whereas those ventilated in excess of their perfusion ($VA/Q > 1.0$) contributed to the arterial–alveolar CO_2 gradient or physiologic dead space.[88,89] On the basis of a normal distribution of VA/Q ratios around a mean of 0.8, Farhi and Rahn computed an average $AaDO_2$ on normal lungs as 8–10 mm Hg.[90] Unfortunately, later experiments determined that the distribution of VA/Q among all alveoli and capillaries was not amenable to a normalized statistical distribution.[91]

Although the ideal homogeneous lung model has seemed a convenient basis for analysis regarding problems of VA/Q distribution, it fails in the analysis of observed VA/Q distribution abnormalities as seen in patients with diffuse pulmonary disease, especially those in whom diffusion and distribution abnormalities coexist. Furthermore, this type of analysis neglects other sources of inhomogeneity such as those created by differences between stroke volume and capillary blood volume and perfusion and diffusing capacity.[92–94]

Practical application of model theory to evaluate the magnitude of venous admixture in any particular situation depends upon the following relationship:

$$\frac{Qs}{Qt} = \frac{CcO_2 - CaO_2}{CcO_2 - CvO_2}$$

where Qs = venous admixture, Qt = cardiac output, CaO_2 = O_2 content of arterial blood, CvO_2 = O_2 content of mixed venous blood, and CcO_2 = O_2 content of end capillary blood. If these determinations are made during ambient-air breathing, the value Qs/Qt will represent the "shunt" flow due to the abnormalities of distribution of VA/Q plus that flow contributed by true venous admixture that totally bypasses the pulmonary capillary bed. If the measurements are repeated during a high-percentage O_2 breathing (70%–100%), the component due to distribution defects is eliminated and the shunt flow is almost entirely due to this anatomic component.

The determination of Cc is not practicable, and so this component is usually estimated from the modified Sackur equation in which

$$\frac{Qs}{Qt} = \frac{(A - a)O_2 \times 0.0031}{[CaO_2 + (A - a)O_2] - CvO_2}$$

where $(A - a)O_2$ = alveolar–arterial O_2 gradient (mm Hg) and 0.0031 = solubility of O_2 (vol%/mm Hg O_2 tension). The derivations and limitations of this method have been summarized by Finley et al.[95]

However, it should be noted that recent studies by Wagner and King and their respective associates have suggested that the $AaDO_2$ determined at 100% O_2 inhalation cannot reliably be used to indicated true venous admixture. It has been pointed out that O_2 per se may alter VA/Q and that areas of poor diffusion may also contribute, especially in severe respiratory failure.[96,97]

Pulmonary Vascular Pressures

Pressures on the right side of the heart can conveniently be measured during right-heart catheterization.[98,99] This may be performed from an antecubital vein or from a femoral vein. With the currently used pressure transducers and recording equipment the major limitation on the fidelity of the recorded pressures is due to the fluid-filled catheter system. The latter not only introduces time delays and phase-angle shifts because of transmission of the pressure waveforms through a 100–125-cm fluid-filled system but also harmonic artifacts produced by catheter movement with the heartbeat. This is particularly troublesome at normal right-heart pressures, but tends to disappear when studies are made in patients with significant pulmonary hypertension. To alleviate these distortions, right-heart pressures have been recorded with catheter-tip manometers, but these have limitations related to cost, ease of manipulation, and calibration; also, until recently there was limited ability to use such catheters for sampling blood or injecting indicator dyes or angiographic contrast media.

The reference level for an externally placed manometer is usually 5–6 cm below the angle of Louis of 10–12 cm above the table top in the supine position. This level is assumed to represent the level of the right atrium. Under many circumstances these reference points are quite similar, and it is not surprising that data obtained from many different laboratories do not differ significantly either among patients with normal hemodynamics or those with myocardial or valvular heart disease. However, some uncertainties may be introduced when these reference levels are applied to patients who are severely dyspneic from cardiac or pulmonary disease. In patients with large hearts or unusual chest configurations it may be difficult to refer external reference positions to the level of the right atrium. Despite these limitations the intrapatient variation is only a few millimeters of mercury. Interpatient variation is of similar or greater magnitude, but is also dependent upon other factors such as the degree of anxiety and sedation.

Right Atrial Pressure

The mean pressure is normally 5 mm Hg or less. In general the a wave dominates the tracing and may vary from 3 to 7 mm Hg, whereas the v wave is usually less, averaging about 5 mm Hg.

Right Ventricular Pressure

The upper limits of normal are frequently given as < 30 mm Hg systolic and 5 mm Hg end-diastolic pressure. It should be appreciated, however, that the 30 mm Hg systolic pressure frequently represents catheter-induced artifact related to an underdamped system having a low resonant frequency.

Pulmonary Artery Pressure

The contours of the pulmonary artery tracing should theoretically mirror that recorded from the ascending aorta, but its magnitude is only one-fifth to one-sixth that of the systemic circulation. When recording through a catheter-tip transducer, the pulmonary artery pressure base has a rapid smooth systolic rise, a rounded peak, and a sharp dicrotic incisura that is followed by a gradual decrease in pressure throughout diastole. With use of conventional catheter gauge systems, however, high-frequency artifacts usually obscure the various components unless critical damping is employed. In contrast, in pulmonary hypertensive states the recordings more closely resemble those obtained using a catheter-tip manometer system.

In general, normal pulmonary artery pressures are less than 30 mm Hg systolic, 12 mm Hg diastolic, and 20 mm Hg mean. The pressure measurements should be recorded over at least two respiratory cycles, since intrathoracic pressure is lower during inspiration than during expiration. In a group of 52 normal subjects the average systolic and diastolic pressures were 22 and 10 mm Hg, respectively.[100] The average mean pressure in 15 normal subjects studied in our laboratory was 13.3 mm Hg (SEM ± 0.6).[76] In many patients the initial pulmonary artery pressure recorded during the cardiac catheterization procedure may be elevated. This is usually followed by a gradual decline as anxiety and apprehension diminish.

Pulmonary Arterial Wedge Pressure

This pressure may be recorded by advancing an end-hole cardiac catheter to a terminal branch of the pulmonary artery.[101,102] Use of a Swan-Ganz flow-directed balloon catheter facilitates the ease of measurement and more frequently provides a clearly pulsatile tracing in which both a and v waves are discernible. The contours of pulmonary wedge and direct left atrial pressures may show phasic differences, but the mean pressures recorded from each site are nearly identical in resting man and during hypoxemic stress.[103] On the other hand, in rare circumstances in which there is isolated pulmonary veno-occlusive disease or cor-triatriatum the pulmonary wedge pressure may be significantly greater than left atrial pressure. With the exception of these conditions, the mean left atrial pressure is almost identical to the mean wedge pressure, and the two pressures may be used interchangeably in studies of the human pulmonary circulation.

Values for left atrial mean pressure in normal subjects are less than 12 mm Hg and average 7.9 to 9 mm Hg.[104,105] In our own experience the average value was 7.6 mm Hg (SEM ± 0.7).[76] Usually the "a" wave amplitude is less than that of the "v" wave by an average of 2–3 mm Hg.

It is unlikely that either the pulmonary wedge pressure or the left atrial pressure reflects true pulmonary capillary pressure. Brody and associates measured pulmonary capillary pressure in isolated lung lobes of dogs.[106] The mean midcapillary pressure was 9.8 mm Hg, compared to an average pulmonary artery wedge pressure of 15 mm Hg and an average pulmonary venous pressure of 6.8 mm Hg.

While pulmonary capillary pressure probably lies somewhere between pulmonary artery diastolic pressure and left atrial mean pressure, it has been estimated from the pulmonary artery wedge pressure in supine normal subjects. Since the upright posture is of greater physiologic significance, attention should also be directed to the variation in pulmonary wedge pressure (or pulmonary capillary pressure) under such circumstances. West has emphasized the importance of the gravitational effect in the lung 30 cm high. Capillary pressure at the base of the upright human lung has been estimated at 13 mm Hg at rest, increasing to 22.3

mm Hg with exercise.[107] Since human plasma has a colloid-osmotic pressure nearly equal to that, leakage of fluid at this site may occur. In contrast, pressure obtained at the top of the lung will be much less and may be negative by 3–4 mm Hg, since the wedge pressure or capillary pressure will be overcome by 7.4–9 mm Hg negative hydrostatic pressure due to the fact that the apex of the lung is 10–12 cm above the level of the pulmonary veins and left atrium.

Pulmonary Venous and Left Atrial Pressures

In the past, measurements of pulmonary venous and left atrial pressure have been limited to catheter passage through an atrial septal defect; or the left atrial pressure was sometimes obtained directly by a variety of techniques including punctures under direct vision at thoracotomy, through the posterior chest wall, via the left bronchus, or by suprasternal puncture. Most of these techniques were hazardous and limited to brief diagnostic procedures, precluding any prolonged observations or hemodynamic investigations.

The development of the transseptal left atrial puncture technique alleviated many of the limitations of other procedures and permitted several hours of hemodynamic observations to be made with less risk to the patient and without undue discomfort.[108–110] The normal values and the contour of the left atrial pressure pulse have been well documented.[104,105] Mean pressures average 8–9 mm Hg, while the heights of the a and v waves average 10–11 and 13–14 mm Hg, respectively. The average mean pressure difference between the left and right atrium varies from 4 to 6 mm Hg.

Types of Pulmonary Circulatory Pressure

Considerable confusion has arisen in the interpretation of pulmonary vascular pressures, pressure differences, and changes in pressure with respect to blood flow because of failure to clearly define the type of pressure measured and its relationship to atmospheric, alveolar tissue, and pleural pressures. Intravascular pressure is the actual luminal pressure within a vessel at a specific locus with respect to atmosphere. Transmural pressure is the difference between the intravascular pressure and the tissue pressure surrounding the vessel. Changes in pleural pressure associated with the respiratory cycle affect all extraalveolar vessels but not capillaries exposed to alveolar pressure. The transmural pressure of the capillaries is determined from the difference of the estimated intracapillary pressure and alveolar pressure. Transmural pressure of all other pulmonary vessels is calculated as the difference between intraluminal pressure and intrapleural pressure.[111] Pleural and pericapillary pressures are probably not identical. The latter may be somewhere between pleural pressure and alveolar pressure and is in all probability negative with respect to the atmosphere.[49–54] In clinical situations it is not possible to routinely measure intrapleural pressure. When pulmonary artery pressure and left atrial pressure are normal or proportionately elevated, an index of transmural pressure may be roughly approximated by averaging the mean pulmonary artery and mean left atrial pressures.[76] The driving pressure is the difference between intravascular pressures at a proximal site and a distal site in a vessel and is responsible for the flow of blood between these two points. As emphasized by Fishman, as long as alveolar pressure is less than left atrial pressure the measurement of the driving pressure across the lungs can be determined from the simultaneous measurement of intravascular pressures in the pulmonary arteries and pulmonary veins or left atrium, irrespective of the intrathoracic pressure.[16]

Pulmonary Resistances

In the past flow through the lungs was considered to be steady, as determined from time-averaged Fick determinations or from indicator dilution techniques. Since measurement of pressures could also be time-averaged, it was expedient to use the analogy of Ohm's law and equate resistance (R) to the pressure drop across the lung (ΔP) divided by the flow (Q). This analysis would be quite acceptable if the fluid were Newtonian (uniform and without particulate matter), if it were flowing in a laminal manner through rigid tubes, and if it were nonpulsatile. None of these conditions are met in the mammalian circulation. While certain calculations can be made using these assumptions, their physiologic meaning is open to question. The traditional application of Poiseuille's law follows, with units expressed in the metric centimeter-gram-second (cgs) system.

$$Q = \frac{\Delta P \pi r^4}{8nl}$$

where Q = flow of fluid (ml/sec), ΔP = pressure differential required to drive fluid through a tube (dynes/cm), r = radius of tube (cm), l = length of tube (cm), and n = coefficient of viscosity of the fluid (poises). The resistance offered to flow is a function of pressure difference (ΔP) and flow:

$$R = \frac{\Delta P}{Q}$$

Pulmonary vascular resistance (PVR) is

$$\text{PVR} = \frac{\text{PAm} - \text{LAm}}{Q_p} \quad \text{or} \quad \frac{\text{PAm} - \text{PWm}}{Q_p}$$

where PAm = mean pulmonary pressure, LAm = mean left atrial pressure, PWm = mean pulmonary wedge pressure, and Q_p = rate of pulmonary flow (liters/min). To express PVR in fundamental units of the metric cgs system with the pressure values in millimeters of mercury and flow values in milliliters per second, the equation is multiplied by a constant of 80. This is the product of the following: $(1.36 \times 980 \times 60)/1000 = 80$, where 1.36 = conversion factor changing millimeters of mercury to centimeters of H_2O, 980 = gravitational acceleration factor, 60 = 60 sec, and 1000 = 1000 ml. For example, in a system where PAm = 30 mm Hg, LAm = 15, and Q_p= 5 liters/min. PVR would equal (30 − 15)/5, or 3 arbitrary resistance units. This would equal 240 dyne-sec-cm^{-5} when multiplied by 80.

In both the pulmonary and systemic circulations the criteria for the use of Poiseuille's equation are not met. The pulmonary vascular bed is a nonlinear, distensible, and frequency-dependent system in which the non-Newtonian fluid flows in a pulsatile manner. The system more closely resembles an alternating-current electrical system that has inductance and capacitance as well as resistance. The vectorial sum of these components constitutes impedance, which has a certain phase angle. It would therefore be expected that calculated resistance varies with pressure and flow but not in a linear manner. In fact, increasing flow through a distensible tube causes an increase in radius because of increased transmural pressure. Thus plots of driving pressure difference and flow would not be linear and would not pass through the origin.[111,112] As the vessel dilates the pressure drop at two points along its course decreases and the calculated resistance to flow also decreases. Not only the physical meaning of PVR but its physiologic meaning becomes quite limited. For example, when pulmonary artery pressure is low or when the difference between pulmonary artery pressure and left atrial pressure is small, minor errors in pressure measurements may result in a large error in calculated PVR.[113] Other problems are to determine whether or not an observed change in PVR represents a change in vessel caliber or whether a change in vascular caliber is due to passive effects or active vasomotor changes.

Several observations will serve to illustrate these points. Increased ventilatory activity accompanying hypoxia, exercise, or hypercapnea complicates both pressure and flow measurement. In addition, because of vascular gradients throughout the upright lung, vascular pressures are influenced to a greater or lesser extent by alveolar pressure. Under circumstances of high alveolar pressure in excess of pulmonary venous pressure, the left atrial pressure or pulmonary wedge pressure may not accurately reflect pulmonary venous pressure. Furthermore, as emphasized by Fishman, at low pulmonary flow rates anomolous viscosity may be mistaken for a change in vascular caliber, i.e., a change in PVR.[16] At the other end of the scale, with relatively high rates of flow a change in caliber may be inferred when only a change in diameters of vascular channels has taken place and when recruitment of parallel vascular channels has occurred in conjunction with augmented pulmonary blood flow.[114]

In the final analysis, when considering the relationships between pressure and flow, passive or mechanical effects must be differentiated from active or vasomotor effects. The conditions enumerated are stringent and have been summarized by Fishman.[16] The passive mechanical factors include (1) changes in circulatory effects such as a change in pulmonary or left atrial blood flow, (2) changes in pulmonary blood volume and in the bronchial circulation, (3) respiratory changes that alter alveolar and intrathoracic pressure, and (4) changes that alter or change the magnitude of interstitial fluid. Once these conditions have been evaluated, factors that may affect vasomotor activity in the pulmonary circulation can be considered. These include (1) the effects of the sympathetic nervous system and circulating catecholamines and (2) critical closing pressures in vessels, chemoreceptor activity, and local reflexes. Vasomotion is probably most manifest in the pulmonary arteries and least in pulmonary capillaries and veins.

In practice the passive circulatory and respiratory effects have been well documented in animal studies.[113,115–125] Briefly, a rise in either pulmonary artery pressure or left atrial pressure is followed by passive dilatation of the vessels, a decrease in artery-to-left-atrial pressure gradient, and a decline in calculated pulmonary vascular resistance. Respiratory effects are more complex. The pulmonary vascular resistance is increased at both large and small lung volumes, but the mechanisms are quite different. At large lung volumes increases in alveolar pressure are transmitted to alveolar vessels, decreasing the caliber because of a reduction in vascular distending pressure. At the opposite extreme of small lung volumes the geometry of extraalveolar vessels is altered by interstitial forces, resulting in increased resistance to flow.

The original recognition of vasomotor properties within the lung rested upon graphic analyses of pressure–flow curves. Once a pressure–flow relationship was established for an individual situation at rest, and for response to an intervention such as exercise, it became possible to construct similar pressure–flow relationships in the same subject under the same stimulus but with an altered inspired O_2 tension. Thus Fishman was able to demonstrate an increase in pulmonary vascular resistance at rest and during exercise when the subject was switched from ambient-air inhalation to a gas mixture containing 12% O_2 in nitrogen.[1] This is the classic representation of pulmonary vasomotor activity in the intact human subject.

Pulmonary Blood Volume

In 1960 the use of transseptal left-heart catheterization combined with right-heart catheterization made possible a better understanding of the interrelationships among pressure, flow, and volume in the pulmonary vascular bed. Determination of the pulmonary blood volume from sequential indicator dilution curves inscribed from a brachial artery following separate pulmonary artery injection and left atrial injection of indicator gave a capacitance estimate of the pulmonary vascular bed.[126–128] Thus it was found that pulmonary blood volume (PBV) was an additional variable that could be included in concepts of active and passive pulmonary vasomotor activity.

PBV is defined as that volume between the main pulmonary artery and an indeterminate portion of the left atrium and includes pulmonary arterial, pulmonary capillary, and pulmonary venous volumes. It may be determined, under steady-state conditions, from the following relationship:

$$\mathrm{PBV} = \overline{\mathrm{CI}}(\mathrm{Tm}_{\mathrm{PA-BA}} - \mathrm{Tm}_{\mathrm{LA-BA}})$$

where PBV = pulmonary blood volume (ml/m^2), $\overline{\mathrm{CI}}$ = mean cardiac index derived from PA and LA injections (ml/sec/m^2), $\mathrm{Tm}_{\mathrm{PA-BA}}$ = mean transit time from main PA to a systemic artery, and $\mathrm{Tm}_{\mathrm{LA-BA}}$ = mean transit time from left atrium to a systemic artery.

Alternative methods include single injection of indicator into the inferior vena cava with simultaneous sampling from the main pulmonary artery and left atrium[128] or simultaneous injection of different indicators into the pulmonary artery and left atrium.[129] The precordial radioisotopic techniques proposed by Donato, Giuntini, and their collaborators give values for PBV that are quite comparable to that estimated from pulmonary artery and left atrial dilution curves.[130–133] These values for patients with normal hemodynamics have ranged from 211 to 313 ml/m^2 (SEM = 10–16). In contrast, other precordial methods have yielded consistently larger values due to variable inclusions of both right- and left-heart volumes.

In man the PBV can be further compartmentalized by measuring pulmonary capillary volume (V_c). This volume is most conveniently estimated by the single-breath carbon monoxide method determined at two levels of inspired oxygen tension.[134–136] The average value for normal subjects is 54 ml/m^2 (range 39–66) supine and 31–45 ml/m^2 sitting.[135,136] These values approximate 20% of the total PBV.

The liquid volume in the lung may also be partitioned into intravascular and extravascular compartments by the method of Chinard and Enns.[137] This method depends upon indicator dilution principles by which tritiated water (THO) after a single injection into the main pulmonary artery or right ventricle penetrates the pulmonary capillary extravascular space during its primary circulation through the lung. The dilution volume of this indicator is compared to that of a nondiffusible indicator such as radioiodinated human serum albumin (RISA) that is injected simultaneously. Dilution curves are constructed from analyses of brachial artery blood samples collected at 1- or 2-sec intervals. The pulmonary extravascular volume (PEV) is then calculated from the following relationship:

$$PEV = CI\,(Tm_{THO} - Tm_{RISA})$$

where CI = cardiac index (ml/m^2/sec), Tm_{THO} = mean transit time for tritiated water (sec), and Tm_{RISA} = mean transit time for RISA (sec).

Determinations of these volume components are important, since they add capacitance parameters not provided by measurements of pressure and flow given earlier. However, they fail to discriminate between arterial and venous volumes either for the whole lung or by region.

These techniques have limitations and potential errors that must be known and minimized; they include timing and rate of injection, positions of the catheters, mixing of indicator within the channels in the pulmonary vascular bed in proportion to flow through them, and adequacy of the sampling system. These factors have been summarized in detail elsewhere.[71,72,76] Thus interpretations of changes in these parameters during interventions or between patients must be made with caution.

Pressure–Volume Relationships

The pulmonary vasculature is considered to be highly distensible. A variety of studies of the entire vascular tree and of the pulmonary-venous-to-left-atrial portion suggest that when beginning from a collapsed state, increments in volume are accommodated with small increments in pressure. However, as the total volume becomes larger each increment in unit volume results in larger and larger pressure increases, until pressures rise precipitously for a given increase in volume.[138]

Pressure is always greater during filling than during removal, and pulmonary veins and arteries are more distensible than capillaries.[138,139] Although the elastic properties of the pulmonary vascular bed are important in defining its distensibility, other factors must be considered.[16] These include tone of vascular smooth muscle, amount of interstitial fluid, perivascular tissue pressures, and alveolar surface tension. Since these factors may predominantly affect small arterials and veins; interstitial fluid pressure is likely to affect small- and medium-sized arteries and veins in dependent portions of the lung), an increase in PBV will be apportioned unequally among the various pulmonary vascular segments. Distensibility has been clearly demonstrated to occur on the arterial side and in the capillaries.[140,141] However, others have emphasized that recruitment of previously closed vascular channels is of equal or even greater importance.[142,143]

It has been suggested that the simultaneous measurement of PBV and PEV might serve to distinguish between the phenomena of increased distensibility and recruitment.[142–145] If recruitment alone were operative, PBV and PEV would be expected to rise and fall together in response to any intervention that augmented pulmonary capillarity. The findings of Luepker et al., in which these volumes were measured before and during 10 min of supine exercise in cardiac patients, support this contention.[145] That is, the ratio of PEV to PBV remained constant between the control and exercise periods. We have made similar observations in our laboratory both with exercise and during atrial tachypacing.[146] However, when left atrial pressure is elevated for a considerable period of time there is suggestive evidence that the ratio of PEV to PBV either increases during the intervention or fails to return to control values during the recovery period. Under these circumstances, factors other than recruitment are in operation, and a true increase in PEV is strongly suggested.

Interpatient observations of the behavior of PEV and PBV provide additional evidence that distensibility and absolute increases in extravascular water are important in the human lung. Studies by Luepker and associates, as well as those from our laboratory in patients with normal hemodynamics, valvular heart disease, and coronary artery disease, indicate relatively little change in PBV as left atrial and/or left ventricular end-diastolic pressure rises and as functional disability increases, but a progressive rise in PEV with increasing left atrial hypertension, functional impairment, and x-ray evidence of pulmonary vascular congestion[147,148] (Figs. 18-4 and 18-5). Thus the ratio of PEV to PBV tends to increase, suggesting that the extravascular volume is more sensitive to pressure elevation within the pulmonary circulation and that the extravascular compartment is a distensible component more sensitive to pressure changes than is PBV (Fig. 18-6).

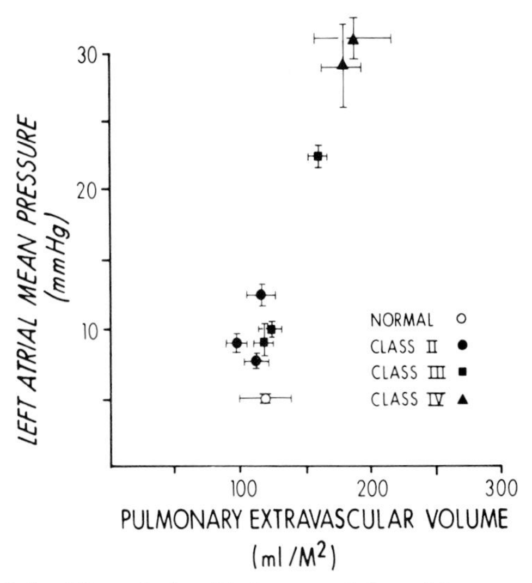

Fig. 18-4. The relationship between left atrial pressure (LA) and pulmonary extravascular volume (PEV) in normal subjects and those with progressive functional impairment. Those without mitral valve obstruction in functional classes II and III cluster at a LA of 10 mm Hg or less and a PEV 100–120 ml/m². Patients with mitral stenosis in functional class II have a normal PEV but LA pressures of 13 mm Hg. Those in functional class III have LA pressures of 23 mm Hg and PEV of 155 ml/m². Class IV patients with aortic valve disease or coronary artery disease have the highest LA pressures and PEV. The bars indicate the standard error of the mean for both parameters. [Reproduced by permission from Bloomfield DA (ed): Dye Curves, 1974, and by University–Park Press, Baltimore.]

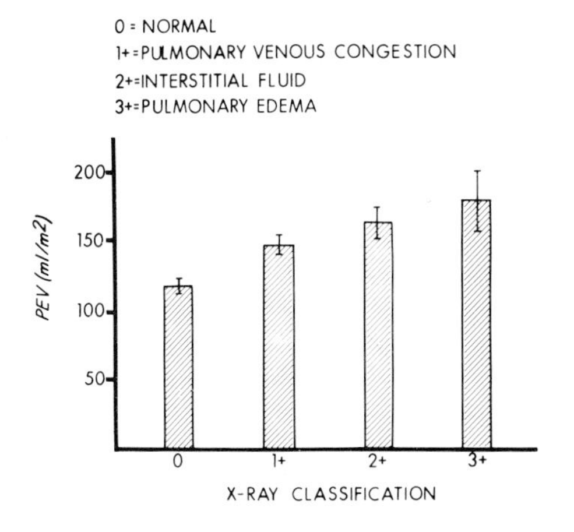

Fig. 18-5. The relationship between pulmonary extravascular volume (PEV) and x-ray evidence of pulmonary congestion. The PEV increases progressively with increasing degrees of pulmonary congestion and edema. The bars represent one standard error of the mean. [Reproduced by permission from Bloomfield DA (ed): Dye Curves, 1974, and by University–Park Press, Baltimore.]

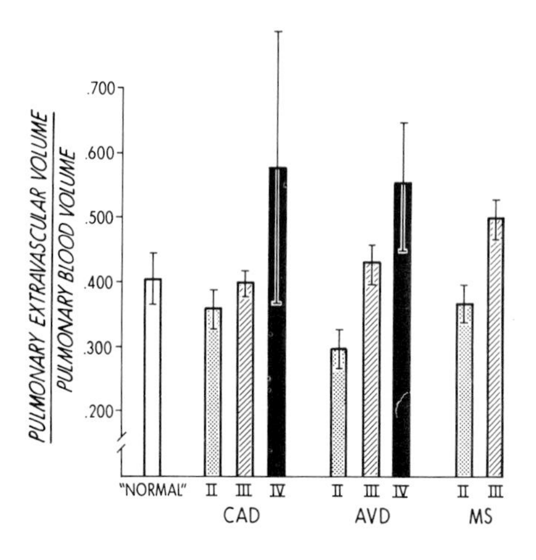

Fig. 18-6. The relationship of pulmonary extravascular volume (PEV) to pulmonary blood volume (PBV) ratio for each functional class of patients. CAD = coronary artery disease; AVD = aortic valve disease; MS = mitral stenosis. Note that the ratio rises as dysfunction increases. This is primarily due to an increase in PEV. The bars represent one standard error of the mean. [Reproduced by permission from Bloomfield DA (ed): Dye Curves, 1974, and by University-Park Press, Baltimore.]

RESPONSES OF THE PULMONARY CIRCULATION TO STRESSES

Exercise

Physical exertion is a universal form of stress; it is usually accomplished in the laboratory with an ergometer or a treadmill. The treadmill provides the most familiar type of cardiovascular stress, but the majority of hemodynamic studies have been conducted using a bicycle ergometer in the supine position.

Cardiac output is usually estimated by the indicator dilution technique or by the Fick principle. While both methods require that a steady state be present for valid measurements, the Fick method tends to become less reliable with heavy work loads (when $\dot{V}O_2$ is over 1000 ml/min/m^2) where a steady state is difficult to achieve. As a rule, when the Fick method is employed the parameters of VE, $\dot{V}O_2$, gas exchange ratio (R), heart rate, and arteriovenous oxygen difference do not become constant at the same time.[149,150] The arteriovenous oxygen difference may become stable within a minute or two, but VE and R may require a much longer time to equilibrate and may never do so at heavy work loads, thus invalidating the method. In contrast, at lighter exercise loads that most patients can tolerate ($\dot{V}O_2 < 400$ ml/min/m^2) these variables become constant within 3 to 5 min. As a general rule it is unwise to make such determinations at less than 5 min of exertion, and it is preferable to make them between 5 and 10 min of exercise. It has frequently been stated that a normal increase in cardiac output is closely related to the increase in minute oxygen consumption. The figures most frequently referred to are a 600- to 800-ml increase in cardiac output per 100 ml increase in oxygen consumption.[151,152] These values depend upon attainment of a steady basal preexercise state. If the latter has been achieved the increment in cardiac output may exceed 1000 ml for a 100-ml increase in oxygen consumption.[1] On the other hand, if there is no steady state the incremental change in cardiac output may be subnormal and misleading.[150] These differences must be taken into account when these parameters are used as determinants of cardiac function.

The relationship between oxygen consumption and cardiac output is usually linear. At rest the values for cardiac output are about 7.5 liters/min with an average oxygen consumption of 250–275 ml/min. When average $\dot{V}O_2$ consumption increases to 3 liters/min the cardiac output is approximately 23 to 24 liters/min.[153] On the other hand, the arteriovenous oxygen difference is distinctly curvilinear during exercise. At rest it averages 35–45 ml/liter, but at 2 liters/min oxygen consumption it increases to 100 ml/liter. At an oxygen consumption of 3 liters/min it increases only to 120–125 ml/liter.[153] Thus there is only a small increase in arteriovenous oxygen difference above 2 liters/min oxygen consumption. Distinct differences are noted between the performance of subjects in the supine position as compared to the sitting position. At all levels of oxygen consumption the arteriovenous oxygen difference is greater sitting than supine.[153]

In normal individuals exercise in the range of oxygen consumption of 700 to 800 ml/min results in an increase of 4 to 6 mm Hg in mean pulmonary artery pressure.[153,154] With more strenuous exercise, pulmonary artery pressure continues to increase in a linear manner with respect to flow, reaching a mean of 23 to 25 mm Hg at cardiac outputs of 20 to 25 liters/min. As milder degrees of exercise are prolonged, the pulmonary artery mean pressure stabilizes and then may begin to decline after 6 or 7 min of exercise.[154] Part of the initial increase in mean pressure is probably related to redistribution of blood centrally as well as to autonomic nervous system "overshoot" during the early exercise period. With termination of exercise the pressure usually falls to levels below that of the control period, as does the cardiac output.[154–156]

Pulmonary wedge pressure responses are less predictable even in normal subjects. Exercise studies in our laboratory suggest that the wedge pressure or direct left atrial pressure shows little change or may fall slightly during moderate exercise (oxygen consumption 800 ml/min).[76] However, Ekelund and Holmgren noted a gradual linear increase in pulmonary wedge pressure from 8 to 13 mm Hg over a cardiac output range from 5 to 23 liters/min.[153] While it is generally agreed that the difference between mean pulmonary artery pressure and pulmonary wedge pressure tends to increase with increasing exercise loads, the calculated pulmonary vascular resistance has been variable. The consensus appears to be that PVR decreases with moderate exercise, since flow increases out of proportion to the increase in gradient between pulmonary artery pressure and pulmonary wedge pressure.[153]

Unlike mean pulmonary wedge pressure and left atrial pressure, which tend to increase as the severity of exercise and the cardiac output increase, right ventricular end-diastolic pressure gradually decreases. However, right ventricular systolic pressure rises from a normal of 22–25 mm Hg to a peak of nearly 50 mm Hg at cardiac outputs of 25 liters/min.[153]

A number of investigators have demonstrated an increase in intrathoracic or central blood volume (CBV) in normal subjects during supine exercise.[157–159] The validity of this technique has been challenged on the basis that local hyperemia and redistribution of arterial blood flow during exercise may cause spuriously high estimates of CBV.[160,161] Central blood volume is a rather nebulous entity and most frequently is defined as that volume between the main pulmonary artery and the systemic sampling site including all temporally equivalent points in the systemic circulation.

Previous studies by Korsgen and associates, as well as studies in our laboratory, clearly indicate that measurements of central and pulmonary blood volumes are not invalidated by local hyperemia and arterial blood distribution.[76,162–164] In our own studies, normal subjects were noted to increase PBV by 69 ml/m^2 with an increase in mean PA pressure of only 3.4 mm Hg and a fall in mean LA pressure of 2 mm Hg. Patients with mild aortic regurgitation behaved in a similar fashion.[76] In contrast, patients with mitral stenosis, mitral regurgitation, aortic stenosis, or primary myocardial disease (either ischemic or nonobstructive cardiomyopathy) had disproportionate increases in both pulmonary artery and left atrial mean pressures, for an increase in PBV that averaged from 40 to 110 ml/m^2. The increase in PBV was accompanied by a steeper rise in pressure[165] (Fig. 18-7). Thus the slopes of these pressure–volume curves were much steeper with respect to the pressure axis (ordinate) than those for normal patients or patients with aortic regurgitation. These data were derived from studies of 54 patients in functional class I to class III. An additional 21 patients with mitral stenosis or primary myocardial disease who were in functional class IV were studied at rest. In these patients PBV was normal (4 patients) or less than normal (17 patients) despite gross elevations in both pulmonary artery and left atrial mean pressures. In these patients the pressure–volume points shifted to the left in comparison to the data obtained in less incapacitated patients who were exercised. These data strongly suggested that the pressure–volume curves in the pulmonary vascular bed were shifted leftward, signifying greater pressure elevations for any in-

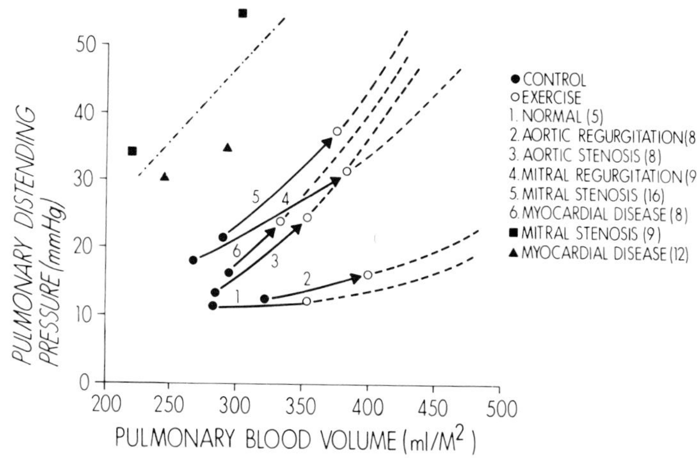

Fig. 18-7. The relationship of pulmonary distending pressure (PD) (the average of the pulmonary artery mean pressure and left atrial mean pressure) to pulmonary blood volume (PBV) at rest (black dots) and during supine leg exercise (circles). Note that the observed curves (solid lines) and their hypothetical extensions (dashed lines) are flat in normals and in patients with aortic regurgitation, but become steeper in those with other valvular or myocardial dysfunction. The solid symbols in the upper left quadrant with the (dot-dash) line represent data obtained at rest in functional class IV patients. The (black box) and (triangle) with PBV below 240 ml/m^2 represent 8 patients with mitral stenosis and 9 patients with primary myocardial disease, respectively. The highest value of PD of greater than 50 mm Hg and a PBV of 290 ml/m^2 is 1 patient with mitral stenosis studied during acute pulmonary edema. The other triangle represents 3 patients with primary myocardial disease in class IV who had normal PBV but an average PD of 34 mm Hg.

crement in PBV (i.e., the compliance of the pulmonary vascular bed was reduced).

The pathologic counterpart of these observations has been well substantiated. With long-standing pulmonary hypertension because of either valvular or myocardial dysfunction there is significant restriction of the pulmonary vascular bed. The exercise response may be viewed as representing different segments of a family of pressure–volume curves analogous to the relaxation pressure–volume curves in the normal dog described by Sarnoff and Berglund.[138] What is more pertinent, perhaps, is to emphasize that chronic postcapillary pulmonary hypertension influences the degree of pulmonary vascular disease. This factor is best appreciated in severe mitral stenosis.[166–169] The pathophysiologic alterations include (1) medial hypertrophy and intimal proliferation and fibrosis in pulmonary arteries and veins, (2) dilatation of proximal pulmonary arteries and narrowing of the distal pulmonary arteries and arterioles, frequently with atheromatous changes, (3) thrombosis and thromboembolism, (4) functional vasoconstriction because of hypoxemia, (5) interstitial and alveolar edema and interstitial fibrosis, and (6) dilatation of lymphatics and increase lymphatic drainage from the lung. The PBV may decrease by any or all of these changes that lead to progressive obliteration of the pulmonary vascular bed and reduction of its volume. The intravascular volume may be redistributed toward the apices of the lungs, partly as a result of increased pulmonary vascular resistance in the basal segments.

Many of the same pathologic changes occur in patients with left ventricular failure and pulmonary venous hypertension. Less prominent changes in the pulmonary vessels and in the interstitial space would be expected in these patients, since the duration of pulmonary venous hypertension is less than in the case of mitral valve disease. Nevertheless the pulmonary vascular volume may be reduced by the same pathophysiologic mechanisms considered above.

It is evident from our observations and those of Varnauskas that an elevated PBV is not a universal accompaniment in patients with chronic pulmonary venous hypertension.[76,129] Pressure elevation, rather than augmented PBV, appears to be the dominant hemodynamic feature, suggesting that the compromised pulmonary vascular bed is operating on a very steep portion of its pressure–volume curve.

From the previous discussion of distensibility it may be argued that the changes in PBV noted with exercise merely reflect recruitment phenomena within the pulmonary vascular bed. The exercise data largely obtained in patients with valvular heart disease in which both PBV and PEV were measured would support this view. That is, observations in a small group of patients suggested that the ratio of PEV to PBV did not change between control and exercise periods.[145] Our findings in 114 patients corroborate these impressions and suggest that failure of the ratio between PEV and PBV to increase would suggest recruitment of previously closed vessels.[146] However, recruitment must become less important as the degree of pulmonary vascular disease increases, since pressure elevations increase more than PBV does. We have also noted that in patients with mitral valve obstruction, atrial tachypacing results in left atrial and pulmonary arterial pressure elevation accompanied by increases in both PEV and PBV but without a change in cardiac output.[146] Under such circumstances any recruitment phenomenon must be accomplished solely by pressure elevation. The ratio of PEV to PBV continues to remain relatively constant during 10 to 15 min of the intervention. By contrast, if left atrial pressure elevation is maintained for 40 to 60 min there is an increase in the ratio of PEV/PBV, especially during the 15 min in the recovery period. These findings strongly suggest a distensibility factor that requires time for its manifestations. These observations in man are consonant with animal studies reported by others.[170-172] We would conclude that the increase in PBV associated with left atrial hypertension during short-term exercise may represent a recruitment phenomenon, but that the increase noted with prolonged atrial pacing more closely reflects an increase in distensibility with less important changes in pulmonary vascular recruitment.

Response to Hypoxemia at Sea Level

In 1946 von Euler and Liljestrand evoked an increase in mean pulmonary artery pressure in anesthesized cats by changing the inhaled gas mixture from room air to one of low oxygen content.[173] Two years later Liljestrand observed that the increase in pulmonary artery pressure was independent of pulmonary blood flow, left atrial pressure, and nervous effects within the lungs.[174] Since that time numerous investigators have noted similar pulmonary hypertensive effects from acute anoxia in experimental animals.[175-179]

The pulmonary circulatory effects of acute hypoxia have also been extensively studied in man.[180-185] These effects include a universal increase in mean pulmonary artery pressure without significant changes in either pulmonary wedge or direct left atrial pressure, little or no change in cardiac output or CBV, but an increase in calculated PVR. It has been suggested that pulmonary vasoconstriction must be involved in these findings, since the mean-PA–mean-LA gradient increases without a change in blood flow[181,183] and that blood flow is diverted from a hypoxic lung to a well-oxygenated lung.[186] It has further been demonstrated that pulmonary vasodilatation is produced by acetylcholine administered during acute hypoxia.[185,187]

The pulmonary hypertensive response to acute hypoxia is related to both oxygen tension and hydrogen-ion concentration. In young calves an increase in mean PA pressure is greater when the alveolar O_2 tension is lower.[188] At relatively high pH the mean PA pressure is less sensitive to hypoxia. However, as hydrogen-ion concentration is increased, the sensitivity of the pulmonary vascular response to hypoxia is greatly enhanced.[189-192]

In studies from our laboratory, inhalation of 12% O_2 in nitrogen in man resulted in a decrease in arterial O_2 saturation from 94% to 72%. Slight increases in heart rate and cardiac index occurred. Mean pulmonary artery pressure increased by 6 mm Hg without a change in mean left atrial pressure or systemic artery pressure. Both CBV and PBV fell significantly. These findings in which mean PA pressure rose but PBV decreased were interpreted as indicative of pulmonary vasoconstriction.[185]

The locus for this action remains uncertain. Present evidence suggests that the primary site is precapillary and includes both pulmonary arteries and arterioles.[191] Alveolar gas may act directly on precapillary vessels, since it has been shown that the former may diffuse through pulmonary tissue into the latter.[193]

Response to High Altitude

Over 40 years ago Hurtado first reported pulmonary hypertension in residents at high altitude.[194] A number of investigators have noted a 50% to 100% increase in mean pulmonary artery pressure in those acclimatized to high altitude, as compared to their sea-level counterparts.[195–198] Despite this, mean pulmonary wedge pressure remained normal, so that the pulmonary hypertension reflected an increase in pulmonary vascular resistance. Pulmonary hypertension was observed not only in adults but also in young children. Penaloza and co-workers found that children born at high altitude had pulmonary hypertension, which did not revert to normal during adulthood.[196] These children differed strikingly from their sea-level contemporaries, who achieved a normal pulmonary artery pressure within 6 months after birth. The pulmonary hypertension of high altitude has both anatomic and functional vasoconstrictive components. Anatomic studies have demonstrated thickening and medial hypertrophy of pulmonary arterioles in both man and animals living at high altitudes.[198–200] Functional vasoconstriction has been demonstrated by a reduction of mean PA pressure during 100% O_2 breathing or following administration of vasoactive drugs such as tolazoline, isoproterenol, and acetylcholine.[200–204] If, on the other hand, natives to high altitudes are exposed to acute hypoxia there is a marked increase in mean PA pressure without significant alterations in cardiac output or mean pulmonary wedge pressure.[204–206] The pulmonary vasoconstriction appeared to be located in the arteries and arterioles.

The studies of Roy and associates indicate a form of successful adaptation to high-altitude hypoxia in normal subjects.[207] In these individuals studied at sea level within 120 hr after return to sea level from high-altitude exposure for 1 to 2 years there was a 60% increase in cardiac index and comparable increases in CBV and PBV despite only a modest increase in mean PA and mean LA pressures. These findings are in keeping with a physiologic adaptation to a generalized increase in blood volume.

RESPONSE TO PHARMACOLOGIC INTERVENTIONS

Digitalis Preparations

Human and animal investigations have demonstrated that digitalis glycosides act on both the myocardium and the peripheral circulation.[208–218] The latter action is quite uniform and results in augmented systemic arterial and venous resistances leading to a decreased venous return. Among normal subjects and patients with cardiopulmonary disease the myocardial effect may vary. However, positive inotropic effects are usually manifested by an increased rate of rise of left ventricular systolic pressure, as well as by augmented myocardial contractility as determined from epicardial strain-gauge arches or changes in force–velocity relationships in the left ventricle or by increased left ventricular stroke work in the presence of a lower left ventricular end-diastolic pressure at rest and during supine exercise.

In normal subjects the cardiac output and pulmonary vascular pressures are affected little by the administration of the drug.[219–221] On the other hand, in patients with cardiomegaly but no clinical evidence of congestive heart failure, digitalis usually caused increases in cardiac output, stroke volume, and left ventricular stroke work with an unchanged or decreased left ventricular diastolic pressure.[213,214,222–224]

In patients with primary pulmonary disease the action of the drug has been variable. In some the cardiac output increased without appreciable changes in the degree of right ventricular and pulmonary hypertension.[225–227] In others a decrease in cardiac output accompanied an increase in mean pulmonary artery pressure.[225,229] A third type of response consisted of a decrease in both cardiac output and pulmonary artery pressure.[225]

In patients with left ventricular failure, cardiac output increased and was accompanied by a fall in mean pulmonary artery, left atrial, and left ventricular diastolic pressures. These changes were not accompanied by changes in right-heart filling pressures.[214,222,227–229]

Studies from our laboratory suggested that acute digitalization has a variable effect on the pulmonary circulation, depending upon pressure changes in the left heart and pulmonary artery.[230] In patients in whom acute digitalization with acetyl strophanthidin resulted in a fall in left ventricular diastolic, left atrial, and pulmonary artery pressures, the cardiac output increased slightly and was accompanied by a significant decrease in heart rate. There was a concomitant significant decrease in both CBV and PBV. In another group of patients the drug caused a slight increase in cardiac output and stroke index without appreciable change in mean pulmonary artery, left atrial, and left ventricular diastolic pressures. In most of these patients the resting pressures were normal to begin with, and no change occurred in either CBV or PBV. These data suggest that the left-heart and pulmonary artery pressures must decrease before changes in CBV and PBV occur. The decrease in PBV is an effect of improved left ventricular performance and suggests that the concordant fall in pressure and volume after acute digitalization is predominantly passive.

Adrenergic Drugs

Epinephrine

Epinephrine stimulates both alpha and beta adrenergic receptors. It is a peripheral vasoconstrictor as well as a powerful myocardial stimulant.[231] In animals intravenous epinephrine usually caused an increase in mean pulmonary artery and mean systemic artery pressures accompanied by an increase in cardiac output.[232,233] A number of explanations for the observed increase in mean pulmonary artery pressure have been suggested, including a rise in left ventricular filling pressure, an increase in pulmonary blood flow, redistribution of the blood volume centrally, and a direct pulmonary vasoconstricting effect. The studies of Borst and associates and of Gaddum and Holtz in isolated lungs suggested a vasoconstricting effect in the pulmonary circulation.[234,235] Despite these vasoconstrictive effects, Feeley and co-workers noted a significant increase in PBV in intact dogs during epinephrine infusion, which was attributed to increased transmural distending pressure.[236] The latter tended to offset vasoconstriction, which was found in the isolated animal preparations.

In man epinephrine infusion results in a rise in mean pulmonary artery pressure, cardiac output, and stroke volume, but little change in pulmonary artery wedge pressure.[237,238]

Norepinephrine

In animals intravenous norepinephrine causes a consistent rise in mean pulmonary artery pressure and an increase in blood flow if the heart rate is accelerated.[233,236,243] Borst and associates felt that norepinephrine caused vasoconstriction.[234]

In human subjects mean pulmonary artery pressure has been observed to increase consistently. Most investigators have interpreted this change as secondary to a rise in left ventricular diastolic and left atrial mean pressures.[243–245] However, Patel and co-workers noted significant increases in PVR and a widened PAm–PWm gradient that was disproportionate to blood flow.[244] They concluded that this finding suggested active pulmonary vasoconstriction. Others, on the basis of an observed increase in PAm pressure prior to changes in LAm pressure, also inferred that the drug caused vasoconstriction.[246]

Isoproterenol

This beta agonist has been extensively studied in animals, normal subjects, and patients with cardiopulmonary disease.[247–254] Isoproterenol causes cardioacceleration as well as a fall in PAm and a reduction in PAm–PWm gradient. The calculated PVR therefore decreases. In anesthetized intact dogs the drug caused a significant increase in PBV both before and after acute pulmonary embolization.[236] Studies in isolated supported lung preparations indicated vasodilator effects on both pulmonary arteries and veins.[255]

In man isoproterenol given sublingually or intravenously has been noted to increase PBV, reduce mean pulmonary artery and left atrial pressures, and decrease end-systolic dimensions of both ventricles.[256,257] Our observations on the effects of intravenous isoproterenol have included study of 35 patients.[254] In 5 patients with myocardial disease and 1 normal subject the administration of the drug resulted in significant increases in cardiac index, heart rate, and stroke index accompanied by a fall in mean left atrial pressure and systemic artery mean pressure. Both PBV and CBV increased. In contrast, in 18 patients with mitral stenosis the positive inotropic effects of the drug were associated with significant increases of mean pulmonary artery and left atrial pressures and a decrease in mean brachial artery pressure. Both PBV and CBV rose significantly. In 11 patients with aortic valve disease the positive inotropic effects of the drug were accompanied by a slight decrease in mean left atrial and brachial artery pressures. However, PBV and CBV did not change. These effects may be explained by the multiple actions of the drug. First, isoproterenol causes pulmonary vasodilation (or increased vascular recruitment) resulting in an increase in volume despite a fall in PAm and LAm pressures (patients with primary myocardial disease or normal function). Second, isoproterenol causes a positive inotropic effect that is responsible for the increase in cardiac output and stroke volume and the decrease in LAm and left ventricular diastolic pressure. If only a fall in left-heart filling pressure occurred, the PBV would be expected to decrease. The fact that PBV did not change significantly in patients with aortic valve disease suggests that opposing effects of pulmonary vasodilatation and increased myocardial contractility were operative, the net result of which was an unchanged PBV. Third, the increased inotropicity in patients with predominant mitral valve obstruction resulted in an increase in cardiac output and in heart rate. These factors increased both mean pulmonary artery and left atrial pressures. In these patients an increase in mitral valve flow is accompanied by a rise in the diastolic pressure gradient between left atrium and left ventricle. Left atrial and pulmonary artery pressures would therefore increase. The increased pressures would tend to increase PBV passively. In these circumstances pulmonary vasodilatation would be obscured by passive increases in PBV. Thus both pulmonary vascular pressures and volume would increase.

In these studies the potential effects of the drug upon bronchomotor tone and intraalveolar pressure were not investigated. Rodbard has suggested that both may influence pulmonary artery pressure.[258]

Aminophylline

The effects of drug infusion have been determined in many patients with cardiovascular and pulmonary diseases.[259–266] The drug has been shown to be an active vasodilator, based upon observations that cardiac output was unaffected or increased and was accompanied by a fall in mean pulmonary artery pressure and the gradient between pulmonary artery and pulmonary wedge pressures. Thus the calculated pulmonary vascular resistance declined.

In our experience in patients with valvular heart disease involving the mitral or aortic valve, infusion of aminophylline caused the cardiac output to change little, but did result in significant falls in both pulmonary artery and left atrial pressures. The calculated pulmonary vascular resistance decreased. Despite the fall in pressures, PBV increased in both groups of patients, suggesting pulmonary vasodilatation. It is likely that the mechanism of action is at least in part a direct one on the pulmonary vessels. However, the role of bronchodilation could not be excluded. Parker and co-workers, in a study of the drug's effect in patients with cor pulmonale, suggested that the fall in pulmonary artery and left atrial pressures was greater than could be accounted for by bronchodilation alone.[264]

Acetylcholine

This drug has a major action on the neuroeffector junctions of the vagus nerve. It was early appreciated that injection of a small dose of the drug into the isolated lung resulted in a decrease in mean pulmonary artery pressure.[267] Twenty years later Harris and associates noted that acetylcholine caused a fall in mean pulmonary artery pressure in man without significant alterations in cardiac output, systemic artery pressure, or indirect left atrial pressure, thus implying that the drug was destroyed or metabolized in the pulmonary circulation.[268,269] Subsequent studies established that in man (1) acetylcholine caused a reduction in mean pulmonary artery pressure with no change in blood flow in normal man,[270] (2) it caused a greater effect on pulmonary artery pressure when administered during inhalation of a low-oxygen mixture,[270] and (3) its effects were more prominent in the presence of pulmonary hypertension secondary to various types of cardiac or pulmonary disease.[76,271–275] In general the fall in mean pulmonary artery pressure was accompanied by a decrease in calculated pulmonary vascular resistance but no change in pulmonary wedge pressure or direct left atrial pressure. Blood flow either remained constant or increased, while PBV and CBV rose. The increase in PBV in the face of a lower mean pulmonary artery pressure suggests pulmonary vasodilatation. Furthermore, the observation of Fritts and associates[270] that the drug counteracted hypoxic pulmonary vasoconstriction was confirmed by the finding that a rise in PBV associated with a fall in mean pulmonary artery pressure occurred despite the continued hypoxic stimulus.[76]

Diuretics

The effectiveness of these drugs in the treatment of acute and chronic congestive heart failure has been well established since the 1940s, beginning with the mercurial diuretics. Early studies noted a rise in cardiac output and a fall in right atrial pressure in a patient in congestive heart failure following a single intravenous injection of a mercurial diuretic. Subsequent studies by Rowe and associates also described a fall in cardiac output as well as in pulmonary artery pressure in hypertensive patients treated with intravenous chlorothiazide.[276] Similar hemodynamic findings have been noted by Rader and by Stampfer and their associates in studies of patients with primary myocardial dysfunction[277] and those with valvular heart disease.[278]

The advent of more rapidly acting diuretics such as ethacrynic acid and furosemide that act primarily to inhibit sodium reabsorption along the ascending loop of Henle and that cause profound diuresis particularly in the presence of congestive heart failure with pulmonary edema has led to more extensive studies of drug effects on both the systemic and pulmonary circulation. Although the drugs are structurally different, they presumably have similar diuretic and hemodynamic effects. The latter, however, are modified by the clinical state of the patient. In general, in patients without clinical evidence of cardiac decompensation, acute administration of the drugs results in a decrease in plasma volume, central venous pressure, and cardiac index, with little or no change in pulmonary artery or systemic artery pressures.[279–281] In the presence of chronic congestive heart failure the fall in plasma volume as well as right atrial, left ventricular filling, and pulmonary artery pressures is accompanied by an increase in cardiac index in some patients and a fall in others.[280–287] In some instances PBV and CBV have been observed to decrease,[282,283] while in other instances no changes in volume have been observed.[287]

In the presence of acute myocardial infarction with heart failure, the mechanisms have been thought to be similar.[284–286,288–290] Tattersfield studied 30 patients with acute myocardial infarction and found a decrease in pulmonary artery pressure, stroke index, and cardiac work 1 hr after drug administration. These effects persisted for an additional 5 hr and were attributed to intravascular volume depletion secondary to diuresis. However, other studies, including our own, suggested that hemodynamic evidence of pulmonary congestion may decrease prior to the diuretic action.[283,286,287,290] In the study of Dikshit and associates of patients with acute myocardial infarction the primary hemodynamic effects of furosemide appeared to be peripheral vascular changes in which venous capacitance increased within 5 min of drug infusion and was accompanied by a concomitant fall in pulmonary artery diastolic pressure.[286] These changes were quickly followed by increases in renal plasma flow and glomerular filtration. Increased venous capacitance and increased renal blood flow probably accounted for the initial central hemodynamic effects of a decrease in pulmonary artery pressure and cardiac output. After 15 min there was a significant diuresis that caused a further reduction in plasma volume and left ventricular filling pressure. Both the early vascular and later diuretic effects of the drug would tend to decrease heart size and myocardial fiber length, resulting in a decreased myocardial O_2 demand.

Our observations in 27 patients with left atrial hypertension tend to confirm and extend these findings.[287] Following intravenous ethacrynic acid or furosemide infusion we have observed consistent decreases in cardiac index, pulmonary artery pressure, and left atrial pressure that begins within 15 min and lasts for 60 to 120 min after infusion. Pulmonary blood volume measured at 20, 40, and 60 min postinfusion did not change significantly despite reduction in cardiac output and significant decreases in pulmonary artery and left ventricular filling pressures. Likewise, pulmonary extravascular volume was unchanged at 1 hr after drug infusion. Nevertheless, plasma volume was significantly lower at 60 min postinfusion.

Summarizing recent observations concerning diuretic action, it may be stated that (1) the cardiac output and right and left ventricular filling pressures are decreased during acute diuresis; (2) the initial effects of furosemide are probably due to increased systemic venous capacitance and renal blood flow and secondarily to altered central hemodynamics as a result of a decrease in circulating plasma volume; (3) the decreases in cardiac output, pulmonary artery pressure, and left atrial pressure are probably not due to a negative inotropic effect of the drugs on the myocardium; (4) the earliest observed decreases in central pressures and in cardiac output probably reflect peripheral venodilatation (furosemide); (5) reduction of the plasma volume is not paralleled by a reduction in pulmonary extravascular volume, suggesting that mobilization and removal of lung water is a delayed phenomenon that requires prolonged (> 2 hr) decreases in pulmonary capillary pressure to be effected; and (6) in patients with chronic postcapillary pulmonary hypertension the fall in pulmonary artery and left atrial pressures without a coincident decrease in PBV is attributable to decreased pulmonary vascular compliance. Under these conditions small decreases in PBV were associated with significant fall in PAm and LAm pressures.

PULMONARY CONGESTION AND EDEMA

As has been pointed out by Robin and associates, the early phylogenetic percursors of mammals forsook a water existence and became atmosphere-dependent over 1 billion years ago.[14] In this evolution the gas-exchange mechanism had to be made dry instead of persistently wet. This required elaborate structural and functional changes in adaptation that have been enumerated previously. These alterations included development of airways and alveoli in close proximity to the pulmonary capillary bed, a system to maintain the geometry of the terminal airways, development of an elaborate but efficient interstitial space to protect the integrity of the alveolar–capillary barrier, and an elaborate lymphatic system to siphon off interstitial fluid and protein, lest it interfere with gas exchange. While the entire system has proved to be efficient, there is an uneven vulnerability to fluid overloads that is determined by anatomic site, geometry, gravitational factors, and the integrity of the capillary–alveolar interface. Failure of one or more of these mechanisms or the inability to compensate for defects in a portion of the system causes pulmonary edema, which may be defined as abnormal accumulation of liquid in the interstitium or air spaces in the lungs.

The etiologic classification of pulmonary edema includes mechanisms of (1) increased pulmonary capillary hydrostatic pressure, (2) altered pulmonary capillary permeability, and (3) mixed causal mechanisms. However, it must be emphasized that in any given condition multiple factors may be causally related, and in other conditions the mechanisms are not known.

Hydrostatic Pulmonary Congestion and Edema

The definition of pulmonary congestion varies, depending upon one's perspective. In adults it is characterized by clinical features of pulmonary dysfunction, namely dyspnea of effort, orthopnea, paroxysmal dyspnea, pulmonary rales, and/or bronchoconstriction; at times it is accompanied by systemic congestion and pleural effusions. It is commonly seen in patients with valvular or myocardial disease, particularly myocardial infarction. In children tachypnea may dominate the clinical picture, whereas signs of pulmonary congestion may not be prominent but may be readily demonstrable radiographically. Radiologic definitions of pulmonary congestion may include "pulmonary vascular engorgement," as seen in patients with left-to-right intracardiac shunts, and "pulmonary edema," a fanlike increase in pulmonary vascular markings described as "bat wing" or "butterfly"[291–293] in which the hilar regions appear edematous whereas the periphery, apices, and bases appear to be spared, as seen in patients with acute left ventricular failure complicating acute myocardial infarction or hypertensive heart disease. These striking anatomic inhomogeneities suggest central impairment of drainage mechanisms as well as greater liberation of water into the central as compared to the peripheral interstitium.[18] Pulmonary congestion also is frequent in mitral or aortic valvular disease. However, in addition to the increase in pulmonary vascular markings there is a cephalad redistribution of pulmonary blood flow that is quite characteristic. Thus pulmonary arteries and veins in the lower lobes are narrowed, while those in the upper lobes are distended. Furthermore, interstitial edema manifested by engorged pulmonary lymphatic and seen as Kerley B lines in the lung bases is characteristic.

In the clinical conditions cited the net effects of gravity and augmented ventilation and the modifying effects of interstitial fluid pressures on regional lung perfusion may become difficult to predict. For example, local hydrostatic capillary pressure at the lung bases may be altered by decreasing precapillary vessel caliber due to increases in local interstitial pressures.[7,32,84] This factor may oppose gravity, which predisposes to dependent edema, because of greater basilar hydrostatic pressure and because of the free communications within the interstitial space with pulmonary lymphatics. The resultant lung response may be further modified by tachypnea, which tends to promote lymphatic drainage.

From a pathologic viewpoint, the lungs in pulmonary edema are filled with fluid that can be easily expressed from the cut surface. However, recognition of early pulmonary edema may be difficult because of the subtle transitions between areas of expanded intravascular and extravascular volumes. The distribution of lung fluid becomes unreliable because of the variable time lag between death and examination. Histologic examination is also compromised by the inability to identify lung water. However, histology is of value in defining the duration of the pulmonary hypertension that precedes pulmonary edema. If the pulmonary edema has been as acute as that observed following acute myocardial infarction, the walls of the pulmonary blood vessels may appear to be nearly normal. On the other hand, chronic pressure elevations as seen in mitral valve disease result in medial hypertrophy and intimal proliferation of the pulmonary arteries and veins, particularly in the lower lobes of the lung.

As has been indicated previously, the major manifestations of hydrostatic congestive heart failure and pulmonary edema associated with left ventricular myocardial disease or mitral valve disease have been respiratory.[294–296] The development of acute pulmonary edema may be an intercurrent or terminal dramatic event.[297–299] These events occur most often with nocturnal recumbency, strenuous exertion, paroxysmal tachycardias, or, at times, rapid intravenous infusions of blood or fluids. Pulmonary function is impaired in these patients and is characterized by a decrease in vital capacity, a relative or absolute increase in residual volume to total lung capacity, and a decrease in pulmonary compliance.[300–303] Pulmonary capillary volume may be increased or slightly decreased, depending upon the underlying cardiac disease and the magnitude of decompensation.[304–306]

Hemodynamically, cardiac output and ventricular stroke work are decreased in heart failure. In the presence of myocardial disease, left ventricular diastolic pressure is elevated and mean left atrial pressure is increased both in the presence of mitral stenosis and left ventricular failure. Pulmonary venous and pulmonary arterial hypertension are present, and the right ventricular filling pressure may or may not be increased depending upon whether the right ventricle remains competent.

The etiologic importance of an increased PBV in pulmonary congestion and in edema associated with mitral valve disease or myocardial disease has been emphasized in the past by many groups.[307–310] The hypothesis of an augmented PBV encroaching upon alveoli, with displacement of air, has also been advanced.[311–313] The supposed dominant role of the pulmonary blood volume depended upon indirect evidence, including clinical examinations, chest roentgenograms, and measurements of pulmonary pressures and central blood volume. The latter was subsequently found to be only slightly elevated in the presence of pulmonary congestion,[314] and others could not correlate diminished vital capacity or lung compliance with elevated pulmonary vascular pressures.[315,316] Into this uncertainty Ebert interjected the explanation that the absence of a large increase in PBV may be the result of altered pressure–volume characteristics of the pulmonary vascular bed.[317] Studies reported from the laboratory of Varnauskas as well as from our own have confirmed the concept that the pulmonary circulation in congestive heart failure is dominated by pressure elevations rather than by an increase in PBV.[76,127,165] These observations were in agreement with the observations of Sarnoff and Berglund, who constructed pressure–volume curves for the isolated canine lung.[138] They observed that in the initial portion of the curve pressures changed little over a large change in volume. However, in a given curve, as volume was increased further, pressures rose rapidly, so that a small increase in volume resulted in a disproportionate increase in pressure.

It seems plausible that in patients with advanced heart disease and chronic pulmonary congestion the normally flat pressure–volume curve was shifted upward and became progressively steeper with respect to the volume axis. Ultimately, in class IV patients the PBV may be normal or only slightly increased, although the mean pulmonary artery and left atrial pressures are markedly elevated. In this situation a further small increment in PBV would result in marked elevation of intrapulmonary pressures (Fig. 18-7). The lines of evidence may be summarized as follows:[165]

1. In patients with normal hemodynamics, supine exercise, atrial pacing, acute transfusion, and phlebotomy alter PAm and LAm pressures little, suggesting that the normal lung operates on a flat portion of its pressure–volume curve.

2. In chronic pathologic states in which pulmonary vascular pressures are elevated, the increase in PBV noted at rest and during these interventions is less than the increments in pressure.

3. In the presence of chronic heart failure secondary to left ventricular dysfunction or mitral valve disease, pulmonary artery and left atrial pressures are very high, but PBV is in the normal range or even below it.

4. An increase in PBV per se does not seem important in producing pulmonary edema in certain conditions. For example, in most patients with atrial septal defects large increases in pulmonary capillary volume[318,319] and presumably in PBV are frequently present with normal pulmonary wedge pressures and without signs or symptoms of pulmonary congestion or edema.

The pathologic expression of the structural changes in the pulmonary vascular bed both from within and from without have been elucidated in previous sections of this chapter.[166–169]

Despite the foregoing emphasis on chronic congestive heart failure, it should not be construed that the PBV may not be significantly increased in acute left ventricular failure. Other clinical situations could arise in which the PBV is acutely elevated. These would include patients with "normal" cardiovascular function as well as patients with "compensated" heart disease who have some elevation of PBV and pulmonary vascular pressures. They may develop acute left ventricular failure after injudicious intravenous administration of fluids or blood. Experimental support for this concept in the dog has been reported by Levine and coworkers.[171]

Extravascular Volumes in Animals

In 1965 Levine and associates attempted to distinguish pulmonary congestion with edema from pulmonary congestion alone based upon clinical observations, pulmonary function abnormalities, and measurements of PBV and pulmonary extravascular volumes (PEV) in dogs.[171] Pulmonary congestion alone was associated with an increase in PBV but not PEV. When pulmonary edema supervened PEV increased significantly while PBV showed no further change. The morphologic correlates of these studies included interstitial edema with swelling of connective tissue around lobar and segmental pulmonary arteries and veins. Electron microscopy revealed dilatation of the interstitium with expansion of collagen-containing areas. However, in areas where alveolar epithelium and capillary endothelium were in intimate proximity, no architectural distortion was noted[28] (Fig. 18-2).

Extravascular Volumes in Man

Assessment of pulmonary extravascular volume has been made by a number of investigators.[145,147,148,165,320–325] The value for PEV in either normal subjects or patients with hemodynamically compensated coronary artery disease (and a presumed normal pulmonary circulation) has varied from 80 to 120 ml/m^2. In two series reporting values for normal subjects the values in milliliters were divided by an assumed body surface area of 1.8m^2.[320,321] The findings in more than 150 patients studied in our laboratory,[148] in which the results in 108 patients have been previously published, indicate the following: There is good correlation between PEV and pulmonary intravascular pressure, particularly when the latter is elevated secondary to mitral valve obstruction. Among patients with coronary artery disease, primary myocardial disease, and aortic valve disease, PEV varied considerably when left atrial mean pressures were normal or nearly so (Fig. 18-4). When the values of PEV were analyzed according to functional disability, those with mild disability (functional classes I and II) had quite normal values for PEV and little or no left atrial hypertension. However, with further functional impairment to class III, both left atrial pressure and PEV were elevated (Fig. 18-4). The latter values were conspicuously higher in patients with mitral valve disease than in those with aortic valve disease or coronary artery disease, suggesting that persistently elevated left atrial and pulmonary capillary pressures are associated with marked increases in lung water. These findings are in substantial agreement with those of McCredie[322] and Luepker and associates.[147] In patients with the most compromised function (class IV), left atrial hypertension and PEV remained linearly related.[320,321]

A relationship between radiographic findings and PEV, PEV/PBV, and left atrial pressure was also identified.[148,325,326] Among patients with normal chest roentgenograms the LAm was normal, the PEV averaged 115 ml/m², and the PEV/PBV averaged 0.37. In patients whose x-rays indicated pulmonary venous congestion and interstitial fluid the LAm averaged 22 mm Hg, the PEV 158 ml/m², and the PEV/PBV 0.50. Radiologic evidence of pulmonary edema was associated with PEV of 180 ml/m², LAm of 28 mm Hg, and PEV/PBV of 0.57. Thus as evidence of fluid accumulation increased the LAm and PEV rose, whereas the PBV did not change, resulting in an increase in the ratio of PEV to PBV. In the presence of normal LAm pressure the PEV/PBV averaged 0.37. With development of left atrial hypertension the ratio increased significantly (Fig. 18-8). It is therefore apparent that the chest roentgenogram reflects primarily an accumulation of lung water and that this parameter must be considered in addition to measurements of PBV and the relationship between flow and pulmonary vascular pressure.

Pulmonary Capillary Pressure in Hemodynamic Cardiac Edema

Although it is common to consider that hydrostatic pressure elevation is always present, this need not necessarily be the case. We have already suggested that regional pulmonary capillary pressure may be quite variable and that there may be time factors associated with the accumulation and removal of both interstitial and intraalveolar edema fluid. The main evidence is the time sequence of removal of edema fluid. Thus in some situations left-sided circulatory pressures may fall at a time when pulmonary edema is still evident.[327] In addition, the rate of removal of edema fluid may be modified by the functional capacity of the pulmonary lymphatics and the magnitude of interstitial hydrostatic and oncotic pressures.[14] Furthermore, hydrostatic pressure elevations may affect endothelial membranes and intercellular junctions to alter capillary permeability.[27]

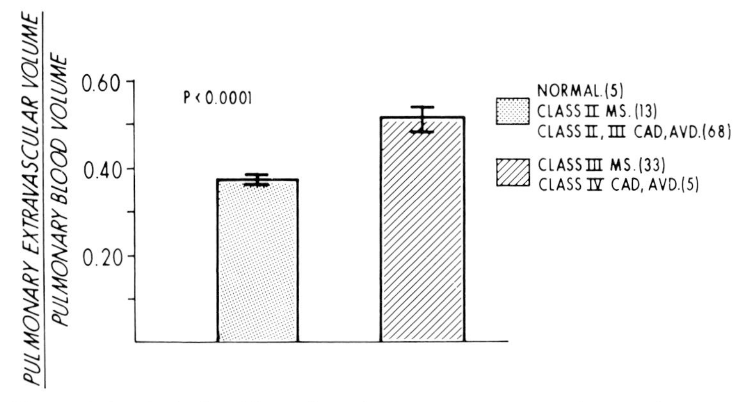

Fig. 18-8. Relationship of pulmonary extravascular volume to pulmonary blood volume in patients with left atrial mean pressure below 13 mm Hg (left) and those with left atrial hypertension (right). The bars indicate one standard error of the mean. [Reproduced by permission from Bloomfield DA (ed): Dye Curves, 1974, and by University–Park Press, Baltimore.]

Noncardiac Hydrostatic Pulmonary Edema

Although diseases of the aortic and mitral valves and of the myocardium account for nearly all hydrostatic causes of pulmonary edema, left atrial myxomas and pulmonary veno-occlusive disease must also be considered. The myoxmas are characterized by symptoms of pulmonary congestion and signs of mitral valve disease, frequently in a clinical setting suggestive of bacterial endocarditis (i.e., fever, weight loss, debility, mitral murmurs, increased sedimentation rate, and peripheral embolic phenomena, but maintainence of sinus rhythm).[328] The last group of diseases includes veno-occlusive disease secondary to increased pulmonary blood flow, congenital stenosis of the pulmonary veins, and acquired pulmonary vein obstruction with mediastinal masses or ganulomata.[14] Veno-occlusive disease is characterized by pulmonary hypertension, impaired right ventricular function, and extensive pulmonary edema without evidence of left-heart disease or intracardiac shunts.[329,330] Pulmonary wedge pressure is frequently normal. The pathology of this disease includes medial and intimal changes in pulmonary arteries and occlusive lesions of pulmonary veins. Pulmonary wedge pressure may be normal, since the measurements may be made in areas downstream from obstruction. In other sectors of the lung the wedge pressure may be elevated if the catheter is lodged upstream from a site of venous obstruction.

Pulmonary Edema with Altered Pulmonary Capillary Permeability

This group of diseases includes those in which there is some form of physical or chemical damage to alveolar epithelium or capillary endothelium or both. Although frequently distinctive, this form of pulmonary edema may accompany hydrostatic pulmonary edema related to left-heart disease or increased intravascular volume. Thus there is frequently overlap in hydrostatic and permeability pulmonary edema.

Etiologic factors including infection, toxic inhalants, circulating toxins, immunologic reactions, radiation fibrosis, disseminated intravascular coagulation, and drowning have been well summarized by Robin et al.[14]

Because of their significant incidence in surgical and medical intensive-care units, two types of altered pulmonary capillary permeability edema deserve special mention.

Oxygen Poisoning

Along with a number of inhaled substances such as oxides of nitrogen, sulfur dioxide, phosgene, and ozone, high concentrations of inhaled oxygen produce direct injury to both capillary and alveolar cells. The direct damage leads to pulmonary edema in which protein content is high and there is subsequent formation of hyaline membranes. Both endothelial and epithelial cells become edematous and may be destroyed. Unlike the type I epithelial cells, the type II pneumocytes, which are metabolically active and contain many organelles, resist destruction by high O_2 exposure. In the monkey they are a source for regeneration of cuboidal alveolar epithelium during the repair phase of oxygen poisoning, which occurs after the fourth day of exposure.[331,332] Observations in man are similar. In the sequence of events of edema formation, interstitial fluid accumulation appears to preceed alveolar edema, just as it does in hydrostatic forms of pulmonary congestion.[333]

The basic mechanism for O_2 toxicity is not clearly established. If only one lung is exposed to high O_2 and the other to room air, the pulmonary pathology is unilateral. The cytotoxicity of oxygen appears to depend upon peroxidation of membrane lipids.[334] Vasoactive substances, proteases, and lysosomal enzymes probably play an important etiologic role.[14]

From a clinical viewpoint the hazards of increased inspiratory O_2 tension begin at concentrations greater than 50%–60% (350–430 mm Hg). In normal volunteer experiments, exposure to O_2 at 1 atm or more is usually terminated with the onset of substernal distress and pain on coughing. Exposure to 100% O_2 at 2 atm causes a reduction in vital capacity and pulmonary compliance after 6 to 11 hr, but without alterations in the pulmonary capillary bed.[335,336] An early effect on 100% O_2 breathing at ambient pressure is a fall in vital capacity, which accelerates after 60 hr of exposure.[337]

In patients who are mechanically ventilated following cardiovascular surgery, or in acute respiratory failure associated with sepsis, pulmonary edema, or pulmonary embolism, subjective symptoms of O_2 toxicity are ordinarily not obvious. Singer et al. studied pulmonary function in patients 24 hr after cardiovascular operations during which they were ventilated with either 100% or 50% O_2.[338] At the end of this period no difference in pulmonary function determined by Q_s/Q_t, dead space, or effective compliance was noted between the two groups. Others have compared the pulmonary function and morphologic changes that develop during mechanical ventilation with air or O_2 in randomly selected patients with irreversible brain damage but normal lungs. The arterial O_2 tension (Pa_{O_2}) measured during O_2 breathing was the most sensitive indicator of changes in pulmonary function. The Pa_{O_2} measured during O_2 breathing fell rapidly in the O_2 group, reaching a low of 120 mm Hg after 50 hr. In contrast, in the air-ventilated group intermittent transient exposure to 100% O_2 resulted in a gradual decline in Pa_{O_2} that did not fall below 300 mm Hg. Radiographic evidence of diffuse pulmonary infiltration and pathologic evidence of pulmonary edema corroborated the physiologic differences noted. The microscopic pictures were similar in the two groups, however.[339]

The pulmonary morphology after more prolonged exposure to O_2 in man was studied by Nash et al.[340] At postmortem examination lungs were edematous. The chronologic response was divided into early exudative and late proliferative phases. The former was dominated by congestion, alveolar hemorrhage, and hyaline membrane formation. The latter, developing in a week to 10 days, was characterized by alveolar and septal edema with proliferation of fibroblasts and hyperplasia of alveolar lining cells, probably type II pneumocytes.

Although some have suggested that mechanical ventilation is of itself[340] important in the genesis of oxygen toxicity, recent experimental evidence in monkeys, rats, and lambs does not support this view.[331,332,341] Mechanical ventilation, instead of aggravating the toxic pulmonary manifestations of O_2, may prevent or delay their onset by reducing gas trapping, thereby preventing absorption atelectasis.[342]

It should not be inferred that the fear of O_2 toxicity should lead to withholding its use. There are many situations such as open-heart surgery, drug poisoning, congestive heart failure, pulmonary infection, neurosurgery, and head trauma in which interstitial and alveolar edema, hemorrhages, and hyaline membranes are the nonspecific responses to these insults. To withhold O_2 therapy from the patient made severely hypoxic by one of these conditions could be fatal long before O_2 toxicity becomes manifest. The physiologic tolerance to acute hypoxemia must be determined for each patient on a day-by-day basis. As a rule it may be assumed that inspired O_2 concentration greater than 50% will lead to serious pulmonary changes if the administration is continued for more than 48 hr. As stated earlier, the higher the O_2 concentration the more rapid will be the onset of O_2 toxicity. The target in O_2 therapy in patients with previously normal lungs is a Pa_{O_2} in the normal range of 70 to 100 mm Hg. As emphasized by Pontoppidan and associates, if the hypoxemia as a consequence of venous admixture is so great that inspired O_2 must be above 50%–60% in order to maintain a normal Pa_{O_2}, it may be desirable to settle for a Pa_{O_2} in the hypoxemic range. This decision will be in large part determined by the cardiovascular status of the patient, by his ability to increase oxygen transport by augmenting cardiac output, and by his tolerance of mild hypoxemia.[342] For a further detailed discussion the reader is referred to this excellent review[342] and to that by Clark and Lambertson.[343]

Adult Respiratory Distress Syndrome

This term has been used to describe many of the pulmonary complications enumerated above in which O_2 therapy and mechanical ventilatory assist play an important role. The litany of possible etiologies is long, probably because we do not understand many of the basic underlying mechanisms, and it includes pneumonia, toxic inhalants, endotoxic shock, hypovolemic shock, postcardiac bypass, disseminated intravascular coagulation, immunologic lung reactions, drug poisoning, pulmonary edema associated with valvular or myocardial disease, and pulmonary embolism. In many general hospitals this syndrome accounts for about 20% of respiratory intensive-care unit admissions and ranks behind only drug ingestion and neuromuscular disease (25% each). The syndrome has been defined by Pontoppidan and associates as a "state in which Pa_{O_2} is below the predicted normal range for the patient's age at the prevalent barometric pressure (in absence of intracardiac right-to-left shunt) or Pa_{CO_2} above 50 mm Hg (not due to respiratory compensation for metabolic alkalemia)."[342]

The major pathophysiologic features are closure of airways or alveoli or both, resulting in abnormal distribution of gases coupled with interstitial lung edema caused by pulmonary capillary hydrostatic pressure elevation or by increased capillary permeability. Both abnormalities are regional, but tend to be greatest at the lung bases. The result is abnormalities of VA/Q and reductions in pulmonary compliance and in functional residual capacity. The reduction in lung volume results in closure of some airways throughout inspiration. The air trapped distally is absorbed and atelectasis results. If perfusion continues, venous admixture (shunting) occurs. Coincidentally, there is impairment of surfactant function, which leads to progressive closure of alveoli.

Pulmonary Edema of Mixed or Unknown Etiology

In this category are such varied and multifaceted conditions as pulmonary parenchymal disease, pulmonary embolism, eclampsia, cardioversion, and cardiopulmonary bypass. Of particular interest are the following conditions:

High-Altitude Pulmonary Edema

The most common clinical profile of this disorder is that of a sea-level-dweller in previous good health who rapidly ascends to high altitude for vigorous work or recreation such as skiing or mountain climbing. After or during exertion he is suddenly seized by acute respiratory distress, followed by the rapid appearance of pulmonary edema that usually responds dramatically to O_2 administration.[344] Pathogenetic mechanisms remain unclear. The most consistent findings have been those of pulmonary artery hypertension associated with normal pulmonary wedge pressure, thus excluding left ventricular failure.[194,345] Mild or moderate hypoxemia does not appear to significantly alter pulmonary capillary permeability.[346,347]

An important etiologic factor appears to be constitutionally increased pulmonary vasoreactivity.[348,349] In addition, there must be regional inhomogeneity of pulmonary capillary pressures, so that some pulmonary-artery–pulmonary-capillary channels are exposed to high pressures[18] and/or that edema formation may occur proximal to vasoconstricted pulmonary arterioles by direct effects upon the pulmonary arterial walls.[351] To account for pulmonary edema that develops in residents of high altitudes after a sojourn at sea level followed by a return to altitude, it has been suggested that pulmonary arteriole medial hypertrophy involutes in a nonuniform manner, while at sea level and upon return to altitude precapillary vasoconstriction is very uneven.[348]

In this condition the major pathophysiologic concept is that nonhomogeneous distribution of pulmonary vascular resistance is central to the development of pulmonary edema.

Overdose with Narcotic Agents

Pulmonary edema has been clearly described as a consequence of heroin overdose.[351] The public health problem from drug abuse has risen rapidly, and other drugs including methadone have been incriminated as a cause of pulmonary edema. The effects of these drugs are related to their pharmacologic effects and cannot be attributed to the vehicle, to adulteration, or to route of administration. The clinical picture is that of marked central nervous system depression, including hypothalamic dysfunction and severe hypoxemia and respiratory depression. At autopsy there is cerebral edema, which further suggests a neurogenic mechanism.[352] Chest x-rays usually reveal patchy pulmonary edema. One must ponder the role of increased capillary permeability either as a direct effect of the narcotic or secondary to hypoxemia, respiratory acidosis, and/or left ventricular failure. According to Fishman, pulmonary capillary hypertension may occur secondary to left ventricular failure (acidosis and hypoxemia) or nonuniform transmission of pulmonary hypertension (due to hypoxemia and acidosis) or impairment of lymphatic drainage (due to respiratory depression).[18]

Functional Disturbances of Pulmonary Edema

To recapitulate, pulmonary edema is manifested by alterations in gas exchange, in the pulmonary circulation, and in the delivery of oxygen to body tissues. Interstitial fluid accumulation and later alveolar edema reduce lung volume and elasticity, which is manifested as a decrease in compliance.[303] Since the distribution of edema is nonuniform, changes in pulmonary function are also heterogeneous, leading to mismatches between alveolar ventilation and corresponding capillary perfusion. The result is pulmonary arteriovenous shunting, giving rise to varying degrees of systemic arterial hypoxemia. If the pulmonary edema becomes alveolar as well, further compromise occurs as a result of inactivation or removal of surfactant. In such alveoli ventilation ceases but perfusion continues which augments pulmonary venous admixtures and intensifies the degree of systemic arterial hypoxemia. Not only does the degree of arterial unsaturation increase, but arterial O_2 tensions generated by 100% O_2 inhalation are considerably less than those normally predicted.[353,354] Initially in pulmonary edema systemic CO_2 tensions are lower and pH is normal or alkalotic, since hyperventilation in areas of lung little affected more than compensate for the areas of decreased or absent ventilation. However, as the edema process becomes more extensive alveolar hypoventilation occurs, resulting in hypercarbia and respiratory acidosis.[355,356]

In the hemodynamic forms of pulmonary edema, right ventricular systolic pressure and hence pulmonary artery pressure must rise to overcome the added resistance to left-heart inflow because of elevation of left ventricular diastolic pressure or atrial mean pressure in the presence of mitral stenosis. The burden on the right ventricle is further compounded by hypoxemia and increased flow resistance to the lower lobes. The former probably involves all pulmonary arteries, while the latter primarily affects the extraalveolar vessels in the lower lobes, as a result of interstitial fluid pressure that changes from negative to positive.

As we have seen, pulmonary venous or left atrial hypertension alters the flow distribution patterns throughout the lung. The normally hypoperfused apices of the upright lung now become hyperperfused as a result of the pulmonary arterial hypertension and the redistribution of blood from the bases toward the apices. These changes are particularly important in the evaluation of mitral or aortic valve disease and recurrent left ventricular failure due to myocardial factors.

Many small distal airways are closed prematurely or become obstructed as the lung volumes fall and the airways are exposed to interstitial pressure. Even in patients with acute myocardial infarction uncomplicated by clinical evidence of heart failure, closing volumes (i.e., the lung volume at which appreciable small-airway closure begins) were found to be elevated.[357] Others have emphasized that in the presence of pulmonary congestion there are major abnormalities in small-airway function that are characterized by increased peripheral airway resistance and premature airway closure.[358,359] The increase in peripheral airway resistance correlates significantly with pulmonary vascular pressures and clinical signs of congestive heart failure. Furthermore, the demonstration of frequency-dependent airway resistance was accompanied by observed elevations of extravascular lung water.[359]

As pulmonary edema increases in severity, bronchial edema, reflex bronchoconstriction, and asthmatic respirations ensue.

The initial insult to the entire organism from pulmonary edema is the development of progressive arterial hypoxemia. The associated hyperventilation secondary to stimulation of lung stretch receptors and anxiety is usually sufficient to cause a respiratory alkalosis that in turn aggravates decreased O_2 delivery to the tissues by increasing O_2 affinity for hemoglobin. However, as pulmonary edema worsens and the situation becomes critical, respiratory acidosis may supervene.[355] This occurs when edema enters the alveoli in amounts sufficient to obstruct terminal bronchioles. Elimination of CO_2 becomes impaired due to ventilation–perfusion abnormalities rather than to impairment of diffusion.[18] At this point ventilation can no longer be maintained at an increased level, either because of physical exhaustion or the concomitant effects of the former plus respiratory depressants such as morphine. Respiratory acidosis is now accompanied by metabolic acidosis as lactate accumulates in the tissues and is less well handled by the liver.

REFERENCES

1. Fishman AP: Respiratory gases in the regulation of the pulmonary circulation. Physiol Rev 41:214, 1961
2. Weibel ER: The ultrastructure of the alveolar capillary membrane or barrier, in Fishman AP, Hecht HH (eds): The Pulmonary Circulation and Interstitial Space. Chicago, University of Chicago Press, 1969, p 9
3. Weibel ER, Gomez DM: The architecture of the human lung. Science 137:3530, 1962
4. Wearn JT, Erstene AC, Bromer AW, et al: The normal behavior of the pulmonary blood vessels with observations on intermittency of the flow of blood in the arterioles and capillaries. Am J Physiol 106:236, 1934
5. Milic-Emili J, Henderson JAM, Dolovich MB, et al: Regional distribution of inspired gas in the lung. J Appl Physiol 21:749, 1966
6. Dollery CT, West JB: Regional uptake of radioactive oxygen, carbon monoxide and carbon dioxide in the lungs of patients with mitral stenosis. Circ Res 8:765, 1960
7. West JB, Dollery CT, Naimark A: Distribution of blood flow in isolated lung; Relation to vascular and alveolar pressures. J Appl Physiol 19:713, 1964
8. Friedman WF, Braunwald E: Alterations in regional pulmonary blood flow in mitral valve disease studied by radioisotope scanning. Circulation 34:363, 1966
9. Gil J: Methods for demonstration of interstitial and alveolar edema by electron microscopy, in Giuntini C (ed): Central Hemodynamics and Gas Exchange. Torino, Minerva Medica, 1971, p 17
10. Klaus M, Reiss OK, Tooley WH, et al.: Alveolar epithelial cell mitochondria as source of the surface active lung lining. Science 137:750, 1962
11. Said SI, Klein RM, Norrell LW, Maddox YT: Metabolism of alveolar cells: Histochemical evidence and relation to pulmonary surfactant. Science 152:657, 1966
12. Carrington CB, Green TJ: Granular pneumocytes in early repair of diffuse alveolar injury. Arch Intern Med 126:464, 1970
13. Green GM, Kass EH: The role of the alveolar macrophage in the clearance of bacteria from the lung. J Exp Med 119:167, 1964
14. Robin ED, Cross CE, Zelis R: Pulmonary Edema. N Engl J Med 288:239, 1973
15. Dunnill MS; An assessment of the anatomical factor in cor pulmonale in emphysema. J Clin Pathol 14:246, 1961
16. Fishman AP: Dynamics of the pulmonary circulation, in Hamilton WF, Dow P (eds): Handbook of Physiology, Section 2, Circulation, vol II. Washington, DC, American Physiological Society, 1963, p 1667
17. Edwards JE: Functional pathology of the pulmonary vascular tree in congenital heart disease. Circulation 15:164, 1957
18. Fishman AP: Pulmonary edema. Circulation 46:390, 1972
19. Drinker CK, Hardenbergh E: Absorption from the pulmonary alveoli. J Exp Med 86:7, 1947

20. Kylstra JA: Lavage of the lung. Acta Physiol Pharmacol Neerl 7:163, 1958
21. Schultz AL, Grismer JT, Wada S, Grande F: Absorption of albumin from alveoli of perfused dog lung. Am J Physiol 207:1300, 1964
22. Chinard FP: The permeability characteristics of the pulmonary blood-gas barrier, in Caro CG (ed): Advances in Respiratory Physiology. London, Edward Arnold, 1966, p 106
23. Meyer EC, Doninguez EAM, Bensch KG: Pulmonary lymphatic and blood absorption of albumin from alveoli: A quantitative comparison. Lab Invest 20:1, 1969
24. Wangensteen OD, Wittmers LE Jr, Johnson JA: Permeability of the mammalian blood gas barrier and its components. Am J Physiol 216:719, 1969
25. Karnovsky MJ: Morphology of capillaries with special reference to muscle capillaries, in Crone G, Lassen MA (eds): Capillary Permeability. New York, Academic, 1970, p 341
26. Staub MC, Nagano H, Pearce ML: Pulmonary edema in dogs, especially the sequence of fluid accumulation in lungs. J Appl Physiol 22:227, 1967
27. Pietra GG, Szidon JP, Levanthal MM, Fishman AP: Hemoglobin as a tracer in hemodynamic pulmonary edema. Science 166:1643, 1969
28. Cottrell TS, Levine OR, Senior RM, et al: Electron microscopic alterations at the alveolar level in pulmonary edema. Circ Res 21:783, 1967
29. Paintal AS: The mechanism of excitation of type J receptors and the J reflex, in Porter R (ed): Breathing. Hering-Breuer Centenary Symposium, Ciba Foundation Symposia. London, J&A Churchill, 1970, p 59
30. Paintal AS: Mechanism of stimulation of type J pulmonary receptors. J Physiol 203:511, 1969
31. Whayne JF Jr, Severinghaus JW: Experimental hypoxic pulmonary edema in the rat. J Appl Physiol 25:729, 1968
32. West JB: Effects of interstitial pressure, in Fishman AP, Hecht HH (eds): The Pulmonary Circulation and Interstitial Space. Chicago, University of Chicago Press, 1969, p 43
33. Lung Morphology: Current Research in Chronic Respiratory Disease: Proceedings of the Eleventh Aspen Emphysema Conference, Aspen Colorado, June, 1968. Washington, DC, U.S. Government Printing Office, 1969, p 199
34. Glazier JB, Hughes JMB, Maloney JE, West JB: Measurements of capillary dimensions and blood volume in rapidly frozen lungs. J Appl Physiol 26:65, 1969
35. Lauweryns JN: The juxta-alveolar lymphatics in the human adult lung: Histologic studies in 15 cases of drowning. Am Rev Respir Dis 102:877, 1970
36. Pump KK: Lymphatics of the human pulmonary alveoli: Preliminary report. Chest 58:140, 1970
37. Tobin CE: Pulmonary lymphatics with reference to emphysema. Am Rev Respir Dis 80:50, 1959
38. Sampson JJ, Leeds SE, Uhley HN, Friedman M: Studies of lymph flow and changes in pulmonary structures as indexes of circulatory changes in experimental pulmonary edema. Israel J Med Sci 5:826, 1969

39. Dumont AE, Claus RH, Reed GE, Tice DH: Lymph drainage in patients with congestive heart failure. N Engl J Med 269:949, 1963
40. Paine R, Butcher HR, Howard FA, Smith JP: Observations on mechanisms of edema formation in the lungs. J Lab Clin Med 34:1544, 1949
41. Uhley H, Leeds SE, Sampson JJ, Friedman M: Some observations on the role of the lymphatics in experimental acute pulmonary edema. Circ Res 9:688, 1961
42. Rabin ER, Meyer EC: Cardiopulmonary effects of pulmonary venous hypertension with special reference to pulmonary lymphatic flow. Circ Res 8:324, 1960
43. Guyton AC, Lindsey AW: Effect of elevated left atrial pressure and decreased plasma protein concentration on the development of pulmonary edema. Circ Res 7:649, 1959
44. Bruner HD, Schmidt CF: Blood flow in the bronchial artery of the anesthetized dog. Am J Physiol 148:648, 1947
45. Weil PE, Salisbury PF, State D: Physiological factors influencing pulmonary artery pressure during separate perfusion of systemic and pulmonary circulation in dog. Am J Physiol 191:453, 1957
46. Liebow AA, Hales MR, Harrison W, et al: Relation of bronchial to pulmonary vascular tree, in Adams W, Veith I (eds): Pulmonary Circulation. New York, Grune & Stratton, 1959, p 79
47. Pietra GG, Szidon JP, Leventhal MM, Fishman AP: Histamine and interstitial pulmonary edema in the dog. Circ Res 24:323, 1971
48. Guyton AC, Granger HJ, Taylor AE: Interstitial fluid pressure. Physiol Rev 51:527, 1971
49. Guyton AC, Taylor AE, Granger HJ: Analysis of types of pressure in the pulmonary spaces, interstitial fluid pressure, solid tissue pressure, and total time pressure, in Giuntini C (ed): Central Hemodynamics and Gas Exchange. Pisa, Minerva Medica, 1971, p 41
50. Levine OR, Mellins RB, Senior RM, Fishman AP: The application of Starling's law of capillary exchange to the lungs. J Clin Invest 46:934, 1967
51. Pearce ME, Wong MJ: Interstitial pressure and compliance changes in experimental pulmonary edema in the dog, in Giuntini C (ed): Central Hemodynamics and Gas Exchange. Pisa, Minerva Medica, 1971, p 181
52. Guyton AC: A concept of negative interstitial pressure based on pressures in implanted perforated capsules. Circ Res 12:399, 1965
53. Meyer BJ, Meyer A, Guyton AC: Interstitial fluid pressure. V. Negative pressure in the lungs. Circ Res 22:263, 1968
54. Courtice FC: Lymph flow in the lungs. Br Med Bull 19:76, 1963
55. Foldi M: Disease of lymphatics and lymph circulation. Springfield Ill., Charles C. Thomas, 1969
56. Staub NC: Steady state pulmonary transvascular water filtration in unanesthetized sheep. Circ Res [suppl 1]:135, 1971
57. Mead J, Takishima T, Leith D: Stress distribution in the lungs: A model of pulmonary elasticity. J Appl Physiol 28:596, 1970
58. Agostoni E, Mead J: Studies of the respiratory system, in Fenn WO, Rahn

H (eds): Handbook of Physiology, section 3, Respiration, vol. I. Washington, DC, American Physiological Society, 1964, p 387
59. Lee GdeJ: Pulmonary edema, in Yu PN, Goodwin JF (eds): Progress in Cardiology, vol 1. Philadelphia, Lea & Febiger, 1972, p 261
60. Clements JA, Tierny DF: Alveolar instability associated with altered surface tension, in Fenn WO, Rahn H (eds): Handbook of Physiology, section 3, Respiration, vol 2. Washington, DC American Physiological Society, 1965, p 1565
61. Taylor AE, Gaar KA Jr: Estimation of equivalent pore radii of the pulmonary capillary and alveolar membranes. Am J Physiol 218:1133, 1970
62. DuBois AB, Marshall R: Measurements of pulmonary capillary blood flow and gas exchange throughout the respiratory cycle in man. J Clin Invest 36:1566, 1957
63. Fick A: Ueber die messung des Blutquatums in den Herzventrikeln. Verh Dtsch Phys Med Ges (Wurzburg) 2:16, 1870
64. Visscher MB, Johnson JA: The Fick principle: Analysis of potential errors in its conventional application. J Appl Physiol 5:635, 1953
65. Hamilton WF: Measurement of the cardiac output, in Hamilton WF, Dow P (eds): Handbook of Physiology, section 2, Circulation, vol I. Washington, DC, American Physiological Society, 1962, p 551
66. Guyton AC, Jones CE, Coleman TG: Circulatory Physiology. Cardiac Output and Its Regulation (ed 2). Philadelphia, W.B. Saunders, 1973
67. Stewart GN: The pulmonary circulation time: Quantity of blood in lungs and output of the heart. Am J Physiol 58:20, 1921
68. Hamilton WF, Moore JW, Kinsman JM, Spurling RG: Simultaneous determination of the pulmonary and systemic circulation times in man and of a figure related to the cardiac output. Am J Physiol 84:338, 1928
69. Dow P: Estimations of cardiac output and central blood volume by dye dilution. Physiol Rev 38:77, 1956
70. Sheppard CW: Mathematical considerations of indicator dilution techniques. Minnesota Med 37:93, 1954
71. Zierler K: Theoretical basis of indicator dilution methods for measuring flow and volume. Circ Res 10:393, 1962
72. Zierler KL: Circulation times and the theory of indicator-dilution methods for determining blood flow and volume, in Hamilton WF, Dow P (eds): Handbook of Physiology, section 2, Circulation, vol I. Washington, DC, American Physiological Society, 1962, p 585
73. Cournand A: Recent observations on the dynamics of the pulmonary circulation. Bull NY Acad Med 23:27, 1947
74. Reeves JT, Grover RF, Filley GF, Blount SG Jr: Cardiac output in normal resting man. J Appl Physiol 16:276, 1961
75. Wade OL, Bishop JN: Cardiac Output and Regional Blood Flow. Oxford, Blackwell, 1962
76. Yu PN: Pulmonary blood volume, in Health and Disease, Philadelphia, Lea & Febiger, 1969
77. Davies LG, Goodwin JF, Steiner RE, Van Leuven BD: Clinical and radio-

logic assessment of the pulmonary arterial pressure in mitral stenosis. Br Heart J 15:393, 1953

78. Doyle AE, Goodwin JF, Harrison CV, Steiner RE: Pulmonary vascular patterns in pulmonary hypertension. Br Heart J 19:353, 1957
79. Steiner RE: Radiological appearance of the pulmonary vessels in pulmonary hypertension. Br J Radiol 31:189, 1958
80. Mattson SB, Carlens E: Lobar ventillation and oxygen uptake in man: Influence of body position. J Thorac Cardiovasc Surg 30:676, 1955
81. West JB, Dollery CT: Distribution of blood flow and ventilation perfusion ratio in the lung measured with radioactive CO_2. J Appl Physiol 15:405, 1960
82. Ball WC, Stewart PB, Newsham LGS, Bates DV: Regional pulmonary function studied with xenon[133]. J Clin Invest 41:519, 1962
83. Dawson A, Kaneko K, McGregor M: Regional lung function in patients with mitral stenosis studied with xenon[131] during air and oxygen breathing. J Clin Invest 44:999, 1965
84. West JB, Dollery CT, Heard BE: Increased pulmonary vascular resistance in the dependent zone of the isolated dog lung caused by perivascular edema. Circ Res 17:191, 1965
85. Rahn H: A concept of mean alveolar air and the ventilation–blood flow relationships during pulmonary gas exchange. Am J Physiol 158:21, 1949
86. Riley RL, Himmelstein A, Motley HL, et al: "Ideal" alveolar air and the analysis of ventilation–perfusion relationships in the lung. J Appl Physiol 1:199, 1949
87. Rahn H, Farhi LE: Ventilation, perfusion and gas exchange—The VA/Q concept, in Fenn WO, Rahn H (eds): Handbook of Physiology, section 3, Respiration, vol I. Washington, DC, American Physiological Society, 1964
88. Severinghaus JW, Stupfel M: Alveolar dead space as an index of distribution of blood flow in pulmonary capillaries. J Appl Physiol 10:335, 1957
89. Riley RL, Permutt S, Said S, et al: Effect of posture on pulmonary dead space in man. J Appl Physiol 14:339, 1959
90. Farhi LE, Rahn H: A theoretical analysis of the alveolar-arterial O_2 difference with special reference to the distribution effect. J Appl Physiol 7:699, 1955
91. Haab P, Piiper J, Rahn H: Attempt to demonstrate the distribution component of the alveolar arterial oxygen pressure difference. J Appl Physiol 2:235, 1960
92. Visser BF, Mass AHJ: Pulmonary diffusion of oxygen. Phys Med Biol 3:264, 1959
93. Piiper J: Variations of ventilation and diffusing capacity to perfusion determining the alveolar arterial O_2 difference: Theory. J Appl Physiol 16:507, 1961
94. Piiper J, Haab P, Rahn H: Unequal distribution of pulmonary diffusing capacity in the anesthetized dog. J Appl Physiol 16:499, 1960
95. Finley TN, Lenfant C, Haab P, et al: Venous admixture in the pulmonary circulation of anesthetized dogs. J Appl Physiol 15:418, 1960

96. Wagner PD, Laravuso RB, Uhl RR, West JB: Distribution of ventilation-perfusion ratios in accurate respiratory failure. Chest [Suppl I] 65:32, 1974
97. King TKC, Weber B, Okinaka A, et al: Oxygen transfer in catastrophic respiratory failure. Chest [Suppl I] 65:40, 1974
98. Cournand A, Ranger HA: Catheterization of the right auricle in man. Proc Soc Exp Biol Med 46:462, 1941
99. Dexter L, Haynes FW, Burwell CS, et al: Studies of congenital heart disease. I. Technique of venous catheterization as a diagnostic procedure. J Clin Invest 26:546, 1947
100. Harris P, Heath D: Human Pulmonary Circulation. Baltimore, Williams & Wilkins, 1962
101. Hellems HK, Haynes FW, Dexter L, Kinney TD: Pulmonary capillary pressure in animals estimated by venous and arterial catheterization. Am J Physiol 155:98, 1948
102. Lagerlof H, Werko L: Studies on the circulation of blood in man. VI. The pulmonary capillary venous pressure pulse in man. Scand J Clin Lab Invest 1:147, 1949
103. Yu PN, Glick G, Schreiner BF, Murphy GW: Effects of acute hypoxia on the pulmonary vascular bed of patients with acquired heart disease, with special reference to the demonstration of active vasomotion. Circulation 27:541, 1963
104. Braunwald E, Brockenbrough EC, Frahm CJ, Ross J: Left atrial and left ventricular pressure in subjects without cardiovascular disease. Circulation 24:267, 1961
105. Samet P, Bernstein WH, Medow A, Levine S: Transseptal left heart dynamics in thirty two normal subjects. Dis Chest 47:632, 1965
106. Brody JS, Stemmler EJ, DuBois AB: Longitudinal distribution of vascular resistance in the pulmonary arteries, capillaries and veins. J Clin Invest 47:783, 1968
107. West JB: Physiological consequences of the apposition of blood and gas in the lung. in DeReuck AVS, Porter R (eds): Development of the Lung. London, Churchill, 1967, p 176
108. Ross J Jr: Transseptal left heart catheterization. Ann Surg 149:395, 1959
109. Brockenbrough EC, Braunwald E: A new technique for left ventricular angiocardiography and transseptal left heart catheterization. Am J Cardiol 6:1062, 1960
110. Ross J: Considerations regarding the technique for transseptal left heart catheterization. Circulation 34:391, 1966
111. Burton AC: The relationship between pressure and flow in the pulmonary bed, in Adams W, Veith I (eds): Pulmonary Circulation. New York, Grune & Stratton, 1959, p 26
112. Haddy FJ, Campbell GS: Pulmonary vascular resistance in anesthetized dogs. Am J Physiol 172:747, 1953
113. Borst HG, McGregor M, Whittenberger JL, Berglund E: Influence of pulmonary arterial and left atrial pressures on pulmonary vascular resistance. Circ Res 4:393, 1956
114. Permutt S, Caldini P, Maseri A, et al: Recruitment versus distensibility in

the pulmonary vascular bed, in Fishman AP, Hecht HH (eds): The Pulmonary Circulation and Interstitial Space. Chicago, University of Chicago Press, 1969, p 375
115. Permutt S, Howell JBL, Proctor DF, Riley RL: Effect of lung inflation on static pressure–volume characteristics of pulmonary vessels. J Appl Physiol 16:64, 1961
116. Howell JBL, Permutt S, Proctor DF, Riley RL: Effect of inflation of lung on different parts of pulmonary vascular bed. J Appl Physiol 16:71, 1961
117. Burton AC, Patel DJ: Effect on pulmonary vascular resistance of inflation of rabbit lungs. J Appl Physiol 12:239, 1958
118. Simmons DH, Linde LM, Miller JH, O'Reilly RJ: Relation between lung volume and pulmonary vascular resistance. Circ Res 9:465, 1961
119. Haddy FJ, Campbell GS: Pulmonary vascular resistance in anesthetized dogs. Am J Physiol 172:747, 1953
120. Whittenberger JL, McGregor M, Berglund E, Borst HG: Influence of state of inflation of lung on pulmonary vascular resistance. J Appl Physiol 15:878, 1960
121. Harasawa M, Rodbard S: Ventilatory air pressure and pulmonary vascular resistance. Am Heart J 60:73, 1960
122. Lenfant C, Howell B: Cardiovascular adjustments in dogs during continuous pressure breathing. J Appl Physiol 15:425, 1960
123. Thomas LJ, Griffo ZJ, Roos A: Effects of negative pressure inflation of the lung on pulmonary vascular resistance. J Appl Physiol 16:451, 1961
124. Patel DJ, Mallos AJ, de Freitas FM: Importance of transmural pressure and lung volume in evaluating effect on pulmonary vascular tone. Circ Res 9:1217, 1961
125. Roos A, Thomas LJ, Negel EL, Prommas DC: Pulmonary vascular resistance as determined by lung inflation and vascular pressures. J Appl Physiol 16:77, 1961
126. Dock, DS, Kraus WL, McQuire LB, et al: The pulmonary blood volume in man. J Clin Invest 40:317, 1961
127. Milnor WR, Jose AD, McGaff CJ: Pulmonary vascular volume, resistance and compliance in man. Circulation 12:130, 1960
128. de Freitas FM, Faraco EZ, Nedel N, et al: Determination of pulmonary blood volume by single intravenous injection of one indicator in patients with normal and high pulmonary vascular pressures. Circulation 30:370, 1964
129. Varnauskas E, Forsberg SA, Widimsky J, Paulin S: Pulmonary blood volume and its relation to pulmonary hemodynamics in cardiac patients. Acta Med Scand 173:529, 1963
130. Donato L, Giuntini C, Lewis ML, et al: Quantitative radiography. I. Theoretical consideration. Circulation 26:174, 1962
131. Donato L, Rochester DF, Lewis ML, et al: Quantitative radiography. II. Technique and analysis of curves. Circulation 26:183, 1962
132. Lewis RL, Giuntini C, Donato L, et al: Quantitative radiography. III. Results and validation of theory and method. Circulation 26:189, 1962
133. Giuntini C, Lewis ML, Sales LA, Harvey RM: A study of the pulmonary

blood volume in man by quantitative radiocardiography. J Clin Invest 42:1589, 1963

134. Oglivie CM, Forster RE, Blakemore WS, Morton JW: A standardized breath holding technique for the clinical measurement of the diffusing capacity of the lung for carbon monoxide. J Clin Invest 36:1, 1957
135. McCredie RM: The diffusing characteristics and pressure volume relationships of the pulmonary capillary bed in mitral valve disease. J Clin Invest 43:2279, 1964
136. Gazioglu K, Yu PN: Pulmonary blood volume and pulmonary capillary blood volume in valvular heart disease. Circulation 35:701, 1967
137. Chinard FP, Enns T: Transcapillary pulmonary exchange of water in the dog. Am J Physiol 178:197, 1954
138. Sarnoff SJ, Berglund E: Pressure volume characteristics and stress relaxation in the pulmonary vascular bed of the dog. Am J Physiol 171:238, 1952
139. Engleberg J, DuBois AB: Mechanics of pulmonary circulation in isolated rabbit lungs. Am J Physiol 196:401, 1959
140. Caro CG, Saffman PG: Extensibility of blood vessels in isolated rabbit lungs. J Physiol 178:193, 1965
141. Glazier BJ, Hughes JMB, Maloney JE, West JB: Measurements of capillary dimensions and blood volume in rapidly frozen lungs. J Appl Physiol 26:65, 1969
142. Permutt S, Caldini P: Theoretical aspects of the relation between pulmonary blood volume and the measurement of extravascular water, in Giuntini G (ed): Central Hemodynamics and Gas Exchange. Torino, Minerva Medica, 1971
143. Maseri A, Caldini P, Harwood P, et al: Determinants of pulmonary vascular volume recruitment versus distensibility. Circ Res 31:218, 1972
144. Permutt S, Caldini P, Maseri A, et al: Recruitment versus distensibility in the pulmonary vascular bed, in Fishman AP, Hecht HH (eds): The Pulmonary Circulation and Interstitial Space. Chicago, University of Chicago Press, 1969
145. Luepker R, Liander BO, Korsgen M, Varnauskas E: Pulmonary intravascular and extravascular fluid volumes in exercising cardiac patients. Circulation 44:626, 1971
146. Austin SM, Schreiner BF, Shah PM, Yu PN: Acute effects of increase in pulmonary vascular distending pressures on pulmonary blood volume and pulmonary extravascular fluid volume in man. Circulation, February 1976 (in press)
147. Luepker R, Liander BO, Korsgren LA, Varnauskas E: Pulmonary extravascular and intravascular fluid volume in resting patients. Am J Cardiol 28:295, 1971
148. Schreiner BF, Murphy GW, Shah PM, et al: Pulmonary extravascular volume in cardiac patients. Circulation [Suppl II] 44:39, 1971
149. Donald KW, Bishop JM, Wade OL: A study of minute to minute changes of arterio-venous oxygen, content difference, oxygen uptake and cardiac output and rate of achievement of a steady state during exercise in rheumatic heart disease. J Clin Invest 33;1146, 1954

150. Donald KW, Bishop JM, Cumming G, Wade OL: The effect of exercise on the cardiac output and circulatory dynamics of normal subjects. Clin Sci 14:37, 1955
151. Dexter L, Whittenberger JL, Haynes FW, et al: Effect of exercise on circulatory dynamics of normal individuals. J Appl Physiol 3:439, 1951
152. Freedman ME, Snider GL, Brostoff P, et al: Effects of training on response of cardiac output to muscular exercise in athletes. J Appl Physiol 8:37, 1955
153. Ekelund LG, Holmgren A: Central hemodynamics during exercise. Circ Res [Suppl I] 20, 21:33, 1967
154. Sancetta SM, Rakita L: Response of pulmonary artery pressure and total pulmonary resistance of untrained convalescent man to prolonged mild steady state exercise. J Clin Invest 36:1138, 1957
155. Slonim NB, Ravin A, Balchum OJ, Dressler SH: The effect of mild exercise in the supine position in the pulmonary arterial pressure in five normal human subjects. J Clin Invest 33:1022, 1954
156. Barrett-Boyes BG, Wood EH: Hemodynamic response of healthy subjects to exercise in the supine position while breathing oxygen. J Appl Physiol 11:129, 1957
157. Roncoroni AJ, Armendia P, Gonzalez R, Taquini AC: "Central" blood volume in exercise in normal subjects. Acta Physiol Lat Am 9:55, 1959
158. Braunwald E, Kelly ER: The effects of exercise on central blood volume in man. J Clin Invest 39:413, 1960
159. Levinson GE, Pacifico AD, Frank MJ: Studies of cardiopulmonary blood volume. Measurement of total cardiopulmonary blood volume in normal subjects at rest and during exercise. Circulation 33:347, 1966
160. Marshall RJ, Shepherd JT: Interpretation of changes in "central blood volume" and slope volume during exercise in man. J Clin Invest 40:375, 1961
161. McIntosh HD, Gleason WL, Miller DE, Bacos JN: A major pitfall in the interpretation of "central blood volume." Circ Res 9:1223, 1961
162. Oakley C, Glick G, Luria MN, et al: Some regulatory mechanisms of the human pulmonary vascular bed. Circulation 16:917, 1962
163. Schreiner BF, Murphy GW, Glick G, Yu PN: Effect of exercise on the pulmonary blood volume in patients with acquired heart disease. Circulation 27:559, 1963
164. Korsgren M, Kallay K, Varnasuskas E: Pulmonary and "central" blood volume during reactive hyperaemia in the legs. Clin Sci 32:415, 1967
165. Schreiner BF, Murphy GW, Kramer DH, et al: The pathophysiology of pulmonary congestion. Prog Cardiovasc Dis 14:57, 1971
166. Harris P, Heath D: Human Pulmonary Circulation. Baltimore, Williams & Wilkins, 1962, chapter 19
167. Parker F, Weiss S: The nature and significance of the structural changes in the lungs in mitral stenosis. Am J Pathol 12:573, 1936
168. Heath D, Edwards JE: Histological changes in the lung in diseases associated with pulmonary venous hypertension. Br J Dis Chest 53:8, 1959
169. Daley R, Goodwin JP, Steiner RE: Clinical Disorders of the Pulmonary Circulation. Boston, Little, Brown, 1960, p 136

170. Levine OR, Mellins RB, Senior RM, Fishman AP: The application of Starling's law of capillary exchange to the lungs. J Clin Invest 46:934, 1967
171. Levine OR, Mellins RB, Fishman AP: Quantitative assessment of pulmonary edema. Circ Res 17:414, 1965
172. Pearce ML, Wong MJ: Interstitial pressure and compliance changes in experimental pulmonary edema in the dog, in Giuntini C (ed): Central Hemodynamics and Gas Exchange. Torino, Minerva Medica, 1971
173. von Euler US, Liljestrand G: Observations on the pulmonary arterial blood pressure in the cat. Acta Physiol Scand 12:301, 1946
174. Liljestrand G: Regulation of pulmonary arterial blood pressure. Arch Intern Med 81:162, 1948
175. Atwell RJ, Hickman JB, Pryor WW, Page EB: Reduction of blood flow through the hypoxic lung. Am J Physiol 166:37, 1951
176. Duke HN: The site of action of anoxia on the pulmonary blood vessels of the cat. J Physiol 125:373, 1954
177. Rahn H, Bahnson HT: Effect of unilateral hypoxia on gas exchange and calculated pulmonary blood flow in each lung. J Appl Physiol 6:105, 1953
178. Nahas GG, Visscher MB, Mather GW, et al: Influence of hypoxia on the pulmonary circulation of non-narcotized dogs. J Appl Physiol 6:467, 1953
179. Rivera-Estrada C, Saltzman PW, Singer D, Katz LN: Action of hypoxia in pulmonary vasculature. Circ Res 6:10, 1958
180. Motley AL, Cournand A, Werko L, et al: The influence of short periods of induced acute hypoxia upon pulmonary artery pressure in man. Am J Physiol 150:315, 1947
181. Doyle JT, Wilson JS, Warren JV: The pulmonary vascular responses to short term hypoxia in human subjects. Circulation 5:263, 1952
182. Fishman AP, McClement J, Himmelstein A, Cournand A: Effects of acute anoxia on the circulation and respiration on patients with chronic pulmonary disease studied during the "steady state." J Clin Invest 31:770, 1952
183. Yu PN, Beatty DC, Lovejoy FW, et al: Hemodynamic effects of acute hypoxia in patients with mitral stenosis. Am Heart J 52:683, 1956
184. Fishman AP, Fritts HW, Cournand A: Effects of acute hypoxia and exercise in the pulmonary circulation. Circulation 22:204, 1960
185. Yu PN, Glick G, Schreiner BF, Murphy GW: Effects of acute hypoxia on the pulmonary vascular bed in patients with acquired heart disease. Circulation 27:541, 1973
186. Himmelstein A, Harris P, Fritts HW, Cournand A: Effect of severe unilateral hypoxia on the partition of pulmonary blood flow in man. J Thorac Cardiovasc Surg 36:369, 1958
187. Fritts HW, Harris P, Clauss RH, et al: Effect of acetylcholine on the human pulmonary circulation under normal and hypoxic conditions. J Clin Invest 37:99, 1958
188. Rudolph AM, Yuan S: Response of the pulmonary vasculature to hypoxia and H+ ion concentration changes. J Clin Invest 45:399, 1966
189. Enson Y, Giuntini C, Lewis ML, et al: The influence of hydrogen ion con-

centration and hypoxia on the pulmonary circulation. J Clin Invest 43:1146, 1964
190. Harvey RM, Enson Y, Betti R, et al: Further observations on the effect of hydrogen ion on the pulmonary circulation. Circulation 35:1019, 1967
191. Kato M, Staub N: Response of small pulmonary arteries to unilabor hypoxia and hypercapnea. Circ Res 19:426, 1966
192. Sackner MA, Will DH, DuBois AB: The site of pulmonary vasomotor activity during hypoxia or serotonim administration. J Clin Invest 45:112, 1966
193. Sobol BJ, Bottex G, Emirgil C, Gissen H: Gaseous diffusion from alveoli to pulmonary vessels of considerable size. Circ Res 13:71, 1963
194. Hultgren H, Glover RF: Circulation adaptation to high altitude. Ann Rev Med 19:119, 1968
195. Rotta A, Caneda A, Hurtado A, et al: Pulmonary circulation at sea level and at high altitudes. J Appl Physiol 9:328, 1956
196. Penaloza D, Sime F, Banchero N, Gamboa R: Pulmonary hypertension in healthy man born and living at high altitudes. Med Thorac 19:257, 1962
197. Vogel JHK, Weaver WF, Rose RL, et al: Pulmonary hypertension on exertion in normal man living at 10,150 feet (Leadville, Colorado). Med Thorac 19:461, 1962
198. Naeye RL: Hypoxemia and pulmonary hypertension. Arch Pathol 71:447, 1961
199. Arias-Stella J, Saldana M: The terminal portion of the pulmonary arterial tree in people native to high altitudes. Circulation 28:915, 1963
200. Grover RF, Reeves JT, Will DH, Blount SG: Pulmonary vasoconstriction in steers at high altitude. J Appl Physiol 18:567, 1963
201. Hultgren H, Kelly J, Miller H: Effect of oxygen upon pulmonary circulation in acclimatized man at high altitude. J Appl Physiol 20:239, 1965
202. Grover RF, Vogel JHK, Voight GC, Blount SG: Reversal of high altitude pulmonary hypertension. Am J Cardiol 18:928, 1966
203. Vogel JHK, Cameron D, Jamieson G: Chronic pharmocologic treatment of experimental hypoxic pulmonary hypertension. Am Heart J 72:50, 1966
204. Tenney SM: Physiological adaptations to life at high altitude. Mod Concepts Cardiovasc Dis 31:713, 1962
205. Grover RF: Pulmonary circulation in animals and man at high altitudes. Ann NY Acad Sci 127:632, 1965
206. Hultgren H, Janis B, Marticorena E, Miller H: Diminished cardiovascular response to acute hypoxia at high altitude. Circulation [Suppl II] 36:146, 1967
207. Roy SB, Bhatia ML, Gadhoke S: Response of pulmonary blood volume to 64 to 114 weeks of intermittent stay at high altitudes. Am Heart J 74:192, 1967
208. Cotten MdeV, Stopp PE: Action of digitalis on the non-failing heart of the dog. Am J Physiol 192:114, 1958
209. Ross J, Waldhausen JA, Braunwald E: Studies on digitalis. I. Direct effects on the peripheral vascular resistance. J Clin Invest 39:930, 1960

210. Rodman T, Pastor BH: The hemodynamic effects of digitalis in the normal and diseased heart. Am Heart J 65:564, 1963
211. Braunwald E, Bloodwell RD, Goldberg LI, Morrow AJ: Studies in digitalis. IV. Observations in man on the effects of digitalis preparations on the contractility of the nonfailing heart and on the total vascular resistance. J Clin Invest 40:52, 1961
212. James TN, Nadeau RA: The chronotropic effects of digitalis studied by direct perfusion of the sinus node. J Pharmacol Exp Ther 139:42, 1963
213. Mason DT, Braunwald E: Studies on digitalis. IX. Effects of ouabain on the nonfailing human heart. J Clin Invest 42:1105, 1963
214. Murphy GW, Schreiner BF, Bleakley PL, Yu PN: Left ventricular performance following digitalization in patients with and without heart failure. Circulation 30:358, 1964
215. Weissler AM, Gamel WG, Grode HE, et al: The effect of digitalis on ventricular ejection in normal human subjects. Circulation 29:721, 1964
216. Weissler AM, Kamen AR, Bornstein RS, et al: Effects of deslanoside on the duration of phases of ventricular systole in man. Am J Cardiol 15:153, 1965
217. Braunwald E, Mason DT, Ross J: Studies of cardiocirculatory actions of digitalis. Medicine 44:233, 1965
218. Williams MH, Zohman LR, Ratner AC: Hemodynamic effects of cardiac glycosides on normal human subjects during rest and exercise. J Appl Physiol 13:417, 1958
219. Selzer A, Hultgren HM, Ebnother CL, et al: Effect of digoxin on the circulation in normal man. Br Heart J 21:335, 1959
220. Dresdale DT, Yuceoglu YZ, Michtom RJ, et al: Effects of lanatoside C on cardiovascular hemodynamics. Acute digitalizing doses in subjects with normal hearts and with heart disease without failure. Am J Cardiol 4:88, 1959
221. Maseri A, Bianchi R, Giusti C, et al: Early effects of digitalis on central hemodynamics in normal subjects. Am J Cardiol 15:162, 1965
222. Lagerlof H, Werko L: Studies on circulation in man; the effects of cedelanid on cardiac output and blood pressure in the pulmonary circulation in patients with compensated and decompensated heart disease. Acta Cardiol 4:1, 1949
223. Harvey RM, Ferrer MI, Cathcart RT, Alexander JK: Some effects of digoxin on the heart and circulation in man. Digoxin in enlarged hearts not in clinical congestive failure. Circulation 4:336, 1951
224. Selzer A, Malmborg RO: Hemodynamic effects of digoxin in latent cardiac failure. Circulation 25:695, 1962
225. Ferrer MI, Harvey RM, Cathcart RT, et al: Some effects of digoxin upon the heart and circulation in man. Digoxin in chronic cor pulmonale. Circulation 1:161, 1950
226. Mounsey JPO, Ritzmann LW, Selverstone NJ, et al: Circulatory changes in severe pulmonary emphysema. Br Heart J 14:153, 1952
227. Stead EA, Warren JV, Brannon ES: Effects of lanatoside C on the cir-

culation of patients with congestive heart failure. Arch Intern Med 81:282, 1948
228. Harvey RM, Ferrer MI, Cathcart RT, et al: Some effects of digoxin upon the heart and circulation in man. Digoxin in left ventricular failure. Am J Med 7:439, 1949
229. Ferrer MI, Conroy RJ, Harvey RM: Some effects of digoxin upon the heart and circulation in man. Digoxin in combined (left and right) ventricular failure. Circulation 21:372, 1960
230. Murphy GW, Schreiner BF, Yu PN: Effects of acute digitalization on the pulmonary blood volume in patients with heart disease. Circulation 43:145, 1971
231. Goldberg LI, Bloodwell RD, Braunwald E, Morrow HG: The direct effects of norepinephrine, epinephrine and methoxamine on myocardial contractile force in man. Circulation 22:1125, 1960
232. Rose JC, Fries ED, Hufnagel CA, Massullo EA: Effects of epinephrine and norepinephrine in dogs studied with a mechanical left ventricle. Demonstration of active vasoconstriction in the lesser circulation. Am J Physiol 182:197, 1955
233. Duke HN, Stedeford RD: Pulmonary vasomotor responses to epinephrine and norepinephrine in the cat. Circ Res 8:640, 1960
234. Borst HG, Berglund E, McGregor M: The effects of pharmacologic agents on the pulmonary circulation in the dog. Studies on epinephrine, norepinephrine, 5-hydroxy-tryptamine, acetylcholine, histamine, and aminophylline. J Clin Invest 36:669, 1957
235. Gaddum JH, Holtz P: The location of the action of drugs on the pulmonary vessels of dogs and cats. J Physiol 77:139, 1933
236. Feeley JW, Lee TD, Milnor WR: Active and passive components of pulmonary vascular response to vasoactive drugs in the dog. Am J Physiol 205, 1193, 1963
237. Witham AC, Fleming JW; The effects of epinephrine on the pulmonary circulation in man. J Clin Invest 30:707, 1951
238. Nelson RA, May LG, Bennett A, et al: Comparison of the effects of pressor and depressor agents and influences on pulmonary and systemic pressures of normotensive and hypertensive subjects. Am Heart J 50:172, 1955
239. Shadle OW, Moore JC, Billig DM: Effects of *l*-arterenol on "central blood volume" in the dog. Circ Res 3:385, 1955
240. Levy MN, Brind SH: Influence on *l*-norepinephrine upon cardiac output in anaesthetized dogs. Circ Res 5:85, 1957
241. Patel DJ, Mallos AJ, de Freitas FM: Importance of transmural pressure and lung volume in evaluating drug effect on pulmonary vascular tone. Circ Res 9:1217, 1961
242. Fowler NO, Holmes JC: Pulmonary presser action of *l*-norepinephrine and angiotension. Am Heart J 70:66, 1965
243. Fowler NO, Westcott RN, Scott RC, McQuire J: The effect of norepinephrine upon pulmonary arteriolar resistance in man. J Clin Invest 30:517, 1951

244. Patel DJ, Lange RL, Hecht HH: Some evidence for active constriction in the human pulmonary vascular bed. Circulation 18:19, 1958.
245. Regan TJ, Defazio V, Binak K, Hellems HK: Norepinephrine induced pulmonary congestion in patients with aortic valve regurgitation. J Clin Invest 38:1564, 1959
246. Bousvaros GA: Effects of norepinephrine on human pulmonary circulation. Br Heart J 24:738, 1962
247. Lands AM, Howard JW: A comparative study of the effects of *l*-arterenol epi- and iso-propylarterenol on the heart. J Pharmacol Exp Ther 106:65, 1952
248. Weissler AM, Leonard JJ, Warren JV: The hemodynamic effects of isoproterenol in man, with observations on the role of the central blood volume. J Lab Clin Med 53:921, 1959
249. Dodge HT, Lord JD, Sandler H: Cardiovascular effects of isoproterenol in normal subjects and subjects with congestive heart failure. Am Heart J 60:94, 1960
250. Gorton R, Gunnells JC, Weissler AM, Stead EA: Effects of atropine and isoproterenol on cardiac output, central venous pressure, and mean transit time of indicators placed at three different sites in the venous system. Circ Res 9:976, 1961
251. Whalen RE, Cohen AI, Sumner RG, McIntosh HD: Hemodynamic effects of isoproterenol infusion in patients with normal and diseased mitral valves. Circulation 27:512, 1963
252. Lee TD, Roveti GC, Ross RS: The hemodynamic effects of isoproterenol on pulmonary hypertension in man. Am Heart J 65:361, 1963
253. Cox AR, Cobb LA, Bruce RA: Differential hemodynamic effects of isoproterenol on mitral stenosis and left ventricular diseases. Am Heart J 65:802, 1963
254. Schreiner BF, Murphy GW, James DH, Yu PN: Effects of isoproterenol on the pulmonary blood volume in patients with valvular heart disease and primary myocardial disease. Circulation 37:220, 1968
255. Brody JS, Stemmler EJ: Differential reactivity in the pulmonary circulation. J Clin Invest 47:80, 1968
256. McGaff CJ, Roveti GC, Glassman E, Milnor WR: The pulmonary blood volume in rheumatic heart disease and its alteration by isoproterenol. Circulation 27:77, 1963
257. Harrison DC, Glick G, Goldblatt A, Braunwald E: Studies on cardiac dimensions in intact unaneasthetized man. IV. Effects of isoproterenol and methoxamine. Circulation 29:186, 1964
258. Rodbard S: Bronchomotor tone. Am J Med 15:356, 1953
259. Howarth S, McMichael J, Sharpey-Schafer EP: Circulatory action of theophylline ethylene diamine. Clin Sci 6:125, 1947
260. Werko L, Lagerlof H: Studies on the circulation of blood in man. VII. The effect of a single intravenous dose of theophylline diethamolamine on cardiac output, pulmonary blood volume and systemic and pulmonary

blood pressures in hypertensive cardiovascular disease. Scan J Clin Lab Invest 2:181, 1950
261. Schuman C, Simmons HG: Cardiac asthma: Its pathogenesis and response to aminophylline. Ann Intern Med 36:864, 1952
262. Sweet HC, Peden BW, Kistner WF, Mudd JG: The effect of aminophylline on the emphysematous patient. Missouri Med 55:1079, 1958
263. Maxwell GM, Crumpton CW, Rowe GG, et al: The effects of theophylline ethylene-diamine (aminophylline) on the coronary hemodynamics of normal and disease hearts. J Lab Clin Med 54:88 1959
264. Parker JO, Kekar K, West RO: Hemodynamic effects of aminophylline in cor pulmonale. Circulation 33:17, 1966
265. Parker JO, Kelly G, West RO: Hemodynamic effects of aminophylline in heart failure. Am J Cardiol 17:232, 1966
266. Murphy GW, Schreiner BF, Yu PN: Effects of aminophylline on the pulmonary circulation and left ventricular performance in patients with valvular heart disease. Circulation 37:361, 1968
267. Alcock P, Berry JL, Daly IdeB: The action of drugs on the pulmonary vessels of dogs and cats. Q J Exp Physiol 25:369, 1935
268. Harris P, Fritts HW, Clauss RH, et al: Influence of acetylcholine on human pulmonary circulation under normal and hypoxic conditions. Proc Soc Exp Biol Med 93:77, 1956
269. Harris P: Influence of acetylcholine on the pulmonary arterial pressure. Br Heart J 19:272, 1957
270. Fritts HW, Harris P, Clauss RH, et al: The effect of acetylcholine on the human circulation under normal and hypoxic conditions. J Clin Invest 37:99, 1958
271. Wood P, Besterman EM, Towers MK, McIlroy MB: The effect of acetylcholine on pulmonary vascular resistance and left atrial pressure in mitral stenosis. Br Heart J 19:279, 1957
272. Soderholm B, Werko L: Acetylcholine and the pulmonary circulation in mitral stenosis. Br Heart J 21:1, 1959
273. Chidsey CA, Fritts HW, Zocche GP, et al: Effect of acetylcholine on the distribution of pulmonary blood flow in chronic pulmonary emphysema. Med Cardiovasc 1:15, 1960
274. Soderholm B, Werko L, Widimsky J: The effect of acetylcholine and gas exchange in cases of mitral stenosis. Acta Med Scand 172:95, 1962
275. Irnell L, Nordgren L: Effects of acetylcholine infusion on the pulmonary circulation in patients with bronchial asthma. Acta Med Scand 179:385, 1966
276. Rowe GG, Castillo CA, Crosley AP, et al: Acute systemic and coronary hemodynamic effects of chlorothiazide in subjects with systemic arterial hypertension. Am J Cardiol 10:183, 1962
277. Rader B, Smith WW, Berger AR, Eichna LW: Comparison of the hemodynamic effects of mercurial diuretics and digitalis in congestive heart failure. Circulation 29:328, 1964
278. Stampfer M, Epstein SE, Beiser GD, Braunwald E: Hemodynamic effects

of diuresis at rest and during intense upright exercise in patients with impaired cardiac function. Circulation 37:900, 1968

279. Kim KE, Onesti G, Moyer JH, Swartz C: Ethacrynic acid and furosemide diuretic and hemodynamic effects and clinical uses. Am J Cardiol 27:407, 1971
280. Lal S, Mortagh JG, Pollock AM, et al: Acute hemodynamic effects of frusemide in patients with normal and raised left atrial pressure. Br Heart J 31:711, 1969
281. Ramirez A, Abelman WH: Hemodynamic effects of diuresis by ethacrynic acid, in normal subjects and in patients with congestive heart failure. Arch Intern Med 121:320, 1968
282. Samet P, Bernstein WH: Acute effects of intravenous ethacrynic acid upon cardiovascular hemodynamics. Am J Med Sci 225:78, 1968
283. Bhatia ML, Singh I, Manchanda SC, et al: Effect of frusemide on pulmonary blood volume. Br Med J 2:551, 1969
284. Scheinman M, Brown M, Rappaport E: Hemodynamic effects of ethacrynic acid in patients with refractory acute left ventricular failure. Am J Med 50:291, 1971
285. Sjogren A: Left heart failure in acute myocardial infarction. Acta Med Scand [Suppl 510, Part III] 1970, p 53
286. Dikshit K, Vyden JK, Forrester JS, et al: Renal and extrarenal hemodynamic effects of furosemide in congestive heart failure after acute myocardial infarction. N Engl J Med 288:1087, 1973
287. Austin SM, Kramer DH, Shah PM, et al: Acute hemodynamic effects of furosemide and ethacrynic acid in man. Circulation, February 1976 (in press)
288. Tattersfield AE, McNicol MW: Diuretics in acute myocardial infarction. Clin Sci 38:32, 1970
289. Davidson RM, Goldman J, Whalen RE, Wallace A: Hemodynamic effects of furosemide in acute myocardial infarction. Circulation [Suppl IV] 44:156, 1971
290. Biagi RW, Bapat BN: Frusemide in acute pulmonary edema. Lancet 1:849, 1967
291. Prichard MML, Daniel PM, Adrian GM: Peripheral ischemia of the lung: Some experimental observations. Br J Radiol 27:93, 1954
292. Herrnheiser G, Hinson KFW: Anatomical explanation of the formation of butterfly shadows. Thorax 9:198, 1964
293. Fleischner FG: The butterfly pattern of acute pulmonary edema. Am J Cardiol 20:39, 1967
294. Peabody FW: Clinical studies on respiration. III. A mechanical factor in the production of dyspnea in patients with cardiac disease. Arch Intern Med 20:433, 1917
295. Altshule MD, Zamcheck N, Iglaner A: The lung volume and its subdivisions in the upright and recumbent positions in patients with congestive heart failure: Pulmonary factors in the genesis of orthopnea. J Clin Invest 22:805, 1953

296. Donald KW: Disturbances in pulmonary function in mitral stenosis and left heart failure. Prog Cardiovasc Dis 1:298, 1958
297. Stead EA: Edema and dyspnea of heart failure. Bull NY Acad Sci 28:159, 1952
298. Richards DW: The nature of cardiac and pulmonary dyspnea. Circulation 7:15, 1953
299. Comroe JH: Dyspnea. Mod Concepts Cardiovasc Dis 25:347, 1956
300. Binger CAL: The lung volume in heart disease. J Exp Med 38:445, 1923
301. Brown CC, Fry DL, Ebert RV: The mechanics of pulmonary ventilation in patients with heart disease. Am J Med 17:438, 1954
302. Frank NR, Lyons HA, Siebens AA, Nealon TF: Pulmonary compliance in patients with cardiac disease. Am J Med 22:516, 1957
303. Sharp JT, Griffith GT, Bunnell IL, Greene DG: Ventilatory mechanics in pulmonary edema in man. J Clin Invest 37:111, 1958
304. McNeill RS, Rankin J, Forster RE: The diffusing capacity of the pulmonary membrane and the pulmonary capillary blood volume in normal subjects and in cardiopulmonary disease. Clin Sci 17:465, 1958
305. Palmer WH, Gee JBL, Mills FC, Bates DV: Disturbances of pulmonary function in mitral valve disease. Can Med Assoc J 89:744, 1963
306. McCredie RM: The pulmonary capillary bed in various forms of pulmonary hypertension. Circulation 33:854, 1966
307. Wood P: Diseases of the Heart and Circulation (ed 2). Philadelphia, J.B. Lippincott, 1956, p 270
308. Rushmer RF: Cardiovascular Dynamics (ed 2). Philadelphia W.B. Saunders, 1961, p 463
309. Harrison TR, Adams RD, Bennett IL, et al: Principles of Internal Medicine (ed 4). New York, McGraw-Hill, 1962, p 98
310. Beeson PB, McDermott W: Cecil-Loeb Textbook of Medicine (ed 11). Philadelphia, W.B. Saunders, 1963 p 620
311. Drinker CK, Peabody FW, Blumgart HL: The effect of pulmonary congestion on the ventilation of the lungs. J Exp Med 35:77, 1922
312. Dow P: The venous return as a factor affecting vital capacity. Am J Physiol 127:793, 1939
313. Glaser EM, McMichael J: Effect of venesection on the capacity of the lungs. Lancet 2:230, 1940
314. Rapaport E, Kuida H, Haynes FW, Dexter L: The pulmonary blood volume in mitral stenosis. J Clin Invest 35:1393, 1956
315. Saxton GA, Rabinowitz M, Dexter L, Haynes FW: The relationship of pulmonary compliance to pulmonary vascular pressures in patients with heart disease. J Clin Invest 35:611, 1956
316. White HC, Butler J, Donald KW: Lung compliance in patients with mitral stenosis. Clin Sci 17:667, 1958
317. Ebert RV: The lung in congestive heart failure. Arch Intern Med 107:450, 1961
318. Bucci G, Cork CD, Hamann JF: Studies of respiratory physiology in children. VI. Lung diffusion capacity of the pulmonary membrane and

pulmonary capillary volume in congenital heart disease. J Clin Invest 40:1431, 1961
319. McCredie RM, Lovejoy FW, Yu PN: Pulmonary diffusion capacity and pulmonary capillary blood volume in patients with intracardiac shunts. J Lab Clin Med 63:914, 1964
320. Lillienfield L, Freis E, Partenope E, Morowitz H: Transcapillary migration of lung water and thiocyanate ion in the pulmonary circulation of normal subjects and patients with congestive heart failure. J Clin Invest 34:1, 1955
321. Ramsey LH, Puckett W, Jose A, Lacy WW: Pericapillary gas and water distribution volumes of the lung calculated from multiple indicator dilution curves. Circ Res 15:275, 1964
322. McCredie RM: Measurement of pulmonary edema in valvular heart disease. Circulation 36:381, 1967
323. McCredie RM: Pulmonary edema in lung disease. Br Heart J 32:66, 1970
324. Schreiner BF, Schnitzler RN, Yu PN: Pulmonary extravascular volume in man. Clin Res 18:522, 1970
325. Yu PN: Lung water in congestive heart failure. Mod Concepts Cardiovasc Dis 40:27, 1971
326. Schreiner BF, Shah PM, Yu PN: unpublished observations
327. Lassers BW, George M, Anderton JL, et al: Left ventricular failure in acute myocardial infarction. Am J Cardiol 25:511, 1970
328. Greenwood WF: Profile of atrial myxoma. Am J Cardiol 21:367, 1968
329. Stovin PGI, Mitchinson MJ: Pulmonary hypertension due to obstruction of the intrapulmonary veins. Thorax 20:106, 1965
330. Heath D, Scott O, Lynch J: Pulmonary veno-occlusive disease. Thorax 26:663, 1971
331. Casarett LJ: Toxicology: The respiratory tract. Ann Rev Pharmacol 11:425, 1971
332. Kapanci Y, Weibel ER, Kaplan HP, Robinson FR: Pathogenesis and reversibility of the pulmonary lesions of oxygen toxicity in monkeys. II. Ultrastructure and morphometric studies. J Lab Invest 20:101, 1969
333. Kistler GS, Caldwell PRB, Weibel ER: Development of fine structural damage to alveolar and capillary lining cells in oxygen poisoned rat lungs. J Cell Biol 32:605, 1967
334. Mengel CE, Kann HE Jr: Effects of in vivo hypoxia on erythrocytes. III. In vivo peroxidation of erythocyte lipid. J Clin Invest 45:1150, 1966
335. Fisher AB, Hyde RW, Ricardo JM, et al: Effect of oxygen at 2 atmospheres on the pulmonary mechanics of normal man. J Appl Physiol 24:529, 1968
336. Ricardo JM, Hyde RW, Fisher AB, et al: Alterations in the pulmonary capillary bed during early O_2 toxicity in man. J Appl Physiol 24:537, 1968
337. Caldwell PRB, Lee WL, Schildkraut HS, Archibald ER: Changes in lung volume, diffusing capacity and blood gases in men breathing oxygen. J Appl Physiol 21:1477, 1966
338. Singer MM, Wright F, Stanley LK, et al: Oxygen toxicity in man: A prospective study in patients after open heart surgery. N Engl J Med 283:1473, 1970

339. Barber RE, Lee J, Hamilton WK: Oxygen toxicity in man: A prospective study in patients with irreversible brain damage. N Engl J Med 283:1478, 1970
340. Nash G, Blennerhassett JB, Pontoppidan H: Pulmonary lesions associated with oxygen therapy and artificial ventilation. N Engl J Med 276:368, 1967
341. deLemos R, Wolfsdorf J, Nachman R, et al: Lung injury from oxygen in lambs, the role of artificial ventilation. Anesthesiology 30:609, 1969
342. Pontoppidan H, Geffin B, Lowenstein E: Acute respiratory failure in the adult. 3 parts. N Engl J Med 287:690, 743, 799, 1972
343. Clark JM, Lambertson CJ: Pulmonary oxygen toxicity: A review. Pharmacol Rev 23:37, 1971
344. Hecht HH: A sea level view of altitude problems. Am J Med 50:703, 1971
345. Hultgren HN, Lopez CE, Lundberg E, Miller H: Physiologic studies of pulmonary edema at high altitude. Circulation 29:393, 1964
346. Nicoloff DM, Ballin HM, Visscher MB: Hypoxia and edema of the perfused isolated canine lung. Proc Soc Exp Biol Med 131:22, 1969
347. Goodale RL, Goetzman B, Visscher MB: Hypoxia and iodoactic acid and alveolo-capillary barrier permeability to albumen. Am J Physiol 219:1226, 1970
348. Viswanathan R, Jain SK, Subramanian S: Pulmonary edema of high altitude. III. Pathogenesis. Am Rev Respir Dis 100:342, 1969
349. Hultgren HN, Grover RF, Hartley LH: Abnormal circulatory responses to high altitude in subjects with a previous history of high-altitude pulmonary edema. Circulation 44:759, 1971
350. Whayne TF, Severinghaus JW: Experimental hypoxic pulmonary edema in the rat. J Appl Physiol 25:729, 1968
351. Steinberg AD, Karliner JS: The clinical spectrum of heroin pulmonary edema. Arch Intern Med 122:122, 1968
352. Cerebral edema seen in many "sudden death" heroin victims. JAMA 212:967, 1970
353. Storstein O, Rasmussen K: The cause of arterial hypoxemia in acute myocardial infarction. Acta Med Scand 183:193, 1968
354. Sutherland PW, Cade JF, Pain MCF: Pulmonary extravascular fluid volume and hypoxemia in myocardial infarction. Aust NZ J Med 1:141, 1971
355. Avery WG, Samet P, Sackner MA: The acidosis of pulmonary edema. Am J Med 48:320, 1970
356. Aberman A, Fulop M: The metabolic and respiratory acidosis of acute pulmonary edema. Ann Intern Med 76:173, 1972
357. Hales CA, Kazemi H: Small-airways function in myocardial infarction. N Engl J Med 290:761, 1974
358. Milic-Emili J, Ruff F: Effects of pulmonary congestion and edema on small airways. Bull Physiopathol Respir 7:1181, 1971
359. Interiano B, Hyde RW, Hodges M, Yu PN: Interrelation between alterations in pulmonary mechanics and hemodynamics in acute myocardial infarction. J Clin Invest 52:1994, 1973

Kevin M. McIntyre
Arthur A. Sasahara

19

Pathophysiology of Acute and Chronic Cor Pulmonale

Of the many disease entities that affect the heart, the internist and even the cardiologist is least familiar with the entity of cor pulmonale, or pulmonary heart disease. Even the term cor pulmonale is misused. For the purposes of this communication, cor pulmonale or chronic cor pulmonale is defined as alteration in structure and function of the right ventricle resulting from disease affecting the structure or function of the lung or its vasculature, except when this alteration results from disease of the left side of the heart or congenital heart disease.[1] Acute cor pulmonale is defined as acute right heart strain or overload resulting from the pulmonary hypertension of massive pulmonary embolism.

The impact of both diseases in terms of morbidity and mortality is considerable, although definitive data on the prevalence of cor pulmonale and acute cor pulmonale are not available. The lack of uniform diagnostic criteria and reporting, particularly in chronic cor pulmonale, in office visits and hospital admissions has been principally responsible for the poor statistics. Nevertheless, estimated figures from indirect means attest to the increasing role of disability and death caused by these two diseases, particularly chronic cor pulmonale. Because of the difficulties in diagnosis of cor pulmonale, aside from an episode of right-heart failure or examination on the autopsy table, an estimation of the prevalence of chronic obstructive pulmonary disease deaths provides an index to the prevalence of chronic cor pulmonale. It has been estimated that over 25,000 deaths each year are caused by chronic obstructive pulmonary disease (COPD) and that this figure is doubling every 5 years.[2] From various reported series the incidence of chronic cor pulmonale averages about 8%–10% of all reported heart disease.[3,4]

ACUTE COR PULMONALE

The term acute cor pulmonale was first used by McGinn and White in their classical description of the electrocardiographic and clinical consequences of massive obstruction of the pulmonary artery by thromboembolism[5] (Figs. 19-1 and 19-2). They proposed that the term be used to distinguish "the immediate result of a high degree of occlusion of the pulmonary artery" from the "progressive enlargement of the right side of the heart secondary to certain pulmonary diseases"—a syndrome that was well recognized at that time. The clinical symptoms of acute cor pulmonale as described by the original authors included retrosternal chest pressure, apprehension, fever, tachycardia, tachypnea, and marked dyspnea, collapse, sweating, pallor, and fall in blood pressure. Physical

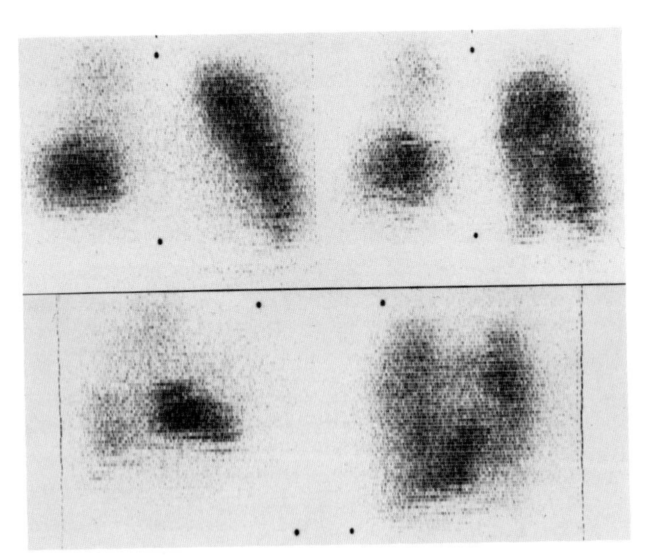

Fig. 19-1. Four-view perfusion lung scans of patient with acute cor pulmonale (massive pulmonary embolism). Approximately 50% of the total pulmonary vasculature is without perfusion. Top left, anterior view; top right, posterior; bottom left, right lateral; bottom right, left lateral view. (Reprinted by permission from Sasahara et al: Therapeutische Umschau/Revue Therapeutique, Band 27, Heft 1, 1970.)

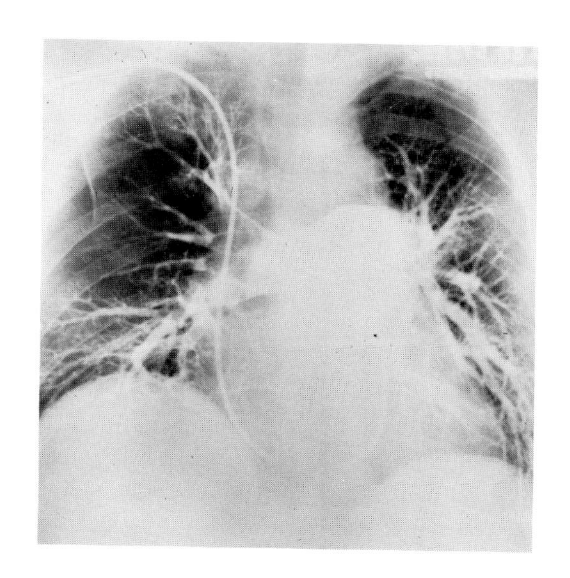

Fig. 19-2. Pulmonary angiogram of same patient as in Fig. 19-1 showing large saddle embolus and smaller emboli on the right, as well as multiple filling defects on the left. (Reprinted by permission from Sasahara et al: Therapeutische Umschau/Revue Therapeutique, Band 27, Heft 1, 1970.)

findings commonly observed by McGinn and White in the original series included cyanosis, enlargement of neck veins, accentuation of the pulmonary second heart sound, gallop rhythm over the right ventricle, palpable dilatation of the pulmonary artery, and occasionally a pericardial friction rub. The authors pointed out that a number of these findings could be transient. They included among the characteristic ECG changes the now classical S1-Q3-T3 pattern and right axis deviation. The pathophysiologic basis for acute cor pulmonale, as originally stated, was "dilatation and partial failure of the chambers of the right side of the heart."

While the description of McGinn and White has effectively endured over 40 years of voluminous writing on the subject of pulmonary embolism (PE) and right-heart failure, additional insight into the hemodynamic consequences of acute pulmonary arterial obstruction has been provided by cardiac catheterization measurements, most of which have been made over the past 10 years since the premortem diagnosis of PE could be confidently established by angiography. It is now clear that hemodynamic measurements made after PE do not necessarily equate to the hemodynamic response to PE—only in patients free of preembolic cardiopulmonary disorders can the hemodynamic status after PE be correctly equated to the hemodynamic response. In this group of patients the hemodynamic response appears to be governed primarily by the magnitude of embolic obstruction. In patients with prior cardiopulmonary disease the hemodynamic abnormalities observed after PE include any such abnormalities that existed before PE (e.g., congestive heart failure) plus any additional abnormalities caused by embolism. The extent to which each of these several processes contributes to the hemodynamic abnormalities may be difficult to determine. Since PE is the most important cause of acute cor pulmonale, we will briefly review the pathophysiology of thromboembolism.

Pulmonary Thromboembolism

Acute PE may be the most common form of acute pulmonary disease among hospitalized patients and has been found in some studies to be the most common cause of sudden unexpected death among hospitalized patients.[6] Effective therapy for PE now exists. Recognition, by far the major problem in pulmonary embolism today, should be enhanced by an understanding of the pathophysiology.

Incidence, Predisposing Factors, and Origin of PE

Postmortem incidence of PE has ranged from 10% to 80%. The incidence is far higher among patients with strong predisposing factors such as trauma, congestive heart failure, or advancing age.[7–9] The incidence of leg vein thrombosis among patients with PE has ranged from 80% to 100%.[9–12] Studies employing ^{125}I-labeled fibrinogen, impedance phlebography, and contrast phlebography have strongly supported the thesis that the deep veins of the legs are the source of the vast majority (> 90%) of PE;[13–15] and in the absence of another clearly recognizable source of PE such as pelvic thrombophlebitis, the leg veins should be considered the source. It has been clearly shown that when thrombi are present in the lower extremities there will be clinical evidence of their presence in only about 50% of cases. The biochemical basis for the apparent imbalance between thrombotic and lytic mechanisms that presumably results in abnormal intravascular thrombus formation is not yet clearly understood.

Embolism in the Pulmonary Vasculature

Pulmonary embolism may have any of the following consequences: pulmonary infarction, hemodynamic disturbance of varying severity, or a combination of the two. Pulmonary infarction usually occurs only when both the pulmonary arterial segment is completely occluded at a rather distal level (usually sublobar) and when the collateral (bronchial arteries) circulation is simultaneously impaired by such processes as congestive failure or systemic hypotension.[16] Clinical symptoms and signs of pulmonary embolism, including hemoptysis, pleuritic pain, and roentgenographic densities, may clearly occur after PE without the pathology of pulmonary necrosis.[17] In the latter circumstance, intraalveolar hemorrhage and congestion are usually present, without lung necrosis, a process referred to by Hampton and Castleman as reversible infarction.[17]

Neurohumoral Considerations

A number of mechanisms have been described in explaining the hemodynamic impact of PE. Many of these were evolved prior to the introduction of diagnostic pulmonary angiography in response to a need to explain the association of small PE and sudden death. Abundant experimental evidence had clearly demonstrated that extreme narrowing or obstruction of the pulmonary circulation was essential before heart function was noticeably impaired.

Both a "pulmonocoronary" reflex and reflexly induced spasm of pulmonary arterioles (pulmono-pulmonary reflex) have been hypothesized in cases of sudden death after small pulmonary emboli.[18,19] Investigations have failed to demonstrate such reflexes.[20,21] Impairment of coronary blood flow by either hypotension or coronary artery spasm has been incriminated as a mechanism of sudden death,[22,23] but there has been little factual support for this hypothesis. Gorham offered pathologic evidence that strongly supported the conclusion that pulmonary embolism had to be extensive to cause sudden death and that sudden death after minor embolism could be explained in nearly all cases on a nonembolic basis, such as underlying heart disease.[24]

Since the original hypothesis of Comroe and his associates,[25] reports have reemphasized the bronchopulmonary and pulmonary vasoconstrictive roles of both serotonin[26,27] and histamine.[28] It is now apparent that vasoconstriction does occur after pulmonary embolism. Hypoxemia has been identified as one important factor responsible for pulmonary vasconstriction,[29,30] and acidosis has been shown to exaggerate the extent to which a given level of hypoxemia will produce vasoconstriction.[31,32] The linear relationship shown to exist between the extent of embolic obstruction and the level of postembolic pulmonary hypertension suggests that whatever role a humoral agent may play in the production of pulmonary hypertension the mass of embolized material appears to be the primary factor.[33]

It is important to point out that the magnitude of obstruction of the pulmonary arteries assessed by angiography underestimates the total increase in right ventricular outflow impedance. Vasoconstriction, whether due to hypoxemia or humoral factors, has been shown to be present consistently after PE, but cannot be quantitated by angiography. In ad-

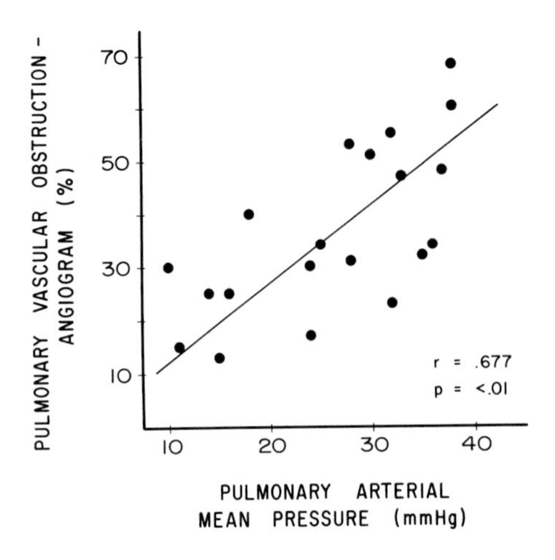

Fig. 19-3. Percent angiographic obstruction—pulmonary arterial mean pressure relationship. Pulmonary vascular obstruction by angiography, expressed in percent on the vertical axis, is shown to correlate closely with pulmonary artery mean pressure in 20 patients free of prior cardiopulmonary disease. The regression line suggests that a normal PAm (< 20 mm Hg) may be anticipated until obstruction exceeds 20% to 30%. On the other hand, no patient had a PAm in excess of 40 mm Hg despite obstruction from 50% to 70%. (Reprinted by permission from McIntyre KM, Sasahara AA: Hemodynamic response to pulmonary embolism in patients free of prior cardiopulmonary disease. Am J Cardiol 28:288, 1971.)

dition, as a result of fragmentation, a substantial amount of embolic obstruction may occur at the level of small arterioles. Since conventional angiography is unable to appreciate detail sufficiently well below the 2-mm level, a significant amount of embolic obstruction may be missed.

Hemodynamic Consequences of PE in Patients Free of Prior Cardiopulmonary Disease (CPD)

The primary determinant of the hemodynamic consequence of PE in patients without prior CPD is the magnitude of embolic obstruction. There may be no hemodynamic disturbance whatsoever after PE, particularly if it is small. Hypoxemia, however, is present in 85% to 90% of cases. Hypoxemia alone is often present with angiographic involvement of 20% or less of the pulmonary vascular bed. Pulmonary hypertension is usually present in addition to systemic arterial hypoxemia when obstruction exceeds 20%–25% (Fig. 19-3). Elevation of right atrial mean pressure (RAm) may be expected when obstruction exceeds 35% or PAm exceeds 30 mm Hg (Fig. 19-4). At this point, perhaps, one might consider the essential ingredient of McGinn and White to be present, i.e., dilatation of the right side of the heart. Cardiac output usually continues to be maintained, however, and may even be increased when RAm is elevated. Finally, usually only after massive obstruction (50% or more by angiography) cardiac output may fall and acute cor pulmonale with right

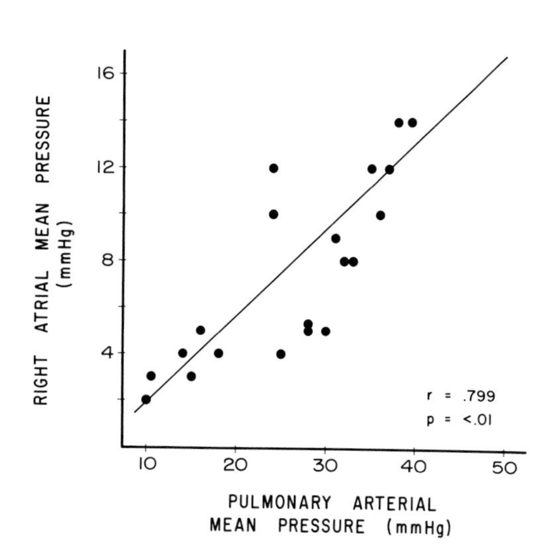

Fig. 19-4. Right atrial mean pressure—pulmonary arterial mean pressure relationship. Right atrial mean pressure was shown to be closely related to the pulmonary arterial mean pressure in 20 patients free of preembolic cardiopulmonary disease. (Reprinted by permission from McIntyre KM, Sasahara AA: Hemodynamic response to pulmonary embolism in patients free of prior cardiopulmonary disease. Am J Cardiol 28:288, 1971.)

ventricular "failure" (in the sense of output impairment) can be said to be present.

The term acute cor pulmonale may be applied with literal accuracy from the time the heart is first affected by PE until it ultimately fails. One may consider, therefore, that there are gradations of severity of acute cor pulmonale that can be identified hemodynamically according to the following categories: (1) increased right ventricular (RV) outflow impedance alone, as evidenced by an isolated increase in RV systolic or PAm pressure; (2) increased RV systolic and diastolic pressure, or RAm; (3) increased RV systolic and diastolic pressures, with impairment of cardiac output. Syndromes 1 and 2 may more accurately be defined as acute compensated cor pulmonale. Alternatively, one may designate syndrome 3 as acute cor pulmonale with right ventricular failure. The term acute cor pulmonale usually implies syndrome 3, in which right ventricular failure has already occurred. In view of the narrow range of adaptability of the normal RV, it may be important from a prognostic and therapeutic viewpoint to be able to distinguish between these pathophysiologic syndromes. The complete syndrome of decompensated acute cor pulmonale, therefore, results from occlusion of the pulmonary circulation sufficient to cause an increase in RV systolic and diastolic pressures, a decline in stroke output of the RV, and a decrease in pulmonary blood flow and left ventricular filling. Systemic blood pressure may be maintained by peripheral arteriolar vasoconstriction until left ventricular filling is critically decreased, at which time hypotension and shock may supervene (acute cor pulmonale, decompensated, with shock). Another manifestation of acute cor pulmonale is sudden cardiac arrest, which is associated with massive PE and appears to result from sudden, virtually complete arrest of pulmonary blood flow and left ventricular filling.

Studies reporting the hemodynamic response to massive PE in patients free of CPD have been remarkably consistent (Table 19-1). It is clear from these studies that the range over which the nonhypertrophied RV can adapt to increases in outflow impedance is very narrow, i.e., from 20 to 40 mm Hg, with 40 mm Hg representing the maximim PAm that the nonhypertrophied RV can generate acutely. Pulmonary hypertension in this group of patients is considered to be very severe when PAm is between 30 and 40 mm Hg. With such a slender margin for adaptability, small increases in outflow obstruction may be catastrophic. This range appears to be so predictable that the finding of a PAm in excess of 40 mm Hg should suggest underlying pulmonary or cardiac disease, or that emboli have been recurrent over a period sufficiently long that right ventricular hypertrophy has occurred. On the other hand, if

Table 19-1
Pulmonary Embolism: Hemodynamics in Previously Normal Patients

	No.	PAm	RAm	PO_2/SAT	CI	HR	% Angiographic Obstruction
Miller and Sutton	23	26.6	9	(50.3)/86.4	—	—	> 50
McDonald et al.							
No shock	20	26.5	9	—	2.7	113	53.4
Shock	3	30.0	12	—	1.4	119	(25–75)
McIntyre and Sasahara[33]	20	26.0	7	64/92.5	3.1	96	37 (13–68)

The observations of Miller and Sutton (Br Heart J 32:518, 1970) were made in a series of 23 patients with massive pulmonary embolism, many of whom required embolectomy. Twenty-one of 23 were free of prior cardiopulmonary disease. The observations of McDonald et al. (Br Heart J 34:356, 1972) were made in a comparable group of patients, but were subdivided by the authors into those with and without shock. Observations by McIntyre and Sasahara were also made in a series of patients free of prior cardiopulmonary disease, but in most patients embolic obstruction by angiography was less than massive, ranging 13%–68%. It is of particular interest to note that the mean pulmonary artery pressure of 26–26.6 mm Hg was virtually identical for each of the large groups observed. Cardiac index was not measured in the series of Miller and Sutton, nor was PaO_2 recorded in that group. The mean PaO_2 (in parentheses) was estimated from the oxygen saturation using standard tables.

PAm is in this range in the absence of significant embolic involvement (i.e., 25% or more) underlying disease should be suspected.

When RAm is elevated as a result of PE, significant obstruction by angiography can be expected. If evidence of significant PE (30% or more) is not found, underlying disease should be suspected. On the other hand, a normal RAm does not exclude PE. When RAm is elevated, cardiac output (CO) may be normal, increased, or depressed. It appears that the RV can compensate to some degree by employing the Frank-Starling mechanism, but this adaptation is very limited. A depression of CO is usually accompanied by tachycardia, the increase in rate only partially compensating for a sharp reduction in stroke output.

Hypotension results from inability of the RV to force blood across the pulmonary circulation, making it unavailable to the LV to maintain systemic pressure and perfusion. Hypoxemia becomes more severe with more extensive obstruction by embolism, increasing RV outflow impedance by causing pulmonary arterial vasoconstriction. Hypoxemia,

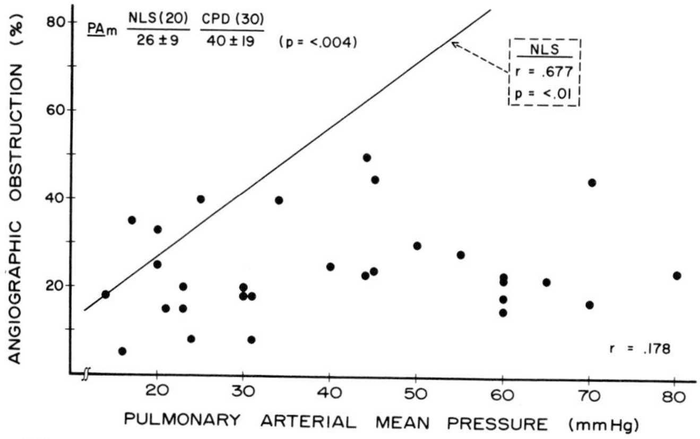

Fig. 19-5. Percent angiographic obstruction—pulmonary arterial mean pressure. The relation between angiographic obstruction (%) and pulmonary arterial mean pressure (mm Hg) in 20 patients free of prior cardiopulmonary disease, represented by the regression line, is contrasted with that of 30 patients with preembolic cardiopulmonary disease, represented by the solid circles. While a close relationship was seen between these parameters in patients free of prior cardiopulmonary disease, no such relationship was found among patients with heart or lung disease prior to embolism. (Reprinted by permission from McIntyre KM, Sasahara AA, Sharma GV: Pulmonary thromboembolism: current concepts. Adv Intern Med 18:199, 1972.)

therefore, can be an important factor in the aggravation of RV loading. When cardiac output falls, tissue ischemia may result in acidosis. It has been shown that hydrogen ion potentiates the vasoconstrictive reaction of hypoxemia. Therefore, while massive embolism is necessary before acute cor pulmonale can be said to be present in this group of patients, a cascade of destructive influences may be precipitated by an additional relatively small increment in RV outflow impedance.

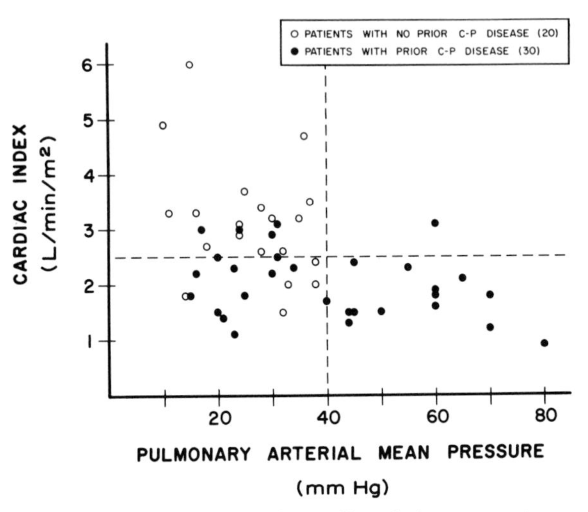

Fig. 19-6. Relationship of cardiac index to pulmonary arterial mean pressure. The postembolic cardiac index was plotted against the pulmonary arterial mean pressure in 20 patients free of prior cardiopulmonary disease (open circles) and in 30 patients with preembolic cardiopulmonary disease (solid circles). Patients free of prior cardiopulmonary disease characteristically had normal or increased cardiac indices, while pulmonary arterial pressures ranged from 10 to 40 mm Hg. No patient in this group had a pulmonary artery mean pressure in excess of 40 mm Hg. In 4 of 5 patients in whom CI was depressed, the PAm was approaching 40 mm Hg. Among patients with prior cardiopulmonary disease, pulmonary arterial mean pressure ranged from normal to 80 mm Hg, exceeding the highest PAm in the former group in most cases. The cardiac index was usually depressed, but the depression of cardiac index bore no apparent relationship to increasing pulmonary pressure. (Reprinted by permission from McIntyre KM, Sasahara AA, Sharma GV: Pulmonary thromboembolism: current concepts. Adv Intern Med 18:199, 1972.)

Hemodynamic Status after PE in Patients with Prior Cardiopulmonary Disease

Among patients with prior cardiopulmonary disease there may be no recognizable relationship between hemodynamic status after PE and the magnitude of obstruction. When judged by the responses observed in normal patients, patients with prior cardiopulmonary disease clearly show a level of pulmonary hypertension that is disproportionate to the degree of embolic obstruction (Fig. 19-5), suggesting that the postembolic PAm in these patients is dependent more on the preembolic state of the pulmonary circulation than on the embolus itself. Depression of CO is common in patients with preembolic heart and lung disease and appears to be independent of both the extent of obstruction and the level of pulmonary arterial pressure (Fig. 19-6). Whereas elevation of total

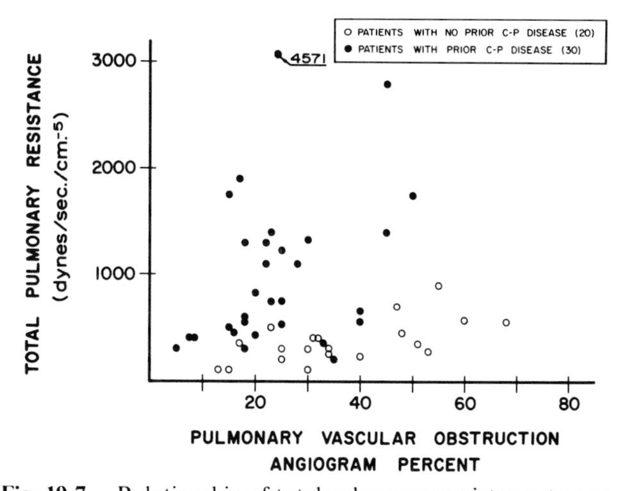

Fig. 19-7. Relationship of total pulmonary resistance to percent angiographic obstruction. Total pulmonary resistance was compared to pulmonary vascular obstruction in 20 patients free of preembolic cardiopulmonary disease (open circles) and in 30 patients with prior cardiopulmonary disease (solid circles). A very predictable relationship was found in the former group, such that the total pulmonary resistance increased linearly with increasing pulmonary vascular obstruction. No such relationship was observed among patients with prior cardiopulmonary disease. The difference in this relationship was attributed to the independent contribution of preexisting processes to the total pulmonary resistance in patients with preembolic heart or lung disease (see text). [Reprinted by permission from Sasahara et al: in Moser K, Stein M (eds): Pulmonary Thromboembolism. 1973, p 144.]

pulmonary resistance is directly proportional to the extent of embolic obstruction in patients previously free of cardiopulmonary disease, total pulmonary. resistance is out of proportion to the angiographic obstruction in patients with prior heart and lung disease (Fig. 19-7). The difference in the total pulmonary resistance between the two groups may represent an approximate measure of the extent to which the preembolic pulmonary vascular status determines the postembolic hemodynamic status. However, recognition of the role of PE and detection of elements of acute cor pulmonale in the setting of prior CPD may be difficult or impossible.

CHRONIC COR PULMONALE

Chronic cor pulmonale develops in the great majority of patients when the responsible bronchopulmonary disease is diffuse, bilateral, and chronic, resulting in severe impairment of oxygenation. Unusually, chronic cor pulmonale may develop simply on the basis of chronic pulmonary vascular obstruction and increased resistance to blood flow, as in recurrent multiple pulmonary embolism, diffuse pulmonary fibrosis, and primary pulmonary hypertension. However, hypoxemia of varying degrees usually occurs in the natural history of these disease, making the development of chronic cor pulmonale more likely.

The most common underlying disease leading to chronic cor pulmonale is chronic obstructive pulmonary disease (COPD). Other diseases such as pulmonary fibrosis, pneumoconiosis, kyphoscoliosis, and tuberculosis, especially following thoracoplasty or resection, may also be responsible. A more complete listing of diseases that may lead to chronic cor pulmonale is tabulated in Table 19-2. Although chronic COPD is three to five times more common in males, the incidence of both sexes with COPD developing chronic cor pulmonale is about 20%.[34] The other 80% of patients with COPD suffer from shortness of breath rather than from chronic cor pulmonale.

Classification of Chronic Cor Pulmonale

Because of the many and varied types of diseases that may lead to chronic cor pulmonale, classifications in the past have been based on causative diseases that affect the bronchopulmonary system, the thoracic cage, or the pulmonary vasculature. However, as advances have been made in both the pathophysiology and treatment of diseases leading to cor pulmonale, it seems more useful to classify chronic cor pulmonale according to the underlying pathophysiologic derangement, which then

Table 19-2
Classification of Chronic Cor Pulmonale

- Disorders in which chronic alveolar hypoventilation predominates
  - Bronchopulmonary disorders
    - Chronic obstructive pulmonary disease
    - Bronchial asthma
    - Cystic fibrosis
  - Thoracic cage disorders
    - Neuromuscular disease
      - Poliomyelitis
      - Muscular dystrophy
    - Thoracic cage deformity
      - Kyphoscoliosis
      - Extreme obesity (Pickwickian syndrome)
    - Diffuse pleural fibrosis
  - Central nervous system disorders
    - Idiopathic alveolar hypoventilation
    - High cervical spinal cord injury
- Disorders in which pulmonary vascular space reduction predominates
  - Obliterative vascular disease
    - Chronic reçurrent pulmonary embolism
    - Primary pulmonary hypertension
    - Arteritides, e.g., polyarteritis nodosa
    - Parasitic occlusion, e.g., schistosomiasis
    - Sickle cell disease
    - Thoracoplasty
    - High altitude
  - Diffuse pulmonary fibrosis
    - Sarcoidosis
    - Pneumoconiosis
    - Chronic diffuse interstitial fibrosis
    - Berylliosis
    - Collagen disease
      - Scleroderma
      - Dermatomyositis
      - Disseminated lupus erythematosus
    - Eosinophilic granulomata or histiocytosis
    - Radiation fibrosis
    - Tuberculosis

permits more rational therapy. Such a classification of chronic cor pulmonale based on principal pathophysiology can be comprised of two large groups: (1) disorders in which chronic alveolar hypoventilation predominate and (2) disorders in which pulmonary vascular space reductions predominate[34,35] (Table 19-2).

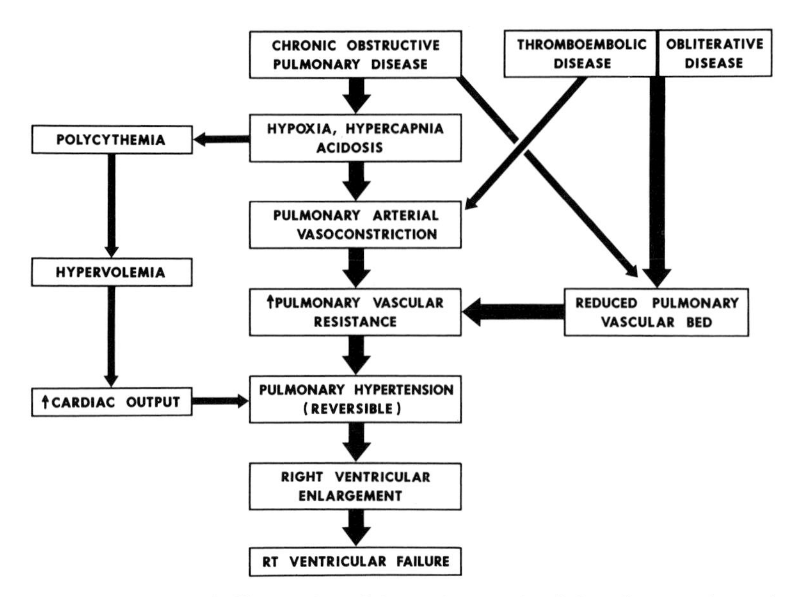

Fig. 19-8. Schematic illustration of the pathogenesis of chronic cor pulmonale.

Pathophysiology of Chronic Cor Pulmonale

The mechanisms leading to chronic cor pulmonale are not entirely known, except for the important role of pulmonary hypertension. This pulmonary hypertension is caused by chronic alveolar hypoventilation or by extensive obliteration of the pulmonary vascular bed or by a combination of both processes (Fig. 19-8). The importance of this distinction rests on the fact that pulmonary hypertension due to obliterative vascular diseases tends to be fixed, whereas when due to hypoventilation the pulmonary hypertension may be potentially reversible. Therefore the nature of the therapy and the expected results depend to a large degree on whether pulmonary vascular obliteration or alveolar hypoventilation is principally responsible for the pulmonary hypertension.

Chronic Alveolar Hypoventilation

The recognition of the importance of alveolar hypoventilation in the pathogenesis of cor pulmonale was a major advance in the understanding of the pathophysiologic changes. These changes include pulmonary hypoxia, arterial hypoxemia, hypercapnea, and respiratory acidosis. The role of these factors independently and in concert upon the pulmonary circulation will be examined in depth.

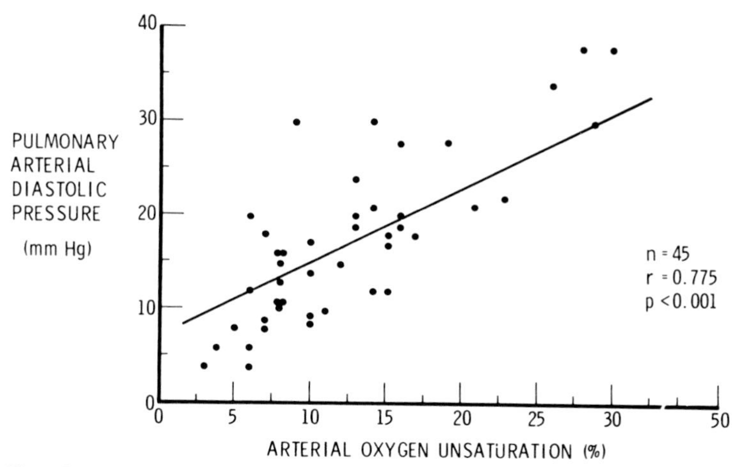

Fig. 19-9. The pulmonary artery diastolic pressure is directly related to the degree of arterial oxygen unsaturation (abscissa). (Reprinted by permission from Harvey RM, Enson Y, Ferrer MI: A reconsideration of the origins of pulmonary hypertension. Chest 59:82, 1971.)

Pulmonary Hypoxia

The role of low ambient oxygen concentration and resulting arterial hypoxemia upon the pulmonary circulation has been extensively investigated for many years.[36,37] A close correlation between the level of pulmonary artery pressure and arterial oxygen unsaturation has been noted in patients with COPD.[34] As the level of arterial saturation decreased, particularly to levels below 80%, the pulmonary arterial pressure rose, the rise being directly related to the oxygen unsaturation (Fig. 19-9). It had also been noted that when normal subjects or patients breathed a 12%–14% oxygen mixture, resulting in significant oxygen unsaturation, pulmonary artery pressure and cardiac output increased (Fig. 19-10). When exercised, a steep rise in pulmonary artery pressure and an increase in cardiac output were noted. When these same subjects breathed ambient air (21% oxygen) and exercised to the same level of cardiac output as during hypoxic (12%–14% oxygen) breathing, the pulmonary artery pressure was significantly less, indicating that for any given level of cardiac output the pulmonary artery pressure was higher during hypoxic breathing than during ambient-air breathing.

Another study pointing to the role of hypoxia on pulmonary circulation vasomotion was reported by Staub,[38] who administered low-oxygen gas to one lung and pure oxygen to the other in an animal model. On section, following rapid freezing, there was marked vasoconstriction

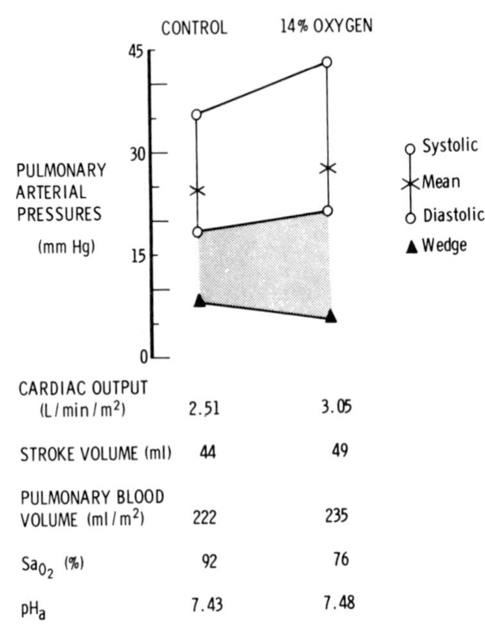

Fig. 19-10. Hemodynamics in a patient with COPD during ambient air and hypoxic breathing. Note the rise in pulmonary artery pressure during hypoxia. (Reprinted by permission from Harvey et al: Chest 59:82, 1971.)

of the arterioles in the hypoxic lung, in contrast to the fully dilated arterioles in the oxygenated lung.

The interest in high-altitude physiology has provided an opportunity to make observations on the adaptations of man and animals to chronic hypoxia. The study of these naturally occurring experimental models has provided valuable information on the integrated responses of the pulmonary vasculature and cardiovascular, hematopoietic, and central nervous systems.

Peñaloza,[39] following up on Rotta's observation,[40] carried out extensive studies in the natives who worked in the tin and copper mines in the Peruvian Andes at an altitude of 15,000 ft (average barometric pressure 444 mm Hg). The pulmonary artery pressure in the healthy natives was approximately twice normal (26 mm Hg); the oxygen saturation averaged 84%; the cardiac output was variable, but tended to be elevated; the hemoglobin and hematocrit were increased, 20 g and 60%, respectively; the blood volume was increased to 95 ml/kg (normal: 80 ml/kg). Interestingly, oxygen inhalation did not lower pulmonary artery pressure, suggesting a fixed hypertension. Histologically, thickening of the precapillary arterioles was found.

Some natives lose acclimatization to high altitude and develop chronic mountain sickness or Monge's disease.[41] Marked alterations in their adaptive mechanisms occur. The arterial saturation decreases further to 70%; the pulmonary artery mean pressure rises to three times normal (36–38 mm Hg); the cardiac output increases an average of 40%; the hemoglobin increases to 24–26 g, the hematocrit up to 80%; the blood volume increases up to 135 ml/kg. The clinical picture consists of dyspnea, fatigue, headache, lightheadedness, and lassitude. The puzzling fact is that patients with chronic mountain sickness do not hyperventilate, reflecting an insensitivity to the rising arterial carbon dioxide tension. It seems likely that in the process of losing acclimatization the respiratory center also becomes involved.

Other studies of high altitude "natives" carried out in this country indicate that the pulmonary hypertension is reversible. Grover reported the hemodynamic findings from 28 healthy adolescents residing in Leadville, Colorado (altitude 10,200 ft).[42] The average mean pulmonary artery pressure was 25 mm Hg, which doubled during exercise. Variability in response to exercise was noted in 7 subjects who developed mean pulmonary artery pressures in excess of 65 mm Hg. One striking example was a 15-year-old girl champion skiier who developed a peak pulmonary artery pressure of 165/95 during exercise. Subsequently she was studied at sea level, where normal hemodynamics and response to exercise were noted.[43] In addition, these adolescent subjects were administered a more hypoxic gas mixture that raised the pulmonary artery pressure in some but not in others, demonstrating the variability of response to hypoxia within the same species.[44]

Animals similarly exposed to high altitude showed great variability in pulmonary artery pressure response between species and within species. In Grover's study,[45] steers taken to 12,700 ft developed average pulmonary artery pressure twice normal within 2 weeks, which tripled in 6 weeks. There was marked variability, the mean pulmonary artery pressure ranging between 55 and 110 mm Hg. In several, the pulmonary artery hypertension was of such severity as to produce right-heart failure, an accumulation of edema fluid presternally, labeled brisket disease.[46] In contrast, lambs did not develop pulmonary artery hypertension at 12,700 ft, indicating a state of hyporeactivity not observed in cattle.

Anatomically, the evidence of right ventricular hypertrophy in animals is more uniform than would be expected on the basis of pulmonary artery hypertension alone.[44] Although lambs did not develop pulmonary artery hypertension at high altitude, the weight of the right ventricle, expressed as a percentage of the total heart weight (RV/T %) increased from 21.7% (sea level) to 26.3% at high altitude. Cats, similarly hyporeactive in their pulmonary artery pressure response, showed an increase of their RV/T % from 24% to 28%.

In man living at high altitude pulmonary artery hypertension (PAH) is accompanied by hypertrophy of the right ventricle and changes in the pulmonary arterial tree. While the relative weight of the right ventricle in the newborn is the same at high altitude as at sea level, postnatal growth is much more rapid at higher altitude. At 1 month the relative weight of the RV is greater at altitude, and at 3 months the maximum degree of right ventricular hypertrophy has occurred, persisting essentially unchanged throughout adult life.[47] The RV/T % at 12,200 ft is approximately 31.5%, compared to 23.7% at sea level.[44]

Histologically there is increased thickness of the muscular media in the terminal portion of the pulmonary arterial tree in natives of high altitudes.[48] In addition, the pulmonary arterial trunk resembles the aorta, where the elastic lamellae are denser, forming a less open branching network—changes compatible with sustained pulmonary artery hypertension existing from infancy. In acquired PAH (e.g., mitral stenosis) this aortic pattern is not observed, although the wall thickness of the pulmonary trunk is increased.[49]

Although the morphometry of myocardial fibers in natives of high altitudes has not been studied, Ishikawa et al. characterized the changes occurring in patients with chronic cor pulmonale due to COPD.[50] They found (1) a direct relationship between myocardial fiber diameter and weight of the heart, (2) a similar proportion of water and phosopholipid content in the myocardium in normal and chronic cor pulmonale hearts, (3) larger septal and right ventricular fibers than left ventricular fibers in pure right ventricular hypertrophy, and (4) a direct correlation between hypoxia and right ventricular fiber diameter.

Hypercapnea and Hydrogen Ion Concentration

Although carbon dioxide retention is common in patients with advanced COPD, its effect on the pulmonary circulation is probably through the acidosis or increased hydrogen ion concentration that it generates, rather than by a direct molecular action (Fig. 19-11). Of importance were the studies of von Euler and Liljestrand, who administered 10% oxygen and 90% nitrogen to their cat model, observing the development of PAH.[36] In addition, the administration of 6.5% carbon dioxide in oxygen also increased pulmonary artery pressure. Subsequently, Liljestrand[51] reemphasized the importance of hydrogen ion concentration as a chemical stimulus for pulmonary artery vasoconstriction. The hypoxemia and retention of carbon dioxide promoted the release of lactic acid, resulting in an increase in hydrogen ion concentration. He also proposed that variability in pulmonary artery pressure response to acutely induced hypoxia or hypercapnea may be

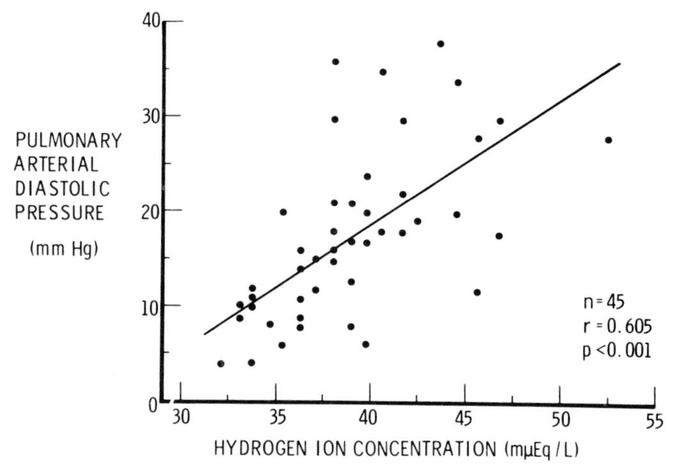

Fig. 19-11. The pulmonary artery diastolic pressure is directly related to the degree of hydrogen ion concentration, shown on the abscissa. (Reprinted by permission from Harvey et al: Chest 59:82, 1971.)

dependent upon net hydrogen ion concentration plus local concentration of hydrogen ion, which in turn is influenced by local variation in ventilation–perfusion ratios. If a segment of diseased lung should have a lower ventilation–perfusion ratio than an adjacent segment, a greater amount of lactic acid would be liberated in the former, increasing the local hydrogen ion concentration and resulting in greater local vasoconstriction.

Subsequently, Enson and his colleagues[52] examined the relationship between hydrogen ion concentration and pulmonary artery pressure and found a highly significant correlation. Acute alkalosis was produced with

	Cardiac Output	Oxygen Sat.	pCO_2	pH	Δ in PA_p
$NaHCO_3$	↑	–	↑	↑	↓
THAM	↑	↓↓	–	↑	↓
HCl	–	sl.↑	–	↓	↑

Fig. 19-12. Relationship between changes in pH and pulmonary artery pressures during various interventions (see text). (Reprinted by permission from Ferrer IM: Disturbances in the circulation in patients with cor pulmonale. Ann NY Bull NY Acad Med 41:942, 1965.)

intravenous sodium bicarbonate and amine buffer, tris-hydroxy-amino-methane (THAM), and acidosis was produced by infusion of dilute hydrochloric acid (Fig. 19-12). Sodium bicarbonate caused a rise in pH and in $Paco_2$, but the change in arterial saturation was minimal. Despite a rise in pulmonary blood flow the pulmonary artery pressure fell. THAM raised pH and cardiac output and lowered oxygen saturation. Despite this unsaturation, the pulmonary artery pressure fell. Dilute hydrochloric acid, in contrast, lowered pH and raised pulmonary artery pressure without altering pulmonary blood flow or $Paco_2$. There was a slight rise in arterial saturation. In view of the importance of both hypoxia and hydrogen ion concentration on pulmonary artery vasomotion, a graph of this relationship was developed that forms the basis for the modern treatment of respiratory failure (Fig. 19-13). At minor degrees of oxygen unsaturation, pulmonary artery pressure is relatively insensitive to hydrogen ion concentration, whereas it is extremely sensitive at high degrees of unsaturation. This protective effect of normal or high pH values (low hydrogen ion concentration) in the face of lowered arterial oxygen saturation is particularly appropriate and important in patients with COPD.

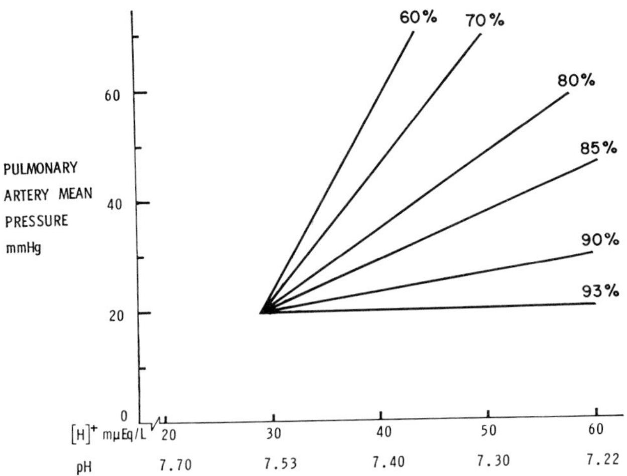

Fig. 19-13. Relationships of arterial oxygen saturation and hydrogen ion concentration with pulmonary artery pressure in 43 patients with COPD. Note the insensitivity of pulmonary artery vasomotion when the saturation is relatively normal and the marked reactivity when the saturation is low. (Reprinted by permission from Ferrer IM: Disturbances in the circulation in patients with cor pulmonale. Ann NY Bull NY Acad Med 41:942, 1965.)

Receptor Sites

The receptor sites for these stimuli have not been totally identified. There is some question whether a common receptor subserves both hypoxia and hydrogen ion concentration and whether the action is upon a muscle cell or is mediated by local parasympathetic ganglia or by molecular receptors. The available evidence now indicates that their site of action is at the level of the muscular arteries, which have a diameter between 100 and 400 μ. Changing alveolar gas tensions are immediately discernible in these precapillary vessels, suggesting a direct pressor response by these vessels.[53,54] Another observation supporting this direct response mechanism is the fact that patients with sympathectomy respond to hypoxia in an identical manner as normals. It is also possible that pulmonary veins may contribute to the vasconconstrictive responses.

Restricted Pulmonary Vascular Bed

When the pulmonary vascular bed is reduced in its capacity by disease processes, there follows a decreased distensibility of the pulmonary vessels as well as an abnormal pressure–flow response. The underlying pathology may be obstruction of the pulmonary vascular bed (vaso-occlusive), as in pulmonary embolism or obliteration of the vascular bed as in COPD.

In the vaso-occlusive process there is a buildup of thromboemboli, which fails to lyse. Although the exact fate of thromboemboli is not known, it has been suggested that massivity of thromboemboli, old thromboemboli, and prior cardiopulmonary and prior pulmonary emboli predispose to reduction of pulmonary vascular space. Tow and Wagner[55] noted that reperfusion following pulmonary embolism was largely dependent upon the size of the initial pulmonary vascular defect estimated by lung scanning (Fig. 19-14). When the initial perfusion defect was less than 15%, 70% of these patients had return of their lung scans to normal at 4 months, while in those with perfusion defects of moderate size (15%–30%) about 40% had returned to normal. In those patients with massive pulmonary embolism (31% or more of the pulmonary vasculature) only 20% had normal scans at 4 months.

In the Urokinase Pulmonary Embolism Trial, massivity was less important as a factor leading to persistence of residual scan defects.[56] More important was the presence of prior cardiopulmonary disease or prior pulmonary embolism. Recent observations suggest that patients with pulmonary embolism treated with conventional heparin therapy fail to clear their pulmonary thromboemboli completely, in contrast to

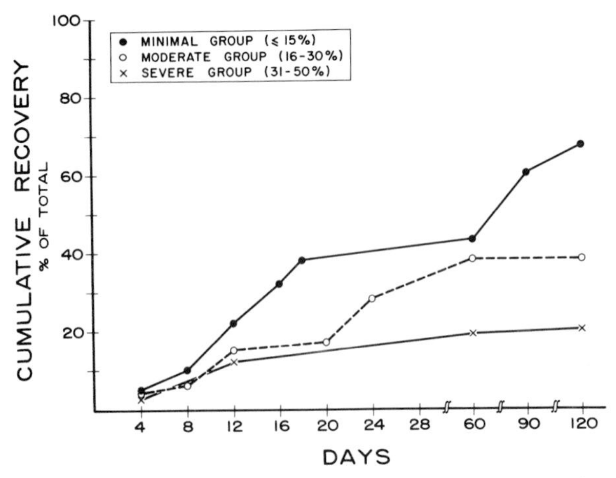

Fig. 19-14. Recovery of lung scan perfusion in patients following pulmonary embolism. The frequency of recovery (ordinate) is related to massivity of embolism (see text). (Adapted from data of Tow DE, Wagner HN Jr: Recovery of pulmonary artery blood flow in patients with pulmonary embolism. N Engl J Med 276:1053, 1967.)

patients treated with thrombolytic therapy.[57] This failure to remove all traces of an embolic event could account for the progressive pulmonary hypertension noted in patients with recurrent pulmonary embolism who develop chronic cor pulmonale. These findings are in consonance with those of Smith and Dexter,[58] who carried out careful postmortem pulmonary artery injection studies, noting the marked reduction in injection mass in those patients with prior pulmonary embolism. Other factors that contribute to this reduced vascular bed are the frequency and massivity of the embolic recurrences.

In primary pulmonary hypertension, a rare entity, there is progressive reduction of the pulmonary vascular bed by thickening of the vascular wall with marked encroachment of vessel lumen. This process eventually leads to thrombotic occlusion of the vessel.

In the obliterative process, such as in emphysema, there may be striking narrowing, distortion, and obliteration of the capillaries and precapillary arterioles.

Infiltrative diseases similarly produce vascular obliteration: sarcoid, berylliosis, silicosis (especially the "complicated" type), asbestosis, scleroderma, etc. The process in these diseases leading to vessel destruction is fibrosis.

Pathophysiology of Restricted Pulmonary Vascular Bed

The reduction in pulmonary vascular space as a factor in the development of PAH, particularly in COPD, is probably not the major factor it was once believed. Numerous lines of evidence have been cited. For instance, in patients undergoing pneumonectomy no significant rise in pulmonary artery pressure has been noted, providing the remaining lung vasculature is not grossly diseased.[59] This 50%–60% loss of pulmonary vascular space is also probably more than is observed in most patients with COPD. The PAH observed in many of the patients with COPD is also reversible, if therapy improves gas exchange. In addition, little correlation between degree of lung destruction and evidence of right ventricular hypertrophy and dilatation at postmortem has been found.[60] However, the importance of pulmonary vascular bed reduction becomes manifest when pulmonary blood flow, for any reason, increases above the level at rest. The compromised vascular reserve, diminished distensibility, and poor storage function all contribute to a disproportionate rise in pulmonary artery pressure when pulmonary blood flow is increased.

Cardiac Output in Chronic Cor Pulmonale

Although controversy currently exists, COPD has been considered a relatively high-output disease since the initial reports of Richards,[61] who found normal to increased outputs in patients with chronic cor pulmonale, in contrast to low output in patients with rheumatic heart disease. Harvey et al.[62] similarly observed higher outputs in their patients with chronic cor pulmonale than in those with hypertensive or coronary artery disease (Fig. 19-15). Forty-three percent of the 79 patients with chronic cor pulmonale had a cardiac index above normal (3.5 liters/min/m^2), while only 5% of the hypertensives had a similar increase. In contrast, 58% of the hypertensives had a cardiac index below normal, whereas only 17% with chronic cor pulmonale had reduced flow.

The mechanism for this increased cardiac output concerns the magnitude of the hypoxemia and polycythemia. The stimulus of hypoxemia upon polycythemia appears to depend upon the severity—arterial oxygen tension below 60 mm Hg stimulating increases in red-blood-cell volume, tensions above 60 mm Hg resulting in little change.[44] This indicates that only those decreases in arterial oxygen tension that produce a fall in arterial oxygen saturation provoked an increase in red-blood-cell mass, implying that red-cell production is regulated by oxygen transport and tissue oxygen tension rather than simply by arterial oxygen tension. In this regard, erythropoietin production and red-blood-cell maturation in

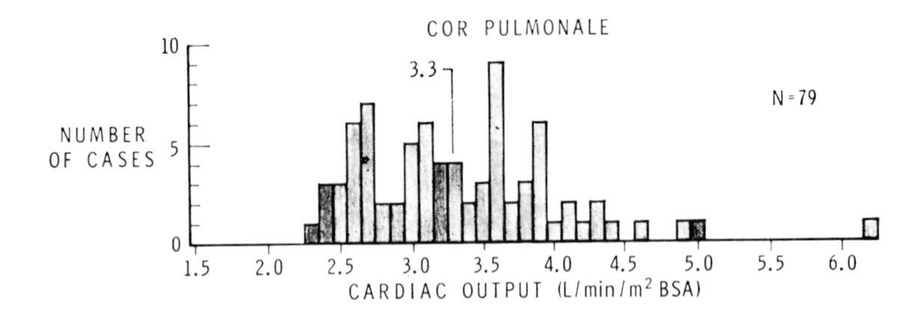

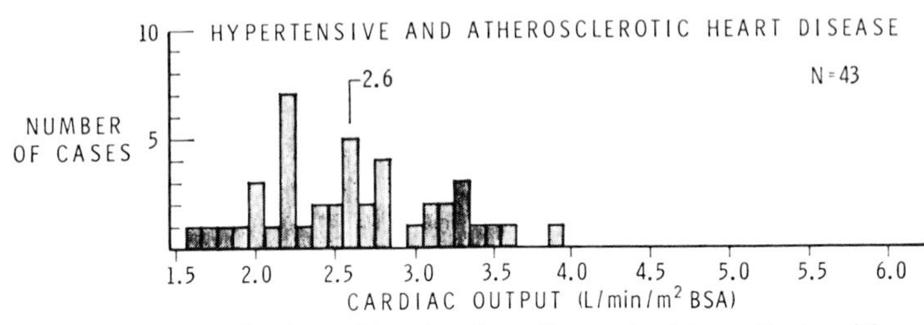

Fig. 19-15. Distribution of levels of cardiac output in patients with cor pulmonale and with hypertensive and atherosclerotic heart disease (see text). (Reprinted by permission from Ferrer IM: Disturbances in the circulation in patients with cor pulmonale. Ann NY Bull NY Acad Med 41:492, 1965.)

the bone marrow increase only when arterial oxygen tension falls below 60 mm Hg.

The consequence of secondary polycythemia is an increase in pulmonary blood volume, which may contribute to the increased cardiac output (Fig. 19-8).[44] The hypervolemia, with its attendant decreased peripheral resistance, contributes to a rapid peripheral runoff, augmenting venous return. Monge reported a pulmonary blood volume of 670 ml/m^2 in Peruvians native to high altitudes, in contrast to a sea-level value of 390 ml/m^2.[44] In those native to high altitude the pulmonary blood volume was 19.4% of total blood volume, in comparison to a value of 15.2% at sea level.[44] There is a suggestion that significant increases in pulmonary blood volume occur only after some 6 months' exposure to hypoxia.

The increase in blood viscosity as a consequence of the increase in red-blood-cell mass does not appear to play any role in the pulmonary hypertension of high altitude.[44] No relationship could be demonstrated between pulmonary artery pressure and hematocrit in the Peruvian or Leadville studies, despite a range in hematocrit from 40% to 78% and a range of pulmonary artery pressure from 14 to 62 mm Hg.

Another observation that may indirectly contribute to the preservation of a normal or slightly increased cardiac output is the reduced peripheral resistance, manifested by lowered blood pressure, observed in patients with COPD and in natives of high altitudes. Galvez reported serial blood pressure determinations in a group of European and North American men living at 12,300 ft in Peru for a period of 2–15 years while maintaining their cultural dietary and occupational habits.[44] He showed a significant decrease in both systolic and diastolic pressures, indicating the effect of hypoxic environment.

Left Ventricle in Chronic Cor Pulmonale

Currently the subject of left ventricular function in COPD and chronic cor pulmonale is under intensive investigation. The topic is not new. Altschule[63] concluded in 1962 that the "whole heart" was involved in patients with COPD and chronic cor pulmonale. This concept was based on the single-syncytium morphology of the heart, and therefore any stress on either side of the heart would be shared by both ventricles. Subsequently, many studies with diverse observations have been recorded.

Necropsy studies, in the main, have involved the left ventricle in patients dying from chronic cor pulmonale. In patients without systemic hypertension during life and without evidence of valvular or coronary artery disease the thickness of the left ventricular wall has frequently been found to be greater than 15 mm, especially where some right ventricular hypertrophy is also present. Also, in congestive cardiomyopathy both ventricles fail in the congestive phase of the disease. Indirectly supporting this "whole heart" concept are the studies of myocardial contractility in which exercise produced similar changes in both ventricles as long as sympathetic fibers of the myocardium were intact. Similarly, in experimental heart failure, profound depletion of myocardial norepinephrine stores in both ventricles occurred, regardless of which ventricle bore the brunt of the hemodynamic load.[64]

Prominent among the factors invoked for the involvement of the left ventricle is hypoxia. Hosono has shown in patients with chronic cor pulmonale that myocardial hypoxia leads to decreased coronary venous saturation, increased myocardial oxygen extraction, excess lactic production, and increased left ventricular residual volume.[65] However, others have cited data that do not support this contention that the left ventricular myocardium suffers from a deficiency of oxygen supply. In one study the oxygen uptake in patients with COPD was normal or slightly increased because of the greater work of breathing. There was also no abnormality of oxygen cost of exercise above resting requirements, indicating an economical performance for muscular work in COPD.[66] The authors believe that the adaptive mechanisms of the body in general tend to assure the tissues of adequate oxygen delivery.

Another factor believed to be responsible for the involvement of the left ventricle in chronic cor pulmonale is the presence of bronchopulmonary anastomoses between the bronchial arterial circulation and the pulmonary veins. Cudkowicz,[67] Roosenberg,[68] and Liebow and his colleagues[69] have demonstrated their existence. In some patients with far-advanced COPD an increased left ventricular output may result from these anastomoses, simulating a left-to-right shunt. The overall importance of these anastomoses, however, remains to be determined.

Davies and Overy,[70] in contrast, found no evidence of left ventricular dysfunction in 12 patients with severe COPD. By carefully subtracting out the large swings in intrathoracic pressure from the ventricular end-diastolic pressure, they found no elevation of left ventricular end-diastolic pressure in any of their patients.

Despite these many studies with contrasting results, the empirical observations of biventricular failure in patients with COPD must be recognized. It seems likely that the combination of hypoxia, increase in $Paco_2$ and hydrogen ion concentration, and the increase in endogenous catecholamine production[42] stimulated by such increases alters the circulatory dynamics of both the right and left ventricles.

REFERENCES

1. Primary prevention of pulmonary heart disease. Report of Inter-Society Commission for Heart Disease Resources. Circulation 41:A17, 1970
2. Bulletin of National Tuberculosis Association. Report of the Task Force on Chronic Bronchitis and Emphysema, May, 1967
3. Chronic Cor Pulmonale. Report of an Expert Committee of the World Health Organization. Circulation 27:594, 1963
4. Gazes PC: Chronic cor pulmonale, in Hurst JW, Logue RB (eds): The Heart (ed 2). New York, McGraw-Hill, 1970, p 1151
5. McGinn S, White PD: Acute cor pulmonale resulting from pulmonary embolism. JAMA 104:1473, 1935
6. McIntyre KM, Levine HJ: Cardiac arrest and resuscitation, in Spitzer S, Oaks WW, Moyer J (eds): Emergency Medical Management. New York, Grune & Stratton, 1971, p 21
7. Sasahara AA, Cannilla JE, Morse RL, et al: Clinical and physiologic studies in pulmonary thromboembolism. Am J Cardiol 20:10, 1967

8. Allison PR, Dunnill MS, Marshall R: Pulmonary embolism. Thorax 15:273, 1960
9. Sevitt S, Gallagher NG: Prevention of venous thrombosis and pulmonary embolism in injured patients. Lancet 2:981, 1959
10. Sevitt S: Venous thrombosis and pulmonary embolism. Am J Med 33:703, 1962
11. Ravdin IS, Kirby CK: Experiences with ligation and heparin in thromboembolic disease. Surgery 29:334, 1951
12. Byrne JJ, O'Neil EE: Fatal pulmonary emboli: A study of 130 autopsy-proven fatal emboli. Am J Surg 83:47, 1952
13. Kakkar VV, Howe CT, Flanc C, Clarke MB: Natural history of deep venous thrombosis. Lancet 2:230, 1969
14. Sharma GVRK, O'Connell DC, Wheeler HB, et al: Deep venous thrombosis as a diagnostic clue to pulmonary embolism. Am J Cardiol 33:170, 1974
15. Wheeler HB, Pearson D, O'Connell D, Mullick SC: Impedance plethysmography. Arch Surg 104:164, 1972
16. Roach HD, Laufman H: Relationship between pulmonary embolism and pulmonary infarction. Surg Forum 5:214, 1955
17. Hampton AO, Castleman B: Correlation of postmortem chest roentgenograms with autopsy findings, with special reference to pulmonary embolism and infarction. Am J Roentgenol Radium Ther Nucl Med 43:305, 1940
18. Hall GE, Ettinger GH: Experimental study of pulmonary embolism. Can Med Assoc J 28:357, 1933
19. Eichetter G: Die Operation der lungen Embolie nach Trendelenburg (Bericht über die bisher bekannt Gewordenen und acht weitere Falle). Chirurg 4:209, 1932
20. Brandfonbrener M, Turino GM, Himmelstein A, Fishman SD: Effects of occlusion of one pulmonary artery on pulmonary circulation in man. Fed Proc 17:19, 1958
21. Hara M, Smith JR: Experimental observations on embolism of pulmonary lobar arteries. J Thorac Surg 18:536, 1949
22. Gray FD Jr: Pulmonary Embolism. Philadelphia, Lea & Febiger, 1966, p 76
23. Katz LN: Pulmonary embolism. Dis Chest 11:249, 1945
24. Gorham LW: A study of pulmonary embolism (II). The mechanism of death based on a clinical pathological investigation of 100 cases of massive and 285 cases of minor embolism of the pulmonary artery. Arch Intern Med 108:189, 1961
25. Comroe JH Jr, Van Lingen B, Stroud RC, Roncoroni A: Reflex and direct cardiopulmonary effects of 5-OH-tryptamine (serotonin); their possible role in pulmonary embolism and coronary thrombosis. Am J Physiol 173:379, 1953
26. Gurewich V, Thomas D, Stein M, Wessler S: Bronchoconstriction in the presence of pulmonary embolism. Circulation 27:339, 1963
27. Thomas D, Stein M, Tanabe G, et al: Mechanism of bronchoconstriction produced by thromboemboli in dogs. Am J Physiol 206:1207, 1964

28. Nadel JA, Colebatch JH, Olsen CR: Location and mechanism of airway constriction after barium sulfate microembolism. J Appl Physiol 19:387, 1964
29. Harvey RM, Enson Y: Pulmonary vascular resistance, in Stollerman GH (ed): Advances in Internal Medicine. Chicago, Year Book Medical Publishers, 1969, p 73
30. Sasahara AA: Pulmonary vascular responses to thromboembolism. Mod Concepts Cardiovasc Dis 36:55, 1967
31. Enson Y, Giuntini C, Lewis ML, et al: Influence of hydrogen ion concentration and hypoxia on the pulmonary circulation. J Clin Invest 43:1146, 1964
32. Harvey RM, Enson Y, Betti R, et al: Further observations on the effect of hydrogen ion on the pulmonary circulation. Circulation 35:1019, 1967
33. McIntyre KM, Sasahara AA: Hemodynamic response to pulmonary embolism in patients free of prior cardiopulmonary disease. Am J Cardiol 28:288, 1971
34. Harvey RM, Ferrer MI: A clinical consideration of cor pulmonale. Circulation 21:236, 1960
35. Hickam JB, Ross JC: Management of the patient with cor pulmonale. Bull NY Acad Med 41:994, 1965
36. von Euler US, Liljestrand G: Observations on the pulmonary arterial pressure in the cat. Acta Physiol Scand 12:301, 1946
37. Motley HL, Cournand A, Werko L, et al: The influence of short periods of induced acute anoxia upon pulmonary artery pressures in man. Am J Physiol 150:315, 1947
38. Staub NC, Storey WF: Relation between morphological and physiological events in lung using rapid freezing. J Appl Physiol 17:381, 1962
39. Peñaloza D, Sime F, Banchero N, et al: Pulmonary hypertension in healthy men born and living at high altitudes. Am J Cardiol 11:150, 1963
40. Rotta A, Canepa A, Hurtado A, et al: Pulmonary circulation at sea level and at high altitudes. J Appl Physiol 9:328, 1956
41. Monge MC: High Altitude Diseases: Mechanism and Management. Springfield, Charles C. Thomas, 1966
42. Grover RF: Pulmonary circulation in animals and man at high altitude. Ann NY Acad Sci 127:632, 1965
43. Grover RF, Vogel JHK, Voigt GC, Blount SG: Reversal of high altitude pulmonary hypertension. Am J Cardiol 18:928, 1966
44. Hultgren HN, Grover RF: Circulatory adaptation to high altitude. Ann Rev Med 19:119, 1968
45. Grover RF, Reeves JT, Will DH, Blount SG: Pulmonary vasoconstriction in steers at high altitude. J Appl Physiol 18:567, 1963
46. Hecht HH, Kuida H, Lange RL, et al: Brisket disease. II. Clinical features and hemodynamic observations in altitude-dependent right heart failure of cattle. Am J Med 32:171, 1962
47. Arias-Stella J, Recavarren S: Right ventricular hypertrophy in native children living at high altitude. Am J Pathol 41:55, 1962

48. Arias-Stella J, Saldena M: The terminal portion of the pulmonary arterial tree in people native to high altitudes. Circulation 28:915, 1963
49. Saldena M, Arias-Stella J: Studies on the structure of the pulmonary trunk. II. The evolution of the elastic configuration of the pulmonary trunk in people native to high altitudes. Circulation 27:1094, 1963
50. Ishikawa S, Fattal GA, Popiewicz J, Wyatt JP: Functional morphometry of myocardial fibers in cor pulmonale. Am Rev Respir Dis 105:358, 1972
51. Liljestrand G: Chemical control of the distribution of the pulmonary blood flow. Acta Physiol Scand 44:216, 1958
52. Enson Y, Giuntini C, Lewis ML, et al: The influence of hydrogen ion concentration and hypoxia on the pulmonary circulation. J Clin Invest 43:1146, 1964
53. Staub NC: Gas exchange vessels in the cat lung. Fed Proc 20:107, 1961
54. Jameson AG: Diffusion of gases from alveolus to precapillary arteries. Science 139:826, 1963
55. Tow DE, Wagner HN Jr: Recovery of pulmonary artery blood flow in patients with pulmonary embolism. N Engl J Med 276:1053, 1967
56. Urokinase Pulmonary Embolism Trial: Perfusion lung scanning. Circulation [Suppl II] 47:II-46, 1973
57. Sharma GVRK, Burleson VA, Roggeveen BB, Sasahara AA: Effects of thrombolytic therapy on pulmonary perfusion and diffusion in pulmonary embolism: Advantages over heparin. Clin Res 19:520, 1971
58. Smith GT, Dexter L, Dammin GJ: Postmortem quantitative studies in pulmonary embolism, in Sasahara AA, Stein M (eds): Pulmonary Embolic Disease. New York, Grune & Stratton, 1965, p 120
59. Cournand A, Riley RL, Himmelstein A, Austrian R: Pulmonary circulation and alveolar ventilation–perfusion relationships after penumonectomy. J Thorac Cardiovasc Surg 19:3, 1950
60. Cromie JB: Correlation of anatomic pulmonary emphysema and right ventricular hypertrophy. Am Rev Respir Dis 84:657, 1961
61. Richards DW Jr: Cardiac output by the catheterization technique in various clinical conditions. Fed Proc 4:215, 1945
62. Harvey RM, Enson Y, Cournand A, Ferrer MI: Cardiac output in cor pulmonale. Arch Krieslaufforsch 46:7, 1965
63. Altschule MD: Cor pulmonale: A disease of the whole heart. Dis Chest 41:398, 1962
64. Braunwald E: Cardiac norepinephrine stores in experimental heart failure in the dog. J Clin Invest 43:2386, 1964
65. Hosono K: Studies on chronic cor pulmonale with special reference to coronary circulation. Jap Heart J 6:318, 1965
66. Filley GF: Pulmonary ventilation and the oxygen cost of exercise in emphysema. Trans Am Clin Climatol Assoc 70:193, 1958
67. Cudkowicz L, Armstrong JB: The bronchial arteries in pulmonary emphysema. Thorax 8:46, 1953
68. Roosenberg JC, Deenstra H: Bronchial-pulmonary vascular shunts in chronic pulmonary affections. Dis Chest 26:664, 1954

69. Liebow AA, Hales MR, Lindskog GE: Enlargement of the bronchial arteries, and their anastomoses with the pulmonary arteries in bronchiectasis. Am J Pathol 25:211, 1949
70. Davies H, Overy HR: Left ventricular function in cor pulmonale. Chest 58:8, 1970

John S. Banas, Jr.

20

Adrenergic Mechanisms and Clinical Heart Disease

The role of the cardiovascular system is to meet the metabolic demands of the body tissues by supplying them with adequate oxygen and nutrients and removing the waste products of metabolism. In order to respond to these constantly changing needs in an integrated, efficient, and immediate manner, the cardiovascular system is supplied with certain control mechanisms. These control mechanisms may be intrinsic or extrinsic to the cardiovascular system. The intrinsic mechanisms are (1) the length–tension relationship of the myocardium that regulates cardiac contraction by the extent of muscle fiber stretch in the resting state, (2) the intrinsic myogenic response of vascular smooth muscle that regulates lumen diameter of blood vessels as a response to the transmural distending pressure, and (3) local metabolic control of vascular smooth-muscle tone. The extrinsic controls are neural, mediated by the autonomic nervous system and a hormonal system that includes such compounds as renin, angiotensin, aldosterone, antidiuretic hormone, and vasoactive substances such as histamine, serotonin, and the prostaglandins. The integrated roles of the intrinsic and extrinsic mechanisms regulating the circulation have been recently reviewed.[1–3] The present discussion will be limited to the role of the autonomic nervous system and its neural mechanism of cardiovascular control via its adrenergic limb and the influence of the sympathomimetic amines and their antagonists on this control system in the normal and diseased heart.

The autonomic nervous system is composed of two divisions—the sympathetic (adrenergic) and the parasympathetic (cholinergic). The heart is supplied with both sympathetic and parasympathetic innervation, whereas the blood vessels are innervated predominantly by the sympathetic division.[1,3,4]

The anatomy of the sympathetic neuroeffector junction consists of the termination of the postganglionic sympathetic neurone and the adjacent effector cell. The postganglionic nerve terminal contains the mechanisms for synthesis, storage, release, and metabolism of the adrenergic neurotransmitter norepinephrine. The fate and metabolism of norepinephrine at the adrenergic nerve terminals have been the subject of extensive study and review.[5–7] Stimulation of the adrenergic nerve fiber releases norepinephrine at the nerve ending. However, norepinephrine release can also be affected by a number of compounds, called the phenylalkyl amines.[8] These amines have vasoactive properties that mimic the effects of adrenergic nerve stimulation and thus have been called the sympathomimetic amines. The sympathomimetic amines that act on the effector cell are called direct, and those whose action is dependent on the release of norepinephrine from the nerve endings are called indirect (Table 20-1). The indirect-acting agents are ineffective when norepinephrine has been depleted from the nerve endings by reserpine or guanethidine.[8–10] Continued administration of the indirect-acting sympathomimetic amines results in a progressively decreasing response. This phenomenon has been called tachyphylaxis and is the result of diminishing norepinephrine release from the adrenergic nerve ending. The prolonged administration of tyramine, an indirect-acting amine, results in a decrease of the total norepinephrine stored. This effect occurs after tachyphylaxis has developed and is associated with the uptake of a hydroxylated metabolic product of tyramine, norsynephrine, which is then bound in the storage vesicles in the adrenergic nerve ending. Thus a "false neurotransmitter" is formed that displaces the normal transmitter norepinephrine.[1,11] A similar process occurs with the use of metaraminol, which also acts as a false neurotransmitter.[12]

The effector cell contains a receptor site that interacts with norepinephrine and other structurally similar chemical agonists (or antagonists) to elicit (or block) a physiologic response characteristic of that particular cell. The concept of a receptor site mechanism for the actions of certain compounds was first described by Langley over 70 years ago.[13] This concept of the adrenergic receptor was elucidated by Dale's studies on the adrenergic blocking activity of the ergot derivatives in 1906.[14] The studies of Ahlquist in 1948[15] established our current classification of two basic adrenergic receptor types, the alpha and beta receptors. In a study of the relative potencies of a series of sympathomimetic amines (agonists) he found that for certain responses such as vascular dilatation, myocardial stimulation, and inhibition of contraction of the uterus,

Table 20-1
Sympathomimetic Amines and Their Effects on the Heart and Blood Vessels

Amine	Action	Receptor Affinity	
		Heart	Blood Vessels
Norepinephrine	Direct	Beta	Alpha
Epinephrine	Direct	Beta	Alpha
Phenylephrine	Direct	±	Alpha
Methoxamine	Direct	0	Alpha
Isoproterenol	Direct	Beta	Beta
Dopamine	Direct	Beta	Alpha Dopaminergic*
Tyramine	Indirect	Beta	Alpha
Mephenteramine	Indirect	Beta	Alpha
Metaraminol	Mixed	Beta	Alpha
Ephedrine	Mixed	Beta	Alpha
Methamphetamine	Mixed	Beta	Alpha

*Dopaminergic refers to the specific receptors located in the renal and mesenteric vasculature that when stimulated by dopamine mediate vasodilatation of these vessels.[112]

isoproterenol was the most potent, followed by epinephrine, with norepinephrine being considerably less active. The receptors mediating these responses were called beta adrenergic receptors. Conversely, responses such as vasoconstriction, excitation of the uterus, dilatation of the pupils, etc., were greater with epinephrine and norepinephrine, with isoproterenol producing a weak response. The receptors responsible for these actions were called alpha adrenergic receptors.

More recently, evidence has been accumulated to suggest that distinct subtypes of beta adrenergic receptors are present in different tissues. Lands and co-workers[16] studied the relative potencies of a series of sympathomimetic amines on several responses considered to be mediated by beta adrenergic receptors. They found a marked correlation in relative potencies for free fatty acid mobilization and for cardiac stimulation and a similar high correlation for bronchodilatation and vasodepression. Based on these results, they proposed two types of beta receptors; the beta-1 adrenergic receptors mediating lipolysis and cardiac stimulation and the beta-2 adrenergic receptors mediating bronchodilatation and vasodepression. Additional studies have revealed that the beta receptors of uterine tissue[17] and those that mediate glycogenolytic effects in liver and muscle[18] appear to be of the beta-2 subtype.

Table 20-2
Physiologic and Pharmacologic Responses of the Beta Adrenergic Receptor Subtypes[16,17,22]

	Receptor Subtype	
	Beta-1	Beta-2
Organ response	Heart: Inotropy Chronotropy	Vascular smooth muscle: Relaxation Bronchial smooth muscle: Relaxation Uterine smooth muscle: Relaxation Liver: Glycogenolysis Muscle: Glycogenolysis
Agonist, selective	—	Soterenol, salbutamol
Agonist, nonselective	Isoproterenol	Isoproterenol
Antagonist, selective	Practolol	Butoxamine
Antagonist, nonselective	Propranolol	Propranolol

Beginning with the work of Powell and Slater in 1958,[19] several new beta adrenergic receptor antagonists have been developed. These compounds have confirmed the existence of different receptor classes as postulated by Ahlquist and the more recent concepts of the beta receptor subtypes. These drugs can selectively block beta receptors of one type more than the other. Butoxamine can selectively block responses mediated by the beta-2 receptors,[20] and practolol is more effective at the beta-1 receptor site.[21] These compounds, such as butoxamine and practolol, have been labeled selective antagonists, whereas propranolol, a drug that blocks both beta-1 and beta-2 adrenergic receptor sites, is called a nonselective antagonist.[22] Furthermore, compounds such as soterenol and salbutamol[23] appear to have selective stimulating actions on the beta-2 receptors and are selective agonists as opposed to isoproterenol, a stimulator of all beta adrenergic receptors and thus a nonselective agonist[22] (Table 20-2). Although the value of selective beta adrenergic agonists and antagonists for clinical use becomes readily apparent, at the present time only the nonselective agonist isoproterenol and the nonselective antagonist propranolol are available for routine clinical use.

Dobutamine, an analogue of dopamine, is a compound that is alleged to be more cardioselective than the existing beta-1 adrenergic receptor agonists. Dobutamine appears to be more selective for inotropy than for chronotropy, with minor effects on systemic blood pressure.[24,25]

If subsequent studies confirm these findings they will provide further evidence for adrenergic receptor subselectivity in the heart.

Thus far, biochemical and electron microscopic studies have not revealed the character of the adrenergic receptor sites. However, recent studies by Lefkowitz and Haber[26,27] have shown that a microsomal fraction of the canine ventricular myocardium concentrates radioactive-labeled norepinephrine; furthermore, this bound norepinephrine can be displaced by beta adrenergic stimulating catecholamines and by propranolol. They suggest that this microsomal fraction contains the beta adrenergic receptor. Despite these very promising studies the adrenergic receptor site has not been convincingly identified. It does appear, however, that adenyl cyclase may represent at least a part of the receptor mechanism.[28] Adenyl cyclase is responsible for the synthesis of adenosine 3′,5′-monophosphate (cyclic AMP) from adenosine triphosphate (ATP). Cyclic AMP acting as a second messenger mediates the effects of many hormones and neurohormones in the body.[29,30] The response of the effector cell to a given hormone is determined by the specificity of the receptor site on the cell, and the second messenger (cyclic AMP) is responsible for mediating the specific cellular response of the effector organ. The role of cyclic AMP as mediator of cardiovascular physiologic responses has been recently reviewed by Kones[31] and Robison and co-workers.[32]

The order of potency of the catecholamines (isoproterenol > epinephrine > norepinephrine) in stimulating adenyl cyclase in tissues such as the heart, liver, skeletal muscle, and uterus is similar to that of a beta receptor. Furthermore, stimulation of adenyl cyclase by catecholamines is blocked by beta but not alpha receptor antagonists.[33] Thus it seems clear that beta adrenergic receptors are functionally linked to adenyl cyclase.

The relationship of alpha adrenergic receptors to adenyl cyclase is not well established. However, in several tissues such as pancreatic islet cells and platelets, alpha adrenergic effects appear to involve reduction in cyclic AMP levels.[34] This finding, however, has not been confirmed in many other tissues, and it will remain for further investigation to delineate the role of the adenyl-cyclase–cyclic-AMP system in the alpha adrenergic receptor sites.

Table 20-3 provides a classification of the adrenergic receptor responses of the cardiovascular system and the mediating receptor types. In the heart, beta-1 adrenergic receptor activation results in an increased heart rate and contractile force, which is mediated by both direct sympathetic nerve stimulation and by circulating catecholamines.[3–5,22,35,36] The vascular system contains both alpha and beta adrenergic

Table 20-3
Classification and Response of Cardiovascular Effector Organs to Adrenergic Receptor Stimuli

Organ	Receptor	Response	Mechanisms
Heart			
Ventricles, atria	Beta-1	↑ Contractility	↑ Rate of isovolumic pressure rise and fall
			↓ Duration of systole; ↑ Peak pressure
			↓ Isovolumic contraction and relaxation time
Pacemaker tissue	Beta-1	↑ Heart rate	↑ Rate of diastolic depolarization (automaticity)
Conducting tissue	Beta-1	↑ Impulse conduction rate	↓ Refractory period in A-V junctional tissue
			↑ Conduction velocity in A-V junctional tissue
Coronary arteries	Alpha	Constriction	Vascular smooth-muscle contraction
	Beta-2	Dilatation	Vascular smooth-muscle relaxation
Metabolism	Beta-2	Glycogenolysis	↑ Cyclic AMP and phosphorylase activation
Peripheral vasculature			
Arterioles, veins	Alpha	Constriction	Vascular smooth-muscle contraction
Arterioles	Beta-2	Dilatation	Vascular smooth-muscle relaxation
Veins	? Beta-2	Dilatation, slight	Vascular smooth-muscle relaxation

receptors. The alpha adrenergic receptors mediating vasoconstriction are present in both arteries and veins; however, vasodilatation mediated by the beta receptors appears to be more convincingly present in the arterial vasculature than in the venous system.[37,38] Vasodilatory response to sympathetic nerve stimulation, as opposed to that of the circulating catecholamines, appears to be mediated via cholinergic or histaminergic receptors.[39,40] Thus arterial and venous dilatation as a result of beta-2 adrenergic receptor activation may be produced by the direct effect of the circulating catecholamines and/or other humoral substances such as the prostaglandins.[3,16,37] There is no convincing evidence to date indicating nerve-mediated vasodilatation via the beta adrenergic receptor.[40] The coronary arteries may be an exception to this statement (vide infra). Glycogenolysis in myocardial tissue is also mediated by the beta adrenergic receptors,[41,42] and recent studies implicate the beta-2 receptor subtype as the probable mediator.[18] Other metabolic effects of adrenergic receptor stimulation have been reviewed by Lucchesi and Whitsitt.[35]

REGULATION OF THE HEART

Cardiac action is regulated by four major factors: (1) ventricular volume (preload), (2) impedance to ejection (afterload), (3) the contractile state of the myocardium (inotropy), and (4) heart rate (chronotropy).[2–3] These four factors are also the major determinants of myocardial oxygen consumption[43] and are constantly under the influence of the adrenergic nervous system either directly (inotropy and chronotropy) or indirectly (preload and afterload) (Fig. 20-1).

Afterload

The ventricular outflow tract, the aortic valve, and the arteriolar vascular bed may act singularly or together to effect impedance to left ventricular ejection (afterload). The ventricular outflow tract may actively cause impedance to left ventricular ejection, as in the patient with idiopathic hypertrophic subaortic stenosis (IHSS). Beta-1 stimulation by exogenous catecholamines (isoproterenol) or endogenous catecholamines (exercise, anxiety, etc.) may increase obstruction to left ventricular outflow in patients with IHSS by augmenting contractility in the region of the hypertrophic muscle and thus increasing afterload. Similarly, the aortic valve, when stenotic, increases intraventricular pressure exponentially with the velocity of flow and thus increases impedance to left ventricular ejection relative to flow across the valve.

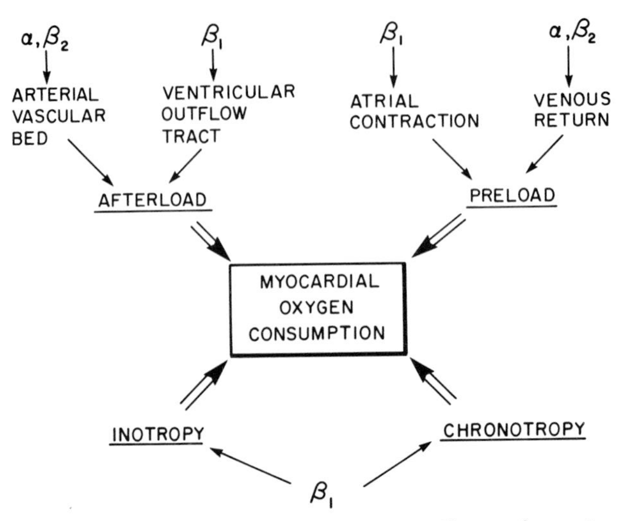

Fig. 20-1. Adrenergic receptors of the cardiovascular system and their proposed roles in the regulation of cardiac functions and myocardial oxygen consumption. α = alpha adrenergic receptor; β_1 = beta-1 adrenergic receptor; β_2 = beta-2 adrenergic receptor.

The arterioles are under constant regulation via stimulation of the alpha and beta-2 adrenergic receptors and by fibers from the dorsal roots and the parasympathetic nervous system. Active vasoconstriction and dilatation via the alpha and beta-2 adrenergic receptors regulate the "runoff" of blood from the aorta, which in turn controls the level of blood pressure and left ventricular impedance for any given rate and volume of ejection. The ventricle responds to these changes in afterload by increasing or decreasing the rate and volume of ejection, which in turn affect both end-systolic and end-diastolic volumes. These changes in afterload and ventricular volume will alter wall stress through the law of Laplace.[43] Thus an inverse relationship exists between ventricular wall stress and the speed and extent of myocardial fiber shortening similar to that demonstrated in isolated muscle.[2,3,43] For example, an increase in afterload results in a decrease in stroke volume, an increase in blood pressure, and an increase in myocardial oxygen consumption. In the ischemic myocardium where coronary blood flow may already be compromised, the potential deleterious effects of raising afterload and increasing myocardial oxygen demands may far outweigh the apparent benefit of maintaining a normal blood pressure. It becomes obvious that afterload must be therapeutically manipulated to provide optimal perfusion pressure to the coronary arteries and the other vital organs while keeping myocardial oxygen demands at a minimum.

Therefore, changes in impedance to left ventricular ejection (afterload) produced indirectly by the regulating action of the adrenergic receptors on the arteriolar bed lead to changes in stroke volume, system blood pressure, and myocardial oxygen requirements.

Preload

Changes in preload (end-diastolic volume) may be effected secondarily by a change in afterload, as described above; however, factors that increase or decrease diastolic inflow to the heart are perhaps of greater importance. Diastolic inflow is regulated by changes in venous flow and venous volume (capacitance). Venous flow may be affected by changes in total blood volume or shifts from the peripheral to the central veins by maneuvers such as elevating the legs to augment venous return or assuming the upright position to decrease venous return to the heart. The use of pharmacologic agents such as the nitrates or their derivatives will decrease venous return to the heart by increasing the volume of the capacitance vessels secondary to their direct action on vascular smooth muscle. Changes in venous capacitance can occur by active venous constriction and dilatation, which is regulated by alpha and beta-2 adrenergic receptor stimulation as described above.

Preload can also be augmented by an increase in the force of atrial contraction, which can provide a substantial increase in end-diastolic volume, particularly in the diseased ventricle. Stimulation of the beta-1 adrenergic receptors in the atrial myocardium mediates the increased force of atrial contraction. Adrenergic receptor stimulation thus alters the diastolic volume of the heart indirectly via its effects on the venous capacitance vessels and by augmentation of atrial contraction.

The changes in diastolic volume of the ventricle can influence stroke volume via the classic Starling effect[44] and by changing fiber length alter the force–velocity relationship so that a greater force will be delivered at any given velocity of shortening.[45] These combined effects altering stroke volume will also influence blood pressure, duration of systole, and rate of rise of ventricular isovolumic pressure and thus profoundly influence myocardial oxygen consumption by changes in wall tension (law of Laplace) and contractile state.[2,43]

Inotropy

Changes in the contractile state of the myocardium are predominantly the result of an intensification of the active state of the muscle. This property of cardiac muscle is determined by the rate of cyclic force-generating processes at the contractile sites, which in turn are dependent on the interaction of the contractile proteins in the presence or absence of ionic calcium.[46,47] The adrenergic neurotransmitter norepinephrine released by stimulation of the cardiac sympathetic nerves and the exogenously administered catecholamines are potent stimulators of the contractile state of the heart. These chemical mediators act on the myocardial beta-1 receptors and through an as yet unidentified mechanism involving cyclic AMP and calcium, intensify the inotropic state of the myocardium. The relationship of beta receptor stimulation, cyclic AMP, and calcium and their postulated effects on the contractile state of the myocardium have been recently reviewed.[4,32,48–50]

Inotropic stimulation in the intact heart causes a more forceful contraction at any given preload and afterload. During the total phase of contraction there is (1) an increase in the rate of isovolumic pressure rise and fall, (2) a decrease in the duration of systole, (3) an increase in peak pressure, and (4) a decrease in isovolumic contraction and relaxation time.[2,4] As a result the stroke volume, the systolic ejection fraction, and the rate of volume ejected are all increased (Table 20-3).

In the intact heart other factors may also affect the inotropic state of the myocardium, such as (1) changes in the speed of the excitation wave both in the conduction system and from cell to cell of the myocardium and (2) regional differences in the development of wall tension.[2] Normally these play an insignificant role in regulating myocardial contractility; however, in the presence of a diseased myocardium they may become increasingly important.[51] Thus cardiac abnormalities that may disrupt the conduction system or damage the myocardial cell may alter the overall contractile state of the myocardium by causing a defect at any point in this chain of events despite adequate stimulation of individual sarcomeres via the adrenergic pathways.

Chronotropy

The autonomic nervous system is of prime importance in regulating heart rate. Stimulation of the beta adrenergic receptors of the heart increases the rate of diastolic depolarization of the specialized automatic cells (automaticity) and enhances conduction velocity and shortens the functional refractory period, particularly in the junctional tissues (Table 20-3).[52–54] The electrophysiologic mechanisms of action of sympathetic stimulation at the membrane level are not well understood. However, recent studies have suggested that catecholamines affect the rate of transmembrane sodium and potassium flux, with the result being a decrease in outward potassium current, which is the apparent cause of a more rapid rate of diastolic depolarization.[4] Activation of the calcium–cyclic-AMP mechanism by sympathetic stimulation is very likely to play a major regulating role in ionic sodium and potassium movement; however, the existing studies are inconclusive.[4,55]

Accelerating the heart rate is a major mechanism for increasing cardiac output at rest and during exercise.[3,35,36] However, this increase in cardiac output occurs only if stroke volume is maintained constant or if it is increased by either an augmentation of preload, a reduction in afterload, or an increase in inotropy. Patients who are unable to increase their heart rates or who have fixed heart rates are unable to augment their cardiac output appropriately, despite the presence of a normal myocardium and where an increase in stroke volume is possible.[56] Braunwald and co-workers[3,57] have studied the roles of the parasympathetic and sympathetic nervous system in the regulation of heart rate in the resting state and during exercise. Their investigation of the baroreceptor reflex arc suggested that in the resting state the decrease in heart rate associated with an elevation in systemic arterial blood pressure was mediated predominantly by the parasympathetic nervous system, whereas cardiac acceleration associated with systemic arterial hypotension was mediated by both the parasympathetic and sympathetic limbs of the autonomic nervous system. During exercise, sympathetic activity appeared to play the dominant role in alteration of the heart rate.

In patients with congestive heart failure, profound defects may exist in the autonomic nervous system. Eckberg[58] studied the mechanisms of heart rate response in normal patients and in patients with organic heart disease by comparing the increase in rate produced by atropine before and after beta adrenergic blockade. They found that the increase in heart rate after the administration of atropine was substantially less in patients with heart disease. These studies imply a profound defect in the parasympathetic control of heart rate in patients with heart disease; and in the presence of heart failure the adrenergic component plays a dominant role in supporting the failing circulation by maintaining or increasing heart rate. Furthermore, when these findings are coupled with the observations of diminished catecholamine stores in the failing myocardium,[59] they suggest a marked defect in autonomic control of the failing heart.

ADRENERGIC MECHANISMS REGULATING CORONARY BLOOD FLOW

Coronary blood flow is regulated by a complex interaction of multiple factors. As in the regulation of cardiac function, the coronary circulation is influenced by both intrinsic and extrinsic (neural) control systems.[60,61] The intrinsic controls include (1) the four major determinants of myocardial oxygen consumption discussed in the previous section and (2) local tissue factors such as changes in tissue po_2. Thus increases or decreases in preload, afterload, inotropy, or chronotropy will directly affect changes in myocardial blood flow. Myocardial hypoxia can alter coronary blood flow dramatically, possibly by stimulating the local release of potent vasodilating substances such as adenosine.[62,63] These intrinsic determinants of myocardial blood flow act independent of the extrinsic (neural) mechanisms and are operative in the denervated heart.[64] Studies of the adrenergic controlling mechanisms of the coronary circulation have been difficult to analyze because of the concomitant effects of alpha and beta adrenergic stimulation on the intrinsic controlling factors mentioned above. Isolated smooth-muscle preparations from coronary arteries of humans, rabbits, dogs, and monkeys have been shown to contain both alpha and beta receptor activity.[65,66] The smaller coronary arteries possessed predominant beta adrenergic activity (vasodilatation) and the larger epicardial vessels demonstrated predominant alpha adrenergic activity (vasoconstriction). These responses could be blocked by using appropriate adrenergic receptor antagonists. However, results of other studies suggest that the vasoconstrictor activity following beta adrenergic blockade with propranolol appeared to be secondary to diminished myocardial oxygen demands.[67] More recently, McRaven and co-workers[68] studied the effects of adrenergic stimulation on coronary vessels in the intact heart using the selective beta adrenergic blocking agent practolol. These results indicated that coronary vasodilatation produced by sympathetic nerve stimulation and the action of norepinephrine was indirect and predominantly secondary to stimulation of the myocardial beta adrenergic receptors. A direct effect causing vasoconstriction was minimal but was thought to be mediated via vascular alpha adrenergic receptors. Isoproterenol, however, had both direct and indirect effects, causing coronary artery vasodilatation with the direct effect on coronary artery beta adrenergic receptors predominating.

At the present time the role of adrenergic receptor activity regulating myocardial blood flow appears to be predominantly secondary to the effects of stimulation of the myocardial beta-1 adrenergic receptors. Thus coronary resistance appears to be mainly regulated by local metabolic factors that remain operative in the presence or absence of the cardiac sympathetic nerves. To date, there is no conclusive evidence that direct neural control via adrenergic receptors in the coronary vessels of man predominates over the intrinsic regulators of myocardial blood flow (Table 20-3).

ROLE OF THE SYMPATHETIC NERVOUS SYSTEM IN CONGESTIVE HEART FAILURE

As mentioned previously, the failing heart is dependent on the support of the adrenergic nervous system, which provides the increase in endogenous catecholamines necessary for inotropic and chronotropic stimulation via the myocardial beta-1 adrenergic receptors.[69] It is well established that adrenergic activity is increased in the presence of congestive heart failure in animals and in man.[70,71] However, whereas endogenous circulating catecholamines are increased in congestive heart failure, there is a reduction in the concentration and content of myocardial norepinephrine stores in the failing heart. This decrease in myocardial norepinephrine content appears to be secondary to impaired activity of tyrosine hydroxylase, the rate-limiting enzyme in the biosynthesis of norepinephrine.[72] In addition, there is a defect in the uptake and binding of norepinephrine by the cardiac sympathetic nerves.[59,73] However, the defect in cardiac contractility in the failing heart is not likely to be a direct consequence of the depleted norepinephrine stores, for the contractile state has been demonstrated to be normal in the catecholamine-depleted heart muscle secondary to surgical denervation.[74] The reduced myocardial content of norepinephrine in the failing heart does appear to be clinically significant, for stimulation of the cardiac sympathetic nerves to the failing heart results in abnormally small increases in heart rate and contractile force.[75] The myocardial beta adrenergic receptors in the failing heart have been demonstrated not only to be intact but to respond in a supersensitive manner to circulating catecholamines. Thus circulating catecholamines either endogenously produced by the adrenal medulla or exogenously administered would appear to be of prime importance in supporting the failing heart and circulation.[76] It is evident that the indiscriminate use of the beta-1 adrenergic blocking agents in the patient with cardiac failure may have deleterious results by blocking the inotropic and chronotropic support of the circulating catecholamines.

ADRENERGIC AGONISTS AND ANTAGONISTS

Knowledge of the adrenergic mechanisms and their roles in the regulation and function of the cardiovascular system is imperative for a better understanding of the clinical pharmacology of the adrenergic receptor agonists and antagonists. The cardiovascular effects of alpha and beta adrenergic receptor stimulation have been reviewed in the previous section and are outlined in Tables 20-2 and 20-3. From the foregoing discussion it should be readily apparent that the sympathomimetic amines (Table 20-1) and the adrenergic receptor antagonists may affect more than one determinant of cardiac function simultaneously. For example, isoproterenol, a nonselective beta adrenergic receptor agent, may increase stroke volume through different mechanisms working in an integrated manner. Following isoproterenol infusion, stroke volume will increase secondary to an increase in inotropy and chronotropy (beta-1 receptor stimulation) despite falls in afterload and preload (beta-2 receptor stimulation). Furthermore, methoxamine, virtually a pure stimulator of the alpha adrenergic receptors (Table 20-1), will increase afterload; however, stroke volume may decrease or increase. The change in stroke volume will be determined by the combined effects of (1) increased impedance to ejection and thus a decrease in muscle fiber shortening, (2) the baroreceptor response of the increased blood pressure on heart rate, and (3) the etiology and severity of the underlying cardiac abnormality.[3]

Adrenergic Agonists

Isoproterenol, norepinephrine, epinephrine, and dopamine are perhaps the most widely used catecholamines clinically. Dopamine has recently been cleared for clinical use and is becoming increasingly popular for the treatment of the failing cardiovascular system. Isoproterenol is a nonselective stimulator of the beta adrenergic receptors (Table 20-2). The major hemodynamic effects of isoproterenol administration are an increase in cardiac output by improving myocardial contractility, increasing heart rate, and lowering systemic vascular resistance (Fig. 20-1 and Table 20-3).[5,77–79] Beta-adrenergic-mediated dilatation of the coronary arteries may occur with isoproterenol through stimulation of the receptors in the coronary vasculature.[65,66,80] However, coronary blood flow is augmented predominantly in response to the marked increase in myocardial oxygen demand produced by isoproterenol.[77,78,81,82] In cardiac disease states where the myocardial demands for oxygen cannot be met by increasing oxygen delivery to the myocardium (coronary heart disease, cardiogenic shock, etc.) isoproterenol may have profound deleterious affects on the heart. In the ischemic myocardium, isoproterenol may exaggerate and extend the ischemic zone despite an unchanged coronary blood flow or an increase in coronary blood flow.[83,84] Furthermore, Saroff[85] has shown that the decay of peak developed tension occurs earlier in the ischemic myocardium when treated with isoproterenol. In patients with cardiogenic shock, isoproterenol has been demonstrated to increase myocardial lactate production secondary to augmented anaerobic glycolysis.[86,87] Isoproterenol and other catecholamines with potent inotropic properties produce myocardial cell necrosis, hemorrhage, and inflammatory cell infiltration despite the presence of a previously normal myocardium.[85,88–90] These histologic changes in the myocardium are most likely to occur after prolonged use or at extraordinarily high doses. The myocardial injury induced by these sympathetic catecholamines and the possible mechanisms by which this might occur are the subjects of an extensive recent review by Haft.[90] The potential deleterious affects on the myocardium by isoproterenol and other catecholamines with direct inotropic properties may well explain the poor clinical response of patients with acute myocardial infarction and cardiogenic shock when treated with these agents.[84,86,91–94] The unwanted effects of augmented myocardial oxygen consumption particularly in the presence of ischemic heart disease has markedly limited the clinical usefulness of a "pure," nonselective, beta adrenergic agonist such as isoproterenol in the treatment of these disorders.

Through its affect on beta-1 adrenergic receptors in the heart (Tables 20-2 and 20-3) isoproterenol will accelerate heart rate by increasing automaticity of pacemaker tissues and by increasing conduction velocity through the atrial–ventricular junctional tissue while shortening its refractory period. Thus isoproterenol continues to be an effective agent in the emergency therapy of symptomatic high-grade atrial–ventricular conduction defects and cardiac standstill and as a prophylactic measure during pacemaker implantation.[95]

Isoproterenol has been reported to be beneficial in the treatment of endotoxic shock[96] and in the low-output syndrome of the postcardiac surgery patient.[97] However, Mueller and co-workers have questioned its value in the latter condition.[98] Based on its potent adrenergic receptor stimulating properties, isoproterenol has also been employed as a provocative test for ischemic heart disease[99] and idiopathic hypertrophic subaortic stenosis.[100]

Alpha adrenergic agonists affect the cardiovascular system predominantly by their vasoconstrictor effects on the peripheral circulation (Fig. 20-1 and Table 20-3). Relatively pure agonists such as methoxamine lack direct myocardial effects, but they may reduce stroke output of the heart by increasing afterload and reducing myocardial fiber shortening, as previously discussed. From this observation it would not appear likely that the use of the relatively pure alpha adrenergic agonists such as methoxamine or phenylephrine (Table 20-1) would be beneficial in the treatment of cardiac failure. Nevertheless, the value of an adequate coronary perfusion pressure, which can be achieved with these agents in the ischemic heart, has been dramatically demonstrated by recent studies in the laboratory and in man.[84,102,103] Methoxamine and phenylephrine in the absence of severe cardiac decompensation are effective in terminating paroxysmal atrial tachycardias via their affect on the baroreceptor reflex arc.[3,95] Phenylephrine is the favored drug in this instance, compared to the other alpha adrenergic receptor stimulating agents, for its side effects are minimal and its cardiovascular effects are relatively short-lived.[95]

Norepinephrine acts on alpha and beta adrenergic receptors (Table 20-1). Through its beta receptor stimulating actions, myocardial contractility will increase,[104] and cardiac output may increase, decrease, or remain unchanged.[105,106] The variable response of stroke volume to infusions of norepinephrine may readily be explained by an understanding of the adrenergic mechanisms involved. Thus the increase in myocardial contractility and heart rate from stimulation of the cardiac beta-1 adrenergic receptors may be offset by the increased afterload (with decreased myocardial fiber shortening) and the decrease in heart rate secondary to stimulation of the baroreceptor reflex arc. The end hemodynamic result will thus be determined to a great extent by the cardiovascular abnormality present at the onset of therapy.

In the presence of cardiogenic shock, norepinephrine produces vasoconstriction in the skin, splanchnic bed, muscle, renal arteries, and peripheral venous bed. On the other hand, in the presence of hypotensive states the coronary and cerebral circulation may be improved.[107–109] It is likely that the increase in coronary blood flow is secondary to the increase in cardiac work produced by the inotropic effects of norepinephrine and by the increased afterload. Despite the complex interactions of the adrenergic mechanisms in the patient with cardiogenic shock, the overall clinical response of patients treated with norepinephrine has been favorable.[109–111]

Dopamine is a unique adrenergic receptor agonist that has recently been made available for clinical use. The therapeutic pharmacology of dopamine and its role in the treatment of congestive heart failure and various shock states have been reviewed by Kones,[31] Tarazi,[111] Goldberg,[112] and Holzer.[113] Dopamine acts on alpha adrenergic receptors causing vasoconstriction of the peripheral resistance and capacitance vessels. It also has beta adrenergic receptor stimulating properties, producing an increase in chronotropy and inotropy. Its unique property is that of selectively stimulating a proposed vascular "dopamine receptor" in the renal and mesenteric arteries (Table 20-1). Through this dopaminergic stimulating effect, dilatation of these vascular beds occurs, and as a result renal and mesenteric blood flows increase. The cardiovascular effects of dopamine appear to be dose-related. In animal studies, doses of less than 10 μg/kg cause predominantly beta adrenergic and dopaminergic stimulating effects, whereas higher doses will produce predominant alpha adrenergic effects.[112,114] These unique hemodynamic properties of dopamine have been confirmed in man.[31,112,113] In patients with cardiogenic shock and low-cardiac-output states, dopamine increased the cardiac output, urinary output, and systemic blood pressure.[113–117]

The dose-related effects of dopamine can be used to advantage in patients who may have a low or high systemic vascular resistance. In the studies of Loeb[116] the mean arterial pressure could be increased by titrating the dose upward, thus providing a greater effect on the alpha adrenergic receptors. In contrast, the systemic vascular resistance may decrease, particularly in patients with high vascular resistance, at lower doses that produce predominant beta adrenergic and dopaminergic receptor stimulation.[31] As a result of the vasodilatory effects on the renal arteries, dopamine may be especially useful in patients where oliguria may be refractory to the more commonly employed sympathomimetic amines.[117]

Coronary blood flow is increased following dopamine administration. Brooks and co-workers[118] demonstrated that in the dog progressive increments in coronary blood flow were proportional to increments in myocardial oxygen consumption. Other investigators have also confirmed the fact that the increase in coronary blood flow following dopamine infusion in the laboratory animal parallels increases in cardiac contractility.[31,112] Despite increases in myocardial oxygen consumption produced by dopamine infusion, the metabolic studies of Winslow and co-workers[119] imply that the balance between oxygen supply and myocardial oxygen demands is less disturbed by dopamine than by either norepinephrine or isoproterenol.

Clinical studies have confirmed the hemodynamic effects of dopamine that one would predict from an understanding of its effects on the adrenergic receptors of the cardiovascular system. The alpha and beta adrenergic effects of dopamine, coupled with its unique vasodilating properties on the renal and mesenteric vasculature, would appear to place dopamine in an ideal-drug category. From the studies reviewed above there is little doubt that dopamine has a hemodynamically beneficial effect in most patients with low-cardiac-output states. However, whether dopamine alone or in combination with other agents or circulatory assist devices[113] will improve survival in cardiogenic shock, remains to be established.

Adrenergic Antagonists

Alpha or beta adrenergic receptor blockade produces hemodynamic and metabolic changes that are in general the reverse of those effects produced by the appropriate adrenergic receptor agonists (Table 20-3).

Alpha adrenergic receptor blockade opposes the effect of sympathetic vasoconstriction and venoconstriction and results in a reduction in peripheral vascular resistance and venous return.[5,120,121] Thus one would expect a fall in systemic blood pressure (afterload) and a decrease in ventricular volume (preload). Secondary cardiovascular effects would then be expected such as a reflex tachycardia via the baroreceptor reflex arc (chronotropy) and some increase in contractility (inotropy) due to the decreased impedance to left ventricular ejection. Directional changes in cardiac output and myocardial oxygen consumption would obviously depend on the underlying state of the myocardium and peripheral vasculature, as described previously.

Phenoxybenzamine and phentolamine are the two most widely used and most potent alpha adrenergic blocking agents available for clinical use. Other agents, such as the phenothiazine drugs chlorpromazine and Thorazine, have much weaker alpha adrenergic receptor blocking properties.[122] Based on their alpha receptor adrenergic blocking effects on the cardiovascular system, these agents have been employed in the treatment of a wide variety of disease states such as pheochromocytoma, peripheral vascular disease, hypertension, circulatory shock, and cardiac low-output syndromes.[5,122–125]

With the exception of pheochromocytoma, the therapeutic value of the alpha adrenergic blocking agents in the other conditions noted has been controversial. Perhaps the greatest controversy has been with regard to the use of these agents in the management of cardiogenic shock. The rationale for their use stems from a basic knowledge of the complex circulatory changes that occur when systemic blood pressure and cardiac output reach critically low levels from whatever cause. Hypotension and a diminished cardiac output will initiate reflex peripheral vasoconstriction and a decrease of blood flow to the peripheral tissues. The diminished peripheral perfusion will promote tissue anoxia, with the subsequent metabolic abnormalities that cause local and systemic acidosis. A series of complex compensatory mechanisms come into play in an attempt to correct these abnormalities; however, without a therapeutic break in the cycle these mechanisms may become self-defeating, and the shock state perpetuates itself.[126,127] The alpha adrenergic receptor antagonists have been used in an attempt at improving tissue perfusion and, as a result, reversing the metabolic abnormalities.

However, any improvement that might result from better tissue perfusion may be far outweighed by the deleterious effects of further decreasing perfusion pressure of the coronary and cerebral arteries. Furthermore, venous return will be diminished and will compromise cardiac output via the Starling mechanism unless strict attention is paid to the replacement of blood volume and the monitoring of the cardiac filling pressures. Despite what seems to be sound physiologic reasoning, the value of alpha adrenergic receptor blockade for the treatment of the circulatory shock states in man remains highly controversial.[125–129]

The disturbing and often life-threatening effects produced by an outpouring of the catecholamines norepinephrine and epinephrine from a pheochromocytoma can be attenuated by the use of the alpha adrenergic receptor and beta adrenergic receptor blocking drugs. Phentolamine and phenoxybenzamine are used to combat the hypertensive effects due to alpha receptor stimulation preoperatively and intraoperatively, particularly during manipulation of the tumor. Propranolol, the beta adrenergic receptor blocking agent, is effective in obliterating the cardiostimulatory effects of norepinephrine and epinephrine secreted by the tumor.[124,125]

Perhaps the greatest advance in recent years in the medical therapy of a wide variety of cardiovascular disorders occurred with the development of the beta adrenergic receptor blocking drugs. At the present time propranolol, a nonselective beta adrenergic receptor antagonist (Table 20-2), is the only beta receptor blocking agent available for clinical use in the United States. An extensive discussion of the clinical pharmacology of propranolol and its use in the treatment of numerous clinical conditions is beyond the scope of this discussion, and the reader is referred to a number of recent reviews.[35,52,78,130–133] Only selected conditions will be briefly discussed to illustrate the use of beta adrenergic receptor blockade in altering the adrenergic mechanisms involved in the pathophysiology of various cardiovascular disorders.

Beta adrenergic receptor blockade by propranolol will attenuate the cardiovascular responses mediated via the beta adrenergic receptors (Fig. 20-1 and Table 20-3), resulting in a decrease in heart rate and a decrease in the force of myocardial contraction. In the normal resting supine man, propranolol was observed to have little effect on ventricular dimensions and myocardial contractility, and it produced only a 10%–15% reduction in resting heart rate and cardiac output.[52,78,134] However, in normal man and in patients with heart disease, propranolol substantially attenuated the increases in cardiac output, heart rate, mean arterial pressure, and left ventricular minute work during exercise.[52,134] These findings suggest that the sympathetic nervous system plays a minor role in cardiovascular stimulation in the resting supine man; however, it assumes a significant role in the cardiovasuclar response to exercise. The extent of the changes in ventricular volume that occur with propranolol during exercise have been variable and undoubtedly depend on a number of factors, including the severity of the exercise performed, whether exercise is performed in the supine or upright position, and the presence or absence of a diseased myocardium.[3,35,36,52,134] Thus, although reduction in heart rate, blood pressure, and contractility tend to lower myocardial oxygen consumption, any increase in ventricular dimensions following propranolol therapy would tend to change myocardial oxygen demand in the opposite direction. Wolfson[135] demonstrated that after propranolol administration to exercising human subjects the increase in myocardial oxygen consumption was markedly attenuated when compared to control observations. Thus in patients with heart disease and in normal subjects propranolol attenuates the chronotropic and inotropic responses to exercise. The increase in cardiac output is also diminished, which in turn contributes to the decreased blood pressure response to exercise. Although ventricular dimensions may increase with exercise, tending to offset the diminished myocardial oxygen demand, the increase is relatively slight and perhaps does not achieve significance unless the patient is exercised to maximum or near maximum levels. The end result of the exercise response in man during propranolol therapy is an increase in the duration of exercise, which is achieved at a lower oxygen cost to the myocardium.

Changes in coronary blood flow following propranolol administration generally parallel the changes in total myocardial oxygen consumption, as discussed previously.[67,68,78] Propranolol has been demonstrated to cause vasoconstriction of the coronary arteries of the normal dog.[61,67] This vasoconstriction was initially thought to be secondary to the unmasking of effects of sympathetic stimulation mediated by alpha adrenergic receptors. However, subsequent studies noted above showed that the increase in coronary resistance following propranolol administration was not the result of direct alpha adrenergic receptor stimulation but was secondary to the decrease in metabolic demands of the heart. Pitt and co-workers[136] demonstrated that propranolol-induced reduction in regional coronary flow occurred only in nonischemic areas of the intact dog heart, but not in the ischemic regions. The potential benefit of this type of response is intriguing to speculate upon. Thus in the ischemic heart vasoconstriction secondary to propranolol therapy may occur due to the unopposed action of circulating catecholamines on the alpha adrenergic receptors in the normal coronary arteries. This preferential vasoconstriction may shunt blood flow via collateral vessels from the nonischemic areas of the heart to the regions of ischemia.

The attenuation by propranolol of the myocardial oxygen demands of the heart at rest and particularly during stress, coupled with its demonstrated effects on the redistribution of coronary blood flow, form the rationale for the use of this beta adrenergic receptor antagonist in the treatment of ischemic heart disease. Classic angina pectoris occurs as a result of an imbalance between myocardial oxygen supply and demand. Of the four determinants of myocardial oxygen consumption (demand), propranolol has its greatest effect on heart rate and contractility (Fig. 20-1). Thus for any level of stress producing angina before beta blockade with propranolol, the chronotropic and inotropic response to the stress will be attentuated by this agent and will allow the same level of stress to be achieved at a lower oxygen cost to the myocardium. The addition of nitroglycerin to the therapeutic regimen for angina will alter the remaining two determinants of myocardial oxygen consumption, afterload and preload (decreasing wall tension by the Laplace principle), through their vasodilating actions on the peripheral arterial and venous beds.[137] Evidence does exist that nitrates also dilate the coronary arteries not involved by significant disease, and it is suggested that this mechanism may increase coronary blood flow to the ischemic areas via the coronary collateral circulation.[138–140] The beneficial effects of the synergism between propranolol and the nitrates have been confirmed by the clinical studies of Goldstein and Epstein.[137]

Based on its aforementioned effects on myocardial oxygen demand and coronary blood flow, it has long been speculated that propranolol may be of value in the therapy of acute myocardial ischemia. In this regard, Becker and associates[83] and Maroko and co-workers[84,102] studied the effects of propranolol in the laboratory and in man. In both instances propranolol decreased the extent and severity of myocardial ischemia, as judged by ST-segment changes from multiple-lead electrocardiographic mapping techniques. Pelides and co-workers[141] noted similar results in patients with acute myocardial infarction after the administration of the selective beta-1 adrenergic receptor antagonist practolol. Mueller and associates[86,142] demonstrated that in patients with acute myocardial infarction the administration of isoproterenol resulted in either an increase in lactate production or a shift from lactate extraction to production. When propranolol was administered to patients with uncomplicated myocardial infarctions, improvement in myocardial oxygenation and metabolism was observed, as well as improvement in the patient's ischemic symptoms.[143]

Thus by pharmacologically manipulating the adrenergic mechanisms regulating cardiovascular functions in patients with ischemic heart disease, one can observe improvement in symptoms, hemodynamics, coronary blood flow, size of the ischemic area, and myocardial metabolism. However, whether the use of the beta adrenergic receptor antagonist in acute or chronic ischemic heart disease can alter mortality has yet to be determined.

Stimulation of the beta-1 adrenergic receptors of the heart increases automaticity of the pacemaker tissue, increases conduction velocity, and decreases the refractory period of the junctional tissues in the conducting system of the heart, as previously described (Table 20-3). The nonselective beta adrenergic receptor antagonist propranolol can block these effects and has been shown to suppress automaticity, to increase the atrial–ventricular refractory period, and to slow atrial–ventricular conduction.[35,52,145,146] The form of propranolol currently available is a mixture of dextro (*d*) and levo (*l*) isomers; *d*-propranolol is said to manifest quinidinelike effects on cellular membranes and contains relatively little beta adrenergic receptor blocking properties, whereas *l*-propranolol has almost no membrane effects but contains the potent beta adrenergic blocking properties.[35,130,145] The reported efficacy of propranolol for the treatment of digitalis-induced arrhythmias is most likely secondary to the membrane effects of the dextro form.[35,52,145] The antiarrhythmogenic properties of *dl*-propranolol are the result of the combination of the properties of the two isomers. Propranolol has been reported to be effective in a wide variety of arrhythmias, and its usage in this context has been the subject of extensive reviews.[35,52,78,130,145] However, from analysis of the available data, those arrhythmias most likely to benefit from beta receptor blockade with propranolol are (1) extrasystoles or rapid atrial arrhythmias produced by anxiety and exertion with or without demonstrable heart disease, (2) atrial flutter and atrial fibrillation, (3) tachycardia associated with the Wolff-Parkinson-White (W-P-W) syndrome, particularly atrial fibrillation in the presence of W-P-W, (4) ectopic atrial or ventricular arrhythmias produced by digitalis excess, and (5) arrhythmias secondary to the release of catecholamines from pheochromocytomas.

The tachyarrhythmias and extrasystoles of anxiety states have been attributed to a heightened sensitivity and/or increased secretion of endogenous catecholamines. Propranolol is occasionally effective in these instances because of its beta adrenergic blocking properties. A direct effect on the reticular formation in the central nervous system has also been suggested as an additional mechanism of action.

Atrial fibrillation and atrial flutter are controlled by propranolol predominantly through its beta blocking effect on junctional tissue of the cardiac conducting system. The rapid ventricular response is thus attenuated by increasing the block to impulses arriving at the atrial–ventricular junction. Rarely will propranolol actually convert atrial flutter or fibrillation to normal sinus rhythm. The indication for the use of propranolol for atrial fibrillation or atrial flutter with a rapid ventricular response may be present when the conventional modalities of therapy such as digitalis or d-c cardioversion are not feasible or are contraindicated. Atrial fibrillation in conjunction with the W-P-W syndrome can often be converted to normal sinus rhythm or the underlying W-P-W rhythm.

Idiopathic hypertrophic subaortic stenosis may be clinically suspected because of symptoms caused by impedance to left ventricular outflow (syncope), arrhythmias (palpitations), or angina secondary to the increase in myocardial oxygen consumption as a result of the hypertrophied muscle and the increase in afterload. The outflow tract obstruction can be increased by the administration of isoproterenol and may be provoked in those patients in whom the obstruction is latent.[100] Propranolol is effective in relieving these symptoms, particularly during exercise when outflow tract obstruction is increased.[147,148] The mechanisms by which propranolol benefits patients with idiopathic hypertrophic suboartic stenosis may be related to a decrease in contractility in the area of left ventricular outflow obstruction, a decrease in myocardial oxygen consumption, and suppression of arrhythmias by a direct antiarrhythmic effect.

Propranolol has been added to the medical management of dissecting aortic aneurysms.[149,150] The mechanism of action is proposed to be secondary to the decrease in the force of myocardial contraction and rate of pressure rise in the aorta. These mechanisms will decrease shear forces at the points of aortic fixation to the pericardium and in the superior portion of the posterior thorax and will decrease the rate of rise of the transmitted arterial pressure pulse. In addition, systemic blood pressure may be reduced secondary to a decrease in cardiac output.[36,151]

Frohlich and associates[152] described a syndrome in patients manifested by emotional outbursts and cardiac awareness associated with a hyperresponsiveness of their heart rates to isoproterenol. Resting hemodynamics were consistent with a hyperkinetic circulation and in many ways resembled the hemodynamics of those patients described by Gorlin in 1962[153] as having the hyperkinetic heart syndrome. Frohlich's patients were mildly hypertensive, and hysteria could be precipitated by the infusion of isoproterenol but not by saline. The entire syndrome could be reversed with oral or intravenous propranolol. The mechanisms of this fascinating syndrome are not clear, but it is suggested that these patients have an unusually increased sensitivity of their beta adrenergic receptors.

Recently there has been a great deal of interest generated in the possible role of the beta adrenergic receptor in the production of malignant hypertension and the use of propranolol as a therapeutic agent.

Subjects with high-renin hypertension demonstrated impressive decreases in their blood pressures when treated with propranolol.[154,155] The response of the systemic blood pressure in this group of hypertensive patients to beta adrenergic receptor blockade does not appear to be due simply to a decrease in cardiac output as previously described.[36,151] Indeed, doses of propranolol as high as 2.0 g and greater have been used successfully and apparently without deleterious effects by Pritchard and Gillam[156] in the treatment of severe hypertension. The mechanism of action by which propranolol exerts its beneficial effects in patients with malignant hypertension is presently unclear.

Adrenergic responses mediated via the beta adrenergic receptors exert important influences on a wide variety of cardiovascular disease states. The fact that these mechanisms can be altered by the use of adrenergic antagonists has contributed greatly to a better understanding of the pathophysiologic mechanisms involved and to the effective therapy of these disorders. Further advances will be made in therapy and in the understanding of the adrenergic mechanisms with the development and use of specific adrenergic antagonists such as those listed in Table 20-2.

Many of the potentially harmful side effects of propranolol will no doubt be obviated by the use of more selective antagonists. Major side effects following the use of propranolol, such as marked bradyarrhythmias, the precipitation or exacerbation of congestive heart failure, heart block, and cardiogenic shock, have been reported in patients treated with this drug.[52,78,130,131] Close clinical observation of the patient, together with a sound physiologic understanding of the action of beta adrenergic receptor blocking mechanisms, will allow one to anticipate clinical situations in which these side effects may occur. Experience in the use of the beta adrenergic receptor blocking agents has made the major side effects a relatively rare occurrence. In patients with obvious or latent cardiogenic shock and overt or latent congestive heart failure, propranolol should be avoided, for the sympathetic nervous system may be exerting maximal affects on the cardiovascular system in order to support the circulation. In patients with a high degree of heart block, severe chronic obstructive pulmonary disease, and acute and chronic asthma, the use of nonselective beta adrenergic receptor blocking agents such as propranolol is contraindicated. Patients with arrhythmias treated by elective d-c cardioversion may develop severe postconversion bradyarrhythmias when pretreated with propranolol. The patient who is undergoing open-heart surgery may experience myocardial failure during or following the procedure when beta blockade is present, presumably due to the removal or attenuation of the supporting sympathetic tone on the cardiovascular system. Except in rare circumstances propranolol should be discontinued 48 hr prior to cardioversion or open-heart surgery.[133] Calcium, atropine, glucagon, the methylxanthines, and digitalis exert their inotropic and chronotropic cardiac effects through pathways that bypass the beta adrenergic receptor and can thus be used to counteract the deleterious effects of propranolol on cardiac function.[3,52,130,157] Temporary pacing can also be effective in treating the severe bradycardias or high-grade atrial–ventricular conduction defects that may occur in susceptible patients following the use of propranolol. Isoproterenol, if infused in sufficiently high doses, can often reverse the beta adrenergic blocking effects produced by propranolol.

The adrenergic nervous system is but one link in the intricate chain of closely interrelated physiologic mechanisms and countermechanisms necessary for the appropriate and efficient response of the cardiovascular system to the constantly changing metabolic requirements of the healthy or diseased human organism. A basic knowledge and a firm understanding of the adrenergic mechanisms and their place in the regulation and control of the normal and abnormal cardiovascular system are essential for better insight into the pathophysiology and treatment of cardiac disease. The adrenergic receptors are the mediators of these mechanisms, whether they are activated by direct stimulation via the sympathetic nerves or whether they are stimulated by endogenous of exogenous sympathomimetic agents. The availability of drugs that can selectively stimulate or block specific adrenergic receptors and receptor subtypes has broadened our understanding of these mechanisms and has provided the physician a valuable therapeutic armamentarium for the treatment of many cardiovascular disorders. This chapter is intended to provide only an introduction to the understanding of the complex functions of the adrenergic nervous system and how these functions may be beneficially altered by drug therapy in the patient with clinical heart disease.

ACKNOWLEDGMENT

The author wishes to express his sincere gratitude to Miss Joan MacDonald for the typing of this manuscript.

REFERENCES

1. Chidsey CA: Neural and hormonal control of the circulation, in Conn HL Jr, Horwitz O (eds): Cardiac and Vascular Diseases. Philadelphia, Lea & Febiger, 1971, p 41
2. Gorlin R, Sonnenblick EH: Regulation of performance of the heart. Am J Cardiol 22:16, 1968

3. Braunwald E: Regulation of the circulation. N Engl J Med 290:1124, 1420, 1974
4. Rolett EL: Adrenergic mechanisms in mammalian myocardium, in Langer GA, Brady AJ (eds): The Mammalian Myocardium. New York, John Wiley & Sons, 1974, p 219
5. Moran NC: Adrenergic receptors, drugs, and the cardiovascular system. Mod Concepts Cardiovasc Dis 35:93, 99, 1966
6. Axelrod J, Weinshilbaum R: Catecholamines. N Engl J Med 287:237, 1972
7. Alousi A, Weiner N: The regulation of norepinephrine synthesis in sympathetic nerves: Effect of nerve stimuli, cocaine and catecholamine releasing agents. Proc Natl Acad Sci USA 56:1491, 1966
8. Trendelenberg V: Supersensitivity and subsensitivity to sympathomimetic amines. Pharmacol Rev 15:225, 1963
9. Burn JH, Rand MJ: The action of sympathomimetic amines in animals treated with reserpine. J Physiol 144:314, 1958
10. Abboud FM: The role of catecholamines in circulatory disease. Adv Intern Med 15:17, 1969
11. Kopin IJ, Fischer JM, Musacchio JM, et al: "False neurotransmitters" and the mechanism of sympathetic blockade by monoamine oxidase inhibitors. J Pharmacol Exp Ther 147:186, 1965
12. Crout JR: Substitute adrenergic transmitters. Circ Res [Suppl 1] 18:120, 1966
13. Langley JN: On the reaction of cells and of nerve endings to certain poisons, chiefly as regards the reaction of striated muscle to nicotine and to curare. J Physiol 33:377, 1905
14. Dale HH: On some physiological actions of ergot. J Physiol 34:163, 1906
15. Ahlquist RP: A study of the adrenotropic receptors. Am J Physiol 135:586, 1948
16. Lands AM, Arnold A, McAuliff JP, et al: Differentiation of receptor systems activated by sympathomimetic amines. Nature 214:597, 1967
17. Lands AM, Luduena FL, Buzzo HP: Differentiation of receptors responsive to isoproterenol. Life Sci 6:2241, 1967
18. Arnold A, McAuliff JP, Colella DF, et al: The β_2 receptor mediated glycogenolytic responses to catecholamines in the dog. Arch Int Pharmacodyn Ther 176:451, 1968
19. Powell CE, Slater IH: Blocking of inhibitory adrenergic receptors by a dichloro analog of isoproterenol. J Pharmacol Exp Ther 122:480, 1958
20. Wasserman MA, Levy B: Selective beta adrenergic receptor blockade in the rat. J Pharmacol Exp Ther 182:256, 1972
21. Dunlop D, Shanks RG: Selective blockade of adrenoceptive beta receptors in the heart. Br J Pharmacol 32:201, 1968
22. Lefkowitz RJ: Selectivity in beta-adrenergic responses. Circulation 49:783, 1974
23. Farmer JB, Levy GP, Marshall RJ: A comparison of the β-adrenoceptor properties of salbutamol, orciprenaline and soterenol with those of isoprenaline. J Pharm Pharmacol 22:145, 1970

24. McRitchie RJ, Vatner SF, Tuttle R, Braunwald E: Cardiovascular effects of dobutamine, a cardiospecific β adrenergic stimulant, in conscious dogs. Circulation [Suppl 4] 48:132, 1973
25. Robie NW, Nutter DO, Moody C, McNay JL: In vivo analysis of adrenergic receptor activity of dobutamine. Circ Res 34:663, 1974
26. Lefkowitz RJ, Haber E: A fraction of ventricular myocardium that has the specificity of the cardiac beta-receptor. Proc Natl Acad Sci USA 68:1773, 1971
27. Lefkowitz RJ, Haber E: Physiological receptors as physicochemical entities. Cir Res [Suppl 1] 21:46, 1973
28. Robison GA, Butcher RW, Sutherland EW: Adenyl cyclase as an adrenergic receptor. Ann NY Acad Sci 139:703, 1967
29. Sutherland EW, Robison GA: Metabolic effects of catechoamines. Pharmacol Rev 18:145, 1966
30. Sutherland EW, Robison GA, Butcher RW: Some aspects of the biological role of adenosine 3′,5′-monophosphate (cyclic AMP). Circulation 37:279, 1968
31. Kones RJ: The catecholamines: Reappraisal of their use for acute myocardial infarction and the low cardiac output syndromes. Critical Care Med 1:203, 1973
32. Robison GA, Butcher RW, Sutherland EW: Cyclic AMP. New York, Academic, 1971, pp 106–119, 145–172, 193–210
33. Ibid pp 151–152
34. Ibid pp 175–176, 229
35. Lucchesi BR, Whitsitt LS: The pharmacology of beta-adrenergic blocking agents. Prog Cardiovasc Dis 11:410, 1969
36. Epstein SE, Robinson BF, Kahler RL, Braunwald E: Effects of beta-adrenergic blockade on the cardiac response to maximal and submaximal exercise in man. J Clin Invest 44:1745, 1965
37. Abboud FM, Eckstein JW, Zimmerman BG: Venous and arterial responses to stimulation of beta adrenergic receptors. Am J Physiol 209:383, 1965
38. Eckstein JW, Wendeling MG, Abboud FM: Forearm venous responses to stimulation of adrenergic receptors. J Clin Invest 44:1151, 1965
39. Folkow B: Nervous control of blood vessels. Physiol Rev 35:629, 1955
40. Glick G, Epstein SE, Wecksler AS, Braunwald E: Physiological differences between the effects of neuronally released and blood borne norepinephrine on beta adrenergic receptors in the arterial bed of the dog. Circ Res 21:217, 1967
41. Hornbrook KR, Brody TM: The effect of catecholamines on muscle glycogen and phosphorylase activity. J Pharmacol Exp Ther 140:295, 1963
42. Kvam DC, Riggilo DA, Lish PM: Effect of some new β-adrenergic blocking agents on certain metabolic responses to catecholamines. J Pharmacol Exp Ther 149:183, 1965
43. Sonnenblick EH, Ross J Jr, Braunwald E: Oxygen consumption of the heart. Newer concepts of its multifactorial determination. Am J Cardiol 22:328, 1968

44. Patterson SW, Starling EH: On the mechanical factors which determine the output of ventricles. J Physiol 48:357, 1914
45. Sonnenblick EH, Braunwald E, Morrow AG: The contractile properties of human heart muscle: Studies on myocardial mechanics of surgically excised papillary muscles. J Clin Invest 44:966, 1965
46. Katz AM: Contractile proteins, in Langer GA, Brady AJ (eds): The Mammalian Myocardium. New York, John Wiley & Sons, 1974, p 51
47. Langer GA: Ionic movements and the control of contraction, in Langer GA, Brady AJ (eds): The Mammalian Myocardium. New York, John Wiley & Sons, 1974, p 193
48. Rasmussen H: Cell communication, calcium ion, and cyclic adenosine monophosphate. Science 170:404, 1970
49. Epstein SE, Skelton CL, Levey GS, Entman M: Adenyl cyclase and myocardial contractility. Ann Intern Med 72:561, 1970
50. Steer ML, Atlas D, Levitzki A: Inter-relations between β-adrenergic receptors, adenylate cyclase and calcium. N Engl J Med 292:409, 1975
51. Herman MV, Heinle RA, Klein MD, Gorlin R: Localized disorders in myocardial contraction: Asynergy and its role in congestive heart failure. N Engl J Med 277:222, 1967
52. Epstein SE, Braunwald E: Beta-adrenergic receptor blocking drugs. Mechanisms of action and clinical applications. N Engl J Med 275:1106, 1175, 1966
53. Wallace AG, Troyer WG, Lesage AM, Zotti EF: Electrophysiological effects of isoproterenol and beta blocking agents in awake dogs. Circ Res 18:140, 1966
54. Daggett W, Wallace AG: Vagal and sympathetic influences on ectopic impulse formation, in Dreifus LS, Likoff W, Moyer JH (eds): Mechanisms and Therapy of Cardiac Arrhythmias. New York, Grune & Stratton, 1967, p 64
55. Brady AJ: Electrophysiology of cardiac muscle, in Langer GA, Brady AJ (eds): The Mammalian Myocardium. New York, John Wiley & Sons, 1974, p 135
56. Segal N, Hudson WA, Harris P, Bishop JM: The circulatory effects of electrically induced changes in ventricular rate at rest and during exercise in complete heart block. J Clin Invest 43:1541, 1964
57. Robinson BF, Epstein SE, Beiser GD, Braunwald E: Control of heart rate by the autonomic nervous system: Studies in man on interrelation between baroreceptor mechanisms and exercise. Circ Res 19:400, 1966
58. Eckberg DL, Drabinsky M, Braunwald E: Defective cardiac parasympathetic control in patients with heart disease. N Engl J Med 285:877, 1971
59. Chidsey CA, Braunwald E: Sympathetic activity and neurotransmitter depletion in congestive heart failure. Pharmacol Rev 18:685, 1966
60. Gorlin R: Regulation of coronary blood flow. Br Heart J 33 [Suppl] 33:9, 1971
61. Berne RM: The coronary circulation, in Langer GA, Brady AJ (eds): The Mammalian Myocardium. New York, John Wiley & Sons, 1974, p 251

62. Berne RM: Cardiac nucleotides in hypoxia: Possible role in regulation of coronary blood flow. Am J Physiol 204:317, 1963
63. Rubio R, Berne RM: Release of adenosine by the normal myocardium in dogs and its relationship to the regulation of coronary resistance. Circ Res 25:407, 1969
64. Kent KM, Cooper T: The denervated heart. A model for studying autonomic control of the heart. N Engl J Med 291:1017, 1974
65. Bohr DF: Adrenergic receptors in coronary arteries. Ann NY Acad Sci 139:799, 1967
66. Zuberbuhler RC, Bohr DF: Responses of coronary smooth muscle to catecholamines. Circ Res 16:431, 1965
67. Whitsitt LS, Lucchesi BR: Effects of propranolol and its stereoisomers upon coronary vascular resistance. Circ Res 21:305, 1967
68. McRaven DR, Mark AL, Abboud FM, Mayer HE: Responses of coronary vessels to adrenergic stimuli. J Clin Invest 50:773, 1971
69. Braundwald E, Chidsey CA, Pool PE, et al: Clinical staff conference. Congestive heart failure; biochemical and physiological considerations. Ann Intern Med 64:904, 1966
70. Chidsey CA, Braunwald E, Morrow AG: Catecholamine excretion and cardiac stores of norepinephrine in congestive heart failure. Am J Med 39:442, 1965
71. Chidsey CA, Harrison DC, Braunwald E: Augmentation of the plasma norepinephrine response to exercise in patients with congestive heart failure. N Engl J Med 267:650, 1962
72. Pool PE, Covell JW, Levitt M, et al: Reduction of cardiac tyrosine hydroxylase activity in experimental congestive heart failure. Circ Res 20:349, 1967
73. Spann JF Jr, Chidsey CA, Pool PE, Braunwald E: Mechanism of norepinephrine depletion in experimental heart failure produced by aortic constriction in the guinea pig. Circ Res 17:312, 1965
74. Spann JF Jr, Sonnenblick EH, Cooper T, et al: Cardiac norepinephrine stores and the contractile state of heart muscle. Circ Res 19:317, 1966
75. Covell JW, Chidsey Ca, Braunwald E: Reduction of the cardiac response to postganglionic sympathetic nerve stimulation in experimental heart failure. Circ Res 19:51, 1966
76. Vogel JHK, Chidsey CA: Cardiac adrenergic activity in experimental heart failure assessed with beta receptor blockade. Am J Cardiol 24:198, 1969
77. Krasnow N: Biochemical and physiologic response to isoproterenol in patients with left ventricular failure. Am J Cardiol 27:73, 1971
78. Pitt B, Ross RS: Beta adrenergic blockade in cardiovascular therapy. Mod Concepts Cardiovasc Dis 38:47, 1969
79. Kassebaum DG, Van Dyke AR: Electrophysiological effects of isoproterenol on Purkinje fibers of the heart. Circ Res 19:940, 1966
80. Pitt B, Elliot EC, Gregg DE: Adrenergic receptor activity in the coronary arteries of the unanesthetized dog. Circ Res 21:75, 1967
81. Clancy RL, Graham TP Jr, Powell WJ Jr, Gilmore JP: Inotropic aug-

mentation of myocardial oxygen consumption. Am J Physiol 212:1055, 1967.
82. Ross J Jr: Clinical implications of cardiac adrenergic mechanisms. Cardiovasc Res [Suppl 3] 4:40, 1970
83. Becker LC, Ferreira R, Thomas M: Effect of propranolol and isopreterenol on regional left ventricular blood flow and epicardial ST segment height in experimental myocardial ischemia. Am J Cardiol 31:119, 1973
84. Maroko P, Braunwald E: Modification of myocardial infarction size after coronary occlusion. Ann Intern Med 79:720, 1973
85. Saroff J, Wexler BC: Isoproterenol-induced myocardial infarction in rats. Circ Res 27:1101, 1970
86. Mueller H, Ayres SM, Giannelli S Jr, et al: Effect of isoproterenol, *l*-norepinephrine, and intraaortic counterpulsation on hemodynamics and myocardial metabolism in shock following acute myocardial infarction. Circulation 45:335, 1972
87. Kuhn LA: Shock in myocardial infarction: Medical treatment. Am J Cardiol 26:578, 1970
88. Mueller RA, Theonen H: Cardiac catecholamine synthesis, turnover, and metabolism with isoproterenol-induced myocytolysis. Cardiovasc Res 5:364, 1971
89. Wexler BC: Acute enzyme and metabolic changes in arteriosclerotic vs. nonarteriosclerotic rats following isoproterenol-induced myocardial infarction. Angiology 22:251, 1971
90. Haft J: Cardiovascular injury induced by sympathetic catecholamines. Prog Cardiovasc Dis 17:73, 1974
91. Cohn JN, Tristani FE, Khatri IM: Studies in clinical shock and hypotension. VI. Relationship between left and right ventricular function. J Clin Invest 48:2008, 1969
92. Smith JJ, Oriol A, March J, McGregor M: Hemodynamic studies in cardiogenic shock: Treatment with isoproterenol and metaraminol. Circulation 35:1084, 1967
93. Gunnar RM, Loeb HA, Pietras RJ, Tobin JR Jr: Ineffectiveness of isoproterenol in shock due to acute myocardial infarction. JAMA 202:1124, 1967
94. Weil MH, Shubin H: Isoproterenol for the treatment of circulatory shock. Ann Intern Med 70:638, 1969
95. Bellet S: Clinical disorders of the heart beat. Philadelphia, Lea & Febiger, 1971, p 934
96. Starzecki B, Spink WW: Hemodynamic effects of isoproterenol in canine endotoxin shock. J Clin Invest 47:2193, 1968
97. Beregovich J, Reicher-Reiss H, Kunstadt D, Grishman A: Hemodynamic effects of isoproterenol in cardiac surgery. J Thorac Cardiovasc Surg 62:957, 1971
98. Mueller HS, Gregory JJ, Giannelli S Jr, Ayres SM: Systemic hemodynamic and myocardial metabolic effects of isoproterenol and angiotensin after open-heart surgery. Circulation 42:491, 1970

99. Combs DT, Martin CM: Evaluation of isoproterenol as a method of stress testing. Am Heart J 87:711, 1974
100. Braunwald E, Ebert PA: Hemodynamic alterations in idiopathic hypertrophic
subaortic stenosis induced by sympathomimetic drugs. Am J Cardiol 10:489, 1962
101. Smith ER, Redwood DR, McCarron WE, Epstein SE: Coronary artery occlusion in the conscious dog. Effects of alterations in arterial pressure produced by nitroglycerin, hemorrhage, and alpha-adrenergic agonists on the degree of myocardial ischemia. Circulation 47:51, 1973
102. Maroko PR, Kjekshus JK, Sobol BE, et al: Factors influencing infarct size following experimental coronary artery occlusion. Circulation 43:67, 1971
103. Epstein SE, Kent KM, Goldstein RE, et al: Reduction of ischemic injury by nitroglycerin during acute myocardial infarction. N Engl J Med 292:29, 1975
104. Goldberg LI, Bloodwell RD, Braunwald E: The direct effects of norepinephrine, epinephrine, and methoxamine on myocardial contractile force in many. Circulation 22:1125, 1960
105. Mueller H, Ayres SM, Grace WJ: Principal defects which account for shock following acute myocardial infarction in man: Implications for treatment. Critical Care Med 1:27, 1973
106. Tuckman J, Finnerty FA Jr: Cardiac index during intravenous levarterenol infusion in man. Circ Res 7:988, 1959
107. Corday E, Williams JH Jr: Effect of shock and of vasopressor drugs on the regional circulation of the brain, heart, kidney, and liver. Am J Med 29:228, 1960
108. Corday E, Vyden JK, Lang TW, et al: Reevaluation of the treatment of shock secondary to cardiac infarction. Dis Chest 56:200, 1969
109. Mueller H, Ayres SM: Systemic and cardiac energetics in acute myocardial infarction and in cardiogenic shock in man: Effects of therapeutic interventions. Bull NY Acad Med 50:341, 1974
110. Kuhn LA; Clinical management of cardiogenic shock. Bull NY Acad Med 50:366, 1974
111. Tarazi RC: Sympathomimetic agents in the treatment of shock. Ann Intern Med 81:364, 1974
112. Goldberg LI: Cardiovascular and renal actions of dopamine: Potential clinical applications. Pharmacol Rev 24:1, 1972
113. Holzer J, Karliner JS, O'Rourke RA, et al: Effectiveness of dopamine in patients with cardiogenic shock. Am J Cardiol 32:79, 1973
114. Beregovich J, Bianchi C, Rubler S, et al: Dose related hemodynamic and renal effects of dopamine in congestive heart failure. Am Heart J 87:550, 1974
115. Rosenblum R, Freiden J: Intravenous dopamine in the treatment of myocardial dysfunction after open heart surgery. Am Heart J 83:743, 1972
116. Loeb HS, Winslow EBJ, Rahimtoola SH, et al: Acute hemodynamic effects of dopamine in patients with shock. Circulation 44:163, 1971

117. MacCannell KL, McNay JL, Meyer MB, Goldberg LI: Dopamine in the treatment of hypotension and shock. N Engl J Med 275:1389, 1966
118. Brooks HL, Stein PD, Matson JL, Hyland JW: Dopamine induced alterations in coronary hemodynamics in dogs. Circ Res 24:669, 1969
119. Winslow E, Loeb H, Rahimtoola SH, et al: Transmyocardial lactate metabolism during treatment of shock with catecholamines. Circulation [Suppl 3] 40:207, 1970
120. Abboud FM, Schmid PG, Eckstein JW: Vascular responses after alpha adrenergic receptor blockade. I. Responses of capacitance and resistance vessels to norepinephrine in man. J Clin Invest 47:1, 1968
121. Abboud FM, Eckstein JW: Vascular responses after alpha adrenergic receptor blockade. II. Responses of venous and arterial segments to adrenergic stimulation in the forelimb of dog. J Clin Invest 47:10, 1968
122. Nickerson M: Drugs inhibiting adrenergic nerves and structures innervated by them, in Goodman LS, Gilman A (eds): The Pharmacological Basis of Therapeutics. New York, Macmillan, 1970, p 549
123. Gitlow SE, Pertsemlidis D, Bertan LM: Management of patients with pheochromocytoma. Am Heart J 82:557, 1971
124. Himathongkam T, Newmark SR, Greenfield M, Dluhy RG: Pheochromocytoma. JAMA 230:1692, 1974
125. Harrison DC, Derber RE, Alderman EL: Pharmacodynamics and clinical use of cardiovascular drugs after cardiac surgery. Am J Cardio. 26:385, 1970
126. Jacobson ED: A physiologic approach to shock. N Engl J Med 278:834, 1968
127. Haddy FJ: Pathophysiology and therapy of the shock of myocardial infarction. Ann Intern Med 73:809, 1970
128. Perlroth MG, Harrison DC: Cardiogenic shock: A review. J Clin Pharmacol Ther 10:449, 1969
129. Misra SN, Kezdi P: Hemodynamic effects of adrenergic stimulating and blocking agents in cardiogenic shock and low output state after myocardial infarction. Am J Cardiol 31:724, 1973
130. Harrison DC: Beta adrenergic blockade, 1972. Pharmacology and clinical uses. Am J Cardiol 29:432, 1972
131. Morreli HF: Propranolol. Ann Intern Med 78:913, 1973
132. Paterson JW, Conolly ME, Dollery CT, et al: Pharmacodynamics and metabolism of propranolol in man. Pharmacol Clin 2:127, 1970
133. Faulkner SL, Hopkins JT, Boerth RC, et al: Time required for complete recovery from chronic propranolol therapy. N Engl J Med 289:607, 1973
134. Sonnenblick EH, Braunwald E, Williams JF Jr, Glick G: Effects of exercise on myocardial force velocity relations in intact unanesthetized man: Relative roles of changes in heart rate, sympathetic activity, and ventricular dimensions. J Clin Invest 44:2051, 1965
135. Wolfson S, Gorlin R: Cardiovascular pharmacology of propranolol in man. Circulation 40:501, 1969
136. Pitt B, Craven P: Effect of propranolol on regional myocardial blood flow in acute ischemia. Cardiovasc Res 4:176, 1970

137. Goldstein RE, Epstein SE: Medical management of patients with angina pectoris. Prog Cardiovasc Dis 14:360, 1972
138. Fam WM, McGregor M: Effect of coronary vasodilator drugs on retrograde flow in areas of chronic myocardial ischemia. Circ Res 15:355, 1964
139. Brachfeld N, Bozer J, Gorlin R: Action of nitroglycerin on the coronary circulation in normal and in mild cardiac subjects. Circulation 19:697, 1959
140. Bing RJ, Bennish A, Bluemchen G, et al: The determination of coronary flow equivalent with coincidence counting technic. Circulation 29:833, 1964
141. Pelides LH, Reid DS, Thomas M, Shillingford JP; Inhibition by β-blockade of the ST segment elevation after acute myocardial infarction in man. Cardiovasc Res 6:295, 1972
142. Mueller H, Ayres SM, Gregory JJ, et al: Hemodynamics, coronary blood flow and myocardial mebabolism in coronary shock; response to *l*-norepinephrine and isoproterenol. J Clin Invest 49:1885, 1970
143. Mueller HS, Ayres SM, Religa A, Evans RG: Propranolol in the treatment of acute myocardial infarction. Effect on myocardial oxygenation and hemodynamics. Circulation 49:1078, 1974
144. Rouse W: Effects of propranolol and ouabain on the conducting system of the heart in dogs. Am J Cardiol 18:406, 1966
145. Gibson D, Sowton E: The use of beta adrenergic receptor blocking drugs in dysrhythmias. Prog Cardiovasc Dis 12:16, 1969
146. Berkowitz WD, Wit AL, Lau SH, et al: The effects of propranolol on cardiac conduction. Circulation 40:855, 1969
147. Cohen LS, Braunwald E: Amelioration of angina pectoris in idiopathic hypertrophic subaortic stenosis with beta adrenergic blockade. Circulation 35:847, 1967
148. Sloman G: Propranolol in management of muscular subaortic stenosis. Br Heart J 29:783, 1967
149. Anaghostopoulos CE, Prabhaker MJS, Kittle FC: Aortic dissection and dissecting aneurysms. Am J Cardiol 30:263, 1972
150. Wheat MW Jr: Treatment of dissecting aneurysms of the aorta: current status. Prog Cardiovasc Dis 16:87, 1973
151. Frohlich ED, Tarazi RC, Dustan HP, Page IH: Paradox of beta-adrenergic blockade in hypertension. Circulation 37:417, 1968
152. Frohlich ED, Tarazi RC, Dustan HP: Hyperdynamic β-adrenergic circulatory state. Arch Intern Med 123:1, 1969
153. Gorlin R: The hyperkinetic heart syndrome. JAMA 182:823, 1962
154. Buhler FR, Laragh JH, Baer L, et al: Propranolol inhibition of renin secretion: A specific approach to diagnosis and treatment of renin dependent hypertensive diseases. N Engl J Med 287:1209, 1972
155. Kincaid-Smith P: Management of severe hypertension. Am J Cardiol 32:575, 1973
156. Pritchard BNC, Gillam PMS: Treatment of hypertension with propranolol. Br Med J 1:7, 1969
157. Williams JF Jr: New developments and therapeutic applications of cardiac stimulating agents. Am J Cardiol 32:491, 1973

J. Michael Criley, Paul A. Lennon
Abdul S. Abbasi, Arnold H. Blaufuss

21
Hypertrophic Cardiomyopathy

M.R.C. was a 16-year-old black male high school athlete who was forced to retire from competitive sports because of severe retrosternal chest pain, dizziness, and diaphoresis on minimal exertion, which had been progressive for the previous 2 years. He was admitted to the Harbor General Hospital Cardiac Care Unit because of an episode of crushing substernal pain of 2 hours duration associated with a rapid and forceful heart beat.

The patient had no history suggestive of acute rheumatic fever, nor had he had recurrent throat infections. His father had a heart murmur, had undergone cardiac catheterization 4 years previously, and had died suddenly at home at the age of 36. His paternal grandmother had been under treatment for "valvular heart disease," and had had congestive failure symptoms for 10 years. Two siblings died at ages 24 hours and 21 years of unknown types of heart disease. A paternal aunt was admitted to another hospital for a "heart attack" at age 36.

On arrival at the hospital he was a well-muscled adolescent male with a pulse rate of 150 (regular), a blood pressure of 90/60, and profuse diaphoresis, and he complained of severe substernal pressure. Limited cardiac examination at that time revealed a left precordial bulge, an active precordium, and a grade III/VI harsh midsystolic murmur along the left sternal border. An electrocardiogram revealed atrial flutter with 2:1 block, marked left ventricular hypertrophy, and deeply inverted left precordial T waves. A 50-W-sec countershock resulted in immediate reversion to sinus rhythm, and his chest pain subsided over the next 30 min.

On cardiac examination the next day he was noted to have a brisk pulse upstroke, a forceful presystolic and systolic apical impulse, an S_4 gallop, and a grade II midsystolic murmur at the left sternal border that did not change appreciably with respiration and was increased slightly by a Valsalva maneuver.

An electrocardiogram, chest roentgenogram, phonocardiogram, and echocardiogram were abnormal (Fig. 21-1A–D), and a diagnosis of hypertrophic cardiomyopathy (HCM) was strongly considered. Cardiac catheterization was

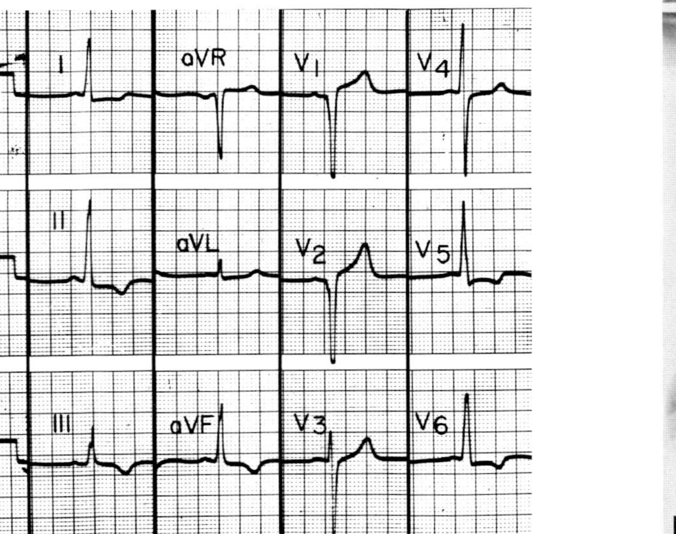

Fig. 21-1A. Electrocardiogram (1/2 standardized) of patient M.R.C. reveals high voltage, prolonged left ventricular activation, and ST-T changes compatible with extreme left ventricular hypertrophy, as well as absence of R waves in V_1 and V_2 suggestive of anteroseptal infarction or fibrosis.

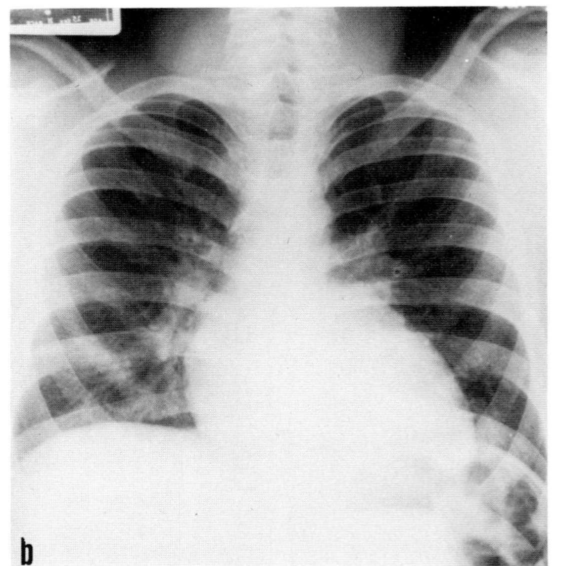

Fig. 21-1B. Chest x-ray (PA projection) of patient M.R.C. demonstrates overall cardiac enlargement principally involving the left ventricle. Analysis of the original film as well as other views suggested left ventricular, left atrial, and right atrial enlargement with increased size of the pulmonary veins.

Fig. 21-1C. Phonocardiogram (apex), apexcardiogram (ACG), and electrocardiogram (ECG, lead a VL) of patient M. R. C. demonstrates a prominent fourth heart sound (S_4) synchronous with a large a wave, a midsystolic murmur (SM), and a late secondary systolic wave (SW).

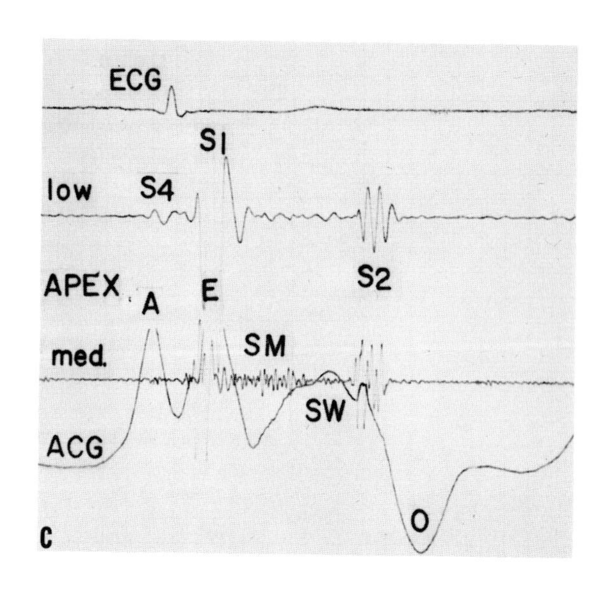

Fig. 21-1D. Echocardiogram of patient M.R.C. reveals a markedly thickened interventricular septum of 2.2 cm and normal thickness (0.8 cm) of the left ventricular posterior wall (LVPW). The anterior mitral leaflet (aML) comes into apposition with the septum in early diastole, and there is a reduced diastolic closing slope of 24 mm/sec and a systolic anterior movement (SAM). These findings are characteristic of HCM with obstruction (see text). (Chord. = chordae tendineae)

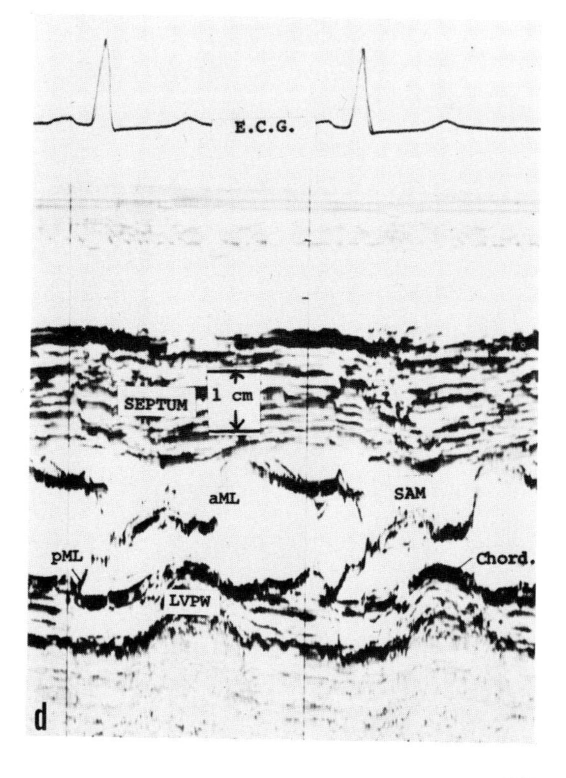

Table 21-1
Clinical and Cardiac Catheterization Data on Patient with HCM and Two Affected Relatives

	Patient M.R.C., Age 16	Father F.C., Age 36	Grandmother J.L.C., Age 56
History	Exertional chest pain, dizziness interfering with athletics	Dyspnea on exertion, fatigue, known to have cardiac enlargement and murmur for 5 years; died suddenly 1 year after study.	Congestive failure treated as "valvular heart disease" for 10 years; recent onset of atrial fibrillation with slow ventricular response, requiring pacemaker 15 months after study
Clinical findings	Brisk pulses; presystolic and late systolic LV impulse, S_4, grade II midsystolic murmur at left sternal border (Fig. 21-1C)	Normal pulses; S_4 and mid–late-systolic murmur at left sternal border and apex	Normal pulses; RV and LV lift, apical S_4 and holosystolic murmur; ejection murmur along left sternal border
EKG	Severe left ventricular hypertrophy with inverted T waves in II, III, aVF, and V_5–V_6 (Fig. 21-1A)	Left atrial and severe left ventricular hypertrophy with QRS duration of 0.11 sec and T inversion in V_4, V_5, V_6	Left atrial and ventricular hypertrophy, recently atrial fibrillation with ventricular response of 40–50
X-ray	Cardiothoracic ratio 54%; 2+ left atrial and 3+ left ventricular enlargement with increased pulmonary venous markings (Fig. 21—1B)	Cardiothoracic ratio 55%; 2+ left atrial and 2+ left ventricular prominence	Cardiothoracic ratio 60%; 3+ left atrial enlargement and 3+ left ventricular enlargement with pulmonary venous congestion

Catheterization data			
Date	1-4-72	9-4-68	1-27-72
Cardiac index (liters/min/m^2)	3.1	2.0	2.1
Right atrial mean pressure (mm Hg)	2	0	13
Right ventricular pressure (mm Hg)	30/0/3	15/0	35/7/10
Pulmonary artery pressure (mm Hg)	30/15	18/6	35/18
Left atrial mean pressure (mm Hg)	16	6	18
Left ventricular pressure (mm Hg)	125/8/16	100/0/8	115/8/18
Aortic pressure (mm Hg)	110/80	100/70	115/70
LV/Ao gradient:			
Rest (mm Hg)	15	0	0
Provoked (mm Hg)	48*	10–18†	0
LVEDP:			
Rest (mm Hg)	16	8	18
Provoked (mm Hg)	13*	40‡	15*
Cineangiography			
Ejection fraction	0.89	0.78	0.59
Mitral regurgitation	2+	0	3+
Ventriculogram	Irregular small thick-walled LV cavity with septal bulge and hyperdynamic contractions	Thick-walled LV with long thin cavity (banana shaped)	Irregular thick-walled enlarged cavity with prominent papillary muscles

*Isoproterenol infusion. †Valsalva—postectopic beat. ‡Angiotensin infusion.

Table 21-2
Clinicopathologic Correlations in Hypertrophic Cardiomyopathy (HCM)

Criteria	Pathology	Clinical Expression	
		History: Patients with manifestations of HCM may be completely asymptomatic	Physical Examination and Phonocardiography: Patients with HCM may have no abnormal findings on examination
Anatomic			
Idiopathic left ventricular hypertrophy	Hypertrophy in absence of discrete cause, rarely symmetric; thick, bizarre-shaped, abnormally oriented myocardial cells in septum and possibly free wall	Angina-like chest pain	
Asymmetrical septal hypertrophy (ASH)	Markedly thickened superior intraventricular septum with bizarre cells (vide supra); abnormal cells in free wall thought to correlate with nonobstruction (Fig. 21-2)	Familial incidence: Murmur, cardiac disability, or sudden death in family members; recent studies suggest that ASH may be transmitted as Mendelian dominant autosomal trait	ASH may produce palpable left parasternal lift in absence of right ventricular hypertrophy
Functional			
Hyperkinesis	Cardiac muscle initially thought to contain excessive norepinephrine and sympathetic nerve endings; however, this finding has not been confirmed	Patients are frequently athletic	Brisk jerky arterial pulse (Figs. 21-7A & 21-8), hyperdynamic double apical impulse; systolic murmur along left sternal border of variable intensity and configuration, which may increase with Valsalva maneuver
Reduced compliance of left or right ventricle	Gross hypertrophy with variable fibrosis	Dyspnea with exertion or tachycardia	Double apical impulse with presystolic expansion and fourth heart sound (Fig. 21-1C); large a wave in jugular pulse (Fig. 21-8)

Table 21-2 *(cont.)*

Criteria	Pathology	Clinical Expression: History: Patients with manifestations of HCM may be completely asymptomatic	Clinical Expression: Physical Examination and Phonocardiography: Patients with HCM may have no abnormal findings on examination
Obstruction	Patients with obstruction thought to have more normal arrangement of muscle fibers in free left ventricular wall, permitting vigorous contraction; otherwise, obstructed ventricles anatomically similar to nonobstructed ventricles; both may have plaque on septal surface thought to be caused by apposition of anterior mitral leaflet and septum and thickened mitral leaflets	Severity of symptoms correlates poorly with presence or magnitude of pressure gradient	Rapid upstroke, bifid arterial pulse, and apical systolic murmur with or without thrill more common in patients with gradients (Fig. 21-8); murmur increases with Valsalva or interventions that increase gradient (see Hemodynamics) (Fig. 21-7A)
Unexplained Dysrhythmia?		Syncope, often related to exertion or tachycardia Death, frequently sudden	

Findings on Investigation: Roentgenography: Patients with HCM may have normal cardiac x-ray	Electrocardiography: Patients with HCM may have normal ECG	Echocardiography	Hemodynamics: Patients with HCM may have no hemodynamic abnormalities	Angiography
Cardiothoracic ratio may be enlarged	ST-T changes and/or increased QRS voltage (Fig. 21-1A); increased ST depression and T inversion often occur with exercise	Increased left ventricular thickness, normal or small cavity size; free wall and septum may be equally thickened, but asymmetrical septal hypertrophy more common (see below)		Irregular, thick-walled, small left ventricular cavity; coronary arteries usually enlarged (Figs. 21-10 & 21-12)

Table 21-2 *(cont.)*

Findings on Investigation				
Roentgenography: Patients with HCM may have normal cardiac x-ray	Electrocardiography: Patients with HCM may have normal ECG	Echocardiography	Hemodynamics: Patients with HCM may have no hemodynamic abnormalities	Angiography
Septal bulge may be prominent on left cardiac border below pulmonary artery	Broad, deep Q waves may simulate myocardial infarction (Fig. 21-4)	Septal thickness $\geq$ 1.6 cm, septal to posterior wall ratio $\geqq$ 1.5 (Fig. 21-3)		Superior septal bulge evident on left anterior oblique or lateral views, well demonstrated by biventricular angiography
		Vigorous excursions of left ventricular posterior wall (but not septum) with high estimated ejection fraction (Fig. 21-6B)	Left ventricular pressure often has notch on upstroke, and arterial pressure has rapid rise (Fig. 21-11); high left ventricular *dp/dt* and high aortic velocity; cardiac output often normal or high	Rapid emptying of ventricle to small end-systolic size (Fig. 21-10)
May have enlarged left atrium and pulmonary venous congestion (Fig. 21-1B)		Reduced E-F slope of anterior mitral leaflet (Fig. 21-1D)	Elevated left ventricular end-diastolic pressure, prominent a wave, slow y descent (Fig. 21-11)	

performed 2 weeks later, and the catheterization data for the patient and two of his affected relatives are listed in Table 21-1. The patient was treated with propranolol 40 mg q.i.d., and his activity was restricted to unlimited walking, with avoidance of running or participation in competitive sports. He continued to have some chest pain on climbing stairs and after unauthorized athletics, but the severity and duration of the pain were less, and he had no recurrence of tachyarrhythmia.

The preceding case report was selected to introduce a discussion of hypertrophic cardiomyopathy (HCM) because the patient illustrates many of the typical features of the disease.

For the purposes of this chapter HCM will be considered as a single disease entity, although it has not been established with certainty that all

Table 21-2 *(cont.)*

Findings on Investigation				
Roentgenography: Patients with HCM may have normal cardiac x-ray	Electrocardiography: Patients with HCM may have normal ECG	Echocardiography	Hemodynamics: Patients with HCM may have no hemodynamic abnormalities	Angiography
Cardiac size correlates poorly with presence or magnitude of gradient	ECG abnormalities correlate poorly with presence or magnitude of gradient	Presence and degree of systolic anterior motion (SAM) (Fig. 21-6B) usually correlates well with presence and magnitude of gradient; however, SAM may occur in patients without gradients and without other evidence of HCM (Fig. 21-6C); left ventricular wall thickness, septal thickness, and left ventricular septal ratio correlate poorly with presence or absence of gradient (Fig. 21-3C)	Intraventricular pressure gradient in left and/or right ventricle; the pressure gradient is greatest in the last half of systole (Figs. 21-5A & 21-11) and may vary spontaneously and/or with physiologic or pharmacologic maneuvers; the postextrasystolic beat may or may not fail to increase aortic pulse pressure despite increase in left ventricular pressure (Brockenbrough-Braunwald phenomenon, Fig. 21-5)	The angiographic site of obstruction has been variously interpreted (Fig. 21-19); most authors feel that lucent line (Fig. 21-10B) on angiograms is site of obstruction, caused by impingement of anterior mitral leaflet against the septum; mitral regurgitation is observed in the majority of cases, thought to be due to distortion of the valve by hypertrophic ventricle and abnormal orientation of papillary muscles

patients thus classified have the same etiology and pathogenesis. Goodwin[1] has subcategorized these patients into HCM with and without obstruction, based on the presence or absence of an intraventricular pressure gradient. The interrelationship of the obstructive and nonobstructive forms has been firmly established within families[2–5] of affected patients and in individuals who have been documented to have a pressure gradient within the ventricle (obstruction) at one time and not at another.[3,6]

Although the patient technically qualifies for the diagnosis of HCM with obstruction, because of the presence of a pressure gradient, albeit small, within the left ventricular cavity, it is not clear what role the obstruction plays in his symptom complex. Similarly, his grandmother had had cardiac symptoms for over 10 years, and his father died suddenly (presumably a cardiac death); yet neither was demonstrated to have significant obstruction.

DEFINITION

In the following section an attempt will be made to define hypertrophic cardiomyopathy based on common criteria found in our cases as well as those reported in the literature. As noted above, it has not been established whether HCM is in fact one disease entity or a conglomeration of patients with idiopathic left ventricular hypertrophy from a variety of causes. Nevertheless, certain characteristic clinical and laboratory findings have emerged that permit classification of patients into the diagnostic category of HCM (Table 21-2).

The anatomic abnormality is left ventricular hypertrophy, involving principally the superior interventricular septum[7] (asymmetrical septal hypertrophy, ASH)[5] in the absence of a definable inciting cause, i.e., hypertension, aortic stenosis, coarctation, etc. However, one of Brock's original cases had a history of systemic hypertension,[8] and it has been alleged by others[9–12] that "functional hypertrophic subaortic stenosis" can coexist with or be caused by coarctation and subvalvular, supravalvular, or valvular aortic stenosis; so rigid exclusion of those cases with discrete lesions that could produce hypertrophy may not be justified.

Electrocardiographic changes of left ventricular hypertrophy occur in the majority of patients,[13,14] and voltages may be massive. Left atrial enlargement and biventricular hypertrophy are commonly seen.[13–15]

Asymmetrical septal hypertrophy (ASH), with a bizarre muscle fiber arrangement (Fig. 21-2A),[7,16–19] has been proposed as a specific marker to identify, by echocardiography (Fig. 21-1D), asymptomatic relatives of patients with known cases of HCM.[5] In our series of 30 patients with HCM studied by echocardiography, all had a septal thickness of $\geq$ 1.6 cm and a septal/posterior wall ratio of $\geq$ 1.5 (Table 21-3 and Fig. 21-3). However, case reports of patients with clinical and hemodynamic features of HCM have described concentric (symmetrical) hypertrophy[8,12,20,21] in operated or autopsy cases studied before echocardiography was available.

Q waves in the electrocardiogram (ECG) simulating myocardial infarction (Fig. 21-4), although with normal coronary arteries, occur in up to half of patients.[6,7,13] These ECG changes have been ascribed to unusual or unopposed septal forces[22–24] and indeed correlate with the angiographic appearance of localized septal hypertrophy.[14]

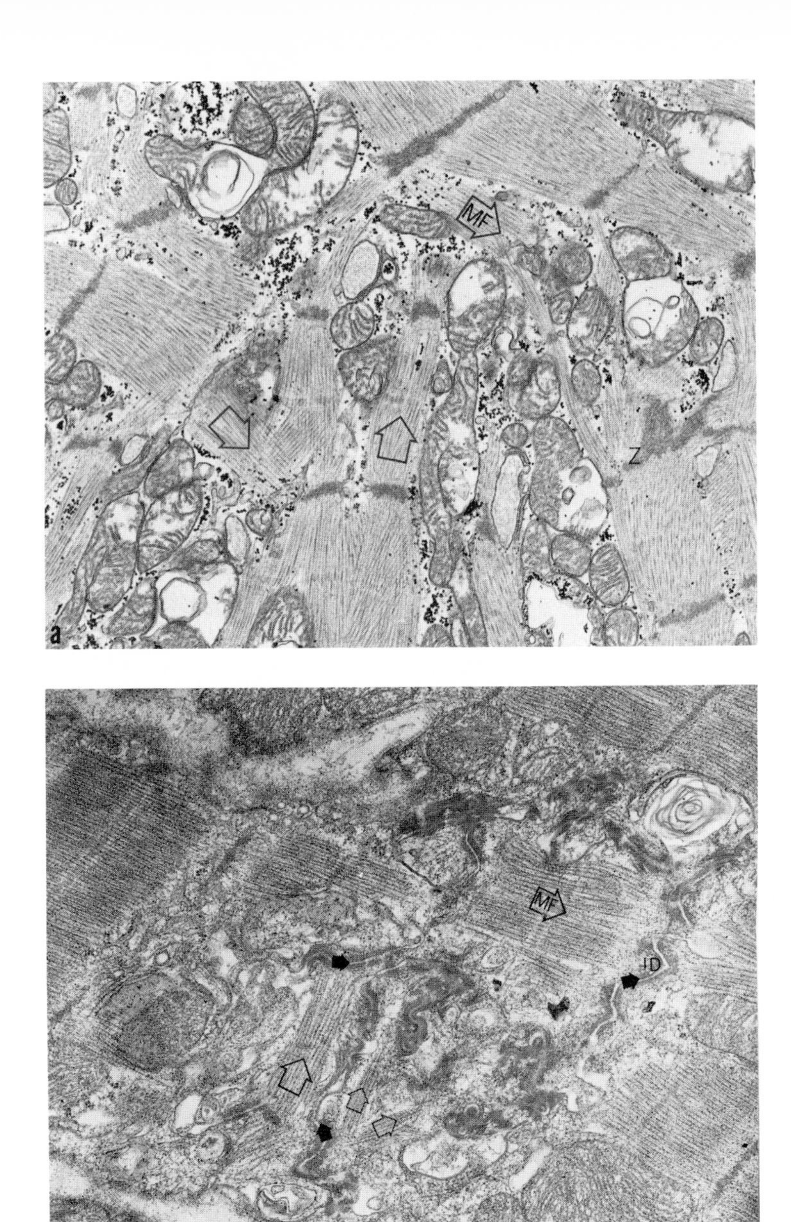

Fig. 21-2A. Electron micrograph of the septum of a patient with hypertrophic cardiomyopathy. Note the multidirectional orientation of the myofibrils (MF, arrows), the widened and split Z bands (Z). (This figure was kindly furnished by Victor J. Ferrans and William C. Roberts, M.D.)

Fig. 21-2B. Electron micrograph of the left ventricle in a dog undergoing 10 weeks of constant infusion of a subhypertensive level of norepinephrine. Note the multiple intercalated discs (ID) oriented at right angles to the major direction of the myofribrils (MF). The myofribrils have a bizarre multidirectional orientation, similar to that seen in HCM.

Table 21-3
Thickness of Septum and Posterior Wall

Group*	Measurement (mm)†		
	Septum	Ventricular Wall	Ratio
Normal	10.3 ± 2.2 (8–12)	8.8 ± 1.6 (6–12)	1.2 ± 0.1 (0.9–1.3)
Aortic stenosis or hypertension (20)	16.6 ± 1.8 (14–20)	14.0 ± 2.2 (11–18)	1.2 ± 0.1 (1.0–1.3)
Hypertrophic cardiomyopathy (30)	22.4 ± 3.5 (16–30)	10.8 ± 1.9 (8–15)	2.1 ± 0.3 (1.5–2.5)

*Figures in parentheses denote number of cases in each group.
†Values are mean and one standard deviation. Figures in parentheses denote range.

A familial incidence of HCM has been reported in 25% to 37% of cases in large series.[13,25] Using echocardiographic ASH as a marker, Clark[5] has postulated that most if not all cases are familial, and he supports a dominant autosomal mode of inheritance initially proposed by Brent.[2]

Intraventricular pressure gradients (Fig. 21-5A) at rest or after provocation, in the left and/or the right ventricle, have been widely used to establish the presence of obstruction in patients with HCM.[1–6,8–11,13–16,19–25] These pressure gradients are provoked or augmented by perturbations that decrease ventricular filling, decrease arterial resistance, or increase the vigor of left ventricular contraction. Conversely, interventions that increase left ventricular filling, increase afterload, or decrease inotropy diminish or abolish the pressure gradients.[3] Since there are nonobstructive mechanisms that can produce pressure gradients,[26–38] it is hazardous to rely on pressure gradients alone for the diagnosis of HCM.

Abnormal mitral valve motion consisting of systolic anterior movement (SAM) (Fig. 21-1D and Fig. 21-6B) of the free edge of the anterior mitral leaflet with approximation to the interventricular septum has been considered by many to be the mechanism of obstruction,[39–42] and when demonstrated by echocardiography or angiocardiography it has been considered diagnostic of HCM with obstruction. The mechanism of the SAM has been variously explained: abnormal insertion of the anterior leaflet,[39] displacement or malorientation of the papillary muscles,[43–45] and the Venturi effect resulting from high-velocity flow through the outflow tract.[46,47] This abnormal systolic anterior movement, however, is neither specific for HCM (Fig. 21-6D) nor universally agreed upon as the mechanism for obstruction in HCM.[3,48] Mitral regurgitation is frequently demonstrated in HCM; it is increased by interventions that increase or provoke intraventricular pressure gradients and decreased by reduction or abolition of the gradient.[49,50]

The systolic murmur characteristic of obstructive HCM[3,51–53] is probably a result of the abnormal valve motion. The murmur usually (but not invariably) begins at an interval after the first sound, is crescendo–decrescendo in character with a midsystolic or late-systolic peak (Figs. 21-5A, 21-6B, 21-7A, and 21-8). It is best heard over the lower left sternal border or apex and more often radiates to the axilla than to the neck. The murmur may be pansystolic, particularly after isoproterenol infusion. It may be quite faint or absent at rest, but is usually augmented by the Valsalva maneuver, vasodilators, and the upright posture. The murmur is not invariably associated with obstruction. Reeve has reported a case in which the characteristic augmentation of the murmur occurred with a Valsalva maneuver in a patient without a pressure gradient.[54]

It is not unusual to hear two murmurs in the same individual—an early peaking systolic ejection murmur over the aortic area and a later peaking systolic murmur over the left precordium. Rarely, a mid-diastolic rumble due to left ventricular inflow obstruction may be audible.[55] The augmentation of the murmur by vasodilators and upright posture and diminution by vasopressors and supine posture are shared by HCM and prolapsed mitral leaflet (Fig. 21-7B);[56] they suggest that the murmur in HCM is caused or contributed to by mitral regurgitation.

Hyperkinesis, as detected by a vigorous apical impulse and rapid arterial pulse upstroke (Figs. 21-5A, 21-7A, and 21-8), is common but not invariably present; and a relatively rapid arterial *dp/dt* tends to differentiate patients with HCM from those with discrete forms of aortic stenosis (Fig. 21-5B)[3] as well as from patients with congestive cardiomyopathy. It has been observed that many patients with HCM have been active in sports[48,57] or have athletic physiques, as did the patient described at the beginning of this chapter. Hyperkinesis is also apparent in cineangiographic studies of left ventricular function that reveal more rapid and more complete emptying of the left ventricle in patients with HCM than in patients with normal ventricles or valvular aortic stenosis.[58,59] Echocardiographic studies also demonstrated vigorous excursions of the posterior wall of the left ventricle (Fig. 12-6B) and high estimated ejection fractions.[60] The intraventricular pressure gradient appears to be associated with hyperkinesis, in that inotropic agents augment the gradient; and patients with more rapid pulse upstroke rates are

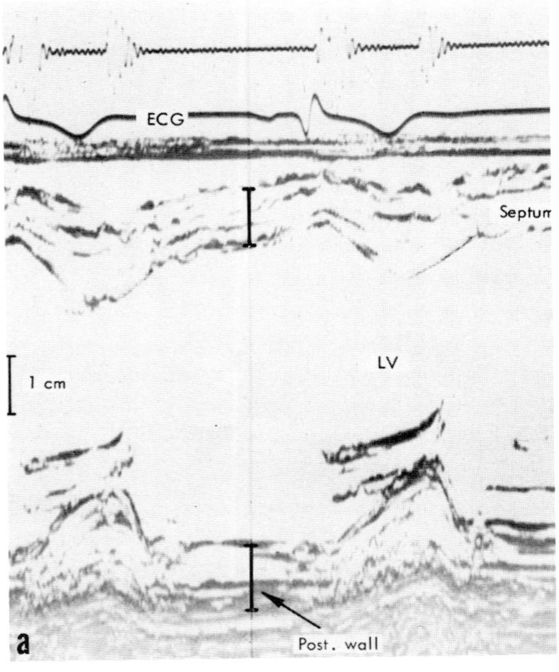

Fig. 21-3A

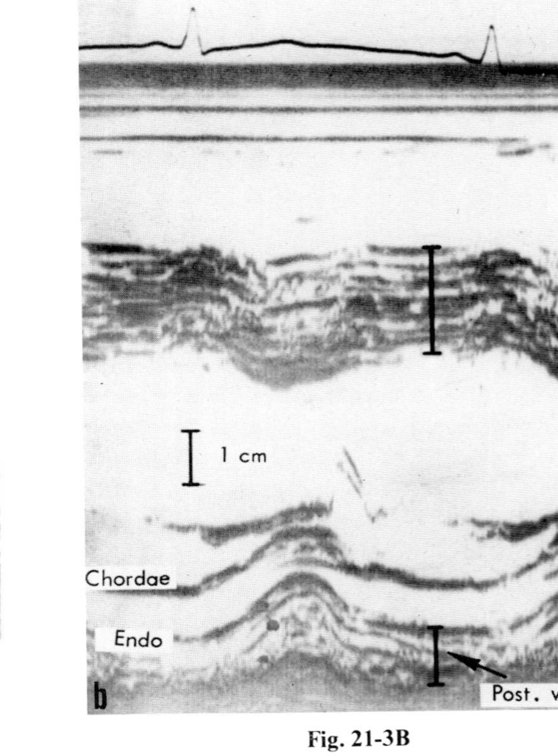

Fig. 21-3B

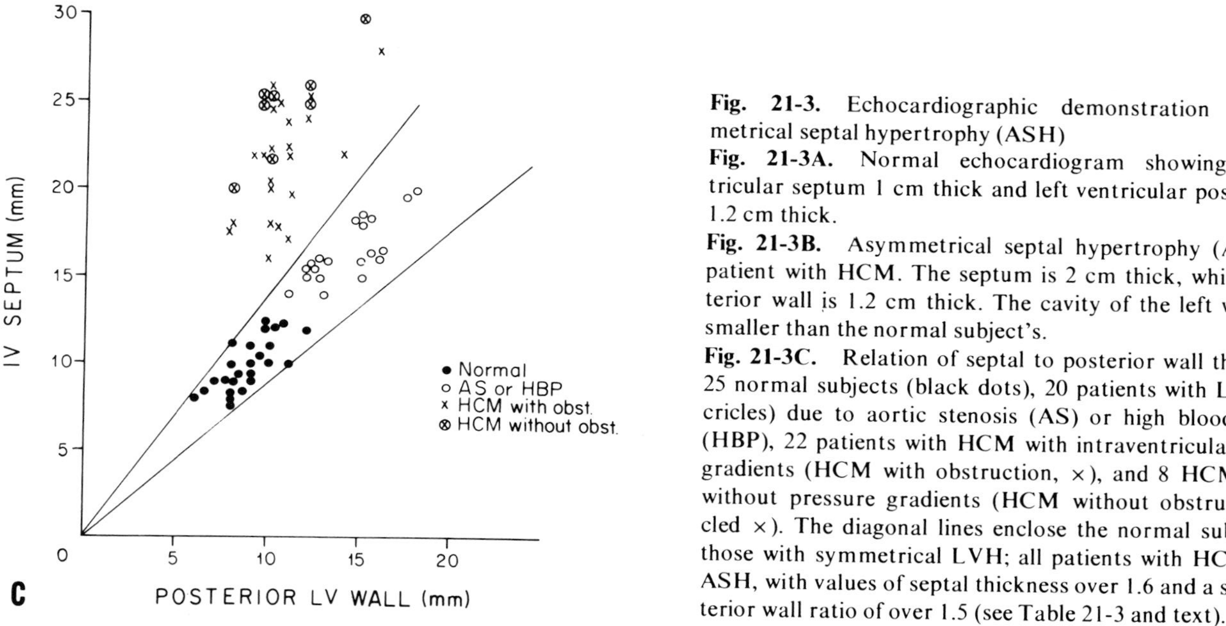

Fig. 21-3. Echocardiographic demonstration of asymmetrical septal hypertrophy (ASH)

Fig. 21-3A. Normal echocardiogram showing interventricular septum 1 cm thick and left ventricular posterior wall 1.2 cm thick.

Fig. 21-3B. Asymmetrical septal hypertrophy (ASH) in a patient with HCM. The septum is 2 cm thick, while the posterior wall is 1.2 cm thick. The cavity of the left ventricle is smaller than the normal subject's.

Fig. 21-3C. Relation of septal to posterior wall thickness in 25 normal subjects (black dots), 20 patients with LVH (open cricles) due to aortic stenosis (AS) or high blood pressure (HBP), 22 patients with HCM with intraventricular pressure gradients (HCM with obstruction, ×), and 8 HCM patients without pressure gradients (HCM without obstruction, circled ×). The diagonal lines enclose the normal subjects and those with symmetrical LVH; all patients with HCM exhibit ASH, with values of septal thickness over 1.6 and a septal posterior wall ratio of over 1.5 (see Table 21-3 and text).

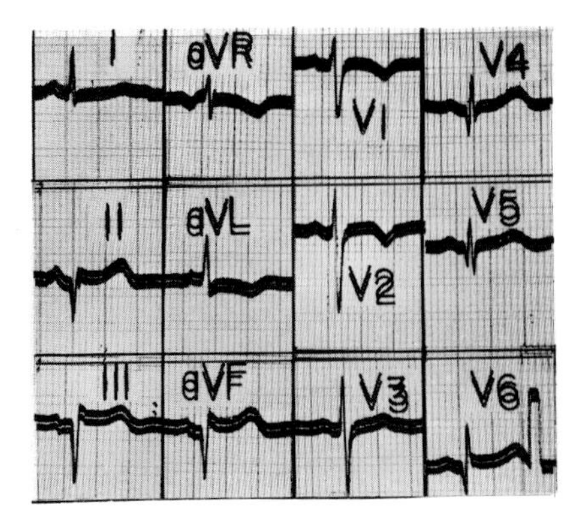

Fig. 21-4. Electrocardiogram in HCM simulating myocardial infarction. The woman from whom this ECG was obtained had no history of acute myocardial infarction, but had clinical and echocardiographic features of HCM. There are deep Q waves in leads II, III, aVF, and V_4–V_6, in the absence of voltage criteria for left ventricular hypertrophy.

more likely to manifest intraventricular gradients. It has been postulated that the contractility of the free wall of the left ventricle is the primary factor responsible for the outflow tract obstruction. If the free wall musculature has the bizarre fiber arrangement characteristic of the septum, the left ventricle contracts poorly and does not produce obstruction, while a more normal muscle fiber distribution permits the vigorous contraction that is apparently responsible for the obstruction.[61] Measurements of flow and velocity in the aortic root in patients with HCM have revealed more rapid ejection in early systole than normal, with ejection of approximately 80% of the stroke volume in the first half of systole,[3,62] even in the absence of a significant pressure gradient.[63]

ETIOLOGY OF HCM

As noted above, at least one-third of reported case of HCM are familial, with an autosomal-dominant mode of inheritance. Meerschwam has found evidence of a skeletal muscle abnormality by electromyography in cases of HCM, suggesting that a generalized myopathy may be present.[48]

We are acquainted with a case of HCM with obstruction in an infant who died of an obscure myopathy, the exact nature of which was undiagnosed even after autopsy and electron microscopy. At the age of 6 months cardiac catheterization demonstrated the typical findings of HCM with obstruction, with an intraventricular gradient of 45 mm Hg in the left ventricle. At autopsy there was gross asymmetrical septal hypertrophy with a measured septal:free-wall ratio of 1.5:1.[64]

HCM has also been noted in association with Friedreich's ataxia,[65] as well as with lentiginosis.[11,66,67] In our laboratory, dogs subjected to

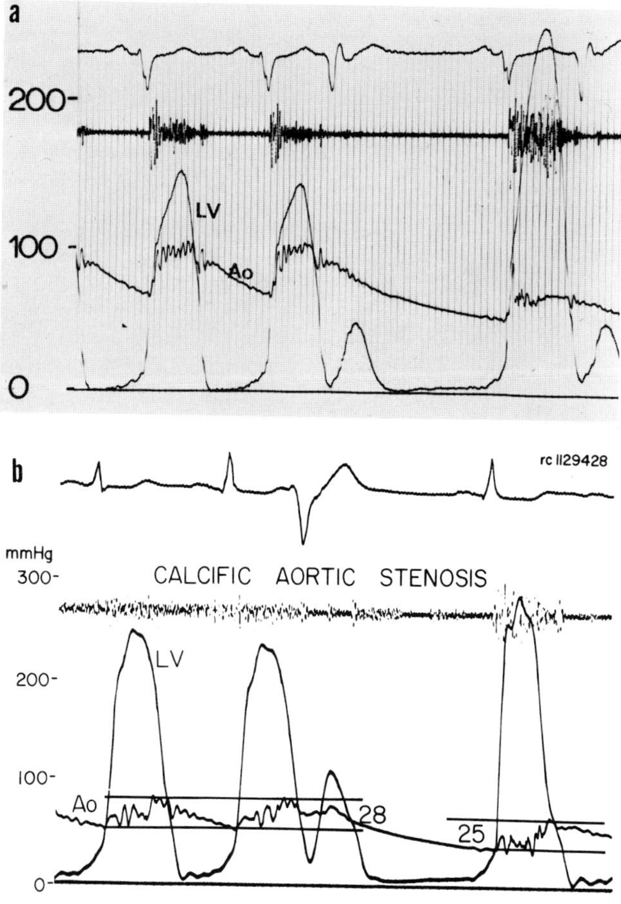

Fig. 21-5. Postectopic beat phenomenon in HCM and calcific aortic stenosis. (A) HCM with obstruction. There is a resting pressure gradient between the left ventricle and aorta demonstrated in the two beats on the left. Following the premature ventricular contraction (center) there is a compensatory pause and a marked increase in left ventricular pressure, pressure gradient, and murmur intensity. The aortic pulse pressure decreases, a phenomenon characteristic of HCM with obstruction. (B) Calcific aortic stenosis with hypertrophic postectopic beat response. A postectopic beat phenomenon characteristic of HCM is recorded in a patient with calcific aortic stenosis. Following the premature ventricular contraction (center), the left ventricular pressure rises to almost 300 mm Hg, while the pulse pressure in the aorta falls from a resting level of 28 mm Hg to 25 mm Hg. The slow pressure rise in the aorta is characteristic of valvular aortic stenosis, and the lesion was confirmed at subsequent surgery. Most patients with discrete aortic stenosis will have an increase in aortic pulse pressure following a premature beat.

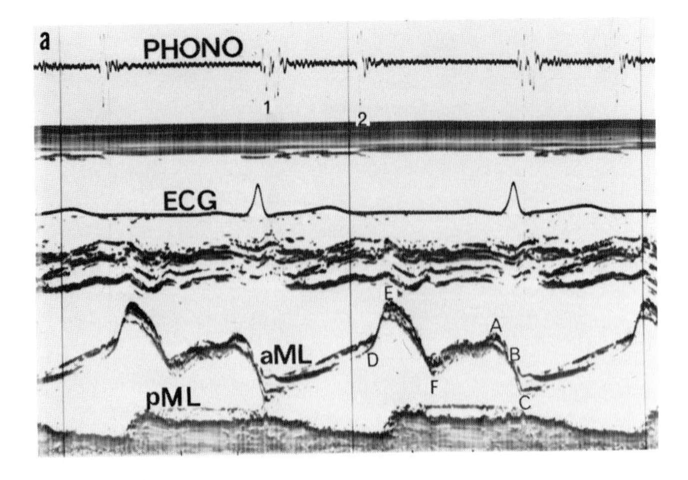

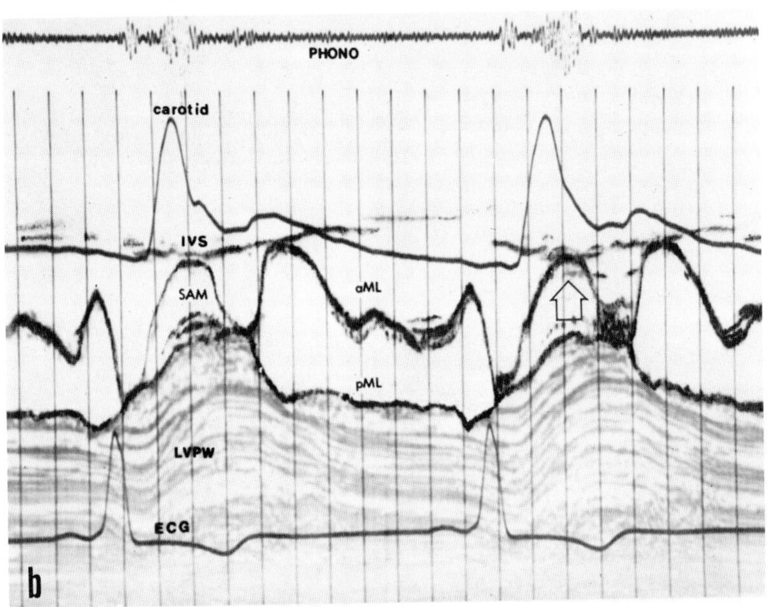

Fig. 21-6A. Normal systolic movement of the mitral valve. In this normal patient the anterior mitral leaflet (aML) moves abruptly toward the interventricular septum (IVS) in early diastole, floats back to a midposition in mid-diastole, and is pushed open again by atrial systole. During ventricular systole the valve remains in a posterior position in apposition to the posterior mitral leaflet (pML).

Fig. 21-6B. SAM in HCM with obstruction. The anterior mitral leaflet swings forward normally in early diastole, but appears to contact the interventricular septum (IVS). It floats back to a midposition more slowly than does the normal, and then is thrust widely open during atrial systole. During systole there is a vigorous excursion of the left ventricular posterior wall (LVPW), and both leaflets are thrown forward with the anterior leaflet once again contacting the interventricular septum. There is an apparent systolic separation of the anterior and posterior leaflets. This SAM is said to be characteristic of HCM with obstruction. (This figure was kindly provided by Richard L. Popp, M.D.)

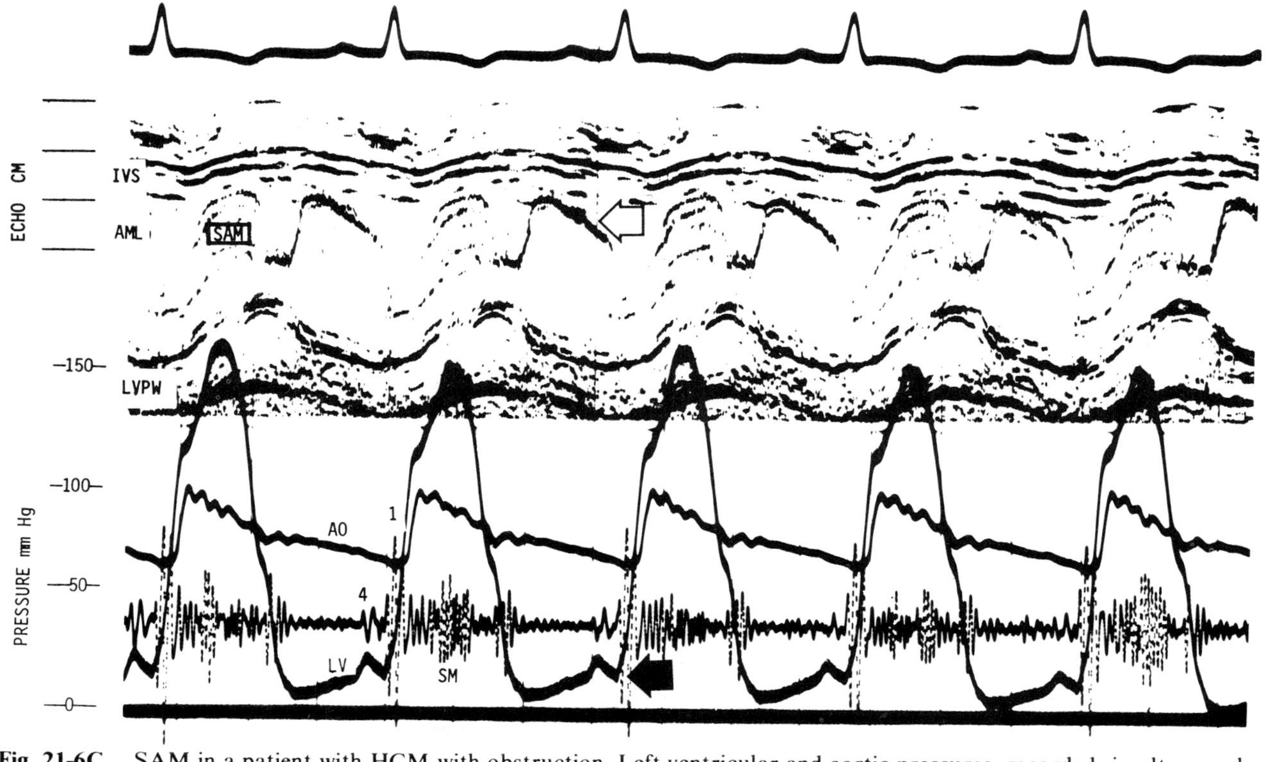

Fig. 21-6C. SAM in a patient with HCM with obstruction. Left ventricular and aortic pressures, recorded simultaneously with the echocardiogram, confirm the pressure gradient of 70 mm Hg. The elevated left ventricular end diastolic pressure (black arrow), indicative of reduced left ventricular compliance, is manifested in the echocardiogram by a reduced diastolic slope of the anterior mitral leaflet (white arrow).

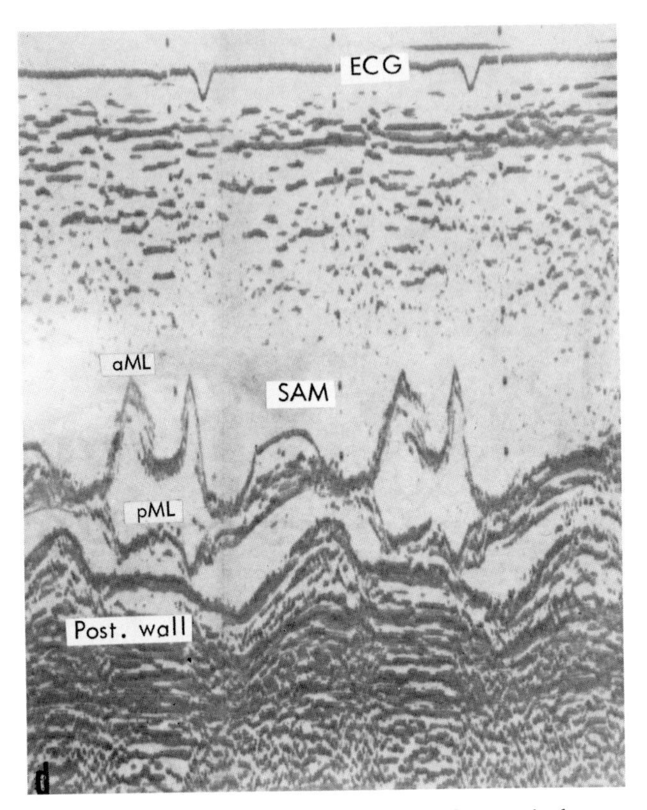

Fig. 21-6D. SAM In a patient with left ventricular aneurysm. Systolic anterior motion suggestive of HCM with obstruction is recorded from a patient with a left ventricular aneurysm. There was no evidence of HCM at cardiac catheterization or at subsequent aneurysmectomy. (This figure was kindly provided by Dr. John H. K. Vogel.)

chronic (3–6 months) nonhypertensive norepinephrine infusions have developed left ventricular hypertrophy, resting intracavitary pressure gradients, and a bizarre muscle fiber arrangement in the ventricle similar to HCM (Fig. 21-2B), suggesting that excessive circulating norephinephrine or enhanced sensitivity to norepinephrine may play a role in HCM.

Subaortic pressure gradients with abnormal systolic position of the mitral valve have been reported in association with several types of congenital heart disease[11,68] (Fig. 21-9), and intraventricular pressure gradients have been recorded after operative relief of discrete forms of aortic stenosis.[9,10,12,69] These pressure gradients may be examples of nonobstructive mechanisms (vide infra) in some cases, since relief of experimental coarctation in dogs results in intraventricular pressure gradients in the absence of obstruction.[37]

APPROACH TO THE PATIENT WITH SUSPECTED HCM

A high index of suspicion is required not to miss the diagnosis of HCM. Table 21-4 summarizes the common presenting features. If the patient is symptomatic, chest pain, dyspnea, and dizziness or syncope are the most common complaints. On the other hand, asymptomatic individuals may be discovered to have an abnormal electrocardiogram or chest x-ray, or may be referred because of a familial incidence of heart disease. The presence of a brisk carotid pulse, hyperdynamic left ventricular impulse, and prominent fourth heart sound should alert the clinician to consider the possibility of HCM and to undertake the various noninvasive procedures that can establish the diagnosis in the majority of instances.

Careful auscultation combined with the Valsalva maneuver as well as postural and/or pharmacologic stimuli can bring out the characteristic murmur or murmurs of HCM. The absence of a murmur does not negate the diagnosis of HCM, however, and should not deter the clinician from proceeding to echocardiography.

Echocardiography has significantly contributed to noninvasive diagnosis of HCM. Echocardiography may also be useful in following the progress of the disease and changes induced by therapy.

Two major echocardiographic criteria for the diagnosis of HCM are (1) systolic anterior motion of the mitral valve (Figs. 21-1D and 21-6B)[70,71] and (2) asymmetrical left ventricular hypertrophy mainly involving the ventricular septum (Fig. 21-3B and Table 21-3).[60,72,73] Three minor criteria are (1) normal to small left ventricular cavity size, with increased ejection fraction (Fig. 21-6B),[60] (2) proximity of the mitral valve to the ventricular septum in diastole (Figs. 21-1D and 21-6B),[74] and (3) reduced early diastolic slope of the mitral valve, suggesting decreased left ventricular compliance (Figs. 21-1D, 21-6B, and 21-6C).[70] These minor criteria are nonspecific; however, they are so commonly present that their absence makes the diagnosis of HCM less likely.

Abnormal systolic anterior motion (SAM) of the mitral valve was the first specific echocardiographic abnormality found in HCM. Normally the mitral valve is seen to open in ventricular diastole and to close

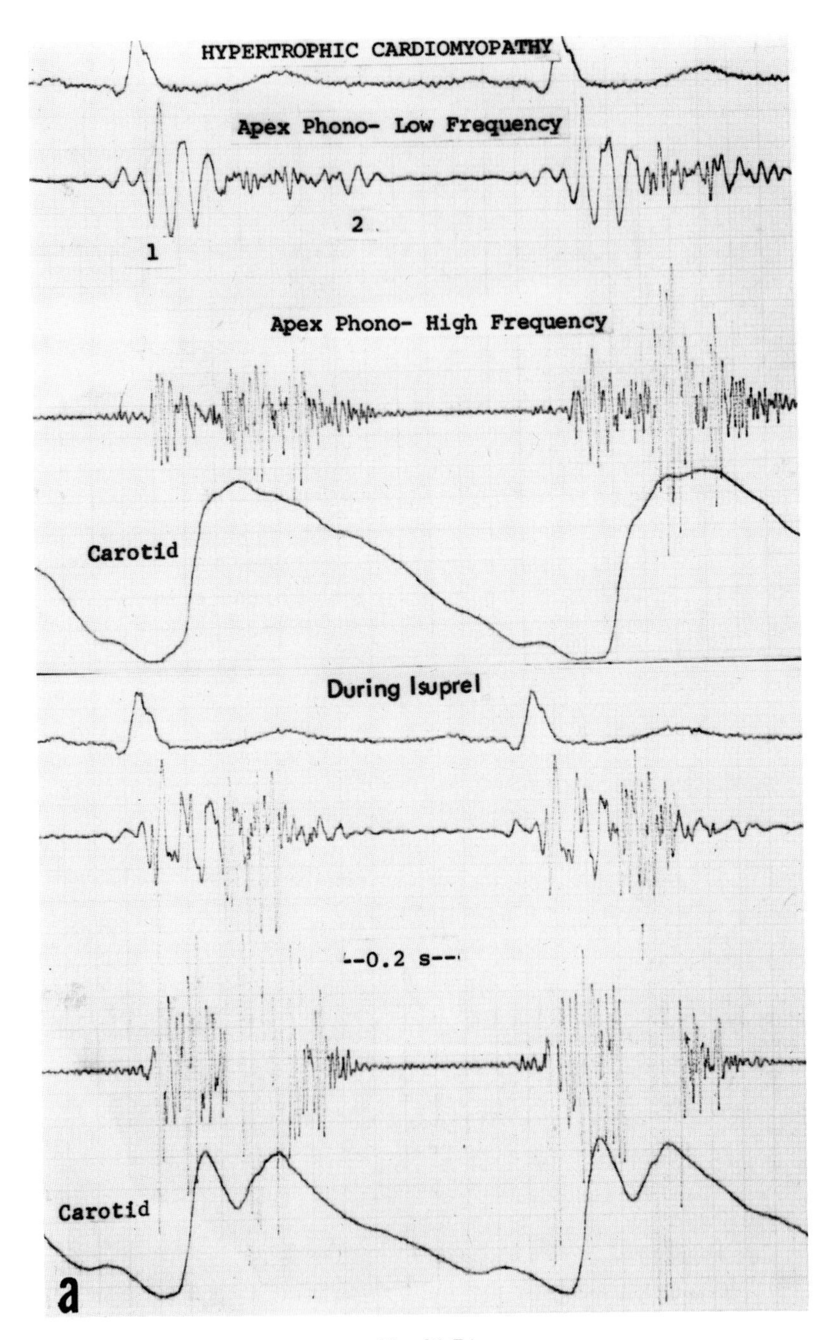
HYPERTROPHIC CARDIOMYOPATHY
Apex Phono- Low Frequency
1
2
Apex Phono- High Frequency
Carotid
During Isuprel
--0.2 s--
Carotid
a

Fig. 21-7A

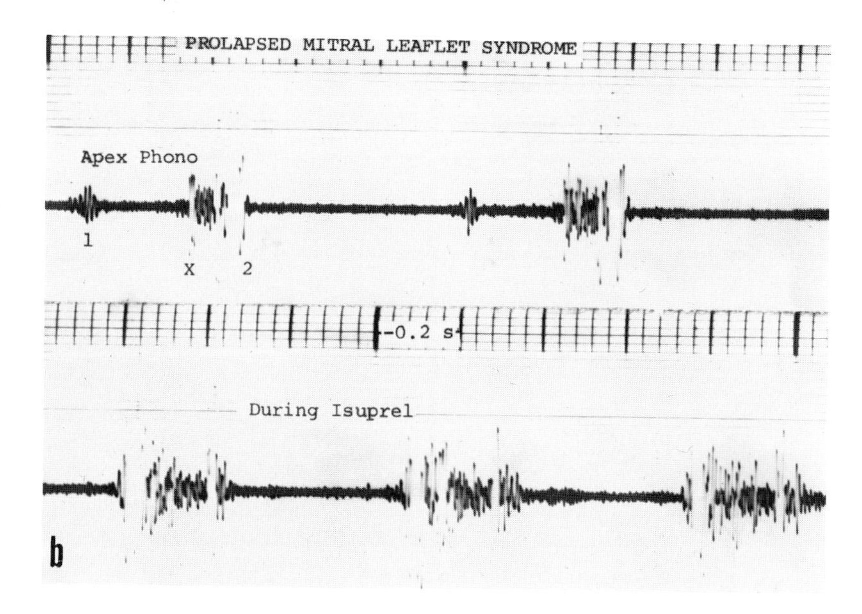

Fig. 21-7A. Phonocardiogram after isoproterenol in HCM. An apex phonocardiogram and carotid arteriogram demonstrate a crescendo-decrescendo murmur with maximal intensity in late systole. During isoproterenol (Isuprel) infusion, the murmur becomes holosystolic and the carotid arteriogram demonstrates a bifid contour.

Fig. 21-7B. Phonocardiogram after isoproterenol in prolapsed mitral leaflet syndrome. The resting apex phonocardiogram reveals a click (×) and late systolic murmur. During isoproterenol infusion the murmur becomes holosystolic. This response of the murmur to isoproterenol is similar to that seen in HCM.

at the beginning of systole (Fig. 21-6A). During ventricular ejection the mitral valve and anulus are displaced slightly anteriorly. In contrast, in HCM there is abnormal SAM of the anterior and sometimes the posterior mitral leaflet toward the ventricular septum (Figs. 21-1D, 21-6B, and 21-6C); SAM is characteristically symmetrical, with the peak at midsystole. If SAM is absent it may often be demonstrated after administration of an agent such as amyl nitrite.[71] It should be pointed out that SAM may be missed unless the beam is pointed inferolaterally to record both the anterior and the posterior mitral leaflets (Fig. 21-1D).[75]

The degree and the duration of the SAM of the mitral valve have been used to estimate the severity of left ventricular outflow obstruction.[75] Because of variation in the size of the SAM in the same individual caused by slight change in the beam direction, and the dynamic

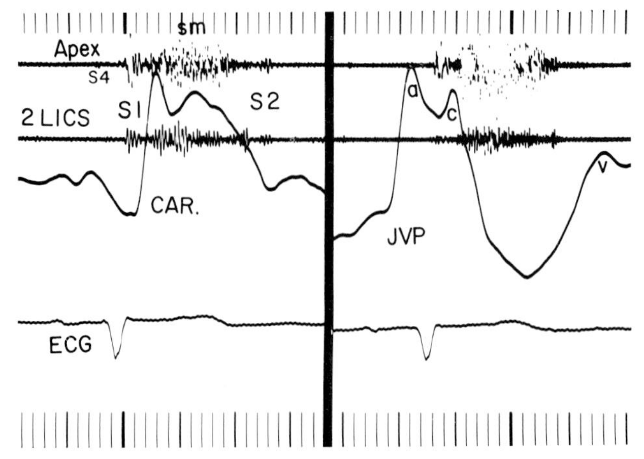

Fig. 21-8. Phonocardiogram, carotid arteriogram, and jugular phlebogram in a patient with HCM. The phonocardiogram reveals a mid–late-systolic accentuation of the systolic murmur (sm) that follows the early systolic dip in the arterial pulse tracing. There is a fourth heart sound (S4) and a prominent jugular a wave.

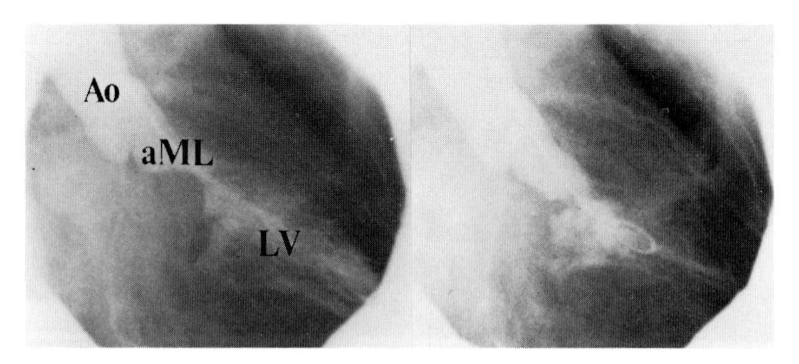

Fig. 21-9. Left ventriculogram in a patient with ostium primum atrial septal defect with abnormal mitral valve insertion. A retrograde left ventriculogram in the right anterior oblique projection on a patient with a 60mm pressure gradient in the outflow tract of the left ventricle in association with an ostium primum atrial septal defect reveals anterior displacement of the entire anterior mitral leaflet (aML) during diastole (left) and a lucent line seen across the outflow tract in systole (right). This abnormal systolic anterior motion is similar to that seen in HCM with obstruction, although there are no other features of HCM in this patient. (This figure was kindly provided by Arnold L. Nedelman, M.D.)

Table 21-4
Principal Presenting Features in
30 Cases of HCM

Feature	Number of Patients
Chest pain	14
Dyspnea on exertion	9
Syncope or dizziness	10
Heart murmur	7
Abnormal ECG	3
Abnormal x-ray	2

Note: Some patients had more than one presenting feature.

and variable nature of the outflow obstruction, the magnitude of any SAM can only be taken as a gross measure of the left ventricular outflow obstruction.[76]

SAM is not always present in HCM, and it sometimes is not demonstrated even after amyl nitrite or Isuprel administration.[60] On the other hand, SAM has been seen in conditions other than HCM, such as ischemic heart disease (Fig. 21-6D), prolapsed mitral leaflet syndrome,[77] and atrial septal defect.[78] Usually in such cases SAM is asymmetrical and small; however, symmetrical and significant systolic motion of the mitral valve occupying more than one-half of the left ventricular outflow has been noted in patients with left ventricular aneurysm (Fig. 21-6D). Thus the absence of SAM does not rule out HCM, and its presence is not specific for the diagnosis.

The second criterion, which is now felt to be more specific for HCM, is asymmetrical left ventricular hypertrophy mainly involving the ventricular septum. Echocardiographic evidence of disproportionate septal thickness has been demonstrated in some patients with pulmonary hypertension;[79,80] however, clinical recognition of pulmonary hypertension in the absence of clinical findings of HCM can usually separate out these cases.

Cardiac catheterization and angiocardiography are not always necessary for confirmation of the diagnosis of HCM, but they are useful in excluding other conditions and in establishing the presence of certain characteristic findings that are often present and are somewhat unique to HCM.

The ventricle in HCM is usually thick-walled and noncompliant, with a small, irregularly shaped cavity that contracts rapidly to a miniscule end-systolic volume (Fig. 21-10).[3] However, some patients who must

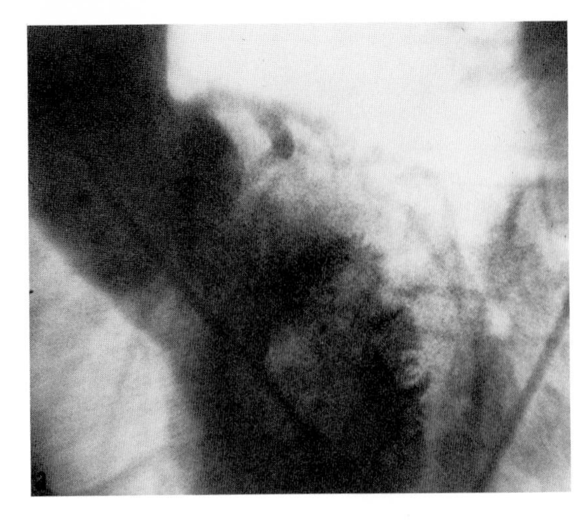

Fig. 21-10A. Left ventriculogram in a patient with HCM with obstruction (gradient = 70 mm Hg). Diastole, LAO. The contour of the left ventricle is relatively normal except for the thickened left ventricular wall and the heavy trabeculations.

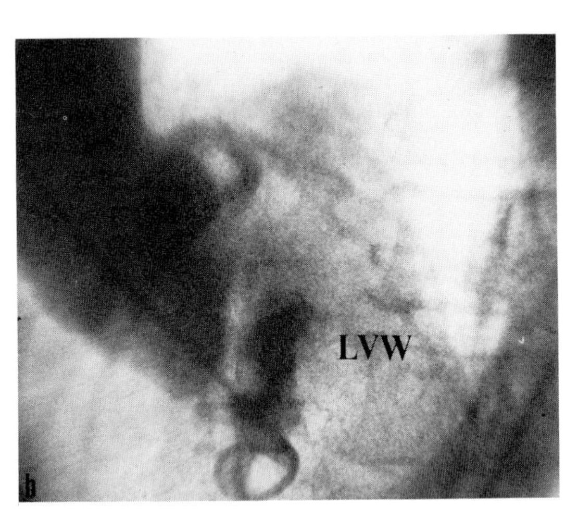

Fig. 21-10B. Left ventriculogram in a patient with HCM with obstruction (gradient = 70 mm Hg). Systole, LAO. The left ventricle has ejected over 90% of its end-diastolic volume, leaving a small volume of contrast medium separated by a lucent line that represents a "sandwich" made up of the anterior and posterior mitral leaflets, which have been thrust forward by the vigorous excursion of the left ventricular wall (LVW).

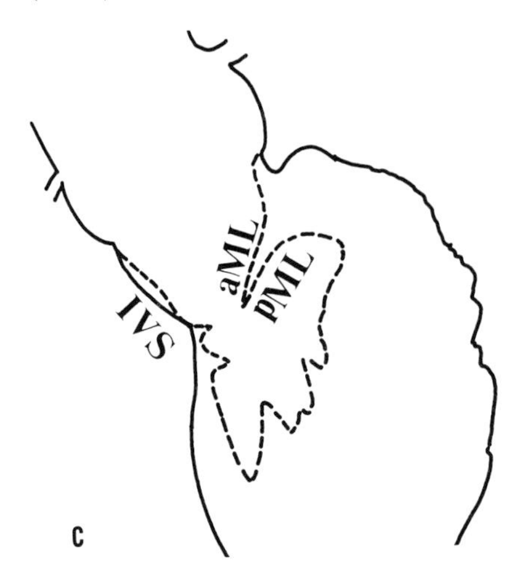

Fig. 21-10C. Left ventriculogram in a patient with HCM with obstruction (gradient = 70 mm Hg). Superimposition of diastole and systole, LAO. The marked degree of systolic emptying can be appreciated by noting the excursion of the left ventricular posterolateral wall (right) between diastole (solid line) and systole (dashed line). The lucent line seen in Fig. 21-10B represents both mitral leaflets. There is relatively little excursion of the interventricular septum (IVS).

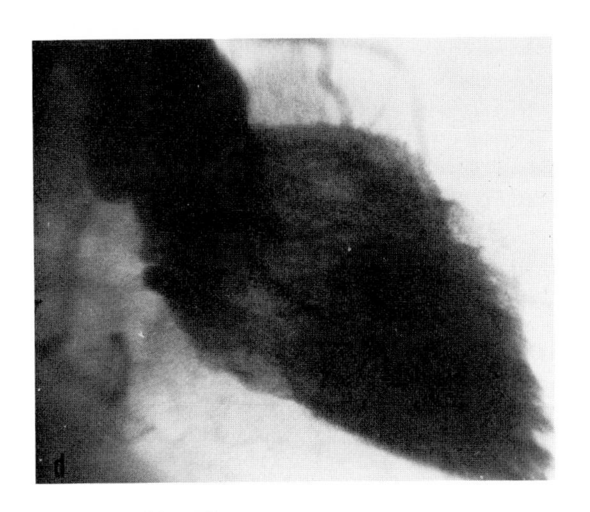

Fig. 21-10D. Left ventriculogram in a patient with HCM with obstruction (gradient = 70 mm Hg). Diastole, RAO. The end-diastolic contour of the left ventricle is not unusual. Frame-by-frame cineangiograms, however, revealed slow filling of this chamber during diastole, with a large increase in volume following atrial systole.

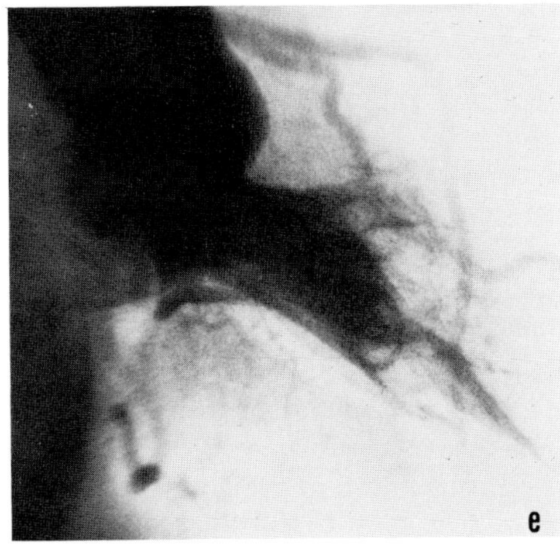

Fig. 21-10E. Left ventriculogram in a patient with HCM with obstruction (gradient = 70 mm Hg). Systole, RAO. The left ventricle has rapidly ejected over 90% of its end-diastolic volume.

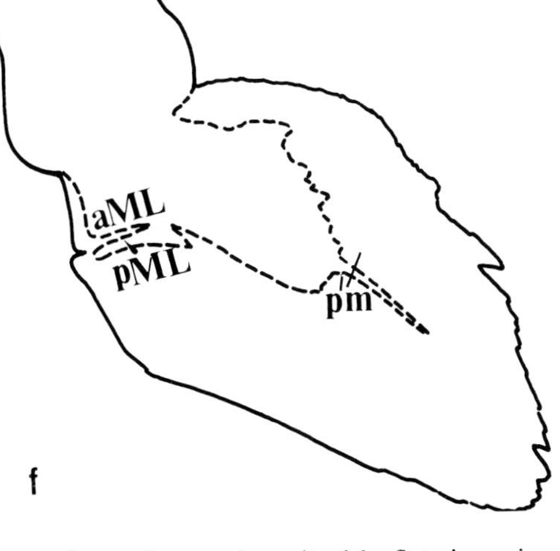

Fig. 21-10F. Left ventriculogram in a patient with HCM with obstruction (gradient = 70 mm Hg). Superimposed diastole and systole, RAO. The remarkable degree of emptying can be appreciated, and the vigorous excursions of the ventricular wall can be appreciated by comparing the diastolic (solid line) and systolic (dashed line) contours of the left ventricle. The papillary muscles (pm) cause indentations in the apical portion of the systolic contour. The "sandwich" made up of the anterior and posterior mitral leaflets is again seen. (The angiocardiograms in Fig. 21-10A–F were kindly provided by Dyrel Faulstick, M.D.)

be assumed to have the same disease (i.e., siblings of patients with typical cases of HCM) may have poorly contracting ventricles suggestive of congestive cardiomyopathy. It is possible that the congestive state is an end stage of HCM, as suggested by Oakley,[6] although one of the present authors has followed such a patient with a stable congestive picture for 11 years who has been symptomatic for 37 years.

The cardiac output is typically normal or high,[3] but decreased values have been reported, particularly in the severely symptomatic stages of HCM.[6,13]

Intracardiac pressures commonly exhibit elevated a waves in both atria in association with increased end-diastolic pressures in both ventricles, reflecting the noncompliance of the hypertrophic ventricular muscle. The y descent of the atrial pressure pulse may be gradual, suggesting inflow obstruction to ventricular filling (Fig. 21-11).[81]

Intracavitary pressure gradients within the right and/or left ventricle, which may be constant or quite varible, have been used as evidence of outflow tract obstruction. The contours of the pressure curves from left ventricle and aorta (Figs. 21-5A, 21-6C, and 21-11) are quite unlike the contours observed in discrete forms of outflow tract stenosis (Fig. 21-5B), in that the left ventricle and aortic pressure in HCM rise together for approximately 100 msec, after which the left ventricular pressure pulse develops a notch and then continues to rise while the pressure in the outflow tract and aorta levels off or falls. The gradient is therefore greater in the last half of systole, while in discrete forms of stenosis the gradient has the greatest magnitude in the first half of systole and is progressively diminished as the aortic pressure gradually rises to a late systolic peak. The upstroke rate (dp/dt) of the aortic or peripheral arterial pressure is rapid,[3] in contrast to discrete forms of stenosis.

A pressure gradient in either ventricle may be present at rest, may appear spontaneously during the course of a hemodynamic study, or may require a provocation to be manifest. Interventions that either decrease ventricular filling, lower arterial resistance, or increase the vigor of ventricular contraction may provoke a pressure gradient that was not initially present or enhance an existing pressure gradient.

Provocation of an interventricular pressure gradient in a patient without other evidence of HCM should not be interpreted as diagnostic of HCM, since nonobstructive gradients have been produced in normal ventricles[26–38] and in patients with other cardiac conditions.[82,83] Furthermore, if extremely unphysiologic stimuli (i.e., Valsalva maneuver during isoproterenol infusion) are required to provoke a pressure gradient in a

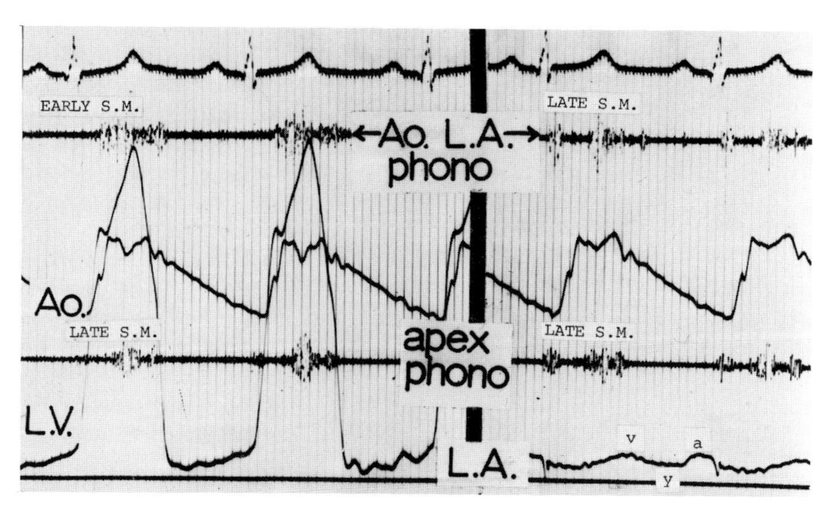

Fig. 21-11. Intracardiac Phonocardiograms in a patient with HCM with obstruction. An intracardiac phonocardiogram is recorded in the aorta (top left) and left atrium (top right) along with pressures in the left heart and an apex phonocardiogram. The pressure contours demonstrate the typical rapid upstroke of the aortic pressure pulse and the midsystolic dip in aortic pressure at the time that the left ventricular pressure continues to rise to a late systolic peak. The aortic phonocardiogram reveals a loud early systolic ejection murmur and soft late systolic vibrations. The atrial phonocardiogram reveals a mid–late-systolic murmur that coincides with the murmur heard at the apex. These recordings would indicate that the mid–late-systolic apical murmur is generated by mitral regurgitation, while the early systolic ejection murmur is due to rapid early systolic ejection of blood to the aorta. Time lines = 0.04 sec.

patient with HCM the diagnosis of obstruction should be made with full knowledge of the fact that hypertrophic, hyperkinetic ventricles are more readily provoked into nonobstructive gradients than are normal ventricles.[35]

The coronary arteries in HCM are usually large and luxuriant,[50] with myriads of left ventricular branches (Fig. 21-12), although the concurrence of coronary atherosclerosis and HCM is well established.[83,84] An interesting finding noted on cine coronary arteriography has been a systolic constriction, particularly in the penetrating septal branches in which these vessels seem to disappear during systole (Fig. 21-12B). It is not known whether this phenomenon is responsible for ischemia, in view of the fact that coronary flow is maximal in diastole.

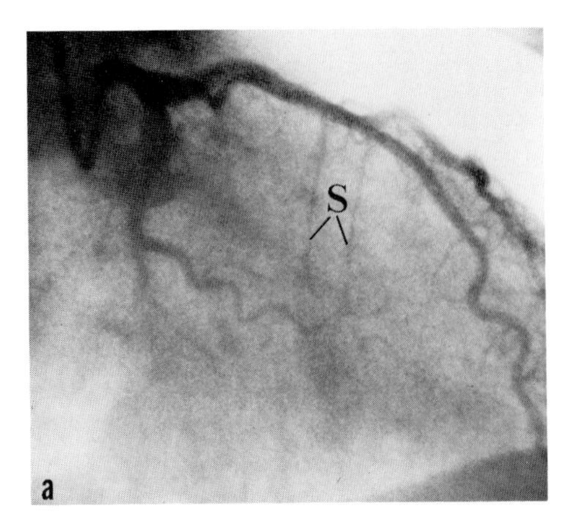

Fig. 21-12A. Left coronary arteriogram in HCM. RAO, left coronary arteriogram, diastole. Selective injection of the left coronary artery reveals large major branches of the coronary artery. Several septal perforating branches (S) are demonstrated coursing at right angles from the anterior descending branch.

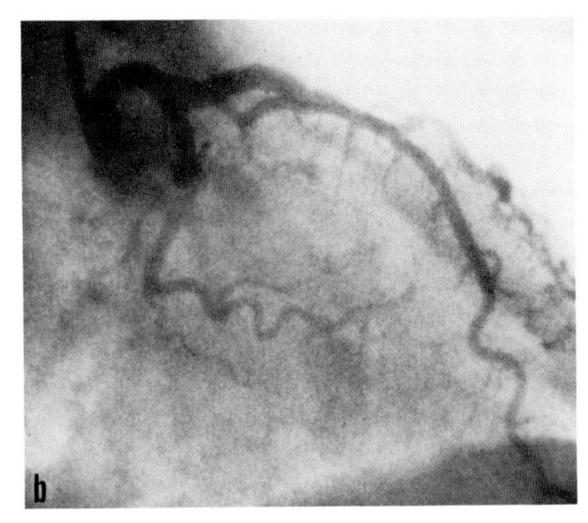

Fig. 21-12B. Left coronary arteriogram in HCM. RAO, left coronary arteriogram systole. During systole the septal branches nearly disappear as a result of the systolic "squeeze" within the septum. These vessels did not fill again until early diastole to mid-diastole of the following cycle. (These angiograms were kindly provided by Luke Chang, M.D.)

DIFFERENTIAL DIAGNOSIS OF HCM

Although many patients with HCM present with clear-cut physical findings of a hyperkinetic, hypertrophic heart with characteristic laboratory findings of HCM, there is a gray zone of overlap with several other conditions with similar findings that can confuse even the astute clinician or laboratory cardiologist. The confusion most often exists between HCM and (1) ischemic heart disease, (2) hyperkinetic heart syndrome, (3) prolapsed mitral leaflet syndrome, (4) mitral regurgitation, (5) mild discrete aortic or subaortic stenosis, (6) hypertensive heart disease, and (7) congestive cardiomyopathy.

Since most patients with HCM present with chest pain or exertional dyspnea (Table 21-4), the murmur may be absent or nonspecific, and the electrocardiogram may exhibit characteristic changes of ischemia or infarction (Fig. 21-4), the initial clinical impression may be ischemic heart disease. Indeed, in an older patient the possibility of HCM and ischemic heart disease coexisting may further cloud the picture. Often only careful echocardiographic, hemodynamic, and angiographic studies can differentiate these two conditions, and indeed even these procedures may lead to further confusion. Systolic anterior motion has been demonstrated in ventricular aneurysm (Fig. 21-6C), and intraventricular pressure gradients have been recorded in patients with coronary artery disease.[83] In these overlap cases we rely heavily on the echocardiographic demonstration of ASH and the angiographic appearance of the ventricle—thick-walled, irregular, hypercontractile ventricle with septal bulge with or without the lucent line (vide infra) of abnormal mitral valve motion—to confirm that HCM is actually present.[43]

The hyperkinetic heart syndrome may simulate many features of HCM: brisk pulses, systolic murmur, active precordium, electrocardiographic evidence of LVH, and intraventricular pressure gradients.[85] The pressure gradients are probably nonobstructive (vide infra). The absence of echocardiographic features of HCM in the patient or his immediate family can help to differentiate this syndrome from HCM.

The murmur of the prolapsed mitral leaflet syndrome (Fig. 21-7B) may sound like HCM and respond to pharmacologic stimuli in a manner indistinguishable from HCM. Patients with the prolapsed mitral leaflet syndrome may also have chest pain and syncope. The presence of a mobile midsystolic click and the echocardiographic[86,87] and angiocardiographic demonstration of billowing of the mitral leaflet or leaflets[88,89] establish the diagnosis of the prolapsed mitral leaflet syndrome.

Although most cases of mitral regurgitation can be clearly differentiated from patients with HCM,[53] it is not uncommon in our experience to discover that HCM is present during cardiac catheterization of a patient previously thought to have rheumatic mitral regurgitation or mitral annular calcification.[90,91] The presence of an S_4, the late systolic timing of the murmur, and augmentation of the murmur with Valsalva maneuver will usually permit correct identification of HCM. In contrast, mitral regurgitation usually has no S_4, a prominent S_3, and a holosystolic murmur that diminishes with the Valsalva maneuver.

Mild discrete aortic or subaortic stenosis may closely resemble many features of HCM, particularly in the young patient in whom the pulse upstroke is not retarded. It is also possible that an actual overlap may exist between discrete lesions of the outflow tract and HCM,[12,59] since intraventricular gradients may be found after surgical relief of

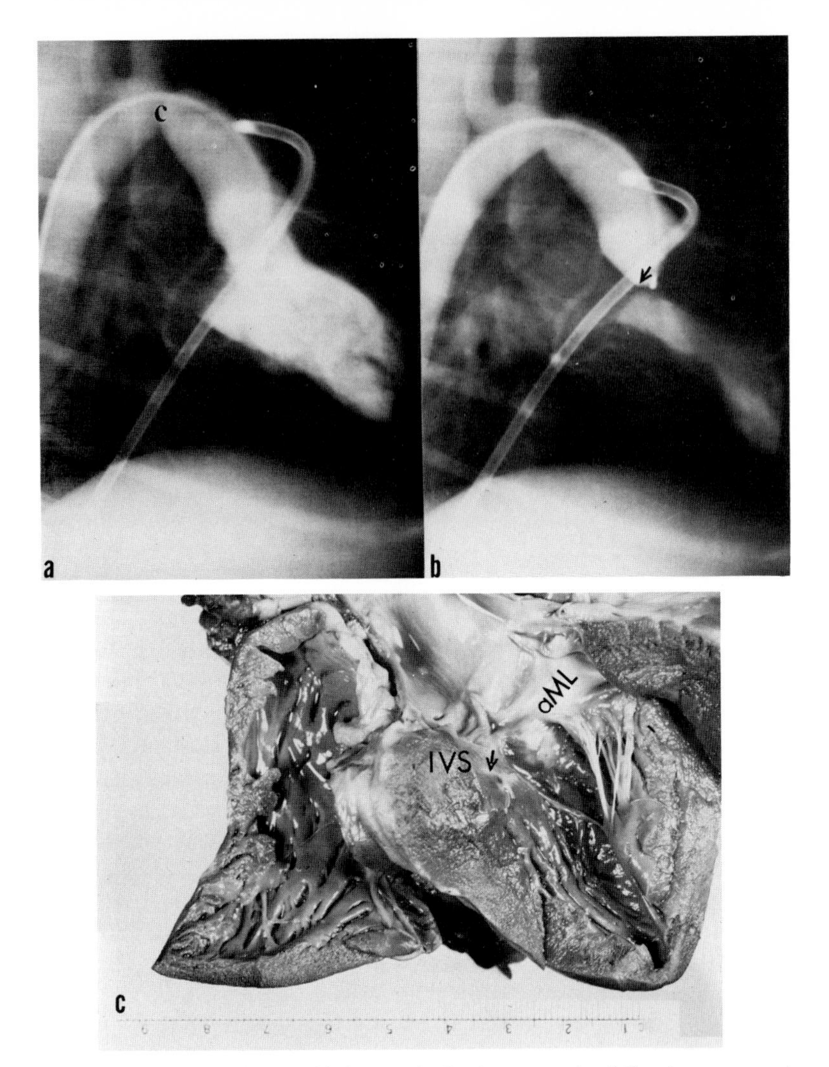

Fig. 21-13A-B. Experimental left ventricular hypertrophy following coarctation of the ascending aorta. (A) Left ventriculogram, RAO projection, diastole. The area of coarction can be seen, and the ventricle has a normal contour. (B) Left ventriculogram, RAO, systole (after isoproterenol infusion). The left ventricle has emptied into a long, thin slitlike chamber. There is a lucent line across the outflow tract (arrow) that appears to result from both the posterior and anterior mitral leaflets. This angiographic appearance is similar to that seen in HCM with obstruction. This animal developed a large pressure gradient after isoproterenol infusion.

Fig. 21-13C. Experimental left ventricular hypertrophy following coarctation of the ascending aorta. Autopsy specimen, left ventricle. There is marked hypertrophy of the left ventricle with asymmetrical septal hypertrophy evident. The interventricular septum (IVS) is 1.8 cm thick, and there is a white plaque (arrow) on the interventricular septum opposite the anterior mitral leaflet (aML). (These figures were kindly provided by Alexis F. Hartmann, Jr., M.D.)

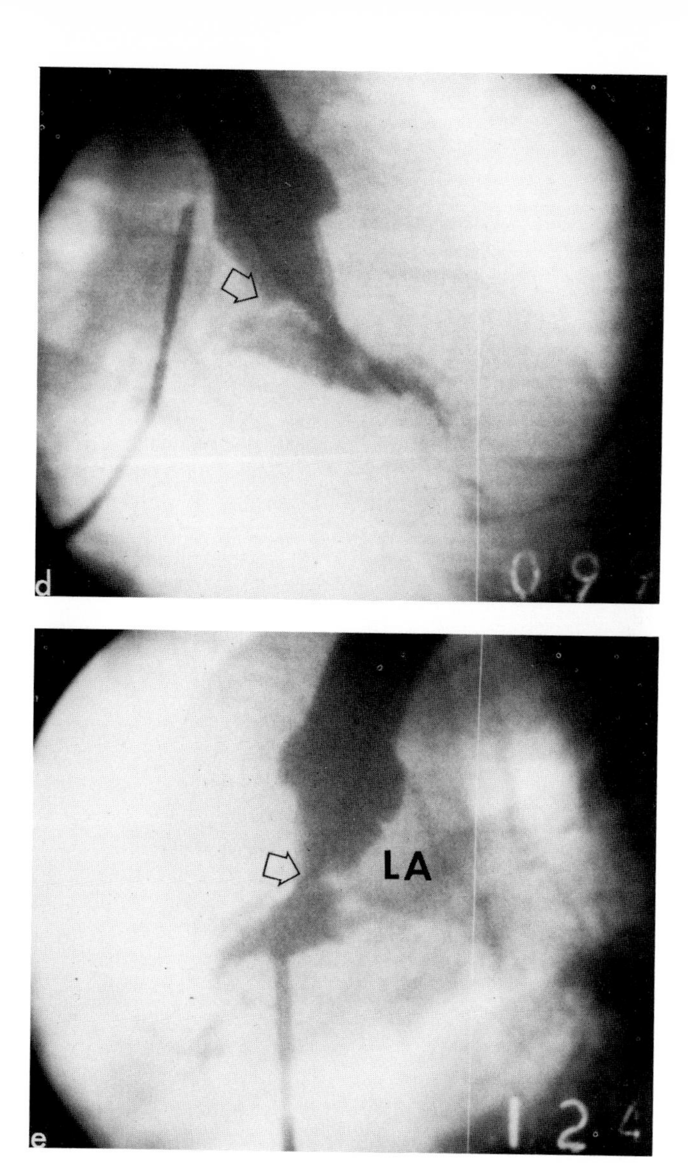

Fig. 21-13 D–E. Experimental hypertrophy following chronic sub-hypertensive norepinephrine infusion. Left ventricular hypertrophy has been induced by 14 weeks of intravenous norepinephrine of 0.67 to 2.0 mgm./24 hours. A pressure gradient of 136 mm Hg was induced by nitroglycerine infusion. (D) Left ventriculogram, RAO projection, systole. A lucent line (arrow), formed by the posterior mitral leaflet from below and the anterior mitral leaflet from above, can be seen across the outflow tract. (E) Left ventriculogram, LAO, systole. The lucent line (arrow) and mitral regurgitation into the left atrium (LA) are demonstrated. The angiographic appearance resembles that seen in hypertrophic cardiomyopathy with obstruction (see Fig. 21-10).

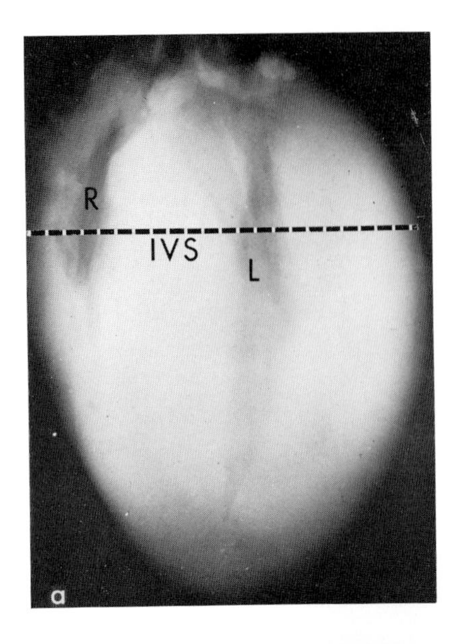

Fig. 21-14A. Concentric hypertrophy in HCM. Soft-tissue x-ray of heart removed at autopsy. The patient was a 17-year-old male who had the characteristic physical and laboratory findings of HCM, as well as an intraventricular pressure gradient. A soft-tissue radiograph has been obtained revealing the markedly thickened intraventricular septum (IVS) and the small right (R) and left (L) ventricular cavities. The dotted line indicates the plane of the section shown in Fig. 21-14B.

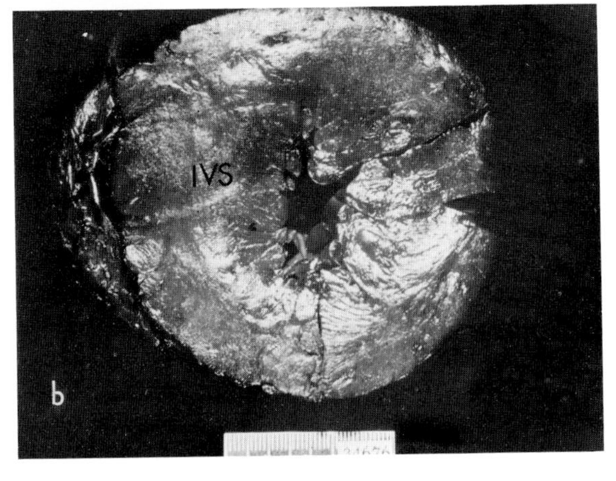

Fig. 21-14B. Concentric hypertrophy in HCM. Cross section of ventricles. The muscle mass is extremely thick, with a small stellate left ventricular cavity (center) and a slitlike right ventricular cavity (left).

these discrete lesions.[10,69,92] It has not been established if these postoperative gradients are due to obstruction. It is possible to produce intraventricular pressure gradients in animals after relief of experimental coarctation,[37,93,94] and these gradients are thought by some to be on a nonobstructive basis.[37] The production of abnormal mitral valve motion in experimental-coarctation animals after isoproterenol stimulus (Fig. 21-13B) provides an additional area of potential confusion and leads to speculation that perhaps any means of developing a hypertrophic-hyperkinetic ventricle could potentially lead to a condition closely simulating HCM with obstruction.

Patients with hypertensive heart disease may resemble patients with HCM in a number of details. Both may have prominent S_4 gallops, forceful apical impulses, and ECG evidence of left ventricular hypertrophy. To further compound the confusion, both conditions may coexist; and indeed Brock[8] first postulated that hypertension may be causative of HCM. There has been no subsequent substantiation of this postulate. Echocardiograms can aid in differentiation, although it might be difficult to categorize a case of apparent HCM with concentric hypertrophy (Fig. 21-14) by echocardiography alone.

The end state of HCM may closely resemble congestive cardiomyopathy (CCM), as demonstrated by the grandmother of the patient described at the beginning of this chapter (Table 21-1). Oakley has described the end-stage ventricle in HCM as "bulky, incompliant and fibrotic,"[6] It would be anticipated that echocardiography might clearly differentiate CCM[95] from HCM by the presence or absence of ASH.

CONTROVERSIES IN HYPERTROPHIC CARDIOMYOPATHY

Although 10 of the 15 commonly used names (Table 21-5) to describe the condition under discussion contain reference to stenosis or obstruction, it is not clear that obstruction plays a major role in the

Table 21-5
Various Names of HCM

1957	Functional obstruction of the left ventricle[8]
1958	Asymmetrical hypertrophy of the heart[7]
1958	Pseudoaortic stenosis[20]
1959	Functional aortic stenosis[130]
1960	Familial muscular subaortic stenosis[2]
1960	Idiopathic hypertrophic subaortic stenosis[131]
1960	Obstructive cardiomyopathy[132]
1961	Diffuse subvalvular aortic stenosis[133]
1961	Hereditary cardiovascular dysplasia[21]
1961	Functional subaortic stenosis[134]
1962	Muscular subaortic stenosis[135]
1964	Subaortic hypertrophic stenosis[136]
1965	Hypertrophic hyperkinetic cardiomyopathy[35]
1970	Hypertrophic cardiomyopathy[1]
1973	Asymmetric septal hypertrophy[5]

symptomatology or mortality associated with HCM. The National Institutes of Health study of the natural history of HCM indicated that patients with large pressure gradients are less likely to die of their disease than those with pressure gradients of less than 30 mm Hg.[13] Similarly, the Hammersmith group has found that those patients without gradients have poorer prognoses than those with gradients,[6] and several studies have indicated no correlation between pressure gradients and symptomatology or mortality.[3,96] It is therefore important to reexamine the evidence that there is indeed a significant obstruction or impediment to outflow of the ventricle in patients with HCM.

EVIDENCE FOR THE PRESENCE OF SIGNIFICANT OBSTRUCTION IN HCM

Clinical evidence for the presence of significant outflow tract obstruction is the similarity in symptoms (angina, syncope, and failure) between HCM and aortic stenosis and the presence of a midsystolic or late-systolic crescendo–decrescendo murmur (Figs. 21-6B, 21-6C, 21-7A, and 21-8) that increases in magnitude with interventions that increase the pressure gradient, i.e., the Valsalva maneuver, isoproterenol (Fig. 21-7A), amyl nitrite, and postextrasystolic beats (Fig. 21-5A).

Hemodynamic evidence of obstruction is the intraventricular pressure gradient in the left (Figs. 21-5, 21-6C, and 21-11) and/or right ventricle. The zone of pressure change between high and low pressure is about 2.5 cm below the aortic valve,[3,8] and transseptal catheterization has localized the high-pressure region to that portion of the ventricle just inside the mitral valve (initial or inflow region) extending to the apex.[41,97] The dynamic nature of the obstruction is attested by its variable magnitude, as exemplified by the Brockenbrough-Braunwald phenomenon in which the postextrasystolic beat has not only a larger gradient (as in discrete forms of stenosis) but a failure of the aortic pulse pressure to rise (Fig. 21-5A).[98]

Angiocardiographic evidence of obstruction falls into two categories. The N.I.H. and Toronto groups[41,43,99] have described a lucent line representing apparent systolic apposition of the anterior mitral leaflet and the bulging septum (Fig. 21-10B), while others have described the ventricle divided into two parts caused by septal apposition with the free wall.[48,100–102] The mitral valve site of obstruction appears to be gaining advocates,[44,45,47] particularly since echocardiographic studies reveal comparable findings. The Toronto group has described the following sequence of events: rapid ejection followed by anterior deviation of the mitral valve to meet the septum, which in turn causes obstruction as well as mitral regurgitation.[99]

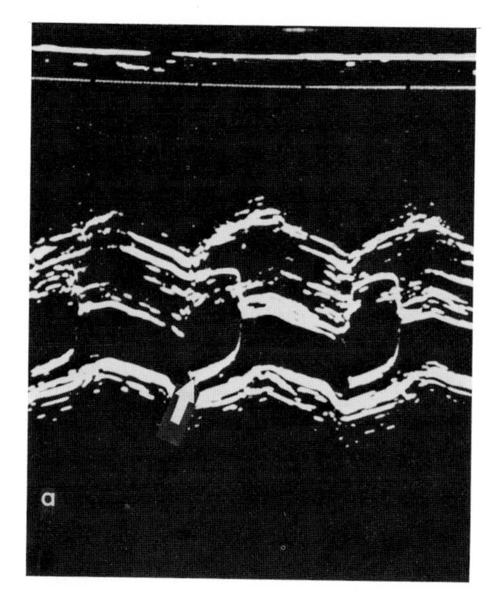

Fig. 21-15A. Echocardiographic demonstration of aortic cusp motion. Normal. The valve cusps swing open in early systole (arrow) and remain widely open throughout the systolic ejection period.

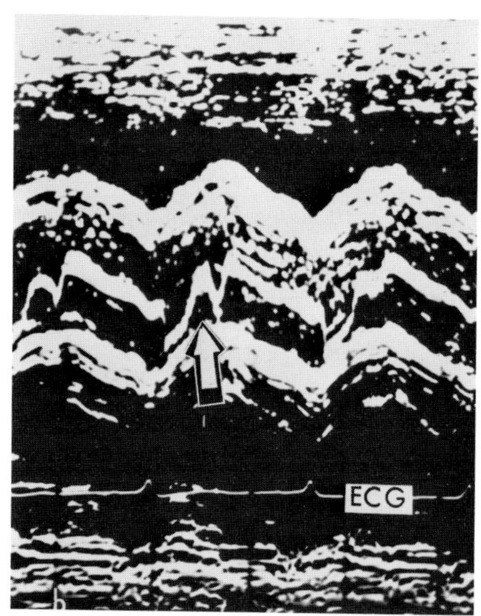

Fig. 21-15B. Echocardiographic demonstration of aortic cusp motion. HCM. Although the aortic valve cusps have opened normally in early systole, there is a transient midsystolic closure motion of the posterior or noncoronary cusp (arrow).

Echocardiographic evidence of obstruction is furnished by systolic anterior movement (SAM) of the anterior mitral leaflet (Figs. 21-1D, 21-6B, and 21-6C) and premature closure of the aortic valve cusps (Fig. 21-15B).[103] In general, SAM has correlated well with pressure gradients, and its absence has correlated well with absence of other evidence of obstruction.[70]

Surgical evidence of obstruction has differed considerably in descriptions by various authors. Although Brock[8,104] failed to find any evidence of obstruction in the beating heart, Morrow and others have vividly described a contraction ring or sphincter approximately 2.5 cm below the aortic valve[3,100] that squeezed tightly around the surgeon's finger.[105,106] Björk[39] furnished the first surgical description of the mitral valve's role in outflow tract obstruction. Relief of symptoms as well as the pressure gradient has followed successful surgery, regardless of whether the operation was designed to incise and/or excise the septal aspect of the contraction ring[100,107–111] or remove the mitral component of the obstruction.[112]

Pathologic evidence of obstruction has also been variously interpreted. The earliest reports of Teare,[7] who first described asymmetrical hypertrophy, and Brock[104] provided no documentation of obstruction at autopsy, and Roberts[113] initially reported that hearts from patients with HCM with and without significant obstruction were anatomically similar. Disproportionate septal thickening and an endocardial plaque opposite the mitral valve were demonstrated in a heart that during life had only a 5 mm Hg intraventricular gradient.[113] Later reports from Roberts and co-workers[61] have indicated that hearts with obstruction have a more normal architectural arrangement of the muscle fibers in the free wall, and therefore the wall is better able to contract vigorously. In the studies by Roberts of the left ventricle in patients who dies of HCM without obstruction the muscle fibers of the free wall, like those of the septum, were short, thick, and disorganized in arrangement, and presumably unable to contract with sufficient vigor to obstruct.

Although the weight of these arguments for significant obstruction in HCM is considerable, there is still ample reason to examine alternative explanations for the various phenomena observed in HCM.

EVIDENCE AGAINST THE PRESENCE OF A SIGNIFICANT OBSTRUCTION IN HCM

Clinical similarities between HCM and aortic stenosis do not prove that there is outflow tract obstruction in HCM. Angina, syncope, failure, and sudden death have been repeatedly seen in patients with HCM without significant pressure gradients,[3,13,96] and can be explained by the thickened, noncompliant ventricles with high energy requirements and resultant excessive myocardial oxygen consumption. The characteristic murmur, although of ejection configuration, has been recorded in patients without pressure gradients.[3,54,114] Intracardiac phonocardiograms have revealed an early systolic murmur in the aortic root at the time of initial rapid ejection and a mid–late-systolic murmur of maximal intensity in the left atrium (Fig. 21-11). Surgical reports have indicated that there is a thrill over the left atrium, but not over the outflow tract and aorta.[106] It is therefore possible that the apical murmur is generated by the mid–late-systolic mitral regurgitation that is seen angiographically in HCM.[99] Indeed, there are many similarities between the murmur of HCM and the murmur of the prolapsed mitral leaflet syndrome, including the response to postural and pharmacologic interventions (Fig. 21-7). In both conditions there is a ventriculo-valvular disproportion; in HCM the end-systolic ventricle is too small for the valve, while in the prolapsed mitral leaflet syndrome the valve is too large for the end-systolic ventricle.

Hemodynamic evidence for obstruction—the pressure gradient—furnished Brock's only evidence that an obstruction was present, since he was unable to find a constricted area in the beating heart or at autopsy.[8,104] However, there are a number of observations dating back to the earliest intracardiac pressure recordings that establish the presence of pressure gradients in normal or unobstructed ventricles.[26–34,36,37] Expermiental studies done in our laboratory with dogs with left ventricular hypertrophy induced by surgical coarctation or long-term norepinephrine infusions have reproduced many of the pressure phenomena observed in HCM with obstruction, including an intracavitary pressure gradient and the Brockenbrough-Braunwald phenomenon (Fig. 21-16), in an unobstructed ventricle. It is postulated that the experimental hypertrophic-hyperkinetic ventricle produces the pressure gradient by cavity obliteration, a vigorous systolic contraction resulting in rapid emptying of the ventricle.[35,36] The failure of the postextrasystolic aortic pressure to rise (Brockenbrough-Braunwald phenomenon) can be explained by either restricted filling of the hypertrophic ventricle, increased mitral regurgitation, decreased aortic impedance (due to a longer diastolic runoff period), or a combination of these factors.

Cavity obliteration results in the generation of a high systolic pressure within portions of the ventricle that are contracting vigorously, and this dynamic pressure rise is not freely transmitted to adjacent portions of the ventricle in which little or no contraction (or possibly expansion) is occurring (Fig. 21-17). The outflow tract or aortic vestibule of the left ventricle[115] is an essentially noncontractile extension of the aorta into the left ventricle, bounded posterolaterally by the anterior mitral leaflet and anteromedially by the muscular and membranous interventricular septum and the atrioventricular septum. This aortic vestibule represents a dead space, while the remainder of the ventricle is capable of virtually complete emptying Figs. 21-10 and 21-17 in little more than half of the

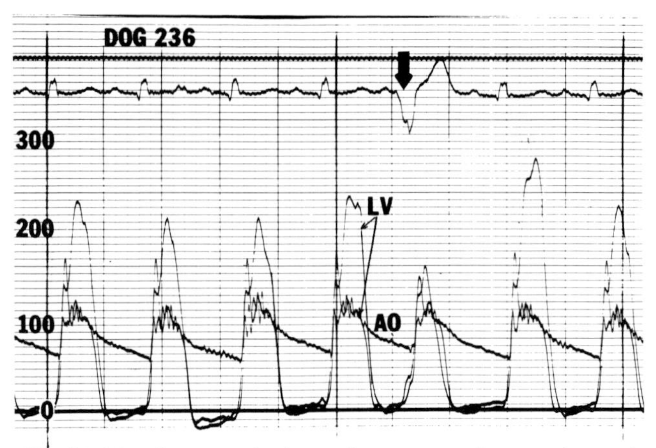

Fig. 21-16. Postectopic beat phenomenon in experimental canine hypertrophy. Experimental hypertrophy has been provoked by a 10-week subhypertensive level of infusion of norepinephrine. The animal had a resting pressure gradient between the left ventricle and aorta of 60 mm Hg, and this gradient could be increased by Isuprel or glucagon infusions. A premature ventricular contraction is induced (arrow), and although there is a higher left ventricular pressure on the postectopic beat, the aortic pulse pressure fails to rise. This finding is said to be characteristic of HCM with obstruction.

systolic ejection period in a hyperkinetic ventricle. In areas that are obliterated early in systole, continued isometric contraction can cause a sustained late systolic pressure rise.

Thus a rapidly emptying ventricle can produce a pressure gradient throughout systole in the absence of any obstruction, and this phenomenon can be demonstrated in hydrodynamic models of the ventricle that are incapable of obstruction.[36] The mechanism of cavity obliteration should be distinguished from catheter entrapment[42,97,116] (Fig. 21-18), in which an artifically high pressure can be recorded by a catheter if it is embedded into the wall of the ventricle and subjected to isometric contraction throughout most of systole. In the latter phenomenon, unlike cavity obliteration, blood cannot be withdrawn or is not ejected through the catheter during systole, and the pressure wave recorded by the entrapped catheter may persist after the intracavitary pressure has declined in late systole.[37,97]

There are many angiographic similarities between the ventricle in HCM with obstruction and the experimental preparation in which hypertrophy and hyperkinesis are induced mechanically or pharmaco-

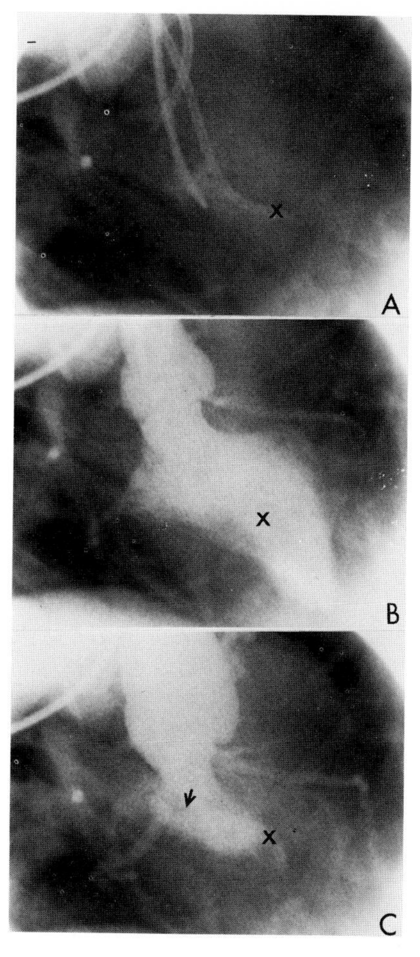

Fig. 21-17. Cavity obliteration. (A) Position of catheters for injection and recording of pressure in the left ventricle and aorta. A multiple-side-hole catheter (×) has been placed in the body of the left ventricular cavity for pressure recording while a second catheter immediately behind is used for an injection. A third catheter (top center) is used to record aortic pressure. A 110-mm pressure gradient was recorded at the time of this angiographic frame between the left ventricular catheter (×) and the aortic catheter. The dog has undergone a coarctation of the ascending aorta 9 months earlier. (B) Left ventriculogram, RAO, diastole. The diastolic contour of the left ventricle is normal, and the position of the catheter is indicated by the ×. (C) Left ventriculogram RAO, systole. Ninety-five percent of the left ventricular volume has been rapidly ejected into the aorta, and the aorta has expanded markedly. A small lucent line (arrow) indicates anterior and superior displacement of the mitral apparatus caused by the vigorous excursions of the ventricle. The × indicates that the left ventricular pressure-recording catheter is not embedded in the myocardial wall. These angiograms demonstrate the phenomenon of cavity obliteration in which the ventricular chamber empties rapidly into a low-resistance bed (low-compliance aorta), causing an intraventricular pressure gradient and abnormal mitral valve motion. There is no catheter entrapment.

logically. In both the ventricle contracts vigorously, ejects blood rapidly, and wrings out the ventricle to a seemingly minimal end-systolic volume (Figs. 21-10 and 21-17). Yet in HCM the ventricle is considered by most investigators to be obstructed, while in the experimental preparation there is considerable evidence that the pressure gradient is not caused by obstruction.[36,37] The angiographic mitral valve abnormality (lucent line)

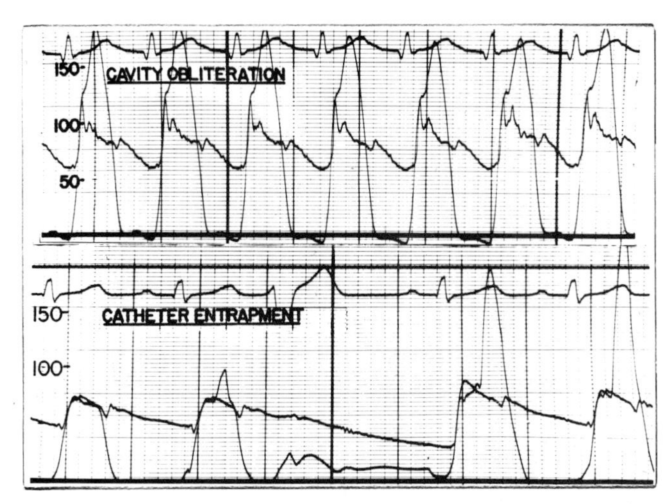

Fig. 21-18. Cavity obliteration versus catheter entrapment. The upper panel demonstrates a pressure gradient obtained in a dog following an 8-week norepinephrine infusion. The left ventricular and aortic pressure contours are typical of HCM with obstruction, and yet this dog was demonstrated to have cavity obliteration (as seen in Fig. 21-17) as well as displacement of the anterior mitral leaflet. In the lower panel the left ventricular catheter has been forcibly advanced into the left ventricular wall causing a premature beat (bottom center) and a bizarre left ventricular pressure contour with a late systolic rise and a decline in left ventricular pressure following the aortic dicrotic notch. This pressure gradient is an artifact caused by entrapment of the catheter tip within the wall of the ventricle.

presumably responsible for obstruction in HCM can be seen in some angiographic studies of experimentally hypertrophic-hyperkinetic dogs (Figs. 21-13 and 21-17)[94] The significance of the abnormal movement of the mitral valve can be interpreted several ways (Fig. 21-19): (1) it is always an obstructive mechanism (Fig. 21-19C); (2) it results from the rapid ejection and wringing out of the ventricle (Fig. 21-19D); (3) in some instances it can obstruct, while in others it is a result of the hyperkinetic contraction of the ventricle and is not necessarily obstructive. We currently favor the third possibility.

Impediment to outflow can be considered a valid working definition of obstruction; yet the majority of studied done in patients with HCM fail to demonstrate that the ventricle is inhibited in its emptying and in fact ejects blood with more velocity than normal[58,62,63] and empties the

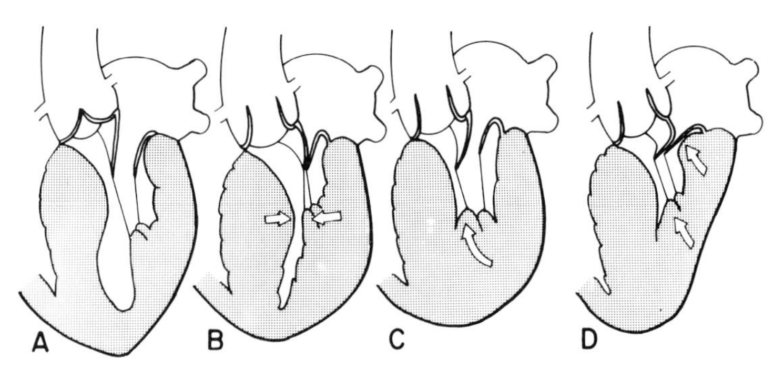

Fig. 21-19. Changing concepts of the site of obstruction in HCM. (A) Diastolic cone. A misinterpretation of angiocardiograms published in the earlier literature[3,128] on HCM suggested that there was an hour-glass narrowing in the middle of the left ventricle, which was in fact caused by the diastolic apposition of the anterior mitral leaflet and the thickened interventricular septum.[35] (B) Contraction ring. A number of reports from the earlier literature[100–102] as well as a recent monograph by Meerschwam[48] indicate that the obstruction is caused by a contraction ring in the middle of the ventricle. Angiographic and diagrammatic demonstrations of this indicate that the obstruction is at the level of the papillary muscles. (C) Obstruction caused by systolic anterior movement. Most authors now support the concept of mitral valvulogenic obstruction of the outflow tract of the left ventricle, the cause of which is thought to be displacement of the anteromedial papillary muscle,[44–47] which permits the leading edge of the anterior leaflet to move across the outflow tract and obstruct the left ventricle. The Venturi effect is also thought to contribute to the tethering of this leaflet into the outflow tract during rapid ejection of blood from the left ventricle.[46,47] (D) Cavity obliteration. This concept of the genesis of the pressure gradient and angiographic features holds that the left ventricle ejects blood rapidly, distorting the mitral valve apparatus but not necessarily obstructing the outflow tract. The posterior left ventricular wall contracts vigorously (upper arrow) forcing the posterior and anterior leaflet "sandwich" forward into the nearly empty ventricle producing the angiographic lucent line. Thus the lucent line and SAM may not necessarily indicate obstruction.

ventricle more completely than normal.[58,59] It has been argued that the rapid and complete emptying is not due to forward ejection, but rather regurgitation into the left atrium after the offending mitral leaflet has swung across the outflow tract and obstructed aortic outflow.[38] This point is partially substantiated by the demonstration that mitral regurgitation is increased by interventions that increase the outflow tract gradient.[49,50] Despite this mitral regurgitation the forward stroke volume is often normal in HCM and is ejected primarily in early systole,[3,62,63] and the midsystolic ventricular size is small.[58] Whether the small midsys-

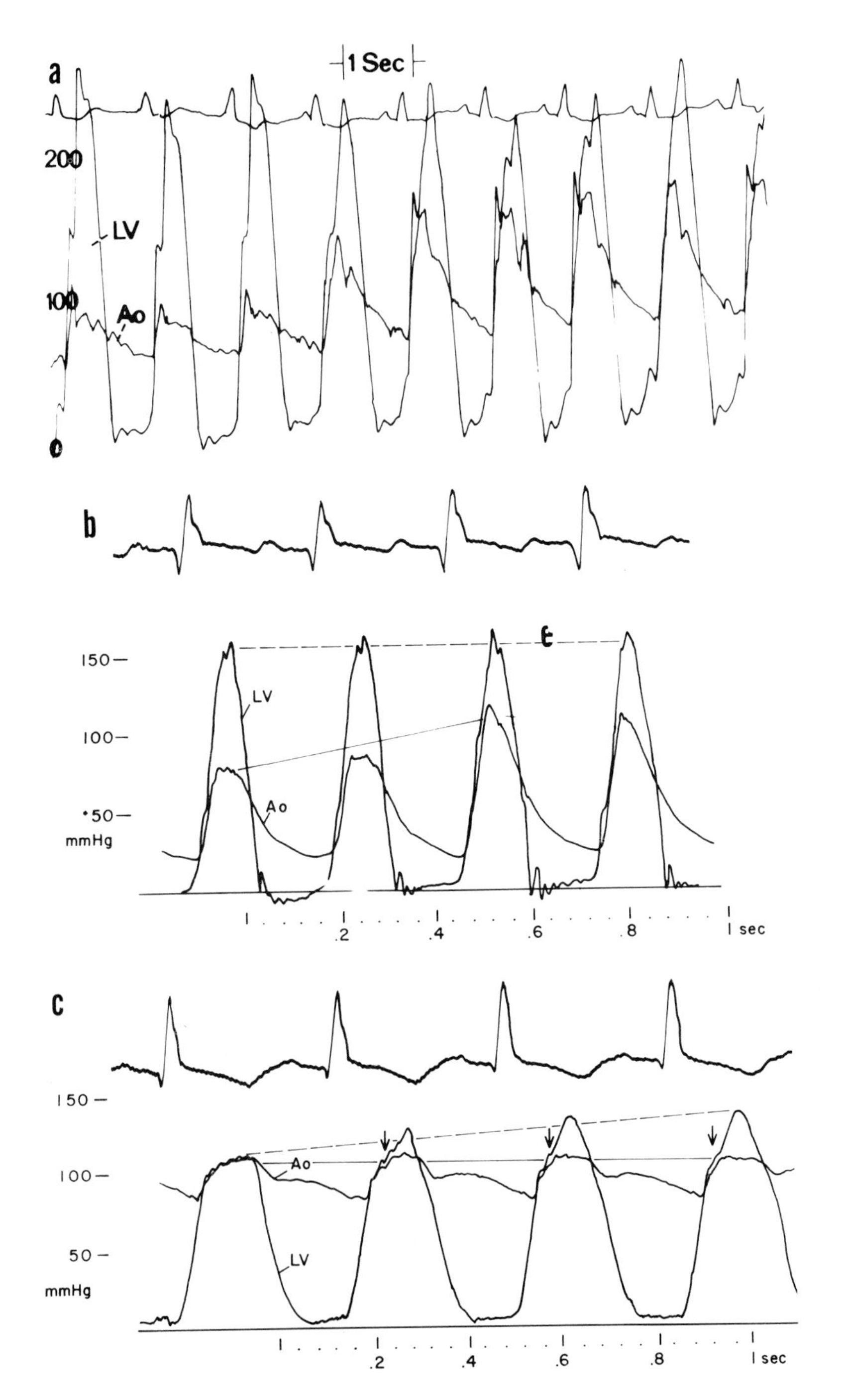
a
1 Sec
200
LV
100
Ao
0
b
150—
100—
50—
mmHg
LV
Ao
.2
.4
.6
.8
I sec
c
150 —
100—
50 —
mmHg
Ao
LV
.2
.4
.6
.8
I sec

tolic volume of blood in the ventricle is inhibited from egress through the outflow tract in late systole or is squeezed through the mitral valve into the atrium because the small ventricular chamber has distorted the mitral valve remains to be settled.

An intriguing phenomenon that is difficult to explain on the basis of obstruction is the wide variation in pressure gradient that often occurs spontaneously with respirations, with release of the Valsalva maneuver, or with certain arrhythmias in patients with HCM[3,117] and experimental animals with pressure gradients (Fig. 21-20). Under these circumstances the developed left ventricular pressure usually does not change as the gradient is diminished, but rather the aortic pressure rises or falls. On the other hand, in an experimental preparation in which variable mechanical obstruction is induced by inflating a balloon in the left ventricular outflow tract (Fig. 21-20C) the left ventricular pressure rises as the obstruction is increased and falls as the obstruction is decreased. Thus it appears in HCM and the experimental animal with nonobstruc-

---

Fig. 21-20A. Variable gradient—A result of variable obstruction? Variation in pressure gradient in a patient with HCM with obstruction during atrioventricular dissociation. During the recording of the left ventricular and aortic pressure, the patient developed a junctional rhythm with A-V dissociation, which then converted to sinus rhythm halfway through the recording. There is some beat-to-beat variation in left ventricular pressure, but ventricular pressure remains more or less constant while the aortic pressure and pulse pressure increase during sinus rhythm to minimize the gradient. If this diminution in pressure gradient were due to diminishing obstruction it would be anticipated that the left ventricular pressure would fall, rather than the aortic pressure rise, as the obstruction is diminished (see Fig. 21-20C). (This figure was kindly supplied by Rex N. MacAlpin, M.D.)

Fig. 21-20B. Spontaneous gradient change with respiration. A marked change in pressure gradient is demonstrated in a dog with experimental left ventricular hypertrophy during inspiration (left) and expiration (right). As in Fig. 21-20A the left ventricular pressure remains constant while the aortic pressure rises. This same respiratory phenomenon has been observed in patients with HCM.[129]

Fig. 21-20C. Increasing gradient due to increasing obstruction. A balloon has been placed in the outflow tract of the left ventricle of a dog and rapidly inflated during systole to simulate a dynamic obstruction. The beginning of inflation is marked with an arrow, and the degree of inflation is increased progressively in the three beats shown on the right. It can be seen that when the degree of obstruction increases the aortic pressure remains more or less constant while the left ventricular pressure rises. This phenomenon, due to controlled variation in the magnitude of obstruction, is markedly different from the phenomenon demonstrated in patients with HCM (Fig. 21-20A) and in hypertrophic dogs (Fig. 21-20B).

tive gradients that the ventricle has a set intracavitary pressure under the circumstances of constant inotropic state and that it transmits more or less of this pressure to the aorta, depending on filling, aortic impedance, etc. However, if the inotropic drive of the ventricle is increased by isoproterenol the left ventricular pressure rises to produce the larger gradient.

Perhaps the most compelling argument that obstruction is not a necessary integral part of the disease process in HCM is the fact that the entire symptom complex (the same massive hypertrophy, the endocardial plaque, and the same clinical findings) can occur in HCM in the absence of a significant pressure gradient. The patient described at the beginning of this chapter exemplifies the severity of symptoms often found in patients with HCM and minimal obstruction.

THE NATURAL HISTORY OF HCM AND ITS THERAPEUTIC IMPLICATIONS

There is general agreement that HCM is a form of cardiomyopathy and that death and disability occur in a significant percentage of patients who are discovered to have the condition. A recent multicenter natural history study of 190 patients with HCM revealed an attrition rate of 3.5% per year, with 26 of the 31 deaths being sudden.[96] The cause or causes of sudden death are unknown, although postexercise asystole has been reported[118] and ventricular fibrillation was observed in one of our patients with HCM following exercise.

We share the view of Oakley and Goodwin, namely that the symptoms of HCM result primarily from poor filling of the thickened diseased ventricle and that outflow tract obstruction does not play a significant role in the majority of patients.[6,119] It is acknowledged that there are abnormalities in contractility of the ventricle—the abrupt, vigorous wringing out of the left ventricle is distinctly abnormal and very probably deleterious. The adverse effects of the contraction abnormalities may in some cases be manifest in true outflow tract obstruction, but in a large number of cases may produce deleterious effects by increasing the myocardial oxygen consumption and impeding the subendocardial perfusion of the left ventricular muscle. It is quite likely that the high pressure and the increased inotropic state of the left ventricle not only cause excess myocardial oxygen consumption but render the heart ischemic and irritable. Therefore we favor the use of propranolol in large doses (0.5 to 3 mg/kg/day) in an attempt to decrease the abnormal contractility.

Although beta adrenergic receptor blockade has not been uniformly successful in either symptomatic or hemodynamic improvement, there is evidence that it does increase the compliance of the left ventricle.[120]

Digitalis glycosides are by no means contraindicated in the treatment of HCM, although hemodynamic studies have indicated that the pressure gradient may be increased,[3,121] and some patients have worsened on this therapy.[3] Digitalis preparations can be quite effective in treating congestive failure or controlling the ventricular response to atrial fibrillation. It is not recommended as primary therapy in HCM, as it is in valvular heart disease, but it can be a useful adjunct.

Atrial flutter, fibrillation, and other tachyarrhythmias are clearly deleterious, as in the patient described at the beginning of this chapter, because of the increased oxygen consumption and decreased filling time of the left ventricle; and antiarrhythmic agents such as quinidine, procaine amide, and diphenylhydantoin are useful in patients with recurrent arrhythmias. Oakley and her co-workers have reported a significant incidence of thromboembolism following the onset of chronic atrial fibrillation, and they recommend that anticoagulation be employed in these instances.[6]

Since infective endocarditis has occurred in HCM,[6] precautions must be taken during dental procedures and potential unsterile surgery to prevent or minimize bacteremia as in patients with valvular heart disease.

Overzealous use of diuretics or fluid restriction are to be avoided. We also recommend abstention from competitive athletics of any type, since syncope and death may follow exertion,[7] although sudden death has occurred at rest or with normal activity as well.

The role of surgery in HCM is not universally agreed upon. Sudden death has been observed after apparently successful surgery, and the combined surgical and late postoperative mortality in the multicenter study did not demonstrate an advantage in life preservation by surgical as compared to medical therapy.[96]

In evaluating the effects of surgery it is difficult to ignore the symptomatic improvement reported by many centers[111,112,122–124] and to speculate on the possible reasons for amelioration. The operations performed by most centers combine various apsects of outflow tract (septal) incision and muscle resection, and the rationale is based on earlier concepts of the nature of the outflow tract obstruction, namely that there is a contraction ring[3,100,105] that closes off the outflow tract. Despite the fact that most medical investigators have changed their concept of this obstruction from the original contraction ring to a faulty mitral valve movement, the operation has not changed appreciably, and for all practical purposes most surgeons are still operating on a contraction ring. On the other hand, Cooley, who has previously championed a variety of approaches to the contraction ring,[123,125] has acknowledged the mitral valvulogenic theory of the obstruction and has recently recommended valve replacement without outflow tract resection.[112] This latter approach is not without its critics.[126]

Oakley has questioned the value of surgery, since the relief of outflow tract obstruction may only simulate the natural progression of the disease, which in her experience progresses to a point of no obstruction and a grave prognosis.[6] It is generally acknowledged that surgery does indeed relieve the gradient, and several mechanisms for this can be proposed: (1) interruption of the contraction ring,[3,100,105] (2) prevention of anterolateral[45] or anteromedial[44] malposition of anterior papillary muscle, or (3) myocardial damage resulting in decreased contractility of the left ventricle. Some investigators have reported an abolition of the abnormal movement of the anterior leaflet concomitant with relief of the gradient,[44,45] while others continue to demonstrate SAM after septal myomectomy through a combined left-ventricular–aortic approach. It has also been reported that an anterior myocardial infarction may result in abolition of the intraventricular pressure gradient.[127]

CONCLUSION

Brock's original description of HCM in 1957 sparked a controversy that has persisted through more than 1000 articles, dozens of scientific meetings, and countless discussions. Brock stated in 1957 that "the functional nature of the obstruction is concealed in death. Indeed it is concealed in life unless we are able to demonstrate its presence by means of a pressure withdrawal record."[8] The mysterious nature of the obstruction was illustrated by Brock's inability to find an obstruction or "any sense of resistance" to the passage of bougies or fully expanded dilators through the valvar and subvalvar regions of the left ventricle in the actively beating heart, despite his repeated demonstration of pressure gradients before and after these manipulations.[8,104]

In the ensuing decade and a half the intraventricular pressure gradient has become the focal point of many hundreds of articles that have sought to define the cause of this elusive obstruction, its location (Fig. 21-19), and the means for relieving it. No small amount of confusion has been generated by the nomenclature applied to patients with HCM (Table 21-5), since most of the names employed emphasize the presence of stenosis or obstruction, while a number of individual patients thus classified have no evidence of impediment to ventricular outflow. Other patients classified as obstructed only develop intraventricular pressure gradients under experimental provocation, while others undergo spontaneous transformations from an obstructed to a nonobstructed state and vice versa. It is also apparent that pressure gradients can exist or can be provoked in ventricles without evidence of HCM or any other evidence of obstruction to outflow.

Based on our own and others' laboratory observations, ASH, SAM, and the postectopic beat phenomenon, considered by others to be specific hallmarks of HCM, may each be produced by diverse etiologies and experimental provocations: discrete outflow tract stenosis, experimental coarctation, chronic nonhypertensive norepinephrine infusions, and ventricular aneurysm.

Despite these pitfalls there is justification for considering HCM a disease entity. There is a condition that is frequently (possibly always) familial in which there is left ventricular hypertrophy that is usually (but not always) asymmetrically distributed and that may or may not be symptomatic.

It is hoped that this communication will have pointed out some of the areas where controversy and confusion still exist. If there is one thing that is clear in HCM, it is that it is not a simple and straightforward disorder.

ACKNOWLEDGMENTS

The authors wish to thank the following individuals for providing the figures for this chapter: Drs. Luke Chang, Dyrel Faulstick, Alexis F. Hartmann, Jr., Rex N. MacAlpin, Arnold L. Nedelman, Richard L. Popp, Victor J. Ferrans, William C. Roberts, John H. K. Vogel; Michael M. Laks and Gerald Adomian for their assistance in obtaining the electronmicrographs of the dog demonstrated in Fig. 21-2B; Dan Garner, James Beazell, and Bruce Ishimoto for their technical assistance; and Mrs. Susan Goldfine for the preparation of this manuscript.

REFERENCES

1. Goodwin JF: Congestive and hypertrophic cardiomyopathies. A decade of study. Lancet 1:731, 1970
2. Brent LB, Aburano A, Fisher DL, et al: Familial muscular subaortic

stenosis. An unrecognized form of "idiopathic heart disease," with clinical and autopsy observations. Circulation 21:167, 1960
3. Braunwald E, Lambrew CT, Morrow AG, et al: Idiopathic hypertrophic subaortic stenosis. Circulation 30 [Suppl 4] 1964
4. Nasser, WK, Williams JF, Mishkin ME, et al: Familial myocardial disease with and without obstruction to left ventricular outflow. Clinical, hemodynamic, and angiographic findings. Circulation 35:638, 1967
5. Clark CE, Henry WL, Epstein SE: Familial prevalence and genetic transmission of idiopathic hypertrophic subaortic stenosis. N Engl J Med 289:709, 1973
6. Oakley CM: Hypertrophic obstructive cardiomyopathy—Patterns of progression, in Wolstenholme GEW, O'Connor M (eds): Hypertrophic Obstructive Cardiomyopathy. London, J&A Churchill, 1971, p 9
7. Teare D: Asymmetrical hypertrophy of the heart in young adults. Br Heart J 20:1, 1958
8. Brock RC: Functional obstruction of the left ventricle. Guys Hosp Rep 106:221, 1957
9. Parker DP, Kaplan MA, Connolly JE: Coexistent aortic valvular and functional hypertrophic subaortic stenosis. Am J Cardiol 24:307, 1969
10. Davies H: Hypertrophic subaortic stenosis as a complication of fixed obstruction to left ventriclar outflow. Guys Hosp Rep 119:35, 1970
11. Brandenburg RO, Tajik AJ, Giuliani ER, et al: Congenital cardiovascular lesions associated with idiopathic hypertrophic subaortic stenosis (IHSS). Circulation [Suppl 2] 1972, p II–134 (abstract)
12. Roberts WC: Valvular, subvalvular, and supravalvular aortic stenosis: Morphologic features. Cardiovasc Clin 5:97, 1973
13. Frank S, Braunwald E: Idiopathic hypertrophic subaortic stenosis. Clinical analysis of 126 patients with emphasis on the natural history. Circulation 37:759, 1968
14. Harmjanz D, Bottcher D, Schertlein G: Correlations of electrocardiographic pattern, shape of ventricular septum, and isovolumetric relaxation time in irregular hypertrophic cardiomyopathy (obstructive cardiomyopathy). Br Heart J 33:928, 1971
15. Hollister RM, Goodwin JF: The electrocardiogram in cardiomyopathy. Br Heart J 25:357, 1963
16. McCallister BD, Brown AL: A fine structure study of idiopathic hypertrophic subaortic stenosis. Am J Cardiol 19:142, 1967
17. Meesen H, Poche R: Beitrage zur pathologischen Anatomie der fallotschen Fehler und zur idiopathischen Herzhypertrophie. Anglo Ger Med Rev 4:73, 1967
18. Poche R: Ultrastructure of cardiac muscle under pathological conditions. Ann NY Acad Sci 156:34, 1969
19. Ferrans VJ, Morrow AG, Roberts WC: Myocardial ultrastructure in idiopathic hypertrophic subaortic stenosis. Circulation 45:769, 1972
20. Bercu BA, Diettert GA, Danforth WH, et al: Pseudoaortic stenosis produced by ventricular hypertrophy. Am J Med 25:814, 1958

21. Pare JAP, Fraser RG, Pirozynski WJ, et al: Hereditary cardiovascular dysplasia. A form of familial cardiomyopathy. Am J Med 31:37, 1961
22. Braudo M, Wigle ED, Keith JD: A distinctive electrocardiogram in muscular subaortic stenosis due to ventricular septal hypertrophy. Am J Cardiol 14:599, 1964
23. Coyne JJ: New concepts of intramural myocardial conduction in hypertrophic obstructive cardiomyopathy. Br Heart J 30:546, 1968
24. van Dam RT, Roos JP, Durrer D: Electrical activation of ventricles and interventricular septum in hypertrophic obstructive cardiomyopathy. Br Heart J 34:100, 1972
25. Emanuel R: The familial incidence of idiopathic cardiomyopathy, in Wolstenholme GEW, O'Connor M (eds): Hypertrophic Obstructive Cardiomyopathy. London, J&A Churchill, 1971, p 50
26. Piper H: Uber die Aorten—und Kammerdruckkurve. Archiv für Anatomie und Physiologie 1913, p 331
27. Wiggers CJ: Studies on the consecutive phases of the cardiac cycle. Am J Physiol 56:415, 1921
28. Hamilton EF, Brewer G, Brotman I: Pressure pulse contours in the intact animal. Am J Physiol 107:427, 1934
29. Gregg DE, Eckstein RW, Fineberg MH: Pressure pulses and blood pressure values in unanesthesized dogs. Am J Physiol 118:399, 1937
30. Gauer OH: Evidence in circulatory shock of an isometric phase of ventricular contraction following ejection. Fed Proc 9:47, 1950
31. Wiggers CJ: Circulatory Dynamics. Modern Medical Monographs No. 4. New York, Grune & Stratton, 1952, p 55
32. Gauer OH, Henry JP: Negative (−G2) acceleration in relation to arterial oxygen saturation, subendocardial hemorrhage and venous pressure in the forehead. Aerospace Med 35:533, 1964
33. Martin AM Jr, Hackel DB, Spach MS, et al: Cineangiography in hemorrhagic shock. Am Heart J 69:283, 1965
34. Morrow AG, Vasko JS, Henney RP, Brawley RK: Can outflow obstruction be induced with the normal left ventricle? Am J Cardiol 16:540, 1965
35. Criley JM, Lewis KB, White RI Jr, Ross RS: Pressure gradients without obstruction. A new concept of "hypertrophic subaortic stenosis." Circulation 32:881, 1965
36. White RI Jr, Criley JM, Lewis KB, Ross RS: Experimental production of intracavity pressure differences. Possible significance in the interpretation of human hemodynamic studies. Am J Cardiol 19:806, 1967
37. Blundell PE, Bedard P, Baron RH, Wigle ED: Nature of intraventricular pressure differences induced by pharmacological agents in dogs. Am Heart J 74:652, 1967
38. Adelman AG, Wigle ED: Two types of intraventricular pressure difference in the same patient. Left ventricular catheter entrapment and right ventricular outflow tract obstruction. Circulation 38:649, 1968
39. Björk VO, Hultquist G, Lodin H: Subaortic stenosis produced by an abnormally placed anterior mitral leaflet. J Thorac Cardiovasc Surg 41:659, 1961

40. Fix P, Moberg A, Soderberg H, Karnell J: Muscular subvalvular aortic stenosis. Abnormal anterior mitral leaflet possibly the primary factor. Acta Radiol [Diag] (Stockh) 2:177, 1964
41. Ross J Jr, Braunwald E, Gault JH, et al: The mechanism of the intraventricular pressure gradient in idiopathic hypertrophic subaortic stenosis. Circulation 34:558, 1966
42. Wigle ED, Auger P, Marquis Y: Muscular subaortic stenosis: The initial left ventricular inflow tract pressure as evidence of outflow tract obstruction. Can Med Assoc J 95:793, 1966
43. Simon AL, Ross J Jr, Gault JH: Angiographic anatomy of the left ventricle and mitral valve in idiopathic subaortic stenosis. Circulation 36:852, 1967
44. King JF, Reis RL, Bolton MR, et al: Superior-to-inferior septal hypertrophy in IHSS: The fundamental determinant of obstruction. Circulation [Suppl 4] 1973, p IV–6 (abstract)
45. Reis RL, Bolton MR, King JF, et al: Anterior-superior displacement of anterior papillary muscle (APM) producing obstruction and mitral regurgitation in IHSS: Operative relief by posterior-medial realignment of APM following ventricular septal myectomy. Circulation [Suppl 4] 1973, p IV–74 (abstract)
46. Wigle ED, Adelman AG, Silver MD: Pathophysiological considerations in muscular subaortic stenosis, in Wolstenholme GEW, O'Connor M (eds): Hypertrophic Obstructive Cardiomyopathy. London, J&A Churchill, 1971, p 63
47. Henry WL, Clark CE, Epstein SE: Mechanism of outflow obstruction in IHSS. Circulation [Suppl 4] 1973, p IV–177 (abstract)
48. Meerschwam IS: Hypertrophic Obstructive Cardiomyopathy. Baltimore, Williams & Wilkins, 1969
49. Wigle ED, Marquis Y, Auger P: Pharmacodynamics of mitral insufficiency in muscular subaortic stenosis. Can Med Assoc J 97:299, 1967
50. Wigle ED, Adelman AG, Auger P, Marquis Y: Mitral regurgitation in muscular subaortic stenosis. Am J Cardiol 24:698, 1969
51. Soulie P, Joly F, Carlotti J: Les stenoses idiopathiques de la chambre de chasse du ventricule gauche (a propos de 10 observations). Acta Cardiol (Brux) 17:335, 1962
52. Nellen M, Beck W, Vogelpoel L, Schrire V: Auscultatory phenomena in hypertrophic obstructive cardiomyopathy, in Wolstenholme GEW, O'Connor M (eds): Hypertrophic Obstructive Cardiomyopathy. London, J&A Churchill, 1971, p 77
53. Lindgren KM, Epstein SE: Idiopathic hypertrophic subaortic stenosis with and without mitral regurgitation. Phonocardiographic differentiation from rheumatic mitral regurgitation. Br Heart J 34:191, 1972
54. Reeve R: Clues to the bedside diagnosis of mild idiopathic subaortic stenosis. JAMA 195:131, 1966
55. Shabetai R, Davidson S: Asymmetrical hypertrophic cardiomyopathy simulating mitral stenosis. Circulation 45:37, 1972
56. Fontana ME, Pence HL, Leighton RF, Wooley CF: The varying clinical

spectrum of the systolic click–late systolic murmur syndrome. Circulation 41:807, 1970
57. Burchell HB: Hypertrophic obstructive type of cardiomyopathy: Clinical syndrome, in Wolstenholme GEW, O'Connor M (eds): Cardiomyopathies. Boston, Little, Brown, 1964, p 29
58. Wilson WS, Criley JM, Ross RS: Dynamics of left ventricular emptying in hypertrophic subaortic stenosis. Am Heart J 73:4, 1967
59. Grant C, Raphael MJ, Steiner RE, Goodwin JF: Left ventricular volume and hypertrophy in outflow obstruction. Cardiovasc Res 4:346, 1968
60. Abbasi AS, MacAlpin RN, Eber LM, Pearce ML: Echocardiographic diagnosis of idiopathic hypertrophic cardiomyopathy without outflow obstruction. Circulation 46:897, 1972
61. Maron BJ, Ferrans VJ, Henry WL, et al: Differences in distribution of myocardial cellular abnormalities in asymmetric septal hypertrophy. Circulation [Suppl 4] 1973, p IV–6 (abstract)
62. Hernandez RR, Greenfield JC Jr, McCall BW: Pressure–flow studies in hypertrophic subaortic stenosis. J Clin Invest 43:401, 1964
63. McGranahan GM Jr, Martin HA, Murgo JP: Simultaneous aortic flow velocity, left ventricular-aortic pressure gradients and mitral motion in hypertrophic subaortic stenosis. Circulation [Suppl 4] 1973, p IV–194 (abstract)
64. Takahashi M, Landing BH: IHSS associated with severe skeletal muscle disease in an infant. (to be published)
65. Gach JV, Andriange J, Franck G: Hypertrophic obstructive cardiomyopathy and Friedreich's ataxia. Report of a case and review of literature. Am J Cardiol 27:436, 1971
66. Somerville J, Bonham-Carter RE: The heart in lentiginosis. Br Heart J 34:58, 1972
67. Hopkins BE, Robinson JS, Taylor RR: Lentiginosis with hypertrophic cardiomyopathy. Proc Cardiac Soc Aust NZ 1973, p 3 (abstract)
68. Shone JD, Sellers RD, Anderson RC, et al: The developmental complex of "parachute mitral valve," supravalvular ring of the left atrium, subaortic stenosis, and coarctation of the aorta. Am J Cardiol 11:714, 1963
69. Binet JP, Langlois J, Leiva-Semper A, David PH: Ventriculomyotomy in hypertrophies of the left ventricle. J Thorac Cardiovasc Surg 56:469, 1968
70. Shah PM, Gramiak R, Kramer DH: Ultrasound localization of left ventricular outflow obstruction in hypertrophic obstructive cardiomyopathy. Circulation 40:3, 1969
71. Popp RL, Harrison DC: Ultrasound in the diagnosis and evaluation of therapy of idiopathic hypertrophic subaortic stenosis. Circulation 40:905, 1969
72. Henry WL, Clark CE, Epstein SE: Asymmetric septal hypertrophy. Echocardiographic identification of the pathognomonic anatomic abnormality of IHSS. Circulation 47:225, 1973
73. Abbasi AS, MacAlpin RN, Eber LM, Pearce ML: Left ventricular hypertrophy diagnosed by echocardiography. N Engl J Med 289:118, 1973

74. Moreyra E, Klein JJ, Shimada H, Segal BL: Idiopathic hypertrophic subaortic stenosis diagnosed by reflected ultrasound. Am J Cardiol 23:32, 1969
75. Henry WL, Clark CE, Glancy DL, Epstein SE: Echocardiographic measurement of the left ventricular outflow gradient in idiopathic hypertrophic subaortic stenosis. N Engl J Med 288:989, 1973
76. Shah PM, Gramiak R, Adelman AG, Wigle ED: Role of echocardiography in diagnostic and hemodynamic assessment of hypertrophic subaortic stenosis. Circulation 44:891, 1971
77. Abbasi AS: unpublished data
78. Tajik AJ, Gau GT, Ritter DG, Schattenberg TT: Echocardiographic pattern of right ventricular diastolic volume overload in children. Circulation 46:36, 1972
79. Brown OR, Harrison DC, Popp RL: Echocardiographic study of right ventricular hypertension producing asymmetrical septal hypertrophy. Circulation [Suppl 4] 1973, p IV–47 (abstract)
80. Goodman DJ, Harrison DC, Popp RL: Echocardiographic features of primary pulmonary hypertension. Am J Cardiol 33:438, 1974
81. Stewart S, Mason DT, Braunwald E: Impaired rate of left ventricular filling in idiopathic hypertrophic subaortic stenosis and valvular aortic stenosis. Circulation 37:8, 1968
82. Krasnow N, Rolett E, Hood WB Jr, et al: Reversible obstruction of the ventricular outflow tract. Am J Cardiol 11:1, 1963
83. Gulotta SJ, Aronson AL, Ewing K: Anginal syndrome in co-existent coronary disease and hypertrophic sub-aortic stenosis. Circulation [Suppl 2] 1972, p II–162 (abstract)
84. Giuliani ER, Tajik AJ, Frye RL, et al: Idiopathic hypertrophic subaortic stenosis (IHSS) and associated coronary artery disease. Circulation [Suppl 2] 1972, p II–156 (abstract)
85. Gorlin R: The hyperkinetic heart syndrome. JAMA 182:823, 1962
86. Kerber RE, Isaeff DM, Hancock EW: Echocardiographic patterns in patients with the syndrome of systolic click and late systolic murmur. N Engl J Med 28:691, 1971
87. Dillon JC, Haine CL, Chang S, Feigenbaum H: Use of echocardiography in patients with prolapsed mitral valve. Circulation 43:503, 1971
88. Barlow JB, Bosman CK: Aneurysmal protrusion of the posterior leaflet of the mitral valve: An auscultatory-electrocardiographic syndrome. Am Heart J 71:166, 1965
89. Criley JM, Lewis KB, Humphries JO, Ross RS: Prolapse of the mitral valve: Clinical and cine-angiocardiographic findings. Br Heart J 28:488, 1966
90. Tajik AJ, Giuliani ER, Frye RL, et al: Mitral valve and/or annulus calcification associated with idiopathic hypertrophic subaortic stenosis (IHSS). Circulation [Suppl 2] 1972, p II–228 (abstract)
91. Krauss KR, Weisinger B, Glassman E: Mitral annular calcification and subaortic stenosis. Circulation [Suppl 2] p II–178 (abstract)

92. Block PC, Powell WJ Jr, Dinsmore RE, Goldblatt A: Coexistent fixed congenital and idiopathic hypertrophic subaortic stenosis. Am J Cardiol 31:523, 1973
93. McLaughlin JS, Morrow AG, Buckley MJ: The experimental production of hypertrophic subaortic stenosis. J Thorac Cardiovasc Surg 48:695, 1964
94. Burford TH, Hartmann AF Jr, Ferguson TB, Ferrier RW: The production of muscular subaortic stenosis in dogs. J Thorac Cardiovasc Surg 54:639, 1967
95. Abbasi AS, Chahine RA, MacAlpin RN, Kattus AA: Ultrasound in the diagnosis of primary congestive cardiomyopathy. Chest 63:937, 1973
96. Shah PM, Adelman AG, Wigle ED, et al: The natural (and unnatural) course of hypertrophic obstructive cardiomyopathy—A multicenter study. Circulation [Suppl 4] 1973, p IV–5 (abstract)
97. Wigle ED, Marquis Y, Auger P: Muscular subaortic stenosis. Initial left ventricular inflow tract pressure in the assessment of intraventricular pressure differences in man. Circulation 35:1100, 1967
98. Brockenbrough EC, Braunwald E, Morrow AG: A hemodynamic technic for the detection of hypertrophic subaortic stenosis. Circulation 23:189, 1961
99. Adelman AG, McLoughlin MJ, Marquis Y, et al: Left ventricular cineangiographic observations in muscular subaortic stenosis. Am J Cardiol 24:689, 1969
100. Morrow AG, Brockenbrough EC: Surgical treatment of idiopathic hypertrophic subaortic stenosis: Technic and hemodynamic results of subaortic ventriculomyotomy. Ann Surg 154:181, 1961
101. Menges H Jr, Brandenburg RO, Brown AL Jr: The clinical, hemodynamic, and pathologic diagnosis of muscular subvalvular aortic stenosis. Circulation 24:1126, 1961
102. Lillehei CW, Levy MJ: Transatrial exposure for correction of subaortic stenosis. JAMA 186:114, 1963
103. Gramiak R, Shah PM, Kramer DH: Ultrasound cardiography: Contrast studies in anatomy and function. Radiology 92:939, 1969
104. Brock RC: Functional obstruction of the left ventricle. (Acquired aortic subvalvar stenosis). Guys Hosp Rep 108:126, 1959
105. Dobell ARC, Scott HJ: Hypertrophic subaortic stenosis: Evolution of a surgical technique. J Thorac Cardiovasc Surg 47:26, 1964
106. Julian OC, Dye WS, Javid H, et al: Apical left ventriculotomy in subaortic stenosis due to a fibromuscular hypertrophy. Circulation [Suppl 1] 1–44 1965
107. Sousa JEMR, Zerbini EJ, Jatene AD, et al: Transaortic infundibulectomy for hypertrophic subaortic stenosis. Am J Cardiol 15:801, 1965
108. Cleland WP: The results of surgical treatment of hypertrophic obstructive cardiomyopathy, in Wolstenholme GEW, O'Connor M (eds): Cardiomyopathies. Boston, Little, Brown, 1964, p 276
109. Bentall HH: The technique of operation for obstructive cardiomyopathy, in

Wolstenholme GEW, O'Connor M (eds): Cardiomyopathies. Boston, Little, Brown, 1964, p 272
110. Wigle ED, Trimble AS, Adelman AG, Bigelow WG: Surgery in muscular subaortic stenosis. Prog Cardiovasc Dis 11:83, 1968
111. Barratt-Boyes BG, O'Brien KP: Surgical treatment of idiopathic hypertrophic subaortic stenosis using a combined left ventricular-aortic approach, in Wolstenholme GEW, O'Connor M (eds): Hypertrophic Obstructive Cardiomyopathy. London, J&A Churchill, 1971, p 150
112. Cooley DA, Leachman RD, Wukasch DC: Diffuse muscular subaortic stenosis: Surgical treatment. Am J Cardiol 31:1, 1973
113. Morrow AG, Roberts WC, Ross J Jr, et al: Obstruction to left ventricular outflow. Current concepts of management and operative treatment. Ann Intern Med 69:1255, 1968
114. Goodwin JF: Disorders of the outflow tract of the left ventricle. Br Med J 2:461, 1967
115. Walmsley R, Watson H: The outflow tract of the left ventricle. Br Heart J 28:435, 1966
116. Wigle ED, Auger P, Marquis Y: Muscular subaortic stenosis. The direct relation between the intraventricular pressure difference and the left ventricular ejection time. Circulation 36:36, 1967
117. Shah PM, Yipintsoi T, Amarasingham R, Oakley CM: Effects of respiration on the hemodynamics of hypertrophic obstructive cardiomyopathy. Am J Cardiol 15:793, 1965
118. Joseph S, Balcon R, McDonald L: Syncope in hypertrophic obstructive cardiomyopathy due to asystole. Br Heart J 34:974, 1972
119. Goodwin JF: Changing concepts of hypertrophic obstructive cardiomyopathy in the last decade, in Wolstenholme GEW, O'Connor M (eds): Hypertrophic Obstructive Cardiomyopathy. London, J&A Churchill, 1971, p 4
120. Webb-Peploe M, Croxson RS, Oakley CM: Beta adrenergic blockade with practolol in hypertrophic obstructive cardiomyopathy. Br Heart J 33:143, 1971 (abstract)
121. Braunwald E, Brockenbrough EC, Frye RL: Studies on digitalis. V. Comparison of the effects of Ouabain on left ventricular dynamics in valvular aortic stenosis and hypertrophic subaortic stenosis. Circulation 26:166, 1962
122. Adelman AG, Wigle ED, Felderhof CH, et al: Natural history and treatment in muscular subaortic stenosis. Am J Cardiol 33:121, 1974 (abstract)
123. Cooley DA, Bloodwell RD, Hallman GL, et al: Surgical treatment of muscular subaortic stenosis. Results from septectomy in 26 patients. Circulation [Suppl 1]: 1–124 1967
124. Morrow AG, Epstein SE, Rodgers BM, Braunwald E: Idiopathic hypertrophic subaortic stenosis: A current assessment of the results of operative treatment, in Wolstenholme GEW, O'Connor M (eds):

Hypertrophic Obstructive Cardiomyopathy. London, J&A Churchill, 1971, p 140

125. Cooley DA, Beall AC Jr, Hallman GL, Bricker DL: Obstructive lesions of the left ventricular outflow tract: Surgical treatment. Circulation 31:612, 1965
126. Roberts WC; Operative treatment of hypertrophic obstructive cardiomyopathy. The case against mitral valve replacement. Am J Cardiol 32:377, 1973
127. Carter WH,Whalen RE, McIntosh HD: Reversal of hemodynamic and phonocardiographic abnormalities in idiopathic hypertrophic subaortic stenosis. Am J Cardiol 28:722, 1971
128. Braunwald E: Obstructive cardiomyopathy (idiopathic hypertrophic subaortic stenosis), in Hurst JW, Logue BB (eds): The Heart. New York, McGraw-Hill, 1970, Fig. 74–5, p 1221
129. Braunwald E, Lambrew CT, Morrow AG, et al: Idiopathic hypertrophic subaortic stenosis. Circulation [Suppl 4] 1964, Fig 43, p IV–47
130. Morrow AG, Braunwald E: Functional aortic stenosis. A malformation characterized by resistance to left ventricular outflow without anatomic obstruction. Circulation 20:181, 1959
131. Braunwald E, Morrow AG, Cornell WP, et al: Idiopathic hypertrophic subaortic stenosis. Am J Med 29:924, 1960
132. Goodwin JF, Hollman A, Cleland WP, Teare D: Obstructive cardiomyopathy simulating aortic stenosis. Br Heart J 22:403, 1960
133. Kirklin JW, Ellis FH Jr: Surgical relief of diffuse subvalvular aortic stenosis. Circulation 24:739, 1961
134. Brachfeld N, Gorlin R: Functional subaortic stenosis. Ann Intern Med 54:1, 1961
135. Wigle ED, Heimbecker RO, Gunton RW: Idiopathic ventricular septal hypertrophy causing muscular subaortic stenosis. Circulation 26:325, 1962
136. Kittle CF, Reed WA, Crockett JE: Infundibulectomy for subaortic hypertrophic stenosis. Circulation [Suppl] 1964, p 29

Leon Resnekov

22
Congestive Cardiomyopathy

An excellent account of cardiomyopathy will be found in the remarkable St. Cyres lecture given by Wallace Brigden in 1956.[1] In it the author was able to clarify some of the confusion surrounding the subject at that time, and his descriptive term cardiomyopathy is now an integral part of medical terminology. Cardiomyopathy, i.e., disease of the heart muscle, is a term now used for a variety of similar and dissimilar conditions, the dominant clinical manifestation of which is cardiac failure. The literature continues, therefore, to be replete with a formidable array of conditions masked under the generic term cardiomyopathy. Largely due to the endeavors of Goodwin and his colleagues at the Royal Postgraduate School of Medicine in London, a working classification has now been suggested in an attempt to bring order into a very confused situation. Figure 22-1, which is a modification of Goodwin's classification, preserves his major divisions of the disease into cardiomyopathies of the congestive, obstructive, and obliterative types.[2]

Cardiomyopathy can be defined as a disorder of heart muscle of unknown cause; it is a primary disease. What previously was frequently termed secondary cardiomyopathy is better defined in terms of the underlying disease process known to be present. For example, the term ischemic cardiomyopathy usually reserved for patients whose severe left ventricular dysfunction and cardiac failure are due to coronary arterial disease, serves little useful purpose and should be abandoned. These patients are best classified as having coronary heart disease with cardiac failure.

In this chapter an attempt will be made to describe the important presenting features, diagnosis, pathophysiology, prognosis, and treatment of congestive cardiomyopathy.

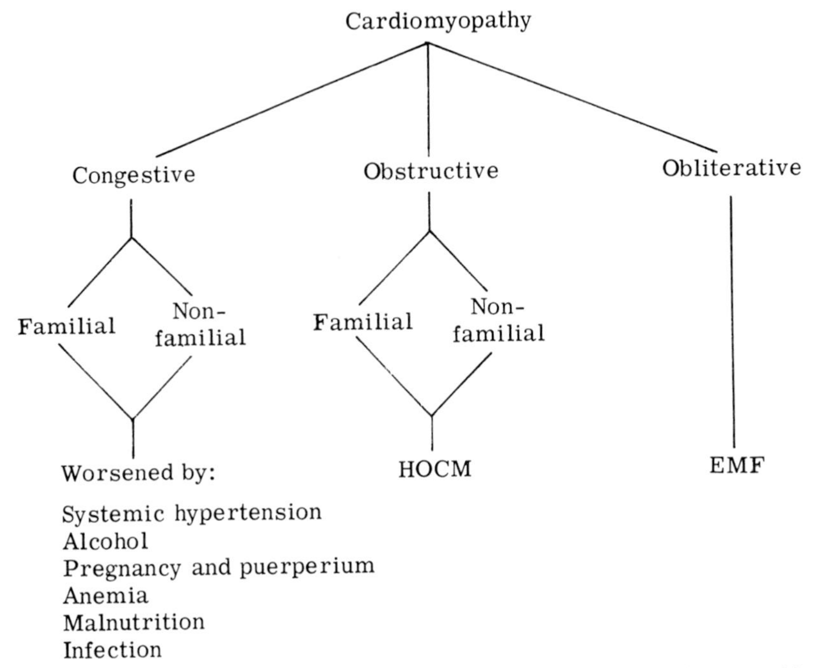

Fig. 22-1. Clinical spectrum of cardiomyopathy. HOCM = hypertrophic obstructive cardiomyopathy, EMF = endomyocardial fibroelastosis.

DEFINITION

Congestive cardiomyopathy may be defined as a generalized myocardial disease predominantly affecting the left ventricle, but ultimately all chambers of the heart are involved. A diminished systolic ejection fraction is mandatory, and dilatation of the heart therefore occurs. It is particularly important that coronary heart disease, systemic hypertension, constrictive pericarditis, and all known systemic diseases and toxic or metabolic causes of congestive heart failure are excluded, since the end result of these conditions is similar to the result of congestive cardiomyopathy.

PATHOLOGY OF THE HEART

As proposed by Hudson,[3] the following features are essential for a pathologic diagnosis of cardiomyopathy: (1) enlargement of the heart due to hypertrophy or dilatation or both; (2) thickening of the endocardium either in a uniform manner or in patches; (3)mural thrombus, frequently of the left ventricle; (4) scars and other lesions in the myocardium, frequently in both ventricles. In addition, there are important negative features: (1) absence of any major coronary arterial disease or congenital arterial lesion; (2) no acquired or congenital primary valvular heart disease; (3) no evidence pathologically of systemic hypertension; (4) absence of any congenital or acquired shunt within or without the heart.

Extensive changes are seen in the myocardium, including lysis of fibers, vacuolation of nuclei of myofibrils, loss of striation, and swelling, with myocytolysis being a constant feature in many reported series.[4] These changes occur most extensively where muscle mass is greatest and are followed by fibrosis in the same region. Mural thrombus is almost inevitably associated.

Maximal endocardial changes are seen in the left ventricle, but the right ventricle is usually involved as well. The earliest changes seem to be smooth-muscle hyperplasia with metachromatic substance between the fibers. The endocardium thickens with smooth-muscle cells and elastic and fibrous tissue, a result of organization of thrombus changes that occur haphazardly throughout the endocardium of the two ventricles. It is frequent for these changes to extend into the myocardium, but only smaller arteries show the typical pathologic pattern; the main coronary arteries and their larger branches are little affected. Changes shown in the small vessels include thickening of the intima and possibly partial or total occlusion by thrombus. Edington and Jackson[5] think that myocytolysis is the earliest pathologic change of cardiomyopathy and that fibrosis, mural thrombus, and thickening of the endocardium occur later. Since all these effects are seen to a greater or lesser extent irrespective of the cause or type of the cardiomyopathy, these authors have suggested a unified concept for the pathology of the disease in general. Others,[6,7] however, have insisted that specific pathologic differences can be demonstrated in fatal cases of cardiomyopathy occurring in different regions of the world, therefore challenging the unified concept. Nevertheless, with the smooth-muscle cell now known to be a multifunctional mesenchyme,[8] it is possible that its hyperplasia could have greater importance in the production of endocardial thickening than was previously appreciated.

ETIOLOGY

According to the definition of congestive cardiomyopathy, its cause remains unknown. Nevertheless, in its diagnosis and differential diagnosis any important known predisposing causes should constantly be borne in mind.

Coronary Heart Disease

Perhaps the most difficult diagnosis of exclusion is coronary arterial disease, since angina pectoris occurs very frequently in patients with congestive cardiomyopathy.[9] Selective coronary arteriography during life and autopsy studies following death may be the only means by which coronary heart disease with heart failure can be differentiated from congestive cardiomyopathy.

Systemic Hypertension

Systemic hypertension causes important diagnostic problems, since it frequently coexists with congestive cardiomyopathy.[10,11] As reported by Kristinsson,[12] 25% of his series of 74 patients with congestive cardiomyopathy had systemic hypertension. Pressure was known to be elevated in about one-third of these before cardiomyopathy was diagnosed; in one-third hypertension became apparent at a later date, and in the remaining third variable blood pressure recordings were obtained. Those initially hypertensive might subsequently become normotensive—a result of worsening ventricular function. Hypertension and cardiomyopathy are known to occur in the same population groups,[13] but the completely dissimilar pathologic effects of the two diseases on the heart and other organ systems make it most unlikely that congestive cardiomyopathy is caused by systemic hypertension. There seems little doubt that hypertension may worsen the prognosis of congestive cardiomyopathy, since it causes an additional load on the already compromised left ventricle. It should, therefore, be treated energetically if associated with cardiomyopathy.

Infections

While certain workers have postulated previous virus infection as an important cause,[14] it now appears more likely that only in a very small number is previous virus infection responsible.[15,16] Viruses thought to have a role in the etiology of congestive cardiomyopathy are: coxsackie virus B, ECHO 9, 6, and 13 strains, and influenza A-2. Goodwin and Oakley[2] were unable to detect evidence of previous virus infection by myocardial biopsy or autopsy electron microscopy in any of their large series of patients with congestive cardiomyopathy. Recurrent infections, bacterial or viral, may worsen the prognosis of congestive cardiomyopathy, particularly if associated with malnutrition or anemia; and any underlying infection should be diligently sought and eradicated as part of the general method of treatment (vide infra).

Alcohol

There does appear to be an important causal relationship to heavy intake of alcohol. The relationship is, however, not simple. Cardiomyopathy of heavy beer drinkers in Quebec and Belgium originally thought to be due to the alcohol content of the beverage is now known to be due to cobalt, which is used as an antifrothing agent in the manufacture of the drink.[17–19] However, ingestion of cobalt in the beer was not by itself the primary cause, since it was only in those individuals who drank beer to such an excess that it was their main source of calories and whose protein ingestion was particularly deficient that eventually developed cardiomyopathy. The second possible cause of cardiomyopathy in heavy drinkers of alcohol is an inadequate thiamine intake as part of the overall picture of poor nutrition. However, it must be admitted that clinically thiamine deficiency, as typified by beriberi, presents completely differently as high-output cardiac failure. Nevertheless, McIntyre and Stanley[20] have described a form of beriberi heart disease that mimics congestive cardiomyopathy as low-cardiac-output failure, and this is sometimes called dry or Shoshin beriberi. Experimental studies have failed to prove conclusively that heavy ingestion of alcohol is a primary cause of myocardial disease. Burch and De Pasquale[21] reported that rats given a quarter of their weight in alcohol for one-quarter of their lives failed to develop cardiomyopathy. While it would be unwise to extrapolate directly from an experiment of that sort to the clinical situation, several groups of investigators[22–24] have documented clinical studies in patients who were known heavy consumers of alcohol and who presented in severe cardiac failure. Very occasionally their cardiovascular status was improved if they abstained completely from further alcohol drinking. No specific pathology of alcoholic cardiomyopathy has been suggested, although Hibbs et al.[25] believe that deposits of neutral lipids throughout the myocardium associated with specific electronmicroscopic changes are diagnostic of the condition. Most pathologists now believe that there are no specific differences by light or electron microscopy or even by histochemical studies when congestive cardiomyopathy is associated with a heavy intake of alcohol. To be sure, alcohol can be shown to have a profound negative inotropic effect on the heart during acute hemodynamic studies,[26] but this does not mean that permanent depression of myocardial function must necessarily accompany chronic ingestion of alcohol. James and Bear[27] suggested that acetaldehyde, a metabolite of alcohol, stimulates the release of norepinephrine from nerve endings in the myocardium, leading eventually to depletion of myocardial norepinephrine stores. Since excess catecholamine is known to damage the heart,[28] and drugs that deplete myocardial catecholamines do not cause cardiac failure, the postulated effects of acetaldehyde seem unlikely. Whether alcohol has any deleterious effect on the enzyme systems of the heart (theoretically an attractive suggestion) remains to be investigated.

Pregnancy and the Puerperium

Considerable speculation continues concerning the nature of congestive cardiac failure in association with pregnancy or the puerperium. Congestive heart failure occurs more frequently at these times in multiparous women, particularly when over the age of 30. Demakis et al.[29] studied 27 patients and showed that persistent enlargement of the heart remaining after the puerperium carries a poor prognosis, the average survival being only 4.7 years and congestive cardiac failure being the cause of death, with pulmonary and systemic embolization additional important causes of fatality; further pregnancies are frequently associated with a worsening of congestive heart failure in this group.[30] In contrast, women whose heart size returns to normal following the puerperium have a much better prognosis; the average survival in this group is 10.7 years, and only minimal symptoms of cardiac decompensation occur until shortly before final deterioration occurs. In this group subsequent pregnancies are common without precipitating further congestive cardiomyopathy. As suggested by Brockington,[31] any associated systemic hypertension may have an important role both in the etiology of cardiomyopathy with pregnancy and in its prognosis, particularly in the black population. That being so, adequate therapy for associated systemic hypertension would appear extremely important.

The condition is much less likely during pregnancy in affluent societies,[10,32] whereas under conditions of poverty multifactorial causes lead to poor nutrition, anemia, and recurrent systemic infections that increase its occurrence.

Inheritance

Emanuel et al.[33] have reported a small series of familial congestive cardiomyopathy. In general, a genetic influence is unusual, although it is interesting to speculate on the possibility of an inherited immunologic defect in these familial cases[34] that causes the heart to be more vulnerable to some as yet unknown acquired cause; but much further study is needed to determine whether there is a true familial pattern to cardiomyopathy and its cause.

ABNORMAL PHYSIOLOGY

Congestive cardiomyopathy presents with ventricular dysfunction and dilatation, reduced contractility, and a diminished ejection fraction. Although the left ventricle is primarily affected, the right ventricle is often involved, particularly in the late stages of the disease. Initially the heart attempts to maintain an adequate function by increasing its diastolic filling. For reasons that are as yet poorly understood, compensatory hypertrophy is inadequate.[11] An inappropriate degree of ventricular dilatation occurs that results in a great increase in wall stress, according to the Laplace relationship. In the early stages of the disease hemodynamics may be normal at rest, although during exercise a widened A-V oxygen difference associated with reduction in the cardiac output is usual. Later, left ventricular end-diastolic pressure and end-diastolic and -systolic volumes are all increased. The stroke volume is reduced, as are stroke work, cardiac index, ejection fraction, and mean systolic ejection rates. Direct measurement of left ventricular contractility results in values lower than normal.

Mitral valvar regurgitation is frequent, since the disease affects not only the papillary muscles but also their insertion into the ventricular myocardium.[35–43] Continuing mitral regurgitation results in further increase in left ventricular end-diastolic pressure. Unlike the chronic left ventricular diastolic dilatation of rheumatic mitral regurgitation, there is in addition differing distensibility and compliance of the left ventricular muscle in congestive cardiomyopathy—a result of associated fibrosis of the ventricular wall. Increased pressure in end-diastole causes a secondary rise in left atrial pressure. Because ventricular compliance is abnormal, a prominent left atrial a wave is usual in sinus rhythm; when mitral regurgitation is severe, the normal x descent is replaced by a large regurgitant wave during systole. Elevation of mean left atrial pressure causes pulmonary arterial hypertension. Once the cardiac output is markedly reduced or recurrent pulmonary emboli occur, pulmonary vascular resistance becomes further elevated.[44–46] A large atrial differential of pressure with higher pressures on the left side is found, mean right atrial pressure often being two or three times lower than the left.[47]

Following the establishment of left ventricular failure, dilatation and hypertrophy of the right ventricle are frequent, eventually leading to high right ventricular end-diastolic pressure and considerable elevation of mean pressure in the right atrium. Later, tricuspid regurgitation may develop, particularly if atrial fibrillation supervenes.[36,46,47] Pulmonary hypertension, increased pulmonary vascular resistance, and reduction in cardiac output and cardiac failure are worse following pulmonary emboli, a common occurrence in late stages of the disease.

The hemodynamic abnormalities of congestive cardiomyopathy associated with pregnancy or the puerperium and with the heavy ingestion of alcohol do not differ in essentials from the description given above, although some modification may be present should there be considerable anemia, intercurrent infection, or systemic hypertension.

DIAGNOSIS

The diagnosis is usually not difficult, since the patient presents with evidence of congestive cardiac failure and severe left ventricular dysfunction. Breathlessness on effort, episodes of paroxysmal cardiac dyspnea at night, physical weakness, recognized dysrhythmias (especially premature beats), and later evidence of right ventricular failure with edema are common. Syncope and dizziness may occur either due to an acute rhythm disturbance or conduction defect or to severe left ventricular dysfunction with an inability to increase the stroke volume and cardiac output normally on effort. Sometimes the presenting symptom is an embolic episode, particularly systemic; but pulmonary emboli are by no means rare, particularly once atrial fibrillation is established. Angina pectoris occurs in 10% of patients and may be a prominent symptom. These patients should be carefully differentiated by the appropriate invasive investigations from those whose cardiac failure is due to coronary arterial disease.

Congestive cardiomyopathy may occur at any age, although its clinical recognition seems most common during the third and fourth decades; there are many documented cases occurring in early childhood, including the neonatal period. The disease affects males and females equally. In this country it has been claimed that it is more common in the black population; but this could be secondary, since it is known that systemic hypertension, so common in the black population, significantly worsens congestive cardiomyopathy.

Physical Examination

The findings reflect the underlying state of the myocardium. At one end of the spectrum there may be clear evidence of severe biventricular dysfunction with associated pulmonary hypertension. In contrast, abnormal physical findings may be few or absent when the disease is in its early stages, with only mild cardiac enlargement on chest radiography with possibly ST-T wave ECG abnormalities to alert the physician to the presence of cardiomyopathy. A fairly common presentation of the patient with moderately severe congestive cardiomyopathy is as follows: (1) normal general appearance with no evidence of central or peripheral cyanosis; (2) mild elevation of the jugular venous pressure with prominence of the a wave in sinus rhythm; (3) normal volume but slow upstroke of the carotid impulse; (4) blood pressure normal or mildly elevated; (5) enlargement of the left ventricle by palpation with an easily felt atrial pulsation in sinus rhythm; (6) right ventricular enlargement appreciated as an abnormal pulsation to the left of the sternum; (7) a prominent atrial sound, normal mitral valve closure, and possibly a pansystolic murmur of mitral regurgitation on auscultation. When left ventricular dysfunction is severe or left bundle branch block is present the second heart sound would be reversed to respiration, pulmonary valve closure preceding aortic. A loud left ventricular filling sound is usual in diastole; (8) later in the disease signs of pulmonary hypertension occur, including a pulmonary ejection click, pulmonary ejection systolic murmur, and increase in intensity of pulmonary valve closure; an immediate diastolic murmur of pulmonary regurgitation is often heard at this stage; (9) finally, evidence of tricuspid regurgitation occurs, particu-

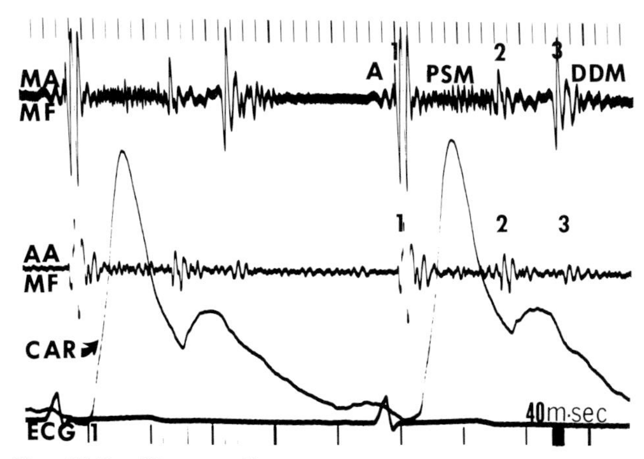

Fig. 22-2. Phonocardiogram, congestive cardiomyopathy. From above downward: mitral area (MA) at medium frequency (MF), aortic area (AA) at medium frequency, carotid (CAR) and ECG lead I. Notice the atrial sound (A), mitral pansystolic murmur (PSM), and loud third heart sound (3) followed by a mitral delayed diastolic murmur (DDM). In this patient mitral regurgitation was associated with cardiomyopathy.

larly when atrial fibrillation is present; the jugular venous pressure is now markedly elevated with a prominent v and rapid y descent and an absent a wave in atrial fibrillation. An inspiratory pansystolic murmur is audible at the lower sternal edge, and pulsation of the liver with or without abdominal ascites may be obvious. Peripheral edema is usual.

Mitral and tricuspid regurgitation, when present, will also be associated with a delayed diastolic murmur incorporated with and following the ventricular filling sound of each of the affected ventricles.

The most characteristic feature of the disease is that apart from obvious enlargement of the ventricles there are very easily appreciated atrial (in sinus rhythm) and ventricular filling sounds (Fig. 22-2).

Electrocardiogram and Vectorcardiogram

It is rare for a completely normal electrocardiogram to be associated with the disease. Although many patients remain in sinus rhythm for several years after the diagnosis is made, the following rhythm disturbances are frequently associated: atrial fibrillation, atrial flutter, and ventricular premature beats, sometimes with runs of ventricular tachycardia. In addition, conduction disturbances of all types and grades are frequent, including the fascicular blocks; eventually complete heart block may occur. In sinus rhythm a broad notched P wave indicative of left atrial enlargement is frequent. Once pulmonary hypertension is present, tall peaking of the P waves in leads II, III, aVF, and V_1 occurs, indicating additional right atrial enlargement. The QRS pattern indicates left ventricular hypertrophy, but abnormal Q and QS patterns not unlike those of myocardial infarction may be recorded and are usually due to fibrosis of the myocardium. T-wave inversions similar to those of myocardial ischemia are frequently recorded. As can be appreciated, the ECG changes are those frequently associated with any form of left ventricular disease with or without associated pulmonary hypertension. Their correct interpretation is made in association with the clinical presentation.

Vectorcardiography provides a more sensitive index of ventricular hypertrophy and is useful in the interpretation of the abnormal QS patterns, but by and large it gives little additional insight into the underlying pathology.

Chest Radiography

Chest radiography is most helpful, since it will provide evidence of the overall size of the heart and show whether specific chamber enlargement is present. In the early stages there is usually mild overall enlarge-

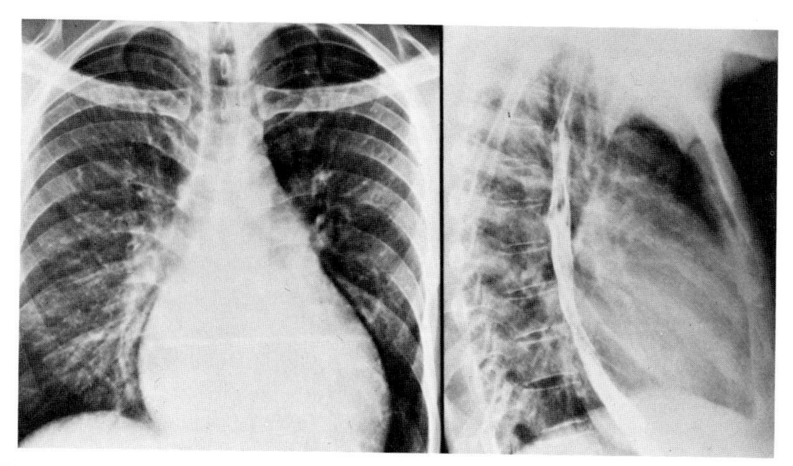

Fig. 22-3. Chest radiograph, congestive cardiomyopathy, posteroanterior and lateral view (barium-filled esophagus). There is overall cardiac enlargement with specific enlargement of the left ventricle and left atrium. Pulmonary venous (upper lobes) and arterial congestion is seen, and Kerley B lines are present at the right base.

ment of the heart with some specific enlargement of the left ventricle. As the disease progresses the heart increases in size, left ventricular enlargement becomes more pronounced, and in addition enlargement of the left atrium is obvious. The pulmonary vasculature shows evidence of congestive failure with pulmonary venous engorgement, particularly to the upper lobes, enlargement of the pulmonary arteries, pulmonary edema, and Kerley B lines due to engorgement of pulmonary lymphatics (Fig. 22-3). Further progression of the disease is associated radiologically with evidence of right ventricular enlargement and bilateral pleural effusions. With the development of severe tricuspid regurgitation, additional enlargement of the right atrium may occur, and pericardial effusions are frequent. Pulmonary emboli will be recognized by the typical segmental changes in the lungs on chest radiography, persistent pleural effusions, and linear atelectasis.

Special Investigations

Echocardiogram

An important noninvasive diagnostic tool is the echocardiogram. By suitably positioning the ultrasound probe, echo signals may be obtained from the ventricular septum, posterior wall of the left ventricle, and

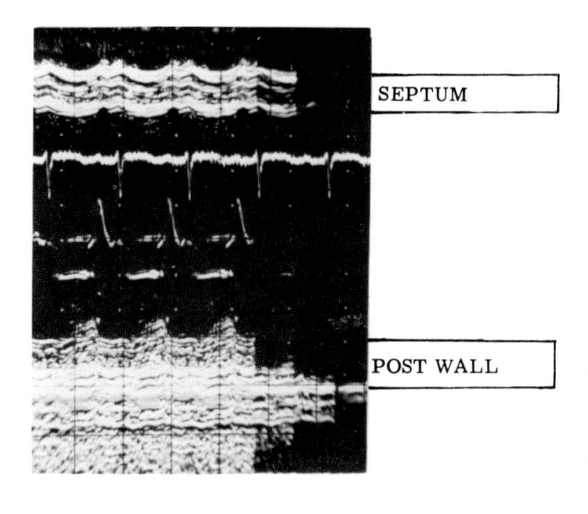

EDd	6.7 cm
EF	39 %
MINOR AXIS, SYSTOLIC CHANGE	11 %

Fig. 22-4. Echocardiogram, congestive cardiomyopathy Echo signals are reflected from the interventricular septum and posterior wall of the left ventricle. There is an increase in the end-diastolic diameter (EDd), systolic change is small, and the ejection fraction (EF) is reduced. This is an example of reflected ultrasound signals from an enlarged decompensated left ventricle (compare Fig. 22-5).

leaflets of the mitral valve. Simultaneous signals from the intraventricular septum and posterior wall of the left ventricle can be used to measure left ventricular dimension.[48] The transverse left ventricular dimension correlates with angiographic major and minor axes of the heart, and there is also a close relationship between the cube of the echocardiographic ventricular dimension and angiographic volume.[49,50] The amplitude of movement of the mitral valve ring during the cardiac cycle with respect to the fixed transducer at the cardiac apex has been used to assess stroke volume as the difference between end-diastolic and end-systolic volumes.[51] Using this technique, left ventricular wall thickness may be measured and the velocity and amplitude of left ventricular motion during the phase of ventricular ejection established.[52] In congestive cardiomyopathy, analysis of the echocardiogram indicates poor contraction of an enlarged left ventricle with an increased ventricular diameter and diastolic volume and reduction of the ejection fraction (Fig. 22-4).

In addition, echocardiography provides a relatively easy method of differentiating hypertrophic obstructive cardiomyopathy from congestive cardiomyopathy. Typical abnormalities of the anterior cusp movement of the mitral valve in hypertrophic obstructive cardiomyopathy have been recorded.[53,54] In addition to the abnormal movement of the mitral valve, the septal echo is prominent, impinging on the mitral valve

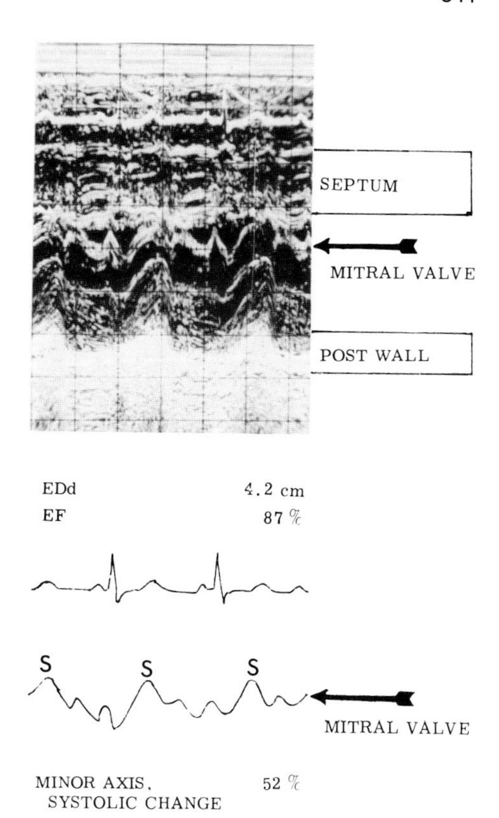

Fig. 22-5. Echocardiogram, obstructive cardiomyopathy. There is marked thickening of the interventricular septum compared to the posterior wall of the left ventricle. The end-diastolic diameter (EDd) is not increased, and the ejection fraction (EF) is high at 87%. Notice the abnormal movement of the anterior leaflet of the mitral valve (S), which abuts on the septum during systole. This is an example of reflected ultrasound signals from a markedly hypertrophied but compensated left ventricle with high ejection fraction and is typical of hypertrophic cardiomyopathy with obstruction to left ventricular outflow (compare Fig. 22-4).

during early diastole. When a true obstructive form of the disease is present, however, the anterior cusp of the mitral valve moves abnormally during systole, showing a strike anterior movement toward the septum (Fig. 22-5).

Cardiac Catheterization and Angiography

Right- and left-heart catheterization, as previously described, provides objective evidence of impaired cardiac function, with a decrease in cardiac output, elevated left and frequently right ventricular end-diastolic and pulmonary arterial wedge pressures, and a large left atrial a wave in sinus rhythm. The differential diagnosis of constrictive pericarditis may be helped by catheterization studies, since the abnormal x descent in right atrial pressure curves, early diastolic dip in right ventricular pressure curves, and a diastolic pressure plateau with equal right and left ventricular end-diastolic and mean pulmonary wedge pressures are said to be specific for constrictive pericarditis. In certain causes of car-

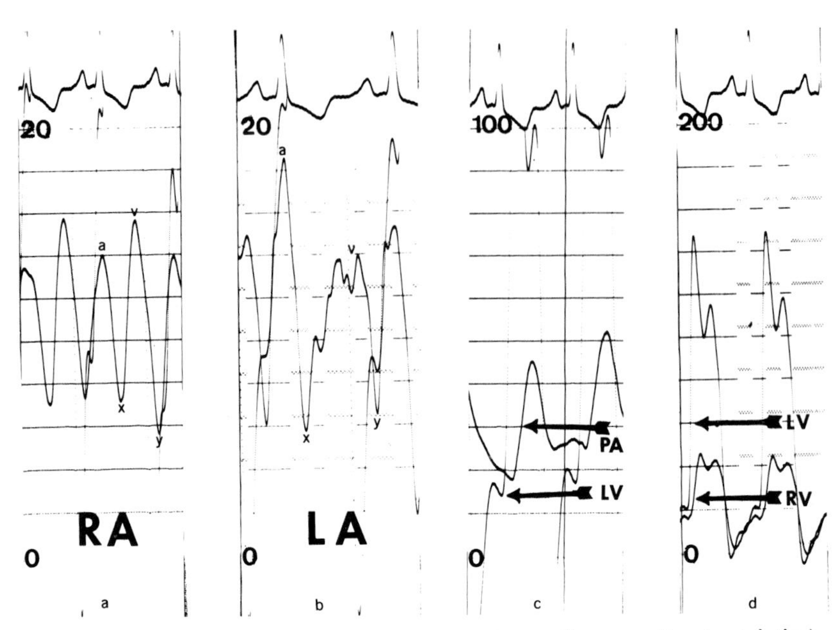

Fig. 22-6. Right and left heart pressures in cardiomyopathy (restrictive). Constrictive pericarditis was not responsible for the hemodynamic abnormalities in this patient, who had restrictive cardiomyopathy. There is elevation of pressure in both atria with a prominent x descent (a and b). Pulmonary arterial end-diastolic pressure is almost the same as that in the left ventricle (c and d). Note the early diastolic dip with subsequent plateau and equal diastolic pressures in both ventricles.

diomyopathy, however, particularly amyloidosis and endomyocardial fibroelastosis, an identical hemodynamic pattern may be found (Fig. 22-6), and the differentiation on hemodynamic grounds alone may be extremely difficult.[36,44,55]

Selective coronary arteriography should be performed in patients with congestive cardiomyopathy who are being investigated by cardiac catheterization as the only method of differentiating with certainty coronary arterial disease. Although the absence of angina pectoris as a symptom is unusual in any patient whose cardiac failure and enlargement of the heart are produced by atherosclerotic disease of the coronary arteries, angina pectoris may occur in patients with congestive cardiomyopathy even in the presence of a completely normal coronary circulation.[56]

Angiography of the left ventricle is useful since it will provide measurements of left ventricular dimensions and volume, will demonstrate and quantitate any mitral regurgitation present, and will give a visual representation of abnormal left ventricular contractions from which the left ventricular ejection fraction can be calculated.

Myocardial Biopsy

Biopsy of the myocardium, although frequently performed, has provided little useful information up to the present time. The techniques that have been used include open thoracotomy, biopsy catheters,[57] or the transthoracic needle.[58]

In a careful histologic, histochemical, and ultrastructural study of the cardiac biopsy material obtained in 53 patients,[59] gross pathologic and histologic differentiation between congestive and hypertrophic obstructive cardiomyopathy was possible, but perhaps surprisingly neither histochemical nor electron microscopic studies revealed any differentiating diagnostic features. Van Noorden et al.[59] indicated that in their opinion cardiac biopsy was therefore not a useful procedure in diagnosing and classifying cardiomyopathy. In this study great care was also taken to search for evidence of a viral origin in every case of congestive cardiomyopathy investigated; none was found.

It may well be that different techniques of electronmicroscopic assessment of the ultrastructure of the myocardium with particular reference to quantitative changes rather than qualitative effects, as proposed by Page,[60] might yield more meaningful data, particularly regarding the cause of dilatation on the one hand and hypertrophy on the other.

PROGNOSIS

Congestive cardiomyopathy is a serious disease. Among 74 patients carefully followed, less than 50 were alive 5 years later.[12] The overall prognosis depends on the severity of the underlying process, and despite the gloomy figures of Kristinsson[12] many patients have benefited quite markedly by intensive therapy for congestive cardiac failure and correction of any associated anemia, infection, or systemic hypertension. The rapidity of their improvement can be most gratifying (Fig. 22-7), and many of these patients are able to return to gainful employment, although their ultimate prognosis must remain guarded. While treatment slows the progression of the disease and often helps to protect the patient against complications, such as emboli or dysrhythmias, the course is usually relentless until death occurs in congestive cardiac failure with or without additional systemic embolization or pulmonary infarction.

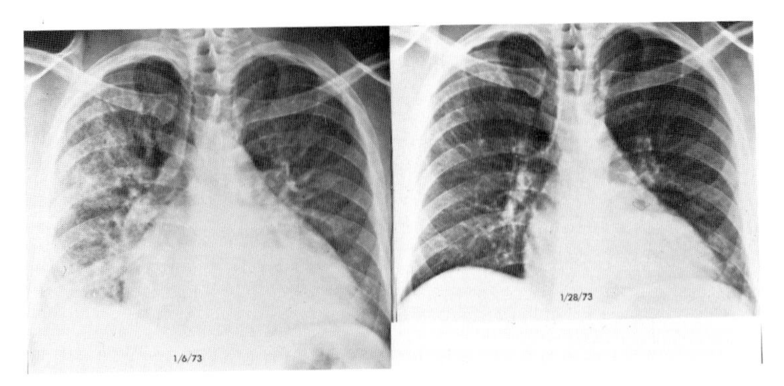

Fig. 22-7. Congestive cardiomyopathy; effect of treatment on heart size. Following 22 days of bedrest, digitalization, and diuretic therapy, there is marked diminution in overall size of the heart as well as clearing of pulmonary edema and absorption of the basal pleural effusion.

TREATMENT

No breakthrough has yet occurred in the treatment of congestive cardiomyopathy, nor have there been documented cures. The mainstays of treatment remain bedrest and subsequent alteration in the amount of physical activity prescribed, digitalization, diuretic therapy, and anticoagulant drugs, in an attempt to protect the patient against systemic or pulmonary emboli. At the same time it is most important to remove any known possible precipitating cause, such as alcohol, and to manage energetically any anemia, systemic hypertension, or infection. A diligent search should be made for any underlying systemic or metabolic disease whose presence may be occult (for example, a collagen disease) and that requires specific therapy.

Bedrest

For many decades bedrest has been a most important form of treatment for any type of cardiac failure. Resting the heart[61] can be assumed to decrease heart rate/volume, arterial blood pressure, rate of work, and power output, as well as heart size. A large heart is at a mechanical disadvantage, since if it is assumed that the heart is a sphere dilatation must increase resting tension (law of Laplace). In addition, myocardial tension does not lessen but increases during systole (a result of the reduced ejection fraction and increased overall size of the heart), and this is associated with an increasing myocardial oxygen consumption.[62] Furthermore, the papillary muscles are disadvantageously positioned when there is a considerable increase in ventricular size. This inefficiency is usually associated with mitral valvar regurgitation, further embarrassing an already overburdened ventricle. As reported initially by Burch and Walsh,[63] prolonged bedrest may have beneficial results in cardiomyopathy. Very impressive reduction in overall size of the heart may be seen, but up to the present time there does not appear to be any objective hemodynamic evidence of permanent improvement, even when periods of bedrest are prolonged to 1 year or longer.

Positive Inotropic Drugs

Digoxin

Once the diagnosis has been made, the patient should be digitalized and maintained on an adequate dose of digoxin. It is unclear whether digitalizing the patient will slow the process of hypertrophy and dilatation that eventually leads to severe cardiac failure, but it is attractive to speculate that this might be so on theoretical grounds. In any case, slowing the heart is likely to significantly reduce myocardial oxygen consumption and therefore to be beneficial.

Glucagon

Lvoff and Wilcken[64] used intravenous glucagon in the management of 50 patients at a dose of 2.5 mg/hr for 1–7 days, controlling the nausea that was produced by antihistamine therapy. Thereafter treatment was continued with routine digitalization and diuretics. Although improvement was claimed for this combined treatment, others have not been able to confirm their results.

Diuretic Therapy

With elevation of ventricular filling pressures (especially of the right ventricle), peripheral edema, and pulmonary congestion or edema, diuretic therapy should be begun and maintained with an appropriate daily or alternate-day dose. Meticulous attention should be paid to preserving electrolyte balance, and adequate replacement of potassium should be undertaken as routine.

Steroids

Steroid therapy was used in 13 patients in large doses (prednisone, 60–80 mg/day) by Kristinsson.[12] Ten of these patients died, and there was no indication of objective benefit as compared to a control group, although some euphoria and increased well-being were commented on by some of the patients. Hemodynamic studies undertaken in 4 of the 13 confirmed objective lack of benefit. Furthermore, as cautioned by Goodwin,[65] an as yet unknown infective process responsible for cardiomyopathy could be significantly worsened by steroid therapy that might cause its dissemination, as happened with the myocardial lesions of coxsackievirus B in mice.[66]

Anticoagulant Therapy

An important consideration in each patient is whether to maintain him on anticoagulant therapy for a long term. As previously described, the incidence of mural thrombus is distressingly high; and as can be predicted, embolism, both systemic and pulmonary, is frequent. On the systemic side this is an important cause of death, since the brain is a favored site. Embolism to other systemic organs is by no means rare. Furthermore, repeated pulmonary embolism and infarction reduce still further the pulmonary vascular bed, increase pulmonary vascular resistance, decrease cardiac output, and hasten the onset of right ventricular failure, atrial fibrillation, and tricuspid regurgitation. Those undertaking invasive circulatory investigation of these patients should be aware of the potential risk of dislodging mural thrombus, especially from the left ventricle but also from the right, and should modify their technique to reduce this hazard. Many physicians believe, therefore, that patients with congestive cardiomyopathy should be maintained on long-term anticoagulant therapy to protect them as far as possible from the distressing and serious complications of embolism, but statistical evidence of the efficiency of such treatment is still awaited.

Pericardiotomy

The concept that the pericardium might exert restriction when considerable enlargement of the heart is present was postulated some time ago.[37] Salisbury et al.[67] showed that experimental acute heart failure could be relieved by pericardiotomy, and Bartle and Herman[68] were able to demonstrate hemodynamically that the pericardium does have a constricting effect in mitral regurgitation of relatively short onset. Accordingly, Goodwin[65] had 3 patients with congestive cardiomyopathy operated on, but was unable to find objective hemodynamic evidence of improvement following pericardiotomy. The possible benefit of such surgery in selected patients who have marked cardiac dilatation is certainly worth further consideration, since a more favorable diastolic fiber length–tension relationship might follow pericardiotomy without causing any increase in contractility.

Vasodilator Therapy

Traditionally, physicians have focused attention on diastolic filling of the ventricle and its contractile state and little attention has been devoted to the role of the arterial system and venous capacitance vasculature in determining performance of the left ventricle when diseased. More recently, dramatic hemodynamic improvement has been shown in patients whose acute myocardial infarction is complicated by severe left ventricular failure when vasodilator drugs have been given.[69] This beneficial effect results from lowering impedance to left ventricular outflow. Although it has been known since 1955 that intravenous sodium nitroprusside is remarkably successful in reducing blood pressure by causing arteriolar dilatation and therefore decreasing systemic vascular resistance, its clinical acceptance has been delayed for many years. Sodium nitroprusside also relaxes venous smooth muscle causing a decrease in venous return and a reduction in ventricular end diastolic pressure. In cardiac failure there is a high impedance during ventricular ejection. Further reduction in stoke volume, a result of ventricular dysfunction, increases the arteriolar resistance still further. In cardiomyopathy, the diseased ventricle has to generate a greater than normal systolic wall tension despite its reduced contractility and a progressive diminution in cardiac output with associated increase in impedance often results.

Treating the patient with vasodilators frequently results in a reduction of left ventricular filling pressure and an increase in cardiac output.[70] The beneficial effects of sodium nitroprusside therapy are produced by its vasodilator properties reducing impedance to left ventricular ejection while at the same time myocardial oxygen consumption is diminished by decreasing both preload and afterload of the left ventricle by its vasodilator effects on systemic arterial and venous smooth muscle.[71]

Vasodilator therapy should be started with intravenous sodium nitroprusside as an infusion of 0.3 to 6 μg/kg per minute. Once the rate of infusion needed has been established, changes in filling pressure of the left ventricle (monitored conveniently by passing a flotation catheter into the pulmonary artery and measuring pulmonary wedge pressure), arterial pressure and cardiac output should be followed. Dramatic improvement is often noted, especially in patients with marked ventricular dysfunction associated with considerable elevation of filling pressure of the ventricle and of peripheral vascular resistance.

Once improvement using intravenous sodium nitroprusside has been demonstrated, chronic vasodilator therapy may be given by mouth. Isosorbide dinitrate 10mg every two to four hours has been recommended and benefit maintained for many months (C. Romero, personal communication).

Further experience with this novel form of treating the chronic congestive cardiac failure of cardiomyopathy is needed to determine the true role of vasodilator therapy, particularly in the long term management.

Although congestive cardiomyopathy is a relentless disease, and treatment up to the present time has been largely disappointing in terms of cure, the patient can usually be materially benefited by a careful medical regimen including bedrest, digitalization, diuretic, and vasodilator therapy, as well as attention to any underlying anemia, systemic hypertension, or infection. Alcohol should be avoided. Occasionally spectacular success follows such careful management, indicating the need for a positive approach to treatment and prognosis, eradication of known precipitating causes, and a continuing diligent search for the etiology of congestive cardiomyopathy.

REFERENCES

1. Brigden W: Uncommon myocardial diseases—The non-coronary cardiomyopathies. Lancet 2:1179, 1957
2. Goodwin JF, Oakley CM: The cardiomyopathies. Br Heart J 34:545, 1972
3. Hudson REB: Pathology of cardiomyopathy. Cardiovasc Clin 4:4, 1972
4. Reddy CRRM, Parvathi G, Rao NR: Pathology of cardiomyopathy from South India. Br Heart J 32:226, 1970
5. Edington GM, Jackson JG: The pathology of heart muscle disease and endomyocardial fibrosis in Nigeria. J Pathol Bacteriol 86:333, 1963
6. Parry EHO, Abrahams DG: The natural history of endomyocardial fibrosis. Q J Med 34:383, 1965
7. Shaper AG: On the nature of some tropical cardiomyopathy. Trans R Soc Trop Med Hyg 61:458, 1967
8. Wissler RW: The arterial medial cell, smooth muscle, or multifunctional mesenchyme. Circulation 36:1, 1967
9. Gau G, Goodwin JF, Oakley CM, et al: Q waves and coronary angiography in cardiomyopathy. Br Heart J 32:554, 1970
10. Kristinsson A, Croxson RS, Everson Pease AG, et al: Investigation, treatment and prognosis of congestive cardiomyopathy, in: Proceedings of the Fifth European Congress of Cardiology, Athens, September 1968
11. Goodwin, JF: Congestive and hypertrophic cardiomyopathies—A decade of study. Lancet 1:731, 1970

12. Kristinsson A: Diagnosis, natural history and treatment of cardiomyopathy. Thesis, University of London, 1969
13. Fodor J, Miall WE, Standard KL, et al: Myocardial disease in a rural population in Jamaica. Bull WHO 31:321, 1964
14. Sainani GS, Krompotic E, Slodki SJ: Adult heart disease due to the coxsackie virus B infection. Medicine 47:133, 1968
15. Bengtsson E: Myocarditis and cardiomyopathy. Cardiologia 52:97, 1968
16. Somerville W: Myocarditis and myocardiopathy, in: Proceedings of the Fifth European Congress of Cardiology, Athens, September 1968, p. 149
17. Kesteloot H, Roelandt J, Willems J, et al: An enquiry into the role of cobalt in the heart disease of chronic beer drinkers. Circulation 37:854, 1968
18. Mercier G, Patrie G: Quebec beer drinkers' cardiomyopathy. Clinical signs and symptoms. Can Med Assoc J 97:884, 1967
19. Morin Y: Quebec beer drinkers' cardiomyopathy. Haemodynamic alterations. Can Med Assoc J 97:901, 1967
20. McIntyre N, Stanley NN: Cardiac beri-beri: Two modes of presentation. Br Med J 3:567, 1971
21. Burch GE, De Pasquale NP: Alcoholic cardiomyopathy. Cardiologia 52:48, 1968
22. Brigden WW, Robinson JF: Alcoholic heart disease. Br Med J 2:1283, 1964
23. Alexander CS: Idiopathic heart disease. I. Analysis of 100 cases with special reference to chronic alcoholism. Am J Med 41:213, 1966
24. Tobin JR, Driscoll JF, Lim MT, et al: Primary myocardial disease and alcoholism—The clinical manifestations and course of the disease in a selected population of patients observed for three or more years. Circulation 35:754, 1964
25. Hibbs RG, Ferrans VJ, Black WC, et al: Alcoholic cardiomyopathy—An electron microscopic study. Am Heart J 69:766, 1965
26. Riff DP, Jain AC, Doyle JT: Acute hemodynamic effects of ethanol on normal human volunteers. Am Heart J 78:592, 1969
27. James TM, Bear ES: Effects of ethanol and acetaldehyde on the heart. Am Heart J 74:243, 1967
28. Van Vliet PD, Burchell HB, Titus JL: Focal myocarditis associated with pheochromocytoma. N Engl J Med 274:1102, 1966
29. Demakis JG, Rahimtoola SH, Sultan GC: Natural course of peripartum cardiomyopathy. Circulation 44:1053, 1971
30. Walsh JJ, Burch GE, Black WC, et al: Idiopathic myocardiopathy of the puerperium (postpartal heart disease). Circulation 32:19, 1965
31. Brockington IF: Postpartum hypertensive heart failure. Am J Cardiol 27:655, 1971
32. Demakis JG, Rahimtoola SH: Peripartum cardiomyopathy. Circulation 44:964, 1971
33. Emanuel R, Withers R, O'Brien K: Dominant and recessive modes of inheritance in idiopathic cardiomyopathy. Lancet 2:1065, 1971
34. Dass KS, Callen JP, Dodson VN, Cassidy JT: Immunoglobulin binding in cardiomyopathic hearts. Circulation 44:612, 1971

35. Massumi RA, Rios JC, Gooch AS, et al: Primary myocardial disease—Report of 50 cases and review of the subject. Circulation 31:19, 1965
36. Dye CL, Genovese PD, Daly WJ, Behnke RH: Primary myocardial disease. II. Hemodynamic alterations. Ann Intern Med 58:442, 1963
37. Goodwin JF: Cardiac function in primary myocardial disorders. Br Med J 1:1527, 1964
38. Yu PM, Cohen J, Schreiner VF Jr, Murphy GW: Hemodynamic alterations in primary myocardial disease. Prog Cardiovasc Dis 7:125, 1964
39. Hamby RI: Primary myocardial disease—A prospective clinical and hemodynamic evaluation of 100 patients. Medicine 49:55, 1970
40. Pietras RJ, Meadows WR, Fort M, Sharp JT: Hemodynamic alterations in idiopathic myocardiopathy including cineangiography from the left heart chambers. Am J Cardiol 16:672, 1965
41. Dodge HT, Baxley WA: Left ventricular volume and mass and their significance in heart disease. Am J Cardiol 23:528, 1969
42. Croxson RS, Raphael MJ: Angiographic assessment of congestive cardiomyopathy. Br Heart J 31:390, 1969
43. Rackley CE, Hood WP Jr, Rollett EL, Young DT: Left ventricular end diastolic pressure in chronic heart disease. Am J Med 48:310, 1970
44. Fowler NO, Gueron M, Rowlands DT Jr: Primary myocardial disease. Circulation 23:498, 1961
45. Yu PN, Schreiner BF, Cohen J, Murphy GW: Idiopathic cardiomyopathy—A study of left ventricular function and pulmonary circulation in 15 patients. Am Heart J 71:330, 1966
46. Storstein O: Hemodynamic studies in primary myocardial disease. Acta Med Scand [Suppl 472] 1967, p 113
47. Hamby RI, Catangay P, Apiado O, Khan AH: Primary myocardial disease. Clinical hemodynamic and angiographic correlates in 50 patients. Am J Cardiol 25:625, 1970
48. Popp RL, Wolfe SB, Hirata T, Feigenbaum H: Estimation of right and left ventricular size by ultrasound—A study of echoes from the intraventricular septum. Am J Cardiol 24:523, 1969
49. Pombo JF, Troy BL, Russell RO Jr: Left ventricular volumes and ejection fraction by echocardiography. Circulation 43:480, 1971
50. Fortuin NJ, Sherman ME, Hood WP Jr, Craige E: Evaluation of left ventricular function by echocardiography. Circulation 42:III-120, 1970
51. Popp RL, Harrison DC: Ultrasonic cardiac echocardiography for determining stroke volume and valvular regurgitation. Circulation 41:493, 1970
52. Kraunz RF, Kennedy JW: Ultrasonic determination of left ventricular wall motion in normal man—Studies at rest and after exercise. Am Heart J 79:36, 1970
53. Shah PM, Gramiak R, Kramer DH: Ultrasound localization of left ventricular outflow obstruction in hypertrophic obstructive cardiomyopathy. Circulation 40:3, 1969
54. Pridie RB, Oakley CM: Mechanism of mitral regurgitation in hypertrophic obstructive cardiomyopathy. Br Heart J 32:203, 1970

55. Burwell CS, Robin ED: Some points in the diagnosis of myocardial fibrosis. Trans Assoc Am Physicians 67:67, 1954
56. Raftery EB, Banks DC, Oram S: Occlusive disease of the coronary arteries presenting as primary congestive cardiomyopathy. Lancet 2:1147, 1969
57. Konno S, Sakakibara S: Intracardiac biopsy. Circulation 30:108, 1964
58. Sutton DC, Sutton GC: Needle biopsy of the human ventricular myocardium—Review of 54 consecutive cases. Am Heart J 60:364, 1960
59. Van Noorden S, Olsen EGJ, Pearse AGE: Hypertrophic obstructive cardiomyopathy—A histological, histochemical and ultrastructural study of biopsy material. Cardiovasc Res 5:118, 1971
60. Page E, McCallister LP: Quantitative electron microscopic description of heart muscle cells—Application to normal, hypertrophied and thyroxin-stimulated hearts. Am J Cardiol 31:172, 1973
61. Burch GE, De Pasquale NP: On resting the human heart. Am J Med 44:165, 1968
62. Sonnenblick EH, Skelton CL: Oxygen consumption of the heart—Physiological principles and clinical implications. Mod Concepts Cardiovasc Dis 40:9, 1971
63. Burch GE, Walsh JJ: Cardiac enlargement due to myocardial degeneration of unknown cause—Preliminary report on effect of prolonged bedrest. JAMA 172:207, 1960
64. Lvoff R, Wilcken DL: Glucagon in heart failure and cardiogenic shock—Experience in 50 patients. Circulation 45:534, 1972
65. Goodwin JF: Treatment of the cardiomyopathies. Am J Cardiol 32:341, 1973
66. Kilbourne ED, Wilson LP, Perrier D: The induction of gross myocardial lesions by a coxsackie pleurodinia virus and cortisone. J Clin Invest 35:362, 1956
67. Salisbury PF, Cross CE, Rieben PA: Acute experimental heart failure relieved by pericardiotomy. Am J Cardiol 13:133, 1964
68. Bartle SH, Herman HJ: Acute mitral regurgitation in man—Hemodynamic evidence and observations indicating an early role for the pericardium. Circulation 36:839, 1967
69. Franciosa JA, Guiha NH, Limas CJ, Rodriguera E, Cohn JN: Improved left ventricular function during nitroprusside infusion in acute myocardial infarction. Lancet 1:650, 1972
70. Majid PA, Sharma B, Taylor SH: Phentolamine for vasodilator treatment of severe heart failure. Lancet 2:719, 1971
71. Miller RR, Visnara LA, Zelis R, Amsterdam EA, Mason DP: Clinical use of nitroprusside in chronic ischemic heart disease. Effects on peripheral vascular resistance and venous tone and on ventricular volume, pump, and mechanical performance. Circulation 51:328, 1975

J. Warren Harthorne
Gerald M. Pohost

23

Electrical Therapy of Cardiac Dysrhythmias

The burgeoning field of biomedical engineering has armed the modern physician with a considerable array of electronic hardware with which to treat a host of medical diseases. It is the purpose of this chapter to review current methods of treating cardiac arrhythmias electrically. Historically the application of an electrical stimulus to regenerate cardiac contraction dates from the early 1800s.[1] Considerable electrophysiologic investigation was carried out in animal laboratories during the 1930s,[2,3] but effective cardiac stimulation of patients was not regularly used until the 1950s.[4] A major impetus to the development of electrical pacing in the control of cardiac arrhythmias was provided by Furman's introduction of temporary transvenous pacing in 1958[5] and by the development of reliable implantable fixed-rate and demand pacing systems. Rather than trying to give an exhaustive treatise on all conceivable types of arrhythmias for which electrical therapy may be beneficial, this chapter will stress the techniques and types of pacing systems available, the physiology of each type of pacing, and the detection of pacemaker malfunction. Cardioversion will then be briefly considered.

ANATOMY OF THE CONDUCTION SYSTEM

A general knowledge of the anatomy and physiology of the conduction system is basic to a proper understanding of pacing techniques (Fig. 23-1). The sino-atrial (S-A) node can be identified in the "roof" of the right atrium adjacent to the superior vena cava as an oblong structure

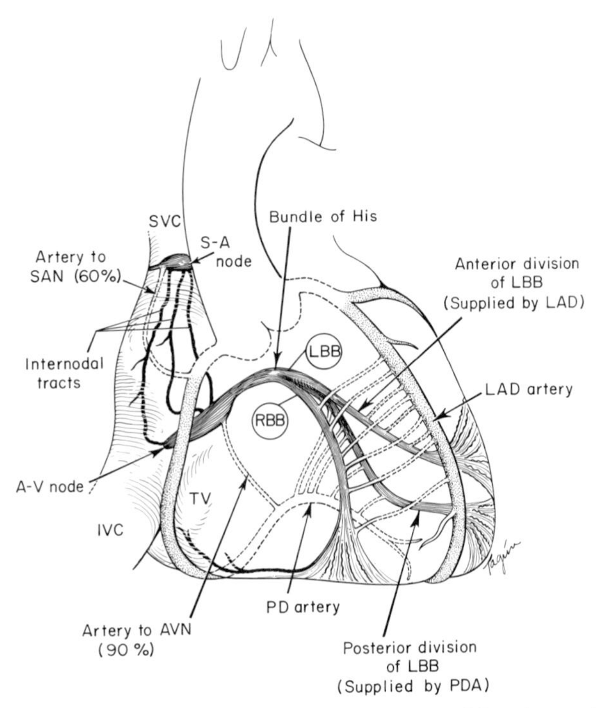

Fig. 23-1. Anatomy of conduction system. The sino-atrial node artery arises as an early branch of the right coronary artery (RCA) in 60% of the population and as a branch of the left coronary artery (LCA) or its circumflex branch (Circ) in 40%. The right coronary artery is dominant in 90% of cases, providing the artery to the atrioventricular node (A-V node) and the posterior descending artery to the inferior myocardium. In 10% of patients with a dominant LCA the artery to the A-V node arises from the circumflex branch. Note that the anterior two-thirds of the ventricular septum with adjacent right bundle branch and anterior division of the left bundle are supplied by perforating septal branches of the anterior descending artery, while the posterior one-third of the septum and posterior division of the left bundle are supplied by perforating septal branches of the posterior descending artery. Not shown is the alternate collateral supply to the A-V node provided by the perforating septal branch of the LAD.

1.5–2.0 cm long. Its arterial blood supply arises (in 55%–60% of patients) as a proximal branch of the right coronary artery, in the remainder of patients as a branch of the left coronary, usually arising from the circumflex. The generation of electrical activity by the S-A node is a complex and not completely understood process. In the absence of disease the S-A node dominates the lower portions of the conduction system with the highest frequency of impulse formation. Many pacemaker cells coexist in close proximity within the S-A node and are quite sensitive and responsive to various neurotransmitter agents. The membrane potential initiated within the pacemaker cells of the S-A node is rapidly transmitted through the atrium by way of three internodal pathways to the atrioventricular node. An ancillary pathway (Bachmann's bundle) transmits the impulse to the left atrium.

The atrioventricular (A-V) node is formed by the convergence of the internodal tracts and lies in the "floor" of the right atrium adjacent to the ostium of the coronary sinus and tricuspid anulus. Its vascular supply derives from the atrioventricular nodal artery, a branch of the distal right coronary in 90% of the population, and this vessel also supplies the bundle of His and portions of the posterior fascicle of the left bundle. The A-V node provides the major delay in impulse transmission to the ventricles through a process of decremental conduction. Below the A-V node, impulse transmission rapidly accelerates and is quickly dispersed through the remainder of the His-Purkinje system. Ischemia of the A-V node due to interruption of the A-V nodal artery, as seen with inferior wall infarction, often culminates in the familiar Wenckebach or Mobitz I block. Escape foci in nonischemic conduction tissue immediately subjacent to the A-V node and the bundle of His prevent extreme bradyarrhythmias. The level at which delay of impulse transmission from atrium to ventricle occurs (i.e., above or below the His bundle) is very important in determining the resultant arrhythmia and the indications for pacemaker insertion. A more refined delineation of this information may be obtained from the technique of recording His-bundle electrograms.

After traversing the A-V node and main bundle, the impulse rapidly advances through the three fascicles of the bundle branch system spreading anteriorly from the left to right and from endocardium to epicardium. The arterial supply of the bundle branches differs from that of structures above. The right bundle branch and anterior division of the left bundle parallel each other on opposite sides of the anterior ventricular septum and are supplied by septal perforating branches of the anterior descending branch of the left coronary artery. Its occlusion, as with an anterior myocardial infarction, may culminate in right bundle branch block (RBBB) and left-axis deviation (LAD), with ventricular activation then tenuously confined to the posterior fascicle. The posterior division of the left bundle is supplied by the A-V nodal artery and perforating septal branches of the posterior descending artery. In the presence of an anterior wall infarction associated with RBBB and LAD, ischemia of the posterior division of the left bundle may culminate in sud-

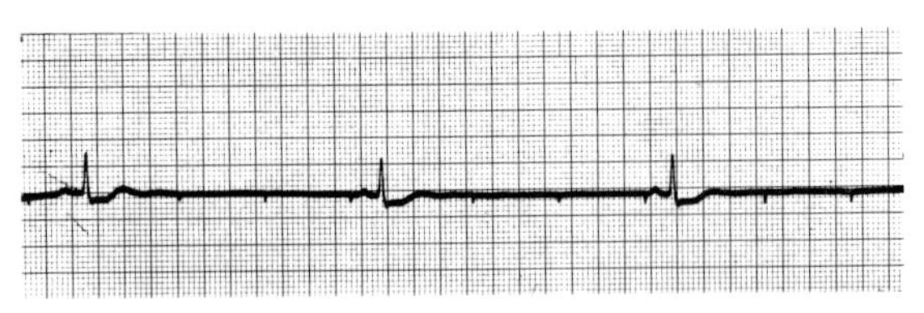

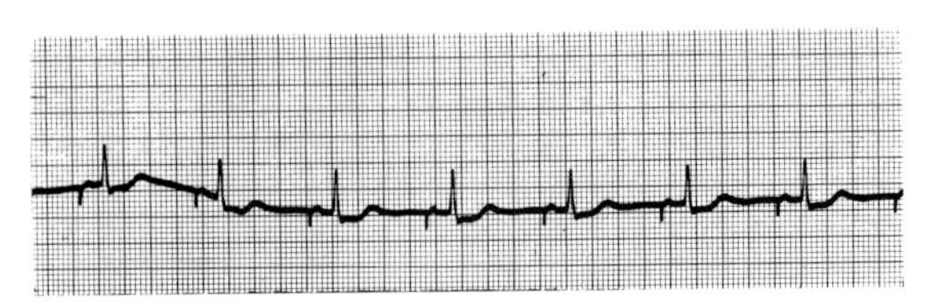

Fig. 23-2. The upper panel demonstrates sinus bradycardia with subthreshold pacemaker impulses superimposed. Increase in the pacing amplitude in the lower panel produces atrial pacing.

den asystolic arrest of the Mobitz II variety. The terminal ramifications of the His-Purkinje network are supplied by deep transmural branches of epicardial coronary vessels.

DISEASE OF THE CONDUCTION SYSTEM

Damage to the conduction system may occur as a consequence of surgical trauma or congenital or acquired disease processes. Inflammatory reactions in myocarditis, degenerative processes associated with fibrotic interruption of conduction pathways (Lenegre's disease), or ischemia due to interruption of arterial blood supply may result in various conduction abnormalities, depending upon the structures involved. Resultant arrhythmias are basically due to defects in impulse generation or impulse transmission. Thus damage to the S-A node, as often occurs with atrial septal defect repair or proximal endarterectomy of the right coronary (when it supplies the S-A nodal artery), or as seen chronically in the so-called sick sinus syndrome, may result in erratic impulse formation. Sinus arrest, sinus pauses, sinus bradycardia, and mixed atrial tachyarrhythmias may supervene, and response to atrial pacing may be beneficial (Fig. 23-2).

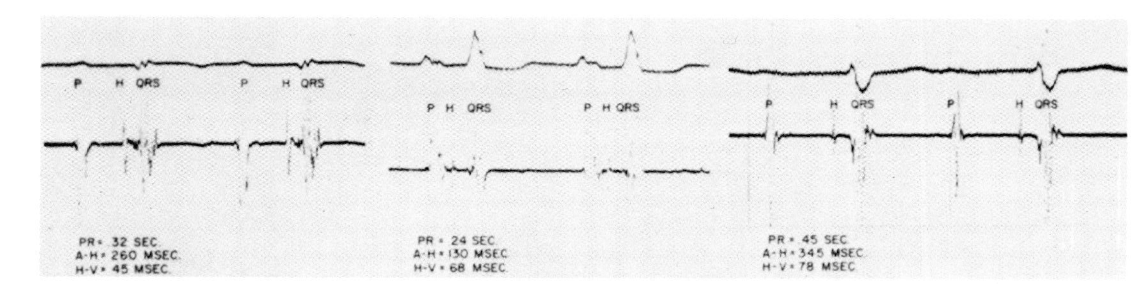

Fig. 23-3. His-bundle and simultaneous surface electrocardiogram showing three forms of first-degree atrioventricular block (above), and below the His spike and a combination of each. (Reproduced by permission from Haft J: The His bundle electrogram. Circulation 47:899, 1973.)

3° AVB (A. Flutter)

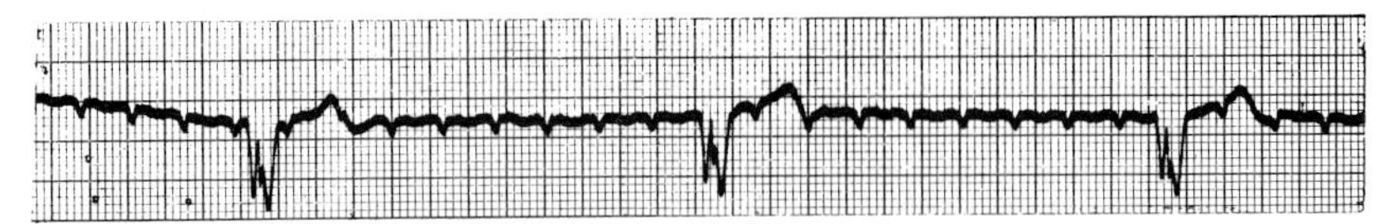

Fig. 23-4. Complete heart block with an idioventricular rhythm and atrial flutter.

Delay in transmission of electrical activity from the S-A node to the ventricle without actual interruption results in prolongation of the PR intervals to varying degree. While it is not possible from the surface EKG to define at what level this delay is imposed (i.e., above or below the His bundle), intracardiac electrograms of His-bundle activation time may prove useful (Fig. 23-3). Extreme vagotonia, digitalis excess, inflammatory diseases, and ischemia of the A-V node are common causes. While the normal PR conduction time is influenced by heart rate, a delay of greater than 0.21 sec is normally considered first-degree A-V block. Further progression of this delay will result in intermittent failure of transmission of atrial activity through the A-V node and bundle of His, causing second-degree A-V block. If this delay in transmission is above the His bundle, as is usually the case in heart block of congenital origin, extreme bradyarrhythmia seldom ensues, despite progression to 3° A-V block, because of escape foci high in the bundle branch system. With block below the His bundle in the bundle branches themselves, Mobitz II A-V block may result; and if complete block occurs unreliably slow escape foci are common. With second-degree block of either variety, varying rates of conduction between atria and ventricles may be seen. Only constant monitoring or His-bundle electrograms can distinguish between Mobitz I and II block. Varying combinations of arrhythmias may coexist within the same patient on a continuous rhythm strip, and panconduction system disease is often present. Thus patients may have evidence of sick sinus disease coexisting with interruption of impulse transmission (Fig. 23-4).

INDICATIONS AND TECHNIQUE OF CARDIAC PACING

The indications for insertion and use of a pacemaker electrode vary according to many factors, including the availability of personnel trained in the technique, the nature of the underlying conduction disturbance and escape mechanism, and the age and general health of the patient. The latter two are perhaps the weakest considerations, and many patients well

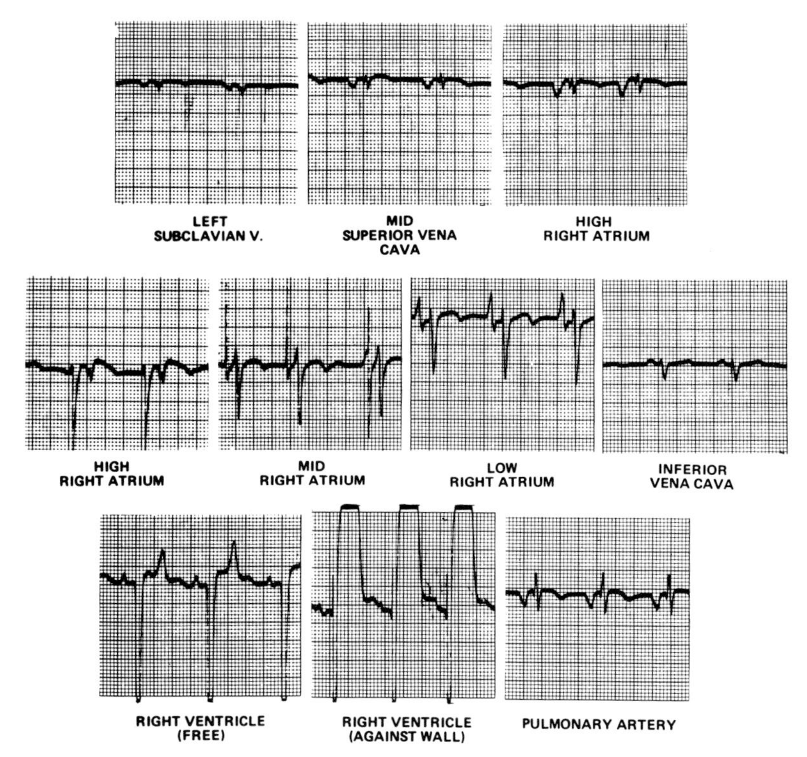

Fig. 23-5. Intracavitary electrocardiogram. Note the dominant intraatrial P complex. An injury current is seen on the ventricular electrogram as the electrode makes contact with the endocardium. (Reproduced by permission from Bing O, McDowell JW, Hantman J, Messer JV: Pacemaker placement by electrocardiographic monitoring. N Engl J Med 287:651, 1972.)

into their eighth, ninth, and tenth decades have had gratifying responses to electrical pacing. Since the indications for temporary and permanent pacing differ somewhat, they will be considered separately.

Temporary Pacing

Insertion of a temporary electrode is performed by several methods, depending upon the urgency of the clinical situation, the skill of the physician, and the equipment used. Blind insertion at the bedside may best be performed via subclavian or internal jugular vein puncture and passage of a small-caliber electrode through a Teflon sheath. EKG monitoring of the intracavitary electrogram (Fig. 23-5) allows recognition of the position of the electrode tip; with experience, reliable ventricular pacing

Table 23-1
Indications for Temporary Pacemakers

Urgency	Clinical Situation	Preferred Method*
Acute	1. Inferior myocardial infarction with:	
(Emergency)	Symptomatic sinus bradycardia unresponsive to atropine	A
	Unstable Mobitz type I block	V or A-V
	2. Anterior myocardial infarction with:	
	RBBB + RAD ± 1° A-V block	V or A-V
	RBBB + LAD ± 1° A-V block	V or A-V
	LBBB ± 1° A-V block	V or A-V
	Progression of above to Mobitz type II block	V or A-V
	3. Postoperative cardiac surgery:	
	Suppression of ventricular irritability	A or V
	Control of intraoperative block	V or A-V
	Aid in rhythm analysis	atrial wire electrogram
Chronic	1. All patients undergoing permanent pacemakers	A, V, or A-V
(Elective)	2. Any bradyarrhythmia in a patient undergoing general anesthesia	A, V, or A-V
	3. Digitalis toxic arrhythmia	A, V, or A-V
	4. Overdrive of ventricular irritability	V or A
	5. Conversion of atrial arrhythmias by rapid atrial stimulation	A
	6. His-bundle electrocardiography	V

*A = atrial pacing, V = ventricular pacing, A-V = synchronous or sequential A-V pacing (in general, demand or noncompetitive pacemakers should be employed).

Table 23-2
Indications for Permanent Pacemakers

Method of Pacing	Clinical Indications	Remarks
Atrial	1. Symptomatic* sinus bradycardia, sinus arrest, sick sinus syndrome	Unlike ventricular pacing system, a fixed-rate generator is used to avert suppression of atrial pacer by its own QRS, or alternatively a generator with a refractory period longer than the PR interval
	2. Suppression of atrial or more commonly ventricular irritability	Variable-rate generator employed and increased to rate required to suppress irritable focus; concurrent drug therapy usually required
Ventricular	1. Symptomatic* A-V block or other bradyarrhythmia	
	2. Suppression of ventricular irritability	Same as 2 above.
Atrioventricular synchronous	1. Physically active young individuals where rate increase in response to physiologic stress is important	Present pervenous method tends to be unreliable due to unstable atrial electrode position
Atrioventricular sequential	1. With bradyarrhythmias where properly timed atrial contraction is important for maximal cardiac performance	

*Symptomatic includes syncope or near syncope, CHF, or low-output symptoms.

can be established in 80% of attempts. Insertion of the electrode by cut-down on the external jugular or basilic venous system and positioning of the electrode under direct fluoroscopic vision, if convenient, is preferred by many institutions. The advantages of one technique over the other are not clearly defined, other than the simplicity of the bedside approach when it is successful versus the somewhat greater success rate of the fluoroscopic method. Most busy catheterization laboratories employ both of the methods. Indications for the use of temporary pacing electrodes may best be itemized in tabular form (Table 23-1).

Permanent Pacing

Since introduction of permanent pervenous pacemaker insertion has reduced the operative mortality to nearly zero, few contraindications to their use exist, and they are widely employed in many types of cardiac disorders where rate control contributes to management. Rather than an exhaustive listing of all clinical conditions under which permanent pacemakers have been employed, a tabular listing of representative uses of each method is presented (Table 23-2).

The vast majority of clinicians dealing with pacemakers now favor the pervenous over the direct epicardial method due to its much lower mortality and the ease with which the procedure can be accomplished. Occasions will continue to arise, however, when use of the epicardial method will be desirable, such as intraoperative heart block complicating cardiac surgical procedures. Although different groups have varying preferences for electrode insertions (jugular, cephalic, axillary) or generator position (prepectoral, subpectoral), the general principles of insertion are as described below.

TECHNIQUE OF PERMANENT TRANSVENOUS PACEMAKER INSERTION

The procedure is usually performed in a cardiac catheterization laboratory or operating room having fluoroscopic facilities. Patients may be premedicated, but general anesthesia is seldom required. Prophylactic antibiotics are recommended. Electrode insertions are made via the cephalic vein or via a deeper branch of the axillary system. No preference is shown toward right or left sides, except in a few instances where the patient's occupation or hobbies require more vigorous use of one arm than the other. Electrode tips are wedged firmly beneath the trabecular system of the right ventricle under fluoroscopic control. When properly positioned, gentle traction on the electrode reveals the tip to be resistant to withdrawal. Endocardial stimulating thresholds are determined in every instance, and the electrode is withdrawn and repositioned until a satisfactory level is attained. Stiffening stylets are removed, and the vein is ligated to prevent backbleeding. Stitches around the electrode are avoided, and the electrode is anchored with a small silastic butterfly. Pulse generators are buried in a pocket over the right or left chest. Wounds may be closed with subcuticular catgut or interrupted nylon sutures. Pressure bandages are applied and remain in place for 7 days. Temporary pacemaker electrodes are usually inserted in every patient prior to the permanent implant to allow intraoperative control of the cardiac rhythm. In each case these are left in place for at least 48 hr after the permanent unit is inserted. Antibiotics are continued for a further 48 hr after removal of the temporary electrode.

Complications of permanent endocardial pacing have been quite uniform from one center to another with rare exception. Intraoperative death has been unusual. The hospital mortality ranges from 1% to 2% due in most instances to acute myocardial infarction or progression of congestive heart failure. Late deaths following successful pacemaker insertion occur in approximately 8% per year follow-up and are usually due to conditions common to an elderly population (e.g., myocardial infarction, CHF, cancer, strokes). Complications other than death involve either the electrode system or the area of the implantation. Early problems of electrode perforation or dislodgement diminish with increasing experience, but result usually in intermittent or complete failure of the pacing system to sense or pace properly. In the case of perforation into the pericardial space, transitory chest pain may occur, and there is often stimulation of the retrosternal or intercostal muscles simulating a cardiac impulse where none exists. Diaphragmatic pacing may also occur. Pericardial tamponade seldom occurs in the absence of anticoagulant therapy.

TYPES OF PACING

For both temporary and permanent use, a variety of pacing systems are available differing in electronic design and the purpose for which they are used.

Asynchronous

The stimulation rate is independent of the electrical or mechanical activity of the heart and in the presence of any endogenous rhythm will cause competition. These units may be used for atrial or ventricular

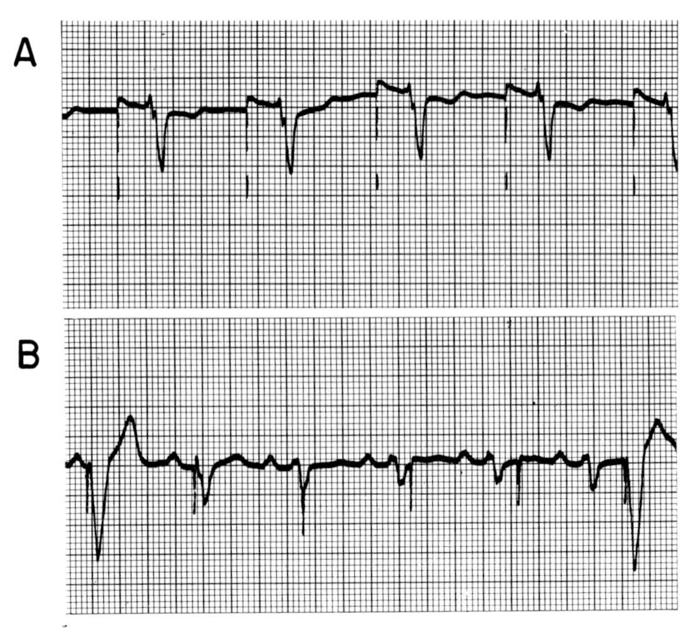

Fig. 23-6. Asynchronous pacing. Panel A shows asynchronous stimulation of the atrium, while panel B shows ventricular pacing. Note the competition between the paced and intrinsic cardiac rhythm in the latter instance.

pacing, but should not be used for ventricular pacing where a low ventricular fibrillation threshold is anticipated (recent myocardial infarction, ischemic heart disease, or recent cardiac surgery) (Fig. 23-6).

Ventricular Inhibited

This pulse generator suppresses its output in response to natural ventricular activity, but produces an asynchronous impulse in the absence of natural activity. Most ventricular inhibited generators have a

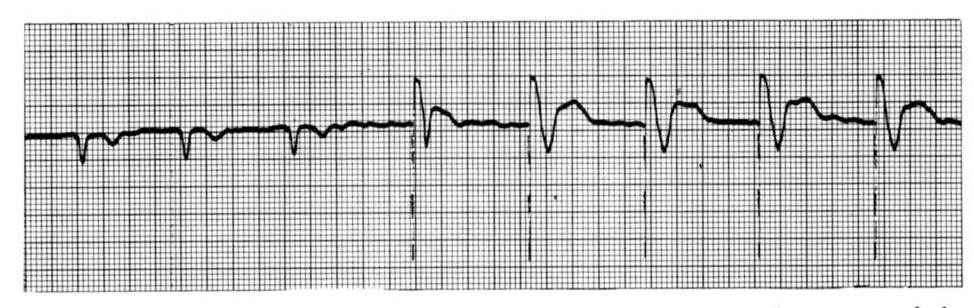

Fig. 23-7. Ventricular inhibited pacing. The first three complexes exceed the preset escape interval of the pacemaker and inhibit its discharge. With further slowing, ventricular pacing ensues.

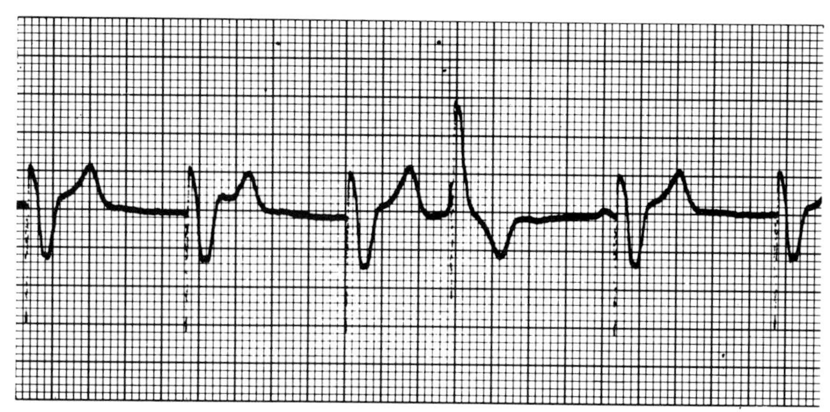

Fig. 23-8. Ventricular synchronous pacing. The rhythm is predominantly paced ventricular. The fourth complex is a premature contraction above the escape rate of the pacemaker and causes instantaneous discharge into its refractory period.

limited maximum sensitivity to avert P-wave or T-wave sensing (1.5–2.0 mV). As a consequence, when the intracardiac electrical impulse recorded across the pacing electrodes is less than that value, competition may result. Conversion to a unipolar configuration by use of a separate indifferent electrode and cathodal stimulation will usually correct this problem (Fig. 23-7).

Ventricular Synchronous or Ventricular Triggered

The output is delivered synchronously with the natural ventricular activity. With this type of system, each naturally occurring QRS has a superimposed pacing artifact just after the inscription of the upstroke. To avert T-wave synchrony these units have an electronic refractory period of approximately 400 msec and may not sense early impulses occurring within that interval (Fig. 23-8).

Atrioventricular Synchronous

The stimulation rate of the pulse generator is directly controlled by the atrial rate by way of a double-electrode system. The mechanism is analogous to a spinal reflex arc in that the atrial electrode serves as an afferent loop through which the pulse generator senses atrial contraction and initiates a pacemaker discharge down the ventricular electrode at an appropriate P-R interval that thus acts as an efferent loop. To avert one-to-one responses to atrial tachyarrhythmias, these pacemaker systems have an electronic refractory period, which results in 2:1 or 3:1 block at higher atrial rates (Fig. 23-9).

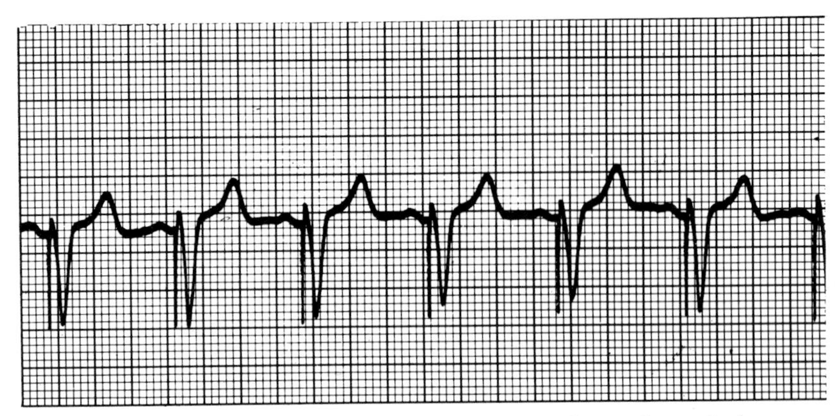

Fig. 23-9. Atrioventricular synchronous pacing. Each paced ventricular complex is preceded by the patient's own P wave, which is sensed by an atrial electrode and triggers the pacemaker to produce ventricular pacing at an appropriate PR interval thereafter.

Atrioventricular Sequential

This type of pacing employs a further modification of the foregoing double-electrode technique in which atrial and ventricular electrodes are each used for active stimulation of atrium and ventricle, respectively, with an appropriate PR interval between the two. This mode of stimulation is most commonly used in the treatment of acute block situations where the contribution of atrial contraction to maintenance of cardiac output is critical. Thus in patients with cardiogenic shock or in the postoperative surgical patient with A-V block this method may improve cardiac performance by 15%–20% (Fig. 23-10).

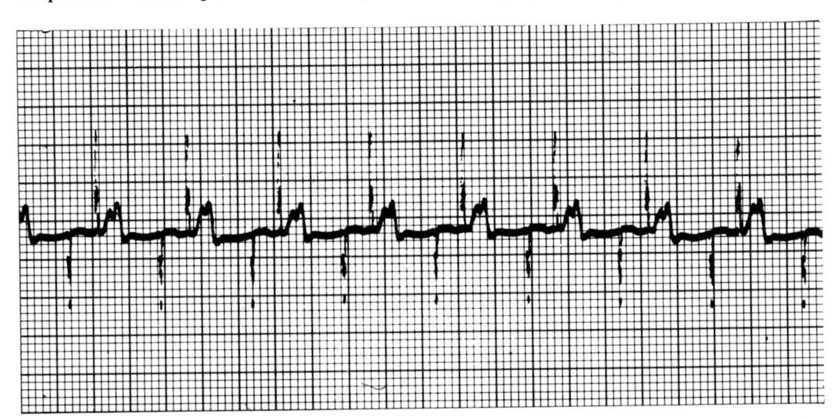

Fig. 23-10. Atrioventricular sequential pacing. The atria and ventricles are both actively stimulated through a double-electrode system with an appropriate PR delay intervening between the stimuli.

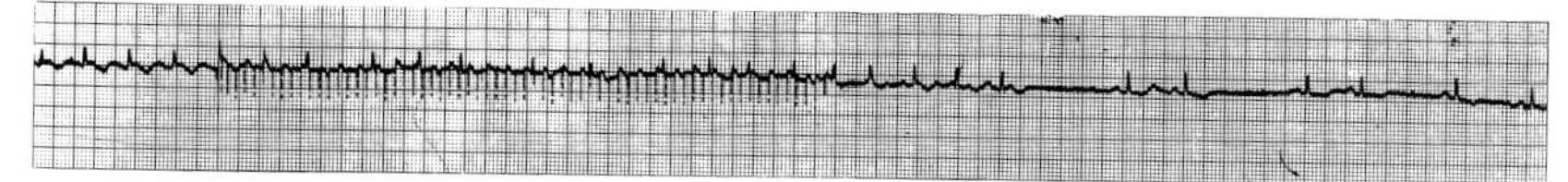

Fig. 23-11. Rapid atrial stimulation. The initial rhythm is atrial flutter with 2:1 block. Stimulation of the atrium at 600 impulses/min results in reversion to sinus rhythm.

Rapid Atrial Pacing

Stimulation of the atrium at rates of 200 to 600 impulses per minute via pervenous and transthoracic atrial electrodes may be useful in conversion of rapid regular supraventricular tachyarrhythmias such as atrial flutter or atrial tachycardia (Fig. 23-11). The technique requires special equipment not generally available and is seldom effective in atrial fibrillation.

THE PHYSIOLOGIC BASIS FOR CARDIAC PACING

The hemodynamic changes induced by heart block and by atrial and/or ventricular pacing should be understood in order that logical clinical judgment can be used in evaluating patients for cardiac pacing and deciding which type of pacemaker to use. Physiologic cardiac impairment in heart block is related to several factors (Table 23-3).

Bradycardia, if severe (HR < 40) or if associated with significant myocardial disease, can lead to symptomatic depression of cardiac function. Cardiac output is diminished, and right and left ventricular end-diastolic pressures and right and left atrial mean pressures become elevated. When myocardial disease coexists these abnormalities may be quite striking and frequently lead to congestive heart failure.

Table 23-3
Factors Implicated in the Physiologic Cardiac Impairment of Heart Block

Decreased heart rate
Inability of heart to respond to stress by increasing frequency
Loss of the dynamic effect of coordinated atrial contraction
Total dissociation
Disturbances of the optimal P-QRS interval
Diminished myocardial performance due to:
Decreased heart rate
Intrinsic myocardial dysfunction

In the presence of complete heart block when the escape pacemaker focus resides in the ventricle (idioventricular rhythm) an increase in cardiac frequency with stress is usually severely limited or not present. Increased circulatory demands imposed upon the heart must then be accomplished by an increase in stroke volume, and the cardiac reserve or extent to which the stroke volume can be augmented is thus extremely important. Increased stroke volume is normally accomplished by improvement in ventricular ejectile ability and increase in ventricular volume through the Starling mechanism. When myocardial disease is present, cardiac volume and diastolic pressure are often maximally increased, and stroke volume cannot be further augmented despite bradycardia. Cardiac output is reduced, and exertional stress results in little if any increase in stroke volume at the expense of large increases in ventricular filling pressures, and further congestion ensues.

When the relationship of atrial contraction to ventricular contraction is disturbed, decreased cardiac performance with diminished cardiac output and elevated ventricular filling pressure may also result. Again, these effects are further augmented with diminishing cardiac reserve. At one end of the spectrum of abnormal atrioventricular synchrony is a deviation from the optimal interval between atrial and ventricular contraction, as in first-degree heart block, while at the other end is total dissociation between atrial and ventricular contraction. Abnormal atrioventricular synchrony is seldom of clinical significance except in the presence of impaired ventricular function. The optimal PR interval for maximum cardiac performance is normally 0.14–0.17 sec (Fig. 23-12). As the interval diverges from this range, cardiac output may fall. More significantly, when the contribution of atrial systole to ventricular filling is totally eliminated, as with idioventricular rhythm and complete heart block, ventricular filling pressures are most abnormal. The influence of atrial systole on cardiac performance may best be appreciated by observing the systemic arterial and jugular venous pulses with complete heart block (Fig. 23-13). When atrial systole occurs at the optimal interval before ventricular systole, pulse pressure and peak systolic pressure are highest, while pulsatile and mean venous pressures are lowest. On the other hand, when atrial contraction occurs during ventricular contraction, while the atrioventricular valves are closed, arterial pressures are lowest and venous pressures highest. The impressively large "cannon" jugular venous wave occurs at this time and may be as high as 20 mm Hg. Occasionally the atrial contraction can generate a pressure higher than the simultaneous ventricular pressure during ventricular systole, and A-V valve regurgitation may result.

The foregoing discussion of A-V block provides a format for clinical selection of one of the various types of pacemakers. Thus a ventricular

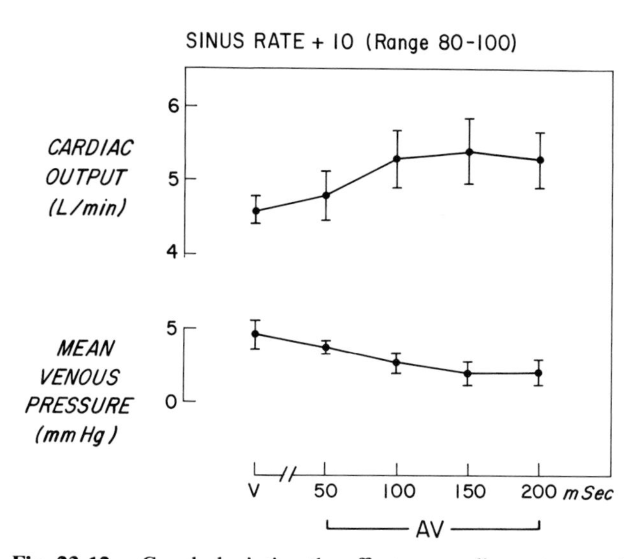

Fig. 23-12. Graph depicting the effect on cardiac output and central venous pressure of varying the PR interval during atrioventricular pacing. (Reproduced by permission of Leinbach RC, Chamberlain DA, Kastor JA, et al: A comparison of the hemodynamic effects of ventricular and sequential atrioventricular pacing in patients with heart block. Am Heart J 78:502, 1969.)

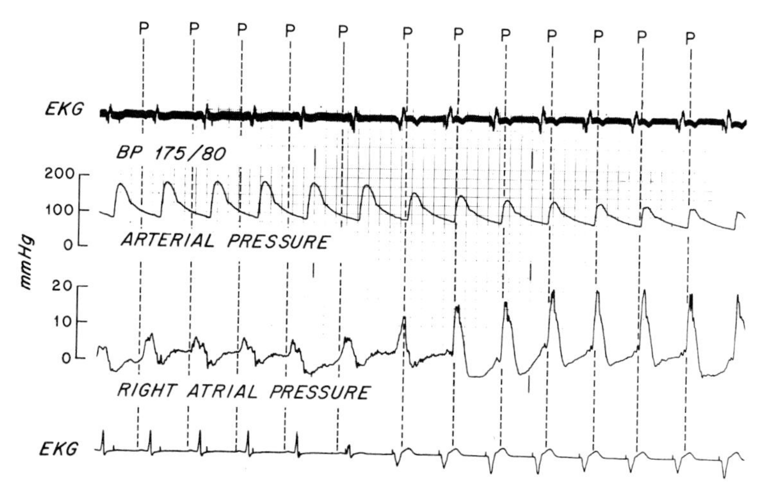

Fig. 23-13. Atrioventricular dissociation produced by fixed-rate ventricular pacing. With the onset of ventricular pacing (complex 6) atrial contraction follows the QRS with resultant "cannon" a waves in the venous pressure and fall in the arterial pressure (P = P wave).

electrode pacemaker is generally used to increase heart rate when cardiac function is adequate without preceding atrial contraction. Atrioventricular synchronous or atrioventricular sequential pacing allows for preservation of the improved cardiac function provided by a normally timed atrial contraction in patients with substantial impairment of cardiac reserve, as with extensive myocardial infarction or cardiomyopathy or in younger individuals where variable physiologic demands are needed. Physiologic testing in the hemodynamics laboratory may be of value in assessing the benefits of atrioventricular pacing. Atrial pacing alone can only be performed in the presence of intact atrioventricular conduction and is less frequently utilized than ventricular pacing due to the relative instability of presently available atrial electrodes. It may, however, be exceedingly useful for overdrive suppression of ectopic arrhythmias, as in patients following myocardial infarction or cardiac surgery and as an aid in the long-term management of these arrhythmias.

Physiologic effects of various modes of cardiac stimulation have been studied with particular reference to the atrial contribution to cardiac output in patients with complete heart block. Ventricular pacing alone produces relatively modest increments in cardiac output with increasing rate, but does result in phasic variations of the arterial and venous pressures due to dissociation of atrial and ventricular contractions. The augmentation of stroke volume mediated by atrial contraction through the Starling mechanism occurs randomly. While the literature is somewhat conflicting, most investigators support the importance of a properly timed atrial contraction to maintenance of cardiac output, although the effect may be less marked in patients with severely diseased myocardial function. The average augmentation of cardiac output by a properly timed atrial contraction in patients with chronic heart block has been found to be 10%–24% in comparison to that observed with ventricular pacing at the same rate.[6,7] Similar findings have been demonstrated in patients with acute block due to myocardial infarction.[8]

Long-term physiologic effects of cardiac pacing have not been widely studied. Young active patients in whom proper physiologic response to stress is desirable (e.g., strenuous physical activity, pregnancy, adolescent growth) may be suitable candidates for atrioventricular synchronous pacemaker systems that provide venticular stimulation in response to the normal variations in sino-atrial discharge produced by such activities. Patients with more diffuse conduction system disease in whom the frequency of sino-atrial discharge is inadequate may benefit from atrioventricular sequential pacemaker systems.

The small number of reports in the literature on the hemodynamic results of cardiac pacing are conflicting. The early rise in cardiac output seen with increased rate may ultimately return to prepaced levels. Furman and Escher,[9] however, found a 21% average early increase in cardiac output with rate increase to 74 per minute and a further average 13% increase at the time of rate restudy, for a total 37% average increase of cardiac output over prepaced levels. While exercise produces little further improvement in patients with fixed-rate pacers, an additional 18%–20% improvement during exercise may be realized with atrial synchronous systems.

POWER SOURCES OF IMPLANTABLE PACEMAKER SYSTEMS

Present implantable pacemaker systems are powered in most instances by the Mallory RM-1 mercury–zinc cell. While theoretical calculation of generator longevity when powered by these cells is 5 years, only slightly less than 2 years are realized in the hostile environment of the human body. Several aspects of this power cell result in premature depletion. The paper separator between the anode and cathode allows the passage of free mercury created as the cell depletes and results in internal short-circuiting with rapid cell depletion. Because hydrogen gas is also produced during the chemical interaction within the cell, this must be vented through a gasket that allows body fluids to enter the power cell, with resultant bridging of the gap between the two poles of the cell. Modification of this basic cell design incorporated in the LeClenche cell employs a horizontal sandwich configuration with improved separator efficiency. Although preliminary tests suggest improved expected pacemaker longevity of 36–40 months, ultimate replacement of the power pack must be anticipated with the attendant hazards of sepsis plus increasing cost of hospitalization. The search for a pacemaker power supply having protracted reliability has been long and arduous and has basically involved five approaches: (1) chemical cell (Hg–Zn, LiI), (2) magnetic-induction system, (3) rechargeable nickel–cadmium cell, (4) biogalvanic cell systems, and (5) nuclear-powered pacemaker.

Chemical Cell

Present chemical sources of energy supply include the Rubin mercury–zinc cell described above and the recently introduced lithium iodide cell. With modifications of battery construction, improved design of separators, hermetic sealing, and external noninvasive electromagnetic circuit programming to minimize current drain per impulse, effective generator function should exceed 4 years.

The recently introduced non-gas-emitting lithium iodide cell can be hermetically sealed without the problems inherent in the use of a "gas getter." In the foregoing chemical cell the anode is lithium and the cathode is an iodine-containing compound. Migrations of ions through lithium iodide electrolyte result in the generation of electricity. Clinical experience to date is inadequate to assess long-term performance, but projected battery life for the 2-amp-hr cell is 7.6 years and for the 4-amp-hr cell is 15.2 years.

A mercury–cadmium cell with a projected life of 40 years is presently under test but has not been put into clinical trial.

Magnetic-Induction System

The heart may be directly stimulated by myocardial electrodes connected with a subcutaneous secondary coil in which impulses are induced by an external overlying primary coil connected to a pulse-forming apparatus. Although the technique theoretically avoids the need for repeated surgery, problems with maintaining proper apposition of the two coils and inconvenience and risk to the patient who is pacemaker-dependent have discouraged its wide application. No such device is commercially available in this country.

Rechargeable Nickel–Cadmium Cell

Elmquist and Senning[10] in 1959 introduced a fixed-rate implantable pacemaker powered by nickel–cadmium cells that could be externally recharged inductively from outside the body by means of a secondary coil and a silicium diode in the pacemaker. Standard Ni–Cd cells function most efficiently at temperatures lower than that of the human body, and premature deterioration of these early cells led to their infrequent use. More recent modifications of the nickel–cadmium cell promise longer generator life. Ten years of effective pulse generator function is presently anticipated, but experience to date is too brief to warrant conclusions.

Biogalvanic Cell Systems

A variety of biologic sources of energy production have been investigated, but most proved impractical. Satinsky et al.[11] in 1965 employed an electrochemical process in the form of a positive platinum electrode and negative steel electrode, and successful stimulation of animals could be achieved. Ultimate deterioration of electrode discs due to oxidative corrosion prevented long-term use. More recent modifications may introduce for the first time a lifetime pacemaker system powered by biogalvanic energy sources.

Nuclear-powered Pacemaker

The use of plutonium 238 as the fuel of a thermoelectric generator has provided a power source sufficient to deliver a power supply of 0.6 V. Amplification in a d-c/d-c converter to 7.5 V provides energy sufficient to stimulate the heart through a standard pacemaker circuit. The isotope used has a half-life of 87 years. The estimated generator life of at least 10 years is roughly double that of current estimates for chemically powered pacemakers.

RESULTS OF CARDIAC PACING

The ultimate benefit to the patient of the various pacing modalities enumerated above is determined by the nature of the underlying disease process for which the technique has been employed. Best results are obtained in patients with the Stokes-Adams syndrome, which carries a 1-year mortality of 50% when treated medically but which can be eliminated by effective pacing. Patients with bradyarrhythmias associated with congestive heart failure fare less well with rate increase and, although initially improved, often go on to die of refractory congestive failure. The longevity of patients in whom pacemakers are used as an adjunct to the suppression of arrhythmias is determined by the combined effectiveness of antiarrhythmic agents and pacing. Considerable dispute exists over the efficacy of pacing in patients with acute myocardial infarction complicated by heart block. Some authors report little influence on ultimate survival figures, despite use of temporary pacing techniques. Regardless of late survival figures in such series, stabilization of rhythm during the early postinfarction phase aids in the early management of such patients and provides a "breathing" period during which more elaborate coronary angiographic study and surgical intervention may be considered.

PACEMAKER FOLLOW-UP AND MANIFESTATIONS OF MALFUNCTION

The vast majority of permanent pacemaker systems presently employed are powered by the Rubin Mallory mercury–zinc cell. Fixed-rate generators with their simple circuit design often provide effective stimulation in excess of 30 months, while most demand systems require replacement in 24 to 30 months. All patients with implanted pacemaker systems must be carefully followed for evidence of premature generator depletion. This follow-up may consist in its simplest form of instructing the patient in counting his pulse each morning with directions to call the responsible physician if it is below the escape rate of the pacemaker. Most pacemaker centers have their pacemaker patients return for regularly scheduled visits during which a general medical exam is performed, EKG recorded, and pacemaker interval determined. Initially such visits are spaced at intervals of 3–4 months, but more frequent examination (every 1–2 months) is appropriate as the time of expected battery depletion draws near. More recently, equipment for telephone transmission of EKG rhythm strips has become available and allows the physician to maintain continued surveillance of pacemaker function without requiring actual hospital visits. Although pulse-wave analysis of pacemaker stimuli provides ancillary information regarding pacer function, it does not contribute significantly to detection of early battery depletion over simple rate determination and is not routinely utilized.

DETECTION OF PACEMAKER MALFUNCTION

It is not possible to present an elaborate review of all of the various manifestations of pacemaker malfunction, which basically involve three aspects of the pacing system.

Battery Depletion

Although theoretical calculation of generator function based on frequency of discharge, current drain, and available power supply suggests an effective function of 5 years, this is seldom realized in practice. As a

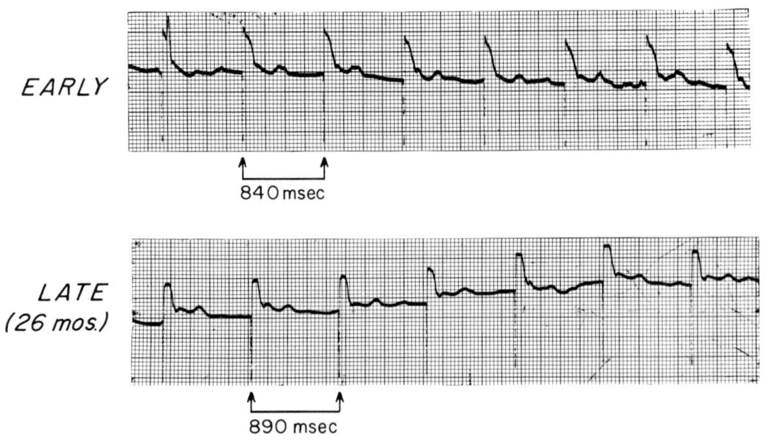

Fig. 23-14. Serial rhythm strips demonstrate slowing of the stimulation rate as evidence of battery depletion.

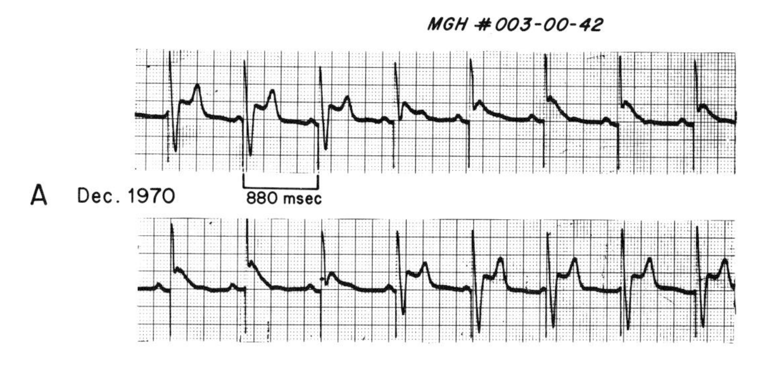

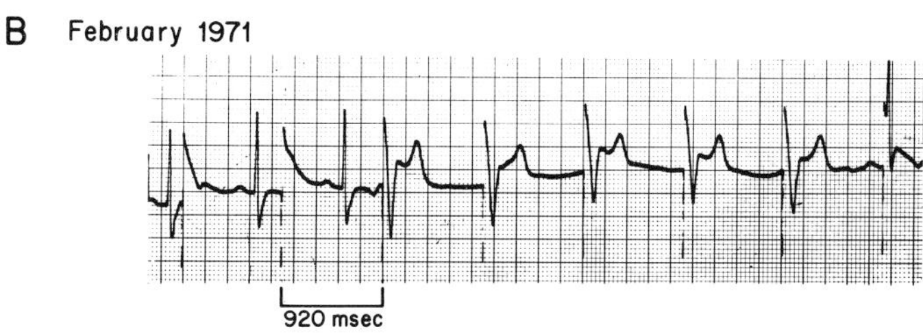

Fig. 23-15. Failing R-synchronous pacemaker with reversion to asynchronous stimulation as evidence of battery depletion.

rule, pacers with simple circuits (i.e., fixed rate) last longest, while more complicated models (demand, A-V synchronous, A-V sequential) show earlier evidence of "premature" battery depletion. Two indicators in serial EKG recordings serve to identify battery depletion. A change in the basic pacing rate (usually slowing) or loss of demand function (Figs. 23-14 and 23-15) dictate the need for generator replacement.

Component Malfunction

This is fortunately one of the least frequent but often most lethal of the problems that beset pacemaker systems. Despite rigid controls and testing procedures, an occasional electronic component may fail, often resulting in sudden cessation of pacemaker function or, rarely, pacemaker "runaway" (Fig. 23-16). The frequency of resultant death due to this cause is unknown, but it seems very infrequent in the authors' experience.

External radiofrequency interference with demand pacemaker circuits has been reported and can be reproduced; it is more common with temporary pacemakers (Fig. 23-17). Implantable pacemaker systems currently in use are undergoing various design changes to avert such undesirable features. In the meantime it is appropriate to advise patients to

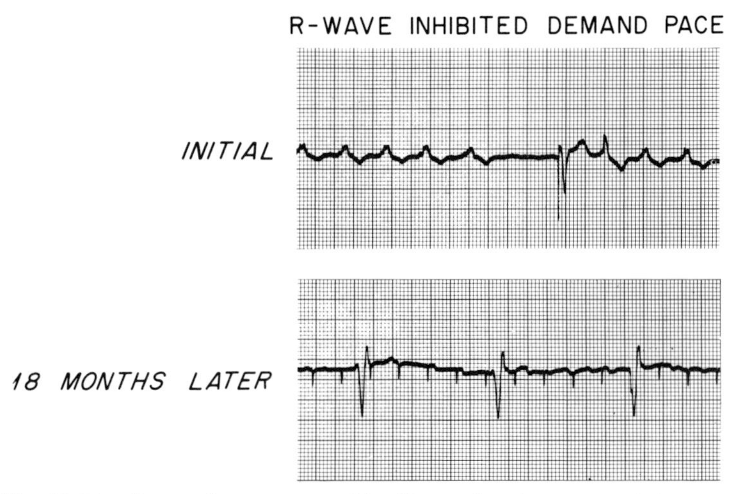

Fig. 23-16. Pacemaker runaway. The illustration demonstrates increase in the pacer discharge rate to over 200 per minute due to a component failure. Stimulation of the heart is prevented by decrease in the energy per impulse as the rate increases. Rarely, stimulation of the heart may occur and culminate in fatal ventricular fibrillation.

avoid unnecessary exposure to strong electromagnetic or ultrasonic energy sources or electrocautery and to report any unusual symptoms of dizziness associated with the use of electromechanical devices.

Electrode Failure

The most obvious problem is a break in electrode continuity with resultant interruption of paced rhythm. This has been infrequent with pervenous systems, but is more common with epicardial pacemakers and

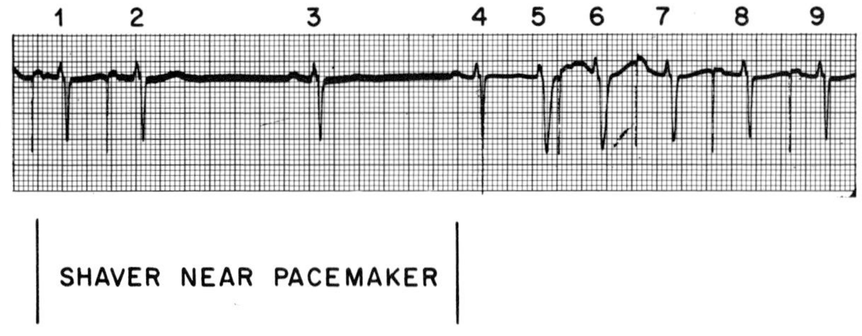

Fig. 23-17. Radiofrequency interference. Note the suppressed output of a temporary demand pacemaker (atrial pacing) when an electric shaver is held in close proximity to the pacemaker. (Reproduced by permission from Crystal RG, Kastor J, De Sanctis RW: Inhibition of discharge of an external demand pacemaker by an electric razor. Am J Cardiol 27:695, 1971.)

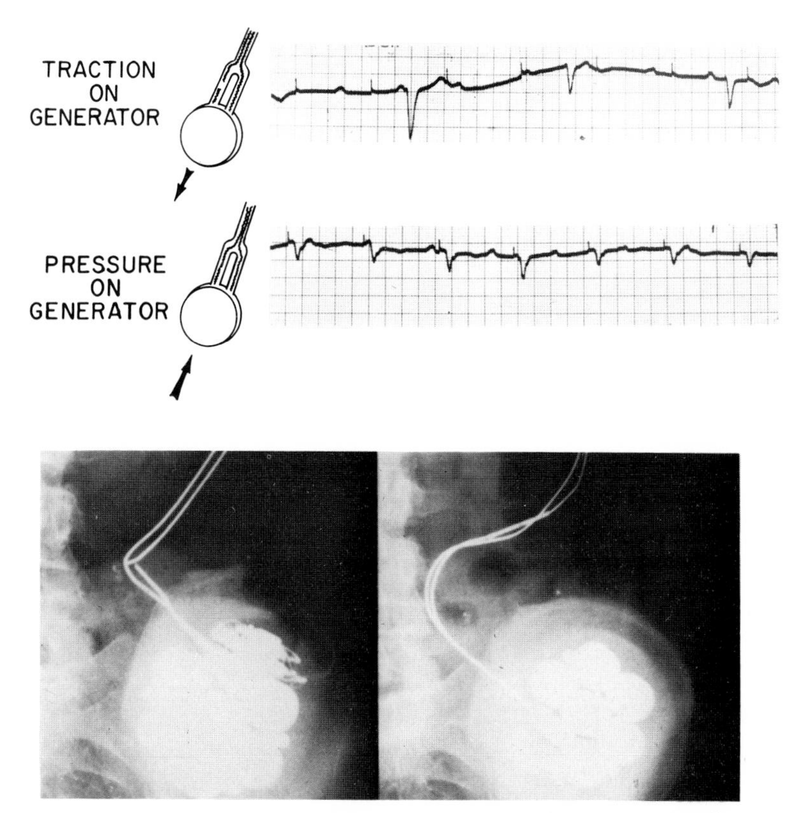

Fig. 23-18. The x-ray demonstrates a broken (epicardial) electrode adjacent to the pulse generator before and after splicing. Failure to stimulate the heart may be intermittent as the broken lead makes and breaks contact.

can often be identified on x-ray (Fig. 23-18). Local lead splicing usually corrects the problem. Perforation of the heart by the electrode tip is confined to pervenous systems, and erratic function ranging from occasional failure to capture the ventricle or sense properly to completely ineffective function. Transitory chest pain and stimulation of intercostal or diaphragmatic muscles may occur. Pericardial tamponade is rare. Revision and repositioning are often necessary to correct the problem. Loss of a stable endocardial position is perhaps the most common electrode problem and diminishes with increasing experience on the part of the physician performing the procedure. Erratic capture, inappropriate demand function, ventricular irritability and unusual mobility of the electrode within the heart on fluoroscopy usually identify the problem. Regardless of the underlying cause of pacemaker malfunction, reexploration and revision are usually indicated to correct it. Instability of

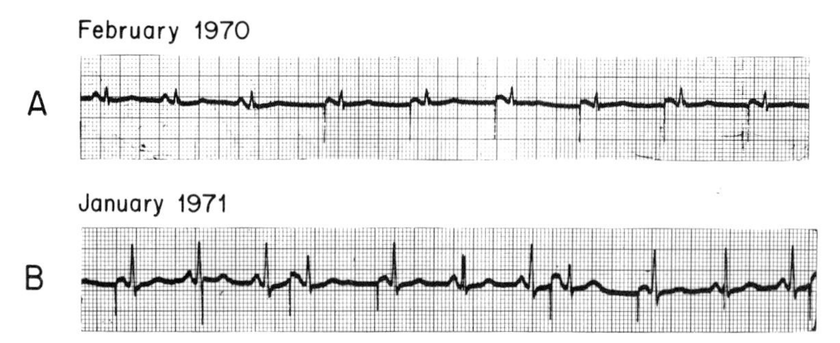

Fig. 23-19. Atrial pacemaker. Faulty demand function due to unsatisfactory lead position. Note competition between the pacemaker and the intrinsic sinus mechanism.

electrode position is particularly common with pervenous atrial pacing systems and often results in erratic demand function (Fig. 23-19).

CARDIOVERSION

External countershock therapy has been used clinically in the treatment of arrhythmias since 1962 when it was introduced by Lown,[12] and it may be performed on an elective or emergency basis. Elective cardioversion is appropriate in the management of those arrhythmias that are not immediately life-threatening, do not revert with pharmacologic therapy, and have a reasonable expectation of preservation of sinus rhythm. Emergency cardioversion is performed for any arrhythmia that is potentially life-threatening, hemodynamically embarrassing, or severely symptomatic and that is unresponsive to pharmacologic management. Ventricular fibrillation is associated with cessation of cardiac output and must be cardioverted immediately. Ventricular tachycardia may lead to ventricular fibrillation, and if not immediately responsive to drug therapy, it should be cardioverted. In the management of all of these arrhythmias, combined antiarrhythmic therapy is indicated to avert a recurrence. Atrial arrhythmias associated with a rapid ventricular response may lead to acute left ventricular failure, accentuation of angina pectoris, or myocardial infarction and thus require emergency conversion.

Complications

Complications of cardioversion include systemic embolism, creation of ventricular fibrillation or other arrhythmia, pulmonary edema, or postconversion electrocardiographic changes. Systemic embolization has been reported in 1%–3% of patients and is more common among patients with a prior history of embolism and with patients with associated mitral valve disease. Anticoagulation of such patients for 10 to 12 weeks prior to cardioversion is recommended. Ventricular fibrillation as a consequence of attempted cardioversion is generally the consequence of an improperly timed shock and requires immediate repeat countershock. Pulmonary edema rarely supervenes following reversion to sinus rhythm and has been attributed to dyssymmetric right and left atrial function with greater right ventricular output than can be accommodated by the left atrium. Finally, electrocardiographic changes including ST- and T-wave abnormalities may appear and rarely may be accompanied by the appearance of new Q waves. Alterations of serum enzyme determinations including SGOT and CPK have generally been attributed to skeletal muscle changes.

Postcardioversion rhythm disturbances vary according to the clinical circumstances. Nodal rhythm and atrial premature beats are common in the immediate postconversion rhythm strips and usually subside promptly. Ventricular irritability may appear, particularly in patients in whom digitalis excess may be a factor. Bradyarrhythmias are not infrequent and require atropine or even electrical pacing. Prior cessation of digitalis preparations and premedication with quinidine can avert many of these ectopic rhythm disturbances.

Contraindications

Cardioversion can be a safe and effective procedure if certain precautions are taken. Conversion of a tachyarrhythmia due to digitalis toxicity may be associated with a high incidence of postconversion arrhythmias, including ventricular fibrillation and bradycardia; digitalis should be discontinued 24 to 48 hr in advance if possible. Preconversion heart block or slow ventricular response in association with atrial fibrillation are usual contraindications to cardioversion unless a temporary pacemaker is in use. Arrhythmias that recur paroxysmally despite pharmacologic measures are usually not suitable for cardioversion unless they are life-threatening. Obvious metabolic derangement such as electrolyte imbalance, impaired oxygenation, or alteration of acid–base balance should be corrected to ensure effective and sustained cardioversion. Hyperthyroid or anemic patients with recurrent supraventricular tachyarrhythmias are best handled pharmacologically until the precipitating or aggravating disorder can be corrected. A recent history of systemic embolization is usually an indication to defer cardioversion until effective anticoagulation can be accomplished. Repeated failure of earlier attempts to retain sinus rhythm is generally a contraindication to repeated cardioversion. Generally speaking, patients with long-standing atrial arrhythmias associated valvular heart disease, cardiomegaly, or congestive failure are least likely to preserve sinus rhythm following conversion and are not ideal candidates. On the other hand, following successful surgical correction of the associated cardiac disorder (e.g., mitral valve replacement, ASD repair) cardioversion attempts may again become advisable.

Probability and Duration of Cardioversion

While the immediate success rate for conversion of an atrial arrhythmia to sinus rhythm approaches 80%–90%, subsequent reversion to the previous rhythm is common. The duration of the preconversion arrhythmia, the cardiac hemodynamic status, the presence of associated valvular lesions, and the relationship of onset of the arrhythmia to stress (e.g., pneumonia) all contribute to the ultimate likelihood of remaining in sinus rhythm. Among patients with atrial fibrillation of less than 1 year's duration, preservation of sinus rhythm for 2 years occurs in only 25%. Those patients without the aforementioned disorders of cardiomegaly, valvular heart disease, hyperthyroidism, and anemia, and in whom a recognizable precipitating stress such as pneumonia or surgery is present, have the greatest likelihood of achieving long-term benefit from reversion to sinus rhythm. Elderly patients are not usually suitable candidates for elective cardioversion because of the risks of general anesthesia and the lack of clinical symptoms in their often sedentary life styles. Most studies to confirm an increase in cardiac output with reversion from atrial fibrillation to sinus rhythm, and patients with symptomatic congestive failure may be expected to improve with successful reversion. As with other therapeutic modalities, one must weight the potential risks against the anticipated benefits.

Technique

Anticoagulation with Coumadin prior to elective cardioversion may reduce the risk of systemic embolization and is especially indicated in those patients with prior documented embolic events. Digitalis is usually discontinued 24 hr before the procedure, and metabolic derangements are corrected. Quinidine is administered for 1 to 2 days before the procedure in a dose of 200–300 mg every 6 hr and may be administered as quinidine lactate 200 mg intramuscularly just prior to the conversion attempt.

The procedure of elective cardioversion is usually performed in an operating room, recovery room, or coronary-care unit, so that accurate monitoring is available. A 12-lead electrocardiogram is obtained before and after conversion. The patient is connected to the electrodes of the synchronizer circuit of the cardioverter, and the electrocardiogram or amplitude of the monitored lead is adjusted to a level adequate for proper sensing by the synchronizing circuit. The synchronizer should be adjusted so that the electrical impulse of the cardioverter falls directly on the R wave of the subject's electrocardiogram. Proper synchrony is confirmed by holding the paddles together, discharging them, and observing the time of occurrence of the discharge spike.

With abundant electrode paste and with the patient anesthetized or heavily sedated the paddles are positioned in such a manner that the discharge impulse between the paddles traverses the heart. Variations in individual patient body shape may require various changes in paddle position before successful conversion is achieved. The hand-held chest paddles are usually placed so that one rests anteriorly over the midsternum and the other in the posterior axillary line facing the anterior paddle. Separate back paddles are available that may be positioned with one just below the scapula and the other over the anterior chest wall. Whatever configuration is chosen, it must be remembered that it is the current density that reaches the heart that determines effective cardioversion, rather than the current applied to the chest wall. No single energy level can be specified as being effective for all patients. In general, one hopes to achieve successful reversion of the arrhythmia at the lowest energy level possible, in the hope of averting electrically induced myocardial damage. Regular supraventricular tachyarrhythmias such as atrial flutter often revert at relatively low energy levels of 25–50 joules, while higher settings of several hundred joules may be required for patients with atrial fibrillation or large body build. The operator predicts an energy level based on familiarity with the foregoing factors and gradually increases the energy level of succeeding shocks until successful reversion is achieved. Rarely, energy levels above the capacity of a single cardioverter will be required and may be achieved by using two converters in series. The electrocardiogram is observed for an interval between shocks to confirm reversion or the appearance of ventricular irritability. Suppression of the latter with intravenous lidocaine is advisable.

In patients where cardioversion constitutes an emergency procedure for therapy of a life-threatening arrhythmia, maximum energy levels are usually selected to ensure prompt reversion. Since the arrhythmia is usually ventricular fibrillation, synchrony with the R wave is often not required.

Postconversion drug suppressant therapy is selected on the basis of the preconversion arrhythmia and will usually include quinidine, Dilantin, digitalis, or Pronestyl in varying combination. Anticoagulation should be continued in those patients with a prior history of systemic embolism.

REFERENCES

1. Aldini G: General Views on the Application of Galvanism to Medical Purposes. London, J. Callow, 1819
2. Hyman AS: Resuscitation of the stopped heart by intracardial therapy. Arch Intern Med 50:283, 1932
3. Gould: (1929); referred to in Hyman AS: Resuscitation of the stopped heart by intracardial therapy. Arch Intern Med 50:283, 1932
4. Weirich WL, Gott VL, Lillehei CW: The treatment of complete heart block by the combined use of a myocardial electrode and an artificial pacemaker. Surg Forum 8:360, 1957
5. Furman S, Robinson G: The use of intracardial pacemaker in the correction of total heart block. Surg Forum 9:245, 1958
6. Leinbach RC, Chamberlain DA, Kastor JA, et al: A comparison of the hemodynamic effects of ventricular and sequential A-V pacing in patients with heart block. Am Heart J 78:502, 1969
7. Samet P, Bernstein WH, Nathan DA, Lopey A: Atrial contribution to cardiac output in complete heart block. Am J Cardiol 16:1, 1965
8. Chamberlain DA, Leinbach RC, Vassaux CE, et al: Sequential A-V pacing in heart block complicating acute myocardial infarction. N Engl J Med 282:577, 1970
9. Furman S, Escher DJW: Hemodynamics of cardiac pacing, in: Principles and Techniques of Cardiac Pacing. New York, Harper & Row, 1970, p 207
10. Elmquist R, Senning A: An implantable pacemaker for the heart, in: Proceedings of the Second International Conference on Medical Electronics London, Illiffe & Sons, 1960, p 253
11. Satinsky V, Dreifus LS, Racine P, et al: Self-energizing cardiac pacemaker. JAMA 35:192, 1965
12. Lown B, Amarasingham R, Neuman J: New method for terminating cardiac arrhythmias. JAMA 192:35, 1962

Norman Brachfeld

24

Diagnostic and Therapeutic Implications of Inadequate Myocardial Oxygenation

Biochemical laboratory techniques for the evaluation of the degree of myocardial ischemia provide valuable tools to aid our understanding of its pathophysiology. The therapeutic and diagnostic values of such observations are obviously maximized by a firm understanding of the basic alterations in cell metabolism induced by myocardial ischemia. A lack of confidence in understanding metabolic interrelationships cannot help but prevent an optimal interpretation of data. Improper application of such information to a given clinical situation may lead to disappointing results and/or confusion. Techniques that should currently be utilized in many clinical situations may thus be restricted to a few highly specialized centers.

In this chapter we will briefly consider some of the distortions of myocardial metabolism induced by ischemia, describe the techniques by which experimental data are obtained, and comment on potential pitfalls in interpretation. The interested reader is referred to additional reviews of current progress.[1–6] The application of such information has improved current diagnostic techniques and has led to the development of several experimental therapeutic modalities that attempt to correct the biochemical abnormalities induced by myocardial ischemia. We will consider those factors in the myocardial cellular environment that support viability and help to maintain cellular homeostasis, those that most

Dr. Brachfeld is a Career Development Awardee of the National Heart Institute (HE-07521).

seriously challenge it, and those that offer the possibility of therapeutic intervention. Many of the changes to be described are quite nonspecific, occurring in response to other types of cellular trauma and with some modifications during physiologic cell death as well.

Metabolism implies living tissue. The processes of ischemic metabolism are dynamic phenomena constantly changing in degree and severity and influenced by multiple hemodynamic and biochemical factors. Myocardial infarction rarely presents initially as a homogeneous localized mass of tissue. This type of injury invariably produces a heterogeneous mixture of progressively changing proportions of necrotic and living tissue. This is true of the zone of infarction as well as of the peri-infarction zone of ischemia.

During the initial 18–20 min that follow cessation of nutritional myocardial blood flow, most of the changes in this "zone of injury" are reversible. Thereafter, progressively more cells become irreversibly injured and eventually necrotic. It is well to emphasize that many studies have demonstrated the initial maintenance of normal and ischemic metabolism, occurring side by side, in tissue suddenly deprived of a primary arterial supply. Even after 45 min of total hypoxia 35%–66% of such cell populations may remain viable.[7]

Significant and measurable changes in carbohydrate, lipid, and protein metabolism and alterations in cytoplasmic, mitochondrial, and lysosomal enzyme systems are demonstrable after the onset of ischemia. These changes establish the basis for several diagnostic approaches to ischemic heart disease and are useful in the evaluation of response to treatment; they may be more sensitive than, and have been shown to chronologically precede, diagnostic changes in the standard ECG. Nevertheless, such changes are not in themselves necessarily diagnostic of irreversible cell injury and have not as yet provided a means of quantitating the severity, degree, or extent of the ischemic injury or the size of the infarct. New techniques have improved our ability to perform extensive acute and chronic metabolic studies in patients suffering from myocardial ischemia or infarction. Data so accumulated are necessarily poorly controlled, are often limited, and have only recently become available.[8]

METABOLIC LABORATORY STUDIES

Our knowledge of the patterns of ischemic myocardial metabolism has been obtained almost exclusively from controlled laboratory experiments performed on the isolated perfused heart of the rat, the guinea pig, and the rabbit or on anesthetized open-chest dogs or pigs. Concepts derived from these studies, although individually valid, are necessarily derivative. They must be applied with caution to man, in whom humoral and nervous influences and often long-standing coronary artery disease have a profound effect on metabolic response. We cannot expect that all such data will exactly parallel metabolic events occurring in the ischemic or infarcting human heart, and we must be prepared to interpret such information critically.

Isolated systems are necessarily denervated; they are perfused with a balanced well-buffered salt solution free of hemoglobin and formed whole-blood elements. The partial pressure of oxygen is invariably either much higher or much lower than that present under in vivo physiologic conditions. Available substrate may be single or multiple, but is usually fixed throughout the experiment. Hemodynamic load, and often performance, is preset by a carefully regulated perfusion apparatus, so that the heart may perform no external cardiac work, may do isovolumic work, or may assume a more physiologic pressure–volume load. Comparisons between such studies may be confusing, since there is an almost linear correlation between substrate utilization and metabolic activity on the one hand and hemodynamic demand on the other. Perfusate solutions may be circulated in either retrograde or antegrade direction. Indeed, coronary flow may be normal or high despite diffuse tissue hypoxia, a state rarely encountered under clinical conditions. The imposed stress of ischemia or hypoxia by intent involves the entire heart, and there is rarely preexisting coronary vascular disease. The design of the experiment itself may so distort the processes being evaluated as to make the broad interpretation of results highly questionable.

The states of ischemia and hypoxia or anoxia differ significantly in both metabolic and histologic expression. Electron micrographs of tissue obtained at the meat market (anoxic) may appear almost normal, while those made from tissue subjected to ischemia show markedly distorted histology. Despite rather marked differences in experimental design, data obtained from these studies are sometimes not evaluated critically. The unwary reader may lump findings together and formulate broad, overly simplified general concepts.

METABOLIC PATHWAYS: AEROBIC GLYCOLYSIS

The initial sequence of reactions in the process of carbohydrate degradation, the so-called Embden-Meyerhof or glycolytic pathway, has been studied in almost all animal preparations. Familiarity with its more general aspects suggests many potential diagnostic and therapeutic applications. Figure 24-1 illustrates its more salient landmarks. It is a highly integrated, self-regulating enzyme system governed by a variety of sensitive allosteric feedback control mechanisms that ensure maximal economy of substrate utilization. The reactions are extramitochondrial and occur in the cytoplasm of the cell.

Several features are of particular interest. When coronary blood flow and oxygen content are normal, glucose passes from the extracellular to the intracellular compartment, is rapidly and irreversibly phosphorylated in the presence of hexokinase, and in the process utilizes a molecule of adenosine triphosphate (ATP). Glycogen stores are protected by feedback inhibition of phosphorylase due to an increase in the concentration of glucose-6-phosphate (G-6-P). Increased concentration of this hexose monophosphate will also suppress hexokinase activity by feedback inhibition if the end product is not utilized. Its hexose isomer, fructose-6-phosphate, consumes an additional molecule of ATP during a phosphorylation regulated by phosphofructokinase (PFK), a major allosteric enzyme. Although several secondary control points are operative, the sensitivity of this multivalent regulatory enzyme to a number of positive or negative end-product modulators has established it as the principal rate-limiting pacemaker of the glycolytic sequence. As is true for most enzymes of this type, the cytoplasmic reaction is essentially irreversible.

Cleavage of fructose-1,6-diphosphate yields two 3-carbon fragments that ultimately produce 4 moles of ATP by cytoplasmic substrate-level phosphorylation. Since 2 moles of ATP are hydrolyzed there is a net glycolytic yield of 2 moles of ATP per mole of glucose consumed. An important step in this sequence is the oxidation of glyceraldehyde 3-phosphate to 1,3-diphosphoglycerate, the sole oxidative reaction (dehydrogenation) of the glycolytic pathway. It conserves the energy of oxidation in the end product, the diphosphate, and subsequently releases it to ADP by substrate-level phosphorylation. Completion of the reaction is dependent upon the presence of the oxidizing agent NAD^+, the obligatory electron acceptor, present in catalytically small amounts. Unless efficient mechanisms for continuous oxidation of NADH are available, feedback inhibition will interrupt the flow of glucose metabolites and terminate the activity of this pathway. During aerobic metabolism regeneration is effected by mitochondrial oxidative phosphorylation. A variety of shuttle mechanisms serve to transport electrons from extramitochondrial to intramitochondrial sites. Thus by the markedly efficient aerobic process of oxidative phosphorylation the carbohydrate fragments produced by the enzymatic interactions discussed above are

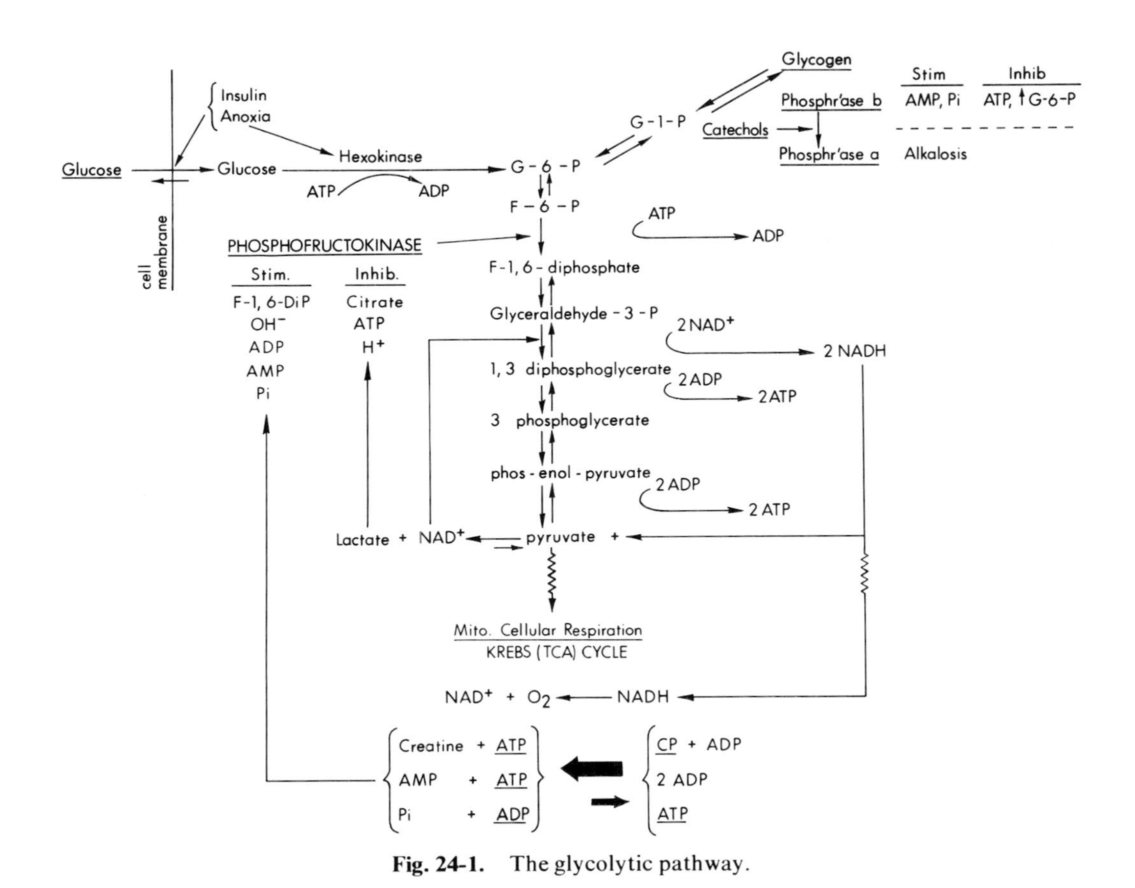

Fig. 24-1. The glycolytic pathway.

ultimately reduced to CO_2 and H_2O after entering the mitochondrion. The net yield of 2 moles of ATP during glycolytic degradation contrasts sharply with the 36 moles of ATP per mole of glucose consumed during mitochondrial metabolism. Thus to maintain the same energy yield, pure glycolytic energy production would have to consume 18 times as much glucose.

The so-called Pasteur effect, which defines the integration of glycolysis and mitochondrial oxidative phosphorylation, is an expression of the inhibition of glucose consumption and glycolysis during cellular respiration. When mitochondrial oxidative phosphorylation is active, the concentration of creatine phosphate (CP) rises, as does the ATP/ADP ratio. This calls forth the negative or inhibitory modulation of PFK activity, and glycolytic flux is markedly reduced. Active mitochondrial production of citrate as a result of the oxidation of free fatty acid (FFA) by the myocardial cell induces negative modulation of PFK activity and explains the preferential utilization of FFA during aerobic myocardial metabolism. Increased availability of acetate, pyruvate, and epinephrine will also enhance the concentration of citrate and will depress PFK activity.

Creatine phosphate cannot provide an immediately utilizable supply of energy for contractile activity, since it is not subject to hydrolysis by actomyosin. It does, however, serve as a readily accessible storage source of energy for regeneration of ATP. The stoichiometry of the reaction in which the enzyme creatine phosphokinase (CPK) catalyzes the reversible transfer of phosphate between CP and ADP markedly favors the formation of ATP. It provides an anaerobic mechanism for the rapid regeneration of ATP.

ANAEROBIC GLYCOLYSIS

When oxygen deprivation is experimentally induced, mitochondrial respiratory metabolism is compromised and oxidative phosphorylation is no longer capable of maintaining CP or ATP stores. This is immediately and radically reflected by a displacement of the glycolytic reactions outlined above. The rate of glycolysis may increase as much as 15- to 20-fold.[9] Nevertheless, in the anoxic isolated perfused rat heart ATP levels begin to fall within 10 sec. At 40 sec after onset 75% of total nucleotides have reached fully reduced levels. Concentrations of ADP, AMP, and Pi rise rapidly to reach a new steady state that exceeds controls by 150%–200%. Creatine phosphate depletion also begins at about 10 sec, is much greater in magnitude, and falls to 50% by 20 sec.[9]

Wollenberger's studies on anoxic canine myocardium showed that changes in the adenylic acid system (ATP, ADP, AMP) were more delayed and were not significant during the first minute of anoxia.[10] He also noted a positive correlation between the rate of increase in orthophosphate (Pi) concentration and the acceleration of glycolysis.

The immediate and marked increase in glycolytic flux induced by anoxia must be supported by an enhancement of glucose consumption, of glycogenolysis, or both. The most severe changes in glycolytic intermediates and enzyme activity are seen during the immediate transition from aerobic to anoxic metabolism.

Morgan et al.[11] have shown in the rat heart that anoxia increases glucose uptake due to an acceleration of transport, an increase in the phosphorylation capacity, and a decrease in the apparent phosphorylation Km. Reduction in end-product inhibition (G-6-P) of hexokinase by an increase in concentration of orthophosphate may also enhance glucose phosphorylation. When insulin was added, transport was accelerated still further, and near-maximal rates of phosphorylation were noted even at very low external glucose concentrations. We have found that reduction of flow in the in vivo dog heart to 36% of control values induced a fivefold increase in glucose extraction and doubled glucose consumption (Fig. 24-2).[12] Oxygen consumption and glucose extraction coefficient showed a striking fixed inverse relationship (Fig. 24-3).

Glycogen, the endogenous carbohydrate store of the heart, does not normally meet myocardial substrate requirements. In the absence of other sources of oxidizable substrate, however, it may account for as much as 25% of myocardial oxygen consumption for a limited period of time.[13] The myocardial free glucose pool supports glycolysis during aerobic perfusion. During severe ischemia this source is unavailable, and glycogen provides the sole substrate support for glycolysis. Glycogenolysis is regulated by the enzyme glycogen phosphorylase normally present almost entirely in the relatively inactive b form, due to low concentrations of its cofactor adenosine 5'-monophosphate and of one of its substrates, orthophosphate. Enzyme activity is further repressed by inhibitory aerobic concentrations of ATP, glucose-6-phosphate, and possibly ADP.[10] When phosphorylated to the a form the enzyme is no longer dependent upon 5'-AMP for activity, nor is it inhibited by either ATP or glucose-6-phosphate.[13] With onset of ischemia or hypoxia the b enzyme activity is increased in response to the fall in ATP and glucose-6-phosphate concentrations and by subsequent elevations of AMP and orthophosphate, the latter serving as substrate for enzyme action. Of greater importance, however, is the rapid and marked increase in the

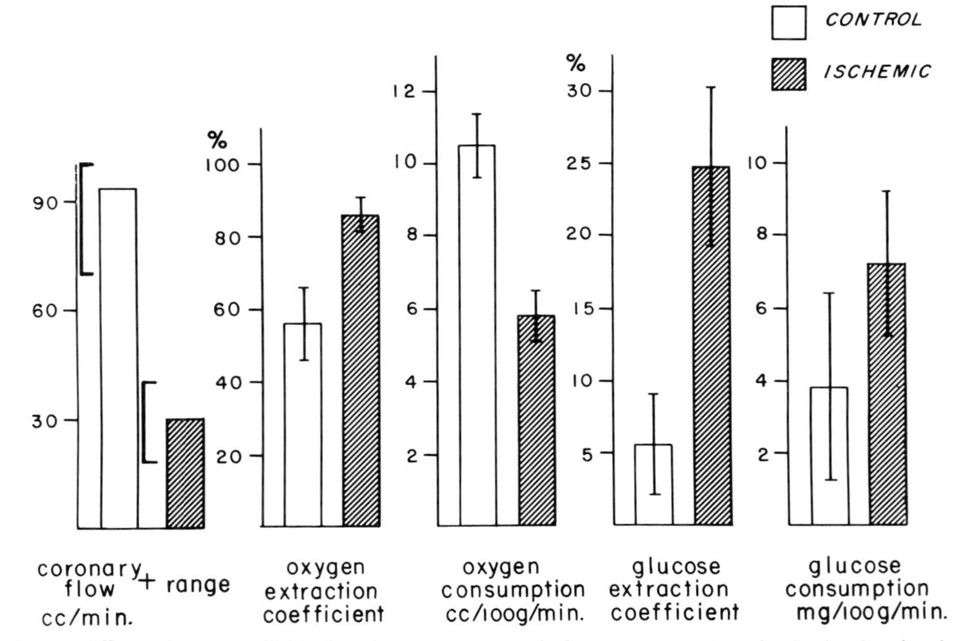

Fig. 24-2. Effect of myocardial ischemia on oxygen and glucose parameters in the in situ dog heart. Left coronary arterial flow reduced to 36% of control.

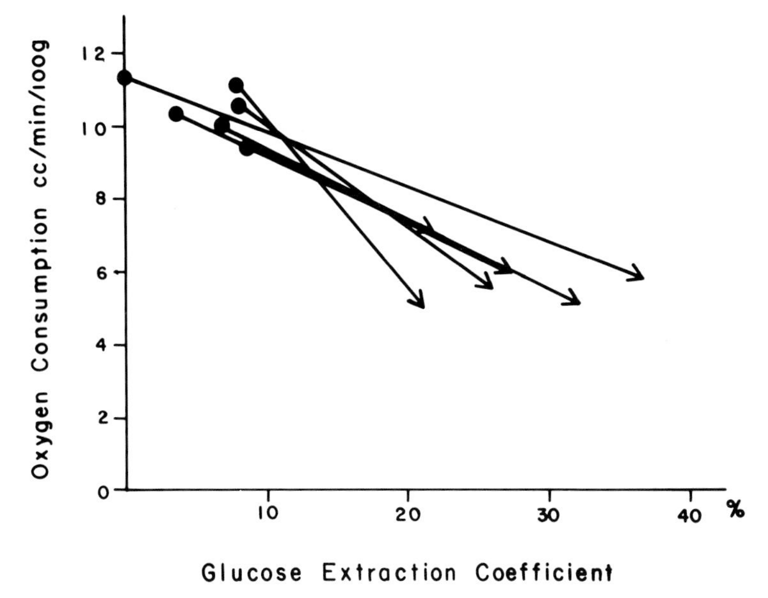

Fig. 24-3. Relationship of myocardial oxygen consumption to percentage of glucose extracted (glucose extraction coefficient) when left coronary arterial flow is reduced 40% in the canine heart.

b-to-a transformation mediated by anoxic stimulation of the enzyme phosphorylase b kinase. Within 17 sec phosphorylase a concentration rises to nearly 60% of total enzyme concentration. Danforth[14] has suggested that the initial increase in transformation may be related to the slight rise in pH that accompanies the early rapid breakdown in creatine phosphate. The mechanism for maintained activation is disputed. Wollenberger[10] suggests that the mechanism is due in part to stimulation of adenyl cyclase and thus of cyclic 3',5'-AMP by cardiac catecholamines. Release of catecholamines from postganglionic sympathetic nerves subjected to ischemia or hypoxia has been demonstrated. Cornblath,[15] however, could not block this conversion by pretreatment with reserpine. Whatever the mechanism, mammalian intracellular (IC) glycogen concentration can be shown to fall to about 70% of control levels within 4 min of onset of anoxia.

The increase in hexokinase and phosphorylase activity enhances the input of carbohydrate substrate into the glycolytic pathway and shifts integration of the rates of glycolysis and respiration to a new metabolic "set." ATP inhibition of PFK activity is most marked at a slightly acidic pH.[14] The initial unregenerated CP breakdown increases intracellular pH slightly. Furthermore, intracellular orthophosphate concentration may rise eightfold during the transition from aerobic to hypoxic cell respiration. Both factors are operative in initiating PFK activity. The latter thereafter serves a more permissive role, since once such activation has begun the reaction tends to be autocatalytic because of the marked antagonism of fructose diphosphate to ATP and citrate inhibition. Enhanced PFK activity means rapid transport of carbohydrate equivalents down the glycolytic pathway. Mitochondrial oxidative phosphorylation is blocked by inadequate oxygenation, and regeneration of NADH is depressed. During aerobic metabolism the isoenzymes of lactic dehydrogenase, MH_3 and H_4, with low affinity for pyruvate as an electron acceptor, allow NADH to be readily oxidized by mitochondria. With hypoxia the M_4 and M_3H isoenzymes appear to be activated, the affinity of pyruvate as electron acceptor increases, and reoxidation of NADH is accomplished by reduction of pyruvate to lactate. This appears to be a relatively efficient emergency measure. Since lactic acid has no other major metabolic pathway open to it, its accumulation does not interrupt subsequent metabolic reactions. Williamson[9] has shown that lactate levels increase within 20 sec and continue to increase at a rate of 60–70 μmoles/g dry weight per minute until a plateau is reached.

Normally glycolysis is an inefficient and quantitatively insignificant method of energy production. During anoxia or severe hypoxia areas of irreversibly damaged myocardium are certainly incapable of maintaining adequate stores of ATP by such means. The variable-sized band of ischemic tissues that surrounds the infarcted myocardium may be extensive, and it leads a precarious existence balanced between potentially reversible changes and terminal cell death. In this zone, glycolytic mechanisms acquire increased importance and have been shown to meet as much as 30% of resting mammalian energy requirements. Although there has been much discussion of the potential importance of this pathway of myocardial energy production, the true significance of its contribution to myocardial energetics has yet to be determined. It will certainly not support normal myocardial wall tension and adequate contractility. Nevertheless, it can maintain low levels of contractility.[16] Agents that inhibit glycolysis inhibit contractile activity. Work in our laboratory and elsewhere[17] has also demonstrated that the presence of glucose in an oxygen-free perfused heart system will support reduced contractile performance and electrical activity. It will prolong the period of reversible injury and enhance recovery when oxygen is readmitted. Thus there is strong evidence that such limited energy production can make a small but perhaps critical contribution to the maintenance of cell viability, to the survival of excitability, and to the functioning of nodal and conducting tissue until a competent collateral circulation is established or vasodilator tonus is induced. It may also play an important role in promoting the recovery of the acutely anoxic heart.

MYOCARDIAL METABOLIC ACIDOSIS

Unfortunately such enhanced glycolytic energy production is not an unmixed blessing. The acidosis that accompanies an accumulation of lactic acid and other reduced metabolites has a deleterious and ultimately lethal action on myocardial function. Recent work[18] has demonstrated that the adverse hemodynamic effects of anoxia may be reproduced by a reduction of pH even when oxygenation is adequate. Depression in contractility occurs despite normal concentrations of high-energy phosphate and may be due to the active competition of hydrogen ions with calcium at excitation–contraction binding sites. In contrast to other tissue, the mammalian myocardium has a relatively poor intrinsic buffering capacity. Myocardial tissue acidosis is accompanied by a reduction in left ventricular contractility and isovolumic work capacity and by refractoriness to vasoactive and inotropic agents. There is a decrement in cardiac pacemaker activity and an increased susceptibility to dysrhythmias. Coronary flow falls in the face of a decrease in pH, accentuating an adverse clinical state, and there may be enhancement of intracapillary blood coagulation. The pH optima of tissue enzyme systems are most frequently found to lie within the normal physiologic range. As the pH falls, such systems are progressively inhibited. A decrease in pH to about 6.6 inhibits the activity of the malate–aspartate shuttle, and this primary mechanism for transporting hydrogen ions between cytosomal and mitochondrial pyridine nucleotides is progressively inactivated. PFK is also vulnerable to pH inhibition. When anaerobiosis is prolonged or severe, inhibition of PFK activity markedly reduces glycolysis, and monophosphate concentrations rise. There is suggestive evidence that at very low pH levels, lysosomal enzymes may be released.[2] Their proteolytic and hydrolytic activity is optimal in the severely acidotic range and they may further destroy the biochemical function of the cell and ultimately lead to autolysis.

DIAGNOSTIC IMPLICATIONS OF METABOLIC CHANGES

The diagnostic and therapeutic implications of these metabolic changes depend upon our ability to sample the venous efflux of the myocardium. Sampling enables us to evaluate change in extraction patterns of common myocardial substrates (glucose, lactate, FFA)[8,19] and can provide valuable clues to the degree of anaerobic metabolism. An increase in the IC orthophosphate level is accompanied by its appearance in abnormal concentrations in venous blood within 2 min.[20] Potassium ion leakage from the damaged cell not only causes ECG abnormalities, but may be reflected by elevations in the coronary sinus samples;[21] potassium is replaced by sodium, chloride, and water. Such electrolyte shifts lead to cell swelling and depression of contractile activity (see below).

BIOCHEMICAL ESTIMATION OF MYOCARDIAL ISCHEMIA

The inability of the tricarboxylic acid (Krebs) cycle pathway to utilize the increased amounts of pyruvate generated by enhanced glycolysis and the accumulation of cytoplasmic NADH forces the conversion of pyruvate to lactate for disposal of H^+ and oxidation of NADH. A reduction in oxygen supply to the myocardial cell causes a shift in its oxidation–reduction system to a more reduced state. The oxidized and reduced forms of nicotinamide adenine dinucleotide (NAD, NADH) and the NAD-linked dehydrogenase systems immediately reflect this change. Our primary concern, however, is with the NAD–NADH redox potential of the mitochondrion, the site of oxidative phosphorylation and the most sensitive index of the adequacy of oxidation. Intramitochondrial redox state is only indirectly coupled to cytoplasmic NAD–NADH ratios, and the latter is grossly reflected by relative shifts in intracellular lactate–pyruvate concentrations.

Such changes in venous lactate–pyruvate concentrations are many stages removed from the critical target site at the mitochondrial level. Venous concentrations represent mixed samples and at best express a net lactate balance. Nevertheless, oxygen deprivation has been qualitatively expressed in terms of a decrease in lactate extraction (percentage lactate extraction), a release of lactate into coronary venous blood samples (lactate production), or an increase in the lactate–pyruvate ratio of blood perfusing the myocardium.

The validity of this type of estimation requires that the muscle cell membrane be freely permeable to lactate and pyruvate and that there be an equilibrium between cytoplasmic and mitochondrial NADH concentration—assumptions that are not entirely justified. Nevertheless, such data are commonly utilized for diagnostic evaluation of suspected

Table 24-1
Significant Factors in Evaluating Myocardial Lactate Data

Nutritional status of the patient
Infusion of glucose, pyruvate, or lactate; a positive linear correlation between lactate extraction and arterial lactate concentration
Arterial lactate elevated with shock, hypoxia
Diabetes
Serum FFA concentration; may suppress extraction of carbohydrate, induce pyruvate efflux
Alkalosis, hyperventilation, infusion of bicarbonate
Catheter placement, inadequate sampling, segmental disease
Sampling site may be proximal to vein draining ischemic area
Drainage from ischemic areas may be meager or absent
Dilution of ischemic drainage by normal venous efflux

myocardial ischemia. If the clinician is aware of potential pitfalls, misinterpretation may be avoided and this extrapolation may be considered valid for all practical purposes.[22] Table 24-1 lists some of the factors of importance in evaluating myocardial venous lactate data. Although patients with myocardial ischemia may show evidence of enhanced glycolysis when coronary reserve is challenged, myocardial glycolysis is accelerated by many factors in the absence of coronary artery disease and despite adequate oxygenation. If the capacity of the hydrogen shuttle is exceeded lactate may be formed from pyruvate.

Coronary sinus studies should not be performed unless the subject is in a steady (basal) nutritional, metabolic, and hormonal state. Sampling after an overnight fast during a period of active metabolism of elevated FFA is accompanied by suppression of circulating glucose, lactate, and pyruvate concentrations, as well as inhibition of pyruvate dehydrogenase activity, so that an increased amount of pyruvate is converted to lactate. Glycolysis is also depressed by citrate-induced inhibition of PFK activity noted during fasting, extraction of carbohydrate is reduced, and myocardial arteriovenous differences are difficult to determine accurately. Starved rats show a decrease in myocardial lactate–pyruvate ratio. Conversely, hyperglycemia induced by glucose infusion will enhance its extraction and both pyruvate and lactate production, prevent steady-state determinations, and induce a transit time artifact.[23] Lactate and pyruvate infusion may similarly distort data without necessarily reflecting changes in oxygenation. Exercise stress tests, once popular for evaluation of coronary insufficiency, are compromised when utilized for estimation of myocardial glycolysis, due to the accompanying increase in arterial lactate concentration, and have been replaced, when biochemical studies are being performed, by pacing-induced stress. Shock and generalized hypoxia from whatever cause are also associated with marked elevations in arterial lactate concentration and are a contraindication for coronary sinus sample studies.

Henderson et al.[24] have postulated compartmentation of tissue lactate and pyruvate and demonstrated differing transmembrane concentration gradients for these metabolites. The rate of pyruvate efflux from the isolated heart was shown to exceed that of lactate by 10-fold, suggesting that a prolonged steady state is required before IC/EC equilibration may be assured. Some investigators have therefore abandoned pyruvate determinations and lactate–pyruvate ratios in favor of lactate assay alone, recognizing that the loss in sensitivity is compensated for by an increased reliability.

The relative inhibition of membrane transport of glucose found in diabetes mellitus reduces glycolytic flux and is associated with a decrease in L/P ratio and with pyruvate accumulation.[22] In studies performed with the isolated diabetic heart, the addition of insulin enhanced glycolytic flow, but pyruvate accumulation persisted. Furthermore, the elevation in circulating FFA and ketone bodies noted in poorly controlled diabetes leads to citrate-induced inhibition of PFK activity and further depresses glycolysis. Both FFA and ketone bodies are extracted from the arterial blood in preference to lactate, and when they are present in significant concentrations they will distort the lactate–pyruvate ratios of the coronary venous samples and cause a misleading drop in the calculated percentage lactate extraction. Enhanced insulin resistance or latent diabetes, frequently expressed in the postcoronary state, must also be considered when such studies are performed.

Other hormones and drugs (growth hormone, cortisol, heparin) modulate glycolytic flux by mobilization of FFA. Perhaps the most common mechanism for mobilization of FFA from adipose depots is that which occurs in response to an increase in catecholamine secretion. Patients undergoing diagnostic pacing studies often demonstrate anxiety symptoms and elevated blood catecholamine levels, especially when paced to the onset of angina. Finally, it must be noted that glucagon has been shown to accelerate conversion of phosphorylase b to a, thus enhancing glycogenolysis and increasing output of lactate.

Scheuer[25] reported that alkalosis increased myocardial lactate production and increased exogenous glucose and endogenous glycogen utilization during normal oxygenation. These effects were noted whether pH changes occurred as a result of lowering pCO_2 or by increasing bicarbonate content. The mechanism appears to be an activation of PFK activity. Huckabee[26] also reported lactate production in hyperventilating human subjects. The Pasteur effect is disrupted by alkalosis, as PFK is relieved from ATP inhibition and the close inverse correlation between aerobic metabolism and glycolysis is disturbed.

During hypoxia, metabolic alkalosis was associated with improved left ventricular pressure and rate of pressure rise; neither oxygen consumption nor lactate production was increased, and lactate–pyruvate ratios were lower than in controls.

Hexokinase and PFK activity may be accelerated in the isolated heart preparation by an acute increase in work.[27] Glucose and glycogen consumptions rise as does output of lactate and pyruvate. Lactate production is similarly doubled when contractility is enhanced by raising the calcium concentration of perfusate solutions.[28] Other more exotic causes of lactate production in the face of normal oxygenation include cardiac transplantation, uncoupling or interruption of the respiratory chain by cyanide, a variety of idiopathic cardiomyopathies, phenformin-induced and idiopathic lactic acidosis, and alcohol ingestion. The latter increases cytoplasmic hydrogen ion content at a rate greater than it can be shuttled to the mitochondrion for oxidation.

Gorlin[23] has emphasized that the diffuse and segmental nature of coronary disease often leads to sampling error due to poor or improper catheter placement. If the catheter tip lies proximal to the vein draining the ischemic zone, evidence of glycolysis may be missed entirely. The nonuniformity of myocardial perfusion is accentuated during ischemia. Poorly perfused areas may drain a meager volume of blood, which is rapidly diluted by mixing with large-volume flow from normal zones. It is therefore recommended that samples be obtained at three venous sites chosen from the great cardiac vein to a point just proximal to the coronary sinus.

Despite great care, one may still fail to obtain evidence of ischemia by reliance on evidence of anaerobic glycolysis alone. It would seem wise to evaluate several parameters simultaneously during pacing-induced stress. Changes in hemodynamics and in the electrocardiogram, evidence of negative potassium and phosphate balance, abnormalities of FFA oxidation and extraction (vide infra), and oxygen desaturation of coronary sinus samples may supply sufficient supplemental information to support a diagnosis. It is evident that such changes are dependent upon viable although ischemic tissue. Infarcted myocardium is by definition necrotic and devoid of metabolic activity. It will not demonstrate glycolytic activity whatever the stimulus.

MYOCARDIAL ISCHEMIA VERSUS MYOCARDIAL HYPOXIA–ANOXIA

The literature makes a poor distinction between anoxia and ischemic hypoxia. In the former, usually confined to isolated heart preparations, oxygen availability is determined by the pO_2 of perfusate, coronary flow is either normal or increased, and the entire heart is subject to the insult. Ischemic in vivo studies, perhaps more physiologic, are usually performed on dogs, pigs, primates, or other large mammals. A localized zone of ischemia is created by ligation, thrombosis, or reduced cannulated flow to a significant portion of the left ventricle. Rovetto et al.[29] compared ischemia to anoxia in the perfused rat heart and noted that although reduced oxygen delivery initially accelerated glycolysis in both preparations the rate was twice as fast in anoxic as in ischemic hearts. During ischemia extracellular lactate concentration exceeded that seen with anoxia by 10-fold. The onset of failure of the preparation correlated closely with the accumulation of tissue lactate. The greater toxicity of ischemia may be due to the inability of this preparation to dispose of end products of glycolysis and to a marked depression of intracellular pH. The earlier onset of acidosis inhibited PFK activity and suppressed glycolytic energy production. Although arterial or perfusate glucose concentration is rarely rate-limiting during the rapid flow associated with hypoxia, its concentration may fall precipitously during ischemia. Under these conditions hypoglycemia introduces another significant variable to glucose utilization.

THERAPEUTIC APPRCACHES

A therapeutic agent can only be effective if there is a means of reaching its target the ischemic cell. Unfortunately most such approaches are hedged by the inability of drugs, oxygen, substrates, etc., to reach the site of injury due to the very hemodynamic conditions that initiated the process. Nevertheless, the defects reviewed above suggest several new approaches to the treatment of myocardial ischemia.

MYOCARDIAL CELL SWELLING IN RESPONSE TO ISCHEMIA

Many workers have noted the persistence of myocardial ischemia and reduced tissue perfusion following restoration of normal proximal flow in the experimental animal. Similar failure of "reflow" to return to normal has been noted in the brain and kidneys.[31] When inadequate oxygenation is prolonged or severe, cell swelling induced by loss of potassium and intracellular accumulation of sodium, chloride, and water may be further accentuated by accumulation of osmotically active particles contributed by autolytic breakdown of protein and tissue metabolites. Myocardial swelling may lead to a palpable firmness of the ventricular wall; pliability is reduced by high membrane tensions and contractility is depressed. Intracellular calcium ion concentration is diluted, and the normal inotropic response to this ion, already burdened by competition with H^+ ion for contractile binding sites, is lessened. Swelling of capillary endothelial cells may cause trapping of formed blood elements and further reduce flow. Thus ischemia may beget further ischemia and impede both substrate and oxygen flux.

An obligatory extracellular hyperosmolal agent, mannitol, improves reflow and prevents closure of regional renal and cerebral vascular beds after temporary arterial occlusion. Willerson et al.[32] infused this agent into the aortic roots of dogs before and during reversible ischemia of the left ventricle. They reported that an increase in

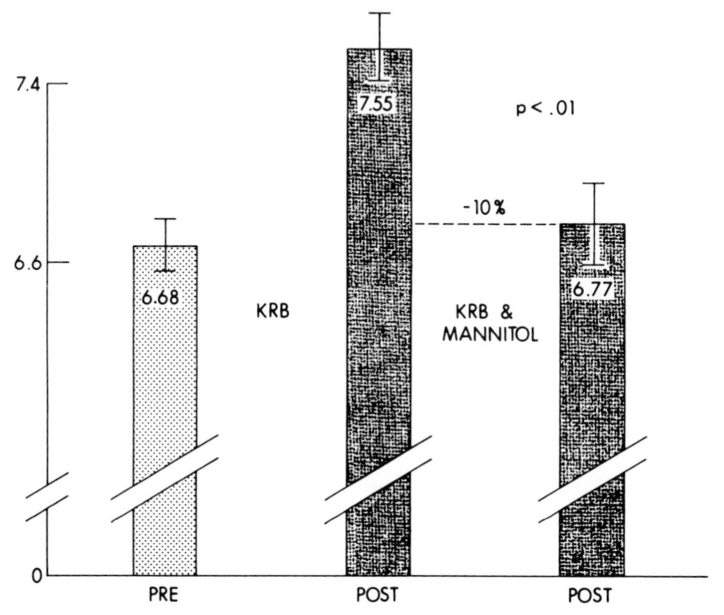

Fig. 24-4. Comparison of myocardial water content (wet/dry weight) during aerobic perfusion (PRE) with Krebs-Ringer bicarbonate (KRB, mOsm = 290) and following 15 min of anoxia + 15 min of recovery with aerobic perfusate (POST). KRB = 290 mOsm, KRB + mannitol = 350 mOsm.

serum osmolality reversed the depressed ventricular function curve, lessened ST-segment elevation, and raised total and collateral coronary blood flow. The improved postischemic myocardial performance persisted after return of osmolality to control levels. Others have shown that mannitol may be effective in reducing infarct size. The mechanism of action of hypertonic mannitol is not entirely clear, but these reports suggest that it prevents closure of capillary beds and decreases cell size, thus enhancing relative intracellular calcium concentration. In our laboratory[33] we have been unable to demonstrate enhancement in energy metabolism or total coronary flow, but recovery of hemodynamic parameters was significantly improved following reoxygenation when mannitol was present. The effect of hypertonic mannitol was related to a direct myocardial osmotic action. The increase in tissue water content expected with prolonged ischemia was less marked, and wet/dry weight ratio fell significantly in its presence (Fig. 24-4). Further effects may be mediated by the known ability of hyperosmolality to alter both contractile-element and series-elastic mechanical characteristics.[34] Unfortunately the clinical applicability of an agent that expands intracellular fluid volume during or following an episode of myocardial ischemia has not as yet been fully determined.

GLUCOSE-INSULIN-POTASSIUM (GIK) INFUSIONS

In 1939 Selye[35] reported that myocardial resistance to cardiotoxic agents was reduced by intracellular hypokalemia and hypomagnesemia. More recently, Sodi-Pallares et al.,[36] Larcan,[37] and others have described a regimen for the early treatment of myocardial ischemia that consists of dietary regulation of sodium, potassium, and water intake and an intravenous infusion of glucose, KCl, regular insulin, and heparin (Sodi-Pallares) or dibasic potassium phosphate, K^+ and Mg^{++} aspartate, cocarboxylase, and cytochrome c (Larcan). Treatment was reported to decrease the frequency, duration, and severity of angina pectoris, dysrhythmias, congestive heart failure, fever, and shock. Electrocardiographic changes rapidly normalized, and disturbances in intracellular water, sodium, and potassium balance were corrected. Response to therapy was attributed to a restoration of intracellular potassium concentration mediated by insulin-stimulated glucose transport and enhanced glycolysis. The impatient advocacy by their respective proponents stimulated an enthusiastic and hopeful reception for these "polarizing solutions."

Earlier we noted that glycolytic activity is near maximal in the insulinized cell during normoglycemia and that high rates of flux have been demonstrated at very low external glucose concentrations. Admittedly a low arterial threshold is an artifact of isolated systems not seen in vivo. Nevertheless, under all but the most severe degrees of hypoglycemia physiologic arterial glucose concentrations are adequate and do not limit glucose transport. Unfortunately clinical trials of this regimen were negated by variations in criteria for patient selection. Data describing the administration and the metabolic, hemodynamic, and electrocardiographic controls were incomplete, and concurrent therapy and subgroup analysis were either inadequate or differed markedly within and/or between studies. A large multicenter clinical trial of 13 hospitals and 986 patients conducted by the Research Council of Great Britain[38] is typical of those reported. It failed to demonstrate a significant difference in mean mortality between control and treated groups.

A recent experimental study by Maroko[39] indicates that cell viability and contractile potential appear to show a favorable response to infusions of GIK or hypertonic glucose when administered after ligation of the anterior descending coronary artery of the dog. It seems doubtful that stimulation of anaerobic glycolysis is responsible for their findings. If one makes the disputable assumption that systemically administered GIK actually does reach the ischemic zone, it is difficult to understand why such solutions have proved ineffective when given by direct intracoronary infusion.

Each component of the solution has rather wide-ranging metabolic effects. When given in combination their actual locus of action is difficult to determine. The rapid introduction of 10%–20% glucose may quickly depress circulating potassium concentration for as long as 6–9 hr and may be severe enough to induce changes in the ECG. There is a simultaneous and quantitatively similar intracellular movement of inorganic phosphate. Other studies indicate that glucose in high concentration may have a direct effect on transmembrane action potential. Insulin has a multivalent effect on cell metabolism and may play a direct and independent role in modulating myocardial ion transport. Insulin may also directly stimulate contractility independent of glucose transport or of the anoxic state.

The histologic lesions associated with well-oxygenated but hypokalemic cells are quite distinct in appearance from those seen after ischemia or anoxia. Reduction in the transcellular potassium gradient by intracoronary infusion of potassium chloride can reproduce the ischemic sequence of injury potential, ectopic beats, and tachycardia despite normal coronary blood flow.[40] Thus electrocardiographic changes reflect shifts in IC/EC potassium concentration and do not necessarily indicate myocardial redox state.

Egress of potassium is a response to ischemia, not the cause of the biochemical lesion. Its leakage from injured cells induces membrane hypopolarization and a fall in transmembrane resting potential, which leads to shifts in the ST segment of the ECG as well as to the onset and potentiation of ectopic foci. Although we may assume that a negative myocardial A-V potassium difference is due to efflux of potassium from ischemic cells, we have no experimental proof that these same injured cells regain potassium when positive balance is restored. The response of the ECG to potassium-oriented therapy may be caused by changes in IC/EC potassium ion gradients of the normal or near-normal cells about the heterogeneous zone of ischemia, rather than by changes in ischemic or necrotic cells.

Reflow experiments suggest that the presence of glucose or potassium is less important than the volume of fluid infused and its osmolal potential. Hypertonic glucose, saline, and sucrose were all equally effective in the treatment of dysrhythmias induced by ischemia and in reduction of potassium efflux.[41]

ROLE OF GLYCOGEN

The in situ heart normally shows little or no dependence upon glycogen as an oxidizable substrate. Preferential utilization of exogenous carbohydrate, free fatty acids, and ketones has been demonstrated by many authors. Unlike liver and skeletal muscle, fasting stimulates myocardial glycogen synthesis mediated by fatty acid mobilization, citrate-induced inhibition of PFK, and increased levels of glucose-6-phosphate. When the isolated heart is perfused with substrate free buffer, glycogenolysis follows utilization of endogenous triglyceride and may account for 15%–20% of oxygen consumption.[13] The acceleration of glycogenolysis noted after an acute increase in heart work, despite the presence of exogenous glucose, appears to be a linear function of orthophosphate concentration. This hypothesis is supported by experiments that have demonstrated a rapid temporary fall in ATP and CP values and an increase in ADP and Pi content under these circumstances.[13,42] Glycogen utilization is again mediated by an increase in PFK activity, a reduction of glucose-6-phosphate concentration, direct stimulation by Pi, and release of phosphorylase inhibition. Thus glycogen utilization in the aerobic heart is regulated by the availability of alternate substrate and by magnitude of the work load.

During anoxia, glycogen plays a much more active role. When the beating isolated rat heart is subjected to an acute work load, the initial peak rates of lactate formation are due to glycogen breakdown. Stores are rapidly depleted.[43]

Glycogen is found in highest concentration in the conduction tissue and ventricular endocardium, areas particularly sensitive to inadequate oxygenation. This distribution offers a teleological emphasis to its importance as an emergency fuel. It is 30% more efficient as a carbohydrate substrate than is glucose and has a net glycolytic yield of 3 moles of ATP per mole of glucose equivalent consumed. It is distributed within the cell as a labile fraction subject to relatively rapid accumulation and depletion and a more stable residual fraction. Glycogen stores in the rat heart may be enhanced by pretreatment with reserpine.[44] Hearts from reserpine-treated rats showed a higher left ventricular pressure rise and lactate output after 2 min of anoxia than did controls. The rate of glycogenolysis correlated well with the initial glycogen concentration, and proportionately more lactate was produced from glycogen than from glucose. Anaerobic ATP production per mole of hexose consumed improved significantly. Similar observations were made in the intact anoxic working canine heart.[45] Glycogen content was elevated by prefeeding with butcher fat and water; anoxia was induced by 95% nitrogen ventilation. During a 5-min period glycogen utilization in the fat-fed dogs exceeded that of controls by 141%. The hemodynamic parameters noted above, as well as cardiac output and left ventricular work, significantly exceed controls. In both studies the degree of glycogen utilization correlated well with the level of stored glycogen and suggested that high levels favor increased glycogenolysis during anaerobic metabolism.

Data obtained from working mammalian perfused hearts indicate a maximal glycolytic flux (glucose + glycogen) of 6.2 times control when heart work is acutely increased and insulin is present.[43] This maximal value of 3.5 μmoles of glucose utilized per minute per gram wet weight is significantly less than the maximal capacity of phosphofructokinase of 42–60 μmoles/min. The increase induced by alkalosis and by anoxia was found to be one-third as great. Hypoxia also fails to stimulate the mammalian heart to its fullest glycolytic potential. If rates achieved during anoxic studies had been reached during mild hypoxia, ATP production would theoretically have been capable of supporting mechanical activity. The observation that the rate of glycogenolysis falls long before glycogen is depleted continues to puzzle investigators.[15] It may well be that only labile glycogen can be rapidly mobilized. Inhibition of glycolysis by the attendant acidosis is only part of the story, since such inhibition would be present to an even greater degree during anoxia, a finding that is incompatible with the observations noted above. Unfortunately we remain unaware of all the factors that limit mammalian glycolytic rate to submaximal levels.

By contrast, the more primitive reptilian heart when performing in a totally anoxic environment demonstrates a unique anaerobic metabolic

Table 24-2
Comparison of (U-^{14}C) Glucose Utilization by Isolated, Perfused Turtle and Rat Hearts in the Aerobic and Hypoxic States

	Heart Rate (beats/min)	Glucose Uptake (μmoles/g wet weight)	$^{14}CO_2$ (μmoles/g wet weight)	Lactate Production (μmoles/g wet weight)
Aerobic				
Rat (A)	280 ± 10	26.8 ± 2.8	2.52 ± 0.27	48.4 ± 10.0
Turtle (B)	39 ± 3	4.8 ± 0.2	0.70 ± 0.07	4.9 ± 0.9
Hypoxic				
Rat (C)	113 ± 7	119.3 ± 7.6	1.69 ± 0.12	223.2 ± 10.8
Turtle (D)	34 ± 3	6.3 ± 0.61	0.20 ± 0.03	36.4 ± 4.5
% Change				
A – C	−60	+345	−33	+361
B – D	−13	+31	−71	+642

All values are means ± SE expressed per 30 min of perfusion to permit comparison between rat and turtle heart experiments; (U-^{14}C) glucose and palmitate were present as substrate; U-^{14}C glucose concentration was 8.4 mM in turtle and 10 mM in rat experiments; palmitate concentration was 0.4 mM in turtle and 0.5 mM in rat experiments; $n = 6$ for each series of experiments.

capacity. Despite PFK tissue activity and maximal rates of lactate production similar to those of the mammal, it can meet total energy requirements by anaerobic glycolysis and produce energy in quantities sufficient to maintain viability and support contractile, electrical, and metabolic activity at near-normal levels. It does so by utilization of glycolytic enzymatic pathways at maximal efficiency.

Studies of aerobic and anoxic metabolism of the turtle heart (*Pseudemys scripta*) performed in our laboratory[46] demonstrated a remarkable adaptability to changes in intracellular oxygen milieu. The reduced cardiac work of the turtle heart is matched by quantitative differences in substrate utilization. Divergence in metabolic demand between these species may be corrected when data are expressed as units per gram wet weight per beat (Table 24-2). During aerobic perfusion the turtle heart preferentially extracted and oxidized free fatty acids despite the presence of available glucose. FFA and glucose metabolism was similar to that seen in the mammalian heart and was adequate for myocardial energy requirements. When hypoxia supervened these similarities in substrate extraction and utilization were disrupted. In the rat an enhanced glycolytic flux was met by an increase in exogenous glucose extraction. Glucose uptake per beat exceeded that of the turtle heart by 10-fold. In contrast, the turtle maintained an endogenous glycogen content 10 times that of the rat heart. It provided a readily available source of metabolizable carbohydrate, and glucose uptake per beat increased insignificantly. The ability of FFA to depress glucose uptake and oxidation in the mammalian heart was not paralleled in studies of turtle myocardial metabolism during oxygen deprivation. FFA utilization did not compete with, but rather enhanced, concomitant carbohydrate metabolism and glycogenolysis and helped to provide sufficient substrate to meet myocardial energy requirements with little change in hemodynamic performance. The concentration of residual glycogen was adequate to maintain stores, and utilization of exogenous glucose was strictly limited. During prolonged perfusions (2 hr or more) such stores were eventually depleted; exogenous glucose supplied a proportionately greater amount of oxidizable carbohydrate, and its metabolic pattern more closely resembled the mammalian. A large store of labile glycogen appears to be the basis for this advantageous metabolic flexibility. It permits normal function under environmental conditions that are lethal to the mammal. In turtle myocardium a relatively low concentration of hexokinase and enhanced glycogen phosphorylase activity combine to support this pattern.

Despite the marked increase in lactate production during active turtle myocardial glycolysis, the extremely high quantities of reptilian

Table 24-3
Carbohydrate Metabolism: Effect of Hypoxia and Palmitate

State	Substrate	Glucose Uptake*	Glycogenolysis†	Lactate Production‡	Total Carbohydrate Utilization§	Fractional Lactate Production¶
Aerobic	Glucose ($n = 4$)	241.1 ± 25.1	81.3 ± 9.2	147.8 ± 10.9	322.4 ± 19.4	0.463 ± .041
	Glucose + palmitate ($n = 5$)	204.8 ± 10.5	53.1 ± 6.8	88.4 ± 9.2	263.8 ± 5.1	0.336 ± .035
	% Change p value	−15% $p < 0.2$	−35% $p < 0.05$	−40% $p < 0.001$	−18% $p < 0.05$	−28% $p < 0.1$
Hypoxia	Glucose ($n = 4$)	570.7 ± 4.4	106.1 ± 8.2	467.0 ± 23.2	676.8 ± 8.2	0.691 ± .039
	Glucose + palmitate ($n = 5$)	298.5 ± 9.9	114.9 ± 9.8	298.6 ± 19.7	413.4 ± 9.6	0.720 ± .032

	% Change	−48%	+8%	−36%	−39%	−4%
	p value	$p < 0.001$	$p < 0.6$	$p < 0.001$	$p < 0.001$	$p < 0.6$
Effect of hypoxia	% Change (glucose)	+137%	+30%	+216%	+110%	+49%
	p value	$p < 0.001$	$p < 0.1$	$p < 0.001$	$p < 0.001$	$p < 0.01$
	% Change (glucose + palmitate	+46%	+116%	+238%	+57%	+114%
	p value	$p < 0.001$	$p < 0.01$	$p < 0.001$	$p < 0.001$	$p < 0.001$

Glucose concentration = 5.5 mM; palmitate concentration = 0.4 mM; data reported ± standard error of the mean.

*Units: μmoles/g dry weight/30 min.

†Decrease in tissue glycogen content (μmoles glucose equivalent/g dry weight/30 min).

‡Units: μmoles glucose equivalent/g dry weight/30 min.

§Glucose uptake + glycogenolysis (μmoles glucose equivalent/g dry weight/30 min).

¶Fraction of total carbohydrate utilized that appeared as lactate.

body buffer help to support tissue pH. A slightly alkalotic control blood pH remains in the alkalotic physiologic range for as long as 3 hr after onset of anaerobiosis.[47] Much of the CO_2 produced resulted from bicarbonate buffering of hydrogen equivalents produced by anaerobic metabolism. Turtles possess a large volume of coelomic fluid with a pH more alkaline than that of plasma and a bicarbonate concentration approximately three times that of plasma. Maintenance of cellular pH thus helps prevent the depression of glycolysis associated with intracellular acidosis in the mammal.

FREE FATTY ACID METABOLISM

Plasma FFA appears to be the major myocardial substrate in the postabsorptive state of normal man. Its oxidation accounts for more than half of the oxygen consumption of the heart.[48] In the aerobic state FFA has been shown to increase myocardial oxygen consumption, preserve or enhance myocardial glycogen stores, and depress glucose uptake. Unfortunately, little is known about the relative effects of FFA and carbohydrate on myocardial hemodynamic performance. The general agreement regarding the importance of carbohydrate to ischemic metabolism is not shared when FFA metabolism is reviewed. Some workers[49–51] report that elevated FFA concentrations increase myocardial oxygen uptake excessively, increase the frequency of dysrhythmias, and depress contractility in the hypoxic heart or papillary muscle. Others[52–54] have been unable to confirm these observations. When palmitate and glucose were added to *moderately* hypoxic rat heart perfusates in our laboratory,[54] normal carbohydrate–FFA relationships persisted (Table 24-3). There was a significant depression of carbohydrate utilization and of lactate production. During aerobic perfusion palmitate depressed glycogenolysis and moderately inhibited glucose uptake. When double substrate was available (glucose + palmitate) and pO_2 was reduced, glucose uptake fell but glycogenolysis remained essentially unchanged. These findings only seem incompatible with our earlier discussion of enhanced glucose extraction and lactate production during marked anaerobiosis when we fail to consider the level of oxygenation provided the cell. In the presence of palmitate peak systolic aortic pressure exceeded controls. There was no disturbance of heart rate, rhythm, or cardiac output at either high or low oxygen levels. Utilization of FFA may support hemodynamic performance by reducing the severity or delaying the onset of tissue acidosis induced by glycolysis or by supplying oxidizable substrate to tissue whose glycolytic potential has been reduced by a fall in intracellular pH. The ATP yield (moles of ATP per gram of substrate consumed) of the complete oxidation of palmitate exceeds that of glucose by 2.5 times and makes it an advantageous fuel when oxygen supply is limited. Regan[52] also noted a fall in glucose uptake and lactate production when FFA levels were moderately elevated in the ischemic canine myocardium by small doses of norepinephrine. These and other studies of rat,[53] dog, and turtle[46] myocardium have shown a persistent uptake and oxidation of FFA during ischemia. Suppression of carbohydrate metabolism by FFA must therefore persist to some degree despite reduction in available oxygen; such correlations are valid only when cellular pO_2 is expressed in relative terms. Ischemic FFA metabolic patterns are incompletely understood, and reports of extraction, oxidation, esterification, and storage as neutral lipid are often contradictory. It is not possible to state whether FFA is "good" or "bad" for the ischemic myocardium, since ischemia cannot be regarded as an all-or-none phenomenon. Myocardial cells within this area reflect an entire spectrum of oxygenation and metabolic rate. Biochemical data expresses the mean response. In general, FFA uptake appears to be fixed or increased during ischemia;[52,55] the efficiency of cellular oxidative phosphorylation determines whether extracted FFA is oxidized or esterified. Storage as neutral lipid, to be expected during extreme anaerobiosis, cannot occur unless a carbohydrate precursor is available to provide the α-glycerophosphate skeleton required for esterification.[55,56]

Reports of a reduction in mechanical myocardial efficiency when FFA was offered as substrate during ischemia[49,50] suggest an uncoupling of oxidative phosphorylation or a direct cellular toxicity. There is inadequate documentation to warrant application of these observations to humans, however. In many of these reports FFA–albumin ratios were abnormally elevated due to the type or amount of FFA used or to relative albumin concentration. In some studies FFA was the sole substrate offered. Unbound FFA in even moderate concentration is indeed toxic to the cell, but this is an artificial laboratory-induced state not seen clinically. Opie[53] points out that an increase in oxygen uptake is not seen when a physiologic concentration of FFA or a perfusate containing mixed substrate is used. Conflicting conclusions regarding the stimulation of dysrhythmias in ischemic canine hearts due to elevation of circulating FFA are undoubtedly due to differences in the preparations studied. Elevation of FFA by heparin administration has not increased the incidence of serious dysrhythmias in patients with recent myocardial infarction.[53]

The state of FFA metabolism may be extremely sensitive to alterations in oxygen availability and may be utilized as a diagnostic aid in cases of atypical angina pectoris. Patients with a normal nutritional blood flow show an enhancement of the myocardial fractional extraction of FFA during a pacing-induced stress. In those with pacing-induced myocardial ischemia there is a reversal of the normal pattern.[57] Extraction falls, but oxidation of the FFA transported into the cell is significantly increased—a finding that is compatible with our discussion.

A unique experimental approach to the problem of increasing anoxic energy production has been suggested by Penney and Cascarano.[58] They attempted to increase substrate level mitochondrial synthesis of ATP by reversed electron flow. Rat hearts were perfused with mixtures of glucose and various Krebs-cycle mitochondrial metabolites. Solutions containing 20-mM glucose + malate + glutamate and 20-mM glucose + oxaloacetate + α-oxoglutarate were evaluated during anoxia. Significant increases in heart rate and glucose uptake were noted with both mixtures. ATP concentration increased during perfusion of the former, and both glycogen and CP increased with the latter. Results were inconsistent, and the mechanism of action is obscure. Oxidation of glucose appeared to be essential. The solutions were markedly hypertonic and contained higher levels of metabolic intermediates than could reasonably be achieved in vivo.

THERAPEUTIC CONCLUSIONS

Whatever drug, nutritional regimen, or hormonal intervention is utilized, an effective collateral flow is required for delivery of treatment to the target site—the ischemic cell. The struggle is to match energy demand with energy supply. Metabolic defects must be corrected. Hyperthyroidism and adrenocortical insufficiency increase oxygen and energy demand and lower cardiac glycogen concentration. Diabetes mellitus inhibits glucose transport and suppresses glycolytic flux. Although hyperglycemia cannot enhance energy production, hypoglycemia is decidedly toxic to the oxygen-depleted cell and is to be avoided. Methods to enhance glycolytic flux are being developed. The factors that prevent hypoxia from stimulating myocardial glycolysis to its maximal potential are unknown. Further study of our reptilian ancestors may yet provide the secret of more efficient utilization of this pathway. Studies of techniques for counteracting the detrimental effects of acidosis on glycolysis and hemodynamics by manipulation of nontoxic buffers are to be encouraged. Alkalotic therapy has proved essential to successful cardiac resuscitation. Although the rate of fall of pH due to lactate accumulation is about 0.5 pH units/15 min, phosphofructinase activity and contractile sites are quite sensitive to slight changes in H^+ ion concentration. An elevation of pH may increase glycolytic flux, enhance excitation-contraction coupling, and increase coronary blood flow. The enhanced performance of hypoxic hearts rendered alkalotic confirms its protective role. This treatment is still experimental and may have profoundly deleterious effects on cardiac function if accompanied by changes in pCO_2. The cell swelling that accompanies the distortions in ion flux of prolonged cellular hypoxia has been challenged by a most provocative therapeutic approach—that of induced extracellular hyperosmolality. Its application to the ischemic human heart is awaited. The experimental use of GIK solutions, hyaluronidase, hydrocortisone, and methylprednisolone has been reported to reduce ischemic damage to the myocardial cell. Their clinical usefulness will undoubtedly be objectively evaluated in the near future. The protective value of increased cardiac glycogen stores prior to the onset of ischemia has been noted in several experimental studies in reptiles and mammals but has not been confirmed clinically. Effective means for elevating myocardial glycogen concentration are cumbersome or may be contraindicated; they include the use of reserpine or beta receptor blocking agents to deplete or interrupt catecholamine stimulation of phosphorylase activity, administration of growth hormone, citrate, ketones, and lipids; starvation; and potentially toxic pharmacologic agents. It is encouraging to note that physical training appears to have a pronounced positive effect on myocardial glycogen concentration. Techniques that carefully control circulating FFA concentrations are currently confined to the experimental laboratory. Initial reports indicate sufficient potential to warrant further investigation. Methods that can effectively increase the energy output of the ischemic cell have yet to be developed. Regardless of how it may be boosted, glycolytic energy production cannot conceivably meet the total energy demands of the normally beating cell. The heterogeneous nature of myocardial ischemia, however, does not require that it do so. If we can accept a decrease in contractility of less than 50% of normal for these cells, anaerobic metabolism is capable of maintaining viability. Current clinical efforts are therefore more pragmatically directed toward reducing the work load of the ischemic myocardium in an effort to meet the deficit and more equitably balance energy demand and supply. The use of beta blocking drugs to reduce heart rate and contractility and furosemide and Arfonad, nitroprusside and nitroglycerine to reduce preload and afterload, respectively, is currently being studied at several centers.

REFERENCES

1. Jennings RB: Symposium on the pre-hospital phase of acute myocardial infarction. Part II: Early phase of myocardial ischemic injury and infarction. Am J Cardiol 24:753, 1969
2. Brachfeld N: Maintenance of cell viability. Circulation [Suppl 4] 39,40:202, 1969
3. Neely JR, Rovetto MJ, Oram JF: Myocardial utilization of carbohydrate and lipids. Prog Cardiovasc Dis 15:289, 1972
4. Scheuer J: Myocardial metabolism in cardiac hypoxia. Am J Cardiol 19:385, 1967
5. Katz AM: Effects of interrupted coronary flow upon myocardial metabolism and contractility. Prog Cardiovasc Dis 10:450, 1968
6. Opie LH: Metabolism of the heart in health and disease. Am Heart J 76:685, 1968
7. Jennings RB, Kaltenbach JP, Sommers HM, et al: Studies of the dying myocardial cell, in James TN, Keys JW (eds): The Etiology of Myocardial Infarction. Boston, Little, Brown, 1963, chap 12
8. Brachfeld N: Myocardial metabolic dysfunction following infarction, in Corday E, Swan HJC (eds): New Perspectives in Diagnosis and Management of Myocardial Infarction. Baltimore, Williams & Wilkins, 1973, chap 3
9. Williamson JR: Glycolytic control mechanisms. II. Kinetics of intermediate changes during the aerobic–anoxic transition in perfused rat heart. J Biol Chem 241:5026, 1966
10. Wollenberger A, Krause EG: Metabolic control characteristics of the acutely ischemic myocardium. Am J Cardiol 22:349, 1968
11. Morgan HE, Henderson MJ, Regen DM, Park CR: Regulation of glucose uptake in muscle. I. The effects of insulin and anoxia on glucose transport and phosphorylation in the isolated perfused heart of normal rats. J Biol Chem 236:253, 1961
12. Brachfeld N, Scheuer J: Metabolism of glucose by the ischemic dog heart. Am J Physiol 212:603, 1967
13. Neely JR, Whitfield CF, Morgan HE: Regulation of glycogenolysis in hearts: Effects of pressure development, glucose and FFA. Am J Physiol 219:1083, 1970
14. Danforth WH: Activation of glycolytic pathway in muscle, in Chance B, Estabrook RW, Williamson JR (eds): Control of Energy Metabolism. New York, Academic, 1965, p 287
15. Cornblath M, Randle PJ, Parmeggiani A, Morgan HE: Effects of glucagon and anoxia on lactate production, glycogen content and phosphorylase activity in the perfused isolated rat heart. J Biol Chem 238:1592, 1963
16. Yang WC: Anaerobic functional activity of isolated rabbit atria. Am J Physiol 205:781, 1963
17. Weissler AM, Kruger FA, Baba N, et al: Role of anaerobic metabolism in the preservation of functional capacity and structure of anoxic myocardium. J Clin Invest 47:403, 1968

18. Williamson JR: personal communication
19. Most AS, Gorlin R, Soeldner JS: Glucose extraction by human myocardium during pacing stress. Circulation 45:92, 1972
20. Opie LH, Thomas M, Own P, Shulman G: Increased coronary venous inorganic phosphate concentrations during experimental myocardial ischemia. Am J Cardiol 30:503, 1972
21. Parker JO, Chiong MA, West RO, Case RB: The effect of ischemia and alterations of heart rate on myocardial potassium balance in man. Circulation 42:205, 1970
22. Opie LH, Mansford KRL: The value of lactate and pyruvate measurements in the assessment of the redox state of free nicotinamideadenine dinucleotide in the cytoplasm of perfused rat heart. Eur J Clin Invest 1:295, 1971
23. Gorlin R: Assessment of hypoxia in the human heart. Cardiology 57:24, 1972
24. Henderson AH, Craig RJ, Gorlin R, Sonnenblick EH: Lactate and pyruvate kinetics in isolated perfused rat hearts. Am J Physiol 217:1752, 1969
25. Scheuer J, Berry MN: Effect of alkalosis on glycolysis in the isolated rat heart. Am J Physiol 213:1143, 1967
26. Huckabee WE: Relationships of pyruvate and lactate during anaerobic metabolism. I. Effects of infusion of pyruvate or glucose and of hyperventilation. J Clin Invest 37:244, 1958
27. Neely JR, Denton RM, England PJ, Randle PJ: The effects of increased heart work on the tricarboxylic acid cycle and its interactions with glycolysis in the perfused rat heart. Biochem J 128:147, 1972
28. Kuhn P, Pachinger O: The effect of calcium on myocardial lactate production under aerobic conditions. J Mol Cell Cardiol 4:171, 1972
29. Rovetto MJ, Whitmer JT, Neely JR: Comparison of the effects of anoxia and whole heart ischemia on carbohydrate utilization in isolated, working rat hearts. Circ Res 32:699–711, 1973
30. Bing OHL, Brooks WW, Messer JV: Protective effect of acidosis on heart muscle viability following hypoxia. Circulation 46:121, 1972
31. Leaf A: Regulation of intracellular fluid volume and disease. Am J Med 49:291, 1970
32. Willerson JT, Powell WP Jr, Guiney TE, et al: Improvement in myocardial function and coronary blood flow in ischemic myocardium after mannitol. J Clin Invest 51:2989, 1972
33. Smithen C, Keller N, Christodoulou J, Brachfeld N: Metabolic and hemodynamic effects of hyperosmolal solutions on recovery from myocardial anoxia. Clin Res 21:451, 1973
34. Wildenthal K, Skelton CL, Coleman HN III: Cardiac muscle mechanics in hyperosmotic solutions. Am J Physiol 217:302, 1969
35. Selye H: Chemical Prevention of Cardiac Necrosis. New York, Ronald, 1959
36. Sodi-Pallares D, Bisteni A, Medrano GA, et al: The polarizing treatment in cardiovascular conditions. Experimental basis and clinical applications, in Bajusz E (ed): Electrolytes in Cardiovascular Disease. Basel, S. Karger, 1966, p 198

37. Larcan A: Pathophysiological basis and practical application of a metabolic therapy of myocardial infarction, in Bajusz E (ed): Electrolytes in Cardiovascular Disease. Basel, S. Karger, 1966, p 198
38. Report of the Medical Research Council Working Party on the Treatment of Myocardial Infarction: Potassium, glucose and insulin treatment for acute myocardial infarction. Lancet 2:1355, 1968
39. Maroko PR, Libby P, Sobel BE, et al: Effect of glucose-insulin-potassium infusion on myocardial infarction following experimental coronary artery occlusion. Circulation 45:1160, 1972
40. Regan TJ, Harman MA, Lehan PH, et al: Ventricular arrythmias and K^+ transfer during myocardial ischemia and intervention with procaine amide, insulin or glucose solution. J Clin Invest 46:1657, 1967
41. Levinson RS, McIlduff JB, Regan TJ: Comparison of polarizing solutions and isovolumic KCl in digitalis induced ventricular tachycardia. Am Heart J 80:70, 1970
42. Opie LH, Mansford KRL, Owen P: Effects of increased heart work on glycolysis and adenine nucleotides in the perfused heart of normal and diabetic rats. Biochem J 124:475, 1971
43. Opie LH: Substrate utilization and glycolysis in the heart. Cardiology 56:2, 1971/1972
44. Scheuer J, Stezoski S: Protective role of increased myocardial glycogen stores in cardiac anoxia in the rat. Circ Res 27:835, 1970
45. Hewitt RL, Lolley DM, Adrouny GA, Drapanas T: Protective effect of myocardial glycogen on cardiac function during anoxia. Surgery 73:444, 1973
46. Brachfeld N, Ohtake Y, Klein I, Kawade M: Substrate preference and metabolic activity of the aerobic and the hypoxic turtle heart. Circ Res 31:453, 1972
47. Robin ED, Vester JW, Murdaugh V, Millen JE: Prolonged anaerobiosis in a vertebrate: Anaerobic metabolism in the freshwater turtle. J Cell Comp Physiol 63:287, 1964
48. Most AS, Brachfeld N, Gorlin R, Wahren J: Free fatty acid metabolism of the human heart at rest. J Clin Invest 43:1177, 1969
49. Henderson AH, Most AS, Sonnenblick EH: Depression of contractility in rat heart muscle by free fatty acids during hypoxia. Lancet 2:825, 1969
50. Kjekshus JK, Mjos OD: Effect of free fatty acids on myocardial function and metabolism in the ischemic dog heart. J Clin Invest 51:1767, 1972
51. Oliver MF, Kurein VA, Greenwood TW: Relation between serum free fatty acids and arrythmias after acute myocardial infarction. Lancet 1:710, 1968
52. Regan TJ, Markov A, Oldewurtel HA, Burke WM: Myocardial metabolism and function during ischemia: Response to *l*-noradrenaline. Cardiovasc Res 4:344, 1970
53. Opie LH: The general and local metabolic response to acute myocardial infarction. Acta Biol Med Ger 28:873, 1972
54. Apstein CS, Gmeiner R, Brachfeld N: Effect of palmitate on hypoxic

myocardial metabolism and contractility, in Bajusz E, Rona G (eds): Myocardiology. Baltimore, University Park Press, 1972, p 126
55. Scheuer J, Brachfeld N: Myocardial uptake and fractional distribution of palmitate-1-^{14}C by the ischemic dog heart. Metabolism 15:945, 1966
56. Wood JM, Hutchings AE, Brachfeld N: Lipid metabolism in myocardial cell free homogenates. J Mol Cell Cardiol 4:97, 1972
57. Brachfeld N, Keller N, Tarjan E, et al: Myocardial metabolism following pacing induced stress. Circulation [Suppl 2] 44:145, 1971
58. Penney DG, Cascarano J: Anaerobic rat heart: Effects of glucose and tricarboxylic acid-cycle metabolites on metabolism and physiological performance. Biochem J 118:221, 1970

Index

a
b
c
6 d
7 e
8 f
9 g
0 h
1 i
8 2 j